13th International Particle Accelerator Conference (IPAC'22)

Journal of Physics: Conference Series Volume 2420

Bangkok, Thailand
12 – 17 June 2022

ISBN: 978-1-7138-6998-6
ISSN: 1742-6588

Printed from e-media with permission by:

Curran Associates, Inc.
57 Morehouse Lane
Red Hook, NY 12571

Some format issues inherent in the e-media version may also appear in this print version.

Institute of Physics
Dirac House, Temple Back
Bristol BS1 6BE UK

Phone: 44 1 17 929 7481
Fax: 44 1 17 920 0979

techtracking@iop.org

Additional copies of this publication are available from:

Curran Associates, Inc.
57 Morehouse Lane
Red Hook, NY 12571 USA
Phone: 845-758-0400
Fax: 845-758-2633
Email: curran@proceedings.com
Web: www.proceedings.com

TABLE OF CONTENTS

Preface

Peer Review Statement

Acceleration and Crossing of Transition Energy Investigation Using an RF Structure of the Barrier Bucket Type in the NICA Accelerator Complex 1
S Kolokolchikov, Y Senichev, A Melnikov, E Syresin

RETRACTED: Investigation of Polarized Proton Spin Coherence Time at Storage Rings .... N/A
A Melnikov, A Aksentyev, Y Senichev, E Syresin

Operational Scenario of First High Luminosity LHC Run 16
R Tomás, G Arduini, P Baudrenghien, O Brüning, R Bruce, X Buffat, R Calaga, F Cerutti, R De Maria, J Dilly, I Efthymiopoulos, M Giovannozzi, P D Hermes, G Iadarola, R Jones, S Kostoglou, B Lindström, E H Maclean, E Métral, N Mounet, Y Papaphilippou, T H B Persson, T Pugnat, S Redaelli, G Sterbini, H Timko, F Van der Veken, J Wenninger, M Zerlauth

Study of Hydrodynamic-Tunnelling Effects Induced by High-Energy Proton Beams in Graphite 24
C Wiesner, F Carra, J Don, I Kolthoff, A Lechner, R Rasile, D Wollmann

Beam-Based Reconstruction of the Shielded Quench-Heater Fields for the LHC Main Dipoles 31
L Richtmann, L Bortot, E Ravaioli, C Wiesner, D Wollmann

A Design of ILC E-Driven Positron Source 38
M Kuriki, H Tajino, S Konno, Z Liptak, T Takahashi, M Fukuda, T Omori, Y Seimiya, J Urakawa, K Yokoya, S Kashiwagi

Beam Loading Compensation of Standing Wave Linac with Off-Crest Acceleration 45
M Kuriki, S Konno, H Tajino, Z Liptak, T Takahashi, M Fukuda, T Omori, Y Seimiya, J Urakawa, K Yokoya, S Kashiwagi

Summary of Numerical Noise Studies for Electron-Ion Collider Strong-Strong Beam-Beam Simulation 51
Y Luo, D Xu, Y Hao, J Qiang, J S Berg, M Blaskiewicz, W Fischer, R Gamage, X Gu, H Huang, G Hoffstaetter, J Kewisch, H Lovelace, C Montag, V Morozov, E Nisson, S Peggs, V Ptitsyn, T Satogata, F Willeke

Fine Decoupling Test and Simulation Study to Maintain a Large Transverse Emittance Ratio in Hadron Storage Rings 56
Y Luo, I Blackler, M Blaskiewicz, W Fischer, A Marusic, C Montag, T Shrey, D Xu

Designing the EIC Electron Storage Ring Lattice for a Wide Energy Range 61
D Marx, JS Berg, Y Cai, BRP Gamage, GH Hoffstaetter, J Kewisch, Y Li, C Montag, V Morozov, Y Nosochkov, V Ptitsyn, D Sagan, MG Signorelli, S Tepikian, J Unger, F Willeke, D Xu

Beam-Beam Interaction for Tilted Storage Rings 68
D Xu, Y Luo, D Holmes, C Montag, F Willeke, Y Hao, J Qiang

Detector Solenoid Compensation in the EIC Electron Storage Ring 76
D Xu, Y Luo, J Scott Berg, H Lovelace, S Tepikian, Q Wu, B Xiao, J Avronsart, H Witte, M Blaskiewicz, C Montag, F Willeke, B R Gamage, V Morozov, Y Hao

Field Enhanced Compact S-Band Gun Employing a Pin Cathode 83
R Bazrafshan, M Fakhari, T Rohwer, K Flöttmann, N H Matlis, F X Kärtner

Beam Orbit Shift Due to BPM Thermal Deformation Using Machine Learning 89
Kemin Chen, Faya Wang, Guanliang Wang, Zhe Wang, Masahito Hosaka, Lei Guo, Wei Xu

Commissioning Simulations for the DIAMOND-II Upgrade 95
H-C Chao, R Fielder, J Kallestrup, B Singh, I Martin

Dependence of CsK2Sb Photocathode Performance on the Quality of Graphene Substrate Film 101
L Guo, K Goto, H Yamaguchi, M Yamamoto, Y Takashima

A Double Dipole Kicker for off and On-Axis Injection into ALBA-II 106
G Benedetti, M Carlà, M Pont

The Status of the in-Vacuum-APPLE II IVUE32 at HZB / BESSY II 112
J Bahrdt, J Bakos, S Gaebel, S Gottschlich, S Grimmer, C Kuhn, S Knaack, F Laube, A Meseck, C Rethfeldt, E Rial, A Rogosch-Opolka, M Scheer, P Volz

Status and Powering Test Results of HTS Undulator Coils at 77 K for Compact FEL Designs 118
S C Richter, A Ballarino, T H Nes, D Schoerling, A Bernhard, A-S Müller

Simulations of the Suitability of a DC Electron Photogun and S-Band Accelerating Structure as Input to an X-Band Linac 125
S D Williams, R P Rassool, S L Sheehy, G N Taylor, M Volpi, R Auchettl, R T Dowd

Achievement of 200,000 Hours of Operation at KEK 7-GeV Electron 4-GeV Positron Injector Linac 132
K Furukawa, M Akemoto, D Arakawa, Y Arakida, Y Bando, H Ego, Y Enomoto, T Higo, H Honma, N Iida, K Kakihara, T Kamitani, H Katagiri, M Kawamura, S Matsumoto, T Matsumoto, H Matsushita, K Mikawa, T Miura, F Miyahara, H Nakajima, T Natsui, Y Ogawa, S Ohsawa, Y Okayasu, T Oogoe, M A Rehman, I Satake, M Satoh, Y Seimiya, T Shidara, A Shirakawa, H Someya, T Suwada, M Tanaka, D Wang, Y Yano, K Yokoyama, M Yoshida, T Yoshimoto, R Zhang, X Zhou

Lifetime Correction Using Fast-Off-Energy Response Matrix Measurements 138
S M Liuzzo, N Carmignani, L R Carver, L Hoummi, T Perron, B Roche, S White

Cooling Challenges in a NEG-Coated Vacuum Chamber of a Light Source 145
S Talebi Motlagh, A Danaeifard, J Rahighi, F Saeidi, F Zamani

Sensitivity of EEHG Simulations to Dynamic Beam Parameters 151
D Samoilenko, W Hillert, F Pannek, S Ackermann, E Ferrari, N Mirian, P Niknejadi, G Paraskaki, L Schaper, F Curbis, M Pop, S Werin

Emittance Feedback for the Diamond-II Storage Ring Using Resonant Excitation 157
Shaun Preston, Teresia Olsson, Beni Singh

Virtual Commissioning of the European XFEL for Advanced User Experiments at Photon Energies Beyond 25 keV Using Low-Emittance Electron Beams 163
Ye Chen, Frank Brinker, Winfried Decking, Matthias Scholz, Lutz Winkelmann, Zihan Zhu

Twin Delayed Deep Deterministic Policy Gradient for Free-Electron Laser Online Optimization 169
M Cai, Z H Zhu, K Q Zhang, C Feng, L J Tu, D Gu, Z T Zhao

Start-To-End Simulations for Bunch Compressor and THz SASE FEL at PITZ 176
A Lueangaramwong, X Li, P Boonpornprasert, M Krasilnikov, F Stephan

Reconstruction and Beam-Transport Study of the cERL Dump Line for High-Power IR-FEL Operation 181
N Nakamura, Y Tanimoto, N Higashi, K Harada, M Shimada, T Uchiyama, T Nogami, A Ueda, S Nagahashi, T Obina, R Takai, H Sagehashi, K Nigorikawa, O Tanaka, R Kato, H Sakai

Tungsten Electron Emitter (TE2) with Direct Heated Cathode by Plasma Stream 187
K I Thoma, T Dönges, M Droba, O Meusel, H Podlech, K Schulte-Urlichs

Studies of a Ka-Band High Power Klystron Amplifier at INFN-LNF 193
M Behtouei, B Spataro, L Faillace, A Leggieri, F Di Paolo, F Marrese, L Valletti, S Fantauzzi, G Torrisi, M Carillo, F Bosco, A Mostacci, L Palumbo, M Migliorati

Photocathode Performance Characterisation of Ultra-Thin MgO Films on Polycrystalline Copper 199
C Benjamin, H M Churn, T J Rehaag, L B Jones, G R Bell, T C Q Noakes

Material Normal Energy Distribution for Field Emission Analyses from Monocrystalline Surfaces 205
J I Mann, Y Li, J K Nangoi, T Arias, J B Rosenzweig

Towards High-Repetition Rate Petawatt Laser Experiments with Cryogenic Jets Using a Mechanical Chopper System 212
M Rehwald, S Assenbaum, C Bernert, C B Curry, M Gauthier, S H Glenzer, S Göde, C Schoenwaelder, U Schramm, F Treffert, K Zeil

Proposal for a Compact Neutron Generator Based on a Negative Deuterium Ion Beam 219
Kouichi Jimbo, Toshiyuki Shirai, Ka-Ngo Leung, Karl van Bibber

Performance of Automated Synchrotron Lattice Optimisation Using Genetic Algorithm 223
X Zhang, S L Sheehy

High Current Heavy Ion Beam Investigations at GSI-UNILAC 228
H Vormann, W Barth, M Miski-Oglu, U Scheeler, M Vossberg, S Yaramyshev

Design and Beam Dynamics Study of Disk-Loaded Structure for Muon LINAC 234
K Sumi, T Iijima, K Inami, Y Sue, M Yotsuzuka, H Ego, M Otani, N Saito, T Mibe, M Yoshida, Y Kondo, K Moriya, Y Takeuchi, Y Nakazawa, H Yasuda

The Zgoubidoo Python Framework for Ray-Tracing Simulations with Zgoubi: Applications to Fixed-Field Accelerators 240
M Vanwelde, C Hernalsteens, R Tesse, E Gnacadja, E Ramoisiaux, N Pauly

Recent Results of Beam Loss Mitigation and Extremely Low Beam Loss Operation of J-PARC RCS 246
P K Saha, K Okabe, T Nakanoya, Y Shobuda, H Harada, F Tamura, H Okita, M Yoshimoto, H Hotchi

RETRACTED: Investigation of Spin-Decoherence in the NICA Storage Ring for the Future EDM-Measurement Experiment .... N/A
A E Aksentev, A A Melnikov, Y V Senichev, V P Ladygin, E M Syresin

Investigation, Simulation and First Measurements of a 2 M Long Electron Column Trapped in a Gabor-Lens Device 259
K I Thoma, M Droba, O Meusel

Status of the Development of the Electron Lens for Space Charge Compensation at GSI 265
K Schulte-Urlichs, M Droba, T Dönges, O Meusel, D Ondreka, H Podlech, P Spiller, K I Thoma

Simulating Quasi-Integrable Optics with Space Charge in the IBEX Paul Trap 271
J A D Flowerdew, D J Kelliher, S Machida, S L Sheehy

Horizontal Beam Response at Extraction Conditions at the Heidelberg Ion-Beam Therapy Centre 277
E C Cortés García, E Feldmeier, Th Haberer

Excitation of the σl_1 = 90° Resonance by the Cavity Rf Accelerating Fields 283
J -M Lagniel

Study of Transverse Resonance Island Buckets at CESR 288
S T Wang, V Khachatryan

Fast Orbit Response Matrix Measurement Via Sine-Wave Excitation of Correctors at Sirius 294
M M S Velloso, F H de Sá, M B Alves

Reverse Bend Option for a 6 GeV Storage Ring Lattice 299
L Hoummi, N Carmignani, L R Carver, F Cianciosi, S M Liuzzo, T Perron, S White

Generalisation and Longitudinal Extension of the Genetic Lattice Construction (GLC) Algorithm 306
S Reimann, M Droba, O Meusel, H J Podlech

Parallelization of Radia Magnetostatics Code 313
A Banerjee, O Chubar, G Le Bec, J Chavanne, B Nash, C Hall, J Edelen

RETRACTED: Quasi-Frozen Spin Concept of Magneto-Optical Structure of NICA Adapted to Study the Electric Dipole Moment of the Deuteron and to Search for the Axion .... N/A
Y Senichev, A Aksentyev, S Kolokolchikov, A Melnikov, V Ladygin, E Syresin, N Nikolaev

Beamline Design and Optimisation for High Intensity Muon Beams at PSI 325
Eremey Valetov

Modeling a Nb3Sn Cryounit in GPT at UITF 331
S Pokharel, A S Hofler, G A Krafft

Review of Alignment and Stability Tolerances for Advanced Light Sources 338
A Khan, S Sharma, V Smaluk

Measurements of Radiation Fields from a Ceramic Break 344
Yoshihiro Shobuda, Takeshi Toyama, Masahiro Yoshimoto, Shuichiro Hatakeyama

Investigation of RF Heating for the Multipole Injection Kicker Installed at SOLEIL 352
A Gamelin, P Alexandre, R Ben El Fekih, J Da Silva Castro, M El Ajjouri, A Letresor, L S Nadolski, R Ollier, S Thoraud, M Sacko, S Taurines

Space Charge Analysis for Low Energy Photoinjector 358
M Carillo, M Behtouei, F Bosco, L Faillace, L Ficcadenti, A Fukasaway, L Giuliano, A Mostacci, M Migliorati, B Spataro, J Rosenzweig, L Palumbo

Modeling and Mitigation of Long-Range Wakefields for Advanced Linear Colliders 365
F Bosco, O Camacho, M Carillo, E Chiadroni, L Faillace, A Fukasawa, A Giribono, L Giuliano, N Majernik, A Mostacci, L Palumbo, J Rosenzweig, B Spataro, C Vaccarezza, M Migliorati

Strong-Strong Simulations of Coherent Beam-Beam Effects in the EIC 372
J Qiang, Y Hao, Y Luo, C Montag, D Xu, F Willeke

The Report of Machine Studies Related to the Vertical Beam Size Blow-Up in SuperKEKB LER 378
S Terui, T Ishibashi, K Ohmi, Y Funakoshi, H Fukuma, R Ueki, Y Ohnishi, M Tobiyama, T Nakamura

Lattice Design of the UVSOR-IV Storage Ring 384
E Salehi, Y Taira, M Fujimoto, L Guo, M. Katoh

A Method for Obtaining 3D Charge Density Distribution of a Self-Modulated Proton Bunch 390
T Nechaeva, L Verra, G Zevi Della Porta, P Muggli

Concrete Shielding Activation for Proton Therapy Systems Using BDSIM and FISPACT-II 395
E Ramoisiaux, C Hernalsteens, R Tesse, E Gnacadja, N Pauly, M Vanwelde, F Stichelbaut

Characterization of the Electron Beam Visualization Stations of the ThomX Accelerator 401
Alexandre Moutardier, Christelle Bruni, Jean-Noël Cayla, Iryna Chaikovska, Sophie Chance, Nicolas Delerue, Hayg Guler, Hugues Monard, Maher Omeich, Scott David Williams

Surrogate-Based Bayesian Inference of Transverse Beam Distribution for Non-Stationary Accelerator Systems 407
H Fujii, N Fukunishi, M Yamakita

A Beam Position Monitor for Electron Bunch Detection in the Presence of a More Intense Proton Bunch for the AWAKE Experiment 413
Collette Pakuza, Philip Burrows, Roberto Corsini, Wilfrid Farabolini, Pierre Korysko, Michal Krupa, Thibaut Lefevre, Stefano Mazzoni, Eugenio Senes, Manfred Wendt

6D Phase Space Diagnostics Based on Adaptively Tuned Physics-Informed Generative Convolutional Neural Networks 419
Alexander Scheinker, Daniele Filippetto, Frederick Cropp

A Machine Learning Approach to Electron Orbit Control at the 1.5 GeV Synchrotron Light Source DELTA 424
D Schirmer

A Data-Driven Anomaly Detection on SRF Cavities at the European XFEL 430
A Sulc, A Eichler, T Wilsken

Online Optimization of NSLS-II Dynamic Aperture and Injection Transient 436
X Yang, G Wang, V Smaluk, L H Yu, T Shaftan, Y Li, F Plassard, S Buda, Y Hidaka, D Durfee, Y Tian, Y Hu, K Ha, A Derbenev, B Bacha, C Danneil, D Padrazo

Dispersion-Free Steering Beam Based Alignment at SwissFEL 443
Eugenio Ferrari, Marco Calvi, Romain Ganter, Christoph Kittel, Eduard Prat, Sven Reiche, Thomas Schietinger

New Generation of Very Low Noise Beam Position Measurement System for the LHC Transverse Feedback 448
Daniel Valuch, Viera Stopjakova

Online Correction of Laser Focal Position Using FPGA-Based ML Models 453
J Einstein-Curtis, S J Coleman, N M Cook, J P Edelen, S Barber, C Berger, J van Tilborg

Commissioning and First Results of an X-Band LLRF System for TEX Test Facility at LNF-INFN 459
L Piersanti, D Alesini, M Bellaveglia, S Bini, B Buonomo, F Cardelli, C Di Giulio, E Di Pasquale, M Diomede, L Faillace, A Falone, G Franzini, A Gallo, G Giannetti, A Liedl, D Moriggi, S Pioli, S Quaglia, L Sabbatini, M Scampati, G Scarselletta, A Stella, S Tocci, L Zelinotti

A Python Framework for High-Level Applications in Accelerator Operations ... 465
J Chrin, V Erçağlar, T Schietinger

Reconstruction of Transverse Phase Space from Transverse Feedback Data for Real Time Extraction of Vital LHC Machine Parameters ... 471
G Kotzian, M Söderén, P S Solvang, D Valuch, V Stopjakova

The New Eddy Current Type Septum Magnet for Upgrading of Fast Extraction in Main Ring of J-PARC ... 477
T Shibata, K Ishii, S Iwata, T Sugimoto, H Matsumoto, K Fan

Spallation Target Optimization for ADS by Monte Carlo Codes ... 483
M Mumyapan, M Ghergherehchi, D H Ha, H Namgoong, J S Chai

Status of the Engineering Design of the IFMIF-DONES High Energy Beam Transport Line and Beam Dump System ... 489
D Sánchez-Herranz, O Nomen, M Sanmartí, B K Singh, F Arranz, I Podadera, C Oliver, P Cara, V Hauer, F Ogando

The Reduction of the Leakage Field of the Injection Septum Magnet in Main Ring of J-PARC ... 495
T Shibata, N Matsumoto, K Ishii, T Sugimoto, H Matsumoto

The New High Field Septum Magnet for Upgrading of Fast Extraction in Main Ring of J-PARC ... 502
T Shibata, K Ishii, S Iwata, T Sugimoto, N Matsumoto, H Matsumoto, K Fan

Role of Surface Chemistry in Conditioning of Materials in Particle Accelerators ... 508
G Sattonnay, S Bilgen, B Mercier, D Longuevergne, S Della-Negra, I Ribaud

Few-Nanosecond Stripline Kickers for Top-Up Injection into PETRA IV ... 513
G Loisch, V Belokurov, F Obier

Ultra-Fast Generator for Impact Ionization Triggering ... 519
A A del Barrio Montañés, V Senaj, T Kramer, Y Dutheil, M Sack

Branch Module for an Inductive Voltage Adder for Driving Kicker Magnets with a Short Circuit Termination ... 525
J Ruf, M J Barnes, Y Dutheil, T Kramer, M Sack

Proposal of a VHEE Linac for FLASH Radiotherapy ... 531
L Giuliano, F Bosco, M Carillo, D De Arcangelis, L Ficcadenti, D Francescone, A De Gregorio, G Franciosini, M Migliorati, A Mostacci, L Palumbo, V Patera, A Sarti, D Alesini, M Behtouei, L Faillace, A Gallo, B Spataro, A Vannozzi, P Cirrone, G Cuttone, G Torrisi, M G Bisogni, J H Pensavalle, F Di Martino, V Favaudon, S Heinrich, A Patriarca

Origin and Mitigation of the Beam-Induced Surface Modifications of the LHC Beam Screens ... 538
V Petit, P Chiggiato, M Himmerlich, S Marinoni, H Neupert, M Taborelli, L Tavian

Evaluation of Geometrical Precision and Surface Roughness Quality for the Additively Manufactured Radio Frequency Quadrupole Prototype ... 544
T Torims, A Cherif, N Delerue, M Foppa Pedretti, D Krogere, T Otto, G Pikurs, M Pozzi, A Ratkus, M Thielmann, M Vedani, M Vretenar, P Wagenblast

Tuning of Superconducting Cavities Using the FFT of Transmitted Power ... 553
E Laface, C G Maiano, P Pierini, M Wang

X-Rays Energy Measurements During the RFQ Conditioning at the European Spallation Source ... 560
E Laface, C G Maiano, R Zeng, O Piquet

Design Studies on a High-Power Wide-Band RF Combiner for Consolidation of the Driver Amplifier of the J-PARC RCS 567
H Okita, F Tamura, M Yamamoto, M Nomura, T Shimada, M Yoshii, C Ohmori, Y Sugiyama, K Hasegawa, K Hara, M M Paoluzzi

CERN Linac4 Chopper Dump: Operational Experience and Future Upgrades 573
C J Sharp, R Franqueira Ximenes, M Calviani, D Grenier, C Y Mucher, E Grenier-Boley, C Torregrosa, J R Hunt, L S Esposito, G Costa, P Andreu Muñoz, A M Krainer

TURBO: A Novel Beam Delivery System Enabling Rapid Depth Scanning for Charged Particle Therapy 580
J SL Yap, S L Sheehy, A F Steinberg, H XQ Norman, R B Appleby

Upgrade of a Proton Therapy Eye Treatment Nozzle Using a Cylindrical Beam Stopping Device for Enhanced Dose Rate Performances 586
E Gnacadja, C Hernalsteens, N Pauly, E Ramoisiaux, R Tesse, M Vanwelde

Achromatic Gantry Design Using Fixed-Field Spiral Combined-Function Magnets 593
R Tesse, C Hernalsteens, E Gnacadja, E Ramoisiaux, N Pauly, M Vanwelde

A Carbon Minibeam Irradiation Facility Concept 600
M Mayerhofer, V Bencini, M Sammer, G Dollinger

Beam Optics Studies for a Novel Gantry for Hadron Therapy 606
E Felcini, G Frisella, A Mereghetti, M G Pullia, S Savazzi, M T F Pivi, E Benedetto

Explorative Studies of an Innovative Superconducting Gantry 612
M G Pullia, E Benedetto, L Dassa, E De Matteis, M Donetti, E Felcini, G Frisella, M Karppinen, C Kurfürst, S Mariotto, A Mereghetti, A Mirandola, A Pella, D Perini, L Piacentini, M T F Pivi, M Prioli, A Ratkus, L Rossi, L Sabbatini, S Savazzi, M Stock, D Tommasini, T Torims, S Uberti, A Vannozzi, J Vilcans, M Vretenar

Study of Coil Configuration and Local Optics Effects for the GaToroid Ion Gantry Design 618
E Oponowicz, L Bottura, Y Dutheil, A Haziot, A Gerbershagen

Slow Extraction Modelling for NIMMS Hadron Therapy Synchrotrons 624
R Taylor, E Benedetto, M Sapinski, J Pasternak

Beam Optics Study for a Potential VHEE Beam Delivery System 631
C S Robertson, P N Burrows, M Dosanjh, A Latina, A Gerbershagen

A Compact Synchrotron for Advanced Cancer Therapy with Helium and Proton Beams 637
M Vretenar, M E Angolettai, G Bisoffi, J Borburgh, L Bottura, K Paļskis, R Taylor, G Tranquille, E Benedetto, M Sapinski

Production of Radioisotopes for Cancer Imaging and Treatment with Compact Linear Accelerators 643
M Vretenar, A Mamaras, G Bisoffi, P Foka

Design of the 590 MeV Proton Beamline for the Proposed TATTOOS Isotope Production Target at PSI 649
M Hartmann, D Kiselev, D Reggiani, J Snuverink, H Zhang, M Seidel

A Novel Intensity Compensation Method to Achieve Energy Independent Beam Intensity at the Patient Location for Cyclotron Based Proton Therapy Facilities 655
Vivek Maradia, David Meer, Damian C Weber, Antony J Lomax, Serena Psoroulas

A Novel Method of Emittance Matching to Increase Beam Transmission for Cyclotron-Based Proton Therapy Facilities: Simulation Study 660
Vivek Maradia, David Meer, Damian C Weber, Antony J Lomax, Jacobus M Schippers, Serena Psoroulas

Ripple Pattern Formation on Silicon Carbide Surfaces by Low-Energy Ion-Beam Erosion 666
Divya Gupta, G.R. Umapathy, Rahul Singhal, Sanjeev Aggarwal

Energy Comparison of Room Temperature and Superconducting Synchrotrons for Hadron Therapy 671
G Bisoffi, E Benedetto, M Karppinen, M G Pullia, M Reza Khalvati, L Rossi, M Sapinski, M Sorbi, U Valente, R van Weelderen, G Venchi, M Vretenar

Study on Construction of an Additional Beamline for a Compact Neutron Source Using a 30 MeV Proton Cyclotron 678
Y Kuriyama, M Hino, Y Iwashita, R Nakamura, H Tanaka

Design of Beam Collection System with High Transmission Efficiency and Tunablity 685
Y Yan, T Yang, D Y Li, H Cheng, Y Z Li, M J Wu, Y L Fang, F N Li, Z Guo, Z X Yuan, K Zhu, C Lin, X Q Yan

Retraction: Investigation of Polarized Proton Spin Coherence Time at Storage Rings (J. Phys.: Conf. Ser.) 690

Retraction: Quasi-Frozen Spin Concept of Magneto-Optical Structure of NICA Adapted to Study the Electric Dipole Moment of the Deuteron and to Search for the Axion (J. Phys.: Conf. Ser.) 691

Retraction: Investigation of Spin-Decoherence in the NICA Storage Ring for the Future EDM-Measurement Experiment (J. Phys.: Conf. Ser.) 692

Author Index

Journal of Physics: Conference Series **2420** (2023) 011001 doi:10.1088/1742-6596/2420/1/011001

Preface

The 13th International Particle Accelerator Conference, IPAC'22, took place at IMPACT Exhibition and Convention Center, Bangkok, Thailand from Sunday, June 12 to Friday, June 17, 2022. IPAC'22 was organized as the first in-person IPAC since the emergence of COVID-19 as a global pandemic and was attended by 742 delegates from 37 countries on all continents. The total includes 118 industry and exhibitor delegates. Hosted by Synchrotron Light Research Institute (Public Organization) - SLRI. SLRI was established in 1996 (originally as National Synchrotron Research Center - NSRC) and is the national synchrotron research institute which also known as "Thai Synchrotron National Lab", situated in Nakhon Ratchasima, Thailand. SLRI performs its main duties on development, operation, and management of synchrotron light source "Siam Photon Source (SPS)" in order to achieve one of its main goals in production of synchrotron light and provide it for user application on research. The Local Organizing Committee (LOC) consisted of 66 staff from SLRI.

There were 169 young scientists or students from all over the world attending the conference. Student grants were offered to 60 of these students thanks to the sponsorship of societies, institutes and laboratories worldwide. The Americas region grants were sponsored by APS-DB, ODU, CLS, SLAC, FRIB, and Jefferson Lab. The Europe region sponsors are EPS-AG, GSI, CNRS/IN2P3, INFN/LNL, CERN, DESY, PSI, CEA Saclay, ELETTRA, CELLS/ALBA, HZB, ESS, Cockcroft Institute, KIT, SOLEIL, ESRF, STFC/DL/ASTeC, MAX IV, and FZJ. The Asian region was sponsored by IPAC'22, SLRI, IPAC'19, ANSTO, SUT, GIT, LION, NSF, iRCT, Ti, Prime Street Advisory, NXPO, PUDITEC, and SPECS. The organizers of IPAC'22 are grateful to all sponsors for their valued support of students.

Welcome remarks were made by Prapong Klysubun (SLRI), Chair of the Organizing Committee (OC). Yoshihiro Funakoshi (KEK), Enrica Chiadroni (INFN/LNF), and David Tarazona (Cornell University) opened the scientific program with presentations, respectively, on "The SuperKEKB Has Broken the World Record of the Luminosity", "Progress Towards Demonstration of a Plasma Based FEL", and "The Accelerator and Beam Physics of the g-2 Experiment". The conference opening ceremony was on Monday afternoon and presided over by H.R.H. Princess Maha Chakri Sirindhorn featuring Special Plenary Talks from Chris Polly (Fermilab) on "Growing Expectations for New Physics" and Prapaiwan Sunwong (SLRI) on "SPS-II: A 4th Generation Synchrotron Light Source in Southeast Asia".

The program was closed with illuminating presentations by Mike Seidel (PSI) on "Towards Efficient Particle Accelerators - a Review"; Manjit Dosanjh (CERN/Oxford University) on "Accelerating the Future: Designing a Robust and Affordable Radiation Therapy Treatment System for Challenging Environments"; and Tomoki Nakamura (Tohoku University) on "Synchrotron Light Illuminates the Origin of the Solar System".

There were 36 invited and 52 contributed oral presentations of very high quality presented during the week. The regional distribution of talks was 16.7 % from Asia, 59.1 % from Europe, and 24.2 % from the Americas. The gender ratio for oral presentations was 71.6 % male and 28.4 % female. The oral presentations were 79.5 % onsite talks and 20.5 % remote talks.

The scientific program was developed by the 16-member IPAC'22 Scientific Program Committee (SPC) comprising 8 leads from Asia and 1 deputy each from the Americas and Europe under the leadership of Hitoshi Tanaka (RIKEN). Valued suggestions for invited talks were contributed by the 90-member Scientific Advisory Board (SAB) representing accelerator laboratories worldwide. Oral sessions were grouped according to the eight IPAC Main Classifications, with Poster Sessions Grouped according to Sub Classifications. The conference program spanned four and a half days, with plenary talks on Monday morning and afternoon in the opening ceremony and Friday morning, and on Thursday afternoon for the Accelerator Awards Session. All other sessions were composed of two invited or contributed talks in parallel.

Four Poster Sessions were held each afternoon on Monday through Thursday, during which 838 posters from 164 institutions were scheduled for presentation. There were 169 students from 81 institutions, representing 17 countries, who attended IPAC'22, with 60 students supported via the student grant program. The Sunday afternoon student poster session was again a successful event and saw 111 student posters presented and was judged by a team of 27 judges, organized by Toshiyuki Mitsuhashi (KEK).

These proceedings contain the reports of 874 total contributions. The regional breakdown is 16 % Asia, 24 % Americas and 60 % Europe. The breakdown by Main Classifications is as follows: MC1 Circular and Linear Colliders with 81 contributions; MC2 Photon Sources and Electron Accelerators with 161 contributions; MC3 Novel Particle Sources and Acceleration Techniques with 60 contributions; MC4 Hadron Accelerators with 75 contributions; MC5 Beam Dynamics and EM Fields with 149 contributions; MC6 Beam Instrumentation, Controls, Feedback, and Operational Aspects with 136 contributions; MC7 Accelerator Technology with 152 contributions; MC8 Applications of Accelerators, Technology Transfer and Industrial Relations with 57 contributions.

The scientific program was supplemented by a variety of special sessions and events.

The Industry Session was held on Wednesday and was in a form of round table session with a moderators, and 6 speakers on the stage based on the topic "Particle Accelerator Technology: From Research to Industry - Present Global, Overview and How to Move Forward". Raffaella Geometrante (Kyma SpA) is the moderator of the session. Presentations were given by Wiboon Rugsancharoenphol (The Federation of Thai Industries) on "Strategy of collaboration with industry in Thailand"; Hans Priem (VDL ETG T&D) on "From big size markets to small size markets"; Enrico Braidotti (CAEN ELS s.r.l.) on "How a small size market company can cross the chasm between a niche market towards wider industrial markets?"; Maurizio Vretenar (CERN) on "The quest for the miniature accelerator: wishful thinking or the key to expanding the particle accelerator market?"; Sandra Biedron (Element Aero and CBB) on "Present status and opportunities for implementing disruptive technologies arising in particle accelerator R&D to industrial market"; and Suzie Sheehy (University of Melbourne) on "Impact of disruptive particle accelerator technologies on human health".

"Lunch and Learn Session" was supported by IPAC'22 for industrial speakers to present and discuss on their innovations and research products as non-commercial talks. This special session allowed accelerator scientists, engineers, students and industrial partners to exchange information and ideas regarding accelerator science and technology through informal interactions.

The industrial exhibition took place from Monday to Thursday. Industrial exhibitors (56 companies) occupied 53 booths at which they presented their high technology products and services to the delegates. Four learned society booths also exhibited during the same 4-day period. This industrial exhibition area featured networking lounges to facilitate discussion and was co-located with the poster areas. It must be acknowledged that the conference would not be possible in its present format without the generous support of the IPAC industry exhibitors and sponsors.

IPAC'22 hosted a range of satellite meetings, including IPACCC meeting, IPAC'22 OC2 meeting, JACoW Stakeholders' meeting, SPC meetings IPAC'23 as well as the film "The fantastic journey of particles in an accelerator" presented by Phu Anh Phi NGHIEM (CAE-IRFU) and the session "Join us for day of education and exploration with Sirepo" by RadiaSoft.

The 2022 Asian Committee for Future Accelerators (ACFA) and IPAC'22 were honored to announce the following award winners during the awards session:

The Xie Jialin Prize for outstanding work in the accelerator field, with no age limit was awarded to Prof. Zhentang Zhao for his significant contributions to the developments of the FEL theory and experiments including the first lasing of the EEHG-FEL in the world and also to the various facility constructions as the main project leader such as SSRF, SDUV-FEL and SXFEL.

The Nishikawa Tetsuji Prize for a recent, significant, original contribution to the accelerator field, with no age limit was awarded to Dr. Xiaobiao Huang for his contribution to the field of particle accelerators, especially in the domain of accelerator design and operation, model-independent beam dynamics analysis, beam-based optimization and control.

The Hogil Kim Prize for a recent, significant, original contribution to the accelerator field, awarded to an individual in the early part of his or her career was awarded to Dr. Daniel Winklehner for his contribution to develop innovative designs of compact, high-current cyclotrons which make new opportunities for high intensity particle physics and industrial applications.

The prizes for best student posters were awarded to Cristhian Eduardo Gonzalez-Ortiz (Michigan State University) for "Third-order Resonance Compensation at the FNAL Recycler Ring"; and Annika Gabriel (University of California and SLAC) for "Temporal and Spatial Characterization of Ultrafast Terahertz Near-Fields for Particle Acceleration".

The proceedings of IPAC'22 are published on the JACoW site (www.jacow.org). The processing of the electronic manuscripts was achieved by IPAC'22 Editorial Team consisting of 28 strong JACoW team members from 18 different institutions prior to, and during the conference. The team, led by Thakonwat Chanwattana (SLRI), Volker RW Schaa (GSI), and Christine Petit-Jean-Genaz (CERN), is a Hybrid Editorial Team includes JACoW experts who also trained less experienced editors, and 6 remote editors who dedicate their time to work remotely regardless of different time zones. In this year, more than 70% of contributions were submitted in a LaTeX format. This is a big improvement in ratio between LaTeX and Word contributions. Some editors have trained themselves during the conference to become LaTeX editors. We can manage the greater number of LaTeX papers also thanks to great supports from remote editors.

The JACoW Collaboration is formed by electronic publishing experts and technicians volunteered by laboratories worldwide. Tasks performed by the proceedings office include: author reception, processing of contributions and transparencies, checking that references are formatted to journal standards, and crosschecking of titles and authors. Setting up the computers and internet network, presentations management and poster session management were a collaborative effort between the LOC and JACoW. Thanks to the work of this dedicated team, a pre-press version with on-time author's submissions that were processed by editors published totaling more than 800 contributions by the last day of the conference. The final version was published at the JACoW site just weeks after the conference. This is yet another impressive record set by the JACoW Collaboration, which is sincerely grateful to the supervisors and institutes of each of the team members for releasing them from their usual duties.

IPAC'22 would also like to acknowledge two brilliant tools, known as JACoW Reference Search tool and JACoW Cat Scan Editor Tool, developed from IPAC'19 by ANSTO in collaboration with JACoW, to assist both authors and editors. These tools allowed the generation of citations of JACoW publications in the corrected format, and the identification of formatting errors in Word DOCX contributions. A significant rise of these tools' usage is a good sign of contributions' quality improvement and efficient contribution processing for the entire JACoW conference community.

IPAC'22 supports light peer review of a limited number of papers to be published as part of the Institute of Physics Journal of Physics: Conference Series. Papers are reviewed by the SAB, while the SPC under the leadership of the Scientific Publication Board Chair, Frank Zimmermann, performed the function of editorial board. Candidate papers were submitted two weeks in advance of the normal deadline to allow a cycle or review, revision, and final review. Over 117 papers were submitted, many were revised, and 111 were approved. Nawin Juntong and Todd Satogata managed the review process within the JACoW SPMS database.

The success of IPAC'22 was achieved thanks to the strong collaboration between the international teams of the OC, SPC and the LOC. Membership of the LOC included the following:
Porntip Sudmuang (SLRI): LOC Chair
Thapakron Pulampong (SLRI): Co-LOC Chair
Prapaiwan Sunwong (SLRI): Scientific Secretariat and SPMS Administration
Siriwan Jummunt (SLRI): Student Program Manager
Thakonwat Chanwattana (SLRI): Proceedings Manager
Natthawut Suradet (SLRI): IT Manager
Athikarn Thongwat (SLRI): IT Manager
Thanapong Phimsen (SLRI): Exhibitor and Sponsor Manager and Poster Presentation Manager
Sasipun Tritan (SLRI): Conference Support Manager
Pinit Kidkhunthod (SLRI): Registration Manager and Facility Tour Manager
and all our supportive LOC members.

It is clear that on this 13th conference of the series that IPAC remains a focus for the coming together of the worldwide accelerator community and the fruitful exchange of ideas, unbounded by nationality, experience or background. Being the first of this conference series to be hosted in the Southeast Asia, it was especially pleasing to see a strong participation from the international community, despite the travel difficulties due to COVID-19 pandemic and Ukraine-Russia conflict, and the extra distance.

The 14th IPAC will return to the European super-region and will take place in Venice, Italy.

Porntip Sudmuang (SLRI) Chair of the IPAC'22 Local Organising Committee

Journal of Physics: Conference Series **2420** (2023) 011002 doi:10.1088/1742-6596/2420/1/011002

Peer Review Statement

All papers published in this volume have been reviewed through processes administered by the Editors. Reviews were conducted by expert referees to the professional and scientific standards expected of a proceedings journal published by IOP Publishing Publishing.

- **Type of peer review:** Single Anonymous
- **Conference submission management system:** Morressier
- **Number of submissions received:** 117
- **Number of submissions sent for review:** 117
- **Number of submissions accepted:** 111
- **Acceptance Rate (Submissions Accepted / Submissions Received × 100):** 94.9
- **Average number of reviews per paper:** 2
- **Total number of reviewers involved:** 191
- **Contact person for queries:**
  Name: Nawin Juntong
  Email: nawin@slri.or.th
  Affiliation: Synchrotron Light Research Institute - Accelerator

Acceleration and crossing of transition energy investigation using an RF structure of the barrier bucket type in the NICA accelerator complex

S Kolokolchikov[1], Y Senichev[1], A Melnikov[1] and E Syresin[2]

[1]Institute for Nuclear Research of the Russian Academy of Sciences, Moscow, Russia
[2]Joint Institute for Nuclear Research, Dubna, Russia

E-mail: kolokolchikov@inr.ru

Abstract. The dynamic of longitudinal motion in Barrier Bucket RF structure is considered. To preserve the stability of the proton beam during the acceleration to the experiment energy it is necessary to cross the transition energy and a rapid jump of transition energy is possible. The influence of the second-order slip factor is taking into account. The dynamic aperture is investigated for various gradients of focusing quadrupoles and corresponding working points which is necessary for transition crossing.

1. Introduction

During particle acceleration to the energy of experiment $E_{exp} = 12.6$ GeV ($\gamma_{exp} = 14.4$) in NICA collider there is a need for the crossing of transition energy $E_{tr} = 5.709$ GeV ($\gamma_{tr} = 7.087$). The slip-factor $\eta = \frac{1}{\gamma_{tr}^2} - \frac{1}{\gamma^2}$, which included in the equation of longitudinal motion, changes the sign, that leads to a loss of stability in the longitudinal plane when γ approaching to γ_{tr}. In order to minimize the reduction of the beam parameters, a rapid jump of transition energy is assumed during the time $t_{jump} \approx 10$ ms [1] with a simultaneous change of the polarity of the RF system to ensure the stability of the beam after the jump. RF structure based on "Barrier Bucket" a feature of the NICA collider accelerating system. A non-zero value of the field between the barriers, ensures the acceleration of the beam. This feature makes the system original.

In this paper, the main features of the dynamics of the longitudinal motion of the beam crossing through the transition energy are considered, taking into account its jump in the RF structure of the "Barrier Bucket" type. Due to the rapid jump of transition energy the time at which the particles are near the zero value of the first order slip-factor is significantly reduced. Obviously, in this case, the second order of the slip-factor begins to play a decisive role in the behaviour of particles inside the barrier bucket and completely determines the stability region near transition. In this case, when energy of particles crosses transition energy, focusing of the beam in the longitudinal plane disappears and the influence of the space charge becomes essential.

Journal of Physics: Conference Series **2420** (2023) 012001 doi:10.1088/1742-6596/2420/1/012001

2. Rapid jump of the transition energy

2.1. Jump schemes

When the particle energy approaches the transition energy, it is assumed to make a rapid jump of the transition energy. This can be achieved by changing the field gradient in the focusing quadrupoles of the arcs, since for periodic structures $1/\gamma_{tr}^2 = \alpha_0 = 1/\nu_x^2$, where α_0 – first order of momentum compaction factor, ν_x – normalized frequency of betatron oscillations (tune) in the horizontal plane. The fast jump of transition energy is proposed to be equal to $2 \cdot \Delta\gamma_{tr} = 0.09$ (from $\gamma_{tr} = 7.087$ to $\gamma_{tr} - 2 \cdot \Delta\gamma_{tr} = 6.997$) [1]. With such a jump, the working point shifts the values from $\nu_x = 9.44$ and $\nu_y = 9.44$ to $\nu_x = 9.365$ and $\nu_y = 9.465$. The maximum change rate of transition energy is limited by the parameters of quadrupoles and their power systems $\dot{\gamma}_{tr} = \frac{d\gamma_{tr}}{dt} = 8.5s^{-1}$. Such a jump can be made during the time:

$$t_{jump} = T_0 \Delta n_{jump} = \frac{2 \cdot \Delta\gamma_{tr}}{\dot{\gamma}_{tr}} = 10.5ms, \quad (1)$$

where $T_0 = 1.7\mu s$ – the time of one revolution period in the ring, $\Delta n_{jump} = 6226$ – the number of revolution periods during which a transition energy jump occurs.

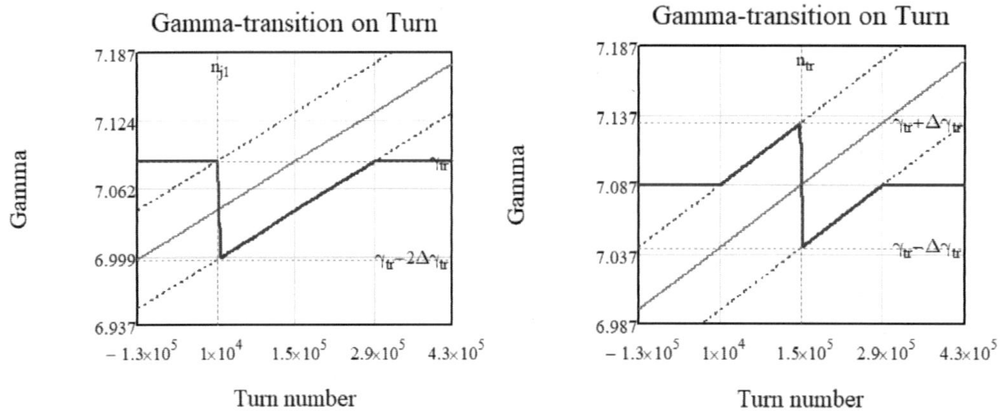

Figure 1. On the left – a schematic diagram of the transition energy γ_{tr} change during the jump and its recovery. On the right – another possible scheme. The blue solid line shows the transition energy γ_{tr}, the solid red line is the energy of particles γ. Blue dashed lines – γ of particles shifted by $\pm\Delta\gamma_{tr}$.

Figure 1 shows different schemes of the transition energy change γ_{tr} together with the beam energy γ at the moment of the jump and after it.

2.2. Dynamic aperture

A change of the transition energy is possible due to a change in the tune of betatron tune in the x-plane, which is achieved by changing the gradient of the focusing lenses in the collider arcs. With the changed parameters of quadrupole lenses, the dynamic aperture was evaluated, it determines the stable area for movement of particles in the transverse plane. The corresponding calculations were carried out using OptiM and MADX programs.

The results shown on figure 2b indicate a working point at $\nu_x = 9.44$ and $\nu_y = 9.44$. For this working point is necessary to use octupole correctors to compensate nonlinear effects. If we assume the fact that it is necessary to reach the low frequency in the horizontal plane, with $\nu_x = 9.3627$ and in the vertical plane $\nu_y = 9.4541$ (see figure 2a). So thus, the transition energy

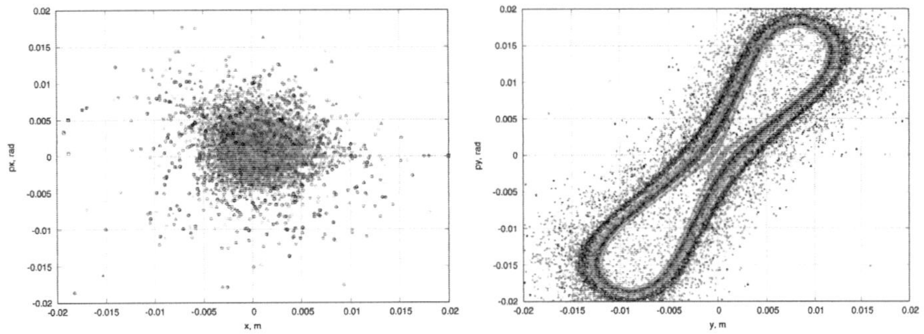

a) $\nu_x = 9.362$ and $\nu_y = 9.454$

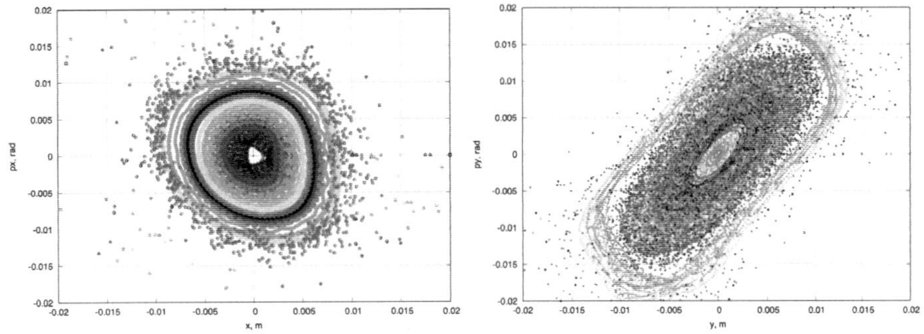

b) $\nu_x = 9.44$ and $\nu_y = 9.44$

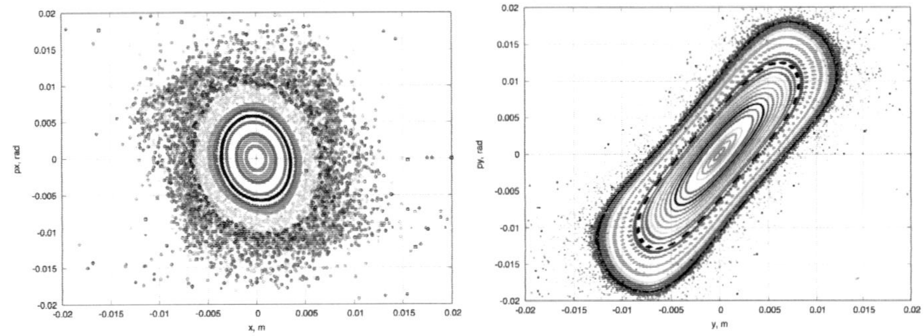

c) $\nu_x = 9.4785$ and $\nu_y = 9.4330$

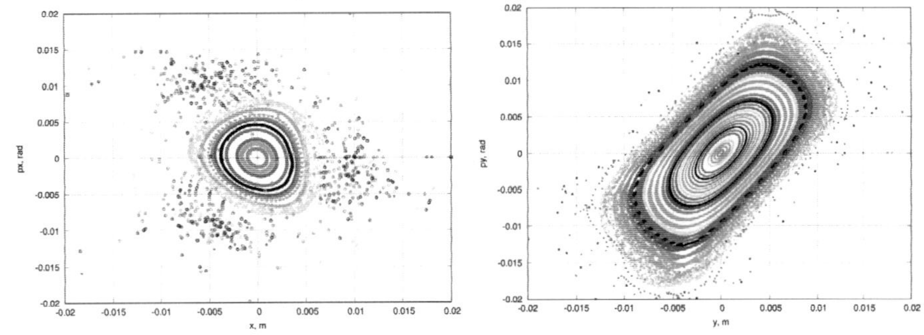

d) $\nu_x = 9.4014$ and $\nu_y = 9.447$

Figure 2. Dynamic apertures (x–plane on the left, y–plane on the right) for the various (a-d) working point with $dp/p = 0$. On the abscissa axis is the coordinate in meters, on the ordinate axis is the relative momentum in radians.

is achieved $\gamma_{tr} - 2 \cdot \Delta\gamma_{tr} = 6.997$, the dynamic aperture in the horizontal plane disappears completely at these values of betatron tunes.

Hence, we considered another option (see figure 1 right): first, we gradually raise the transition energy to $\gamma_{tr} + \Delta\gamma_{tr} \approx 7.13$, then we make a rapid jump down to $2 \cdot \Delta\gamma_{tr}$ up to $\gamma_{tr} - \Delta\gamma_{tr} \approx 7.04$. In this case, the working point changes from $\nu_x = 9.44$ and $\nu_y = 9.44$ up to the value before the jump $\nu_x = 9.4785$ and $\nu_y = 9.4330$ (see figure 2c) and after the jump down to $\nu_x = 9.4014$ and $\nu_y = 9.447$ (see figure 2d). In this case, the dynamic aperture achieved a sufficient value.

3. Barrier bucket RF system

The RF-1 system is used to retain, accumulate and accelerate particles to the experimental energy in the collider rings. Each collider ring has one RF-1 system. During retention and accumulation, 2 pairs of rectangular pulses with opposite signs are generated with the amplitude of each barrier $V_{bb} = \pm 5$ kV (see figure 3). The time duration of a single pulse can vary from $T_{bb} = 10$ till 80 ns. The accumulated particles enclosed between 2 pulses will be inductively accelerated by a constant potential $V_{acc} = 300$ V, which is additionally created also by the RF-1 system [1].

When the energy approaches the transition value, the RF barriers turn off and, after the proton energy becomes greater than the transition energy, the RF barriers turn on and a polarity changes. This is necessary because the slip-factor value changes it's sign after transition energy crossing.

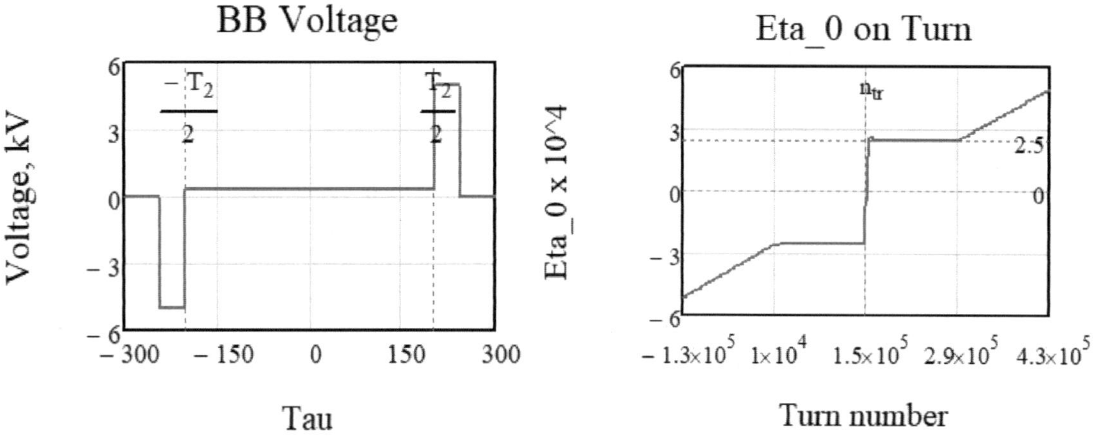

Figure 3. Left – shows the Barrier Bucket RF type; right – scheme of η_0 change during jump.

4. Slip-factor near transition energy

On the one hand, when the slip-factor is purely zero $\eta = 0$ the system is isochronous and with any spread of momentum, the bunch does not increase the length. On the other hand, the following second order of slip-factor begins to play an essential role $\eta = \eta_1\delta$, which distorts the movement and can lead to an increase in momentum spread. And finally, in the absence of focusing in the longitudinal plane, the space charge can introduce large distortions into the phase portrait of the bunch. During acceleration and retention of particles far away from the transition energy, the value of the slip-factor is calculated as (see figure 3 right)

$$\eta = \eta_0 = \alpha_0 - \frac{1}{\gamma_0^2} \tag{2}$$

however, such an expression takes into account only the first order, and in general, the slip-factor depends on $\delta = dp/p$ as [2]:

$$\eta = \eta_0 + \eta_1\delta + \cdots \tag{3}$$

where $\eta_1 = \alpha_1 - \frac{\eta_0}{\gamma_0^2} + \frac{3}{2}\frac{\beta^2}{\gamma^2}$.

Modeling the process of transition energy crossing, the first two orders of decomposition are taken into account. The corresponding momentum compaction factor (MCF) coefficients depend only on the collider magneto-optical structure.

Making a rapid jump of transition energy, parameters $\alpha_0, \alpha_1, \ldots$ also undergo a jump. Value α_0, as we have already said, it is determined by $\alpha_0 = 1/\gamma_{tr}^2$. Coefficient α_1 for different values γ_{tr} is calculated using the program MADX/PTC [3].

5. Longitudinal dynamics

Equations of longitudinal motion in Barrier Bucket in coordinates $(\tau, \Delta E)$ [4]:

$$\frac{d\tau}{dt} = \frac{\eta h}{\beta^2 E_0}\Delta E \quad \frac{d\Delta E}{dt} = \frac{Ze}{A}\frac{\omega_0}{2\pi}Ug(\tau) \tag{4}$$

where E_0 – synchronous particle energy, $Ug(\tau)$ – voltage generated by RF-barriers, $\omega_0 = 2\pi/T_0$, h – harmonic number.

For modelling, in equations (4) it is convenient to switch from the time derivative to the derivative over turn number $t = nT_0$: (also note that for protons $Z/A = 1$)

$$\frac{d\tau}{dn} = \eta\frac{T_0 h}{\beta^2 E_0}\Delta E \quad \frac{d\Delta E}{dn} = V(\tau) \tag{5}$$

5.1. The influence of the second-order slip-factor η_1 on the longitudinal dynamics during the jump

First, let's consider a jump taking into account only the first-order of the slip factor $\eta = \eta_0$. Let's evaluate the value η_0 before the jump itself, due to the symmetry with respect to zero, it will be equal to the value after:

$$|-\eta_{0tr}| = |+\eta_{0tr}| = |\eta_0(\gamma_{tr} - \Delta\gamma_{tr})| = 2.5 \cdot 10^{-4}$$

As it can be seen on figure 4(c,d) there is a jump of slip-factor in a different time for different particles because of dependence of slip-factor on δ. Obviously, after the jump, particles with a negative value of the slip-factor will not be in a stable region, since the polarity of the retaining barriers changes and will tend to leave the phase plane, as it can be seen on figure 4(a, c, d). Also, due to the momentum spread, there is an asymmetry of the phase portrait relate to the zero value of the momentum spread dp/p.

It can be seen that after the jump itself, not all particles are retained inside the separatrix due to the change of the polarity of the RF voltage of retaining barriers. Let's review figure 4a, it represents the blurring of phase space. Before the jump ($\eta_0 < 0$) particles with negative $\delta < 0$ have a slip-factor value η closer to zero than particles with positive $\delta > 0$. Because of this, the phase plane behaves asymmetrically, since the total value of the slip-factor affects the dynamics and change coordinate τ. Thus, the particles accumulate in the area of the left barrier. During the jump, the RF is turned off and does not affect the dynamics. After the jump ($\eta_0 > 0$) the particle distribution is shifted to the left edge and now the opposite is true – for particles with positive $\delta > 0$ the value of the slip-factor η is closer to zero, than for particles with $\delta < 0$. Those particles that turned out to be on the left in the area of the RF barrier (with reversed polarity) due to the proximity to the zero value of the slip-factor, do not have time to stay in the separatrix.

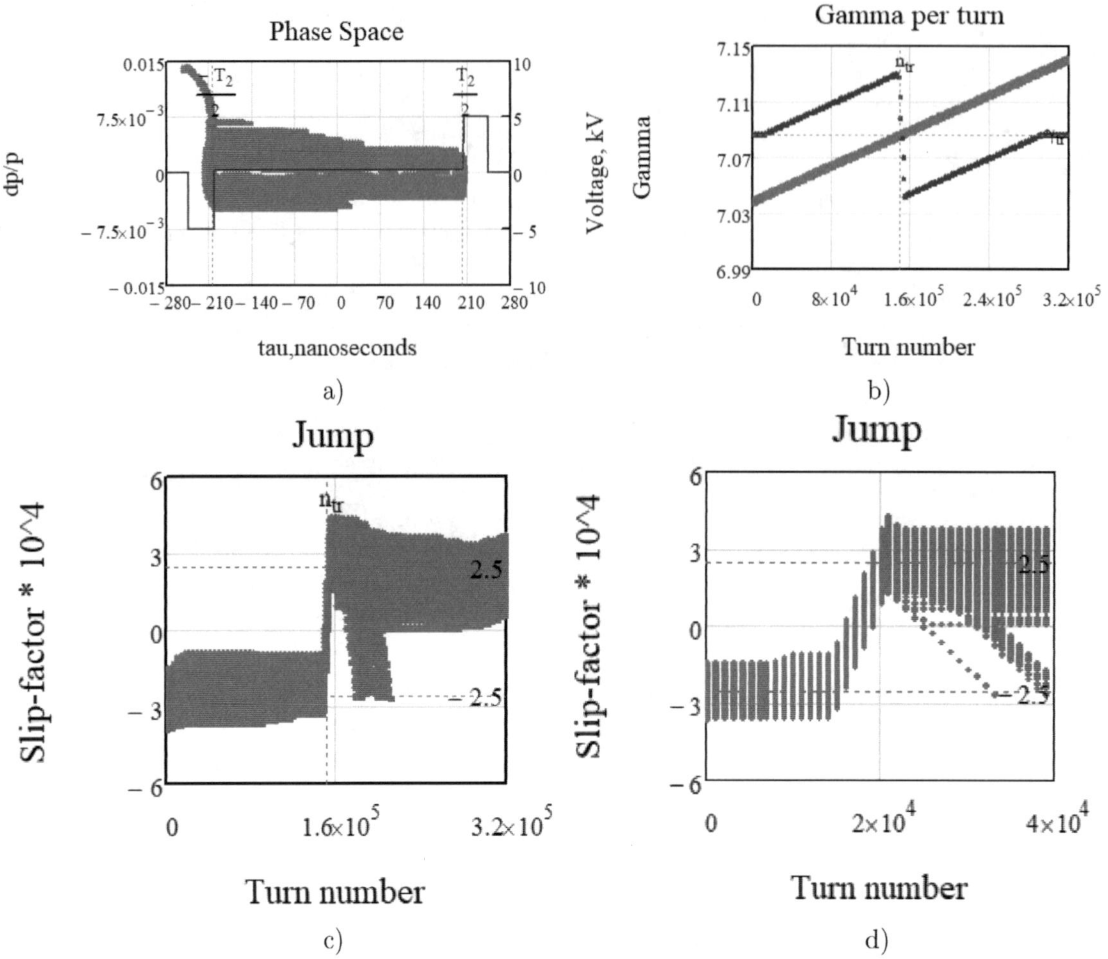

Figure 4. Crossing of the transition energy by a jump taking into account the second order of the slip-factor: a) Blurring the phase space in barrier bucket RF; b) jump of transition energy γ_{tr} (blue dots), γ – gamma particles (red dots); c) Jump of the slip-factor $\eta = \eta_0 + \eta_1\delta$; d) other scale of jump.

6. Conclusion

In this paper, the dynamics of the longitudinal motion of particles in an RF accelerating system of the Barrier Bucket type at a transition energy jump near the transition is considered. The influence of the second order slip-factor during the transition is considered. When the transition energy jumps, the working points changes, dynamic apertures are considered for them.

During modelling and calculations, it was found that with a transition energy jump, the determining factor, which influence on the stability of the particles inside the Barrier Bucket RF system is the second order of slip-factor. It depends on momentum spread and can determine the maximum momentum spread for the transition energy crossing in a jump-scheme without loss. Also, it is necessary to take into account the space charge and impedances. These problems require further research.

References

[1] Syresin E M *et al.* 2021 *Phys. Part. Nucl.* **52** pp 997–1017

[2] Ng K Y 2002 Physics of Intensity Dependent Beam Instabilities Tech. rep. Fermilab-FN-0713
[3] Schmidt F, Skowronski P, Lebedev V and Valishev A 2018 Higher Order Dispersion and Momentum Compaction in MAD-X/PTC using NormalForm Tech. rep. CERN Geneva CERN-ACC-NOTE-2018-006
[4] Stockhorst H, Katayama T and Maier R 2016 *Beam Cooling at COSY and HESR Theory and Simulation – Part 1 Theory* Forschungszentrum Jülich GmbH Zentralbibliothek, Verlag Jülich pp 161–171

Gap in pagination due to unavailable paper.

Pages 8-15

Operational scenario of first high luminosity LHC run

R Tomás, G Arduini, P Baudrenghien, O Brüning, R Bruce, X Buffat, R Calaga, F Cerutti, R De Maria, J Dilly, I Efthymiopoulos, M Giovannozzi, P D Hermes, G Iadarola, R Jones, S Kostoglou, B Lindström, E H Maclean, E Métral, N Mounet, Y Papaphilippou, T H B Persson, T Pugnat, S Redaelli, G Sterbini, H Timko, F Van der Veken, J Wenninger and M Zerlauth

CERN, Geneva, Switzerland

E-mail: rogelio.tomas@cern.ch

Abstract. A new scenario for the first operational run of the High Luminoisty LHC (HL–LHC) era (Run 4) has recently been developed to accommodate a period of performance ramp-up to achieve an annual integrated luminosity close to the nominal HL–LHC design target. The operational scenario in terms of beam parameters and machine settings, as well as the different phases to reach optimal performance, are described here along with the impact of potential delays to key hardware components.

1. Previous operational scenarios

The HL–LHC operational scenario for Run 4 has been developed in [1], considering changes to the hardware configuration and new findings in beam dynamics with respect to the previous scenario presented in [2]. However, very recent changes to the schedule and potential delays in the installation of key hardware components are motivating its revision. This section presents the operational scenario as in [1] while the potential new scenarios are presented in the second section as reported in [3]. The key changes between [2] and [1] follow:

Postponing the installation of sextupoles (MS10) in the dispersion suppressor to after Run 4[1]. It has been verified [4, 5] that beam lifetime due to Dynamic Aperture (DA) without these sextupoles is acceptable for optics with $\beta^* \geq 20$ cm in the two main experimental interaction points, IP1 & IP5, which is the current assumption for Run 4. Figures 1 and 2 show the DA at the start of collisions ($\beta^* = 1$ m) and at the end of the luminosity levelling ($\beta^* = 20$ cm), respectively. Note that at $\beta^* = 15$ cm these sextupoles are mandatory [6].

Reducing the scope of the secondary collimator upgrade, with not all being replaced with low-impedance collimators[1] **[7]**. This decision is driven by cost considerations, with the drawback of a small increase of the machine impedance. The low-impedance collimators are Molybdenum coated Molybdenum Carbide-Graphite composite collimators.

Increasing in the primary collimator gap from 6.7σ to 8.5σ at top energy. To ensure beam stability during the collision adjustment process [8], the impedance of the collimation system is reduced by increasing the collimator gaps. Some key collimator

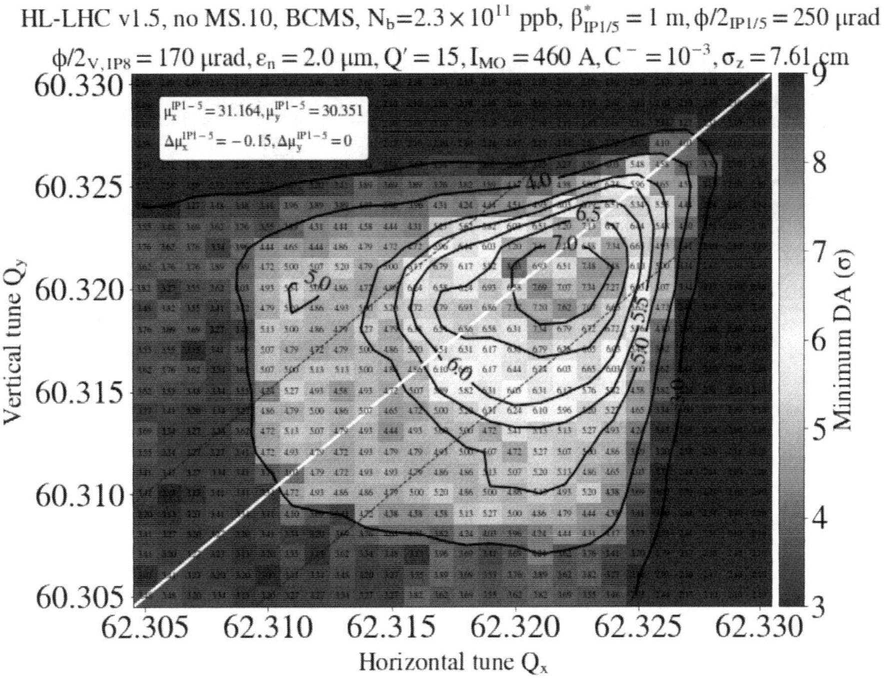

Figure 1. DA including beam-beam interactions at the end of the collision adjustment process versus horizontal and vertical tunes. An optimisation of the betatron phase between IP1 and 5 is applied as $\Delta\phi_x$=−0.15, $\Delta\phi_y$=0. A normalised emittance of 2 µm and bunch intensity of $2.3{\times}10^{11}$ ppb are pessimistically considered to allow for brighter beams.

settings are shown in table 1. The increased gap has the advantage of reducing the halo density at the primary collimator.

Including Hollow Electron Lenses (HEL) in the HL–LHC baseline[1] [9]. The HEL is an advanced tool for active control of the diffusion speed of halo particles, which will serve to mitigate losses from fast processes. Due to resources limitations the HEL will not be ready for Run 4.

Cancelling of the installation of 11T dipoles in LS2 [10]. Due to resources limitations Run 4 will happen without 11 T dipoles and associated new collimators in the IR7 dispersion suppressor [11].

The previous plan for the HL–LHC performance ramp-up allowed an integrated luminosity over 550 fb^{-1} at the end of Run 4. With 160 days of physics, the yearly integrated luminosity is expected to reach 240 fb^{-1}. In the first 1.5 years of Run 4, the bunch intensity is assumed to reproduce that of the Run 3, with minimum β^*=30 cm and full crossing angle of 450 µrad.

Figure 3 shows a schematic view of the Run 4 operational cycle including key beam parameters and expected luminosity. The abrupt jumps in bunch intensity and emittances during the collision adjustment process, just before 2.5 h, correspond to the intensity loss and emittance growth budgets assigned to the interval between injection and the start of collisions. The slow horizontal emittance growth at injection is due to intra-beam scattering (IBS). No

[1] System under review for the new Run 4 scenario

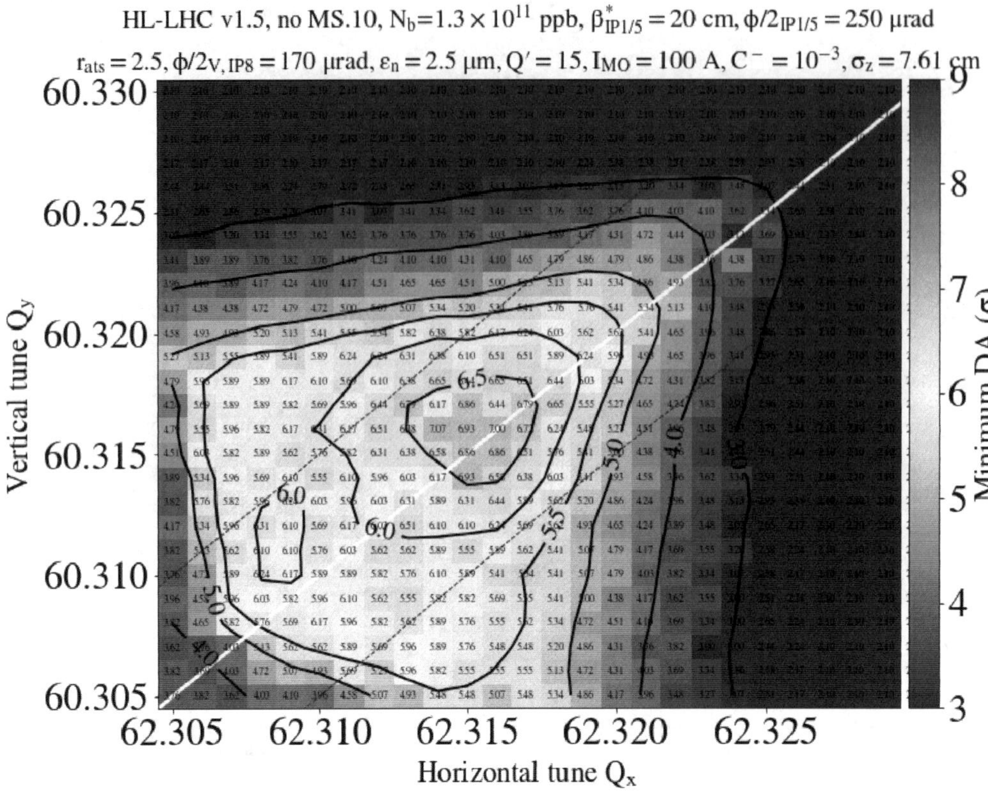

Figure 2. DA at the end of the luminosity levelling versus horizontal and vertical tunes including beam-beam interactions. Target DA is to reach 6σ above the upper blue line [12].

Table 1. Key collimation settings at 7 TeV for an emittance of 2.5 μm and β^{*}=20 cm.

Collimator	Gap $[\sigma]$
TCP/TCPM IR7	8.5
TCSPM/TCSG IR7	10.1
TCLA IR7	13.7
TCP IR3	17.7
TCSG IR3	21.3
TCSP IR6	11.1
TCT H4-V4-H6-V6 IR1&5	13.2
TCDQ IR6	11.1
TCL 4-5-6 IR1&5	16.4

emittance growth from electron cloud, as observed in Run 2, is included in the model. The luminosity starts with a step to $2.5 \times 10^{34}\mathrm{cm}^{-2}\mathrm{s}^{-1}$ followed by a linear ramp to meet cryogenic requirements [13, 14]. The bunch intensity and emittance evolution during physics include burn-off, IBS, synchrotron radiation (SR) damping, and emittance growth from Crab Cavity (CC) noise. Emittance growth from luminosity burn-off has a 1% effect on the integrated luminosity of HL–LHC [15], and it is not included in this report.

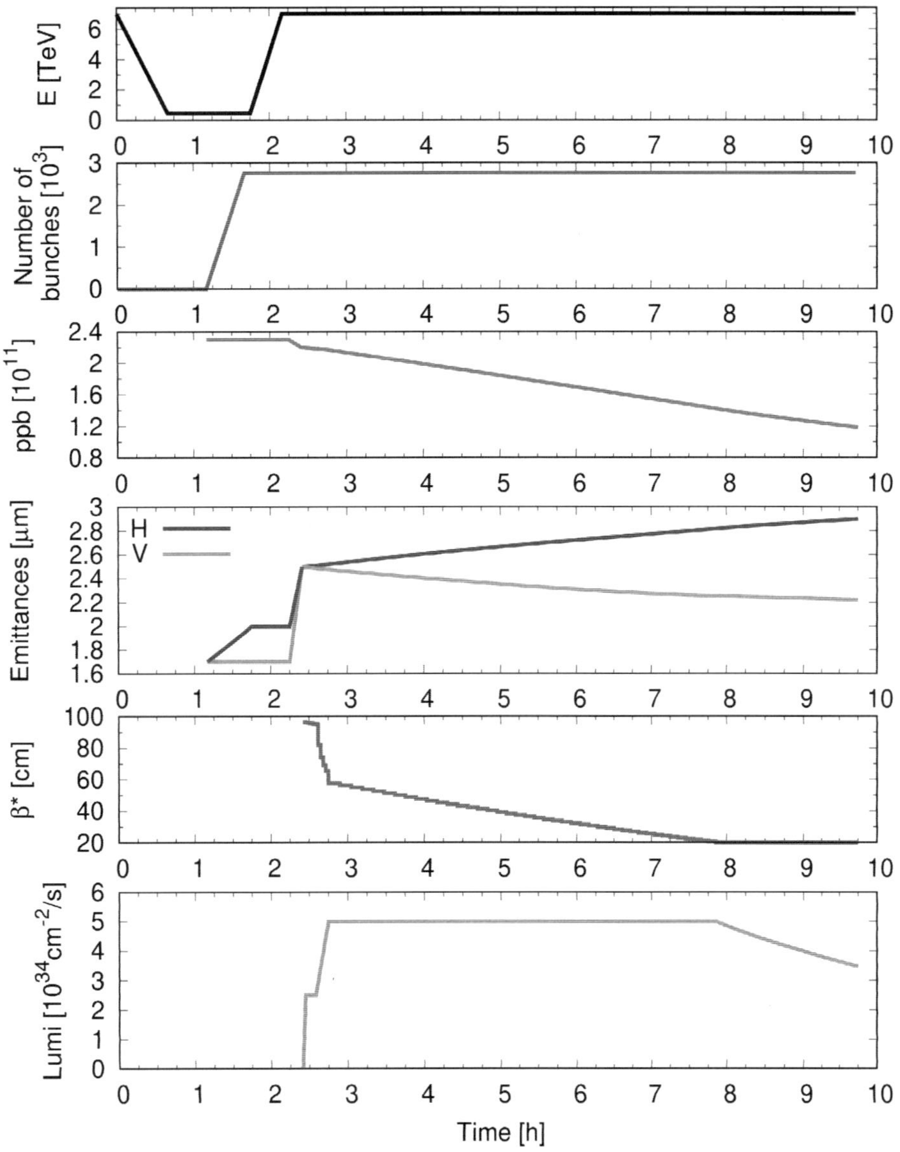

Figure 3. A schematic view of the Run 4 HL–LHC physics cycle showing magnetic cycle, number of bunches, protons per bunch (ppb), transverse emittances (BCMS beam case [12]), and luminosity (top to bottom) versus time until the beam dump.

2. The new Run 4 scenario

Recently, it was decided to extend Run 3 by 1 year and the subsequent long shutdown (LS3) by half a year. This implies a delay of Run 4 by 1.5 years, starting in 2029 with a full-year of operation with about 90 days of physics instead of the 30 days in the previous scenario. Furthermore, it is assumed that Run 4 will last 4 full years (instead of the previous 3.5).

At the moment of writing this report, the hardware configuration of HL–LHC is still being discussed due to the accumulated delays in some components and current political and economic

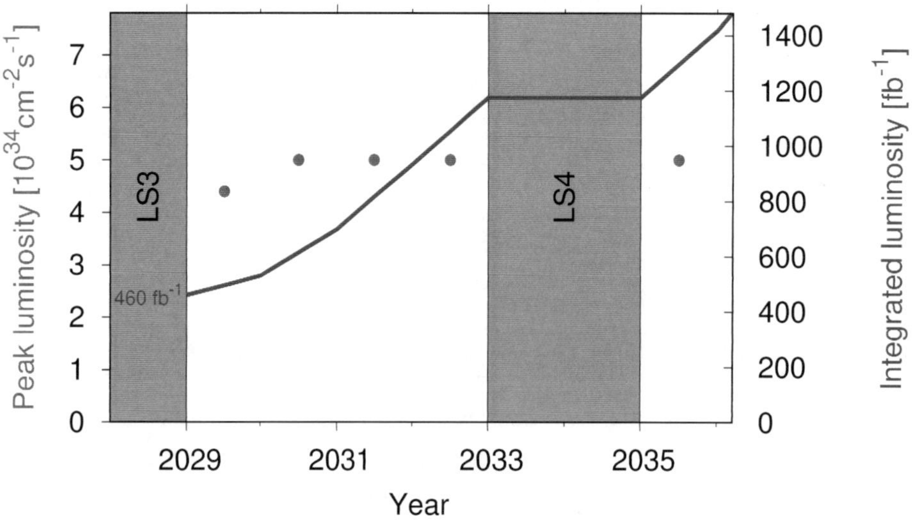

Figure 4. Peak and integrated luminosity during Run 4 (assuming 460 fb^{-1} by the end of Run 3).

uncertainties. In particular, the HEL is unlikely to be ready for the start of Run 4. The increase in collimator gaps by 1.8σ represents an effective mitigation of potential halo issues, since the density of the halo close to the collimator jaws decreases significantly. The potential need for additional mitigation measures in Run 4, for example, to reduce the charge of the bunch, will need to be evaluated using dedicated measurements in Run 3. Beam experiments in Run 3 should also determine if the low impedance collimator upgrade should be carried out in full, or if a reduction in the number of upgraded units would be acceptable.

Assuming no limitations to the beam parameters in Run 4, the potential HL–LHC performance ramp-up is given in figure 4, allowing to integrate luminosity over 715 fb^{-1} during Run 4. This increase compared to the previous Run 4 scenario is a result of the 6 months longer run, the faster intensity ramp-up assumed (1 year instead of 1.5 years) and the slightly larger initial bunch intensity to match the updated Run 3 expectation of 1.8×10^{11} ppb [16]. The minimum β^* in Run 4 is tentatively kept to 20 cm but 15 cm is being considered, however requiring the installation of sextupoles in the dispersion suppressor regions.

3. The first years of operation

The first years of operation of the HL–LHC are expected to take place at a minimum β^*=30 cm, without the use of CCs in physics and with a reduced bunch intensity of 1.8×10^{11} p. This allows reducing the crossing angle to about 450 μrad as validated with the DA simulations shown in figure 5. It is foreseen to steadily reduce the crossing angle during the physics fill as the bunch population decays to maximise performance and reduce the radiation dose to the triplet magnets [17].

The beam-based IR non-linear corrections are expected to require considerable commissioning time and iterations between the different magnet types [18, 19]. Therefore, it is assumed that optics commissioning in the first years will only include magnets up to the octupolar order, leaving the commissioning of the decapolar and dodecapolar correctors for the years with lower β^*. Simulations have confirmed that DA is sufficient at β^*=30 cm without decapolar and dodecapolar IR corrections [20]. Moreover, techniques to speed-up the optics commissioning

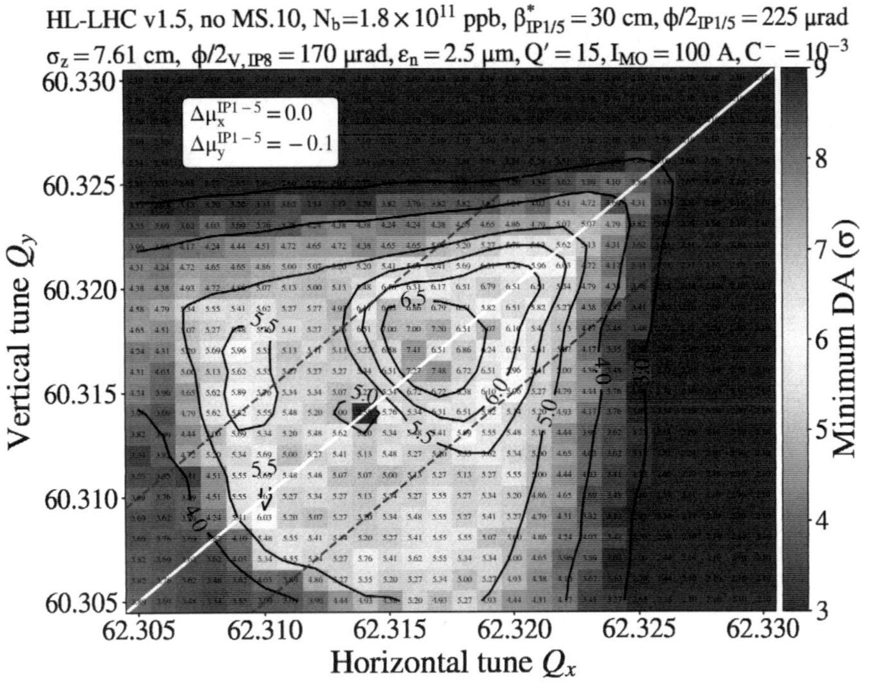

Figure 5. DA after the luminosity ramp-up (β^*=30 cm, full crossing angle of 450 μrad and 1.8×10^{11} ppb) in the first years of operation versus horizontal and vertical tunes including beam-beam interactions and for optimised phase advance between IP1 and IP5.

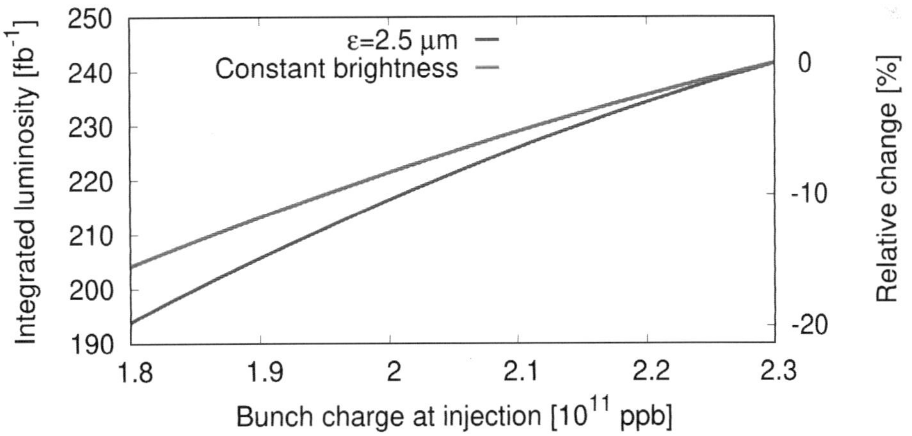

Figure 6. Annual integrated luminosity versus bunch charge at injection in Run 4 (assuming 160 days for physics).

including high-order corrections are being developed [21] and will require dedicated machine experiments in Run 3.

4. Potential intensity limitations

Bunch intensity could be limited in Run 4 due to the absence of the HEL or if RF voltage limitations are encountered at injection. To accommodate the expected larger longitudinal emittance at SPS extraction with 2.3×10^{11} ppb, between 8 and 8.8 MV is required [22], potentially exceeding the RF power capability of the current system with the half-detuning scheme. Acceptable capture losses, injection transients, SPS-LHC energy errors and line-by-line variations of the RF amplifiers may require additional margin. A minimum bunch intensity of 1.8×10^{11} ppb is estimated to be easily achievable. Further studies are ongoing to investigate the maximum bunch intensity feasible with the current RF system. If bunch charge is limited to 1.8×10^{11} ppb at injection, the optimal fill shortens by more than 2 hours and the levelling time by more than 3 hours with respect to the baseline shown in figure 3, and the β^*_{start} is reduced to 45 cm. Annual integrated luminosity (assuming 160 days) is reduced from 242 to 194 fb^{-1}, reducing the expected Run 4 integrated luminosity by about 20% for the case with $\epsilon = 2.5\,\mu\mathrm{m}$. Figure 6 shows the annual integrated luminosity versus bunch charge at injection in the range between 1.8×10^{11} ppb and 2.3×10^{11} ppb. Mitigation measures imply reducing β^* or the crossing angle. The first requires that the MS10 sextupoles are installed before Run 4 to guarantee sufficient lifetime. The latter requires that long-range beam-beam compensators [23], not yet in the baseline, are installed.

5. Summary and outlook

The HL–LHC schedule, hardware configuration and operational scenario are continuously evolving. A new performance ramp-up has been proposed based on the previous scenario to reach the nominal performance of the HL–LHC. Nevertheless, there are important uncertainties in the hardware configuration that need to be clarified before this scenario can be confirmed.

Acknowledgments

Research supported by the HL–LHC project.

References

[1] Arduini G *et al.* 2022 HL–LHC Run 4 proton operational scenario Tech. rep. CERN Geneva CERN-ACC-2022-0001

[2] Métral E *et al.* 2018 Update of the HL–LHC operational scenarios for proton operationTech. rep. CERN Geneva CERN-ACC-NOTE-2018-0002

[3] Tomás R *et al.* 2022 Run 4 performance ramp-up, baseline and alternative scenarios Presented at LHC performance Workshop 2022

[4] Plassard F *et al.* 2021 Sextupole scheme optimization for HL–LHC CERN Geneva CERN-ACC-NOTE-2021-0012

[5] Kostoglou S 2021 Review of the situation without MS10 at the beginning of collisions and at the end leveling Presented at 189th HiLumi WP2 Meeting

[6] Fessia P *et al.* HL–LHC ECR - WP3/WP2/WP15 Deferral of Installation of Lattice Sextupole in Q10 in Points 1 and 5 to LS4 Tech. rep. EDMS No. 2636486 v.0.5, ref. LHC-M-EC-0005 v.0.5

[7] Métral E 2019 Impedance models, operational experience and expected limitations Presented at International review of the LHC collimation system

[8] Buffat X *et al.* 2020 Strategy for Landau damping of head-tail instabilities at top energy in the HL–LHC Tech. rep. CERN Geneva CERN-ACC-NOTE-2020-0059

[9] Bruce R *et al.* Functional Specification HELs - Functional and Operational Conditions Tech. rep. EDMS no. 2514085

[10] HL-LHC: Decision Management Non-installation of 11T in LS2 Tech. rep. EDMS No. 2441244, 2021-01-10

[11] Lindstrom B *et al.* 2022 *Proc. 13th Int. Particle Accelerator Conf. (IPAC'22)* (Bangkok, Thailand) pp 1366–69

[12] Fernandez I Z *et al.* 2020 High-Luminosity Large Hadron Collider (HL–LHC): Technical design report Tech rep. CERN Geneva CERN-2020-010

[13] Claudet S 2019 Request for slow luminosity increase Presented at 160th HiLumi WP2 Meeting

[14] Claudet S 2020 HL–LHC WP9 Cryogenics, Process studies, Coping with Peak Lumi (ramps) Presented at 171st HiLumi WP2 Meeting

[15] Tomás R, Keintzel J and Papadopoulou S 2020 *Phys. Rev. Accel. Beams* **23** 031002

[16] Bernardes A P *et al.* 2022 Updates on LHC Beam Dump Block Readiness for Run3: UD68 Autopsy and HRMT56 Preliminary Results Tech. rep. CERN Geneva LMC #435, Mar 16th

[17] Karatathis N *et al.* 2019 *Proc. 9th LHC Operations Evian Workshop* (Evian Les Bains, France) pp 273–84

[18] Buffat X *et al.* 2022 Optics measurement and correction strategies for HL–LHC Tech. rep. CERN Geneva CERN-ACC-2022-0004

[19] Carlier F *et al.* 2017 Optics Measurement and Correction Challenges for the HL-LHC Tech. rep. CERN Geneva CERN-ACC-2017-0088

[20] Pugnat T 2022 Impact of non-linear correctors on DA for $\beta^* = 30$ cm Presented at Special Joint HiLumi WP2/WP4 Meeting, Apr 5th

[21] Dilly J *et al.* 2022 *Proc. 13th Int. Particle Accelerator Conf. (IPAC'22)* (Bangkok, Thailand) pp 1995–98

[22] Timko H *et al.* 2021 LHC injection loss studies and power limit update Presented at 11th HL-LHC Collaboration Meeting

[23] K. Skoufaris *et al.* 2021 *Phys. Rev. Accel. Beams* **24** 074001

Journal of Physics: Conference Series **2420** (2023) 012004 doi:10.1088/1742-6596/2420/1/012004

Study of hydrodynamic-tunnelling effects induced by high-energy proton beams in graphite

C Wiesner, F Carra, J Don, I Kolthoff, A Lechner, R Rasile and D Wollmann
CERN, CH-1211 Geneva, Switzerland
E-mail: `christoph.wiesner@cern.ch`

Abstract. The design and assessment of machine-protection systems for existing and future high-energy accelerators comprises the study of accidental beam impact on machine elements. In case of a direct impact of a large number of high-energy particle bunches in one location, the damage range in the material is significantly increased due to an effect known as hydrodynamic tunnelling. The effect is caused by the beam-induced reduction of the material density along the beam trajectory, which allows subsequent bunches to penetrate deeper into the target. The assessment of the damage range requires the sequential coupling of an energy-deposition code, like FLUKA, and a hydrodynamic code, like Autodyn. The paper presents the simulations performed for the impact of the nominal LHC beam at 7 TeV on a graphite target. It describes the optimisation of the simulation setup and the required coupling workflow. The resulting energy deposition and the evolution of the target density are discussed.

1. Introduction

In order to evaluate the machine-protection systems for high-energy and high-intensity accelerators, the consequences of beyond-design failures (see p. 445 in [1]) have to be considered. This includes the accidental, direct beam impact in one location. In this scenario, the damage range in the material can be dominated by an effect known as hydrodynamic tunnelling [2, 3]. The effect is caused by the beam-induced reduction of the material density along the beam axis, which allows subsequent bunches to penetrate deeper and deeper into the target. To take into account the changing material densities for the simulation, it is required to sequentially couple an energy-deposition code and a hydrodynamic code. In this generic study, the worst-case scenario of the direct impact of the full nominal Large Hadron Collider (LHC) proton beam at 7 TeV on a graphite target was simulated [4, 5]. The main aim of the study was to establish an efficient coupling workflow based on the FLUKA code [6–8] and the commercially available Autodyn code [9]. To facilitate future comparisons, similar beam parameters and the same equation of state as in a previous study (see Sec. 5.2 in [2], Sec. V in [10] and Chap. 8.3.2 in [11]), which was performed by coupling the codes FLUKA and BIG2 [12], were chosen.

2. Simulation setup and parameters

2.1. Beam parameters

For this study, the nominal LHC beam parameters [1] and a beam size of $\sigma = 0.5\,\text{mm}$ were used. The details of the LHC filling patterns were neglected and a constant bunch spacing of 25 ns was assumed. Table 1 summarizes the main beam parameters.

Table 1. Beam parameters used for the simulations.

Particle species	Protons
Beam Energy	7 TeV
Total number of bunches	2808
Protons per bunch	1.15×10^{11}
Bunch spacing	25 ns
Bunch length	0.5 ns
Filling scheme	None
rms beam size (sigma)	0.5 mm

2.2. Simulation workflow

The simulation workflow for this study consisted of three main steps, which had to be performed in a loop:

(i) Run the FLUKA simulation.
(ii) Import the energy-deposition map into Autodyn and perform the hydrodynamic simulation for a certain number of impacting bunches.
(iii) Update the target densities by interpolating the Autodyn results to the FLUKA regions and create the new FLUKA input file.

The general approach for the coupling simulations is discussed in [13]. For this study, an optimised coupling script [14] was used to update the target geometry of the FLUKA input file. The overall simulation cycle was stopped once it became evident that the propagation speed of the density depletion front in the target reached a near constant value, allowing to extrapolate the total penetration depth after the impact of all the bunches.

2.3. FLUKA setup

The main target and FLUKA simulation parameters are summarised in table 2. The target was modelled as a graphite cylinder with length of 10 m and a radius of 5 cm with axial symmetry. For the initial simulation step, a uniform target density of $2.28\,\mathrm{g\,cm^{-3}}$ was used. Therefore, no segmentation was required and the FLUKA model was implemented with a single region. However, for all subsequent simulation steps, the geometry of the target had to be divided into various regions as only a single density value can be assigned to each FLUKA region.

As a compromise between resolution and FLUKA run-time, a model with 13 400 regions was chosen. This implied a running time of up to seven days using at least 50 000 primary protons per simulation step at typically 25 to 30 CPUs on the available cluster at CERN. This resulted in an acceptable relative statistical error of the energy-deposition density below 1.5 % in the beam-heated region.

To ensure a sufficiently high spatial resolution close to the target axis, where the largest density gradients occurred, a radially adapted region size was chosen. For the 40 innermost regions, a radial bin width of 125 µm was chosen, corresponding to one fourth of the beam sigma of $\sigma = 500\,\mu\mathrm{m}$. This was followed by 20 regions with a 375 µm radial width, 6 regions with a 1.25 mm radial width, and one 30 mm thick verge region to cover the full 5 cm radial target size.

Table 2. Target and FLUKA simulation parameters.

Target length	10 m
Target radius	5 cm
Material	Graphite
Initial target density	2.28 g cm^{-3}
Radial region size	Radially increasing (67 regions in total)
Longitudinal region size	5 cm length (200 regions in total)

2.4. Autodyn setup

Table 3 shows the main Autodyn simulation parameters. The material model used for graphite in the Autodyn simulations [15] consisted in a tabular equation of state (EOS) from the SESAME library. This is the most sophisticated type of EOS, as it can express strong nonlinearities and discontinuities between the different parameters. Since, in this impact scenario, pressures are in the order of GPa, the hydrostatic response of the material, controlled by the EOS, is dominant with respect to the deviatoric component, controlled by strength and failure models. For such reason, no strength/failure model was adopted in the simulation.

Table 3. Autodyn simulation parameters.

Equation of State	Tabular (SESAME) [15]
Strength and failure model	none
Mesh	Eulerian
Longitudinal element size	2.5 cm
Radial element size	125 µm (up to $r = 2$ cm)
Time step	0.5 ns

A Eulerian mesh was used for the model. With respect to a Lagrangian mesh, which was used in previous studies [3], Eulerian methods have the advantage of maintaining a constant element size. This simplifies the FLUKA/Autodyn interaction, and avoids element distortion that, numerically, can lead to high energetic errors, or even to a premature interruption of the simulation. The energy-deposition map per primary proton was loaded from FLUKA into Autodyn and scaled to the required number of impacting protons. Finally, a time step of 0.5 ns was selected in order to be consistent with the bunch length.

3. Results

Table 4 gives an overview of the performed simulation steps for this study. The first step was stopped after the impact of 70 bunches, corresponding to a density change in the most-loaded mesh element of 16 %. For all subsequent steps, a fixed step size of 100 bunches was used, leading to a reduction of the minimum target density by around 50 % per simulation step.

As discussed below and visible from figure 3, the density depletion front at the target axis started to propagate with near constant speed after 4.25 µs, which allowed for stopping the simulations

Table 4. Simulation steps, total number of bunches n_{bun} that have impacted the target at time t_{step}, the resulting peak energy deposition E_{max}, the longitudinal location of the shower maximum $z(E_{\text{max}})$, and the minimum graphite density ρ_{min} in the target.

Step	n_{bun}	t_{step} (μs)	E_{max} (GeV/g/p)	$z(E_{\text{max}})$ (mm)	ρ_{min} (g cm^{-3})
0	0	0	8.4	1350	2.28
1	70	1.75	7.2	1450	1.91
2	170	4.25	4.7	2450	0.96
3	270	6.75	4.1	3175	0.51
4	370	9.25	3.9	3925	0.22
5	470	11.75	3.8	4550	0.11
6	570	14.25	-	-	0.07

after 6 steps and estimating the total damage range by extrapolating to the impact of all 2808 bunches. Note that the simulation loop was stopped after obtaining the Autodyn results for an impact of 570 bunches, and therefore no energy-deposition map is available for Step 6.

For the 6 coupling steps, a total simulation time of around two months was required, dominated by the running time of the FLUKA simulations. This does not count the time for the simulation setup and for post-processing the final results.

3.1. Energy deposition

Figure 1 shows the longitudinal energy density profile close to the target axis as calculated with FLUKA. For the first simulation step with uniform target density, a peak energy-deposition

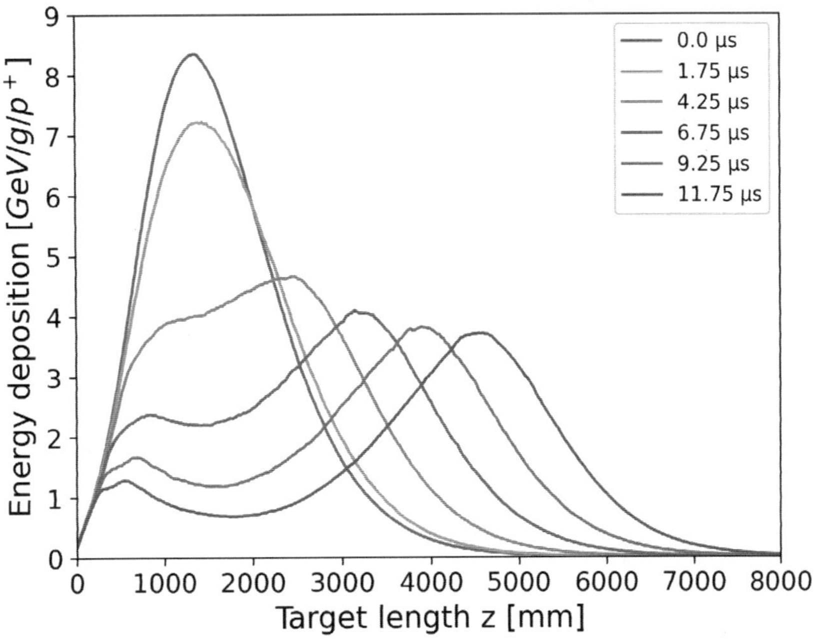

Figure 1. Energy density profiles along the target.

density of 8.4 GeV/g/proton is reached, and the shower maximum at the target axis is located around $z = 1.35\,\mathrm{m}$. Due to the depletion of the target density in the beam-heated region, the peak energy-deposition density is reduced to below $4\,\mathrm{GeV/g/proton}$ after $9.25\,\mathrm{\mu s}$. However, the energy is now deposited much deeper into the target. Consequently, the location of the shower maximum moves downstream and reaches approximately $z = 4.6\,\mathrm{m}$ after the impact of 470 bunches. In addition, a double peak structure becomes visible, which is a consequence of the material-depleted region around 2 m inside the target.

The energy escaping the target volume stayed almost constant during the simulations, changing from 49.4 % after the impact of 70 bunches to 49.7 % after 470 bunches. The simulations showed that a peak pressure of 1.2 GPa was reached after the impact of 170 bunches, while the maximum peak temperature of nearly 8400 K occurred after the impact of 370 bunches [4]. For larger times, the peak pressure decreased because the peak energy-deposition density in the target was reduced due to the hydrodynamic-tunnelling effect.

3.2. Density depletion and damage range

Figure 2 shows the graphite density inside the target after the impact of 570 bunches. The density reduction in the beam-heated region close to the target axis is clearly visible.

Figure 2. Simulated material density inside the first 5 m of the target after the impact of 570 bunches.

Figure 3 shows the evolution of the material density. The minimum density in the beam-heated region decreases from the initial $2.28\,\mathrm{g\,cm^{-3}}$ to less than $1\,\mathrm{g\,cm^{-3}}$ after the impact of 170 bunches ($4.25\,\mathrm{\mu s}$) and further to $0.07\,\mathrm{g\,cm^{-3}}$ after the impact of 570 bunches ($14.25\,\mathrm{\mu s}$).

It can be inferred from the evolution of the density profiles that after $4.25\,\mathrm{\mu s}$ the density depletion front moves along the axis with a nearly constant speed of $25.5 \pm 1.1\,\mathrm{cm\,\mu s^{-1}}$. Here, the speed was averaged over the last four steps. This agrees with the value of $25\,\mathrm{cm\,\mu s^{-1}}$ derived in the previous study [2, p. 77], which used the BIG2 code.

Obtaining the speed of the density depletion front is a key result for the machine-protection assessment because it allows to estimate the total damage range as follows: After $4.25\,\mathrm{\mu s}$, the

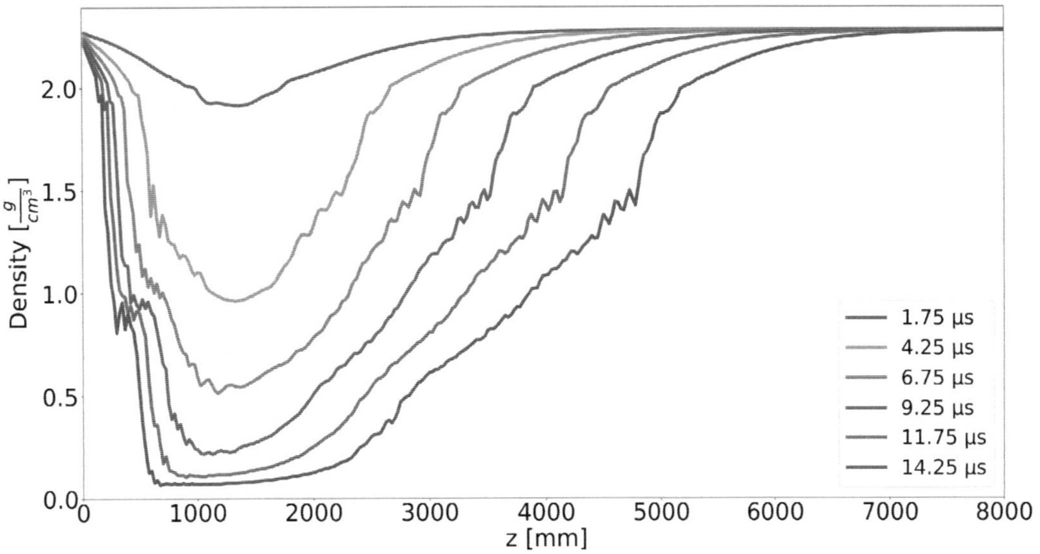

Figure 3. Evolution of the graphite density close to the target axis (i.e. at the innermost mesh line of 125 µm).

density is reduced up to approximately 3.5 m into the target. Afterwards, the depletion front moves with a near constant speed until the total number of 2808 bunches have impacted the target, which corresponds to 70.2 µs, assuming a constant bunch spacing of 25 ns. This gives an additional 16.8 m, and, therefore, a total damage range of the LHC nominal beam in graphite of approximately 20 m.

The result confirms that the full impact of the LHC beam is an unacceptable, beyond-design failure that would lead to significant damage to the installed collimators and absorbers and underlines the necessity for highly redundant and reliable protection and dump systems.

4. Conclusions and outlook

In this study, the spot impact of a proton beam with nominal LHC parameters and a beam size of $\sigma = 0.5$ mm on a cylindrical graphite target was simulated. For this purpose, the energy-deposition code FLUKA was coupled with the hydrodynamic code Autodyn, taking into account the beam-induced density change in the material during the beam impact. In total, 6 simulation steps up to the impact of 570 bunches, corresponding to 14.25 µs, were performed. After 4.25 µs, the density depletion front moved with a near constant speed, allowing to estimate the damage range in graphite for the worst-case scenario of the direct impact of the full nominal LHC beam to be approximately 20 m.

After having successfully established the required coupling workflow, future studies of beyond-design failures using FLUKA and the commercially available code Autodyn are planned. In particular, the effect of the higher intensities of 2.2×10^{11} protons per bunch [16] in the High-Luminosity LHC era on the damage range will be assessed.

Acknowledgments

The authors would like to thank Y. Nie, R. Schmidt and N. Tahir for important discussions and insights about hydrodynamic tunnelling.

References

[1] Brüning O S, Collier P, Lebrun P, Myers S, Ostojic R, Poole J and Proudlock P (eds) 2004 *LHC Design Report, Vol. I The LHC Main Ring* (Geneva, Switzerland: CERN)
[2] Tahir N, Burkart F, Schmidt R, Shutov A and Piriz A 2018 *Nucl. Instrum. Methods Phys. Res., Sect.* B **427** 70–86
[3] Nie Y, Fichera C, Mettler L, Carra F, Schmidt R, Tahir N A, Bertarelli A and Wollmann D 2019 *Phys. Rev. Accel. Beams* **22** 014501
[4] Don J and Wiesner C 2022 MPE Technical Note 2022-01, EDMS Nr. 2739903 (CERN, Geneva, Switzerland)
[5] Rasile R 2022 *Dynamic Response of Advanced Materials impacted by MMOD* Master's thesis La Sapienza University of Rome, Italy, unpublished
[6] Battistoni G *et al.* 2015 *Annals of Nuclear Energy* **82** 10–18
[7] Ahdida C *et al.* 2022 *Front. Phys.* **9**
[8] FLUKA.CERN website `https://fluka.cern/`
[9] Ansys Inc Ansys autodyn `https://www.ansys.com/products/structures/ansys-autodyn`
[10] Tahir N A, Sancho J B, Shutov A, Schmidt R and Piriz A R 2012 *Phys. Rev. Spec. Top. Accel Beams* **15** 051003
[11] Blanco Sancho J 2014 *Machine Protection and High Energy Density States in Matter for High Energy Hadron Accelerators* Ph.D. thesis École Polytechnique Fédérale de Lausanne, Lausanne, Switzerland
[12] Fortov V E, Goel B, Munz C D, Ni A L, Shutov A V and Vorobiev O Y 1996 *Nucl. Sci. Eng.* **123** 169–89
[13] Wiesner C, Carra F, Kruse-Hansen J, Masci M, Nie Y and Wollmann D 2021 *Proc. 12th Int. Particle Accelerator Conf. (IPAC'21)* (Campinas, Brazil) pp 119–122
[14] Kolthoff I 2021 Coupling script for hydrodynamic-tunnelling studies `gitlab.cern.ch/machine-protection/hydrodynamic-tunnelling/coupling-script-lhc-carbon`
[15] Tahir N 2020 Equation of state for nominal graphite Private Comm.
[16] Aberle O *et al.* 2020 *High-Luminosity Large Hadron Collider (HL-LHC): Technical design report* CERN Yellow Reports: Monographs

Beam-based reconstruction of the shielded quench-heater fields for the LHC main dipoles

L Richtmann, L Bortot, E Ravaioli, C Wiesner and D Wollmann
CERN, CH-1211 Geneva, Switzerland
E-mail: Lea.richtmann@aei.mpg.de

Abstract. Small orbit oscillations of the circulating particle beams have been observed immediately following quenches in the LHC's superconducting main dipole magnets. Magnetic fields generated during the discharge into the quench heaters were identified as the cause. Since the resulting, shielded field inside the beam screen cannot be measured in-situ, the time evolution of the field has to be reconstructed from the measured beam excursions.

In this paper, the field-reconstruction method using rotation in normalized phase space and the optimized fitting algorithm are described. The resulting rise times and magnetic field levels are presented for quench events that occurred during regular operation as well as for dedicated beam experiments. Finally, different approaches to model the shielding behavior of the beam screen are discussed.

1. Introduction and motivation

The Large Hadron Collider (LHC) at CERN uses superconducting dipole magnets to guide the particles around their circular trajectory. The energy stored in each of the 8 main dipole circuits, comprising 154 magnets each, is around 1 GJ. Therefore, a sophisticated Quench Protection System (QPS) is installed to avoid damage in case of a resistive transition (quench) in one of the magnet coils [1–3].

An essential part of the QPS are the so-called quench heaters [4] that uniformly heat up the entire superconducting coil after the detection of a quench to avoid a local temperature hot spot. In case of a quench, a nominal current of 75 A [5, p. 174] is discharged through the quench heater strips that are attached to the outer part of each LHC dipole coil. To protect the superconducting coils from beam-induced heating and secondary particles, the beams travel inside a stainless-steel beam screen, which has a co-laminated copper layer with a nominal thickness of 75 µm [5,6].

During Run 2 of the LHC (2015-2018), it was recognized that the magnetic field generated during the discharge of the quench heaters can cause oscillations of the circulating beam [7]. Although this effect is not considered critical for the current LHC operation, the fast development of the orbit perturbation has triggered detailed studies to understand its criticality for High Luminosity (HL)-LHC operation [7,8].

Since the resulting field inside the beam screen cannot be measured in-situ, a method was developed to reconstruct it based on the measured beam excursions [9], which is a crucial input for any shielding model.

Journal of Physics: Conference Series **2420** (2023) 012005 doi:10.1088/1742-6596/2420/1/012005

2. Reconstruction of the quench-heater field

2.1. BPM measurements

In each of the two LHC rings, 516 Beam Position Monitors (BPM) are installed [5, ch. 13]. They measure the beam position averaged over all circulating bunches with a turn-by-turn resolution. Figure 1 shows the beam position measured by one BPM over 50 LHC turns after the firing of the quench heaters in one dipole magnet. The orbit deviation after the quench heater firing, taking effect at turn 14, can clearly be observed. Since the quench heaters produce a horizontal dipole field, the beam is displaced in the vertical plane [7, 8]. In contrast, a reduction of the main dipole field following the quench would displace the beam in the horizontal plane and would occur only on a time scale of tens of milliseconds.

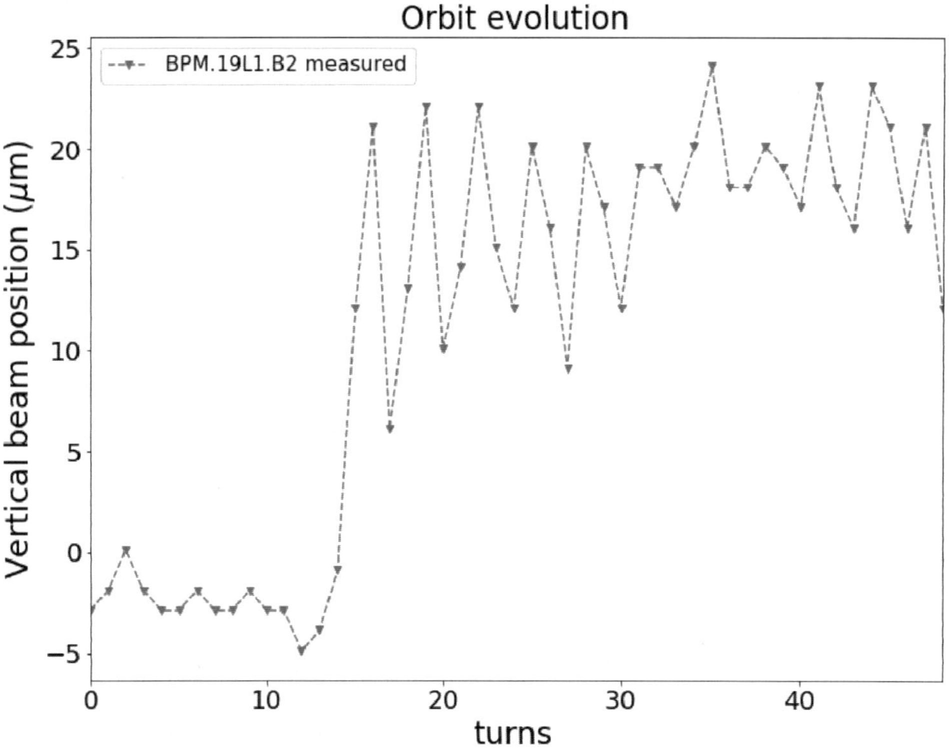

Figure 1. Example of the displacement of Beam 2 measured at one BPM after the firing of the quench heaters (timestamp: 2016-07-12 14:17:13, beam mode: collision, beam energy: 6.5 TeV, 1807 bunches per beam circulating). One LHC turn corresponds to a revolution time of 89 μs.

From the measured beam positions, one can then calculate the change of momentum or angle for each turn, which allows to reconstruct the magnetic field causing this change by using the method described in the following section.

2.2. Field-reconstruction method

For this method [9], it is assumed that the beam is kicked at one single location with a pure dipole field. However, no specific assumption on the time profile of the kick is made, such that this method can also be used for other cases.

In general, the envelope of a particle beam can be described by an ellipse in phase space [10], defined by the position y and the angle y'. By normalizing the position and angle using the twiss

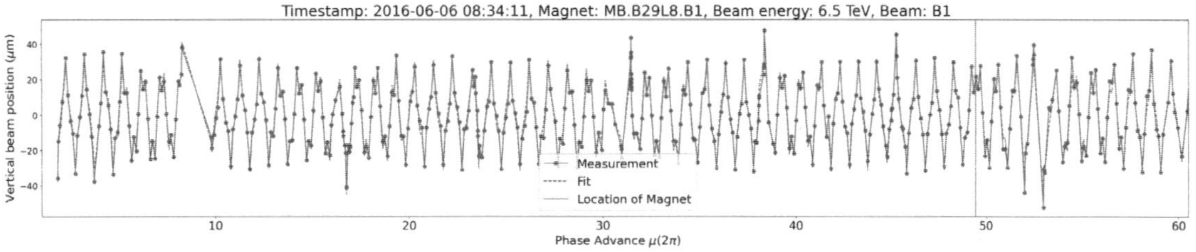

Figure 2. Comparison of the measured (blue) and fitted (red) beam positions around the complete LHC ring five turns after the firing of the quench heaters. The location of the magnet where the kick occurs is indicated with a green line.

parameters [9, p. 6], the ellipse becomes a circle, allowing to calculate the change of the beam coordinates at different locations in the ring by rotation in normalized phase space [10, p. 173].

If a magnetic field provides a kick to the beam at a location M, this causes a change in the momentum or angle. To determine the beam position and angle in the current turn n, one, thus, has to rotate the coordinates from the previous turn $(n-1)$ by the vertical tune Q_y and then add the kick angle that the beam receives in this turn θ^n_{kick} (see Eq. 1).

$$\begin{pmatrix} y^n_{\text{M}} \\ y'^n_{\text{M}} \end{pmatrix} = \begin{pmatrix} \cos(Q_y) & \sin(Q_y) \\ -\sin(Q_y) & \cos(Q_y) \end{pmatrix} \begin{pmatrix} y^{n-1}_{\text{M}} \\ y'^{n-1}_{\text{M}} \end{pmatrix} + \begin{pmatrix} 0 \\ \theta^n_{\text{kick}} \end{pmatrix} \tag{1}$$

To determine the impact of this kick at any other element in the ring, one can rotate the coordinates (y^n_{M}, y'^n_{M}) by the phase advance between the kick location M and the desired element. This means, if position and angle at M from the turn before, as well as the kick angle given in this turn and the phase advance to the desired element are known, one can obtain a function that calculates the position at any desired element, i.e. at the BPMs. This function can then serve as a fitting function.

The beam positions at the BPMs are known from measurements. The phase advance to the kick location and the β function at the BPMs are known from the LHC optics models [11], which are implemented in MAD-X [12]. They are validated with measurements with a remaining β beating below 2 % root mean square (rms) [13]. What remains unknown are the beam position and angle at the magnet and the kick angle θ^n_{kick}. These unknown values are, thus, fitted using the large number of available BPM measurements.

It was, however, observed that a fit leaving all these parameters free led to un-physically large values in position and angle at the magnet. Therefore, an optimised algorithm was developed where the initial values (y, y') at the magnet are fitted for a turn before the kick happens. Afterwards, these coordinates are calculated analytically via rotation by the tune, such that the kick angle remains the only variable to be fitted. This method significantly improved the quality and consistency of the beam-based reconstruction compared to the initial approach.

Figure 2 shows an example of the fitting result. The BPM measurements around the complete ring for one turn are compared with the beam positions as predicted with the fit function, giving an rms deviation of 3 μm.

After obtaining the kick angle, one can transform the coordinates back into un-normalised phase space and calculate the kick field with Eq. 2:

$$B^n_x = \frac{\theta^n_{\text{kick}} \cdot B\rho}{\sqrt{\beta_{\text{M}}} \cdot L_{\text{M}}}, \tag{2}$$

where $B\rho$ is the beam rigidity, β_{M} is the β-function at the magnet, and $L_{\text{M}} = 14.3\,\text{m}$ is the magnetic length [5, p. 164]. As final results, one obtains the time evolution of the field inside the beam screen with a resolution of one turn.

In addition, it was verified that there was no significant impact on the reconstructed field due to the missing synchronisation between the turn-by-turn logging of the BPM data and the firing of the quench heaters [9, p. 29].

3. Field-reconstruction results

Figure 3 shows the reconstructed fields for all known quench events that led to a beam excursion during Run 2 of the LHC at flat-top (triangular markers) and at injection energy (round markers) during proton operation. For easier comparison, the fields are scaled to the same nominal quench heater current of 75 A, and the fields for Beam 2, which have negative sign, are multiplied by (-1).

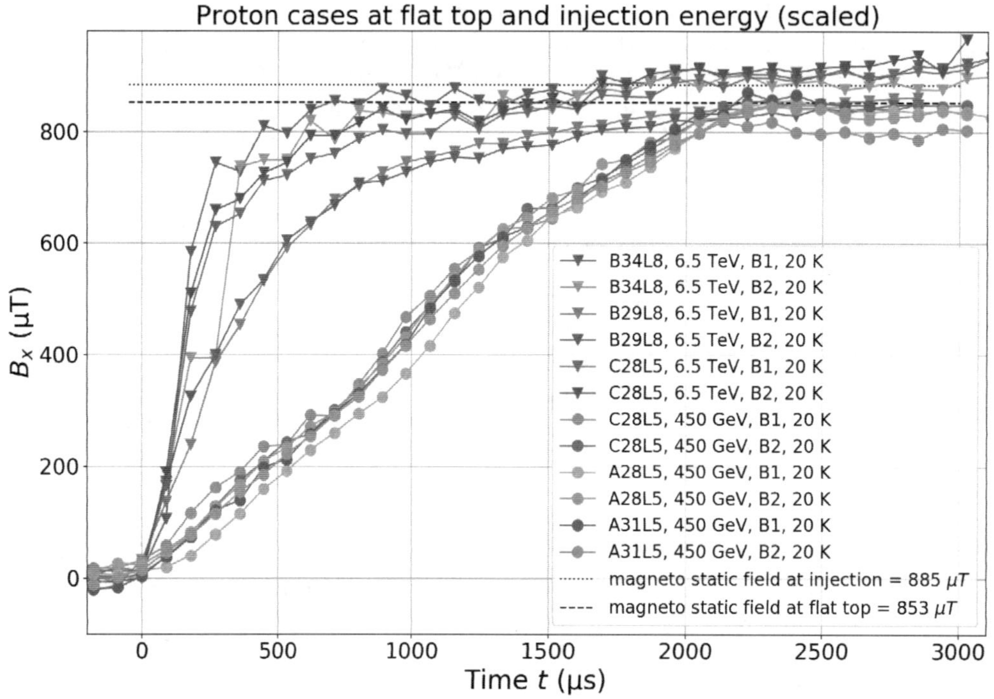

Figure 3. The reconstructed fields for all proton events at flat-top (6.5 TeV) and injection energy (450 GeV). The legend indicates the name of the dipole magnet where the quench heaters fired, the beam energy, the beam (B1/B2) and the temperature of the beam screen.

All reconstructed fields reach within a spread of 10 % the same level. In addition, the field levels agree very well with the expected magneto-static quench heater field (885 μT at injection, 853 μT at flat-top energy), which was independently calculated using the finite-element-method (FEM) code COMSOL [14] including the contribution of the surrounding iron. To quantify the rise times, the reconstructed field $B(t)$ was fitted with a double exponential function:

$$B(t) = a(1 - \exp(-t/\tau_1))(\exp(-t/\tau_2)) \tag{3}$$

with the time constants τ_1 for the first initial rise, τ_2 for the second rise and the amplitude factor a. This ansatz was chosen because the current in the quench heater circuits is described by the same type of function.

For the LHC dipole magnet C28L5, beam-based measurements are available for the following four cases: Case a) quenches during dedicated beam experiments performed at injection energy with the nominal beam screen temperature of 20 K [15], Case b) quench during regular operation at 6.5 TeV with the nominal beam screen temperature, Cases c) and d) quenches during dedicated beam experiments performed with an increased beam screen temperature of 70 K at 3.5 TeV and 6.5 TeV, respectively [16]. Note that the fields for Cases c and d were reconstructed based on the beam positions measured with the ADTObsBox, which allows for a bunch-by-bunch resolution [16]. The reconstructed rise times are shown in table 1. The errors were estimated taking into account the time resolution of the BPMs as well as one standard deviation for the error of the fit algorithm [9, p. 41]

Table 1. Reconstructed rise times τ_1, rise times τ_{sim} simulated with Model III (see below) and nominal copper resistivities $\rho(B,T)$ [17] of the beam screen for four quench events available for dipole C28L5 for Beam 1.

Case	$\rho(B,T)$ ($10^{-10}\,\Omega\,\mathrm{m}$)	τ_1 (µs)	τ_{sim} (µs)
a) 450 GeV, 20 K	2.50	1900 ± 200	2790
b) 6.5 TeV, 20 K	5.28	250 ± 90	1290
c) 3.5 TeV, 70 K	18.20	170 ± 30	380
d) 6.5 TeV, 70 K	19.33	140 ± 30	360

As expected, the reconstructed field rise inside the beam screen is much slower than the rise of the external quench heater field, which has a time constant of $\tau_1 = 7.5\,\mu\mathrm{s}$. It can be observed that the field rises faster at flat-top than at injection energy. This is to be expected because the shielding efficiency of the beam screen depends on its copper resistivity, which changes with the main dipole field (or beam energy) due to magnetoresistivity and with temperature.

4. Shielding models and outlook

It was noticed that the fields reconstructed from the beam position measurements rise faster than expected from FEM simulations. Therefore, several additional models were leveraged to simulate and cross-check the shielding behaviour of the beam screen [9]:

- Model I: A model that calculates analytically the eddy currents for a simplified geometry and takes explicitly the skin effect into account [18].
- Model II: A model of the shielding as a magnetic diffusion process, allowing for an analytical calculation of its time constant.
- Model III: The shielding was calculated with the script used for Model I [18], but using input fields that oscillate at fixed frequencies. From the field attenuation, the transfer function was derived, allowing for a calculation of the time constant by determining the cutoff frequency of the low-pass filter [9, p. 77].

All models predict an increased rise time of the field inside the beam screen when compared to the external quench heater field. It was found that all three models give consistent predictions of the rise time [9]. However, even though the simulated rise times agree better with the measurements than in previous studies, the models still predict more shielding than observed experimentally. Table 1 compares the reconstructed rise times with the ones simulated with

Model III. Compared to the reconstruction, the simulated time constants are 1.5 times longer for Case a, about 2 times longer for Cases c and d, but 5 times longer for the quench events during regular LHC operation (Case b) [9].

As the models yield consistent results among them and as there is no indication that the input parameters are more than 20 % off, it was concluded that the models are incomplete. In particular, all models used so far were 2D models that do not take into account how the eddy current loops close. 3D effects could thus be responsible for the observed discrepancy and should be included in the future.

5. Conclusion

A fully developed method to reconstruct the magnetic fields that are responsible for beam excursions observed after the firing of the quench heaters at the LHC main dipole magnets is presented. The method is based on the measured BPM data and uses rotation in normalized phase space. Since no assumption about the time evolution of the kick is made, this method can be applied generally.

The fields were reconstructed for different quench events. The resulting field levels agree within 10 % between the events and with magnetostatic calculations. Simulations using different 2D shielding models predict rise times that are a factor 1.5 to 5 longer than the ones from the beam-based reconstruction but are consistent among the models.

Acknowledgments

This research was supported by the HL-LHC project. The authors would like to thank M. Morrone for important input and discussions about the beam screen, J. Wenninger for support with the BPM data and M. Maciejewski for help with querying data from the Post Mortem System.

References

[1] Wilson M 1987 *Superconducting Magnets* (Clarendon Press)
[2] Schmidt R, Assmann R, Carlier E, Dehning B, Denz R, Goddard B, Holzer E B, Kain V, Puccio B, Todd B, Uythoven J, Wenninger J and Zerlauth M 2006 *New Journal of Physics* **8** 290
[3] Denz R, Dahlerup-Petersen K and Mess K H 2008 *Proc. 11th Eur. Particle Accelerator Conf. (EPAC'08)* (Genoa, Italy) pp 2422–24
[4] Sonnemann F and Rodriguez-Mateos F 2001 *Proc. Particle Accelerator Conf. (PAC'01)* (Chicago, IL USA) pp 3451–53
[5] Brüning O S, Collier P, Lebrun P, Myers S, Ostojic R, Poole J and Proudlock P (eds) 2004 *LHC Design Report, Vol. I The LHC Main Ring* (Geneva, Switzerland: CERN)
[6] Morrone M, Martino M, De Maria R, Fitterer M and Garion C 2019 *Phys. Rev. Accel. Beams* **22** 013501
[7] Valette M *et al.* 2018 *Proc. 9th Int. Particle Accelerator Conf. (IPAC'18)* (Vancouver, Canada) pp 3115–18
[8] Lindstrom B, Bélanger P, Bortot L, Denz R, Mentink M, Ravaioli E, Mateos F R, Schmidt R, Uythoven J, Valette M, Verweij A, Wiesner C, Wollmann D and Zerlauth M 2020 *Phys. Rev. Accel. Beams* **23** 081001
[9] Richtmann L 2021 *Effect of Quench Heater Discharges on the Circulating Beam in the Large Hadron Collider* Master's thesis Leibniz Universität Hannover, Germany
[10] Wiedemann H 2015 *Particle Accelerator Physics* (New York: Springer International Publishing)
[11] LHC Optics Web: LHC Run II pp physics optics `http://abpdata.web.cern.ch/abpdata/lhc_optics_web/www/opt2018/`
[12] MAD-X, `http://madx.web.cern.ch/madx/`
[13] Persson T, Carlier F, de Portugal J C, Valdivieso A G T, Langner A, Maclean E H, Malina L, Skowronski P, Salvant B, Tomás R and Bonilla A C G 2017 *Phys. Rev. Accel. Beams* **20** 061002
[14] Comsol multiphysics® `www.comsol.com`
[15] Valette M, Wollmann D, Lindstrom B H F, Bortot L, Fernandez Navarro A, Schmidt R and Verweij A 2018 Rep. CERN-ACC-2018-0007 (CERN, Geneva, Switzerland)
[16] Wiesner C, Bortot L, Lindström B, Mentink M, Ravaioli E, Richtmann L, Valette M and Wollmann D 2020 Rep. CERN-ACC-NOTE-2020-0003 (CERN, Geneva, Switzerland)

[17] Simon N, Drexler E and Reed R 1992 Properties of copper and copper alloys at cryogenic temperatures Tech. Rep. MONO-177 National Inst. of Standards and Technology (MSEL) Boulder, CO (United States)
[18] Ravaioli E and Wiesner C 2021 MPE Technical Note 2021-01, EDMS Nr. 2470088 (CERN, Geneva, Switzerland)

Journal of Physics: Conference Series **2420** (2023) 012006 doi:10.1088/1742-6596/2420/1/012006

A design of ILC E-driven positron source

M Kuriki[1], H Tajino[1], S Konno[1], Z Liptak[1], T Takahashi[1], M Fukuda[2], T Omori[2], Y Seimiya[2], J Urakawa[2], K Yokoya[2] and S Kashiwagi[3]

[1] Hiroshima U. ADSE, Higashihirosima, Japan
[2] KEK, Tsukuba, Japan
[3] Tohoku U. ELPH, Sendai, Japan

E-mail: mkuriki@hiroshima-u.ac.jp

Abstract. ILC is an electron-positron linear collider based on Superconducting linear accelerator. Linear collider is the only solution to realize high energy electron-positron collision beyond the limit of synchrotron radiation energy loss by ring colliders. Beam current of injector of linear colliders is much larger than that of ring colliders because the beam is not reusable. Providing an enough amount of particles, especially positron is a technical issue. In this article, we present a design of electron driven positron source for ILC. After optimizations, the system design is established with an enough technical margin, e.g. avoiding potential damage on the production target.

1. Introduction

ILC is an e+e- linear collider with center of mass energy 250 GeV - 1000 TeV [1]. It employs Super-conducting accelerator to boost up the beam up to the designed energy. The beam is accelerated in a macro pulse with 1300 bunches by 5 Hz repetition. The bunch charge is 3.2 nC resulting the average beam current 21 μA. This is a technical challenge, because the amount of positron per second is 40 times larger than that in SLC [2], which was the first linear collider.

The configuration of the positron source is schematically shown in figure 1. The positron generated by electron beam is captured and boosted up to 5 GeV by two linacs. In the E-driven ILC positron source, the drive beam energy is 3.0 GeV and the target is 16 mm thick W-Re alloy rotating with 5 m/s tangential speed. FC (Flux Concentrator) generates a strong magnetic field to compensate the transverse momentum. 36 1.3 m L-band Standing Wave (SW) cavities with 0.5 Tesla solenoid field are placed for positron capture. This section is called as the capture linac. At the downstream, a chicane is placed to removes electrons. The positron booster is composed from 2.0 m L-and and 2.0 m S-band Traveling Wave (TW) cavities. ECS (Energy Compression Section) is composed from 3.0 m L-band TW cavities with chicane.

In E-driven ILC positron source, positrons are generated in a multi-bunch format as shown in figure 2. One RF pulse contains 66 bunches with 80 ns gap. To generate 1312 bunches for positrons of one RF pulse in the main linac, this pulse is repeated 20 times in 64 ms to mitigate the potential damage on the production target. The number of positrons in one RF pulse is 20 times less than those in one RF pulse in the main linac, the instantaneous heat load on the target is also 20 times less [3].

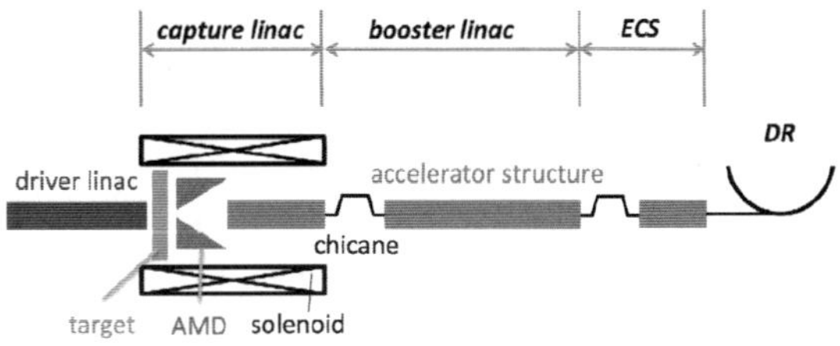

Figure 1. Configuration of E-driven ILC positron source is schematically shown.

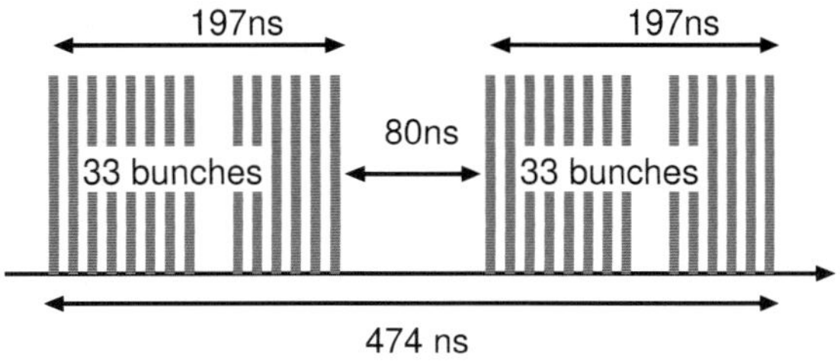

Figure 2. The beam structure in the positron source. Each mini-train contains 33 bunches.

A first simulation was performed by T. Omori [3] only for the capture linac. A simulation with the tracking down to DR (Damping Ring) was made by Y. Seimiya [4], but no beam loading effect was accounted. A new simulation with the beam-loading effect was done by Kuriki and Nagoshi [5, 6]. For those simulations, the peak energy deposition density on the target is kept less than 35 J/g [7], which is considered to be a practical limit of the safety operation.

To obtain uniform intensity positrons over the pulse, the transient variation of the acceleration field by the beam loading has to be compensated so that positrons are accelerated uniformly. Compensation for the transient beam loading by Phase Modulation (PM) on the input RF was proposed by Urakawa [8, 9]. The detail study of the compensation is discussed in Ref. [10, 11]. In this article, we present the electron driven ILC positron source accounting this effect.

2. Subsystems

2.1. Electron driver

3 GeV electron driver is composed from 2600 MHz (S-band) normal conducting TW cavity. The cavity is originally designed for ATF(Accelerator Test Facility) at KEK [12] in 2856 MHz and the parameters are scaled to 2600 MHz. 4 cavities are driven by two klystrons. Accounting 10 % power loss in RF wave guide, the input power to one cavity is 36 MW. The shunt impedance is 57.2 MΩ/m with L=3.228 m and the attenuation τ is 0.57 resulting 0.91 μs filling time. The beam loading compensation was performed by amplitude modulation (AM) [11]. The acceleration voltage per cavity is 31.9 ± 0.4 MV with 0.65 A beam loading current I_B with 36 MW input power [11]. The first two cavities are driven by two klystrons, so that the input power per cavity is 72 MW resulting 43.5 ± 0.6 MV with the same I_B. The electron driver is composed from two cavities as the injector and 100 cavities as the normal section. The total beam energy is 3.3 GeV which has 10% margin.

AM and PM are employed to compensate the beam loading. These modulation on the input RF is made by mixing two RF inputs with PM on the low level RF. The detail is expressed in Ref. [11].

Journal of Physics: Conference Series **2420** (2023) 012006 doi:10.1088/1742-6596/2420/1/012006

2.2. Target and magnetic focusing

Target is 16 mm tungsten-rhenium alloy. To avoid overlapping of the pulse on the target, it is rotated with 5 m/s tangential speed. At the down stream of the target, a flux concentrator as a magnetic focusing to suppress the transverse momentum is placed. The detail of those arrangements is presented in Ref. [6]. The loading on the target is evaluated by PEDD (Peak Energy Deposition Density in J/g); the W-Re destruction limit is 76 J/g [13], and the safe operating threshold is considered to be 35 J/g [6]. The positron production yield is 1.17 as evaluated in 3, the electron bunch charge is 4.1 nC. PEDD for 4.8 nC electron bunch is 33.6 J/g [6], giving 28.7 J/g for 4.1 nC. This is even lower the threshold.

2.3. Capture linac

The capture linac is composed from 36 APS (Alternate Periodic Structure) $\pi/2$ mode SW cavity. The structure is 1.3 m with 21 cells. The shunt impedance is 31.5 MΩ/m. The purpose of the capture linac is to confine the generated positrons in a RF bucket. An effective way for the capturing is the deceleration capture method where positrons are placed on the deceleration phase [14]. The positron is moving to the acceleration phase by phase slipping gradually and the positron is fixed at the acceleration phase. Figure 3 shows the captured positron distribution in the longitudinal phase space z and $\delta = (\gamma - \bar{\gamma})/\bar{\gamma}$ where γ and $\bar{\gamma}$ are Lorentz factor and the average. The positrons after acceleration are distributed along the RF curvature.

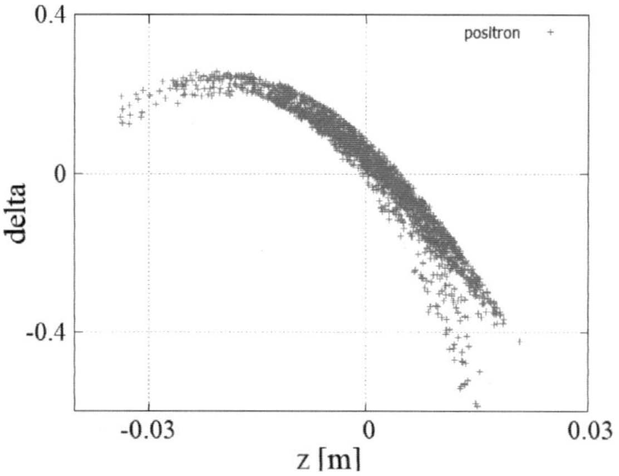

Figure 3. Longitudinal phase space ($z - \delta$) distribution of positron at the end of the capture linac.

The acceleration voltages induced by the input RF and the beam (beam loading voltage) are off-phase with a finite angle, θ in the capture linac. Due to the phase, the beam loading compensation by adjusting the timing doesn't work, because the cancellation between the growth of voltage by RF and voltage by beam is impossible. Instead of the timing adjustment, we employ PM on the input RF. General voltage evolution of SW cavity V is given as

$$\frac{dV(t)}{dt} = -\frac{(1+\beta)\omega}{2Q}\left[V(t) - V_{BL}u(t-t_b) - V_{RF}u(t)\right] \tag{1}$$

where β is coupling to the cavity, Q is Q value, ω is angular frequency, V_{RF} and V_{BL} are the asymptotic voltages by RF and beam, $u(t)$ is a step function. The RF feed is started at $t = 0$ and the acceleration is started at $t = t_b$. Be careful those voltages are complex numbers in general. To obtain an uniform acceleration, the following condition has to be satisfied,

$$V(t_b) - V_{BL} - V_{RF} = 0. \tag{2}$$

If V_{BL} and V_{RF} are in different phases, this condition is never satisfied because $V(t_b)$ and V_{RF} are in phase. PM on V_{RF} is necessary to achieve this as,

$$V_{RF'} = V(t_b) - V_{BL} \tag{3}$$

where $V_{RF'}$ is the asymptotic value of the input RF after PM. V_{t_b} is minimized when $\theta = \pi$ which corresponds to crest acceleration. The maximum beam loading current is expected to be 2.0 A according to a simulation. By assuming 22.5 MW RF input and $\beta = 5$, $V_{RF} = 22.6$ MV and $V_{BL} = 13.7$ MV are expected giving 8.9 MV as the lowest acceleration voltage for each cavity. The total energy after the capture cavity is at least 320 MeV which has an enough margin comparing to the design energy 250 MeV.

2.4. Chicane and booster linac

After the capture linac, a chicane is placed to remove electrons. As a sub-effect, the bunch length is shortened by the momentum compaction factor. The chicane parameter is adjusted as the bunch length after the chicane is minimized as shown in figure 4.

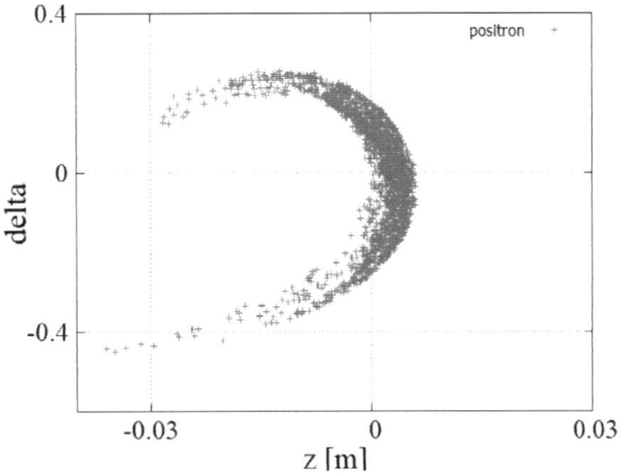

Figure 4. Longitudinal phase space $(z - \delta)$ distribution of positron after the chicane.

The positron booster is composed from L-band [15] and S-band [16] Traveling Wave (TW) cavities. The parameters are summarized in table 1.

Table 1. Parameters of L-band and S-band TW cavities in the booster.

Parameter	L-band	S-band	unit
Shunt impedance	46.5	55.1	MΩ/m
Length	2.00	1.96	m
Aperture (2a)	35.0 39.4	24.3-20.3	mm
Attenuation τ	0.261	0.333	
Filling time t_f	1.28	0.55	μs

Beam loading compensation method is different from that for the capture linac. It is compensated with AM on the input RF [10]. According to Ref. [10], a perfect compensation is possible with two components AM, i.e. AM with a step pulse and a trapezoidal pulse. In this design study, we employ one component AM with a step pulse. E_0 is the initial cavity field, and E_1 is the amplitude of AM. E_1 is determined as

$$E_1 = \frac{rI}{2}\left(\frac{-\omega/Qt_p e^{-2\tau}}{1-e^{-\omega/Qt_p}}+1\right), \tag{4}$$

where t_p is the pulse length. The cavity voltage is same at $t = t_f$ and $t = t_f + t_p$. The initial field E_0 is determined as

$$P_{max} = \frac{L}{r(1-e^{-2\tau})}(E_0 + E_1)^2 \tag{5}$$

where P_{max} is the maximum input RF power.

By assuming 4.8 nC bunch charge, the beam loading current in the booster is 0.78 A. The acceleration voltage per cavity is evaluated as 16.5±0.1 MV for L-band cavity and 29.2±0.6 MV for S-band cavity. Those voltages are evaluated with P_{max} is 22.5 MW for L-band and 36 MW for S-band including 10% power loss by wave guide. Table 2 summarizes the lattice configuration of the booster based on the design in Ref. [4]. As the energy at the booster entrance, 250 MeV is assumed.

Table 2. Lattice configuration of the booster. 4Q+1L means the lattice is composed from four quadrupoles and one L-band cavity, etc. Energy gain is the acceleration energy in the section and Energy is the beam energy at the end of the section in MeV.

Lattice	unit #	# of cavity	Energy gain	Energy
4Q+1L	3	12	198	448
4Q+2L	15	60	990	1438
4Q+4L	18	72	1188	2626
4Q+4S	25	100	2860	5486

2.5. Energy compression section

The purpose of ECS to compress the energy spread of positron after the booster, down to ±0.75%, within DR (damping ring) acceptance. ECS is composed from three chicane (18.6m, 55.8 m in total) and 4 L-band 3m TW cavities (4Q+4L lattice, 22.4 m) driven by 4 klystrons. Other two L-band 3m TW cavities (2L, 6.4 m) driven by two klystrons are set for the beam loading compensation.

3. Positron yield evaluation

The positron yield is defined as the number of positrons obtained in DR dynamic aperture (acceptance) per electrons on the production target. The DR acceptance is [1]

$$\gamma A_x + \gamma A_y < 0.07 \text{ m}, \tag{6}$$

$$\left(\frac{z}{0.035\ m}\right)^2 + \left(\frac{\delta}{0.0075}\right)^2 < 1.0. \tag{7}$$

Figure 5 shows the longitudinal phase space distribution after ECS with DR acceptance (solid circle). Figure 6 shows positron yield as a function of R_{65} of ECS. R_{56} is optimized for each R_{65}. Yield is 1.17 and uniform in region of $-1.2 < R_{65} < -1$.

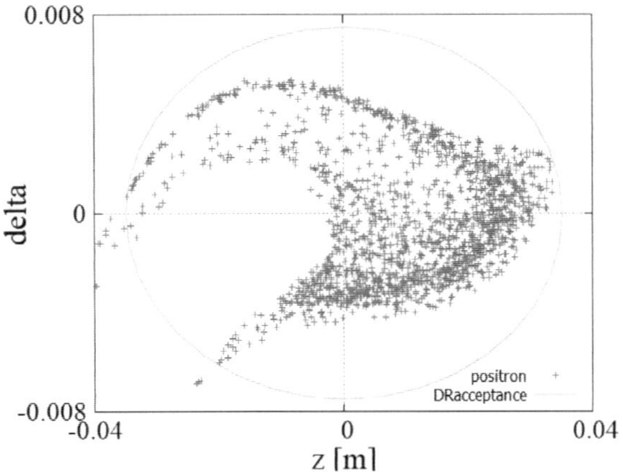

Figure 5. Longitudinal phase space ($z - \delta$) distribution of positrons after ECS section. R_{56} of ECS is -1.04. The solid circle shows DR acceptance.

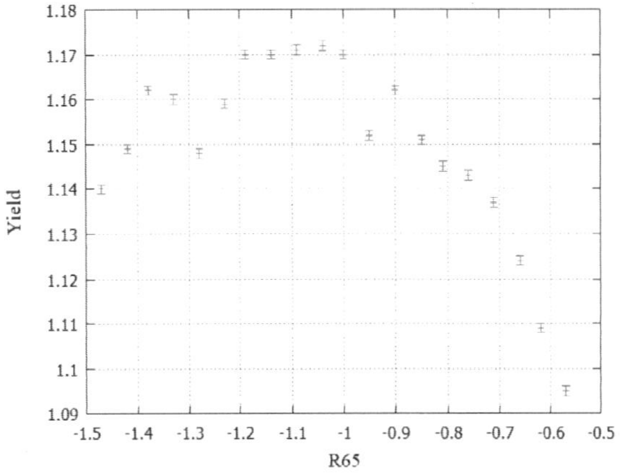

Figure 6. Positron yield as a function of R_{65} of ECS. R_{56} is optimized for each R_{65}. Yield is uniform $-1.2 < R_{65} < -1$.

4. Summary

According to detail studies of the beam loading compensation for SW and TW cavities, the acceleration field is determined. Based on those parameters, the number of RF units and the lattice are fixed and the ILC E-driven positron source design is established. The evaluated positron yield was 1.17 giving the electron intensity on the target satisfying the safety condition.

Acknowledgments

This work is partly supported by Grant-in-Aid for Scientific Research (B) and US-Japan Science and Technology Cooperation Program in High Energy Physics.

References

[1] ILC collaboration 2013 ILC Technical Design Report Tech. rep. KEK-Report
[2] SLC Design Report Tech. rep. SLAC-R-714, 1984
[3] Omori T *et al.* 2012 *Nucl. Instrum. Methods Phys. Res., Sect.* A **672** 52
[4] Seimiya Y *et al.* 2015 *Prog. Theor. Exp. Phys.* **2015** 103G01
[5] Kuriki M *et al.* 2016 in *Proc. 28th Linear Accelerator Conf. (LINAC'16)* (East Lansing, MI USA) pp 430–33

[6] Nagoshi H and Kuriki M *et al.* 2019 *Nucl. Instrum. Methods Phys. Res., Sect.* A **953** 163134
[7] Takahashi T 2018 *Proc. of Int. Workshop on Future Linear Colliders (LCWS2018)* (Arlington, Texas USA)
[8] Kuriki M *et al.* 2021 *Proc. 12th Int. Particle Accelerator Conf. (IPAC'21)* (Campinas, Brazil) pp 1368–71
[9] Kuriki M *et al.* 2021 *Proc. 18th annual meeting of Particle Accelerator Society of Japan (PASJ2021)* (Takasaki, Japan) pp 315–19
[10] Kuriki M 2018 *Proc. 9th Int. Particle Accelerator Conf. (IPAC'18)* (Vancouver, Canada) pp 307–10
[11] Kuriki M *et al.* 2022 *Proc. 13th Int. Particle Accelerator Conf. (IPAC'22)* (Bangkok, Thailand) pp 1893–96
[12] ATF Design and study report Tech. rep. KEK Internal 95-4, 1995
[13] Ecklund S 1981 Positron target materials tests Tech. rep. SLAC-CN-128
[14] James M *et al.* 1991 *Nucl. Instrum. Methods Phys. Res., Sect.* A **307** 207
[15] Saito K *et al.* 2010 *Proc. 7th annual meeting of Particle Accelerator Society of Japan (PASJ2010)* (Himeji, Japan) pp 532–36
[16] Matsumoto S, Higo T, Kakihara K, Kamitani T and Tanaka M 2014 *Proc. 5th Int. Particle Accelerator Conf. (IPAC'14)* (Dresden, Germany) pp 3872–74

Beam loading compensation of standing wave linac with off-crest acceleration

M Kuriki[1], S Konno[1], H Tajino[1], Z Liptak[1], T Takahashi[1], M Fukuda[2], T Omori[2], Y Seimiya[2], J Urakawa[2], K Yokoya[2] and S Kashiwagi[3]

[1] Hiroshima U. ADSE, Higashihiroshima, Japan
[2] KEK, Tsukuba, Japan
[3] Tohoku U. ELPH, Sendai, Japan

E-mail: mkuriki@hiroshima-u.ac.jp

Abstract. In E-Driven positron source of ILC, the generated positron is captured by a standing wave cavity. Because the deceleration capture method is employed, the positron is off-crest over the linac. Because the beam-loading is expected to be more than 1A in a multi-bunch format, the compensation is essential to obtain uniform intensity over the pulse. A conventional method for the compensation controlling the timing doesn't work because RF and Beam induced field are in different phase. In this manuscript, we discuss the compensation with the off-crest acceleration case. A simple phase modulation on the input RF is a solution.

ILC is an e+e- linear collider with CME 250 GeV - 1000 TeV [1]. It employs Super-conducting accelerator (SCA) to boost up the beam up to the designed energy. The beam is accelerated in a macro pulse with 1300 bunches by 5 Hz repetition. The bunch charge is 3.2 nC resulting the average beam current 21 μA. This is a technical challenge, because the amount of positron per second is more than 40 times larger than that in SLC [2].

1. Introduction

In the ILC positron source, the generated positron is captured in a RF bucket by a linac based on a standing wave (SW) linac with APS (Alternate Periodic Structure) cavity. The capture linac is composed from 36 1.3 m L-band APS cavities with 0.5 Tesla solenoid field.

In E-Driven ILC positron source, positrons are generated in a multi-bunch format as shown in figure 1. It contains 66 bunches with 80 ns gap. To generate 1312 bunches for positrons in one RF pulse in the main linac, the positron generation is repeated 20 times in 64 ms. Because the positron is generated over 64 ms, the instantaneous heat load on the target is much suppressed [3, 4].

The generated positron has a large spread in both longitudinal and transverse momentum space. Capturing the positron in an RF bucket for further acceleration is the role of the capture linac. Deceleration capture was proposed by M. James et al. [5] for better capture efficiency. In this method, the positrons are placed on a deceleration phase initially and move to the acceleration phase by phase slipping.

This deceleration capture cause a difficulty on the beam loading compensation, because the cavity field induced by the beam (beam loading field) and the cavity field by RF (acceleration

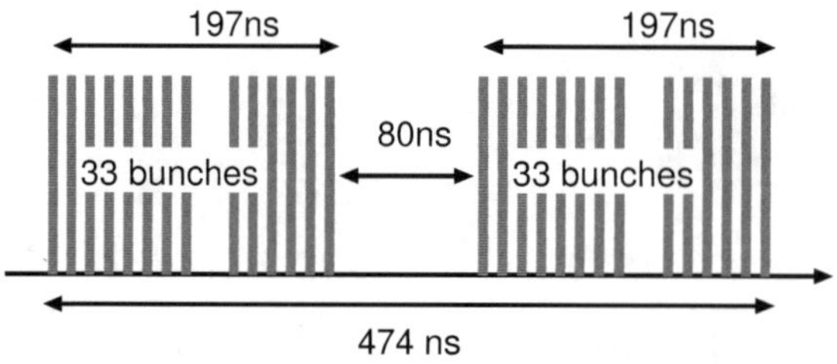

Figure 1. The beam structure in the positron source. Each mini-train contains 33 bunches. Each pulses contain 2 or 1 mini-trains.

field) has a finite angle. Because the conventional theory of the compensation assumes the crest acceleration, the conventional method doesn't work in this case.

2. Beam loading compensation with a standing wave linac

The acceleration voltage by a standing wave RF accelerator with the beam loading is

$$\begin{aligned} V(t) &= \frac{2\sqrt{\beta P r L}}{1+\beta}\left(1-e^{-\frac{t}{\tau}}\right) \\ &-\frac{IrL}{1+\beta}\left(1-e^{-\frac{t-t_b}{\tau}}\right)e^{\imath\theta} \end{aligned} \tag{1}$$

where β is coupling beta, P is input RF power, r is shunt impedance, L is structure length, $\tau = 2Q/\omega/(1+\beta)$, I is beam loading current, t_b is timing to start the beam acceleration, and θ is relative phase of the beam center to the RF. Here, we omit the RF oscillation term, $e^{\imath\omega t}$. If $\theta = 0$ and t_b is adjusted properly as $t_b = -\ln\left(\frac{I}{2}\sqrt{\frac{rL}{\beta P}}\right)$, $V(t)$ can be a constant.

If θ is a finite value, there is no solution in this framework. To compensate the voltage variation by the beam term, RF term should contain imaginary part, i.e. phase modulation (PM) is introduced. Figure 2 shows the phase diagram of the beam loading compensation with PM. V_{RF} is the amplitude of the asymptotic value of the cavity field by the input RF, ϕ is PM angle on the input RF, V_B is the asymptotic value of the beam loading voltage, θ is the beam phase, V_C is the cavity voltage when the beam acceleration is started. If the sum of $V_{RF}e^{\imath\phi}$ and $V_Be^{\imath\theta}$ is equal to V_C, the cavity field is kept as a constant at V_C.

$$V_{RF}e^{\imath\phi} + V_Be^{\imath\theta} = V_C \tag{2}$$

From this condition, ϕ is determined as

$$\phi = \sin^{-1}\left(-\frac{V_B}{V_{RF}}\sin\theta\right) \tag{3}$$

and V_C is

$$V_C = \sqrt{V_{RF}^2 + V_B^2\sin^2\theta} + V_B\cos\theta \tag{4}$$

t_b is obtained as

$$t_B = -\tau\ln\left(1 - \frac{V_C}{V_{RF}}\right) \tag{5}$$

If $\theta > \pi/2$, the solution is different, because V_B gives acceleration. In this case, V_C is

$$V_C = V_{RF}e^{\imath\phi} + V_Be^{\imath\theta} \tag{6}$$

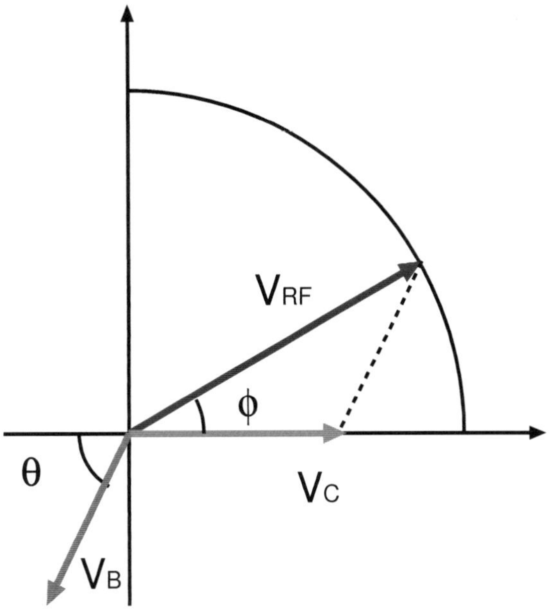

Figure 2. Phase diagram of V_{RF} (blue arrow), V_B (red arrow), and V_C. PM angle ϕ should determined as the $V_{RF}e^{\imath\phi} + V_Be^{\imath\theta} = V_C$.

and is larger than V_{RF}, but it is impossible. Instead, we change the RF input power as V_{RF1} after the acceleration is started. In this case

$$V_C = V_{RF1}e^{\imath\phi} + V_Be^{\imath\theta} \leq V_{RF} \tag{7}$$

if we take the equal as the maximum acceleration,

$$V_C = V_{RF}, \tag{8}$$

and

$$V_{RF1}^2 = V_{RF}^2 + V_B^2 - 2V_BV_{RF}\cos\theta. \tag{9}$$

The PM angle ϕ is

$$\phi = \sin^{-1}\left(-\frac{V_B}{V_{RF1}}\sin\theta\right). \tag{10}$$

3. Beam loading compensation at the pulse gap

ILC pulse has a gap in a pulse and the duration is 80 ns as shown in figure 1. In the gap, the balance between the RF term and the beam term isn't maintained because the beam term amplitude becomes zero. We want keep V_C without V_B in the gap. The solution is very simple, the input RF power should be modulated as giving V_C as the asymptotic value. The input power P_1 should be

$$P_1 = \frac{(1+\beta)^2(\sqrt{V_{RF}^2 - V_B^2\cos^2\theta} + V_B\sin^2\theta)^2}{4\beta rL} \tag{11}$$

At the gap end, amplitude and phase of the input RF should be back to values according to equations (2) and (3), respectively. The beam voltage and the counter part by the input RF are growing again, but they are cancelled to each other. This cancellation is maintained if the input RF is switched in a same manner.

4. Input RF modulation

The cavity voltage variation including the pulse gap is perfectly compensated by the input RF phase and amplitude modulation as demonstrated in the previous section. Here, we consider how the modulation is implemented.

A direct method to implement AM and PM on the input RF is AM and PM on the input RF signal to klystron. This method is not ideal because the response (amplification) is not linear and changing the input RF signal amplitude causes a beat-wave.

Instead of the direct modulation, we consider a combination of two input RF signals with a constant RF amplitude. The phase of each input RF signals are modulated. If the phase modulation for the two klystrons are same sign, the combined RF is

$$V_{RF}e^{\imath\phi} + V_{RF}e^{\imath\phi} = 2V_{RF}e^{\imath\phi}. \tag{12}$$

This is a phase modulation with ϕ. If the sign are opposite,

$$V_{RF}e^{\imath\phi} + V_{RF}e^{-\imath\phi} = 2V_{RF}\cos\phi. \tag{13}$$

This is an amplitude modulation with $\cos\phi$. If we want the phase and amplitude modulation simultaneously, the modulation should be

$$V_{RF}e^{\imath(\phi_1+\phi_2)} + V_{RF}e^{\imath(\phi_1-\phi_2)} = 2V_{RF}e^{\imath\phi_1}\cos\phi_2. \tag{14}$$

resulting phase modulation with ϕ_1 and amplitude modulation with $\cos\phi_2$.

If the phase modulation to RF is instantaneous, the above conclusion is realized. In reality, however, the klystron is driven by a cavity, which has a finite time constant; even if the phase of the RF input is modulated, the modulation of the klystron output appears with a delay. To evaluate the effect of the delay of PM, we perform the numerical calculation. RF voltage (asymptotic value) $\tilde{V_{RF}}$ in the cavity by PM with no delay is expressed as

$$\tilde{V_{RF}} = V_{RF}\left\{u(t) - u(t-t_b)\right\} + V_{RF}u(t-t_b)e^{\imath\phi}, \tag{15}$$

where $u(t)$ is the step function, t_b is time to start the beam acceleration.

If there is a delay with τ as the time constant, it becomes as

$$\begin{aligned} \tilde{V_{RF}} \;=\;& V_{RF}\left\{u(t) - u(t-t_b)\right\} \\ &+V_{RF}e^{-\frac{t-t_b}{\tau}}u(t-t_b) \\ &+V_{RF}u(t-t_b)e^{\imath\phi}\left(1-e^{-\frac{t-t_b}{\tau}}\right). \end{aligned} \tag{16}$$

It shows the RF input is gradually modulated.

We calculate the cavity response with this modulation. As the klystron Q value, we assume 2000. It can be compared with Q value of APS cavity, 25000. Klystron response is 12 times faster than that of APS cavity. The time constant τ is 0.24μs in this case. This Q value of klystron is quite ordinal and further optimization for a faster response is possible, but we assume this value anyway. The cavity voltage is calculated using a coupled pendulum model developed by T. Shintake [6].

Figures 3 and 4 show the cavity voltage evolution with zero delay of the klystron output. Figures 3 and 4 show the real part and the imaginary part, respectively. We start the beam acceleration at the filling time (kink in figure 3), the real part is kept as a constant. The input RF power is 22.5 MW and the beam current is 1 A. The beam angle θ is $\pi/6$.

Figures 5 and 6 show the cavity voltage evolution with delay of the klystron output according to equation (16). Figures 5 and 6 show the real part and the imaginary part, respectively. The conditions are same as those for figures 3 and 4 case. The real part is almost identical, but the imaginary part (figure 6) has a small dip. The amplitude is -0.15 MV. This value should be compared to the real part amplitude, 16 MV. The dip of the imaginary component cause the phase fluctuation of 9.4 mrad, whose impact on the acceleration is limited.

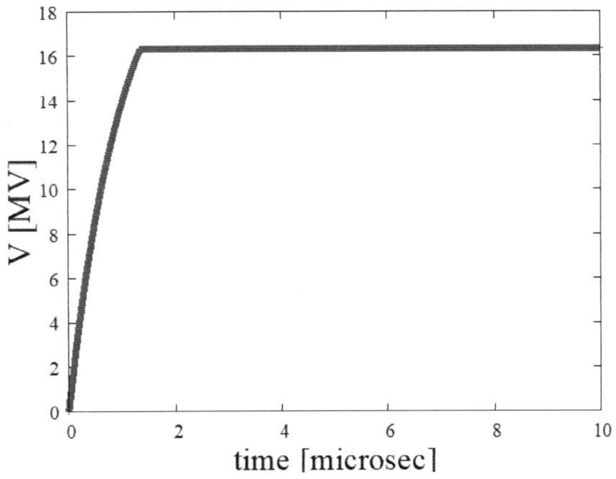

Figure 3. Real part of the cavity voltage evolution with 22.5 MW RF power with 1 A beam loading current starting at the filling time. PM with no delay is applied.

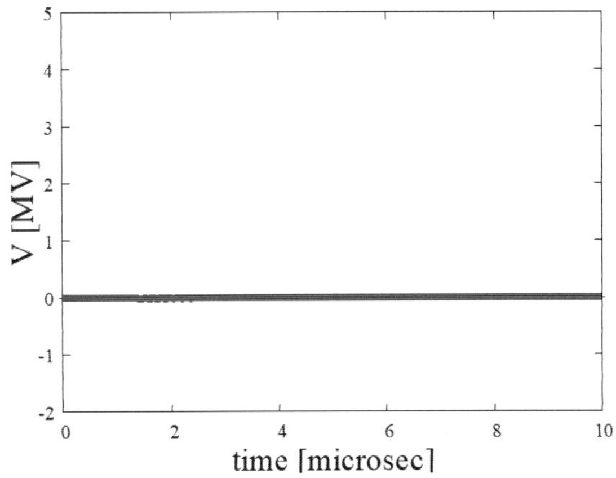

Figure 4. Imaginary part of the cavity voltage evolution with 22.5 MW RF power with 1 A beam loading current starting at the filling time. PM with no delay is applied.

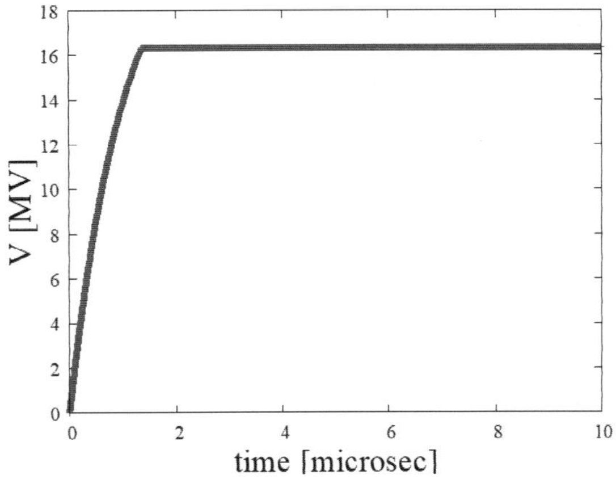

Figure 5. Real part of the cavity voltage evolution with 22.5 MW RF power with 1 A beam loading current starting at the filling time. PM with delay is applied.

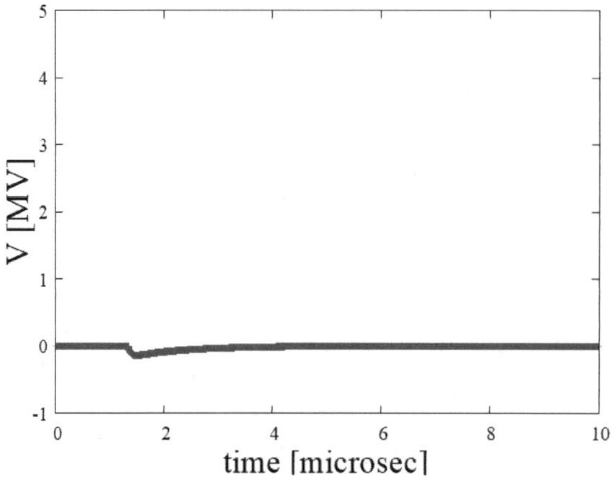

Figure 6. Imaginary part of the cavity voltage evolution with 22.5 MW RF power with 1 A beam loading current starting at the filling time. PM with delay is applied.

5. Conclusion

We consider the beam loading compensation of the positron capture linac for ILC E-Driven positron source. Due to the heavy beam loading, its compensation is essential to obtain an uniform intensity positron pulse. We consider PM and AM on the input RF to compensate the beam loading with a finite beam angle comparing to RF input. It works well including the gap in a pulse. AM and PM can be implemented by combining two RF with PM on the input signal. It was confirmed that the effect of the finite time response of the klystron output to the input low level RF PM was very limited.

Acknowledgments

This work is partly supported by Grant-in-Aid for Scientific Research (B).

References

[1] ILC Technical Design Report Tech. rep. KEK-Report 2013-1(2013)
[2] SLC Design Report Tech. rep. SLAC-R-714(1984)
[3] Omori T *et al.* 2012 *Nucl. Instrum. Methods Phys. Res., Sect.* A **672** 52
[4] Seimiya Y *et al.* 2015 *Prog. Theor. Exp. Phys.* **2015** 103G01
[5] James M *et al.* 1991 *Nucl. Instrum. Methods Phys. Res., Sect.* A **307** 207
[6] Shintake T 1996 *Proc. of Joint US-CERN-Japan International School*,ISBN981-02-3838-X(1996)

Summary of numerical noise studies for Electron-Ion Collider strong-strong beam-beam simulation

Y Luo[1], D Xu[1], Y Hao[2], J Qiang[3], J S Berg[1], M Blaskiewicz[1], W Fischer[1], R Gamage[5], X Gu[1], H Huang[5], G Hoffstaetter[4], J Kewisch[1], H Lovelace III[1], C Montag[1],V Morozov[5], E Nisson[5], S Peggs[1], V Ptitsyn[1], T Satogata[5] and F Willeke[1]

[1]Brookhaven National Laboratory, Upton, NY, USA
[2]Facility for Rare Isotope Beams, Michigan State University, East Lansing, MI, USA
[3]Lawrence Berkeley National Laboratory, Berkeley, CA, USA
[4]Cornell University, Ithaca, NY, USA
[5]Thomas Jefferson National Accelerator Facility, Newport News, VI, USA

E-mail: `yluo@bnl.gov`

Abstract. The Electron-Ion Collider (EIC) presently under construction at Brookhaven National Laboratory will collide polarized high energy electron beams with hadron beams with design luminosities up to $1\times10^{34}\text{cm}^{-2}\text{s}^{-1}$ in center mass energy range of 20-140 GeV. We studied the planned electron-proton collisions using a Particle-In-Cell (PIC) based Poisson solver in strong-strong beam-beam simulation. We observed a much larger proton emittance growth rate than in weak-strong simulation. To understand the numerical noise and its impact on strong-strong simulation results, we carried out extensive studies to identify all possible causes for artificial emittance growth and quantify their contributions. In this article, we summarize our study activities and findings. This work will help us better understand the simulated emittance growth and the limits of the PIC based strong-strong beam-beam simulation.

1. Introduction

The Electron-Ion Collider (EIC) presently under construction at Brookhaven National Laboratory will collide polarized high energy electron beams with hadron beams with design liminosities up to $1 \times 10^{34}\text{cm}^{-2}\text{s}^{-1}$ [1]. Both weak-strong and strong-strong models are used for the EIC beam-beam interaction simulation studies [2, 3]. For the weak-strong model, the electron bunch is assumed as a rigid 6-d Gaussian charge distribution. The proton bunch is represented with macro-particles. The beam-beam kick to proton macro-particles are calculated analytically. For the strong-strong model, each bunch is represented with typically 0.5 to 1 million macro-particles. Particle-in-cell (PIC) method and Fast Fourier Transformation (FFT) are used to solve 2-d Poisson equation on rectangle grids. The charge of each macro-particle is deposited onto nearest 9 grids and the beam-beam kick to each macro-particle is interpolated from the potentials on those 9 nearest grids.

In the following, we present simulation results for the collision between 275 GeV protons and 10 GeV electrons [4]. At this collision mode, the horizontal and vertical beam sizes are 95 μm and 8.5 μm at IP. The beam-beam parameter is 0.012 for the protons and 0.1 for the electrons, both reach their highest values in the EIC. We pay more attention on the proton bunch's emittance growth, especially in the vertical plane. We normally extrapolate beam size growth rate from a linear fitting in the tracking turns to %/hour. From strong-strong simulation, the simulated proton's vertical beam size growth rate is normally larger than 1000%/hour. However, from weak-strong simulation, it is less than 5%/hour.

Strong-strong simulation is subject to numerical noises due to limited macro-particle numbers, transverse grids, longitudinal slices, and the algorithm itself. To better understand the simulated growth rates and the numerical noises in the EIC strong-strong beam-beam simulation, we carried out extensive studies to identify all possible causes for the artificial emittance growth and quantify their contributions.

2. Converging study

Analytically, the numerical noise introduced artificial emittance growth rate in the strong-strong beam-beam simulation is inversely proportional to the number of macro-particles and proportional to the square of the beam-beam parameter. Figure 1 shows one example of simulated proton's vertical beam size growth rate versus the number of macro-particles of electron bunch. Increasing the macro-particles of electron bunch will reduce the simulated proton's emittance growth rate. If fitting the simulated growth rates with a function a/M_e^p, where M_e is the number of macro-particles of electron bunch, we found p is close 0.5 instead of 1 as predicted. We also found that fitting with a function $a/M_e + b$ can better match the simulation results. For the CDR design parameters, b is about 300%/hour.

Figure 2 shows the simulated proton beam size growth rates as function of proton's beam-beam parameter. In this study, we adjusted proton's beam-beam parameter by scaling down electron beam's bunch intensity while keeping the tunes of proton bunch center unchanged. The simulated proton's beam size growth rates can be fitted well with a function $a \times \xi_{bb}^q$, with q about 2.75 for both planes. With a lower beam-beam parameter about 0.008, we re-scanned the growth rate's dependence on the electron bunch's macro-particle number and obtained b between 50-100%/hour.

Sufficient number of longitudinal slices of electron bunch in strong-strong simulation is also important to reduce the simulated proton beam size growth rate. For EIC, due to electron beam's large beam-beam parameter and near-integer tunes, its vertical beam size vaies dramatically near IP.

We normally use 128×128 transverse grids for PIC Poisson solver. With 256×256 grids, the simulated proton growth rates will be smaller too. However, with increased grids, we need to increase the number of macro-particles at the same time and the tracking time will be significantly increased.

3. Turn-by-turn dipole motion

In strong-strong model, the distribution of macro-particles varies turn by turn (TBT). With typically 0.5 to 1 million macro-particles per bunch, the RMS variations of electron bunch center and transverse beam sizes are about 0.2-0.3% of their RMS beam sizes at IP.

In the spectrum of proton bunch's center motion, the electron's imprints at electron's tunes are clearly seen. The amplitudes of electron tune peaks scale with $1/\sqrt{M_e}$. The 'noise' of other frequencies in the spectrum also drops with increased macro-particle number but does not scale as $1/\sqrt{M_e}$.

The impact of dipole moment of electron bunch can be eliminated by introducing virtual symmetric macro-particles in the PIC Poisson solver. For example, for one macro-particle at (x, y), we can introduce virtual macro-particles at $(-x, -y)$, or at $(x, -y)$ and $(-x, -y)$. Strong-strong simulation shows that this approach can eliminate the dipole moment and moderately reduce the proton's beam size growth rate. The growth rates with this method are comparable

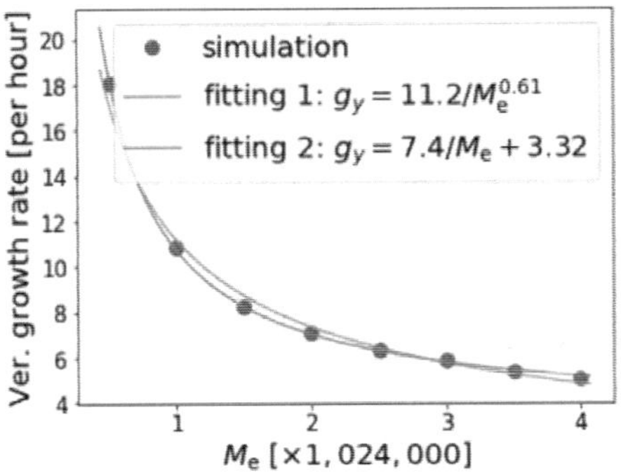

Figure 1. Simulated growth rate of proton vertical beam size versus number of macro-particles of electron bunch.

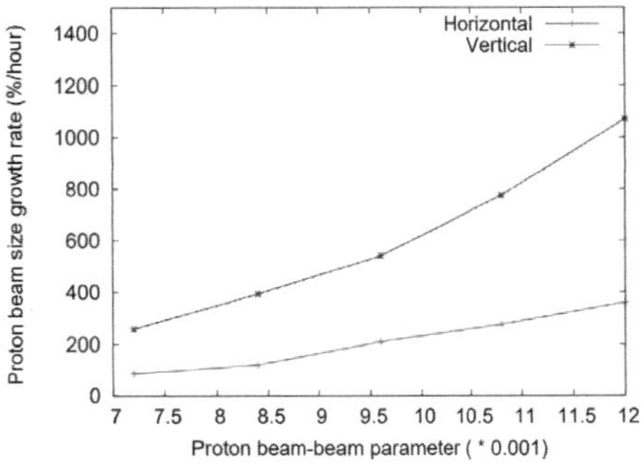

Figure 2. Simulated growth rate of proton transverse beam size versus proton's beam-beam parameter.

to directly increasing electron bunch's macro-particle number by a same amount of virtual macro-particles.

Knowing the levels of RMS TBT variations in electron bunch's center position and beam sizes, we could estimate their contributions by introducing the same levels of artificial random noises into weak-strong simulation. Figure 3 shows the simulated proton's vertical beam size growth rates with different levels of random noises. With a typical relative 0.3% TBT variations in the weak-strong simulation, the proton vertical beam size growth rate due to the noises is between 100-400%/hour.

From above studies, we learned that the TBT variation in the electron bunch's center position can generate artificial proton beam size growth. However, it can not explain majority of the simulated proton beam size growth we observed in the strong-strong simulation which is typically more than 1000%/hour.

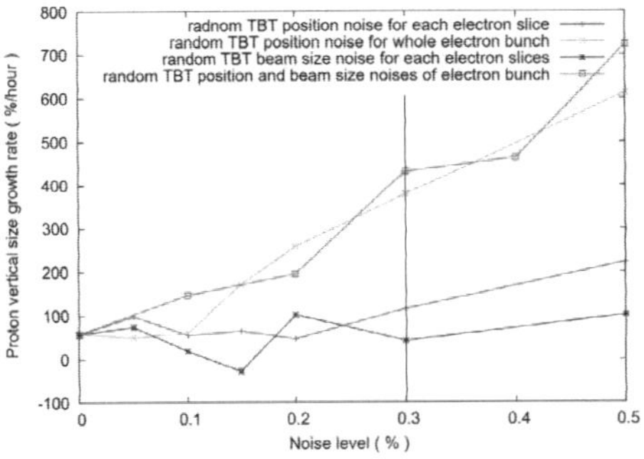

Figure 3. Simulated growth rates from weak-strong simulation with TBT noises added to the electron bunch's centroid and beam sizes.

4. Contribution to growth

To identify the sources of proton beam size growth in the strong-strong simulation, we study the dependence of beam size growth rate on the amplitudes of macro-particles [5]. We group macro-particles according to their initial longitudinal action and transverse action in units of σ. We use calculated geometric beam size growth rate as an observable to measure stability of each group of macro-particles.

Figure 4 shows the proton's vertical beam size growth rate as function of longitudinal amplitude of macro-particles. Three models are used. Frozen model is an extended weak-strong simulation after a strong-strong simulation with electron's charge distribution frozen. From the plot, the vertical beam size growth rate increases with the longitudinal action for all models. However, the strong-strong model gives a much faster growth rate than other two models and its beam size growth starts at a very small longitudinal action.

Figure 5 shows the vertical beam size growth rate as function of transverse amplitude of macro-particles. From the plot, the growth rates from strong-strong model are much larger than other two models. Strangely, the macro-particles with a transverse amplitude less than 1.5σ has a very big growth rate in the strong-strong model, which is not seen in the weak-strong models. This means that macro-particles in the transverse bunch core contribute most of the artificial emittance growth rate in the strong-strong beam-beam simulation.

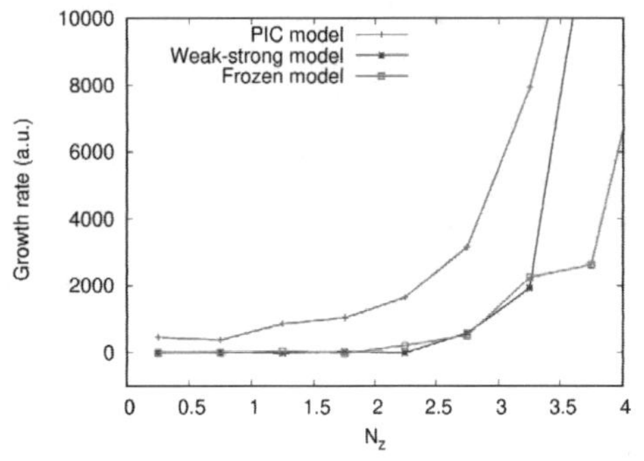

Figure 4. Comparison of growth rates of macro-particles versus their longitudinal actions for three models.

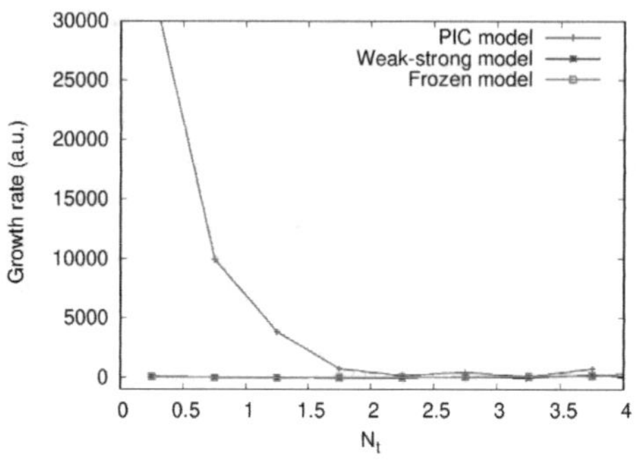

Figure 5. Comparison of growth rates of macro-particles versus their transverse actions for three models.

5. Noise from PIC solver

In the strong-strong simulation, the beam-beam kick is calculated through PIC based Poisson solver. Here we calculate RMS variation of beam-beam kick for a 4-D Gaussian macro-particle distribution [6]. As we know, the emittance growth rate of macro-particles are proportional to the square of the variation in the beam-beam kick. In the following, we use 1000 distributions of macro-particles with known beam sizes to estimate the variation in beam-beam kick in PIC method.

Both round and flat beams are used for comparison. The round beam's transverse sizes are (77 µm, 77 µm), which is close to the RHIC case. The flat beam's transverse sizes are (77 µm, 7.7 µm), which is close to the EIC case. Figure 6 shows the RMS variation of vertical beam-beam kick for both beams. Horizontal axis is radial amplitude in σ. From the plot, the variation of beam-beam kick is larger for small transverse amplitudes for both cases. However, the variation is about 2-3 times larger for the flat beam than the round beam.

Figure 7 shows the beam-beam kick variation on the horizontal and vertical axes for the flat beam. The horizontal axis is x/σ_x or y/σ_y in units of σ. From the plot, the beam-beam kick variation is always larger in the vertical direction than in the horizontal direction. And the variation in the vertical direction stays at a high level even for large amplitudes.

From these studies, we learn that PIC based Poisson solver tends to generate a larger variation (noise) in the beam-beam kick for macro-particles in the bunch core, especially in the vertical plane for a flat beam. This explained why we observed much faster beam size growth rates for macro-particles in the bunch core in the strong-strong simulation.

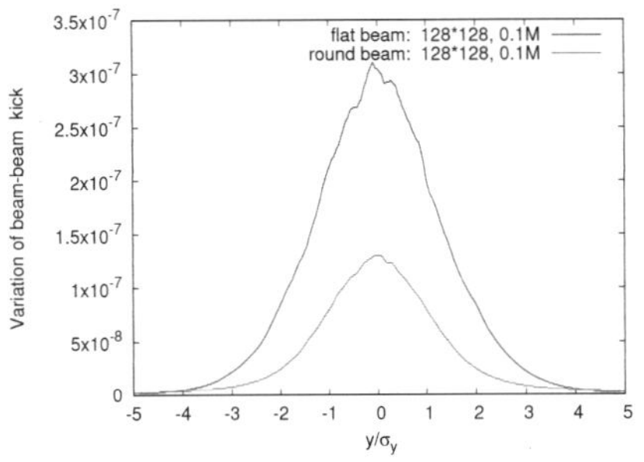

Figure 6. Comparison of beam-beam kick variations for a round beam and a flat beam.

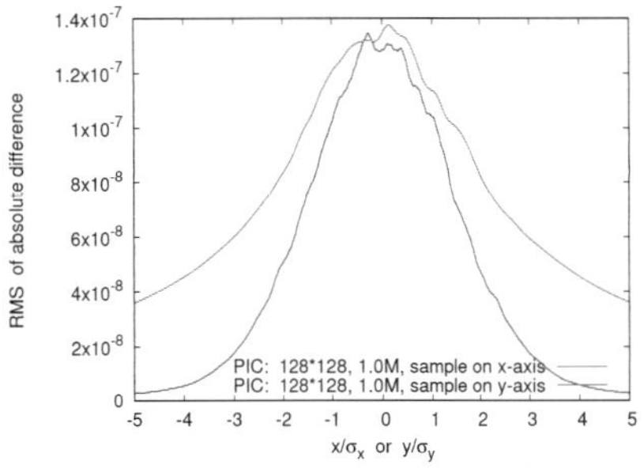

Figure 7. Comparison of beam-beam kick variations on horizontal and vertical axes for a flat beam.

6. Conclusion

Strong-strong beam-beam simulation based on PIC Poisson solver is subject to large numerical noises, which can be reduced by increased macro-particle number, transverse grids, and longitudinal slices. Turn-by-turn variation of bunch center can increase the simulated proton emittance growth rate but not the main contributor to the artificial emittance growth. We found that macro-particles in bunch core contribute most of the artificial emittance growth in the strong-strong simulation. The reason is found that PIC based Poisson solver generates a larger numerical variation in the beam-beam force calculation for macro-particles in the bunch core, especially in the vertical plane for a flat beam.

References

[1] Montag C *et al.* 2022 *Proc. 13th Int. Particle Accelerator Conf. (IPAC'22)* (Bangkok, Thailand) pp 1954–57
[2] Qiang J, Furman Miguel A and Ryne Robert D 2004 *J. Comput. Phys.* **198** 278
[3] Luo Y 2015 *Nucl. Instrum. Methods Phys. Res., Sect.* A **801** 95
[4] Luo Y *et al.* 2021 *Proc. 12th Int. Particle Accelerator Conf. (IPAC'21)* (Campinas, Brazil) pp 3808–11
[5] Luo Y *et al.* 2022 *Proc. 13th Int. Particle Accelerator Conf. (IPAC'22)* (Bangkok, Thailand) pp 1924–26
[6] Luo Y *et al.* 2022 *Proc. 13th Int. Particle Accelerator Conf. (IPAC'22)* (Bangkok, Thailand) pp 1939–41

Journal of Physics: Conference Series **2420** (2023) 012009 doi:10.1088/1742-6596/2420/1/012009

Fine decoupling test and simulation study to maintain a large transverse emittance ratio in hadron storage rings

Y Luo, I Blackler, M Blaskiewicz, W Fischer, A Marusic, C Montag, T Shrey and D Xu

,

Brookhaven National Laboratory, Upton, NY, USA

E-mail: `yluo@bnl.gov`

Abstract. In previous and existing hadron storage rings, the horizontal and vertical emittances are normally the same or very close. For the Hadron Storage Ring (HSR) of the Electron-Ion Collider (EIC), the design proton transverse emittance ratio is 10:1. To maintain this large emittance ratio, we need to have an online fine decoupling system to prevent transverse emittance exchange. For this purpose, we carried out fine decoupling experiments in the Relativistic Heavy Ion Collider (RHIC) and reviewed its previous operational data. Analytical prediction and numerical simulation are preformed to estimate how small the global coupling coefficient should be to maintain a 10:1 emittance ratio.

1. Introduction

The Electron-Ion Collider (EIC) presently under construction at Brookhaven National Laboratory will collide polarized high energy electron beams with hadron beams, reaching luminosities up to $1 \times 10^{34}\mathrm{cm}^{-2}\mathrm{s}^{-1}$ in the center-of-mass energy range of 20-140 GeV. To achieve high luminosities, the EIC adopts crossing angle collision with more than 1000 bunches per ring, flat beams with small β^*s at the interaction point (IP), and strong hadron cooling [1].

For the collision between 275 GeV protons and 10 GeV electrons, the beam-beam parameters for both beams reach their highest values in EIC. The beam-beam parameters are 0.012 for protons and 0.1 for electrons. The beam sizes at interact point (IP) are (95 μm, 8.5 μm). The β^*s in the Hadron Storage Ring (HSR) are (80 cm, 7.2 cm). The un-normalized proton transverse emittances are (11.3 nm, 1.0 nm).

For previous or existing hadron colliders, the horizontal and vertical emittances are normally the same or close to each other. To achieve about a 10:1 emittance ratio in HSR, we will pre-cool the proton bunches at lower beam energies and employ strong hadron cooling at store energy to maintain this ratio. As we know, betatron coupling will mix transverse emittances. Therefore, fine decoupling is required in HSR to stop emittance transfer from the horizontal plane to the vertical plane. For HSR we may not have vertical cooling at store energy. If so, we also need to prevent any sizable emittance growth in the vertical plane.

The HSR of EIC will re-use the existing RHIC arcs. For RHIC, we developed a feedback system to continuously monitor and correct the global betatron coupling [2, 3]. This system

Journal of Physics: Conference Series **2420** (2023) 012009 doi:10.1088/1742-6596/2420/1/012009

was proved robust and routinely operated at injection, on the energy ramp, and at store. For HSR, considering the large transverse emittance ratio, we need to make sure that current RHIC decoupling feedback is sufficient for HSR's fine decoupling requirement.

In the following, we will first present experimental results from previous years and in 2021 to demonstrate fine decoupling and possible 10:1 transverse emittance ratio in RHIC. Then we will analytically estimate the emittance exchange due to global coupling and its requirement for HSR. In the end, we will carry out multi-particle simulation to determine and verify the global decoupling requirement.

2. Results from previous years

To demonstrate if we can achieve a large transverse emittance ratio in RHIC with current decoupling feedback, we performed several beam experiments in RHIC. The earliest one was done in 2017. In this experiment, we used Au (gold) ion beam. We filled the Yellow ring with nominal intensity bunches ($2x10^9$/bunch) for the first half of the bunch train and then half intensity bunches for the remainder. Intra-beam scattering (IBS) will generate longitudinal emittance growth and horizontal emittance growth through horizontal dispersion in the ring. IBS growth rate is related to bunch intensity and emittances. For RHIC, there is no vertical dispersion in the design lattice. We ramped Au ion bunches to 100 GeV/n and observed emittance growth for about 2 hours at store.

Figure 1 shows the emittance measurement data with Ion Profile Monitor (IPM). In the plot, there are two horizontal and two vertical emittance measurement data for Yellow ring at each time point, one for full bunch intensity and one for half bunch intensity. With global decoupling in Yellow ring, we observed a much slower vertical emittance growth than horizontal. After about 2 hours at store, the horizontal emittance reached double the vertical emittance. The vertical emittance increased slightly by 20-30% in 2 hours. At the end of experiment, we pushed the tune settings closer, from (0.2365, 0.2268) to (0.2340, 0.2291) in steps, we observed horizontal emittance transferred to vertical emittance.

A second experiment was done in 2018, where we focused on the minimum transverse tune split needed to stop the transverse emittance exchange. For this experiment, we also used Au ion beam at 100 GeV/n but with stochastic cooling. We deliberately switched off the horizontal stochastic cooling and kept the vertical plane cooling on. We adjusted the tune split by increasing the horizontal set tune to check at what tune split the horizontal emittance will not get cooled by the vertical cooling through betatron coupling. Figure 2 shows the measured tunes and emittances during this experiment.

The tune split before any tune change was about 0.001, as shown at 22:05:00. We lost tune tracking between 22:08:00 and 22:21:00. At about 22:20:00, as shown by a red arrow in the plot, the horizontal emittance began to grow. Then the tune meter was tuned back and the tune split was measured as 0.013. The minimum tune split ΔQ_{min} or the global coupling coefficient $|C^-|$ under the experimental condition was probably about 0.001-0.0015. This hints that we need to have the global coupling coefficient $|C^-|$ at least be 10 times less than the transverse set tune split to stop transverse emittance exchange.

3. Beam experiments in 2021

In 2021 we re-initiated a similar experiment with a main purpose to demonstrate a transverse emittance ratio 10:1 in RHIC and to test the capability of existing RHIC decoupling system. This experiment should have been be done at store energy with stochastic cooling as in 2018. Unfortunately stochastic cooling was not set up in 2021 since this run was dedicated for polarized proton operation. To increase the IBS growth rate in the horizontal plane to obtain a larger transverse emittance ratio, we carried out this experiment at injection with 7.3 GeV Au/n beam.

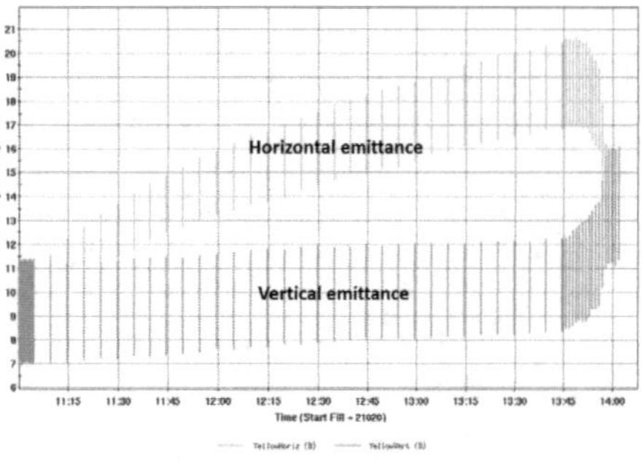

Figure 1. Emittance measurements at store in 2017 beam experiment. Transverse emittance ratio 2:1 was obtained.

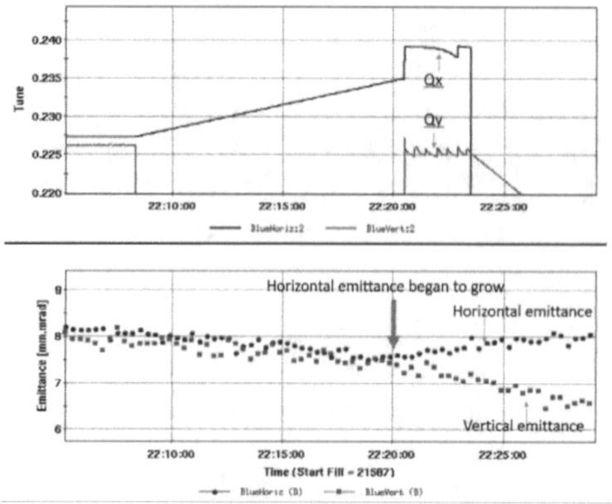

Figure 2. Tunes and emittances at store in 2018 beam experiment. When tune split larger than 0.013, the emittance exchange between the transverse planes stopped.

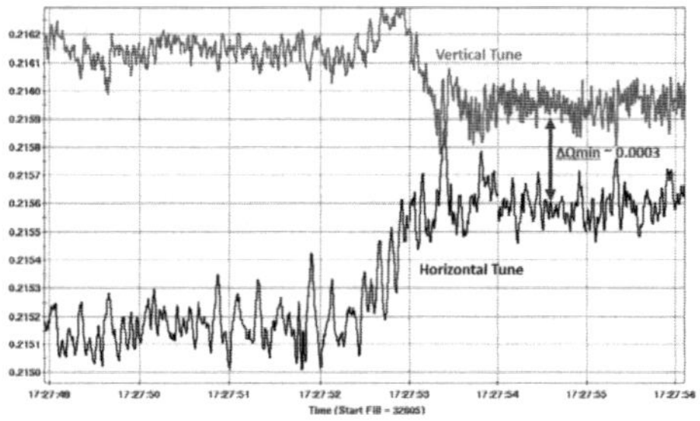

Figure 3. The minimum tune split ΔQ_{min} or $|C^{-}|$ about 0.0003 obtained in 2021 experiment.

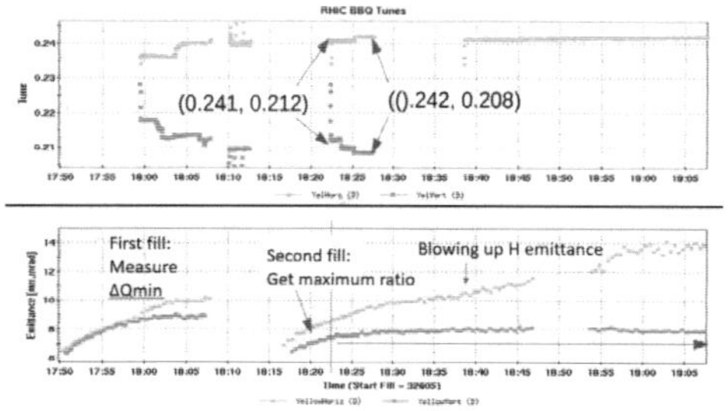

Figure 4. Measured emittances and tunes in 2021 beam experiment. Maximum 1.8:1 transverse emittance ratio was obtained at injection.

Before the experiment, we decoupled the Yellow ring with decoupling feedback and measured the minimum tune split ΔQ_{min} or global coupling coefficient $|C^{-}|$. Figure 3 shows that $\Delta Q_{min} = 0.0003$ can be obtained with current decoupling feedback and with some manual decoupling. This is the smallest ΔQ_{min} we have observed in RHIC. Then we injected 12 new bunches into the well decoupled ring to study how large emittance ratio we could obtain.

During this process, the vertical emittance basically stayed unchanged and horizontal emttance continued growing due to IBS. To increase the transverse emittance ratio in a short period of time, we decided to blow up the horizontal emittance with additional measures. We used horizontal offset injection, horizontal kick from base band tune meter (BBQ), and simply consecutive single kicks for tune measurement (Artus). Figure 4 shows the measured emittances and tunes in this experiment.

For this experiment, we were only able to demonstrate a maximum emittance ratio about 1.8:1 in 45 minutes at injection. With a very large horizontal emittance, we observed a very bad beam lifetime due to limited physical aperture. Again, this experiment should be done at store energy with stochastic cooling. We will come back to re-do it in 2023 RHIC Au run.

4. Analytical prediction

Here we present analytical prediction of transverse emittance exchange due to global betatron coupling. Based on Guignard's Hamiltonian perturbation theory [4], the horizontal and vertical emittances with betatron coupling are

$$\epsilon_x = \epsilon_{x,0} - \frac{|C^-|^2/2}{\Delta^2 + |C^-|^2}(\epsilon_{x,0} - \epsilon_{y,0}), \tag{1}$$

$$\epsilon_y = \epsilon_{y,0} + \frac{|C^-|^2/2}{\Delta^2 + |C^-|^2}(\epsilon_{x,0} - \epsilon_{y,0}), \tag{2}$$

where $\epsilon_{x,0}$ and $\epsilon_{y,0}$ are the emittances without any coupling, Δ is the fractional lattice tune split.

For the HSR of EIC, we need to control vertical emittance increase, say less than 5%. Based on above equations, with uncoupled transverse emittances (11 nm, 1 nm), we need to have $|C^-/\Delta| < 10$ to achieve that goal. For the original proton design tunes (0.310, 0.305) for HSR, we need to have $|C^-|$ less than 0.0005 which is reaching the best we can do with RHIC global decoupling feedback. To minimize the vertical emittance growth rate due to beam-beam interaction, we moved the proton design tunes to (0.228, 0.210) [5]. For the new design proton tunes, we only need to correct the global decoupling coefficient $|C^-| < 0.0018$ to have less than 5% of vertical emittance increase. In the RHIC routine operation, we are able to operate with $|C^-|$ less than 0.001. This means that RHIC global decoupling feedback meets the HSR's global decoupling requirement.

5. Numerical simulation

Next we carry out numerical simulation using SimTrack [6] to determine the decoupling requirement. First, we simulate with RHIC's Blue ring lattice for 100 GeV proton beam. We set the lattice tunes the same as HSR's tunes (0.228, 0.210). We use the existing skew quadruopole families to introduce global coupling in this simulation study. The initial horizontal and vertical emittances are set to be 9.5 nm and 1 nm. The emittance is about 9.5 nm in both transverse planes for the RHIC 255 GeV proton beam at store. We ramp up skew quadrupoles to their full strengths in 1000 turns and continue tracking for another 500 turns. In this study, we used 2000 macro-particles.

Figure 5 shows the evolution of both horizontal and vertical emittances for 10 sets of final skew quadrupole strengths, which corresponds $|C^-|$ from 0.001 to 0.018. From the simulation results, if we plan to have vertical emittance increase less than 5% from its original value 1 nm, the global coupling coefficient should be less than 0.002, which is in good agreement to our above analytical prediction.

Then we move to simulation with the HSR lattice. Here we use an older version of HSR lattice labeled as Version 2020Dec08. For this study, we randomly assign roll angles to all quadrupoles to generate global coupling. The RMS roll angles we used in the simulation study are 140 μrad, 100 μrad, 50 μrad, and 10 μrad, which correspond to $|C^-|$= 0.0065, 0.0045, 0.0022,and 0.00045. The evolution of vertical emittance is shown in Figure 6. Based on the simulation results, with design tunes (0.228, 0.210), the vertical emittance will not grow more than 5% if the global coupling coefficient is smaller than 0.0022, which is in agreement with our early analytical prediction too.

6. Conclusion

To be able to maintain a large 10:1 horizontal to vertical emittance ratio for the HSR of EIC, we carried out beam experiments in RHIC to test the capability of the existing global decoupling feedback system and to demonstrate 10:1 emittance ratio. In these experiments,

Journal of Physics: Conference Series **2420** (2023) 012009 doi:10.1088/1742-6596/2420/1/012009

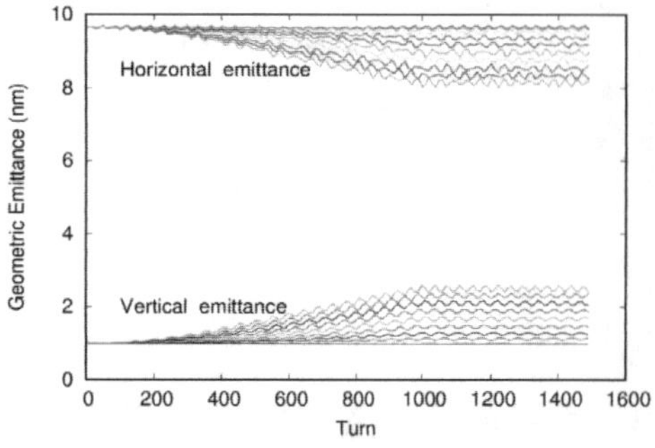

Figure 5. Evolution of horizontal and vertical emittances from numerical simulation done with RHIC lattice.

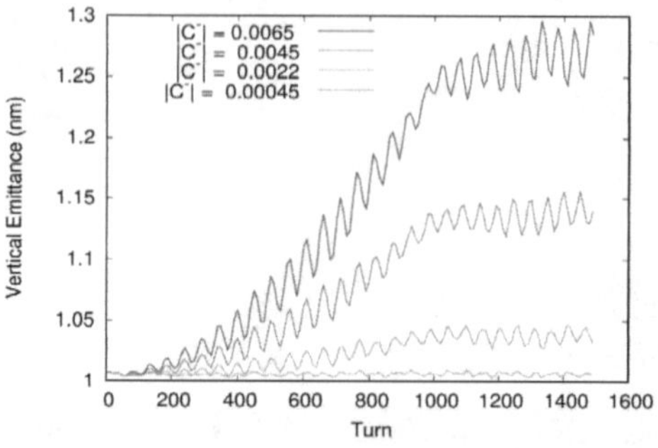

Figure 6. Evolution of vertical emittance from numerical simulation done with HSR lattice.

we demonstrated that we could reach a best global decoupling with ΔQ_{min} or global coupling coefficient $|C^-|$ less than 0.0003. However, we failed to demonstrate larger than 2:1 emittance ratio due to non-optimum experimental condition in 2021. From analytical calculation and numerical multi-particle tracking, we showed that with the new HSR design tunes (0.228, 0.210), we can have the proton's vertical emittance increase less than 5% if the global coupling coefficient is less than 0.002. For RHIC routine operation, we can easily correct global coupling less than 0.001 with the existing RHIC global decoupling feedback. We plan to re-do this beam experiment in 2023 RHIC Au run with stochastic cooling at store.

References

[1] Montag C *et al.* 2022 *Proc. 13th Int. Particle Accelerator Conf. (IPAC'22)* (Bangkok, Thailand) pp 1954–57

[2] Luo Y, Cameron P, DellaPenna A, Hoff L, Marusic A, Peggs S, Schultheiss C and Jones R 2006 *Phys. Rev. Spec. Top. Accel Beams* **9** 124001

[3] Cameron P, DellaPenna A, Hoff L, Luo Y, Marusic A, Schultheiss C, Tepikian S, Gasior M, Jones R and Tan C. Y 2006 *Phys. Rev. Spec. Top. Accel Beams* **9** 122801

[4] Guignard G 1995 *Phys. Rev.* E **51** 6104

[5] Luo Y *et al.* 2021 *Proc. 12th Int. Particle Accelerator Conf. (IPAC'21)* (Campinas, Brazil) pp 3808–11

[6] Luo Y 2015 *Nucl. Instrum. Methods Phys. Res., Sect.* A **801** 95

Designing the EIC electron storage ring lattice for a wide energy range

D Marx[1], JS Berg[1], Y Cai[4], BRP Gamage[5], GH Hoffstaetter[2], J Kewisch[1], Y Li[1], C Montag[1], V Morozov[3], Y Nosochkov[4], V Ptitsyn[1], D Sagan[2], MG Signorelli[2], S Tepikian[1], J Unger[2], F Willeke[1] and D Xu[1]

[1] Brookhaven National Laboratory, Upton, NY, USA
[2] Cornell University, Ithaca, NY, USA
[3] Oak Ridge National Laboratory, Oak Ridge, TN, USA
[4] SLAC, Menlo Park, CA, USA
[5] Thomas Jefferson National Accelerator Facility, Newport News, VA, USA

E-mail: dmarx@bnl.gov

Abstract. The Electron-Ion Collider (EIC) will collide electrons with hadrons at center-of-mass energies up to 140 GeV (in the case of electron-proton collisions). A 3.8-kilometer electron storage ring is being designed, which will store electrons with a range of energies up to 18 GeV for collisions at one or two interaction points. At energies up to 10 GeV the arcs will be tuned to provide 60 degree phase advance per cell in both planes, whereas at top energy of 18 GeV a 90 degree phase advance per cell will be used, which largely compensates for the horizontal emittance increase with energy. The optics must be matched at three separate energies, and the different phase-advance requirements in both the arc cells and the straight sections make this challenging. Moreover, the spin rotators must fulfill requirements for polarization and spin matching at widely different energies while satisfying technical constraints. In this paper these challenges and proposed solutions are presented and discussed.

1. Introduction

The Electron-Ion Collider (EIC) [1, 2] is a new machine currently being designed that will collide polarized electrons with polarized hadrons (protons up to heavy ions) for the purpose of investigating the structure and properties of nucleons. It will be built at Brookhaven National Laboratory in the 3.8-kilometer tunnel that currently houses the Relativistic Heavy Ion Collider (RHIC) [3–5]. Two electron rings will be built in this tunnel: a rapid cycling synchrotron (RCS) [6] for accelerating electrons to collision energies and an electron storage ring (ESR) for colliding electrons with hadrons from a separate storage ring. Figure 1 shows a schematic of the EIC. Collisions will occur at a range of center-of-mass energies between 29 GeV and 140 GeV, providing luminosities up to $10^{34}\,\mathrm{cm}^{-2}\mathrm{s}^{-1}$. These luminosities are considerably greater than those achieved at HERA [5, 7], an electron/positron-proton collider that operated until 2007 at DESY, and, although the top center-of-mass energy will be less than HERA's 318 GeV, the EIC will be optimized for a much wider range of energies.

The ESR lattice consists of six arcs with straight sections in between, labeled according to the numbers on a clock face. The baseline design for the EIC includes a single interaction point

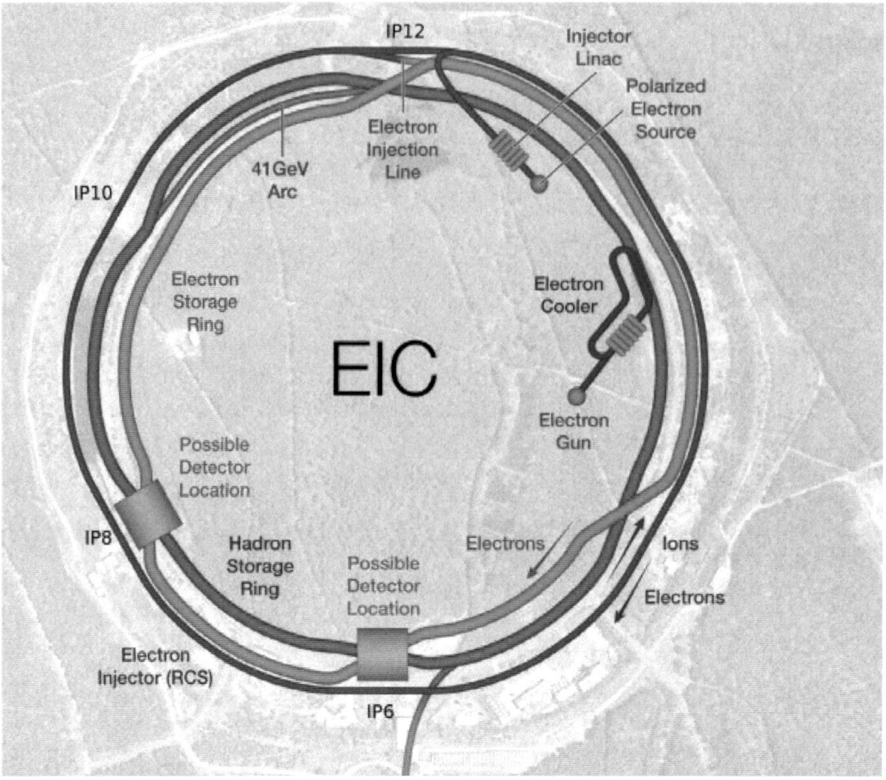

Figure 1. Schematic of EIC (not to scale).

(IP), called IP6, where the beams will collide. A second IP and detector in the neighboring straight section, IP8, may be included in a future upgrade. Superconducting RF cavities will be installed at IP10, and the beam will be injected into the ring at IP12.

In order to achieve the desired range of center-of-mass energies, three values for the electron beam energy are considered: 5-6, 10, and 18 GeV. This introduces several challenges in the lattice design. As the horizontal equilibrium emittance increases with energy, this needs to be compensated to satisfy aperture requirements. Superbends in arc cells and changes to the arc phase advance are used to keep the horizontal emittance within an acceptable range. The nonlinear chromaticity correction scheme also needs to be altered for different energies, as the different arc phase advances call for different sextupole configurations.

One of the key requirements for the experimental program is a high level of longitudinal spin polarization. Beams are injected with vertical spin, which is the stable spin direction for the ring. On each side of the IPs, a solenoid-based spin rotator rotates the spin vector into or from the longitudinal direction. As the spin-rotation angles depend on energy, designing these spin rotators for operation at a wide range of energies is challenging.

Table 1 shows the main lattice parameters at different energies. In the following sections, the design challenges arising from the wide energy range are described.

2. Damping and emittance

The horizontal equilibrium emittance in an electron ring is given by the equation [8]

$$\epsilon_x = \frac{C_q}{J_x}\gamma^2\frac{\langle G^3\mathcal{H}\rangle}{\langle G^2\rangle}, \tag{1}$$

Table 1. ESR 1IP lattice parameters at various energies. The optics depend on energy, and the parameters for 5 and 10 GeV are from an earlier version of the lattice.

Parameter	5 GeV	10 GeV	18 GeV
Arc phase	60°	60°	90°
Superbends	On	Off	Off
Hor. emit. (nm)	26	22	27
Energy spread	0.050%	0.053%	0.095%
Tunes, Q_x/Q_y	40.12/37.1	40.12/37.1	52.12/44.1
Nat. chrom.	-80/-93	-75/-90	-93/-93

where $C_q = 3.84 \times 10^{-13}$ m, J_x is the horizontal damping partition number, γ is the Lorentz factor, G is the inverse of the bending radius, and $\mathcal{H} = \gamma_x \eta^2 + 2\alpha_x \eta\eta' + \beta_x \eta'^2$, where $\alpha_x, \beta_x, \gamma_x$ are the Courant-Snyder parameters and η is the horizontal dispersion. The dependence on γ^2 means that the emittance would change by an order of magnitude if the same lattice were used at all energies. In order to compensate this effect, $\mathcal{H}$ may be changed by adjusting the optics of the lattice. At lower energies up to 10 GeV the phase advance in the arc cells is 60°, while at 18 GeV it is set to 90°, which largely compensates for the emittance increase with energy.

Superbends are used in the arcs, consisting of two long dipoles with a shorter one in between, as shown in figure 2. At energies above 10 GeV, all three magnets produce a field in the same direction with the effect of a single dipole. At lowest energy, however, the central dipole has reversed field, which produces additional damping. This allows a larger beam-beam tune shift to be supported [9], while increasing the horizontal equilibrium emittance [10, 11].

Fine-tuning the emittance may be achieved in operation by adjusting the damping partition numbers by means of changing the frequency to produce a small radial shift [12].

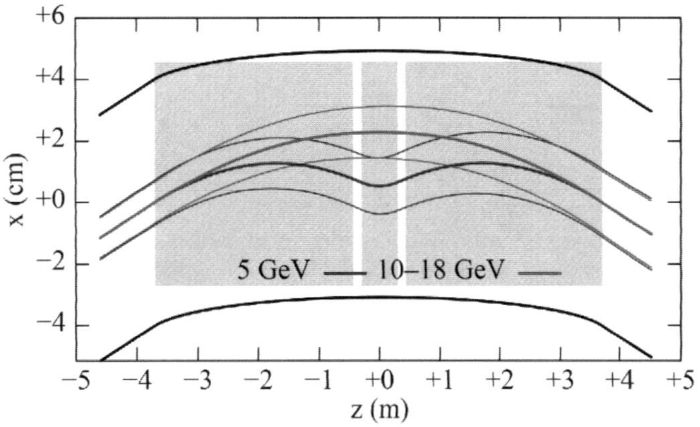

Figure 2. Illustration of superbends in the arcs at various energies. The beam envelopes shown correspond to 10σ for the 60° arc-cell phase advance.

3. Sextupole configuration

The ESR nonlinear chromaticity correction scheme is based on multiple families of -I sextupole pairs in the arcs. One of the consequences of changing the phase advance of the arc cells is that this affects the sextupole family arrangement. The chromatic beta-beat wave propagates at twice the betatron phase advance, which means the simplest family arrangement is three families per plane at 60° arc-cell phase advance and two families per plane at 90°. The sextupole wiring scheme will be designed to allow switching between these two configurations. Chromaticity correction at 18 GeV represents the most challenging case due to the larger energy spread at higher energies, which leads to a greater momentum acceptance requirement ($10\sigma_\delta$ is required for sufficient lifetime), as well as stringent phase-advance requirements caused by using just two families of sextupoles per plane (per arc). More details on the chromaticity-correction scheme are presented in [13].

4. Spin rotators

Spin rotators will be located on both sides of the IP to rotate the spin vector from the vertical direction in the arcs to the longitudinal direction at the IP and back [14]. Designing spin rotators for a broad range of energies is a challenge. The lowest energies of 5 to 6 GeV preclude the use of spin rotators based purely on dipoles, as were used at HERA, due to the excessively large aperture required to accommodate orbit excursions at these energies. Instead, the spin rotators will be composed of a combination of solenoids, dipoles, and matching quadrupoles [15–17]. In general, the spin precesses around an axis parallel to the field direction, namely vertical in dipoles and longitudinal in solenoids. A solenoid followed by dipoles can therefore rotate the spin direction from vertical in the arc to radial after the solenoid, and finally to longitudinal after the dipoles. The configuration is mirror symmetric about the IP to rotate the spin from longitudinal back to vertical on the other side.

As the spin configuration needs to rotate the spin at three different energies and the dipole angles cannot be changed in operation, the solenoid-bend structure is repeated, as shown in figure 3, which offers the required flexibility. However, the lengths of the two solenoids, as well as the bending angles, are different in the two cases to optimize to the requirements at various energies.

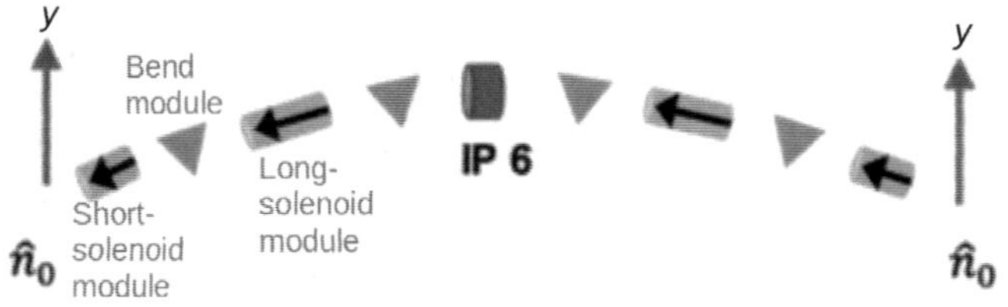

Figure 3. The spin rotators are composed of short- and long-solenoid modules, as well as bend modules, to rotate the spin from vertical in the arcs to longitudinal at the IP and back to vertical on the other side.

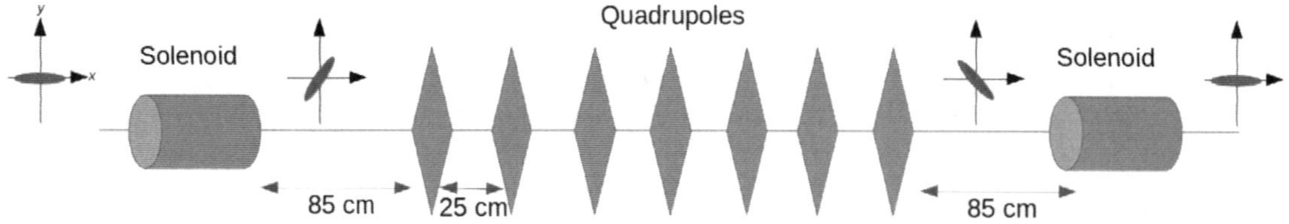

Figure 4. Each solenoid module consists of two half-solenoids with five or seven quadrupoles in between.

Furthermore, each solenoid is actually split into two halves with a set of quadrupoles in between. This unit, collectively called a solenoid module, is shown in figure 4. This configuration allows decoupling between the horizontal and vertical planes, as well as horizontal spin-matching conditions, to be fulfilled. There are four superconducting half-solenoids per spin rotator: two long and two short. An alternative scheme involving skew quadrupoles outside of the half-solenoids was also considered and determined to be less desirable due to the added complexity in dealing with coupling from these skew quadrupoles.

The short-solenoid module sits directly next to the arc with two quadrupoles for matching in between. Between the short- and long-solenoid modules is a bending section consisting of dipoles and quadrupoles for matching. The long-solenoid module leads onto the IR section, containing quadrupoles and dipoles for matching various constraints, including a fixed dipole angle to the IP to complete the spin rotation.

Ideally zero dispersion in both short- and long-solenoid modules would be achieved by suppressing the dispersion at the end of the arcs. The challenge is matching dispersion in as few arc cells as possible, which is necessary because changes in the β and dispersion functions will ruin the sextupole pairing. Previous studies showed that pairing the sextupoles is particularly important for maximizing the dynamic aperture of the 90° lattice. For this reason the preferred solution is to leave the 90° arcs unchanged to the end and to bring the dispersion down in the short-solenoid module and bend module. The dispersion is then zero in the long-solenoid module. The implication of this scheme is that the short solenoids have to be turned off at top energy, which does not allow full longitudinal spin matching.

At intermediate energy of around 10 GeV and lower, the phase advance of the arc FODO cells is 60°, and the short solenoids must be turned on. It would require very large strengths to match the dispersion and its derivative to zero by varying only two quadrupoles, although it can be done with reasonable strengths if all quadrupoles in the last two arc cells are used. As a compromise, it was decided to vary only the quadrupoles in the last arc cell, which is sufficient to lower the dispersion to less than about 0.2 m in the short-solenoid module. The bend section between the solenoid modules can then be used to bring down the dispersion to zero in the long-solenoid module.

The bend angles must be set to provide the required spin rotation at all energies. The spin rotates in a dipole with bend angle θ by $\psi = a\gamma\theta$ with respect to the beam axis, where a is the anomalous magnetic moment of an electron [15]. The dipoles between the end of the spin rotator and the IP must rotate the spin vector by a full $\pi/2$ angle at the maximum energy of 18 GeV, as only the long solenoids are turned on, which requires a bending angle of 38.79 mrad. At minimum energy, the total bend angle from the beginning of the spin rotator to the IP should result in a $\pi/2$ rotation. This would result in a 97.81 mrad bend angle between the solenoid modules at the nominal minimum energy of 5 GeV. At an energy of 6 GeV, this angle is reduced to 77.57 mrad. The current lattice has this latter bend angle, and 6 GeV is taken as the minimum energy, while a study is ongoing to assess the feasibility of adjusting the geometry to achieve longitudinal polarization down to 5 GeV. The total bend angle is repeated in all the other straight sections for symmetry reasons.

The spin-rotation angle in a solenoid is related to its length, L, and strength, K_s, by $\phi = (1 + a)K_sL$ [14]. At top and bottom energies, only one solenoid module is used, which must provide $\pi/2$ spin rotation, whereas at 10 GeV both are used to rotate the spin, with the spin-rotation angles shown in table 2. The energies have been adjusted slightly to ensure half-integer spin tunes ($a\gamma$).

Matching the solenoid modules to achieve decoupling and horizontal spin matching by optimizing the quadrupole strengths is a key challenge described in [14].

Table 2. Solenoid module spin-rotation angles at various energies. ϕ_1 is the angle between the short- and long-solenoid modules, and ϕ_2 is the angle between the long-solenoid module and IP.

Energy (GeV)	ϕ_1 (rad)	ϕ_2 (rad)
17.846	0	$\pi/2$
9.915	0.7111	1.7193
5.949	$\pi/2$	0

5. IR matching

Optimizing the beam-beam interaction at different energies results in different required β^* parameters [18]. They depend on both the electron and hadron energies. In addition, the detector solenoid strength does not scale with energy, so its effect will be particularly significant at lower energies and will require compensation [19].

6. Conclusion

The ESR is designed to operate at a large range of energies. There are many physical effects, such as radiation, spin polarization, and beam-beam interactions, that mean that a simple scaling of the magnet strengths is not sufficient. Instead, the lattice design at different energies is radically different, with different cell phase advances, superbend settings, sextupole configurations, spin-rotator settings, and β^* parameters. Significant progress has been made in the lattice design at various energies, in particular the spin-rotator design, which is still evolving. In the future, the possibility of achieving longitudinal spin polarization at 5 GeV energy will be studied.

Acknowledgments

Work supported by Brookhaven Science Associates, LLC, under Contract No. DE-SC0012704, by Jefferson Science Associates, LLC, under Contract No. DE-AC05-06OR23177, by UT-Battelle, LLC, under contract No. DE-AC05-00OR22725, and by SLAC under Contract No. DE-AC02-76SF00515 with the U.S. Department of Energy.

References

[1] Beebe-Wang J *et al.* 2021 Electron-Ion Collider: Conceptual Design Report Tech. rep. Brookhaven National Laboratory, Jefferson Lab
[2] Montag C *et al.* 2022 *Proc. 13th Int. Particle Accelerator Conf. (IPAC'22)* (Bangkok, Thailand) pp 1954–57
[3] Montag C 2019 *AIP Conference Proceedings* **2160** 040006
[4] RHIC Configuration Manual, 2006 Tech. rep. Brookhaven National Laboratory
[5] Hübner K, Ivanov S, Steerenberg R, Roser T, Seeman J, Oide K, Mess K H, Schmüser P, Bailey R and Wenninger J The largest accelerators and colliders of their time *Particle Physics Reference Library : Volume 3: Accelerators and Colliders* ed Myers S and Schopper H (Springer)
[6] Ranjbar V H, Lovelace H, Meot F and Lin F 2022 *Proc. 13th Int. Particle Accelerator Conf. (IPAC'22)* (Bangkok, Thailand) pp 2439–41
[7] An Assessment of U.S.-Based Electron-Ion Collider Science Tech. rep. National Academies of Sciences
[8] Peggs S and Satogata T 2017 *Introduction to Accelerator Dynamics* (Cambridge University Press)
[9] Guignard G 1981 *IEEE Trans. Nucl. Sci.* **28** 2506–08
[10] Nishimura H and Robin D 1999 *Proc. 18th Particle Accelerator Conf. (PAC'99)* (New York, NY USA) pp 203–5
[11] Robin D 2003 *Proc. 20th Particle Accelerator Conf. (PAC'03)* (Portland, OR USA) pp 224–28
[12] Minty M G, Akre R, Decker F J, Frisch J, Kuroda S and Zimmermann F 1998 *Proc. 17th Int. Conf. on High Energy Accelerators (HEACC'98)* (Dubna, Russia)
[13] Nosochkov Y M *et al.* 2022 *Proc. 13th Int. Particle Accelerator Conf. (IPAC'22)* (Bangkok, Thailand) pp 1950–53

[14] Marx D *et al.* 2021 Task Force Report: ESR Linear Lattice Design Tech. Rep. BNL-222485-2021-TECH, EIC-ADD-TN-025 Brookhaven National Laboratory
[15] Barber D *et al.* 1985 *Particle Accelerators* **17** 243–62
[16] Emma P 1994 A spin rotator system for the NLC Tech. rep. SLAC NLC Note 7
[17] Ptitsin V and Shatunov Y 1997 *Proc. 17th Particle Accelerator Conf. (PAC'97)* (Vancouver, Canada) pp 3500 02
[18] Qiang J *et al.* 2022 *Proc. 13th Int. Particle Accelerator Conf. (IPAC'22)* (Bangkok, Thailand) pp 1942–45
[19] Xu D and Willeke F J 2022 *Proc. 13th Int. Particle Accelerator Conf. (IPAC'22)* (Bangkok, Thailand) pp 1576–82

Journal of Physics: Conference Series **2420** (2023) 012011 doi:10.1088/1742-6596/2420/1/012011

Beam-beam interaction for tilted storage rings

D Xu[1], Y Luo[1], D Holmes[1], C Montag[1], F Willeke[1], Hao Y[2] and Qiang J[3]

[1]Brookhaven National Laboratory, Upton, NY, USA
[2]Michigan State University, East Lansing, MI, USA
[3]Lawrence Berkeley National Laboratory, Berkeley, CA, USA

E-mail: dxu@bnl.gov

Abstract. In the Electron-Ion Collider (EIC) design, to avoid vertical orbit bumps in the Electron Storage Ring (ESR) at some crossing points with Hadron Storage Ring (HSR) to preserve the electron polarization, we plan to tilt the ESR plane by 200 μrad with an axis connecting IP6 and IP8. In this article, we study the beam-beam interaction when two rings are not in the same plane. The Lorentz boost formula is derived and the required vertical crabbing strength is calculated to compensate the dynamic effect. The beam-beam simulations are performed to validate the theory.

1. Introduction

A large crossing angle in the interaction region (IR) is necessary for fast separation of two colliding beams in ring-ring type colliders to achieve high collision rates, IR background minimization, and overall detector component and IR magnet arrangements. Crab crossing is an effective way to restore the head-on collision for circular colliders [1]. The crab crossing scheme was first successfully implemented at KEKB-factory [2]. The crab cavities, which are key components in the crab crossing scheme, were also demonstrated for the hadron beam in CERN's Super Proton Synchrotron (SPS) recently [3]. Other dynamics effects about the crab cavities can be found in [4–6].

The Electron-Ion Collider (EIC) project adopts the local scheme to achieve the desired luminosity 10^{34} cm^{-2}s^{-1} [7], as shown in figure 1 . In the local scheme, a pair of crab cavities are installed at both sides of the interaction point (IP). The upstream crab cavity tilts the beam in x - z plane, and the downstream crab cavity rotates the beam back. In the rest of the rings, both planes stay unaffected.

EIC rings consist of six arcs separated by six straight sections, which are labelled according to the hour markings of a clock from IR2 to IR12. A dedicated IR has been designed at 6 o'clock, and a potential second IR at 8 o'clock is reserved for future upgrade. To resolve the interference between rings, transfer lines, cooler ERS in IR2, the Electron Storage Ring (ESR) is proposed to be tilted by $\sim$ 200 μrad to avoid vertical bends. Figure 2 illustrates the concept. The rotating axis goes through both interaction points: IP6 and IP8. Although the rotation angle is small, the longitudinal axis is changed, and the dynamic effect has to be studied.

This paper derives the Lorentz Boost formula for the tilted storage rings. The beam-beam simulation is also presented to demonstrate if the tilted scheme works or not. A summary is given at last.

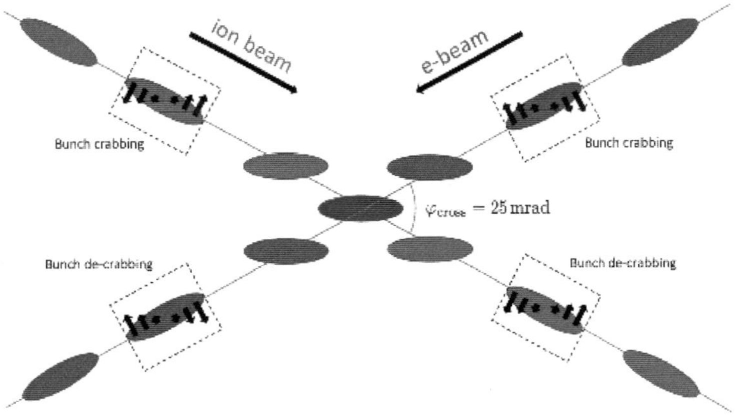

Figure 1. EIC local crabbing compensation scheme [7].

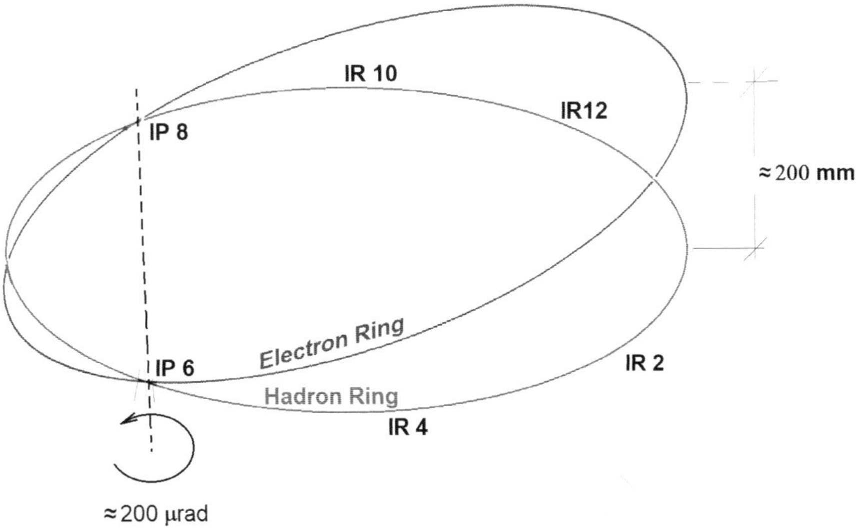

Figure 2. Schematic diagram of tilted ESR in EIC.

2. Lorentz boost

When two beams collide with a crossing angle, it is convenient to consider the beam-beam interaction in the boost frame [8], as shown in figure 3. Let $(\mathbf{E}_x, \mathbf{E}_y, \mathbf{E}_z)$ be the basis vectors for the Frenet-Serret frame of the electron ring, and the left-handed $(\mathbf{I}_x, \mathbf{I}_y, \mathbf{I}_z)$ for the ion ring. The vertical axes, $\mathbf{E}_y$ and $\mathbf{I}_y$, are parallel with each other before rotating the ESR frame. $(\mathbf{X}, \mathbf{Y}, \mathbf{Z})$ are the basis vectors of the boost frame. In the boost frame,

$$\begin{aligned} \mathbf{N} &= (\sin\psi, 0, \cos\psi)^{\mathrm{T}}, & \mathbf{E}_y &= \mathbf{I}_y = (0,1,0)^{\mathrm{T}} \\ \mathbf{E}_z &= (\sin\theta_c, 0, \cos\theta_c)^{\mathrm{T}}, & \mathbf{I}_z &= (\sin\theta_c, 0, -\cos\theta_c)^{\mathrm{T}} \\ \mathbf{E}_x &= (\cos\theta_c, 0, -\sin\theta_c)^{\mathrm{T}}, & \mathbf{I}_x &= (\cos\theta_c, 0, \sin\theta_c)^{\mathrm{T}} \end{aligned} \tag{1}$$

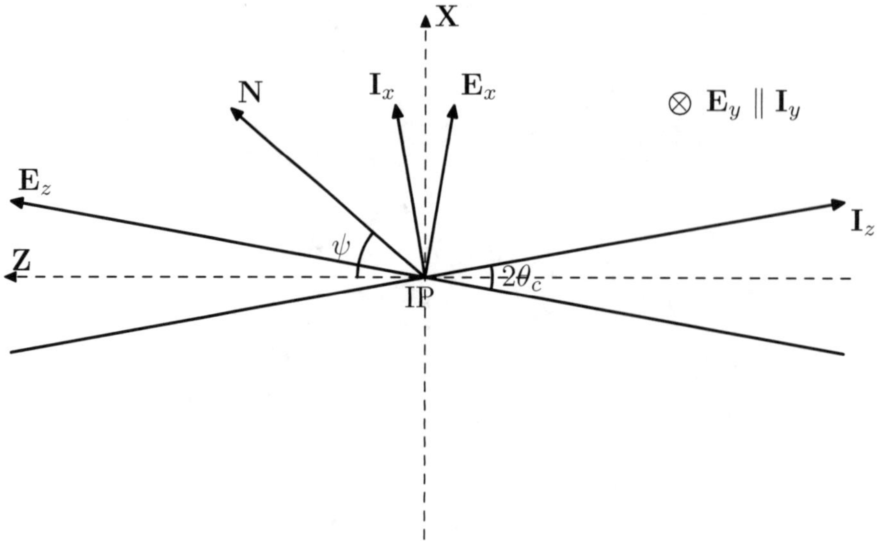

Figure 3. The boost frame for two beams colliding with a large crossing angle. **N** points from IP6 to IP8. $\theta_c = 12.5$ mrad is the half crossing angle, and $\psi = \pi/6 + \theta_c$.

where θ_c and ψ are defined in figure 3, and the superscript "T" denoting the transformation of a vector.

The rotation around **N** can be described by a matrix

$$R(\mathbf{N}, \phi) = \begin{bmatrix} \cos\phi + (1-\cos\phi)\sin^2\psi & -\sin\phi\cos\psi & (1-\cos\phi)\sin\psi\cos\psi \\ \sin\phi\cos\psi & \cos\phi & -\sin\phi\sin\psi \\ (1-\cos\phi)\sin\psi\cos\psi & \sin\phi\sin\psi & \cos\phi + (1-\cos\phi)\cos^2\psi \end{bmatrix} \tag{2}$$

After the ESR is rotated, the new basis vectors are

$$\left(\mathbf{E}'_x, \mathbf{E}'_y, \mathbf{E}'_z\right) = R\left(\mathbf{E}_x, \mathbf{E}_y, \mathbf{E}_z\right) \tag{3}$$

$$\mathbf{I}'_x = \mathbf{I}_x, \quad \mathbf{I}'_y = \mathbf{I}_y, \quad \mathbf{I}'_z = \mathbf{I}_z \tag{4}$$

The new crossing angle is determined by their longitudinal axes,

$$\begin{aligned} \cos\left(2\theta'_c\right) &= -\mathbf{E}'_z \cdot \mathbf{I}'_z \\ &= \cos\left(\psi+\theta_c\right)\cos\left(\psi-\theta_c\right) \\ &\qquad + \cos\phi\sin\left(\psi+\theta_c\right)\sin\left(\psi-\theta_c\right) \\ &= \cos\left(2\theta_c\right) - (1-\cos\phi)\sin\left(\psi+\theta_c\right) \\ &\qquad \times \sin\left(\psi-\theta_c\right) \end{aligned} \tag{5}$$

Substituting the numeric value

$$\theta_c = 12.5 \text{ mrad}, \quad \phi = 200 \text{ µrad}, \quad \psi = \pi/6 + \theta_c \tag{6}$$

the new crossing angle is

$$\Delta\theta_c = \theta'_c - \theta_c \approx \frac{\phi^2}{32\theta_c} = 1 \times 10^{-4} \text{ mrad} \tag{7}$$

It turns out the change of the crossing angle is negligible due to the tilted ESR.

The new boost frame is

$$\mathbf{Z}' = \frac{\mathbf{E}'_z - \mathbf{I}'_z}{2\cos\theta'_c}, \quad \mathbf{X}' = \frac{\mathbf{E}'_z + \mathbf{I}'_z}{2\sin\theta'_c}, \quad \mathbf{Y}' = \frac{\mathbf{E}'_z \times \mathbf{I}'_z}{\sin(2\theta'_c)} \tag{8}$$

From equation (3), equation (4), and equation (8), $\mathbf{E}'_y$ and $\mathbf{I}'_y$ are no longer parallel with $\mathbf{Y}'$. To reuse the Hirata map, we need to rotate the vertical axes around their longitudinal axes first. The rotation angles are

$$\cos\chi_{\mathrm{e}} = \mathbf{E}'_y \cdot \mathbf{Y}' = \frac{-\cos(\psi+\theta_c)\sin(\psi-\theta_c) + \sin(\psi+\theta_c)\cos(\psi-\theta_c)\cos\phi}{\sin(2\theta'_c)} \tag{9}$$

$$\cos\chi_{\mathrm{i}} = \mathbf{I}'_y \cdot \mathbf{Y}' = \frac{\sin(\psi+\theta_c)\cos(\psi-\theta_c) - \cos(\psi+\theta_c)\sin(\psi-\theta_c)\cos\phi}{\sin(2\theta'_c)} \tag{10}$$

Therefore, the new Lorentz Boost map is concatenated by the rotation map and the Hirata map,

$$\mathcal{L}' = \mathcal{R}(\chi) \circ \mathcal{L}(\theta'_c) \tag{11}$$

where $\mathcal{R}$ is a pure rotation around the longitudinal axis, and $\mathcal{L}$ is the well-know Hirata map in [8].

With the numeric values in equation (6), the pure rotation angles are

$$\chi_{\mathrm{i}} \approx \chi_{\mathrm{e}} \approx -4 \text{ mrad} \tag{12}$$

which have to be considered in beam-beam simulation.

Due to the rotation in equation (12), the $y-z$ coupling is introduced in the boost frame. To prevent the luminosity loss, we have to eliminate the y dependence on z. It can be achieved by introducing the vertical crabbing.

Borrowing the concept from [9], the crab dispersion is defined as

$$\zeta = \left(\frac{\partial x}{\partial z}, \frac{\partial p_x}{\partial z}, \frac{\partial y}{\partial z}, \frac{\partial p_y}{\partial z}\right)^{\mathrm{T}} \tag{13}$$

The crab dispersion caused by the upstream crab cavity is

$$(0, \lambda_x/\Lambda_x, 0, \lambda_y/\Lambda_y)^{\mathrm{T}}$$

where $\Lambda_x = \sqrt{\beta_x^*\beta_{x,c}}, \Lambda_y = \sqrt{\beta_y^*\beta_{y,c}}$ are used to normalize the crab cavity strength, $\lambda_{x,y}$ the relative crab cavity kick strength.

Assuming the upstream crab cavity is placed at $\Psi_x = \Psi_y = -\pi/2$ phase advance to IP. After propagated to IP, the crab dispersion turns into

$$\zeta_u = (\lambda_x, 0, \lambda_y, 0)^{\mathrm{T}} \tag{14}$$

Taking the linear term, the Hirata map $\mathcal{L}(\theta'_c)$ is equivalent to a crab dispersion

$$\zeta_L \approx \left(\theta'_c, 0, 0, 0\right)^{\mathrm{T}} \tag{15}$$

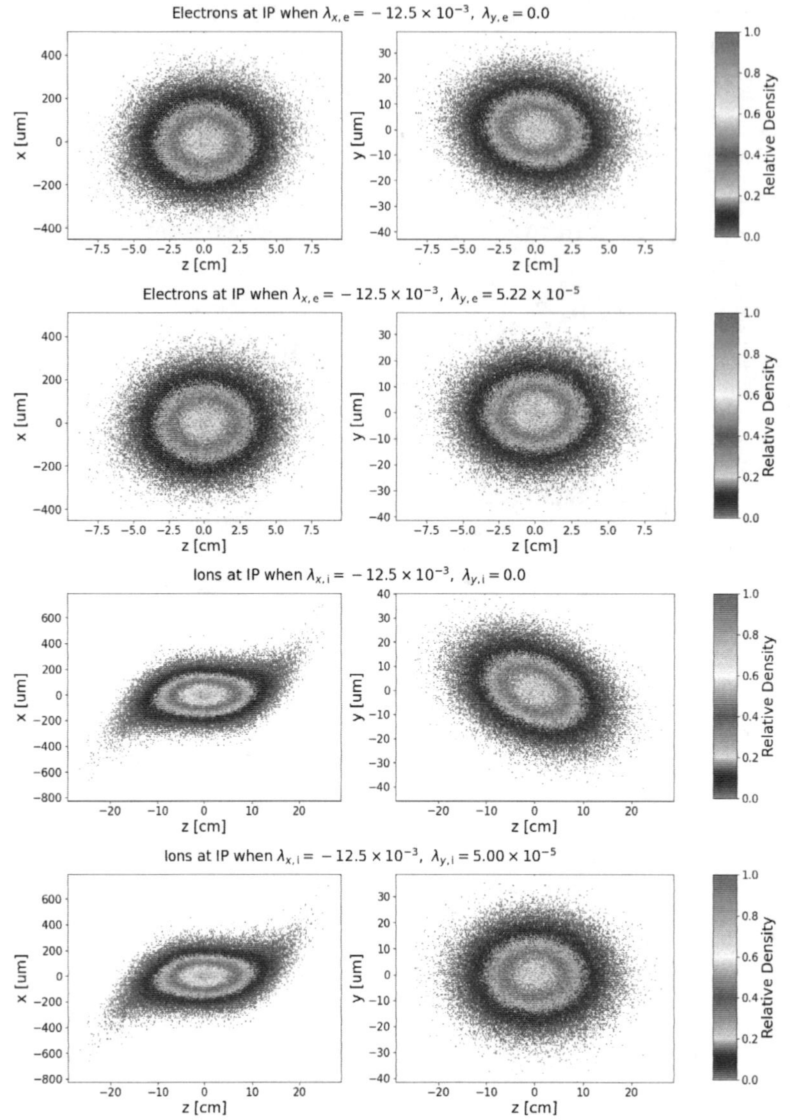

Figure 4. Electron and ion beam distribution in the boost frame.

To provide an effective head-on collision in the boost frame, the crab dispersion in equation (14) and equation (15) have to cancel with each other,

$$R(\chi)\,\zeta_u + \zeta_L = 0 \tag{16}$$

where $R(\chi)$ is the 4-by-4 rotation matrix

$$R(\chi) = \begin{bmatrix} \cos\chi & 0 & \sin\chi & 0 \\ 0 & \cos\chi & 0 & \sin\chi \\ -\sin\chi & 0 & \cos\chi & 0 \\ 0 & -\sin\chi & 0 & \cos\chi \end{bmatrix} \tag{17}$$

The required crab cavity can then be resolved,

$$\lambda_x = -\theta_c' \cos\chi, \quad \lambda_y = -\theta_c' \sin\chi \tag{18}$$

The strength of the downstream crab cavity can be determined similarly.

3. Beam-beam simulation

Figure 4 compares the beam distribution in the boost frame. With appropriate vertical crabbing, the y dependence on z in both electron beam and ion beam is removed.

Figure 5 shows the strong-strong simulation results by BeamBeam3D [10]. With vertical crabbing, the beam size evolution in both ESR and HSR overlaps with the reference curve. Because of the radiation damping and shorter bunch length, the electron beam is still stable even without vertical crabbing. However, when there is no vertical crabbing for the ion ring, the longitudinal motion is coupled to the vertical plane, and the vertical size grows much faster which has to be corrected.

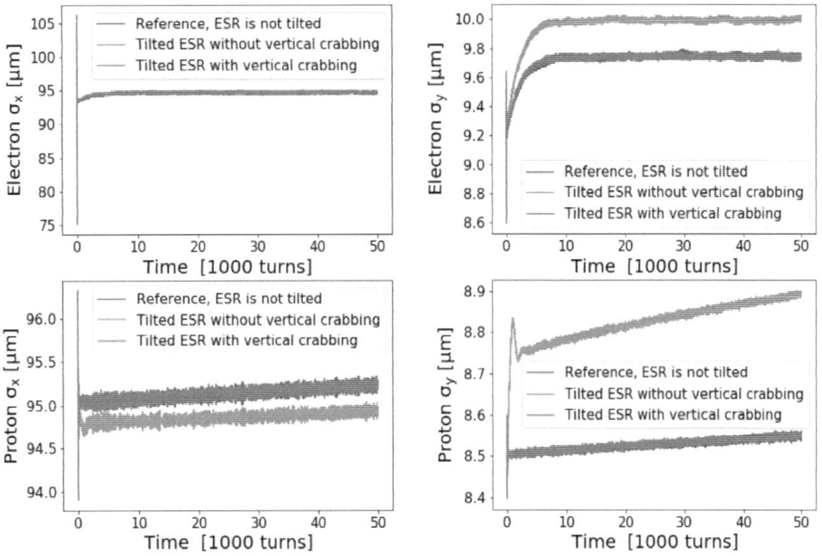

Figure 5. Beam size evolution by strong-strong simulation. The vertical crabbing strength is determined by equation (18).

Figure 6 presents the beam size evolution with vertical crabbing only in ion ring. It shows that the tilted ESR is acceptable for the electron beam. The consequence is a slight luminosity loss.

It is well-known that the strong-strong simulation is prone to numeric noise [11]. Therefore, the weak-strong simulation is performed to cross check the vertical crabbing compensation. The similar physical model is used in weak-strong simulation [12], except that the electron beam is assumed as a fixed Gaussian distribution, and the beam-beam kick is calculated by Bassetti-Erskine formula [13].

Figure 7 presents the weak-strong simulation results. Since the vertical plane is problematic, only the vertical size evolution is shown here. EIC adopts the 2nd-order harmonic crab cavity to mitigate the higher-order synchro-betatron resonances [14]. The left plot shows the tracking result without 2nd-order harmonic crab cavity, and the right plot shows the tracking data with 2nd-order harmonic crab cavity. With vertical crabbing and 2nd-order harmonic crab cavity, the vertical growth caused by tilted ESR is compensated. Without vertical crabbing, no matter if 2nd-order harmonic crab cavity is present, the vertical size growth caused by tilted ESR is unacceptable.

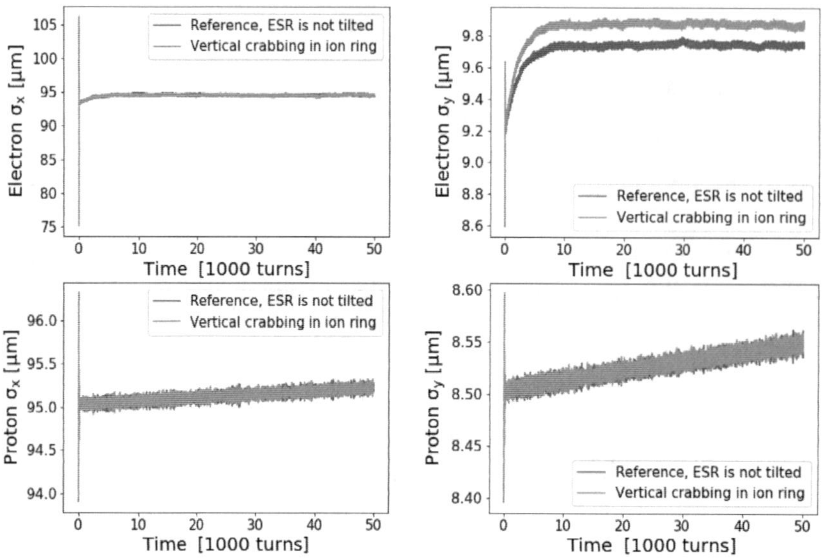

Figure 6. Beam size evolution by strong-strong simulation. Only ion ring is corrected, and there is no vertical crabbing in ESR.

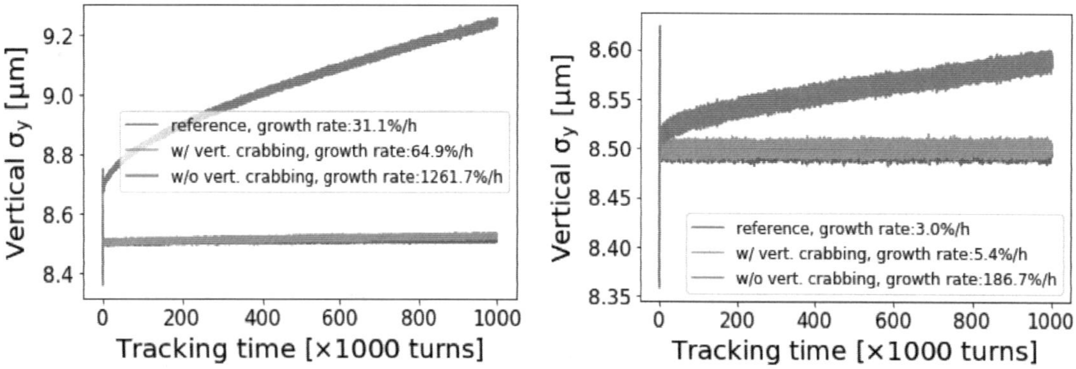

Figure 7. Beam size evolution by weak-strong simulation, left: without 2nd-order harmonic crab cavity, right: with 2nd-order harmonic crab cavity. The last 60% tracking data is fitted with the tracking time $\sigma_y = kt + \sigma_0$, and the growth rate is defined as k/σ_0.

4. Conclusion

We derived the Lorentz Boost map for the tilted ESR scheme in EIC. The required vertical crabbing strength is calculated to compensate the tilted effect on beam-beam interaction. The strong-strong and weak-strong simulation show that the vertical crabbing is necessary in the ion ring to avoid significant vertical size growth. A realistic scheme to introduce vertical crabbing is described in [15].

Acknowledgment

Work supported by Brookhaven Science Associates, LLC under Contract No. DE-SC0012704 with the U.S. Department of Energy.

References

[1] Oide K and Yokoya K 1989 *Phys. Rev.* A **40** 315
[2] Oide K *et al.* 2007 *Proc. 22nd Particle Accelerator Conf. (PAC'07)*, (Albuquerque, NM USA) pp 27–31
[3] Calaga R *et al.* 2021 *Phys. Rev. Accel Beams* **24** 062001
[4] Sun Y-P *et al.* 2009 *Phys. Rev. Spec. Top. Accel Beams* **12** 101002
[5] Ohmi K *et al.* 2011 *Phys. Rev. Spec. Top. Accel Beams* **14** 111003
[6] Xu D *et al.* 2021 *Phys. Rev. Spec. Top. Accel Beams* **24** 041002
[7] Willeke F and Beebe-Wang J 2021 Electron Ion Collider Conceptual Design Report 2021 Tech. rep. United States (doi:10.2172/1765663)
[8] Hirata K 1995 *Phys. Rev. lett.* **74** 2228
[9] Yi-Peng Sun *et al.* 2010 *Phys. Rev. Spec. Top. Accel Beams* **13** 031001
[10] Qiang J *et al.* 2004 *J. Comput. Phys.* **198** 278
[11] Kesting F and Franchetti G 2015 *Phys. Rev. Spec. Top. Accel Beams* **18** 114201
[12] Furman M *et al.* 1995 Comparisons of Beam-Beam Simulations Tech. rep. LBNL No. LBL-37680, CBP Note-152
[13] Bassetti M and Erskine G A 1980 Closed expression for the electrical field of a two-dimensional Gaussian charge Tech. rep. CERN Geneva CERN-ISR-TH/80-06
[14] Xu D *et al.* 2021 Study of Harmonic Crab Cavity in EIC Beam-Beam Simulations Tech. rep. Brookhaven National Lab. BNL-222291-2021-COPA
[15] Xu D *et al.* 2022 *Proc. 13th Int. Particle Accelerator Conf. (IPAC'22)* (Bangkok, Thailand) pp 1972–75

Detector solenoid compensation in the EIC electron storage ring

D Xu[1], Y Luo[1], J Scott Berg[1], H Lovelace III[1], S Tepikian[1], Q Wu[1], B Xiao[1], J Avronsart[1], H Witte[1], M Blaskiewicz[1], C Montag[1], F Willeke[1], B R Gamage[2], V Morozov[3] and Y Hao[4]

[1]Brookhaven National Laboratory, Upton, NY, USA
[2]Jefferson Lab, Newport News, VA, USA
[3]Oak Ridge National Lab, Oak Ridge, TN, USA
[4]Michigan State University, East Lansing, MI, USA

E-mail: dxu@bnl.gov

Abstract. The Electron-Ion Collider (EIC) uses crab cavities to restore the geometrical luminosity loss associated with the large crossing angle. Due to space limitations, the detector solenoid cannot be compensated locally. This paper presents the lattice design to compensate the detector solenoid effects without interfering with the crab cavities. Skew quadrupoles are employed to avoid additional crab cavities. The correction scheme is checked by beam-beam simulation.

1. Introduction

A large crossing angle in the interaction region (IR) is necessary for fast separation of two colliding beams in ring-ring type colliders to achieve high collision rates. Crab cavities, first proposed for linear colliders [1], can compensate for the geometrical luminosity loss induced by crossing angle. This idea was later expanded to include circular colliders [2].

The Electron Ion Collider (EIC) adopts the local crabbing scheme to achieve the desired luminosity [3], as shown in figure 1. In the local crabbing scheme, a pair of crab cavities are installed at both sides of the IP. The upstream crab cavity tilts the beam in $x-z$ plane, and the downstream crab cavity rotates the beam back. In the rest of the rings, both planes stay unaffected.

In the ideal local crabbing scheme, the two crab cavities, located at the location with the betatron phase advance of $\pm\pi/2$ from IP, create desired crab "bump" between them. However, the detector solenoid, a 4 m long solenoid with an integrated strength of up to 12 $\mathrm{T}\cdot\mathrm{m}$ in EIC, would destroy the ideal configuration. Due to the IR space limitation, the anti-solenoid between the crab cavity and the detector solenoid, is not an option. Therefore, the vertical crabbing — the coupling of the vertical and longitudinal motion — is introduced, and has to be corrected in case of a significant luminosity loss.

This paper will first introduce the concept of crab dispersion. With the help of the crab dispersion, two methods are proposed to correct the vertical crabbing caused by the detector solenoid. The beam-beam simulations are performed to check the correction effect.

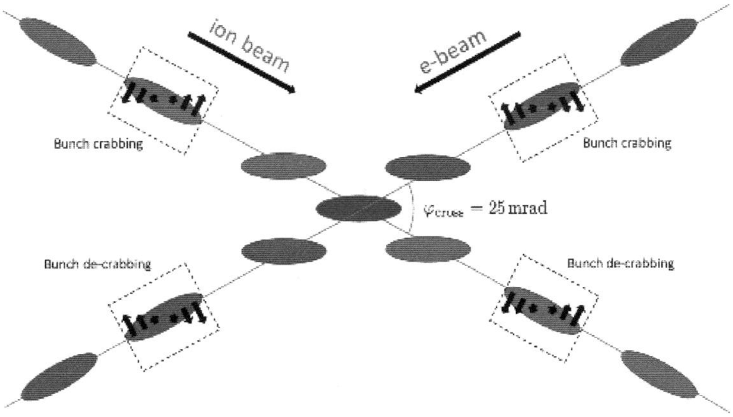

Figure 1. EIC local crabbing compensation scheme [3].

2. Crab dispersion

The Hamiltonian of a thin crab cavity is

$$H = \left(\frac{x\lambda_x}{\Lambda_x} + \frac{y\lambda_y}{\Lambda_y}\right)\frac{\sin(k_c z)}{k_c} \tag{1}$$

where x, y, z are the horizontal, vertical, and longitudinal coordinates when a test particle pass through the thin crab cavity, $\lambda_{x,y}/\Lambda_{x,y}$ denoting the horizontal and vertical kick strength with $\Lambda_x = \sqrt{\beta_x^*\beta_{x,c}}, \Lambda_y = \sqrt{\beta_y^*\beta_{y,c}}$, and k_c the wave number. Here we use the lattice related parameters $\Lambda_{x,y}$ to normalize the crab cavity strength for simplicity.

Following the definition in [4], the crab dispersion is defined as the transverse dependence on longitudinal coordinate z,

$$\zeta = \left(\frac{\partial x}{\partial z}, \frac{\partial p_x}{\partial z}, \frac{\partial y}{\partial z}, \frac{\partial p_y}{\partial z}\right)^{\mathrm{T}} \tag{2}$$

Assuming the lattice is symmetrical around IP, the crab dispersion from the upstream and downstream crab cavities is,

$$\begin{aligned}\zeta_b &= (0, \lambda_{b,x}/\Lambda_x, 0, \lambda_{b,y}/\Lambda_y)^{\mathrm{T}} \\ \zeta_a &= (0, \lambda_{a,x}/\Lambda_x, 0, \lambda_{a,y}/\Lambda_y)^{\mathrm{T}}\end{aligned} \tag{3}$$

where the subscript "b" denotes before IP, and "a" denotes after IP. Projecting the crab dispersion back to IP, the crab dispersion before and after the collision is

$$\begin{aligned}\zeta_b^* &= \begin{bmatrix}\lambda_{b,x}\sin\Psi_x \\ \lambda_{b,x}\cos\Psi_x/\beta_x^* \\ \lambda_{b,y}\sin\Psi_y \\ \lambda_{b,y}\cos\Psi_y/\beta_y^*\end{bmatrix} \\ \zeta_a^* &= \begin{bmatrix}(\lambda_{b,x}-\lambda_{a,x})\sin\Psi_x \\ (\lambda_{b,x}+\lambda_{a,x})\cos\Psi_x/\beta_x^* \\ (\lambda_{b,y}-\lambda_{a,y})\sin\Psi_y \\ (\lambda_{b,y}+\lambda_{a,y})\cos\Psi_y/\beta_y^*\end{bmatrix}\end{aligned} \tag{4}$$

where $\Psi_{x,y}$ is the horizontal/vertical phase advance from the upstream crab cavity (IP) to IP (downstream crab cavity).

To create an effective head-on collision in the boost frame and avoid the crab dispersion leaking out of IR, ζ_b^* and ζ_a^* should fulfill the following conditions,

$$\zeta_b^* = (\theta_c, 0, 0, 0)^{\mathrm{T}}, \qquad \zeta_a^* = (0, 0, 0, 0)^{\mathrm{T}} \tag{5}$$

where θ_c is the half crossing angle.

A possible solution to equation (4) is

$$\begin{aligned} \Psi_x &= \pi/2 \\ \lambda_{b,x} = \lambda_{a,x} = \theta_c, &\qquad \lambda_{b,y} = \lambda_{a,y} = 0 \end{aligned} \tag{6}$$

Taking the detector solenoid into consideration, and projecting the crab dispersion at IP,

$$\begin{aligned} \zeta_b^* &= \begin{bmatrix} \lambda_{b,x} \cos\phi \sin\Psi_x + \lambda_{b,y} \sin\phi \sin\Psi_y \\ \lambda_{b,x} \cos\phi \cos\Psi_x/\beta_x^* + \lambda_{b,y} \sin\phi \cos\Psi_y/\beta_y^* \\ \lambda_{b,y} \cos\phi \sin\Psi_y - \lambda_{b,x} \sin\phi \sin\Psi_x \\ \lambda_{b,y} \cos\phi \cos\Psi_y/\beta_y^* - \lambda_{b,x} \sin\phi \cos\Psi_x/\beta_x^* \end{bmatrix} \\ \zeta_a^* &= \begin{bmatrix} (\lambda_{b,x} \cos(2\phi) - \lambda_{a,x}) \sin\Psi_x \\ +\lambda_{b,y} \sin(2\phi) \sin\Psi_y \\ (\lambda_{b,x} \cos(2\phi) + \lambda_{a,x}) \cos\Psi_x/\beta_x^* \\ +\lambda_{b,y} \sin(2\phi) \cos\Psi_y/\beta_y^* \\ (\lambda_{b,y} \cos(2\phi) - \lambda_{a,y}) \sin\Psi_y \\ -\lambda_{b,x} \sin(2\phi) \sin\Psi_x \\ (\lambda_{b,y} \cos(2\phi) + \lambda_{a,y}) \cos\Psi_y/\beta_y^* \\ -\lambda_{b,x} \sin(2\phi) \cos\Psi_x/\beta_x^* \end{bmatrix} \end{aligned} \tag{7}$$

where 2ϕ is the rotation angle by the detector solenoid. As a demonstration, the solenoid focusing effect, which will be considered in the following sections, is ignored here. Substituting equation (7) into equation (5), a possible solution is

$$\begin{aligned} \Psi_x &= \Psi_y = \pi/2 \\ \lambda_{b,x} &= \lambda_{a,x} = \theta_c \cos\phi \\ \lambda_{b,y} &= -\lambda_{a,y} = \theta_c \sin\phi \end{aligned} \tag{8}$$

However, the conditions in equation (8) are very hard to match. The required vertical crab cavity voltage is extremely large due to the small β_y^*. Multiple crab cavities may conflict with the impedance budget. As a result, the equation (8) is unrealistic in the actual lattice design.

3. Correction by skew quadrupoles

The equation (5) gives 8 constraints. Without introducing additional crab cavities, there are only 2 knobs provided by the crab cavity voltage. The other 6 knobs can be found from the appropriate transfer matrix between the crab cavities and IP.

Let M be the 4-by-4 transfer matrix from the upstream crab cavity to IP. If M has the following form

$$M = \begin{bmatrix} m_{11} & m_{12} & m_{13} & m_{14} \\ m_{21} & 0 & m_{23} & m_{24} \\ m_{31} & 0 & m_{33} & m_{34} \\ m_{41} & 0 & m_{43} & m_{44} \end{bmatrix} \tag{9}$$

tuning the crab cavity voltage so that

$$\zeta_b = (0, \theta_c/m_{12}, 0, 0)^{\mathrm{T}}$$

it is easy to show

$$\zeta_b^* = M\zeta_b = (\theta_c, 0, 0, 0)^{\mathrm{T}}$$

i.e. the beam is correctly crabbed for crab crossing collision.

Let N be the 4-by-4 transfer matrix from the upstream crab cavity to the downstream crab cavity. If N has the following form

$$N = \begin{bmatrix} n_{11} & 0 & n_{13} & n_{14} \\ n_{21} & n_{22} & n_{23} & n_{24} \\ n_{31} & 0 & n_{33} & n_{34} \\ n_{41} & 0 & n_{43} & n_{44} \end{bmatrix} \tag{10}$$

tuning the downstream crab cavity so that

$$\zeta_a = (0, -\theta_c n_{22}/m_{12}, 0, 0)^{\mathrm{T}}$$

Then we have

$$\zeta_a^* = N\zeta_b + \zeta_a = (0, 0, 0, 0)^{\mathrm{T}}$$

i.e. there is no crab dispersion leaking out of IR.

We can put skew quadrupoles between the crab cavities and IP to adjust the linear transfer matrix. Noticing that M and N are $x-y$ coupled, more skew quadrupoles are needed out of the crab bump to finish the decoupling. The $x-y$ decoupling scheme for EIC can be found in [5].

4. Matching for EIC electron storage ring

We added skew components to the existing straight quadrupoles in IR6, and matched the linear transfer matrix to equation (9) and equation (10). Correction results are shown in figure 2. After compensation, the linear transfer matrix from the upstream crab cavity to IP is

$$M = \begin{bmatrix} -0.001553 & 9.394 & -0.06912 & 2.809 \\ -0.1064 & -0.3939 & 0.002898 & -0.1178 \\ 0.03014 & -2.790\text{e}{-10} & -0.1585 & 2.347 \\ 0.03214 & -4.362\text{e}{-11} & 0.07506 & -7.419 \end{bmatrix}$$

Except m_{22}, m_{32} and m_{42} are well matched to 0. The non-zero m_{22} is acceptable as the beam-beam has large tolerance on $\partial p_x/\partial z$ (see beam-beam simulations below). The linear transfer matrix from the upstream crab cavity to the downstream crab cavity is

$$N = \begin{bmatrix} 0.8591 & -3.662\text{e}{-9} & -4.449\text{e}{-10} & 6.804\text{e}{-9} \\ 0.0002619 & 1.164 & -0.006533 & 0.2299 \\ -0.1832 & 2.289\text{e}{-9} & 1.495 & -19.97 \\ 0.02776 & -3.066\text{e}{-11} & -0.2572 & 4.104 \end{bmatrix}$$

The N matrix is well matched.

The maximum skew component strength used in our compensation is 1.2 T/m which is below the limitation of 1.5 T/m from the magnet design group.

The chromatic effect may harm the compensation effect. To check the chromatic effect, we express the transfer matrix as a polynomial of momentum spread δ,

$$M = M_0 + M_1\delta + M_2\delta^2 \cdots \tag{11}$$

and the inverse matrix is expressed as

$$\begin{aligned} M^{-1} = M_0^{-1} &- M_0^{-1}M_1M_0^{-1}\delta \\ &- M_0^{-1}\left(M_2 - M_1M_0^{-1}M_1\right)M_0^{-1}\delta^2 + \cdots \end{aligned} \tag{12}$$

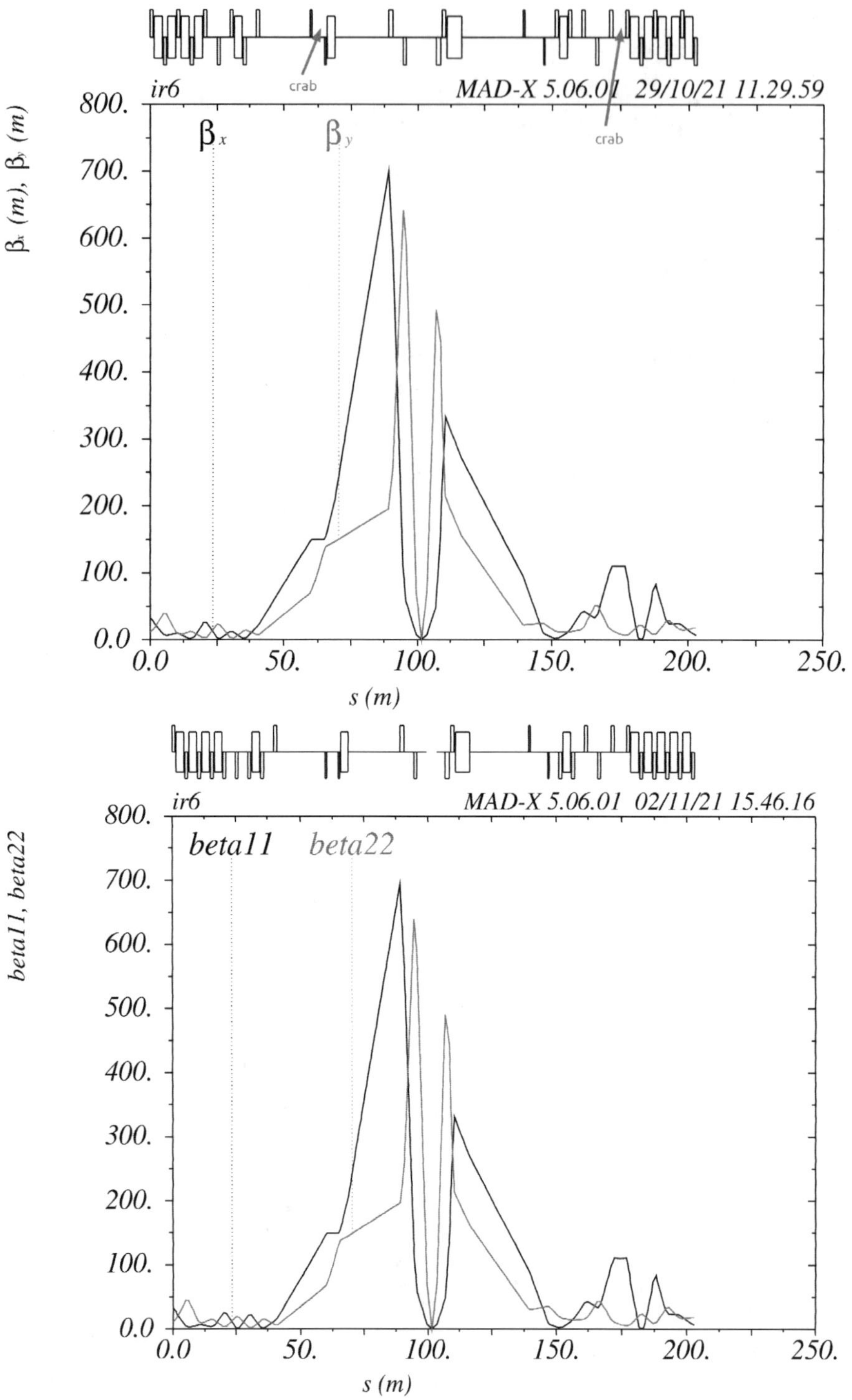

Figure 2. Ripken coupled β functions after coupling compensation for ESR detector solenoid, top: with detector solenoid off, bottom: with detector solenoid on and compensation by skew quadrupoles.

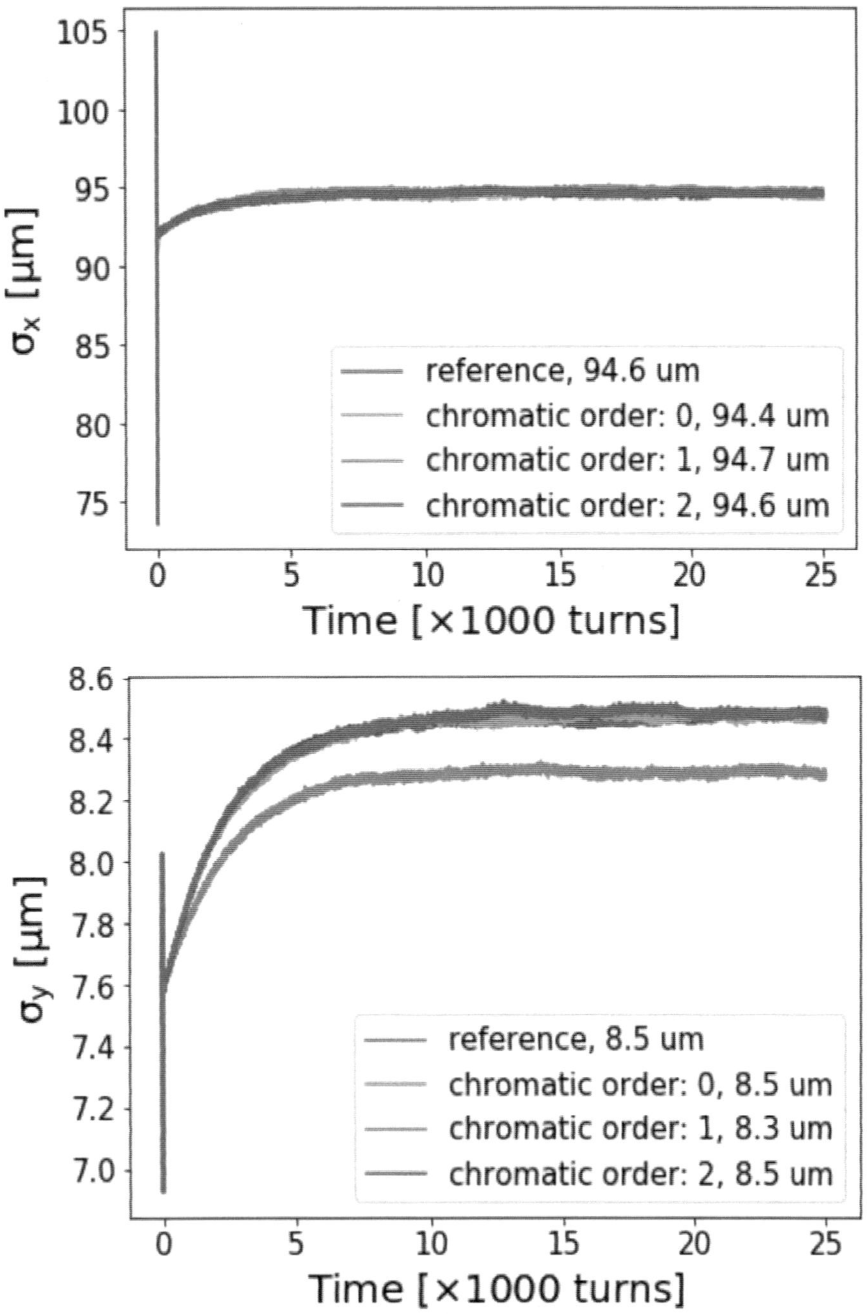

Figure 3. Weak-strong simulation with detector solenoid compensated. The blue curve is a reference with detector solenoid off. The linear transfer matrix is truncated at different orders of δ, as shown in equation (11) and equation (12).

Weak-strong simulation is a widely used approach in beam-beam study [6, 7]. We use a self-written weak-strong code to check the correction result. In the simulation, the ion beam is rigid with a horizontal centroid as

$$x_i = -\theta_c \left[\frac{4}{3} \frac{\sin(k_{c,i}z)}{k_{c,i}} - \frac{1}{3} \frac{\sin(2k_{c,i}z)}{2k_{c,i}} - z \right] \tag{13}$$

where $k_{c,i}$ is the wave number of the crab cavities in the ion ring. A second order harmonic crab cavity is used to flatten the ion bunch [8]. The ion bunch is cut into multiple slices. Each slice is represented by a 2D Gaussian distribution in $x - y$ plane. The weak electron beam are simulated by a number of macro particles. The one-turn map at IP is represented by the linear betatron map. The transfer maps between crab cavities and IP, including the detector solenoid and quadrupoles, are expressed by equation (11) truncated upto different orders. The crab cavity kick is from the Hamiltonian in equation (1). The beam-beam kick is calculated by the Bassetti and Erskine formula [9]. The effects of radiation damping and quantum excitation are represented by a lumped element [10].

The simulation results are shown in figure 3. The orange curve shows the beam size evolution after the detector solenoid is compensated, while the blue curve displays the tracking data with the solenoid off. Both of them overlap with each other. It turns out that our compensation scheme works successfully. The red and green curves show the tracking when equation (11) and equation (12) are truncated upto δ and δ^2. As M and M^{-1} are symplectic upto the truncated order, it is understandable the green cuve reaches a different equilibrium size in the vertical plane. The red curve overlaps with the reference line too, which turns out the chromatic effect is negligible.

5. Summary

The detector solenoid in EIC Electron Storage Ring introduces vertical crabbing which has to be compensated. This paper proposes two different methods to compensate the detector solenoid with the help of crab dispersion. The second method is applied to the ESR lattice, and the weak-strong simulation validates the compensation effect.

Acknowledgment

Work supported by Brookhaven Science Associates, LLC under Contract No. DE-SC0012704 with the U.S. Department of Energy.

References

[1] Palmer R 1988 Energy scaling, Crab crossing and the pair problem Tech. rep. Stanford Linear Accelerator Center SLAC-PUB-4707
[2] Oide K and Yokoya K 1989 *Phys. Rev.* A **40** 315
[3] Willeke F and Beebe-Wang J 2021 Electron Ion Collider Conceptual Design Report 2021 Tech. rep. United States (doi:10.2172/1765663)
[4] Yi-Peng Sun *et al.* 2010 *Phys. Rev. Spec. Top. Accel Beams* **13** 031001
[5] Gamage B R *et al.* 2021 *Proc. 12th Int. Particle Accelerator Conf. (IPAC'21)* (Campinas, Brazil) pp 1439–42
[6] Papaphilippou Y *et al.* 1999 *Phys. Rev. Spec. Top. Accel Beams* **2** 104001
[7] Luo Y *et al.* 2012 *Phys. Rev. Spec. Top. Accel Beams* **15** 051004
[8] Xu D *et al.* 2021 Study of Harmonic Crab Cavity in EIC Beam-Beam Simulations Tech. rep. Brookhaven National Lab. BNL-222291-2021-COPA
[9] Bassetti M and Erskine G A 1980 Closed expression for the electrical field of a two-dimensional Gaussian charge Tech. rep. CERN Geneva CERN-ISR-TH/80-06
[10] Qiang J *et al.* 2002 Parallel simulation of beam-beam interaction in high energy accelerators Tech. rep. LBNL LBNL-50786

Field enhanced compact S-band gun employing a pin cathode

R Bazrafshan [1,4], M Fakhari [1], T Rohwer [1], K Flöttmann [2], N H Matlis [1] and F X Kärtner [1,3,4]

[1] Center for Free-Electron Laser Science, Deutsches Elektronen-Synchrotron DESY, Hamburg, Germany
[2] Deutsches Elektronen-Synchrotron DESY, Hamburg, Germany
[3] The Hamburg Centre for Ultrafast Imaging, Universität Hamburg, Germany
[4] Physics Department, Universität Hamburg, Germany

Email: reza.bazrafshan@desy.de

Abstract. S-band RF-guns are highly developed for production of low emittance relativistic electron bunches, but need powerful klystrons for driving. Here, we present the design and first experimental tests of a compact S-band gun, which can accelerate electrons up to 180 keV powered by only 10 kW from a compact rack-mountable solid-state amplifier. A pin-cathode is used to enhance the RF electric field on the cathode up to 100 MV/m as in large-scale S-band guns. An electron bunch is generated through photoemission from a flat copper surface on the pin ex-cited by a UV laser pulse followed by a focussing solenoid producing a low emittance bunch with 0.1 mm mrad transverse emittance for up to 100 fC bunch charge. We are currently in the conditioning phase of the gun and first experiments show good agreement with simulations. The compact gun will serve three purposes: (i) it can be used directly for ultrafast electron diffraction; (ii) as an injector into a THz booster producing 0.3 MeV to 2 MeV electron bunches for ultrafast electron diffraction; (iii) The system in (ii) serves as an injector into a THz linear accelerator producing a 20 MeV beam for the AXSIS X-ray source project.

1. Introduction

Photocathode rf guns have been proven to be an excel-lent electron source for very high-quality beams required for vacuum ultra-violet (VUV) and X-ray FELs [1]. In photocathode rf guns, electron beams are generated at the photocathode by the drive-laser pulses and accelerated immediately by the rf field. Since the electron bunches after emission from the photocathode are of very high intensity and have close to zero a velocity, high space charge forces quickly deteriorate beam quality. Thus, a high acceleration field at the cathode is of great importance for reaching a high electron beam quality. The beam quality can be further optimized by means of three-dimensionally (3D) shaped laser pulse with a well-defined profile [2]. The expansion and non-linearity of the electron bunch in the phase space can be minimized with the optimized initial electron distribution. The initial profile of the beam after extraction from the cathode undergoes a modification under the rf and the solenoid fields as well as the space charge force. Due to the strong accelerating field in the cavity, electrons can be field-emitted from the copper cavity surface and the photo-cathode. Any dusts on the front surface of the cathode may be a strong field emitter as well causing increased dark current. Dark current at the gun can be accelerated together with the electron bunch, thus the dark current dynamics must be investigated. During gun operation, multipacting peaks are found at the beginning and/or at the end of the rf pulse. The multipacting depends on the solenoid field profile. The amount of dark current and multipacting can be reduced by careful conditioning of the gun.

The concept of the AXSIS project at the Deutsches Elektronen-Synchrotron (DESY) is the development of a compact X-ray light source using terahertz acceleration [3]. Although the final AXSIS machine is expected to be powered exclusively by THz radiation, intermediate phases may resort to a radio-frequencies (RF), i.e. 3 GHz. S-band RF-guns are highly developed for production of low emittance relativistic electron bunches, normally need powerful klystrons which are bulky and expensive. Here, we present the design and first experimental tests of a compact electron source which can accelerate electrons up to 180 keV by only 10 kW from a compact rack mountable amplifier.

2. RF gun design

The AXSIS RF gun is an S-band compact RF gun with a frequency of 2.998 GHz, combining both acceleration and compression of the electron bunch in a single device. By implementing a pin-shaped photocathode with a flat 0.8 mm tip [4,5], strong enhancement of the RF-field strength in the vicinity of the cathode is achieved which substantially reduces the power requirements, and hence the cost and complexity of the device. A solenoid is also used just after the gun to focus the electron beam transversely.

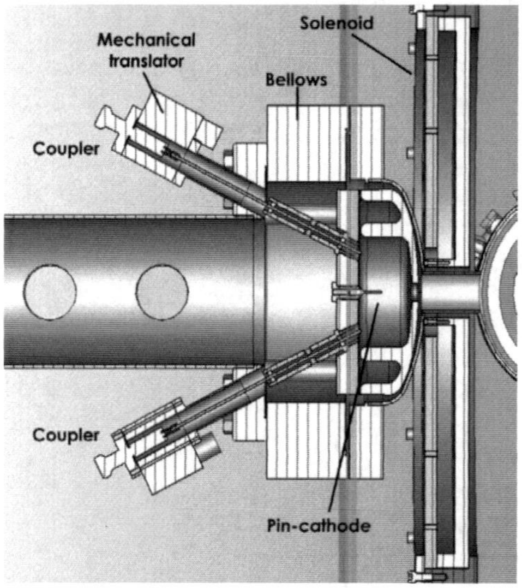

Figure 1. Sectional view of AXSIS RF gun (first prototype).

Figure 1 shows the designed structure. The electrons are generated via photoemission from a circular area with a diameter of 0.8 mm at the top of the stud. The 10 kW input power is injected into the cavity with an input antenna via a coaxial feedthrough shown in figure 1. A second antenna is used as the pick-up for measuring the electric field inside the cavity. For tuning the cavity, this gun consists of two parts: the anode part is fixed, while the cathode part can be moved by a mechanical translator giving us also the possibility of tuning the cavity by adjusting the distance between the cathode and anode. In addition, the cavity design employs a choke filter which confines the field, and also facilitating access to the interior of the cavity and tuning the cavity for optimal coupling.

3. Beam dynamics

A half-cell pillbox cavity employing a metallic pin-cathode is designed to be used as the electron injector. The electric field distribution in the cavity is computed with CST Microwave Studio Eigen-mode solver [6] and is shown in figure 2. The electrons are injected in the negative phase of the RF signal. Therefore, a negative energy chirp is introduced to the bunch which leads to longitudinal compression. The ASTRA particle tracking solver is used to investigate the beam dynamics in the gun [7].

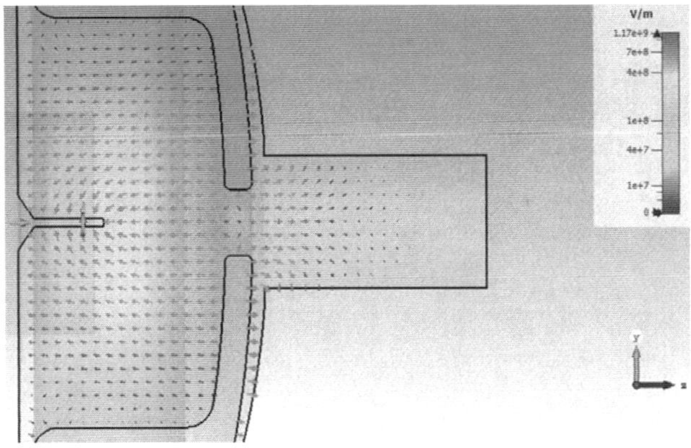

Figure 2. Electric field distribution.

The simulation results are shown in figure 3. The cavity field, reaches over 110 MV/m using only a 10 kW peak power solid-state amplifier, to both accelerate the electrons from rest up to 180 keV kinetic energy and impart an approximately linear velocity chirp onto the bunches with 0.1 mm length at the beginning near the cathode, allowing them to compress via velocity bunching to 10 μm corresponding to 30 fs rms durations at a distance 8 cm after the cathode. In order to focus the beam transversely to provide the required spatial resolution we designed a solenoid to be installed directly after the cavity. According to figure 3(c) this solenoid focuses the beam to 100 μm rms spot size.

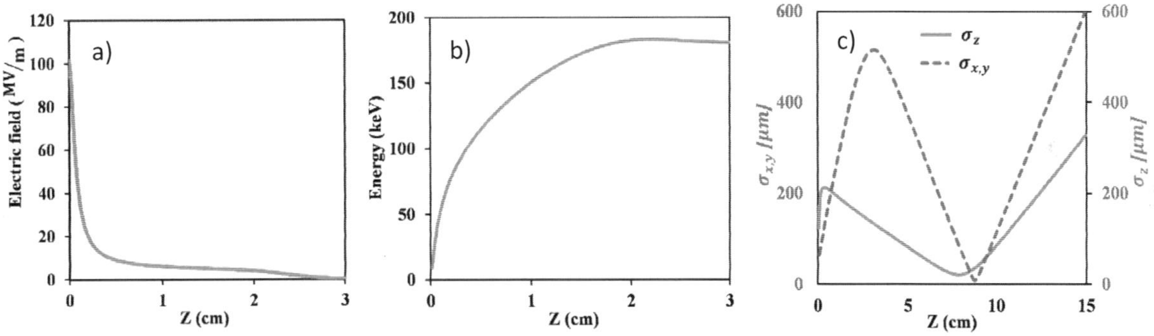

Figure 3. a) Field enhancement at the cathode vs propagation distance (Z), b) Acceleration dynamics of the electron bunch vs Z, c) Focusing and compression of the bunch vs Z.

4. RF gun performance

The first prototype of the RF gun based on the above design was constructed and is currently being commissioned. In the initial design, in order to access the cavity from behind, a chamber connected to the cathode part. Using a vector network analyzer, the frequency response of the cavity was determined and a loaded quality factor of 6,000 was measured with a reflection (S_{11}) reaching below -50 dB. Although the RF gun demonstrated a satisfactory frequency response in agreement with the design, RF leakage into the back chamber resulted in arcing when power above 8 kW was applied. In addition, high levels of secondary electron emission from multipacting was observed. To address these issues, a second prototype was designed and fabricated with shorter couplers surrounded by RF fingers (figure 4). The revised prototype demonstrated excellent performance, in good agreement with simulation, as well as near complete suppression of multipacting.

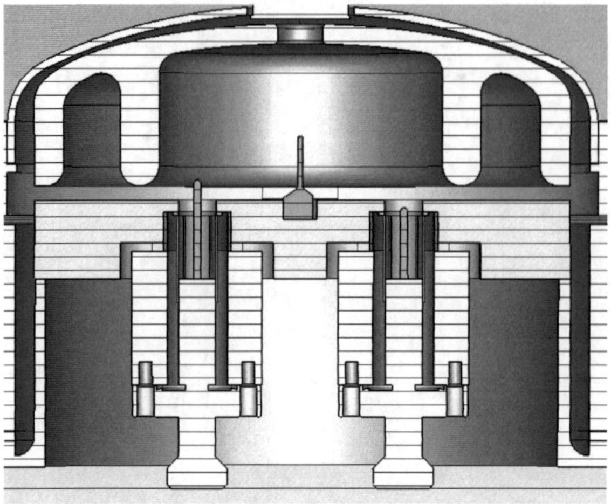

Figure 4. Second prototype design of developed compact RF gun.

Figure 5a shows the frequency response of the cavity with a resonance at 2.998 GHz. Conditioning of the gun using the 10 kW amplifier enabled strong suppression of multipacting and reduction of dark charge to below 400 fC per RF pulse of 5.5 μs (figure 5b) at a repetition rate of 12.5 Hz, it can be reduced by increasing the repetition rate and continuing conditioning. A combination of a Faraday Cup and a YAG scintillator (30 cm away from the cathode) were used to measure the charge and field-emitted beam shape (figure 5c). The resulting transmitted power (P_{21}) of 10 kW signal and negligible reflected power (P_{11}) were measured using the probe and feed antenna respectively (figure 6).

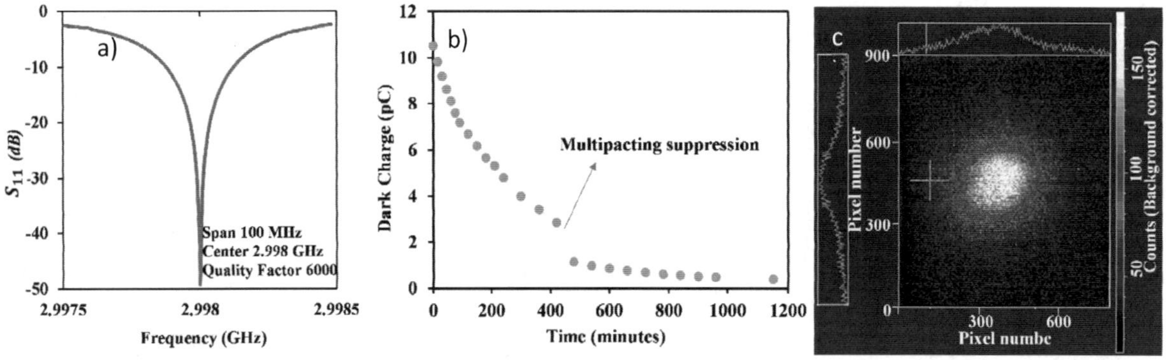

Figure 5. a) Frequency response of the RF Cavity. b) Reduction of dark charge and multipacting during conditioning measured with a Faraday-cup. c) Field-emitted beam detected with a YAG scintillator.

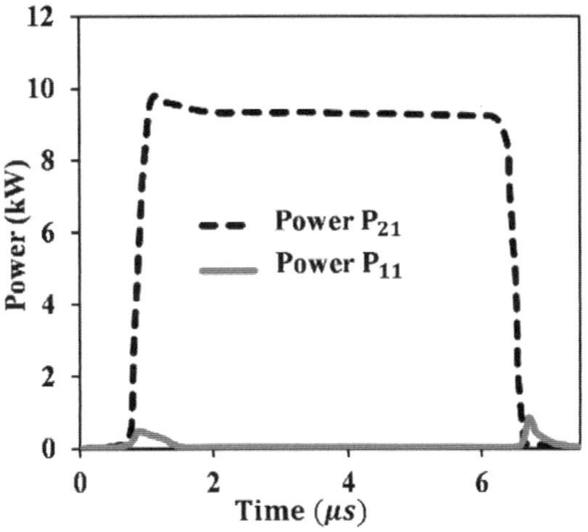

Figure 6. Coupled and reflected RF power.

5. Conclusion

A compact RF gun has been designed and fabricated (figure 7) and has undergone initial testing demonstrating a high cavity quality factor of 6,000 and low levels of multipacting and dark charge (400 fC). Based on simulation, this performance, which matches or slightly exceeds expectation, will allow operation of the gun at an energy of up to 180 keV low emittance electron bunches, powered by only 10 kW solid-state amplifier as the power source. A solenoid is used to focus the beam transversely. The rms longitudinal and transverse sizes of the electron bunch at the sample interaction point would be 10 mm and 100 mm respectively, which fit excellent for the purposes of the UED and AXSIS applications. Following completion of the conditioning phase, photo-triggered electron generation using a UV laser on the photocathode will follow, with first results expected in the next couple of months. The next round of testing will target characterization of the energy, energy spread and emittance of the beam.

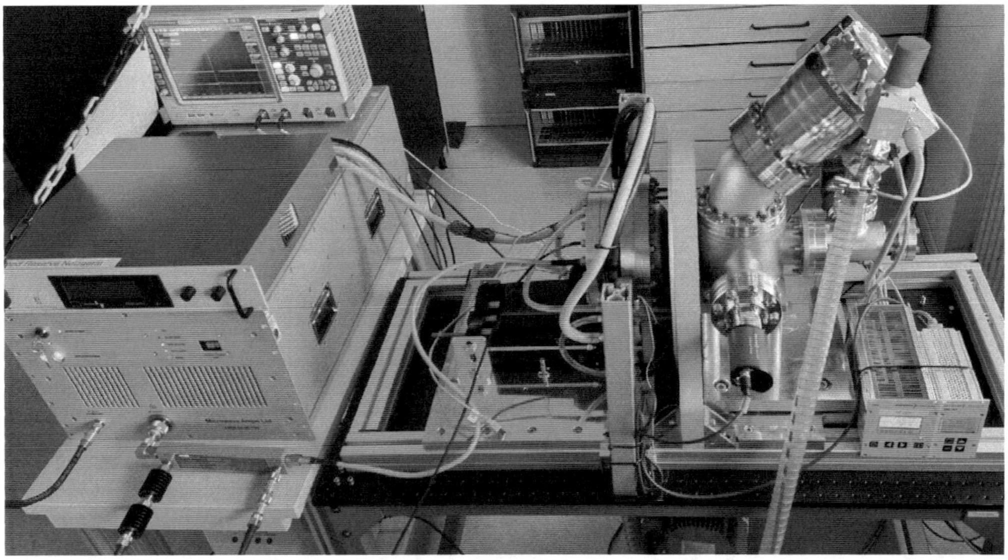

Figure 7. Compact RF gun setup with a compact amplifier.

Acknowledgements

The authors are very grateful for the help and support in electron beam conditioning and diagnostic from Dr. Hossein Delsim-Hashemi and Dr. Matthias Hoffmann.

References

[1] Ayvazyan V *et al.* 2002 *Phys. Rev. Lett.* **88** 104802
[2] Hua Y L *et al.* 2003 *Phys. Rev. Lett.* **91** 074801
[3] Kaertner F X *et al.* 2016 *Nucl. Instrum. Methods Phys. Res., Sect.* A **829** 24
[4] Daoud H, Floettmann K and Dwayne Miller R J 2017 *Struct. Dyn.* **4** 044016
[5] German patent application DE 10 2016 105 443.7. 23. March 2016
[6] CST, Computer Simulation Technology, `https://www.cst.com/`
[7] Flöttmann K 2017 ASTRA: a space charge tracking algorithm, user manual

Beam orbit shift due to BPM thermal deformation using machine learning

Kemin Chen[1], Faya Wang[1], Guanliang Wang[1], Zhe Wang[1], Masahito Hosaka[1], Lei Guo[2] and Wei Xu[1]

[1] NSRL, University of Science and Technology of China, Hefei, Anhui 230029, China
[2] Nagoya University, Furo, Chikusa, Nagoya, Aichi 464-8601, Japan

E-mail: wxu@ustc.edu.cn

Abstract. Stabilizing beam orbit is critical for advanced synchrotron radiation light sources. The beam orbit can be affected by many sources. To maintain a good orbit stability, global orbit feedback systems (OFB) has been widely used. However, the BPM thermal deformation would lead to BPM misreading, which can not be handled by OFB. Usually, extra diagnostics, such as position transducers, is needed to measure the deformation dependency of BPM readings. Here, an alternative approach by using the machine operation historic data, including BPM temperature, insertion device (ID) gaps and corrector currents, is presented. It is demonstrated at Hefei Light Source (HLS). The average orbit shift due to BPM thermal deformation is about 34.5 μm/°C (horizontal) and 20.0 μm/°C (vertical).

1. Introduction

Synchrotron radiation source has many advanced characteristics such as high brightness, transverse coherence, good time structure, etc, which is developed as one of the most powerful scientific tools over decades [1]. The beam orbit stability is required to be better than 10% of the beam size for modern light sources [2, 3]. Many sources could lead to beam orbit shifts. Global orbit feedback system has been widely used to suppress them. However, the BPM thermal deformation would lead to BPM misreading, instead of direct orbit disturbance, which could not be effectively corrected by OFB. Significant beam orbit shift due to BPM misreading has been observed in sources like APS [4], KEKB [5], and HLS [6].

Dedicated diagnostics is needed to measure and correct BPM thermal deformation, for example, the work carried out at HLS [6]. It is a straight forward method and generally provides good results. However, few downsides are there like the cost of hardwares and the need of machine study time, which actually make it difficult for wide implementation. In this paper, we will present an alternative approach by using machine learning, which is successfully demonstrated at HLS-II by predicting effective beam orbit shift accompanied by direct corrector current. It could be integrated to orbit feedback system to improve beam orbit stability.

In the following, we first introduce the theoretical model of beam orbit change due to BPM vacuum chamber thermal deformation, then a surrogate model built using a neural network (NN). Finally results from the surrogate model are presented.

2. Theoretical model

Thermal effects from synchrotron radiation, parasitic heat induced by machine impedance and tunnel temperature variation can cause BPM vacuum chamber mechanical deformation, which leads to its geometric center shift in global reference coordinate and results in BPM misreading. BPM reading, $\vec{u}$, can be written as

$$\vec{u} = \vec{R} - \vec{r}, \tag{1}$$

where $\vec{r}$ and $\vec{R}$ are BPM center and beam position respectively in global coordinate.

In principle, BPM center shift does not affect beam orbit directly, except when OFB is functioning, where the beam experiences extra kicks from correctors. Accordingly, a BPM reading changes as,

$$\Delta\vec{u}_i = \Delta\vec{R}_i - \Delta\vec{r}_i = (\frac{\partial\vec{R}_i}{\partial\vec{r}_i} - \overline{\overline{I}}) \cdot \Delta\vec{r}_i, \tag{2}$$

where $\Delta\vec{r}_i$ is BPM center shift due to heating, $\Delta\vec{R}_i$, beam position change at BPM due to OFB, $\partial\vec{R}_i/\partial\vec{r}_i$, Jacobian matrix and $\overline{\overline{I}}$, a unit 2 by 2 tensor.

With an ideal OFB, beam orbit is corrected back to its original, it then yields

$$\Delta\vec{u} = (\Delta\vec{u}_1, ..., \Delta\vec{u}_N) = 0, \tag{3}$$

$$\Delta\vec{r} = \Delta\vec{R}|_{\Delta\vec{u}=0} = \mathcal{M} \cdot \Delta\Theta, \tag{4}$$

where $\Delta\Theta$ is corrector kick strength change, and $\mathcal{M}$ is machine orbit response matrix (ORM), which ideally depends only on machine optics and is given by

$$\mathcal{M}_{ik} = \frac{\partial u_i}{\partial\theta_k} = \frac{\sqrt{\beta_i\beta_k}}{2\sin\pi\nu}\cos[\nu(\phi_i - \phi_k + \pi)], \tag{5}$$

where (β_i, ϕ_i) and (β_j, ϕ_j) are the beta function and phase advance at position i and k and ν is the betatron tune.

The Θ only responses to the BPM readings, which depends on the beam reference orbit and real beam orbit change. The change of reference orbit is caused by the BPM misreading. The real beam orbit shift can be generated by the gap variation of IDs. The physics model of the corrector strength can be described as a function of the ID gaps and BPM temperature by

$$\Theta(\vec{g}, \vec{T}) \approx \Theta(\vec{g}, \vec{T_0}) + \frac{\partial\Theta}{\partial\vec{T_0}}\Delta\vec{T}, \tag{6}$$

where $\vec{T}$ is the BPM vacuum chamber temperature, $\Delta\vec{T}$ is the BPM temperature variation and $\vec{g}$ is the ID gap.

Therefore, beam orbit shift due to BPM thermal deformation is

$$\Delta\vec{r} = \mathcal{M}\frac{\partial\Theta}{\partial\vec{T_0}}\Delta\vec{T} = \mathcal{M}\mathcal{N}\Delta\vec{T}, \tag{7}$$

where $\mathcal{N}$ is the temperature response matrix (TRM) of corrector strength at given ID gaps.

Technically, TRM could be obtained with implementation of dedicated diagnostics [6], which is difficult for a large storage ring with hundreds of BPMs. Alternatively, we will use a neural network model to obtain the TRM.

Journal of Physics: Conference Series **2420** (2023) 012014 doi:10.1088/1742-6596/2420/1/012014

3. Neural network surrogate model for TRM

Neural networks could efficiently find the function between inputs and outputs, which handles regression and classification problems very well, especially for nonlinear problems with large quantity data [7]. Therefore, a neural network is applied to this work.

According to theoretical analysis above, the BPM temperature, IDs gap and correctors current data from the historic data for the entire year of 2019 are used to study the thermal effects. We take ID gaps (see figure 1) and temperature (see figure 2) as inputs and corrector currents (see figure 3) as outputs, and use the about 80 % of the processed data as training dataset to train NN models and the rest data as testing dataset to test models. Then get an NN surrogate model which describes the mapping function,

$$\Theta(\vec{T}, \vec{g}) = f(\vec{T}, \vec{g}). \tag{8}$$

The TRM can be obtained using this surrogate model, while the ID gaps are fixed. This process is similar to the measurement of the ORM in a real storage ring. A disturbance of the BPM temperature ΔT (0.1 °C) is applied to each BPM in turn. The new temperature $\vec{T}'$ is used as the inputs to the model to get the corresponding corrector currents

$$\Theta'(\vec{T}', \vec{g}) = f(\vec{T}'(T_1, ..., T_i + \Delta T, ..., T_N), \vec{g}). \tag{9}$$

Finally, the TRM is calculated as

$$N_{ij} = \frac{\partial \theta_j}{\partial T_i} \approx \frac{\theta_j' - \theta_j}{\Delta T}, \tag{10}$$

where N_{ij} is the term of TRM $\mathcal{N}$, θ_j is the j-th term of Θ and T_i is the i-th term of $\vec{T}$.

For more accurate results, the predictions with NN multiple models are combined, which is called the model ensemble. Averaging usually works well for a wide range of problems. Here we train several different single models by changing the activation function, the number of layers and the number of neurons in each layer. The results from different models are averaged as the ensemble result in this work.

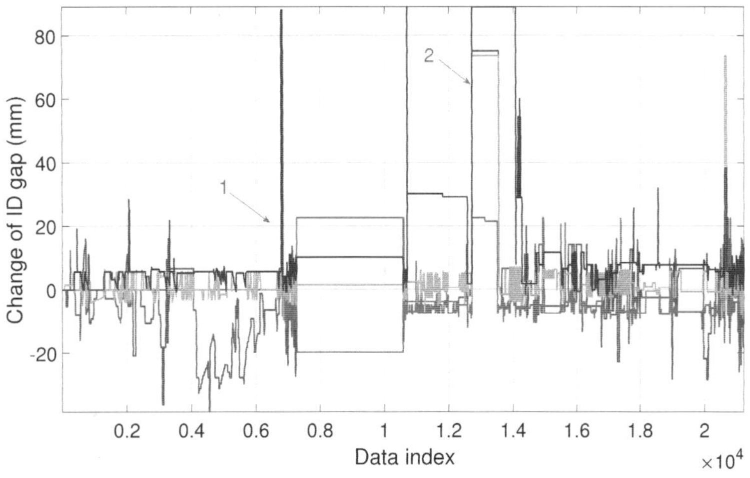

Figure 1. Change of ID gaps.

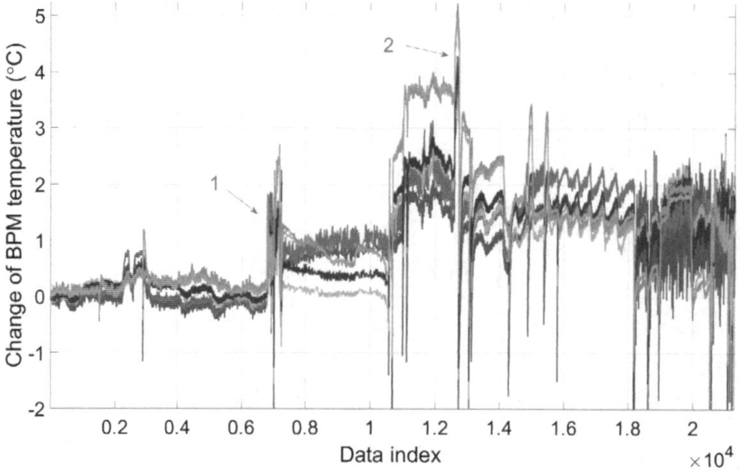

Figure 2. Change of BPM vacuum chamber temperature. Temperature of six BPMs are shown.

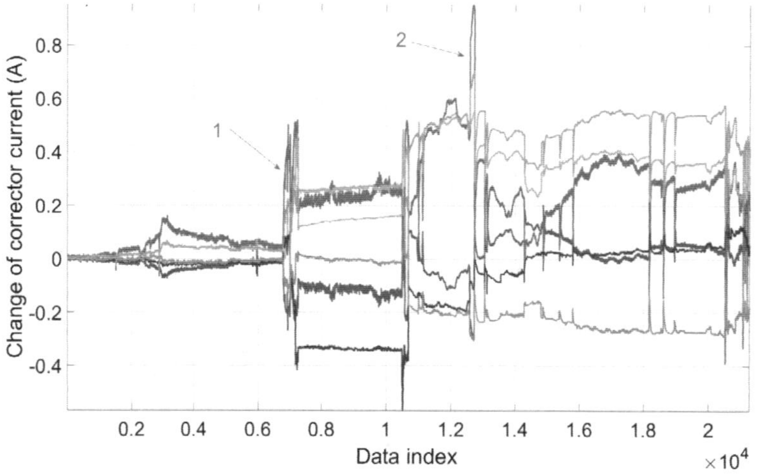

Figure 3. Change of corrector currents. Currents of six correctors are shown.

4. Data preparation

The orbit feedback system of the HLS-II storage ring is composed of 32 sets of BPMs and 32 sets of correctors [8]. There are two temperature sensors attached to each BPM, for a total of 64. The raw data are needed to be processed properly before training as following.

1. Select the data of user mode (the data with beam current at 355 mA - 360 mA). This is because the factors that affect the beam orbit are relatively fewer in the user mode.
2. Discard the data which are out of the normal range.
3. Remove 4 sets of BPM temperature related to broken sensors.
4. Apply interpolation to form consistent data frequencies.
5. Normalize the data.

Comparing the data (see figures 1, 2, and 3), we can see that they have a strong correlation, especially at positions 1 and 2 in the figures, which is consistent with theoretical analysis above.

5. Results

After training, several single models and an ensemble model are obtained. The figure 4 shows the predictions of one horizontal corrector current from testing dataset. From the results, we can see that the prediction of the ensemble model is better than that of a single model, and the real values are almost within the error range of its predicted values, which indicates good performance of this neural network model.

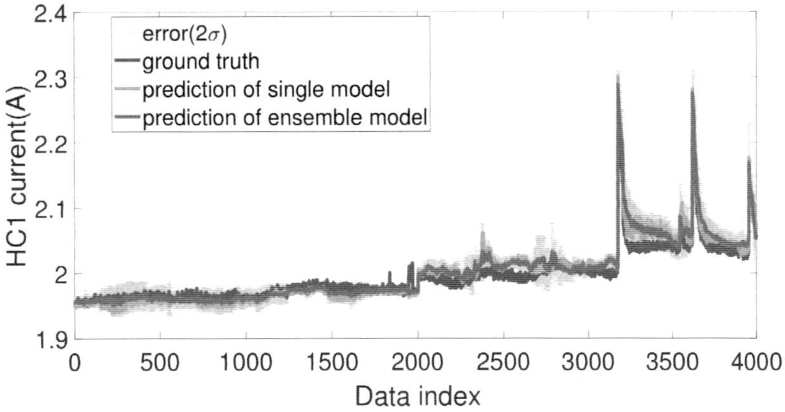

Figure 4. Predictions of the models. The result of one corrector current is shown in the figure, and the rest of the results are similar. The real values are almost within the error range of its predicted values.

By changing the BPM temperature sequentially, the beam orbit shift due to BPM thermal deformation could be obtained using the NN model and the equations (7), (9), and (10). From the results (see figure 5 and figure 6), the BPM temperature variation has a great impact on the beam orbit, especially in some locations, which is related to the lattice optics and the BPM mechanical structure.

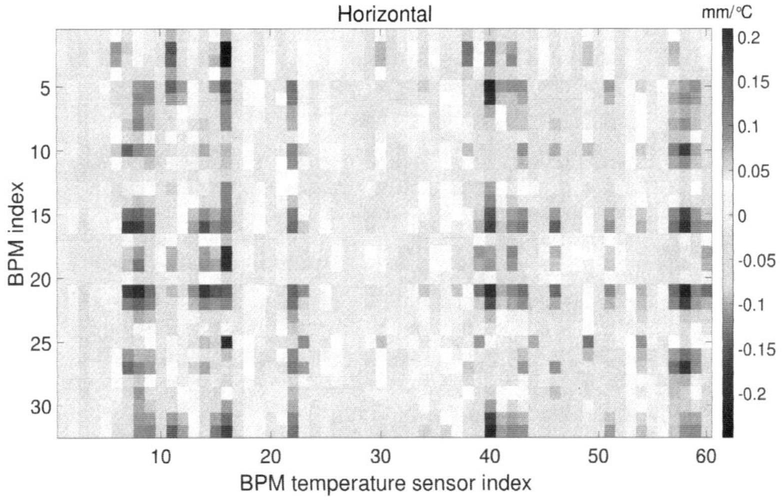

Figure 5. Response of horizontal BPM readings to BPM temperature. The average orbit shift due to BPM temperature is about 34.5 μm/°C.

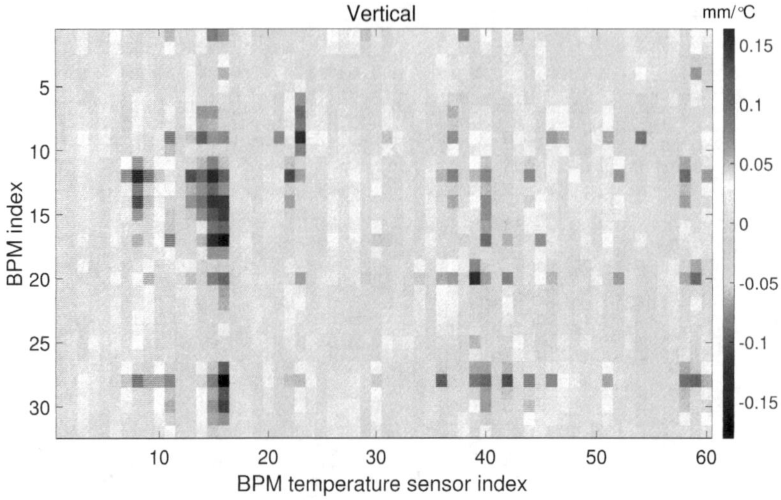

Figure 6. Response of vertical BPM readings to BPM temperature. The average orbit shift due to BPM temperature is about 20.0 μm/°C.

6. Summary

In this paper, we report our study on the beam orbit dependency of BPM temperature. An NN surrogate model is introduced using the historic operation data of the HLS-II storage ring. The average orbit shift due to BPM thermal deformation is about 34.5 μm/°C and 20.0 μm/°C in horizontal and vertical plane respectively.

The neural network only needs machine data, which is easy to implement for storage ring based light sources. Besides, this model could be used for the compensation of beam reference orbit to improve beam orbit stability.

Acknowledgments

The authors would like to thank Prof. Gongfa Liu for providing the historical operation data of HLS-II. This work is supported by the National Natural Science Foundation of China (No. 11975227).

References

[1] Leemann S C, Liu S, Hexemer A, Marcus M A, Melton C N, Nishimura H and Sun C 2019 *Phys. Rev. Lett.* **123** 194801
[2] Decker G 2015 *Proc. 7th Eur. Workshop on Beam Diagnostics and Instrumentation for Particle Accelerator Conf. (DIPAC'05)* (Lyon, France) pp 233–37
[3] Rehm G 2013 *J. Phys.: Conf. Ser.* **425** 042001
[4] Emery L 2001 *Proc. Particle Accelerator Conf. (PAC'01)* (Chicago, IL USA) pp 1276–78
[5] Tejima M *et al.* 2005 *Proc. Particle Accelerator Conf. (PAC'05)* (Knoxville, TN USA) pp 3253–55
[6] Li J W, Sun B G, Cao Y, Xu H L, Lu P, Li C, Xuan K and Wang J G 2010 *AIP Conf. Proc.* **1234** 768–71
[7] Neural network models in R `https://www.datacamp.com/community/tutorials/neural-network-models-r`
[8] Xu W *et al.* 2017 *Proc. 8th Int. Particle Accelerator Conf. (IPAC'17)* (Copenhagen, Denmark) pp 2719–22

Commissioning simulations for the DIAMOND-II upgrade

H-C Chao, R Fielder, J Kallestrup, B Singh and I Martin
Diamond House, Harwell Science and Innovation Campus, Fermi Ave, Didcot OX11 0DE, UK
E-mail: hung-chun.chao@diamond.ac.uk

Abstract. The Diamond-II [1] storage ring, compared to Diamond, improves the natural emittance from 2.7 nm to 160 pm and the beam energy from 3 to 3.5 GeV. The number of straight sections is also doubled from 24 to 48 thanks to the modified hybrid six-bend-achromat lattice. To reduce the impact on the existing science program, the dark time period must be minimised. To assist in this aim, storage ring commissioning simulations have been carried out to predict and resolve possible issues. These studies include beam commissioning starting from on-axis first-turn beam threading up to beam based alignment and full linear optics correction with stored beam. The linear optics corrections with insertion devices are also included. The machine characterisations at different stages are compared. Considerations on realistic chamber limitations, error definitions and some commissioning strategies are also discussed.

1. Introduction

To predict and resolve possible problems during the storage ring commissioning, start-to-end beam commissioning simulations have been carried out. The simulations help to validate the specification and error tolerances defined for the machine components. The code Simulated Commissioning toolkit [2] based on the Accelerator Toolbox [3] which is well integrated with Matlab Middle Layer [4] scripts is chosen for the simulation. The commissioning of the storage ring can be divided into different phases:

(i) Achieve a stored beam, starting with beam-threading and on-axis injection,
(ii) Achieve beam accumulation with off-axis injection,
(iii) Correct the linear optics, optimise the injection efficiency and the beam lifetime,
(iv) Commissioning of all insertion devices (IDs), harmonic cavity, etc.

The commissioning is simulated from the beam threading step to linear optics correction including all ID effects. Forty random machines are simulated to get the statistics.

The structure of this paper is organised as follows. Firstly, the error sources are defined. Then the early stage commissioning procedures including the beam threading and the RF tuning are described. Next is the linear lattice correction including the coupling correction and the alleviating of beta-beating caused by IDs. Finally a short conclusion is given.

2. Error definitions

The error tolerences are developed in consultation with the Diamond-II Magnet and Engineering groups. They are tabulated in table 1 and listed in table 2. The systematic and random multipolar field errors are included up to 15-th order. All the errors used here are populated in a normal distribution with truncation at two standard deviations.

Table 1. Magnet and girder errors.

Element	**Misalignment**		**Relative Field**	
	Offset (µm)	Roll (µrad)	Main Field (%)	Secondary Field (%)
Girder	150/150§	150		
DL Magnet†	100	100	0.05	
Gradient Dipole	50	100	0.05	0.10
Anti-bend Quad	35	100	0.05	0.10
Pure Quad	35	100	0.10	
Sextupole	35	100	0.10	
Octupole	35	100	0.10	
CM	150	150	0.10	
CM‡				0.10
Skew Quad‡				0.50

§ Independent offsets at each end of the girder
† Longitudinal varying dipole
‡ Embedded in sextupole trim coils

Table 2. BPM, RF and other error sources.

BPM Errors		
Initial Offsets	500	µm
Roll	10	mrad
Calibration	5	%
Noise (Turn-by-Turn)	60	µm
Noise (Closed Orbit)	1	µm
RF Errors		
Frequency	100	Hz
Voltage	0.5	%
Phase Offset	90	°
Injected Beam Jitter		
Transverse Displacement	100	µm
Transverse Divergency	10	µrad
Energy Deviation	0.1	%
Phase Shifts	0.1	°
Other Errors		
Circumference	1	µm
Injection Pulsed Magnet Jitter	0.024	%

Aside from these errors, the additional field errors induced by the correctors in the sextupoles are also included in the commissioning studies. The lowest order components, being normal and skew decapolar terms, are linearly proportional to the corrector strengths with a factor of 1.35×10^7. No errors are included in the ID kick-maps [5].

Realistic physical aperture limits are imposed by the round chambers of 20 mm inner diameter in general and the narrow-gap vacuum chambers in some straights since some IDs will be in place on day one of commissioning. All collimators are left open until the commissioning of IDs.

3. Commissioning simulations

3.1. Beam threading and RF tuning

In the beam threading process the sextupoles are turned off to avoid extra sources of errors. During this step the orbit correction is iteratively applied to globally flatten the trajectory. The threading is then progressively extended from one-turn to multi-turns, during which the trajectory correction gradually improves. After multi-turn threading is achieved, the sextupoles are ramped to full strength in multiple steps. Some additional dipole errors are introduced during this process so further global orbit corrections are applied as required. The orbit tuning is continued until the beam survives a few hundred turns and the betatron tune can be detected. A simple quadrupole scan then can be performed using the beam transmission rate or the injection efficiency as an objective.

The next step is to turn on the main RF cavities leaving the harmonic cavity off. The applied RF errors result in a mismatch in the longitudinal phase space, so the RF phase and frequency are adjusted at this stage to remove the unwanted synchrotron oscillations.

The beam-based RF tuning procedures to find and set the optimal phase and frequency are applied iteratively until the beam is centred in the real RF bucket. After this process is complete, another quadrupole scan is performed followed by chromaticity adjustment to maximise the beam survival time. Up to here the first phase beam commissioning is complete.

3.2. Moving towards off-axis injection

From this moment we are entering the second phase where we switch from on-axis to off-axis injection for beam accumulation.

The commissioning procedure in this phase comprises iteratively applying closed orbit corrections, tune correction, beam-based alignment (BBA), and the linear optics correction by Linear Optics from Closed Orbits (LOCO) [6]. The closed orbit correction convergence is controlled by the α parameter in the Tikhonov regularisation [7], which is a parameter used to regularise the corrector strengths.

The BBA follows a procedure modified from the one in used at Diamond, which is originally developed at the ALS [8]. To reduce the commissioning time, the BPMs have been grouped into 48 subsets based on the girders they sit on. During the commissioning the BBA process visit all groups three times.

The commissioning steps are listed and elaborated below.

(i) First Tuning
This is an optional intervention step in case of the low beam transmission at the end of Phase 1. Some interventions, for example, further quadrupole scans or a simplified LOCO correction using a reduced number of quadrupole families can be applied to improve the dynamic aperture (DA). Out of the 40 random machines simulated only one special case needs manual intervention.

(ii) First BBA
This is a coarse BBA step performed with relaxed corrector strengths ($\alpha = 50$). A subsequent orbit correction sequence is launched with descending α until the maximum corrector strength reaches 0.5 mrad. Figure 1 shows the improvement of orbits as the BBA progresses.

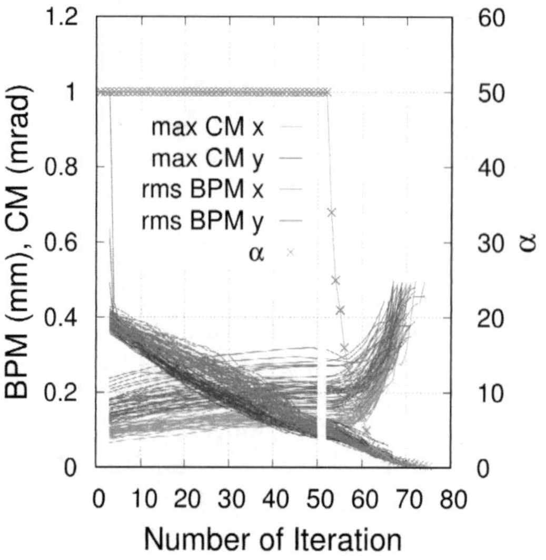

Figure 1. The change of the closed orbits and corrector strengths in the first BBA step. Before the iteration number 50, the BPM reading values decrease as the new BPM offsets are updated closer toward quad centres. After that, the BPM readings continue to decrease but the corrector strengths begin to increase rapidly in line with the decreasing α.

(iii) First LOCO
The goal of this step is to provide a coarse linear lattice correction. For this, LOCO is configured to use just the pure quadrupoles magnets.

(iv) Second BBA and LOCO
We then refine the BBA measurements by launching the BBA procedure again with smaller α and the CM strength limit 0.7 mrad. This is the preparation for the next LOCO with the dispersion included in the response matrix measurement.

After this stage the DA is large enough for off-axis injection [9]. We are now ready to move on to the next phase.

3.3. Coupling correction
In the third phase of commissioning, the goal is to establish the nominal machine optics and thereby improve the injection efficiency and beam lifetime. The first objective is to further correct the linear optics including the coupling. Therefore, all skew quadrupoles are now used. The rest of the field gradient tuning knobs (in strong gradient dipoles and anti-bend quadrupoles) are also used. The simulation is extended to a third iteration of BBA and LOCO.

In summary, the DAs and their improvements at each LOCO step are shown in figure 2. The average rms beta-beating is controlled to 1.0% horizontally and 1.7% vertically, as shown in the top plot of figure 3.

By controlling the vertical emittance to 8 pm-rad, which is the diffraction limit of 10 keV photons, the average Touschek lifetime (estimated at 0.6 nC bunch current and neglecting collective effects) is brought back to 2.25 hours, close to the 2.4 hours derived from the ideal lattice.

3.4. Insertion device commissioning
After the nominal machine optics without ID have been reached, we can move on to the commissioning of the IDs. Here we focus on the commissioning step to alleviate the beta-beating

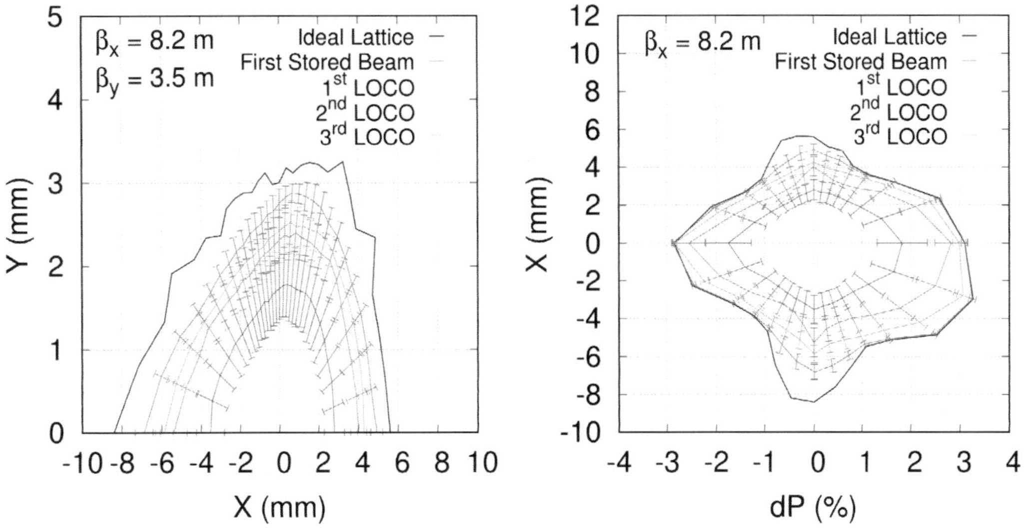

Figure 2. The on- and off-momentum dynamic apertures at different commissioning stages. The averages are connected by solid lines and the error bars are shown in dashes.

induced by all IDs. Further in-depth studies of ID impacts are described in [10].

Multiple LOCO steps are applied to recover the linear optics to the ideal case. The first LOCO step is a coarse correction by 6 families of quadrupoles, while the rest of the LOCO steps perform fine-corrections using all possible tuning knobs. As a result, the mean rms beta-beating at BPMs of all these 40 random machines can be controlled to the values 1.5% horizontally and 2.6 % vertically. The distributions of the RMS beta-beating at different commissioning stages are shown in the bottom plot of figure 3.

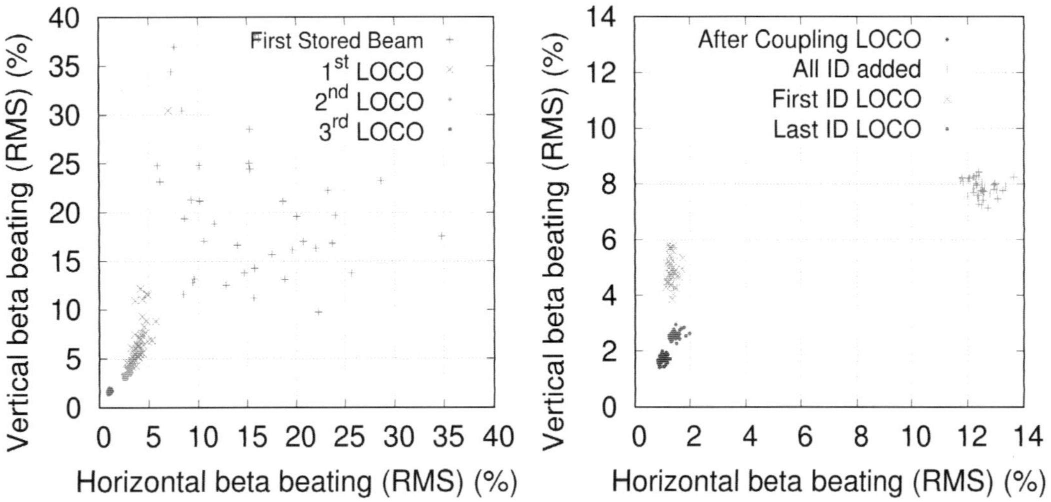

Figure 3. The distributions of the rms beta-beating at different stages. (Left: without ID. Right: with ID)

4. Conclusion

With reasonable error definitions, the start-to-end storage ring commissioning simulations of 40 random machines are carried out with mature techniques that have been used for decades in the synchrotron light source community. The performance of the storage ring is restored close to the ideal values. This gives us good confidence for the future commissioning of Diamond-II storage ring.

Acknowledgments

The authors would like to thank Simone Liuzzo and Thorsten Hellert for the discussions.

References

[1] Martin I P S, Chao H-C, Fielder R T, Ghasem H, Kallestrup J, Olsson T, Singh B and Wang S-W 2022 *Proc. 13th Int. Particle Accelerator Conf. (IPAC'22)* (Bangkok, Thailand) pp 1491–94
[2] Hellert T, Amstutz P, Steier C, and Venturini M 2019 *Proc. 10th Int. Particle Accelerator Conf. (IPAC'19)* (Melbourne, Australia) pp 1441–44
[3] Terebilo A 2001 *Proc. Particle Accelerator Conf. (PAC'01)* (Chicago, IL USA) pp 3203–05
[4] Portmann G J, Corbett W J, and Terebilo A 2005 *Proc. Particle Accelerator Conf. (PAC'05)* (Knoxville, TN USA) pp 4009–11
[5] Elleaume P 1992 *Proc. 3rd Eur. Particle Accelerator Conf. (EPAC'92)* (Berlin, Germany) pp 661–64
[6] Safranek J, Portmann G, and Terebilo A 2002 *Proc. 8th Eur. Particle Accelerator Conf. (EPAC'02)* (Paris, France) pp 1184–86
[7] Amstutz P and Hellert T 2019 *Proc. 10th Int. Particle Accelerator Conf. (IPAC'19)* (Melbourne, Australia) pp 353–56
[8] Portmann G, Robin D, and Schachinger L 1995 *Proc. Particle Accelerator Conf. (PAC'95)* (Dallas, TX USA) pp 2693–95
[9] Kallestrup J, Ghasem H, and Martin I P S 2022 *Proc. 13th Int. Particle Accelerator Conf. (IPAC'22)* (Bangkok, Thailand) pp 2606–09
[10] Singh B, Fielder R T, Ghasem H, Kallestrup J, Martin I P S, and Olsson T 2022 *Proc. 13th Int. Particle Accelerator Conf. (IPAC'22)* (Bangkok, Thailand) pp 539–42

Dependence of CsK_2Sb photocathode performance on the quality of graphene substrate film

L Guo[1], K Goto[1], H Yamaguchi,[2] M Yamamoto[3] and Y Takashima[1]

[1] Nagoya University, Furo, Chikusa, Nagoya, Aichi 464-8601, Japan
[2] Los Alamos National Laboratory, P.O. Box 1663, Los Alamos, New Mexico 87545, U.S.A.
[3] High Energy Accelerator Research Organization, 1-1 Oho, Tsukuba, Ibaraki 305-0801, Japan

Email: l.guo@nusr.nagoya-u.ac.jp

Abstract. A high-performance photocathode is required to advanced accelerators and electron microscopes. In particular, the CsK_2Sb photocathode is of interest because it has features such as low emittance, excitability with visible light, and high quantum efficiency (QE). Generally, the CsK_2Sb photocathode is produced by depositing a cathode element on a substrate, so that the cathode performance strongly depends on the surface condition of the substrate. We have found graphene as reusable substrate, which has the property of being chemically inactive. In this study, graphene film quality dependence of CsK_2Sb photocathode performance was evaluated. Specifically, CsK_2Sb photocathode was deposited using different quality graphene film substrates and their QE values and uniformity were compared. The quality of graphene films was analyzed using X-ray Photoelectron Spectroscopy (XPS) and X-ray Absorption Spectroscopy (XAS). We found that the graphene film can be cleaned by heating at 500 deg. The QE of the cathode on a good quality graphene film was higher and more uniform than that on a poor quality graphene film.

1. Introduction

Cesium potassium antimonide (CsK_2Sb) is one of the highest performing alkali (multi) antimonide photocathodes, achieving quantum efficiency that exceeds 10% by a green laser (532 nm) [1, 2]. It is also realized of record-high beam current of 60 mA in a DC injector with 30 hours 1/e lifetime [3].

Generally, CsK_2Sb photocathode is produced by depositing the cathode element on a substrate, so that the cathode performance strongly depends on the substrate material, including crystallinity1,4, surface state (contamination, roughness, and surface orientation) [1, 4-6] and dopant types [2]. If the substrate after cathode production is used again by heat cleaning, the performance of the reproduced cathode will drastically deteriorate. Therefore, since the cathode performance is prioritized, the substrates are generally not reusable. On the other hand, the exchange of substrate takes an enormous amount of time and labor. This hinders basic research and practical application of photocathodes. As an effectual method to deal with this problem, we have coated graphene on the substrate as a protective layer [7].

In this study, we investigated the QE and its mapping using different quality graphene film substrates to understand the effect of graphene film quality on the photocathode performance. For this purpose, we used XPS and XAS to evaluate the quality of graphene films. The CsK_2Sb photocathode was fabricated onto them, and their performance was compared.

2. Evaluation of graphene quality

Graphene is produced on a Cu substrate by the chemical vapor deposition (CVD) method and transferred to Si substrate by a polymer support [8]. However, the Polymethyl methacrylate (PMMA) support remains on the graphene surface after transfer and must be removed. Considering the photocathode formation environment, the heat cleaning method under ultrahigh vacuum is used for the graphene-coated substrate.

We prepared two kinds of samples of graphene-coated Si substrate (Gra/Si), which one is Si coated with good quality graphene film, the other is Si coated with poor quality graphene film. They were analyzed of XPS and XAS under different heat cleaning condition. Measurements were performed at beamline (BL7U) of Aichi Synchrotron Radiation Center, Japan.

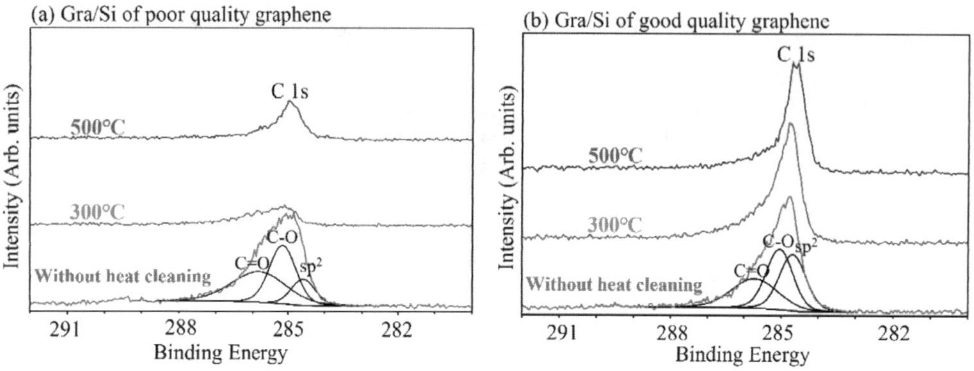

Figure 1. The XPS results of C 1s peaks for under different heat cleaning conditions. Red, green, and blue lines are for cleaning condition of without heat cleaning, 300°C, and 500°C, respectively. (a) is Gra/Si of poor quality graphene. (b) is Gra/Si of good quality graphene. The fitting for the sp^2, C-O, and C=O peaks are shown in the black lines below in the spectrum.

Table 1. The composition % of sp^2, C-O and C=O for heat cleaning condition of without heat cleaning, 300°C, and 500°C, respectively.

Gra/Si of poor quality graphene	
Cleaning condition	sp^2, C-O, C=O (%)
Without heat cleaning	11, 67, 22
300°C	30, 52, 18
500°C	43, 45, 12
Gra/Si of good quality graphene	
Cleaning condition	sp^2, C-O, C=O (%)
Without heat cleaning	18, 45, 37
300°C	66, 18, 16
500°C	88, 7, 5

Figure 1 shows the XPS results of C 1s peaks for graphene of two samples under different heat cleaning conditions ($h\upsilon$ = 650eV). The information depth of the analyzed region is about 0.5–1 nm, which is determined by the electron inelastic mean-free path length and the take-off emission angle of 45°. The spectra were fit by the Doniach–Sunjic function after subtracting a Shirley background. We used CasaXPS software for the analysis after converting the kinetic energy of data to binding energy. Table 1 summarizes the composition percentage of sp^2, C-O and C=O for the two samples under heat cleaning condition of without heat cleaning, 300°C, and 500°C. It is clear from figure 1 that the C 1s spectrum contains three peaks at about 284.6, 285.1, and 286.0eV, corresponding to the sp^2, C-O, and C=O, respectively. For the two samples, C 1s peak without heat cleaning has a wide peak width due to the three components of sp^2, C-O and C=O, and it became sharp with heat cleaning. It shows that the PMMA support was removed by heating and the graphene film was observed.

C 1s peak for Gra/Si of poor quality graphene became weak after PMMA removal by heat cleaning (figure 1a). The structure of graphene was found to be few. On the other hand, C 1s peak for Gra/Si of figure 1.

Good quality became sharp as in the previous study [8] after PMMA removal by heat cleaning (figure 1b). It is clear that quite a good quality of graphene was prepared. The XPS spectrum of graphene did not contain any elements other than C and O, indicating the absence of impurities.

We then performed XAS to confirm the quality of graphene. Figure 2 shows the XAS results of the C K-edge under different heat cleaning conditions for the Gra/Si of two types of quality graphene film. π at 285eV and σ at 292eV. The σ and π band carbon structures for Gra/Si of poor quality graphene was weak after 500°C heat cleaning (figure 2a). On the other hand, a strong σ and π band carbon structures for Gra/Si of good quality graphene after 500°C heat cleaning is shown in figure 2b. Since the shapes after 500°C and 600°C heat cleaning is almost the same, they show sufficient detergency at 500°C. It is shown that heating above 500°C was required to obtain a clean graphene surface. From these results, the quality of the graphene film could be confirmed.

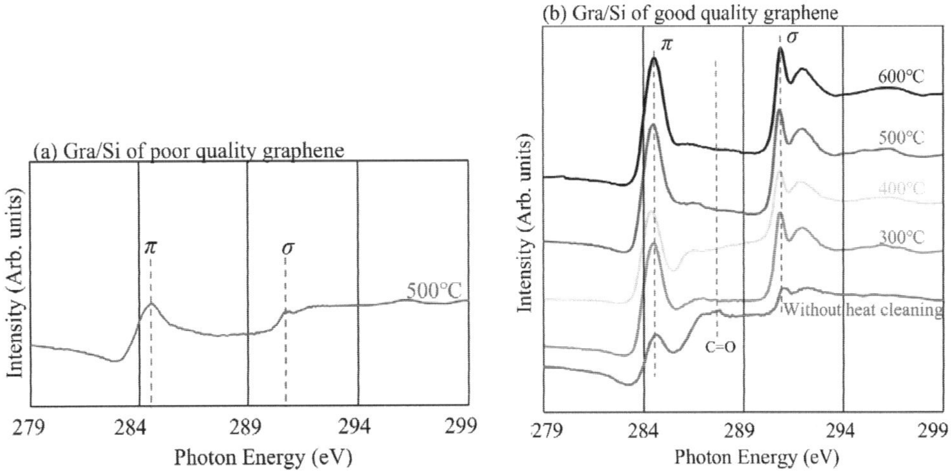

Figure 2. The XAS results of the C K-edge for the Gra/Si substrates of two types of quality graphene film under different heat cleaning conditions. Red, green, yellow, blue, and black lines indicate heat cleaning condition of without heat cleaning, 300°C, 400°C, 500°C, and 600°C, respectively.

3. QE measurement

After confirming the quality of graphene by XPS and XAS, we deposited CsK_2Sb photocathode using the Gra/Si substrates of two types of quality graphene film. The experimental setup and cathode deposition techniques were described in Ref. [1, 2, 7]. Figure 3 shows that spectral response for a CsK_2Sb photocathode as a function of the photon energy on two kinds of these substrates. The laser spot size is about ϕ0.4 mm and the cathode was biased at −100 V. The typical photocurrent was 25 μA. The photocurrent can be detected with a resolution of 0.1 nA. The error was calculated as the standard deviation for the eight measurements. The QE on Gra/Si of good quality graphene showed a satisfactory QE as high as 20% at 400 nm with a relatively small error bar, and it on Gra/Si of poor quality graphene was low 6% at 400 nm with a significant increase in the error bar.

The straight-line part from 2.0 eV to 3.0 eV can be fitted in a linear relationship, and X with Y = 0 is used as a work function. The work function on Gra/Si of poor quality graphene was about 2.1 eV, not at 1.9 eV of the CsK_2Sb photocathode. It suggests that the cathode is not properly formed. On the other hand, the result on Gra/Si of good quality graphene is consistent with a typical spectral response of CsK_2Sb photocathode [2, 5-7]. The photocathode performance of the Gra/Si of good quality graphene was higher than that of Gra/Si of poor quality graphene. This is the experimental evidence that the CsK_2Sb photocathode performance depends on the quality of graphene.

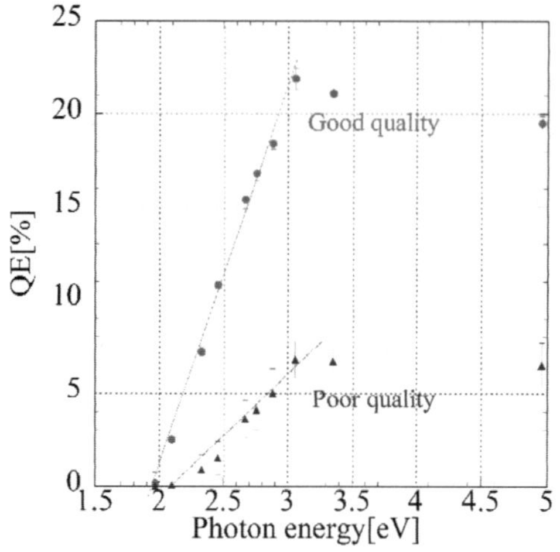

Figure 3. The Spectral response for a CsK_2Sb photocathode on different quality graphene film substrates. Red circles, and blue triangles are for the good quality graphene and poor quality graphene substrates, respectively. The red and blue dotted lines show the fitting results in the straight line from 2.0eV to 3.0eV.

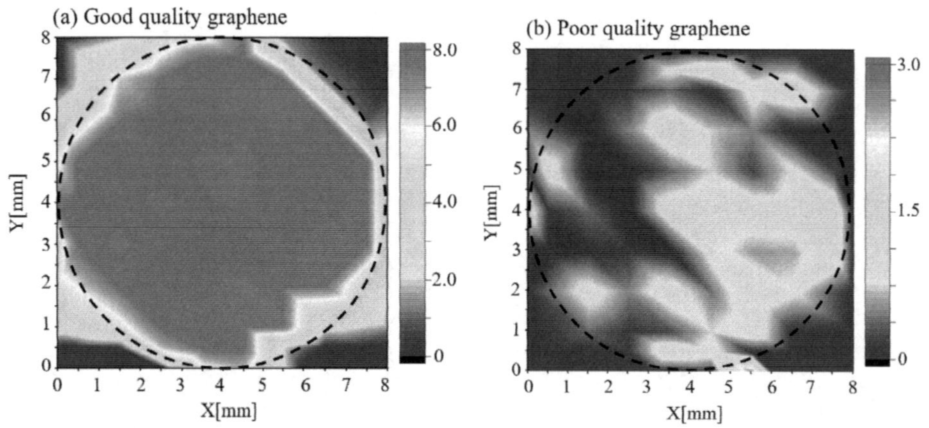

Figure 4. The QE mapping for a CsK_2Sb photocathode on different quality graphene film substrates. (a) The QE on good quality graphene substrate. (b) The QE on poor quality graphene substrate. The black dotted circle indicates the evaporation region of the CsK_2Sb.

We also used 532 nm laser ($h\upsilon$ = 2.33 eV) to measure QE mapping on the two kinds of Gra/Si substrates in figure 4. The value on the X and Y axis is the movement amount of the XY-stage used for measurement. Figure 4a shows a high and uniformly QE of the Gra/Si substrate of good quality graphene. The dispersion of QE within the deposition range is 5% or less. On the other hand, a low and island-shaped QE on the Gra/Si substrate of poor quality graphene is shown in figure 4b. It is shown that the quality of graphene affects not only the maximum of QE but also the uniformity of QE.

The reason for the low photocathode performance on poor quality graphene is that amorphous carbon contaminates the substrate surface and interferes with the formation of the photocathode.

4. Summary

We studied the graphene quality dependence of CsK_2Sb photocathode performance. We used two kinds of Gra/Si substrates, one is Si coated with good quality graphene film, the other is Si coated with poor quality graphene film and evaluated the qualities of graphene by XPS and XAS. We found that heating above 500°C was required to obtain a clean graphene surface. We fabricated CsK_2Sb cathodes on them and compared their performance. We found that the cathodes on good quality graphene had significantly better performance than on poor quality graphene. This showed that the cathode performance depends strongly on the quality of graphene.

Acknowledgments

The authors would like to thank researchers Nakatake and Takakura of Aichi Synchrotron Radiation Center BL7U for their support of XPS and XAS measurement.

The work was financially supported by the High Energy Accelerator Research Organization (KEK), Japan for the Japanese team and the U.S. Department of Energy (DOE) Office of Science for the U.S. team under the U.S.-Japan Science and Technology Cooperation Program in High Energy Physics.

References

[1] Guo L *et al.* 2017 *Prog. Theor. Exp. Phys.* **2017** 033G01
[2] Guo L and Katoh M 2019 *Phys. Rev. Accel. Beams* **22** 033401
[3] Dunham B *et al.* 2013 *Appl. Phys. Lett.* **102** 034105
[4] Sommer A H 1968 *Photoemissive Materials* (New York: Krieger)
[5] Yamaguchi H *et al.* 2019 *Phys. Status Solidi* A **216** 1900501
[6] Yamaguchi H *et al.* 2018 *Adv. Mater. Interfaces* **5** 1800249
[7] Guo L *et al.* 2020 *Appl. Phys. Lett.* **116** 251903
[8] Reina A *et al.* 2008 *J. Phys. Chem.* C **112** 46

Journal of Physics: Conference Series **2420** (2023) 012017 doi:10.1088/1742-6596/2420/1/012017

A double dipole kicker for off and on-axis injection into ALBA-II

G Benedetti, M Carlà and M Pont

ALBA Synchrotron, Carrer de la Llum 2-26, 08290, Cerdanyola del Vallès, Barcelona, Spain

E-mail: gbenedetti@cells.es

Abstract. Injection into the ALBA-II storage ring will be performed off-axis in a 4 meters straight section with a single multipole kicker. We present a novel topology for the coils of the injection kicker, named double dipole kicker (DDK). The resulting magnetic field is the superposition of two opposite dipoles, generated by four inner and four outer conductor rods. When the eight rods are powered, the dipole term cancels and the remaining multipole field is used for off-axis injection. Alternatively, when the four inner rods are switched off, an almost pure dipole is produced, that is useful for on-axis injection during the commissioning. A prototype of DDK is presently under design to be installed and tested in the existing ALBA storage ring. The positioning of the rods is calculated in order to maximize the kick efficiency in mrad/kA and minimise the disturbance to the orbit and the emittance of the stored beam. A metallic coating with optimised thickness along the inner ceramic vacuum chamber should provide compensation for the eddy currents induced field in order to minimize the disturbance to the stored beam while ensuring sufficiently low heat dissipation by the beam image currents.

1. Introduction

ALBA started to study a pulsed multipole kicker for single turn off-axis injection in the previous years [1–3]. The proposed design was a so called non-linear kicker (NLK), whose coils topology was first conceived at BESSY-II [4] and then adopted and tested in other machines [5, 6]. The studies initiated in 2020 for the ALBA upgrade to a new low emittance ring have lead to a different design of the kicker which fits better the new needs of the injection into a ring with a smaller physical aperture and a very compact arrangement of the magnets. In the kicker presented in this paper, the coils have a different topology consisting of four inner rods and four outer rods which produce two opposite dipole fields cancelling each other along the longitudinal axis of the magnet and a non-linear field off axis [7]. The novel pulsed magnet, named double dipole kicker (DDK), has two important advantages: first, despite the small vertical aperture, the peak of the non-linear field can be generated at the position of the injected beam, and second, switching on either the inner or the outer coils an almost pure dipole is produced, that is very useful for on-axis injection in the first turn commissioning of the upgrade ring. In this paper the magnetic concept of the DDK and the parameters to design the prototype to be installed and tested in the existing ALBA storage ring are presented.

2. Double dipole kicker topology

We propose a new idea for the topology of the kicker coils. The topology designed in [1, 3] is changed to a different one shown in figure 1.

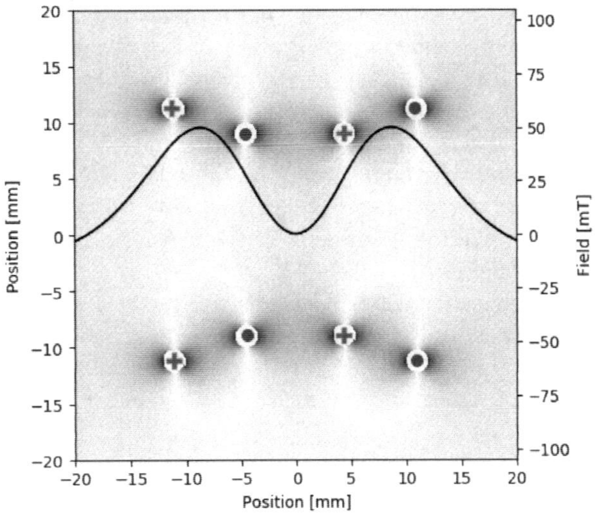

Figure 1. Cross section of the coils topology for the proposed DDK prototype for the ALBA storage ring (the red cross and the blue circle represent the direction of the current carried by the rods) and 2-D magnetic field (colour code) and the field along the horizontal mid plane (black solid line). The coils geometry produces a null field at the center where the stored beam is passing through. The minimum vertical aperture among the rods of ± 9 mm and the field reaches its peak value at the injected beam position $x = 8.7$ mm.

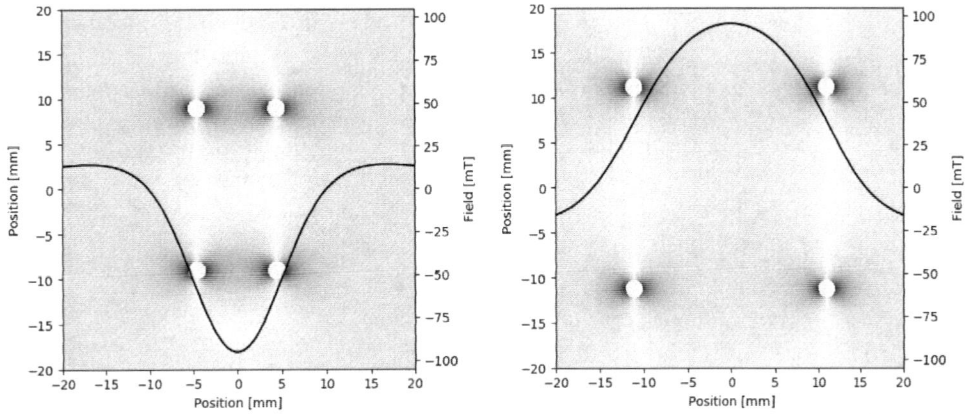

Figure 2. In the DDK, when either the four outer (left plot) or inner (right plot) rods are switched off, a dipole field is produced around the center.

The main difference of this field profile with respect to that of the NLK previously designed is that it results in a narrower plateau and a non zero crossing. On the other hand, the attractive feature of the DDK is that by switching off either inner or outer coils a dipole field is produced around the center as shown in figure 2.

The first motivation to chose such kicker design is that a dipole kicker would be very useful for on-axis injection in the first turn commissioning of the upgrade storage ring of ALBA. The upgrade lattice design, ALBA-II [8], has a very compact arrangement of the magnets and an injection straight section of only 4 meters, with no room for a second injection kicker. Hence, combining two magnets, the dipole and the multipole, in one kicker would be very useful for the injection into ALBA-II. The second advantage of the DDK topology is that the horizontal position where the peak of the field is located is not limited any more by the vertical aperture of the vacuum chamber ($h = \pm 6$ mm). This means that the DDK can have a peak at the position of the injected beam trajectory $x = 8.7$ mm, which was not possible with the NLK design for ALBA [3]. That makes the injection scheme more tolerant to injection trajectory errors, which are mainly due to the septum field fluctuations [2].

Finally, in the DDK geometry needed for ALBA the inductance of the coils is almost halved (0.9 µH for inner plus outer coils and even less for the two separated circuits) with respect to the NLK (1.7 µH [3]) and consequently the required voltage for the power supply is significantly reduced.

3. Tolerances for transparent injection

The standard condition on the orbit stability for transparent injection for the users is that it has to be less than 10% of the beam size: $\Delta x < 10\%\,\sigma_x$ and $\Delta x\prime < 10\%\,\sigma\prime_x$. For ALBA this implies that the defect field at the stored beam position must be $\Delta B_x < 13$ µT and $\Delta B_y < 70$ µT and the gradient must be less than $\frac{\partial B_y}{\partial x} < 0.36$ T/m.

In order to guarantee top-up transparency in ALBA, the mechanical tolerance on the rods positions should be ± 1 µm in both directions, which is unrealistic with the available technologies. Realistic mechanical tolerances with the existing technology can be as small as ± 20 µm, which can produce a displacement of the position of the minimum field of the order of $\Delta x \sim \Delta y \sim \pm 200$ µm.

After the kicker assembly, the profile of the magnetic field will be measured (both with magnetic and beam-based measurements) and the stored beam position should be aligned in order to make it pass through the position where the kick on the beam has its minimum (null gradient), where still a field of the order of $\Delta B_{x,y} \sim \pm 200$ µT is expected. The proposed solution in order to make null field at the position of the stored beam and ensure transparent top-up consists of powering the inner and outer coils with two independent power supplies, with very stringent specifications on the matching between the two pulses. The 10% rule for the orbit stability determines that the integral of the difference between the current of the two pulses has to be less than 0.12% of the pulse value.

4. Field error due to finite propagation speed

The signal propagation through the magnet coils also plays a role in terms of field error. In the previous discussion the implicit assumption that at any time the current flowing in all the conductors had the same magnitude was made, this indeed is not the case when considering transient effects in conjunction to parasitic capacitance between conductors. An exact analysis of the effect would require a detailed 3-D electromagnetic simulation of the magnet geometry. Nevertheless an approximated solution can be derived by assuming the current pulse to propagate through the magnet conductors at constant speed, which is assumed equal to the speed of an electromagnetic wave in the dielectric material that surrounds the coils. In the proposed design a non negligible part of the field is inside the ceramic vacuum chamber, most likely of alumina, that has a propagation speed three times slower than vacuum. It can be safely assumed that the actual speed lies in between what observed in a conductor immersed entirely in alumina and the case of vacuum only.

The connection order of the rods affects the overall field distortion induced by this delayed propagation effect. Furthermore by employing two independent pulser units to power the inner and outer coils it is possible to partially compensate the field distortion, resulting in a remarkable reduction of the effect [7].

The overall effect on the stored beam can be estimated from the mean of the absolute value of the field error along the pulse. Table 1 shows the mean field error for the single and double power supply configurations, both for vacuum only and alumina only magnet.

Table 1. Average field error along the kicker axis due to finite propagation speed through the coils surrounded by vacuum or ceramic.

	Single power supply		Double power supply	
	Vacuum	Alumina	Vacuum	Alumina
ΔB_x	160 µT	475 µT	50 µT	155 µT
ΔB_y	210 µT	630 µT	0	0

5. Coating effects of the ceramic chamber

A second source of perturbation of the stored beam are the eddy currents in the metallic coating (titanium) of the ceramic chamber of the kicker that are produced by the time dependent field. Typical thickness of the coating to ensure an acceptable power dissipation is of the order of 1 µm [3, 5], however the coating thickness can be also optimized in order to minimize the perturbation on the stored beam.

The magnet field is generated by a current pulse that is assumed to be a semi-sinus of time length $t_p = 1.75$ µs, i.e. less than two revolution times. The eddy currents in the coating produce a time dependent induced field that perturbs the profile of the design magnetic field.

With a diameter for the conductor rods of 2 mm, a rectangular-like cross section of the vacuum chamber with full aperture $H \times V = 30$ mm $\times$ 12 mm has been chosen (figure 3) and the eddy currents have been calculated for a 2-D field (no dependence on the longitudinal coordinate z) by solving Maxwell's induction law and imposing the continuity equation for the induced current density inside the coating, which means that the integral of the eddy currents through a transverse coating cross section has to be zero.

The two distributions of eddy currents on the surface of the top/bottom walls and on the sides walls of the vacuum chamber (regions respectively in blue and red in figures 3) produce opposite contributions to the induced field at the center $(x, y) = (0, 0)$. Applying a uniform coating thickness of 1 µm on all the surface, the induced field at $(x, y) = (0, 0)$ would have a value $B_y^{eddy} = 0.75$ mT much larger than the tolerance for transparent top-up ($\Delta B_y < 70$ µT).

The induced field at the center can be made zero by employing on the top and bottom surfaces a metallic coating five times thicker than on the sides surfaces. We expect this to be realized by masking the side surfaces after reaching the nominal thickness in that part of the ceramic chamber. Discussions with the manufacturer will take place in due time. In addition we do not expect that a 2 µm step should have any influence on the behavior of the coating neither on the beam dynamics. With this solution, the perturbation on the stored beam due to the eddy currents is minimized and in principle zero (figure 4), while the attenuation and delay of the field at the position of the injected beam are even smaller than for a uniform coating of 1 µm thickness [7].

6. DDK magnetic specifications

The design parameters chosen for the ALBA DDK prototype are summarized in tables 2 and 3.

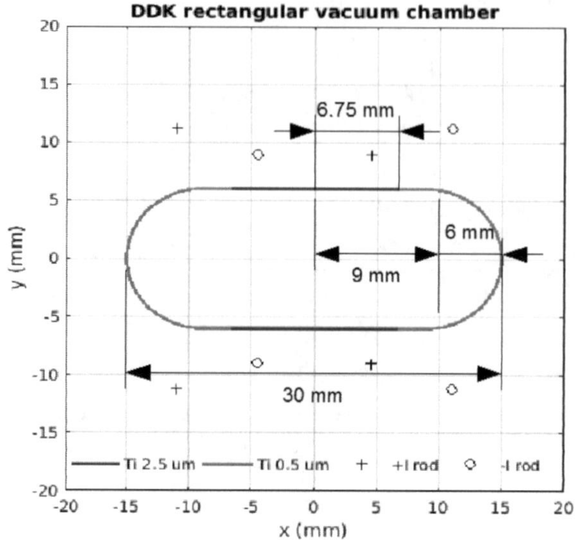

Figure 3. Cross section of the ALBA prototype DDK and the aperture of its ceramic vacuum chamber. The black circles and crosses indicate the positions of the kicker rods. The solid line in blue and red shows the inner aperture of the ceramic chamber where the titanium coating is applied.

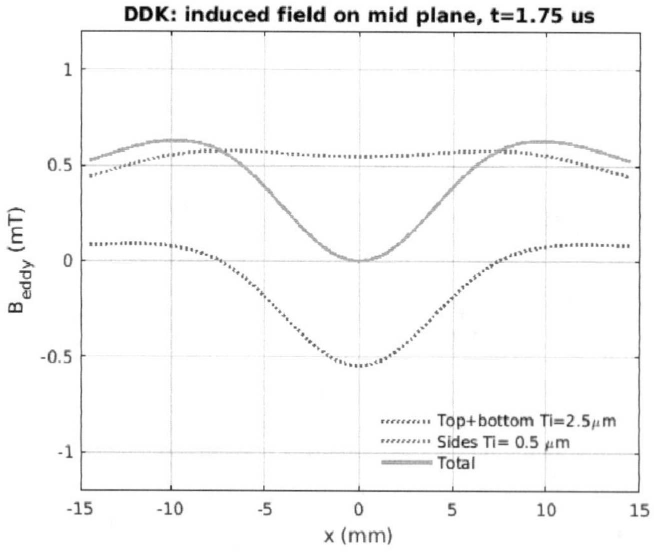

Figure 4. The induced field on the horizontal mid plane (yellow line) and the contributions from the top and bottom coated surface (blue dot line) and from the sides surface (red dot line) of the ceramic chamber. The coating is made of titanium. Applying a thickness of 2.5 µm on the top/bottom surface and 0.5 µm on the sides surface, the field at $(x, y) = (0, 0)$ is zero.

Table 2. Main parameters of the DDK prototype for ALBA.

Inner rod position	x_i, y_i	4.50, 9.00	mm
Outer rod position	x_e, y_e	11.00, 11.20	mm
Diameter of the rods		2	mm
Field at x = 8.7 mm	B_y	50	mT
Magnetic length	L	300	mm
Total coils inductance	L_{tot}	0.90	µH
Inner coils inductance	L_{in}	0.83	µH
Outer coils inductance	L_{out}	0.55	µH
Mutual inductance	M_{inout}	-0.24	µH

Table 3. Main parameters of the half-sinus current pulse.

Pulse duration	t_p	1.75	µs
Peak current	I_0	2675	A
Nominal repetition rate	f_{rep}	3.125	Hz

7. Conclusions

The DDK magnetic design concept for a prototype, to be installed in the existing ALBA storage ring in the next two years, has been completed. The tests and performances with beam of this first kicker should validate this novel design. The experience and results acquired from this prototype will be the basis for the design of a second DDK magnet with specifications for the injection into ALBA-II.

References

[1] Ramos R 2017 Development of a pulsed multipole injection magnet Tech. rep. ALBA Accelerator Division internal report
[2] Benedetti G, Iriso U, Pont M, Ramos D and Ahmadi E 2018 *Proc. 9th Int. Particle Accelerator Conf. (IPAC'18)* (Vancouver, Canada) pp 4030—32
[3] Bueno A 2021 Design specifications for the ALBA non-linear kicker Tech. rep. ALBA Accelerator Division internal report
[4] Atkinson T, Dirsat M, Dressler O and Kuske P 2011 *Proc. 2nd Int. Particle Accelerator Conf. (IPAC'11) (San Sebastián, Spain)* pp 3394—96
[5] Alexandre P *et al.* 2021 *Nucl. Instrum. Methods Phys. Res., Sect.* A **986** 164739
[6] Liu L, Alves M B, de Sá F H, Oliveira A C S and Resende X R 2021 *Proc. 12th Int. Particle Accelerator Conf. (IPAC'21)* (Campinas, Brazil) pp 13-–18
[7] Benedetti G and Carlà M 2022 Magnetic design concept of a double dipole dicker for the injection into ALBA Tech. rep. ALBA Accelerator Division internal report
[8] Benedetti G, Carlà M, Iriso U, Martí Z and Pérez F 2021 *Proc. 12th Int. Particle Accelerator Conf. (IPAC'21)*(Campinas, Brazil) pp 2762—65

The status of the in-vacuum-APPLE II IVUE32 at HZB / BESSY II

J Bahrdt, J Bakos, S Gaebel, S Gottschlich, S Grimmer, C Kuhn, S Knaack, F Laube, A Meseck, C Rethfeldt, E Rial, A Rogosch-Opolka, M Scheer and P Volz

Helmholtz-Zentrum Berlin für Materialien und Energie, Wilhelm-Conrad-Röntgen-Campus, Albert-Einstein-Straße 15, 12489 Berlin, Germany

Email: johannes.bahrdt@helmholtz-berlin.de

Abstract. At BESSY II, two new beamlines for RIXS and for X-Ray-microscopy need a short period variably polarizing undulator. For this purpose, the first in-vacuum APPLE II undulator worldwide is under construction. The parameters are as follows: period length $\lambda_0 = 32mm$, $\# \, periods = 78$, $minimum \; gap = 7mm$. The design incorporates a force compensation scheme as proposed by two of the authors at the SRI 2018. All precision parts of the drive chain are located in air. New transverse slides for the transversal slit adjustment have been developed and tested. Optical Micrometers measure the gap and phase positions, similar to the system of the CPMU17 at BESSY II. They provide the signals for motor feedback loops. A new UHV-compatible soldering technique, as developed with industry, relaxes fabrication tolerances of magnets and magnet holders and simplifies the magnet assembly. A 10-period prototype has been setup for lifetime tests of the new magnetic keeper design. The paper summarizes the status of the undulator IVUE32.

1. Introduction

Photon beams with variably polarized light were already produced and used by the users in the 1980s at BESSY, the predecessor of BESSY II. BESSY II is a dedicated machine for soft X-ray photon production of arbitrary polarization with specific undulators. The low electron-beam energy of 1.7 GeV limits the fundamental photon energy range, and requires the usage of higher harmonics, e.g.: the 3rd harmonic for the transition metals and the 5th harmonic for the rare earth M-edges. Shorter periods lower the harmonic number and thus, enhance the brightness and the degree of polarization. Therefore, HZB launched the development of an in-vacuum APPLE II several years ago and published the design considerations and the mechanic layout at the IPAC 2018 [1].

The IVUE32 will serve two beamlines: A RIXS-beamline will use vertical linearly polarized photons in the range of 250-1000 eV for an efficient suppression of angle dependent optical effects. An X-ray microscope will operate in the range of 250-2500 eV, utilizing all polarization modes and the undulator harmonics 1–7.

In this paper, we focus on the new parts of the device and refer to [1] for the general design and the strategies, which did not change significantly. Meanwhile, the cast iron pieces and all components for the gap and phase motion are already fabricated. The bellows dimensions permit an inclined mode operation, which requires a longitudinal range of ±16 mm. Besides the inclined operation, the larger phase range (in the elliptical mode only ±8mm are required) bears more flexibility in the design of the RF-shielding system.

2. Transverse slides

After the IPAC 2018, we added transverse slides for the horizontal adjustment of the in-vacuum sub-girders. The slides employ two functions: i) the correction of geometric fabrication errors, ii) the correction of systematic field errors, where both errors may stay in relation, respectively. The slide adjustment provides an accuracy of 20 µm over a range of a few 0.1 mm. The slides consist of two halves, which are preloaded with spherical washers. Adjustment plates with a specific thickness define the final position. A Dicronite-coating (http://www.dicronite.de/) guarantees a smooth sliding.

Only recently, we did a thorough analysis of systematic field errors and related phase errors of the IVUE32, and we derived the tolerances of the in-vacuum girder bending. The analysis proved the viability of the slides for a precise horizontal slit adjustment [2].

In many cases, an analytic derivation of the phase error from a systematic field error is simple [2–4]. Usually, the probability distribution $P(\Phi)$ of a systematic phase error is non-Gaussian. Often, an analytic evaluation of this distribution for a strictly monotonic function is possible [5]. As an example, we illustrate the procedure for a linear taper, where s is the longitudinal coordinate and Φ^{-1} the inverse function of Φ:

$$\Phi(s) = s^2 \qquad s = \Phi^{-1}(\Phi) = \sqrt{\Phi}$$
$$P(\Phi) = \partial\,(\Phi^{-1}(\Phi))/\partial\,\Phi = 1/\left(2\sqrt{\Phi}\right)$$

If the phase function $\Phi(s)$ is not strictly monotonic, it can be split in piece-wise strictly monotonic segments $\Phi_i(s)$, which can be treated with the procedure above, and a weighted sum of the contributions $P_i(\Phi_i)$ yield $P(\Phi)$. Figure 1 gives a few examples for $\Phi(s)$ and the related probability function $P(\Phi)$.

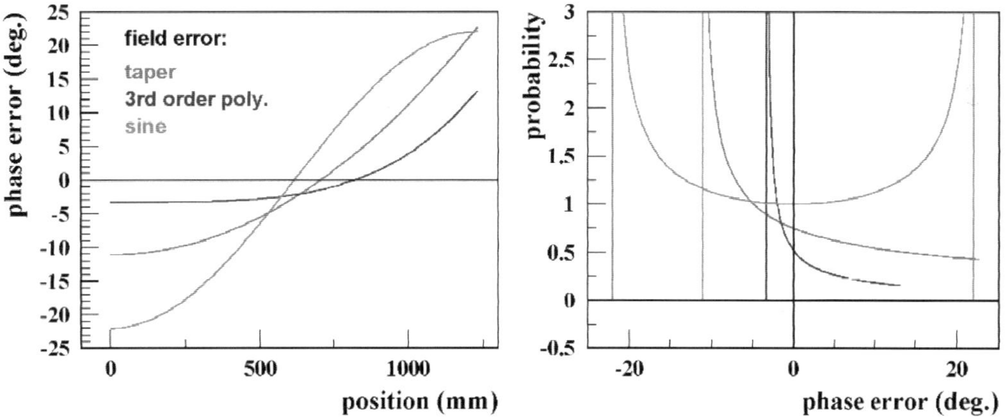

Figure 1. Examples of various systematic phase errors and their probability distributions as discussed in [2].

The rms-values of these phase distributions were derived analytically for several error types in [2–4]. An offset is subtracted from the phase distributions. The probability distributions are scaled vertically for better visibility.

3. RF-shielding system

Different from the first ideas of a split liner as described in [1], the shielding of the magnet structure follows closely the strategy in planar IVUs and CPMUs, where a 100 µm CuNi foil is adopted. In this case the foil must be split longitudinally in order to permit the phase motion of the magnets. The foils are canted at the undulator center, and the short part is hidden in the slit between the magnet rows. Wakefields of geometric errors of the foil are expected to be low. Detailed simulation will start soon. The complete RF-shielding system is divided into four independent groups as depicted in figure 2. A separation of the length adaption and the height adaption in to two independent units helps a lot. The RF-shielding system consists of the following groups:

- 8 foil tensioning systems
- bridging of small slits
- 8 height adaption units
- 8 length adaption units
- canted CuNi-foils on movable and fixed rows

The foil tensioning systems as well as the pieces for bridging small slits with CuNi-foil and RF-springs are adopted from the BESSY II CPMU17 [6]. The height adaption will be realized with a sliding, solid, indirectly cooled Cu-block, similar to the solution at NSRRC [7]. The other two components still need some development effort. The length adaption has to cover a range of $\pm\lambda_0/2 = \pm 16mm$. This is four times the range, which must be covered in the CPMU17 during cool-down. CPMUs in other laboratories demonstrated a safe handling of a much larger thermal shrinkage during cool-down even with an extended length and with Al-girders [7] instead of 1.4429 (CPMU), and thus, this issue is not regarded as a severe engineering challenge.

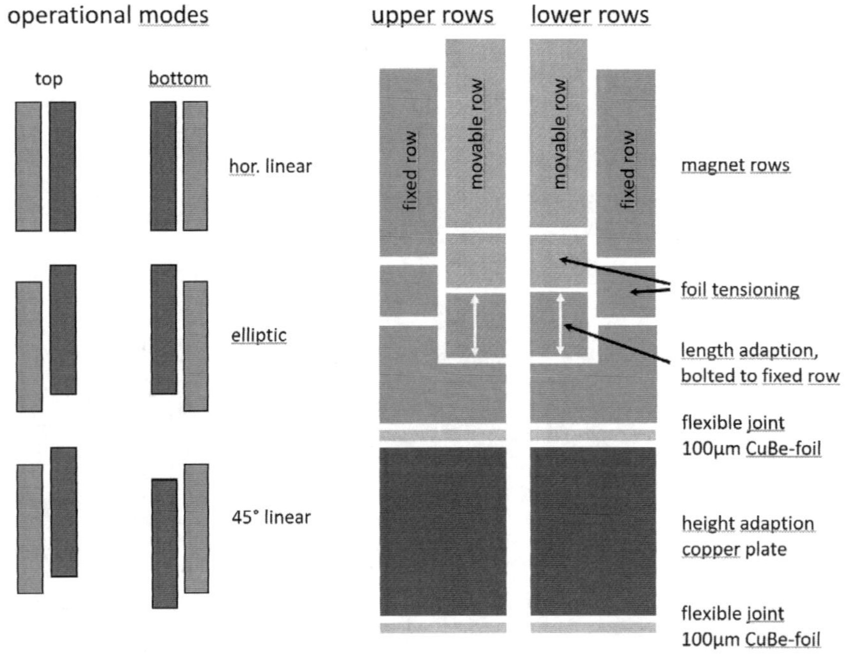

Figure 2. Independent components of the IVUE32 RF-shielding system.

The canted CuNi-foil needs some investigation in order to guarantee small geometric errors. The tolerances of these errors still have to be defined based on CST-simulations.

4. UHV-compatibility

The undulator must meet the UHV-requirements of the BESSY II machine. The chamber must be heated up to 200°C, whereas the magnets and specifically the magnet soldering must stay below 100°C. For this purpose, an integrated water cooling of the four magnet subgirders was developed (figure 3). Each sub-girder employs two long holes. They are short-cut on one end, and at the other end two stainless steel water pipes are welded. The pipes go to the outside without further welding or soldering. The design guarantees a stressless operation of the water pipes inside the chamber during gap and phase motion. The bellows compensate for vertical and longitudinal motions. Figure 3, top shows the water cooling for two diagonal rows. The cooling of the opposite rows is connected to the opposite side of the chamber.

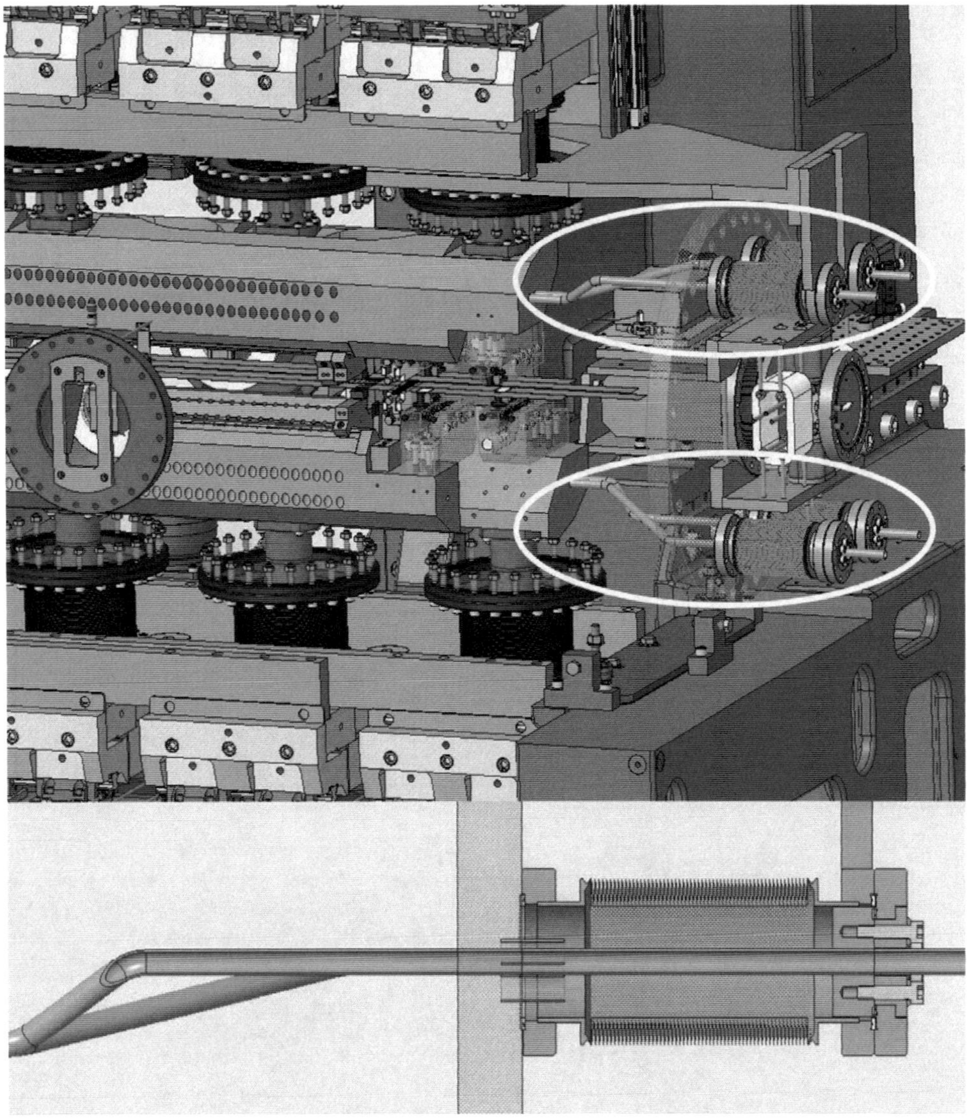

Figure 3. Water cooling of the four magnet sub-girders (yellow ellipses, top). A specific flange and a double-Cu-sealing (bottom) avoids critical couplings inside the UHV, which may leak.

5. Testing of a transverse slide

An accurate horizontal alignment of the in-vacuum sub-girders will be done with five transverse slides per sub-girder. The procedure will be based on magnetic field measurements with an in-vacuum Hall-probe bench. The slides will be used for a phase shimming as well [2]. The transverse slides (figure 3, top, yellow parts) define the parallelism of neighboring sub-girders. The slide stiffness is a critical number, which must be measured. The first slide has already been fabricated. Measurements with a specific measurement setup (figure 4) are on-going. We aim for the determination of the spring constant (μm/kN) of the system slide/column, and the non-linearity. Both numbers will be compared to the FEM-results. After the successful test of the prototype, the other slides will be fabricated on a short time scale, since they are already pre-machined.

Figure 4. Test setup for the prototype transverse slide.

6. 10-period prototype and assembly of full-scale girders

One A- and one B-magnet (horizontal and vertical easy axis orientation) are soldered to a pair. This procedure replaces the magnet gluing of in-air APPLE II devices. The pair will be either soldered or clamped into the keeper. The soldering and clamping of the magnets will be checked carefully with a 10-period lifetime-test setup (figure 5), applying 850.000 full load cycles to the critical joints. The field integrals are measured regularly during the test. Loose connections will immediately be detected via the wire measurements.

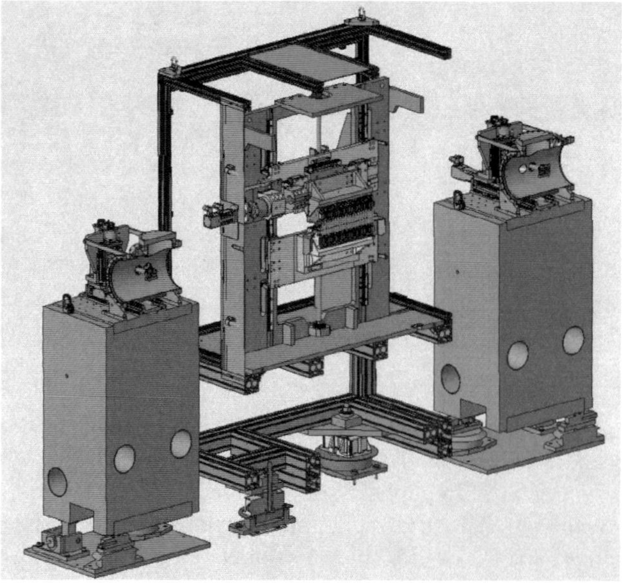

Figure 5. 10-period prototype. Half the structure is cut away for better visibility.

We established a strategy for a successive assembly of the four sub-girders. In the 1st step, the upper and the lower two sub-girders are assembled individually to pairs (figure 6, top). The two sub-girders are shifted already by $\lambda_0/4$ for minimum relative forces [8]. The assembly setup is located at a measurement bench, where the virtual shimming of complete keepers happens (functional magnets and compensation magnets installed). In the 2nd step, both sub-girder halves are combined in a similar procedure as for conventional in-air APPLE II (figure 6, bottom). In the 3rd step, the complete package rolls into the vacuum chamber, and each sub-girder is bolted individually to five columns after separation from the package.

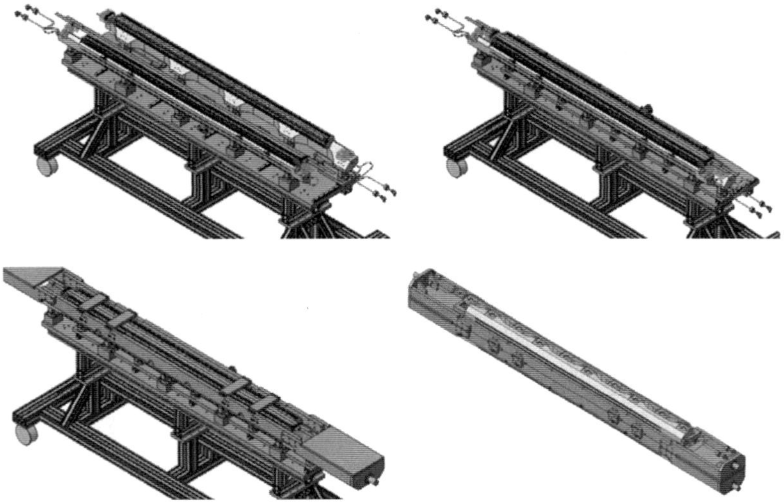

Figure 6. Successive assembly of the four sub-girders from top left to bottom right.

Acknowledgments

The work is supported by ATHENA, a project of the Helmholtz Association.

References

[1] Bahrdt J *et al.* 2018 *Proc. 9th Int. Particle Accelerator Conf. (IPAC'18)* (Vancouver, Canada) pp 4114–16
[2] Bahrdt J *et al.* 2022 *Proc. 14th Int. Conf. Synchrotron Radiation Instrum. (SRI2021)* (Hamburg, Germany) to be pulished in *J. Phys. Conf. Ser.*
[3] Li Y, Faatz B and Pflueger J 2008 *Phys. Rev. Spec. Top. Accel Beams* **11** 100701
[4] Li Y, Ketenogly B and Pflueger J *Phys. Rev. Spec. Top. Accel Beams* **18** 060704
[5] Titze M 2019 *Space Charge Modeling at the Integer Resonance for the CERN PS and SPS* Ph.D. thesis Humboldt University Berlin
[6] Bahrdt J *et al.* 2019 *Proc. 10th Int. Particle Accelerator Conf. (IPAC'19)* (Melbourne, Australia) pp 1415–18
[7] Huang J -C *et al.* 2017 *Phys. Rev. Spec. Top. Accel Beams* **20** 064801
[8] Bahrdt J and Grimmer S 2019 *AIP Conf. Proc.* **2054** 030031

Status and powering test results of HTS undulator coils at 77 K for compact FEL designs

S C Richter[1,2], A Ballarino[1], T H Nes[1], D Schoerling[1], A Bernhard[2] and A-S Müller[2]

[1] CERN, Geneva, Switzerland
[2] Karlsruhe Institute of Technology (KIT), Karlsruhe, Germany

E-mail: sebastian.richter@cern.ch

Abstract. The production of low emittance positron beams for future linear and circular lepton colliders, like CLIC or FCC-ee, requires high-field damping wigglers. Just as compact free-electron lasers (FELs) require high-field but as well short-period undulators to emit high energetic, coherent photons. Using high-temperature superconductors (HTS) in the form of coated ReBCO tape superconductors allows higher magnetic field amplitudes at 4 K and larger operating margins as compared to low-temperature superconductors, like Nb-Ti.

This contribution discusses the development work on superconducting vertical racetrack (VR) undulator coils, wound from coated ReBCO tape superconductors. The presented VR coils were modularly designed with a period length of 13 mm. Powering tests in liquid nitrogen of multiple vertical racetrack coils were performed at CERN. The results from the measurements are presented for three VR coils and compared with electromagnetic simulations.

1. Introduction

Undulator magnets play a major role in accomplishing the next generation of compact and highly brilliant light sources, as they are used in synchrotrons and free-electron lasers as the main photon source with energies up to hard x-rays. When making the facilities more compact, short periods and high field undulators are crucial for delivering the needed high energetic photons for lower beam energies.

To achieve higher magnetic flux densities on the beam axis, the low-temperature superconductor (LTS) Nb-Ti was used to replace permanent magnets. An improvement of the resulting brilliance of high energy x-rays made it the state-of-the-art technology for superconducting undulators, e.g. Nb-Ti undulators operate successfully in KARA at KIT and at Argonne National Laboratory [1–3], whereas permanent magnet wigglers are used to lower the beam emittance, e.g. in PETRA III at DESY [4]. The CLIC wiggler design (3 T, 12 mm gap, 50 mm period) benefits from LTS as demonstrated with a Nb-Ti prototype at KARA, KIT [2]. An HTS version might be able to deliver around 4 T peak fields [5].

Consequently, employing high-temperature superconductors (HTS) like ReBCO (rare-earth barium copper oxide), to superconducting undulators is an attractive choice for the next upgrade. First, operations with larger temperature margins or at higher temperatures than 4 K would relax cryogenic requirements, thus reducing costs compared to LTS technologies. Second, a significant enhancement of the parameter space in terms of B_0, λ_u and gap could be shown in previous studies due to the ReBCO's high B_{c2} field at low temperatures (e.g., YBCO: $B_{c2,\perp}(4.2\,\mathrm{K}) \approx 100\,\mathrm{T}$) [6–8].

Other studies have already shown that ReBCO may be feasible for HTS undulator coils with $\lambda_\mathrm{u} = 16\,\mathrm{mm}$ to reach high current densities in the order of $2\,\mathrm{kA/mm^2}$ at 4.2 K [9]. Yet, undulator periods smaller than 15 mm and quench protection stay challenging for these high currents. With our dry-wound, non-insulated (NI) design, we always give the current the option to circumvent normal conducting zones.

This paper presents and discusses the latest powering test results at 77 K of our three HTS VR undulator coils and briefly compares them to electromagnetic simulations, performed with Opera and J_c fits done at CERN [10–12].

2. Experimental setup

CERN and KIT are both contributing to the EU-financed CompactLight (XLS) study that explores feasibilities for a compact FEL, hence high on-axis fields for a given λ_u of 13 mm are part of the investigations [13]. Consequently, we decided to adopt this value to our modular coil design. All three presented VR undulator coils were wound with a 4 mm wide coated ReBCO tape superconductor onto a D-shaped copper winding body with an iron core (see figure 1). For VR coils #1 and #2, we used tape from Bruker HTS with a total thickness of 100 µm, including a 50 µm substrate and a total copper stabilizer thickness of 45 µm. For VR coil #3, 45 µm thick tape from SuperPower Inc. with a 30 µm substrate and a total copper stabilizer of 10 µm was used.

Each VR coil represents one period and thus consists of two sub-coils and two iron poles. The voltage was monitored via voltage taps on the inner- and outermost turn of each sub-coil. Two Hall probes were placed in a fixed position at a distance $d = 3.5\,\mathrm{mm}$ from the iron pole face as displayed in figure 2. More design-related aspects can be found in table 1. The complete design and manufacturing process were already described in previous work [14].

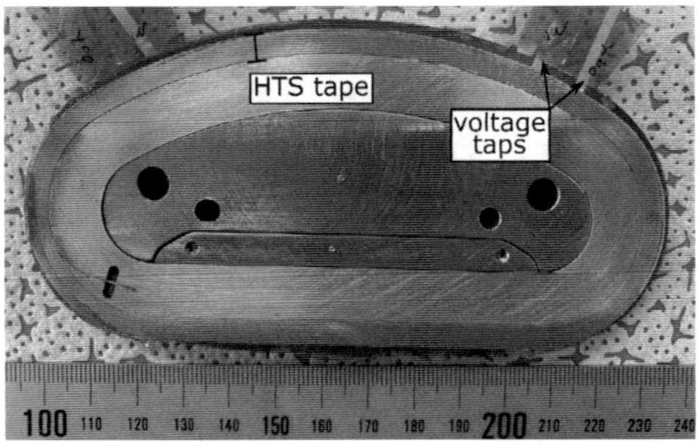

Figure 1. VR coil #3 with its four voltage taps. Scale in mm.

Figure 2. Cross-section of our VR coil design with the copper winding body, iron core and poles, sub-coil 1 and 2.

3. Results and discussion

All powering tests and measurements were performed at 77 K, in liquid nitrogen. The first current ramps were used to investigate the coil's behavior up to the region of the expected critical current I_c. Next, we investigated the voltage decay, thus the charging of the coils by applying current step functions. Finally, we tested the limits of our NI design and went with current steps well above the critical current and measured the radial resistance R_R.

Table 1. Design parameters of the VR undulator coils.

$\lambda_u = 13\,\mathrm{mm}$	**VR coil #1,#2**	**VR coil #3**
Sub-coil x-section	4 mm × 5 mm	4 mm × 4.6 mm
HTS conductor	Bruker HTS	SuperPower
with dimensions	4 mm × 100 µm	4 mm × 45 µm
Number of turns	51	102
$J_{e,sim}(0.6\,\mathrm{T}, 77\,\mathrm{K})$	75 A/mm^2	222 A/mm^2
$I_{c,sim}(0.6\,\mathrm{T}, 77\,\mathrm{K})$	30 A	40 A
$B_0(I_{c,sim}, 3.5\,\mathrm{mm})$	0.14 T	0.28 T

The measured critical current of each sub-coil was derived from U-I measurements following an electric field criterion of 1 µV/cm and the length of the outermost turn (~30 cm). This assumption gives us a threshold value of 30 µV. The I_c of one VR coil is defined by min($I_{c,\text{sub-coil1}}$, $I_{c,\text{sub-coil2}}$). The Hall probe output has an accuracy of ±2%.

3.1. Current ramps

Figure 3 shows exemplarily the measured voltage for each sub-coil of a VR coil when ramping the current with a ramp rate of 0.5 A/s up to plateaus of 10 A, 20 A, 30 A and 35 A. The VR coil's asymmetry lets us expect a different critical current per sub-coil due to the magnetic flux distribution in the superconductor [12]. With increasing current, the voltage is also expected to rise: most of the current will first penetrate the NI coil radially before settling into the superconducting path, which can then be detected as voltage decay. When the I_c of the superconductor is reached locally, the current is forced to flow radially and/or in the copper matrix of the coated super-conductor, which can be measured as an elevated asymptotic voltage plateau level after a decay time. As described above, the sub-coil's I_c is reached, once this constant voltage hits 30 µV. Similar voltage responses were found for all three VR coils.

3.2. Current steps and overflow

All three VR coils were powered gradually up to 112 A with a current step function of 2 A, followed by 300 s decay time before the next step. The decayed voltage and the Hall probes' signal are shown in figure 4 for the tested VR coils. Measured values are displayed in table 2.

VR coils #1 and #2 performed qualitatively as expected. The magnetic field B_0 increased up

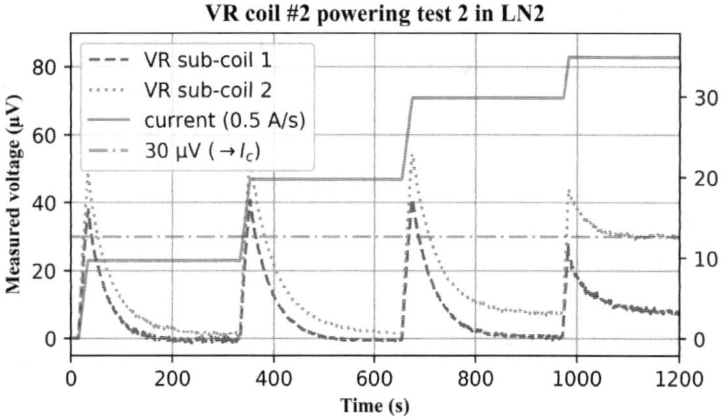

Figure 3. Voltage responses of each sub-coil for VR coil #2 caused by current ramps with 0.5 A/s ramping speed.

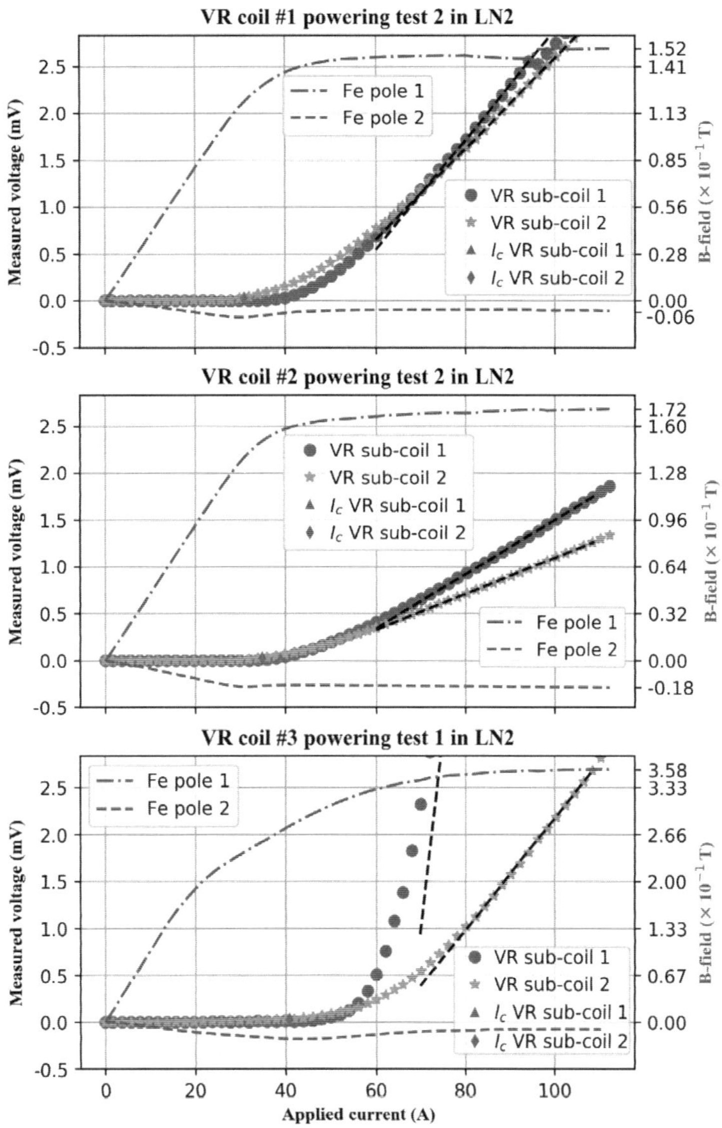

Figure 4. Measured voltages and magnetic pole fields of VR coil #1 (top), VR coil #2 (center) and VR coil #3 (bottom) for a current step function: 2 A step, 300 s hold. Black dashed lines represent linear fits of the radial resistance.

to I_c and reached the region of our predicted values for 30 A (6% and 2% disparity, respectively). After having reached I_c, B_0 began to stabilize and the sub-coils became increasingly resistive. With every current step more turns reached their local I_c and the V-I curve converges to Ohm's law, representing the radial resistance R_R across each sub-coil. Differences between the values within a VR coil can be seen as irregularities in the winding process. The difference between VR coil #1 and #2 can be explained by different winding tensions (30 N vs. 25 N) during manufacturing, which influences the contact resistance between tapes.

Sub-coil 2 in VR coil #3 performed as expected and described above, whereas sub-coil 1 showed a strong R_R. Yet, the region of calculated B_0 could be reached for 40 A with a disparity of 2%. Twice the number of turns let us expect a higher R_R compared to the previous coils.

Table 2. Measured parameters of the VR undulator coils.

	VR coil #1		VR coil #2		VR coil #3	
Sub-coil	**1**	**2**	**1**	**2**	**1**	**2**
I_c (A)	40	30	39	35	47	41
$B_0(I_{c,sim})$ (cT)	1.29	0.16	1.35	0.18	2.75	0.24
R_R (µΩ)	58	48	29	19	400	58
τ (s)	26	42	32	46	(4)	90

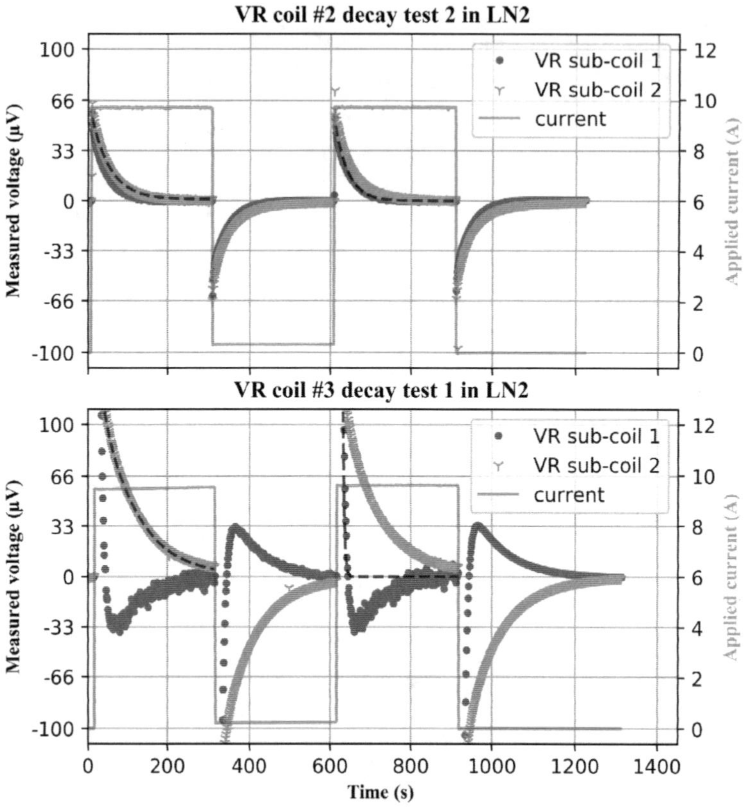

Figure 5. Measured decaying voltages of VR coil #2 (top), which also represents VR coil #1, and VR coil #3 (bottom). Black dashed lines represent exponential fits of the decay.

However, one order of magnitude discrepancy together with the non-saturated B_0 behavior for Fe pole 1 might indicate potential damage to the superconductor, which needs to be further investigated.

3.3. Effective time constants

When varying the undulator field for tuning to different photon energies of users' demands, it is crucial to know how fast an undulator can be charged or discharged. This can be indicated by the effective time constant τ, when measuring the above-mentioned voltage decay time after a current step, thus the time the current needs to enter or exit the superconducting path. The voltage decay can be approximated as exponential decay [14, 15]: $U \approx a \cdot e^{-t/\tau}$, with $\tau = L/R$

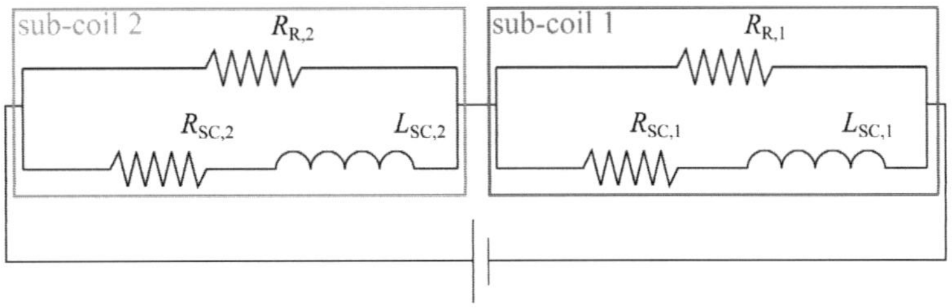

Figure 6. Simplified circuit diagram of one NI VR coil. Inside each sub-coil, the current can select between the radial resistive path $R_{\mathrm{R,i}}$ and the superconducting, inductive path $R_{\mathrm{sc,i}}$. Both sub-coil's inductances may interact.

, inductance L and constant a.

For all three VR coils, a current step to a plateau of 10 A for 300 s was applied, followed by a step to a plateau of 0.5 A for another 300 s, etc. The measured voltage is shown in figure 5. VR coils #1 and #2 showed similar, expected behaviors: an exponential decay with similar τ_i for sub-coil 1 and 2, respectively, as noted in table 2. Sub-coil 2 in VR coil #3 performed as described above. Twice the number of turns leads to a longer (dis-)charging time, here roughly doubles it.

As noticed before, VR coil #3's sub-coil 1 behaves differently. In the VR coil's series circuit, both sub-coils influence each other's (dis-)charging response, especially when one operates differently (see figure 6). Caused by the noticed defects in sub-soil 1, its voltage decay time was drastically reduced. For this reason, the slower decaying voltage of sub-coil 2 pulled/pushed current out/in of the sub-coil 1, which can be detected as a change of sign in the voltage signal.

4. Conclusion

We powered three non-insulated VR undulator coils at 77 K, in liquid nitrogen, each consisting of two sub-coils. All performances reached their design current of 30 A and 40 A with a magnetic peak field in the gap of 0.135 T and 0.275 T, respectively. This was verified in multiple electrical and thermal cycles. Moreover, all coils showed stable performances, even for currents beyond the critical current ($\sim$300% I_c), without any superconductor degradation. Measured time constants around 45 s for two of our three coils may be suitable for tunable undulators.

Summarized, we successfully demonstrated the application of ReBCO coated superconductors to undulator coils with $\lambda_u = 13\,\mathrm{mm}$ and showed the stability and magnetic field performance of a non-insulated undulator coil design for low magnetic fields (<1 T) at 77 K.

Next, we plan to power our HTS VR undulator coils at 4.2 K to investigate operations at high current densities ($\sim 2.3\,\mathrm{kA/mm^2}$) and to explore the magnetic field limits.

Acknowledgments

The authors would like to acknowledge C. Fernandes, J. Mazet, F. Garnier, F.-O. Pincot, P.-A. Contat, N. Bourcey, S. Clement, P. Martin Vazquez and J. C. Perez for support in the manufacturing process. G. Lenoir, P. Koziol and C. Barth are thanked for supporting the powering tests.

This work has been supported by the Wolfgang Gentner Program of the German Federal Ministry of Education and Research.

References

[1] Boffo C *et al.* 2016 *IEEE Trans. Appl. Supercond.* **26** pp 1–4
[2] Bernhard A *et al.* 2016 *Proc. 7th Int. Particle Accelerator Conf. (IPAC'16)*, Busan, Korea pp 2412–15
[3] Ivanyushenkov Y *et al.* 2015 *Phys. Rev. Spec. Top. Accel Beams* **18** 040703

[4] Tischer M *et al.*, *Proc. Particle Accelerator Conf. PAC'05* (Knoxville, TN USA), pp 2446–48
[5] Richter S C 2019 HTS undulator development Presented at CLIC Workshop 2019 (CERN, Geneva)
[6] Richter S C *et al.* 2019 High-temperature superconducting undulators for compact free-electron lasers *Verhandlungen der Deutschen Physikalischen Gesellschaft* (Mueche2019issue) **1**
[7] Nguyen F *et al.* 2019 XLS - D5.1: Technologies for the Compact Light Undulator *Zenedo* (doi:10.5281/zenodo.5024409)
[8] Golovashkin A *et al.* 1991 *Physica C* **185-159** pp 1859-1860
[9] Kesgin I, Kasa M, Ivanyushenkov Y and Welp U 2017 *Supercond. Sci. Technol.* **30** 04LT01
[10] Opera 2020, Simulation Software https://www.3ds.com/de/produkte-und-services/simulia/produkte/opera/
[11] Fleiter J and Ballarino A 2014 Parameterization of the critical surface of REBCO conductors from Fujikura Tech. rep. CERN Internal CERN Notes EDMS: 1426239
[12] Danial M and Nugteren J van 2017 Parameterization of the critical surface of REBCO conductors from Bruker Tech. rep. CERN Internal CERN Notes CERN-STUDENTS-Note-2017-013
[13] CompactLight project (XLS) 2022 http://www.compact-light.eu/
[14] Richter S C *et al.* 2022 *IEEE Trans. Appl. Supercond.* **32** pp 1–5
[15] Nes T H *et al.* 2022 *IEEE Trans. Appl. Supercond.* **32** pp 1–6

Simulations of the suitability of a DC electron photogun and S-band accelerating structure as input to an X-band linac

S D Williams[1], R P Rassool[1], S L Sheehy[12], G N Taylor[1], M Volpi[1], R Auchettl[2] and R T Dowd[2]

[1] The University of Melbourne, 3010 Parkville, Australia
[2] AS - ANSTO, 3168 Clayton, Australia

E-mail: scottdw@student.unimelb.edu.au

Abstract. Work has been underway for some time to design a compact electron beamline utilising X-band linear accelerating structures in the new Melbourne X-band Laboratory for Accelerators and Beams (X-LAB). The original design utilised an S-band RF photogun as an input to a pair of high gradient X-band linear accelerating structures, but we have been motivated to investigate an alternative starting section to allow for initial testing. This will utilise a DC photogun and S-band accelerating structure similar to those used at the Australian Synchrotron. Simulation results incorporating space charge of a beamline composed of a DC photogun, S-band accelerating structures, and two high gradient X-band structures will be presented. These simulation results will be optimised for minimum emittance at the end of the beamline.

1. Introduction

As part of the arrival of the X-Box test stand at the University of Melbourne we have been planning a beamline that can be installed in the University of Melbourne X-LAB accelerator hall utilising X-band RF accelerating cavities of a similar design to the structures used by the compact linear collider [1, 2].

For optimal beam capture and acceleration an RF photogun is planned as an injector, but in this document we investigate the possibility of integrating a relatively low-cost alternative for use during commissioning, that is a 100 keV DC photogun. However, the use of non-relativistic electrons ($\beta \approx 0.55$) will cause acceptance, capture, and acceleration issues as the X-band structures are matched to $\beta \approx 1$.

To alleviate this, we have investigated the possibility of using an S-band RF buncher available locally to increase the energy of the bunch to an satisfactory level for acceptance.

2. X-band RF accelerating structure acceptance

The X-band structure in question is a TW X-band structure operating at 11.9942 GHz similar to the CLIC T24 structure [3]. Following simulation in CST Microwave Studio [4] its fieldmap is exported for further use in simulations, and can be seen in figure 1. At this stage in Melbourne X-LAB planning it is estimated to be fed 20 MW of RF power, yielding an average accelerating gradient of 70 MW m^{-1}, with a maximum gain per structure of approximately 17.5 MeV.

To demonstrate the acceptance issues we shall show results from single particle tracking simulations using the code ASTRA [5]. We use single particle simulations for this section due to their simplicity, and ease of cross checking by numerical integration. Although we neglect whole bunch dynamics at this stage it is very unlikely for a whole bunch to pass through without issues if a single particle can't.

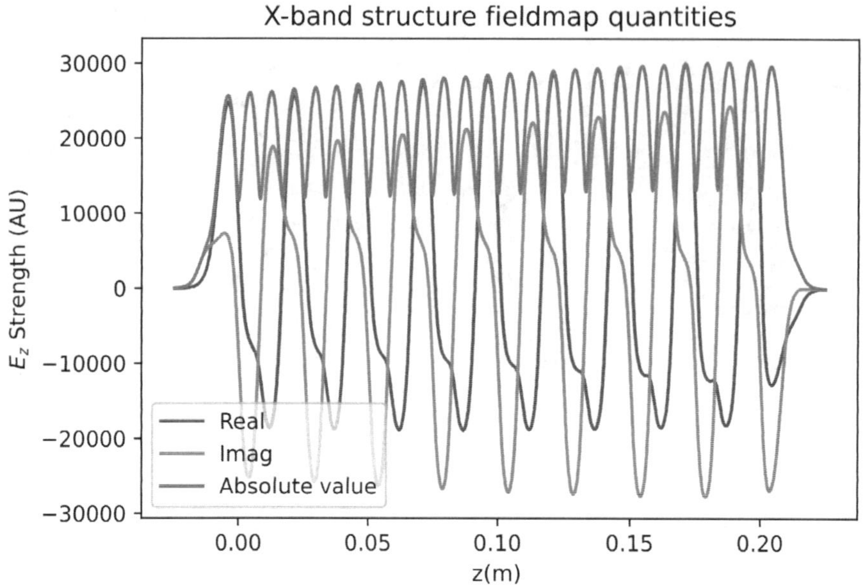

Figure 1. X-band structure fieldmap used.

To investigate acceptance we simulate a single on axis electron multiple times, for different initial Kinetic Energies (KE) and structure phases in increments of 1°. A particle is deemed to have passed though if it reaches the other end of the accelerating structure, and its longitudinal momentum p_z has not dropped below 200 keV $c_0{}^{-1}$ (KE 37.75 keV). This threshold was chosen after inspection of multiple simulations to give good separation between simulations where the particle passes through or is brought to near halt.

In figure 2 we present a combined plot showing the number of simulations deemed to have passed according the previous criteria for a given initial KE, and the maximum energy gain for a given initial KE. The range of phase acceptance (acceptable starting phases), in blue, is calculated by counting the number of simulations where a particle is transmitted for the given initial energy. We see that we need an initial energy of approximately 150 keV to have any phase range acceptable for transmission, and then as KE increases the entire phase range is allowed for transmission. We specifically note that this is for transmission only, not acceleration; a 1 GeV particle would be able to pass through for all phases, but only roughly half would result in any nett energy gain. The orange scatter plot demonstrates the maximum energy gain, out of all phases, for a given initial KE, and we see we can achieve near maximal energy gain above initial KEs of 500 keV.

In figure 3 we show the momentum change of a single particle just as it enters the accelerating structure for different initial KE, where the phasing of the X-band structures has been set for maximum energy gain or greatest depth for the KE=0.1 MeV case. We see that for an initial KE less than or equal to 0.5 MeV there is a initial decrease in momentum the particles are accelerated, whereas for the case of KE=1000 MeV acceleration starts almost immediately. For

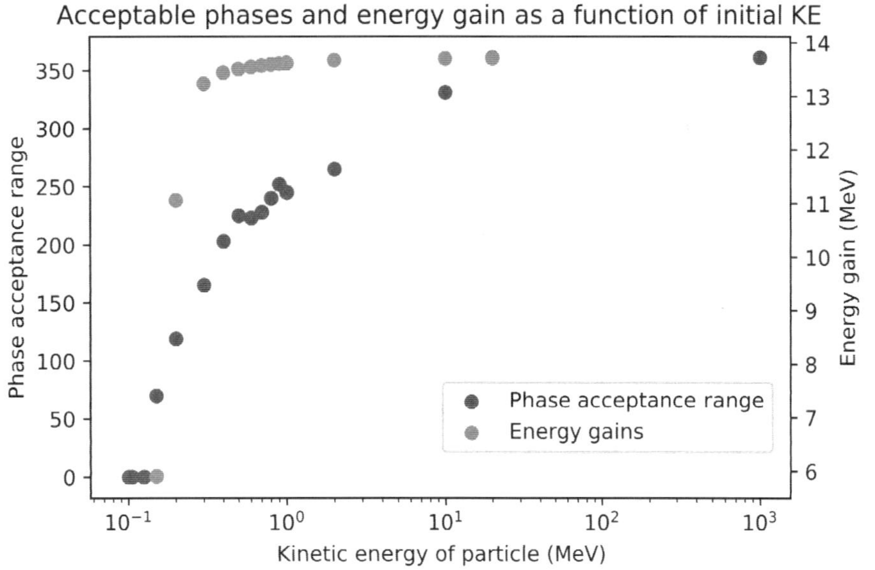

Figure 2. Acceptable for transmission phase range, and maximum energy gain for a given initial kinetic energy.

less than KE<0.5 MeV we view this as a case of phase slippage, where the particles arrive ahead of the accelerating crest in the non-beta matched first cell.

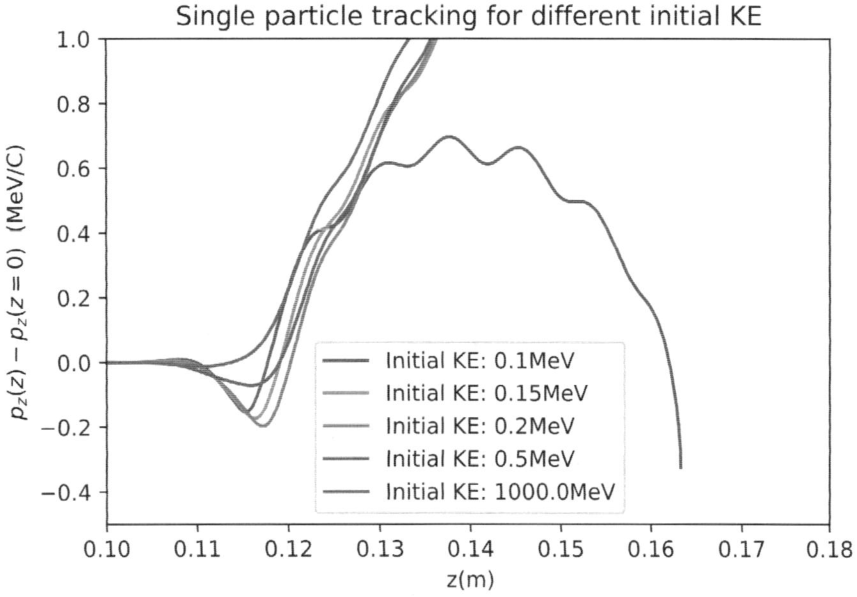

Figure 3. Trajectory of single particles for different initial energies.

To summarise; if we were to use a 100 keV DC photogun we would need an extra energy boost to at least 500 keV, thus motivating the use of an additional accelerating section before the X-band structures. Additionally, concerns about bunch lengthening also motivate the use of that section to also re-bunch the beam.

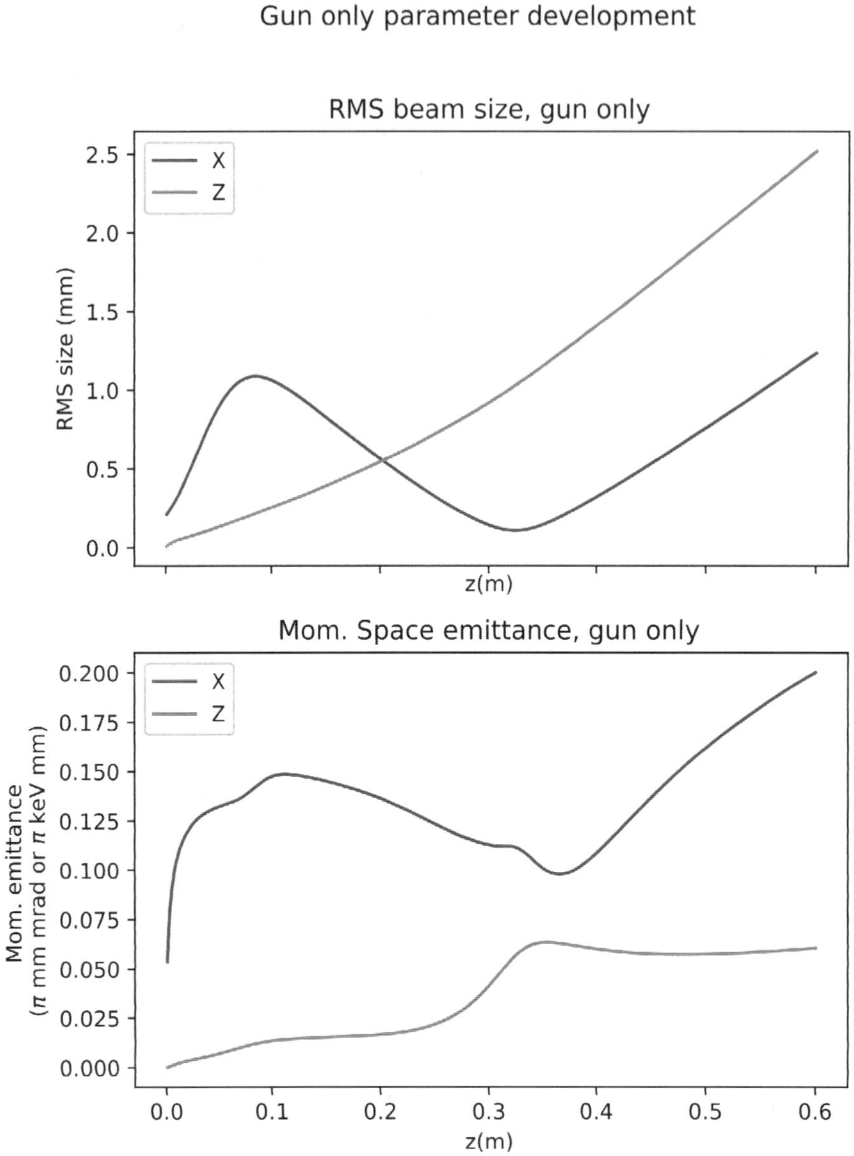

Figure 4. Gun only RMS and emittance development.

3. DC photogun simulation

In the following simulations we consider a 100 keV DC photogun with a focusing solenoid magnet with peak field 0.042 T. We note in contrast to the first section we will now be considering multi-particle simulations with space charge effects present. The magnetic field has been optimised to minimise the transverse emittance of the bunch downstream of the gun. The initial bunch has the parameters in table 1.

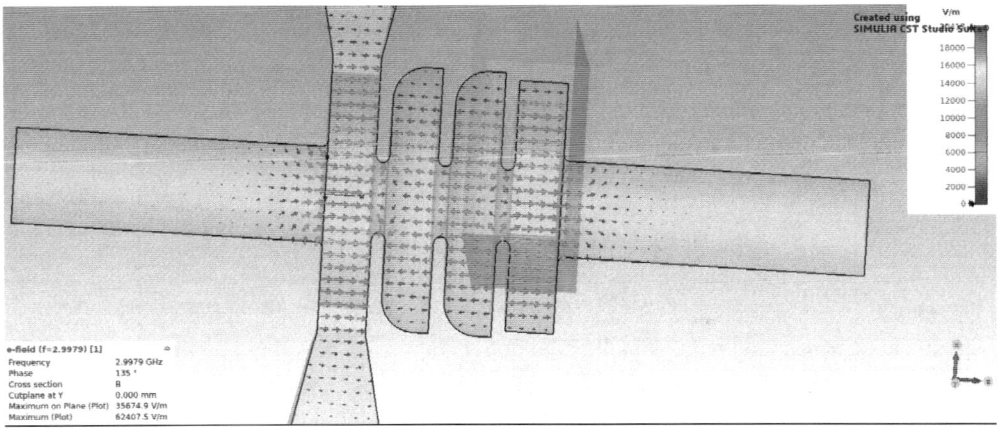

Figure 5. 3D model of buncher.

Table 1. Initial conditions and simulation parameters.

Parameter	Value
Bunch charge	1 pC
Number of macro particles	10000
Transverse shape	2D Gaussian cutoff at $2\sigma_x$
Transverse dimension (σ_x)	0.25 mm
Longitudinal shape	Gaussian in time
Pulse Length (σ_t)	100 fs

Plots of how the beam size and emittance develops can be seen in figure 4, and were generated by the code ASTRA.

We note that although this configuration results in low transverse emittance bunches that focus to an RMS size of 100 µm, we have a growing Z RMS. Also, the Z RMS at the focal point of approximately 1 mm would correspond to a phase difference of 14.4° at X-band wavelengths, strongly motivating rebunching the beam.

4. S-band buncher

To provide the energy boost required for acceptance of the bunch into the X-band accelerating structures we investigate the use of an S-band buncher similar to those already present at the Australian Synchrotron. This was remodelled using CST Microwave Studio and the fieldmap exported for use in the code ASTRA. The CST model can be seen in figure 5.

This is a four cell constant impedance S-band (2.9979 GHz) travelling wave operating in $\frac{2\pi}{3}$ mode with an average gradient of $7\,\mathrm{MV\,m^{-1}}$ and a total cell length of 80 mm. This structure has a design phase velocity of $0.6\,c_0$, and and is thus adequately matched to a kinetic energy 100 keV ($\beta \approx 0.55\,c_0$).

5. Proposed beamline configuration

We present a configuration of the DC photogun, S-band buncher, and two X-band accelerating structures in figure 6 that produces a low and stable emittance bunch with kinetic energy of

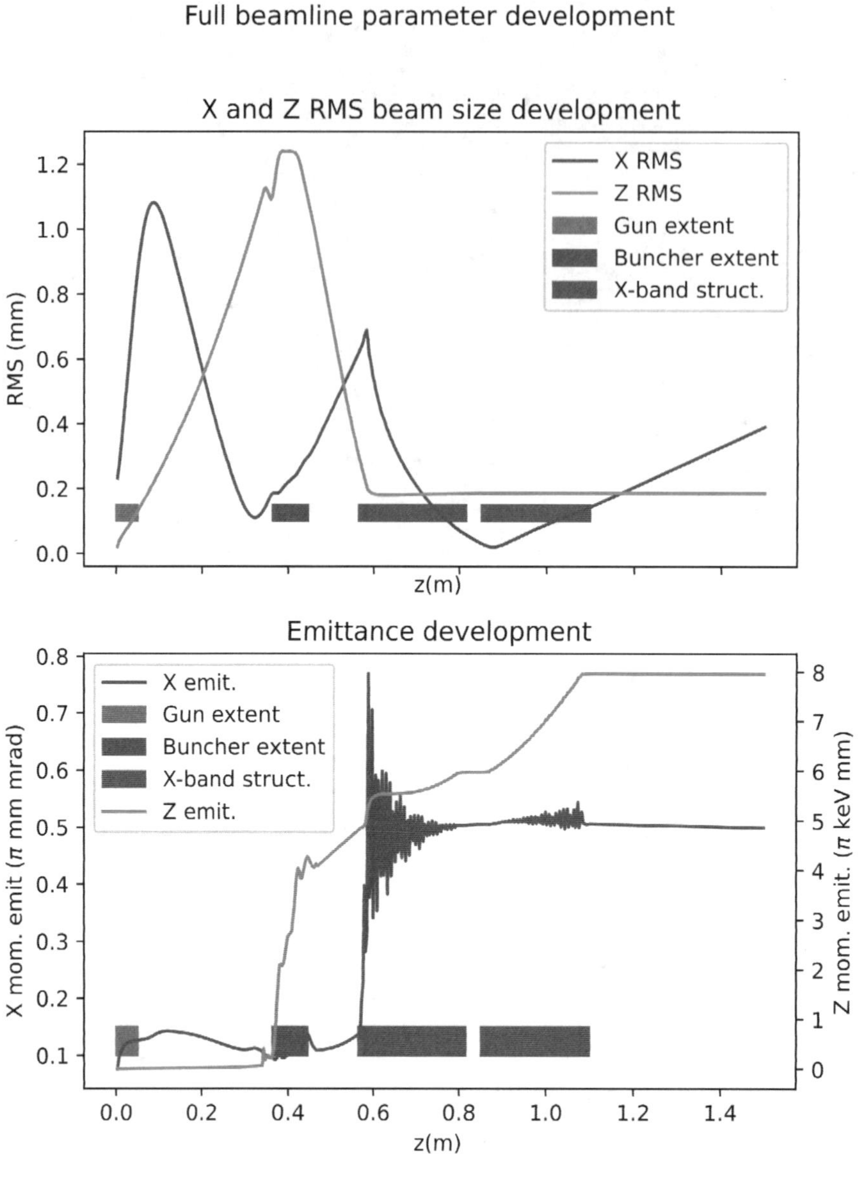

Figure 6. Development of bunch RMS and emittance through beamline.

approximately 27.29 MeV. This configuration was found by iterative optimisation of the position and phase of the buncher and X-band accelerating structures through successive simulations in ASTRA with space charge effects present.

6. Discussion

We have been able to demonstrate that the addition of the S-band buncher to the DC photogun is able to boost the energy of the bunch to the point where it can be accepted and accelerated by the X-band structures.

With regards to the suitability of this beamline, two future applications have been discussed so far; one is for dosimetry studies involving biological, and industrial samples, and other was as an electron source for an Inverse Compton Scattering light source. In the first case, as long as the beam size is amenable to the sample, the main performance characteristics are energy, bunch charge, and repetition rate. Although we will not discuss the repetition rate of the entire line, the beamline presented here would be appropriate for accelerating bunches.

With regards to using the beamline as an electron source suitable for an Inverse Compton Scattering light source, one of the design goals is to minimise the transverse emittance so that we may focus the bunch to a small interaction point, thus maximising scattering. The transverse emittance here of approximately 0.5 π mm mrad is comparable to what may have been achieved with an RF photogun, but with a smaller bunch charge (1 pC here, compared to 100 pC elsewhere).

In both cases this would be sufficient to commission the beamline, to test downstream diagnostics and feedback systems, systems for laser alignment, and to perform early test experiments.

We also note that this investigation has not considered the practical implications of two separate devices versus a single one, nor of working with two separate frequencies and klystron feeds. A possible future development here is to investigate the use of a custom designed and appropriately matched X-band boost cavity instead of the S-band buncher presented here.

7. Conclusion

We have been able to demonstrate issues with the X-band structures in accepting 100 keV electrons. Although far from ideal, we have also shown that the addition of an extra bunching cavity gives an extra energy boost that allows for acceptance. Although the bunch charge will be low it is possible to generate and preserve low emittance bunches for early commissioning and testing.

References

[1] Volpi M *et al. Proc. 12th Int. Particle Accelerator Conf. (IPAC'21)* (Campinas, Brazil) pp 3588–91
[2] Volpi M *et al. Proc. 13th Int. Particle Accelerator Conf. (IPAC'22)* (Bangkok, Thailand) pp 724–27
[3] Higo T *et al. Proc. 1st Int. Particle Accelerator Conf. (IPAC'10)* (Kyoto, Japan) pp 3702–04
[4] Dassault Systemes 2021 CST STUDIO SUITE `https://www.cst.com`
[5] K Floettmann 2021 ASTRA (A Space charge TRacking Algorithm) `https://www.desy.de/~mpyflo/`

Achievement of 200,000 hours of operation at KEK 7-GeV electron 4-GeV positron injector linac

K Furukawa[1,2], M Akemoto[1], D Arakawa[1], Y Arakida[1], Y Bando[2], H Ego[1,2], Y Enomoto[1,2], T Higo[1], H Honma[1], N Iida[1,2], K Kakihara[1], T Kamitani[1,2], H Katagiri[1], M Kawamura[1], S Matsumoto[1,2], T Matsumoto[1,2], H Matsushita[1], K Mikawa[1], T Miura[1,2], F Miyahara[1,2], H Nakajima[1], T Natsui[1,2], Y Ogawa[1], S Ohsawa[1], Y Okayasu[1,2], T Oogoe[1], M A Rehman[1], I Satake[1], M Satoh[1,2], Y Seimiya[1,2], T Shidara[1], A Shirakawa[1], H Someya[1], T Suwada[1,2], M Tanaka[1], D Wang[1], Y Yano[1], K Yokoyama[1,2], M Yoshida[1,2], T Yoshimoto[1,2], R Zhang[1,2] and X Zhou[1,2]

[1] High Energy Accelerator Research Organization (KEK), Tsukuba, Ibaraki, 305-0801, Japan
[2] Graduate University for Advanced Studies (SOKENDAI), Tsukuba, Ibaraki, 305-0801, Japan

E-mail: `kazuro.furukawa@kek.jp`

Abstract. KEK electron positron injector LINAC initiated the injection operation into Photon Factory (PF) light source in 1982. Since then for 39 years, it has served for multiple projects, namely, TRISTAN, PF-AR, KEKB, and SuperKEKB. Its total operation time has accumulated 200 thousand hours on May 7, 2020. We are extremely proud of the achievement following continuous efforts by our seniors. The construction of the injector LINAC started in 1978, and it was commissioned for PF with 2.5 GeV electron in 1982. In parallel, the positron generator linac was constructed for the TRISTAN collider project. The slow positron facility was also commissioned in 1992. After the KEKB asymmetric-energy collider project was commissioned in 1998 with direct energy injections, the techniques such as two-bunch acceleration and simultaneous injection were developed. As the soft structure design of the LINAC was too weak against the great east Japan earthquake, it took three years to recover. Then the construction and commissioning for the SuperKEKB project went on, and the simultaneous top-up injection into four storage rings contributes to the both elementary particle physics and photon science.

1. Introduction

In the experimental particle physics research in the 1970s, there was growing expectation for the world-class Japanese domestic collider after the successes in the world-level theoretical research. On the other hand, in the field of synchrotron radiation science, there were demands for the establishment of a synchrotron radiation research institute with a dedicated accelerator. In order to realize both of those accelerator projects, a 400-m electron linear accelerator capable of 2.5 GeV direct energy injection was constructed in 1982.

Since then, the injector had been in operation for 38 years in 2020, supporting the consecutive accelerator projects of Photon Factory (PF), TRISTAN, KEKB, PF-AR, and SuperKEKB as

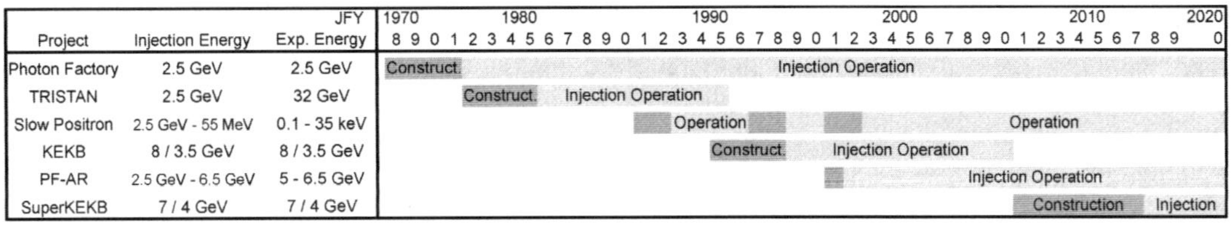

Figure 1. Electron positron accelerator projects operated by the injector LINAC with their injection beam energy and experimental beam energy.

depicted in figures 1, 2, 3 and 5. It has finally achieved 200,000 hours of operation in May 2020. The brief operation history would be given.

2. 2.5 GeV injection for PF/TRISTAN

At the time of the initial construction since 1978, several university-based accelerator facilities had achieved research successes in Japan. It was fortunate that a new linear electron accelerator had been built to share resources for advanced scientific research in both particle physics and synchrotron radiation science, which were based on the accumulated experiences at universities. The situation might be similar to recent J-PARC's focus on both nuclear and particle physics, as well as neutron and muon science. It seems that there are not many accelerator projects in the world that have multiple disciplines from the beginning. Those Japanese accelerators are quite unique that deliver beams for multiple science fields.

While many researchers of synchrotron radiation research wanted the institute independent from particle physics, the accelerator science played a role as a bridge between those two. One hundred sixty of 2-m-long S-band traveling-wave accelerating structures with $2\pi/3$ mode quasi-constant gradient were installed, and driven by forty 20-MW klystrons to achieve 2.5 GeV electron and positron beams [1]. 2.5 GeV direct energy injection was made for PF and 2.5 GeV electron positron beams from the injector were accelerated through the TRISTAN synchrotron chain for the 32 GeV collision as depicted in figure 2.

When the synchrotron radiation facility, PF, was completed, there must be an enthusiastic response to the dedicated synchrotron radiation facility that would promote research in

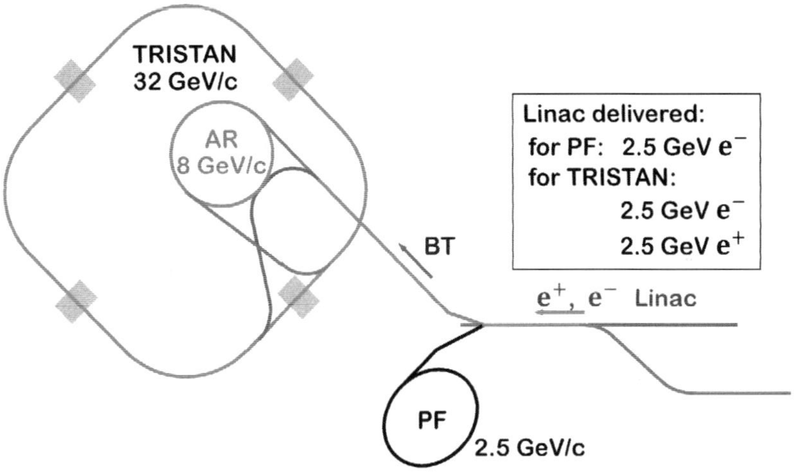

Figure 2. Injection for TRISTAN and PF.

synchrotron radiation science. The TRISTAN project also began to take the lead in high-energy experimental physics, after Japanese theorist had contributed to the worldwide development of particle physics theory. It was realized recently that the specifications of the electric power facilities built at that time secured the expansion possibility even for the fields of nuclear physics and ultraviolet synchrotron radiation science. Such design facilitated the upgrades towards KEKB and SuperKEKB later.

3. 8-GeV upgrade for KEKB

Since the construction of TRISTAN, a linear collider was expected as the next particle physics experiment project. Instead of immediately promoting a large scale collider, however, the construction of a B Factory, KEKB, was carried out to experimentally confirm the remaining issues of CP violation in particle theory proposed by Makoto Kobayashi and Toshihide Maskawa. Since the injection performance was important to achieve the integral performance as a factory, there were concerns compared to SLAC/PEP-II, which was planning the B-factory in parallel.

The direct energy injections for the asymmetric energy collider rings were realized by an energy upgrade from 2.5 GeV to 8 GeV by introducing SLED microwave pulse compressors and adding 50% more accelerating structures that extends the length of the injector up to 700 m in a shape of a letter 'J' as in figure 3. It was a substantial and difficult construction because the injector continued the PF injection [2,3]. It made maximum use of the opportunities of the longest shutdown of nine months in 1997 during the construction.

The success of the KEKB operation was largely due to the online operating environment of the SAD beam simulation software and the EPICS control framework which were developed during the injector commissioning period [4].

KEKB and PEP-II projects developed the friendly competition, and they compared the performance of the accelerators in those control rooms everyday [5]. The performance of the KEK injector was gradually improved during the project step-by-step [6].

The positron beam injection was enhanced by employing dual bunch acceleration. The separation of two bunches was chosen to be 96 ns because of the frequency relationship between the injector and the ring [7].

Then later, it also succeeded in continuous injection by switching the injection every several minutes while the Belle collision detector was in operation. In 2008, the injection was further

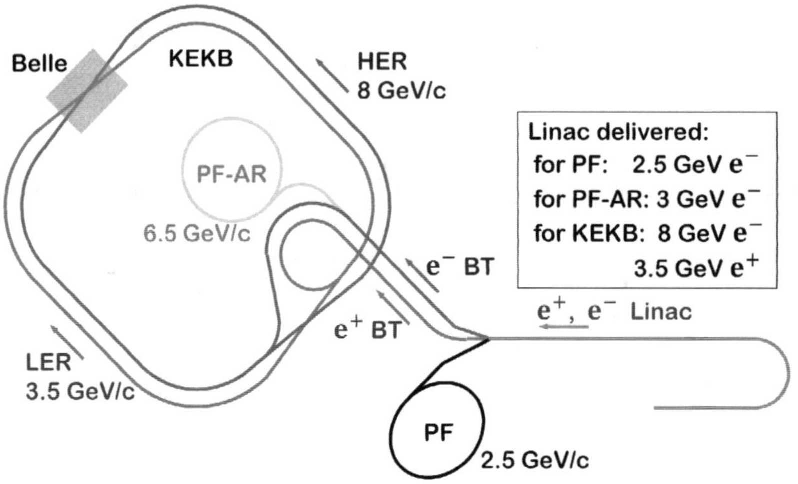

Figure 3. Injection for KEKB, PF and PF-AR.

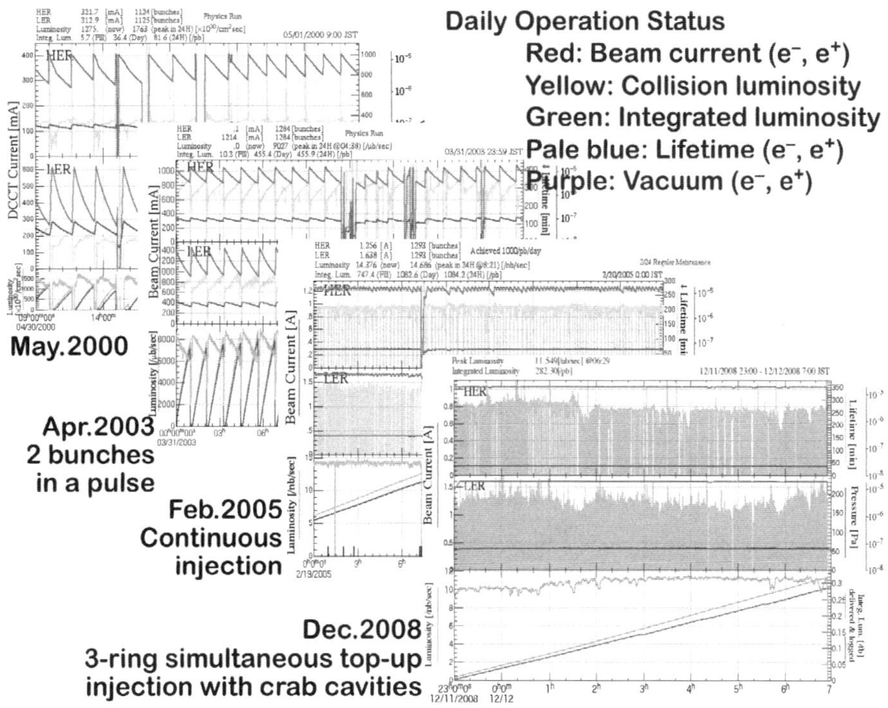

Figure 4. Progress of the daily operation during the KEKB project, with the stored beam current in red and collision luminosity in yellow.

improved by the simultaneous top-up injection in to three storage rings except PF-AR in pulse-to-pulse modulation switching beams with distinct properties every 20 ms employing event-based control architecture [8] as in figure 4.

4. Slow positron and others

The injector facility usually contributes by injecting beams into the downstream storage rings. However, there are cases where experiments are performed using direct electrons. In the past at KEK LINAC, there have been experiments at the beam dump for a search for an axion like particle, beam irradiation calibration of tile fiber calorimeter detectors for the SSC project, positron beam channeling experiments, and positron production using electron channeling radiation. As a result of the positron generation experiment, a crystal target was actually applied to the positron generation for KEKB injection in 2006, and the positron generation efficiency was successfully improved by about 25% [9].

Another ongoing project is the research experiments for material science and particle physics by slow positrons. Between injections to the TRISTAN project slow positrons were generated at the end of LINAC. While the experimental setup had to be moved twice due to the change in the beam operation of the main LINAC, now a dedicated 50 MeV linac is used for slow positron experiments sharing the same resources as the main LINAC [10]. It is valuable to produce scientific results within the LINAC building.

5. SuperKEKB and simultaneous top-up injection

After the KEKB project was completed in 2010 the injector, whose girders were built with the concept of a flexible structure, suffered a serious damage due to the Great East Japan Earthquake in 2011. While light source injections were recovered in 3 months, it took three years to start the

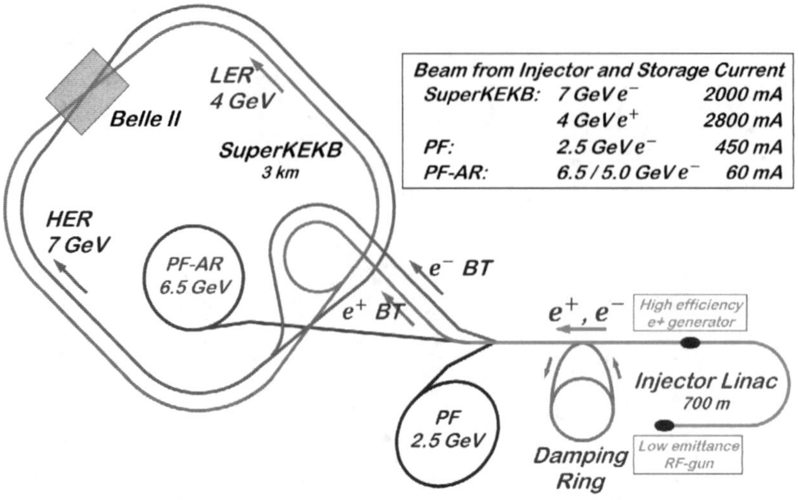

Figure 5. Injection for SuperKEKB as well as PF and PF-AR.

beam test for SuperKEKB as the alignment was severely deteriorated. Actually, the alignment requirement became much tight in order to suppress the wakefield effect in the accelerating structure. While the light source injections were maintained even during the upgrade towards SuperKEKB, the longest construction period of five months was secured in 2017.

The energy enhancement was the main subject for the previous KEKB injection upgrade, on the other hand the SuperKEKB project required quality improvements of the injection beam for the electron positron collision with the novel nano-beam scheme [11] as in figure 5. A beam emittance of about 20 mm.mrad was required with a charge of 4 nC per bunch. The positron capture section was reconstructed [12] in order to increase the positron charge employing a flux concentrator pulsed solenoid, large aperture S-band structures (LASs), long DC solenoid magnets and a hundred of quadrupole focusing magnets. A damping ring was newly constructed to damp the positron emittance. For the electron beam, a new high-current and low-emittance RF gun was developed with an iridium-cerium photocathode and a quasi-travelling-wave side-coupled cavity (QTWSC) [13].

More than hundred pulsed focusing and corrector magnets were newly installed [14]. Pulsed low-level RF controls were also developed. Based on those equipment upgrades [15], the simultaneous top-up injections for four storage rings of SuperKEKB HER, LER, PF ring and PF-AR were realized in 2019, which contributed to the considerable improvement of experimental efficiency [16].

The continuous operation of the electron-positron injector has supported the history of advanced research at KEK by providing beams for various projects as shown in figure 1.

At 8:50 a.m. on May 7, 2020, 200,000 hours of operation was accomplished as shown in figure 6, since the PF injection operation started in 1982. Authors are very proud to have reached the milestone of 200,000 hours of operation by inheriting the operation results accumulated by our predecessors. Although no special event was held due to the effects of COVID-19, a toast was made with morning coffee via video conference. In the statistics, the partial failure rate indicates that a certain device has failed but the beam injection operation was still possible due to redundancy, and the injection stop rate indicates that the failure was so serious that the beam injection was impossible. At the beginning of each program, the failure rate increases before understanding the equipment characteristics, but it tends to stabilize as countermeasures are taken.

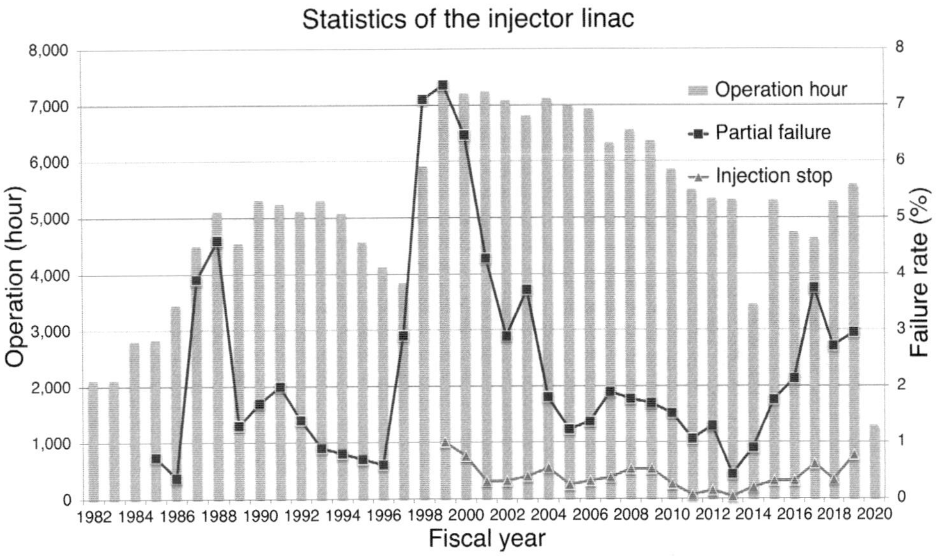

Figure 6. Yearly operation hours of the injector LINAC with failure rates.

6. Conclusion

The KEK electron-positron injector LINAC has been able to achieve 200,000 hours of operation while supporting experiments in multiple fields of photon science and particle physics.

References

[1] Tanaka J 1980 *Nucl. Instrum. Methods* **177** 101
[2] Abe I *et al* 2003 *Nucl. Instrum. Methods Phys. Res., Sect.* A **499** 167
[3] Akemoto M *et al* 2013 *Prog. Theor. Exp. Phys* **2013** 03A002
[4] Flanagan J W *et al* 1998 *Proc. Int. Computational Accelerator Physics Conf. (ICAP'98)* (Monterey, USA) F-Tu09
[5] Seeman J 2008 *Proc. 11th European Particle Accelerator Con. (EPAC'08)* (Genoa, Italy) pp 946–50
[6] Furukawa K *et al* 2018 *Proc.62nd ICFA Advanced Beam Dynamics Workshop on High Luminosity Circular e+e− Colliders (eeFACT2018)* (Hong Kong, China) pp 121–25
[7] Furukawa K *et al* 2001 *Proc. 8th Int. Conf. on Accelerator and Large Experimental Physics Control Systems (ICALEPCS2001)* (San Jose, USA) pp 266–68
[8] Furukawa K *et al* 2009 *Proc. 12th Int. Conf. on Accelerator and Large Experimental Physics Control Systems (ICALEPCS2009)* (Kobe, Japan) pp 765–67
[9] Suwada T *et al* 2007 *Phys. Rev. Spec. Top. Accel Beams* **10** 073501
[10] Hyodo T *et al* 2018 *AIP Conf. Proc.* **1970** 040004
[11] Akai K, Furukawa K and Koiso H 2018 *Nucl. Instrum. Methods Phys. Res., Sect.* A **907** 188
[12] Enomoto Y *et al* 2021 *Proc. 12th Int. Particle Accelerator Conf. (IPAC'21)* (Campinas, Brazil) pp 2954–56
[13] Yoshida M *et al* 2014 *Proc. 27th Int. Linear Accelerator Conf. (LINAC'14)* (Geneva, Switzerland) pp 21–25
[14] Enomoto Y *et al* 2018 *Proc. 29th Int. Linear Accelerator Conf. (LINAC'18)* (Beijing, China) pp 609–14
[15] Furukawa K *et al* 2018 *Proc. 9th Int. Particle Accelerator Conf. (IPAC'18)* (Vancouver, Canada) pp 300–03
[16] Furukawa K *et al* 2020 *Proc. 17th Annual Meeting of the Accelerator Society of Japan (PASJ2020)* (Online, Japan) WEOOP01 pp 1–6

Lifetime correction using fast-off-energy response matrix measurements

S M Liuzzo, N Carmignani, L R Carver, L Hoummi, T Perron, B Roche and S White

ESRF, Grenoble, France

E-mail: `simone.liuzzo@esrf.fr`

Abstract. Following the measurements done at MAX-IV [1], we try to exploit for the ESRF-EBS Storage Ring (SR) off-energy response matrix measurement for the optimization of Touschek lifetime. The measurements performed with fast AC steerers on- and off-energy are analyzed and fitted producing an effective model including quadrupole and sextupole errors. Several alternatives to extrapolate sextupoles strengths for correction are compared in terms of lifetime. For the time being none of the corrections could produce better lifetime than the existing empirically optimized set of sextupoles.

1. Introduction

The ESRF-EBS SR Touschek lifetime is optimized routinely using an empirical approach [2]. In the recent paper by Olsson et al. [1] it was shown that it is possible based on on- and off-energy orbit response matrix measurements to determine a set of sextupoles able to improve the lifetime, by correcting the off-energy optics of the SR.

Having seen the promising results presented, the Nonlinear optics from off-Energy Closed Orbit (NOECO) technique shown in [1] is applied to the EBS SR. This paper reports about the initial simulations and the preliminary results obtained during the first two dedicated experimental shifts.

2. NOECO correction

The NOECO correction exploits the dependence on sextupole strengths of the off-energy orbit response matrix. A set of normal sextupole correctors $K^{\text{cor}}_{\text{sext}}$ is computed based on two off-energy orbit response matrices (ORM) and dispersion (η_h) measurements by solving the following system of equations using a SVD pseudo-inverse:

$$V = \frac{(ORM, \alpha_\eta\, \eta_h)_{+\delta} - (ORM, \alpha_\eta\, \eta_h)_{-\delta}}{\partial\delta} \tag{1}$$

$$V_{\text{measured}} - V_{\text{model}} = \frac{\partial V}{\partial K_{\text{sext}}} K^{\text{cor}}_{\text{sext}} \tag{2}$$

where V is a vector including the variation of all the ORM elements and horizontal dispersion with respect to an energy deviation δ and α_η is a weight to determine the relative balance among ORM and η_h correction (initially set to give similar amplitude to the dispersion response compared to the ORM and then empirically tuned). Only the diagonal blocks (horizontal

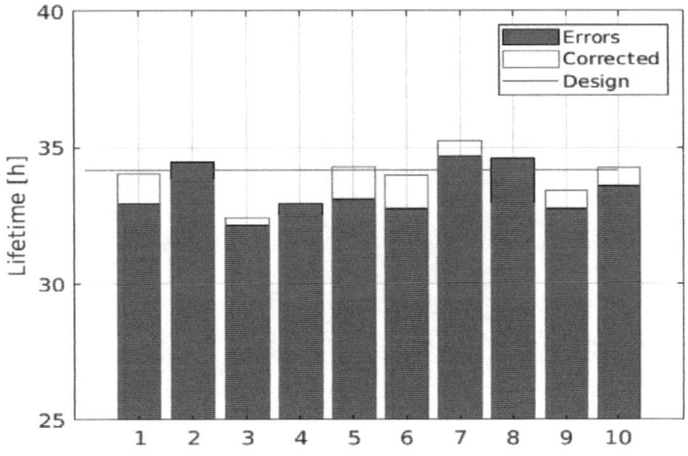

Figure 1. Lifetime before and after NOECO correction for 10 simulated lattices including errors and corrections with identical α_η and number of singular values. Three seeds out of ten require ad-hoc tuning of α_η and number of singular values in order to improve the lifetime.

response to an horizontal steerer and vertical response to a vertical steerer) of the ORM are used. The Jacobian $\partial V/\partial K_{\text{sext}}$ is computed numerically using the Matlab [3] version of Accelerator Toolbox (AT) [4]. In the analysis all the BPMs are considered equal and no weight factor is used. For the details of the correction we refer to the original paper.

3. Simulations

Simulations of the NOECO correction were done in AT using lattices with realistic errors and corrections based on a commissioning-like simulation loop [5]. For most of the seeds considered the computed lifetime improved after NOECO correction, as shown in figure 1. With a change of the singular values or of the weights α_η, seeds 2, 4 and 8 had their simulated lifetime improved compared to the one without NOECO sextupole correction. The local momentum acceptance for the first cell of the SR is compared in figure 2 before and after correction to the one without any errors. The NOECO correction effectively restores the ideal momentum acceptance. The injection efficiencies contextually computed for each seed do not show significant variations ($<1\%$). The sextupole errors present in these lattices are taken from a random Gaussian with standard deviation of 0.35% of the main sextupole field as defined by the magnet design simulations [6]. The sextupole corrections strengths obtained with NOECO are of the same amplitude but uncorrelated. The corrections are small compared to the main magnet gradients, and well within the available power supply limits. An energy shift of 0.33 % is used in these simulations.

4. Measurements

Given the promising results in simulations, two dedicated machine time shifts were devoted to the test with beam of the NOECO correction.

The capability to measure off-energy ORM and dispersion was included in the operation optics correction application [7]. Due to the large number of ORM necessary, the ORM measurement was done using the Fast Orbit Feedback AC steerers.

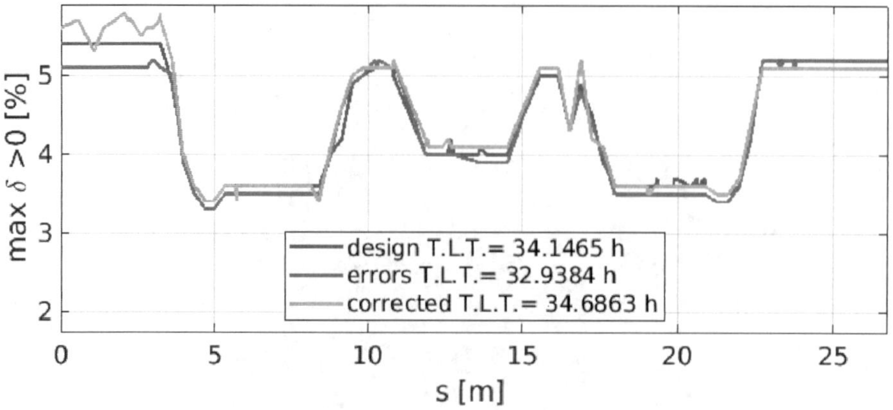

Figure 2. Local momentum acceptance for the first cell of the EBS SR for the design lattice without errors, a lattice with errors and the same lattice with sextupole corrections computed using NOECO.

4.1. Fast orbit response matrix

Fast ORM are measured exciting sequentially the 96 AC steerers of the ring with a sine wave signal and detect the resulting orbit variation amplitude at the same frequency with the 320 BPMs in the ring. The excitation frequency is 21 Hz such that it is well below the cut-off frequency of the steerers limited by the screening effect by the vacuum chamber. The average orbit is kept constant by running the fast orbit feedback (FOFB) with reduced PID parameters to have a slow time response (of the order of 1 second) as described also in [7]. Measuring two ORM (horizontal and vertical) takes approximately 5 minutes, which has to be compared to approximately 1 hour required to measure the response matrix with the legacy method using DC kicks.

4.2. Measurement procedure

All measurements were preceded by an optics correction (fitting of the measured ORM to the theoretical one [7]) to determine the quadrupole (19 per cell) and skew quadrupole (9 per cell) strengths for the model and by a chromaticity measurement. The optics correction guarantees that the off-energy measurements are minimally polluted by initial optics modulation. The measured chromaticity is used to match the reference lattice chromaticity in order to provide the correct tunes off-energy. Variations of frequency of 100 Hz were applied in order to obtain the necessary $\delta = 0.33\%$ energy deviation, as used in simulations. Larger frequency shifts of 200 Hz were also tested without an evident gain in correction performance.

All BPMs not providing appropriate signals (bad electronics, cabling issues, large rotations or scale factors, ...) were excluded from the retrieved data. Finally 310 out of 320 BPMs are used to fit all the 192 sextupoles for the NOECO correction. The ORM repeatibility was tested in the past and showed variations of the computed correction below 1 % when comparing two consecutive measurements. Considering the BPMs resolution of 80 nm at 10 kHz and the magnets power supplies stability, the resolution of a fast ORM measurement is estimated to be better than ~0.04 µmrad^{-1}. The expected maximum amplitude of $V \cdot \delta$ is for comparison 83 µmrad^{-1}.

Once the two ORM and dispersion datasets at $+\delta$ and $-\delta$ to compute V_{measured} (see equation (1)) were collected a Matlab script would compute the Jacobian based on the present SR optics. The same script would analyze the data collected to apply the NOECO correction to finally provide a set of sextupoles corrections to be set to the SR.

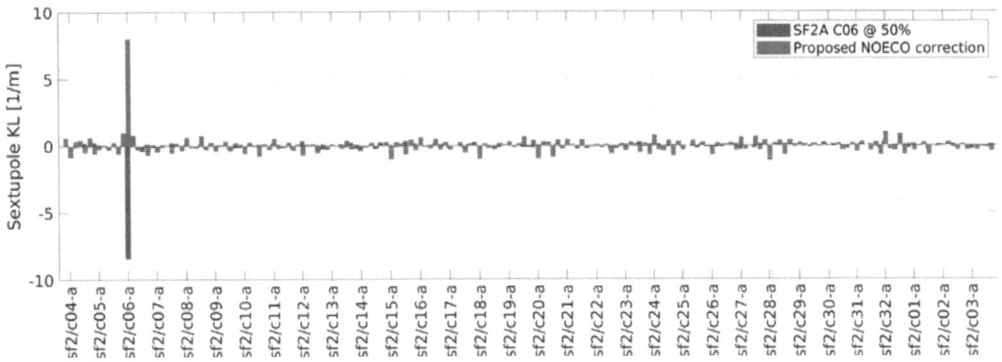

Figure 3. Sextupole pattern set in the SR for test of the NOECO correction scripts. The single magnet mistuned is correctly identified by the correction algorithm.

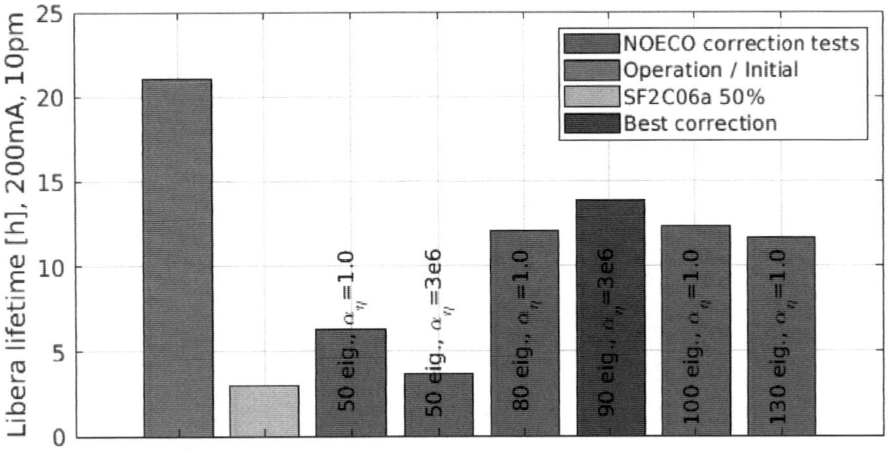

Figure 4. Beam lifetime for the operation setting (red), with a sextupole mistuned (yellow) and for several NOECO correction parameters (blue). The best set of correction parameters is shown in purple.

4.3. Tests of NOECO correction

The first test was reduced to a single source of error, localized in a SF focussing sextupole. A random magnet was selected and set to $\sim$ 50% of its nominal strength. The pattern of sextupoles correction proposed by NOECO was clearly indicating the specific sextupole as the source of discrepancy in the measurement, as shown in figure 3.

The α_η weight and the number of singular vectors used for pseudo-inversion were varied to find the optimal ones. Settings with fixed chromaticity were also produced. Figure 4 shows the lifetime measured at high current and with fixed vertical emittance at ϵ_v =10 pm, normalized to 200 mA.

The mistuned sextupole has a significant impact on the beam lifetime, reducing it from 21 h to 3 h. The NOECO correction restored in this test the lifetime to 14 h. This is positive, as it proves that the algorithm can be used to spot sextupole issues. However, the lifetime after correction is not completely recovered. In fact all other sextupoles apart the mistuned one are modified and by a larger amount compared to the existing corrections, thus erasing their positive effect on lifetime [2]. This may be also due to lack of reiteration of the correction, to a poor measurement (for example not excluding some bad signal BPM) or to a lack in the nonlinear

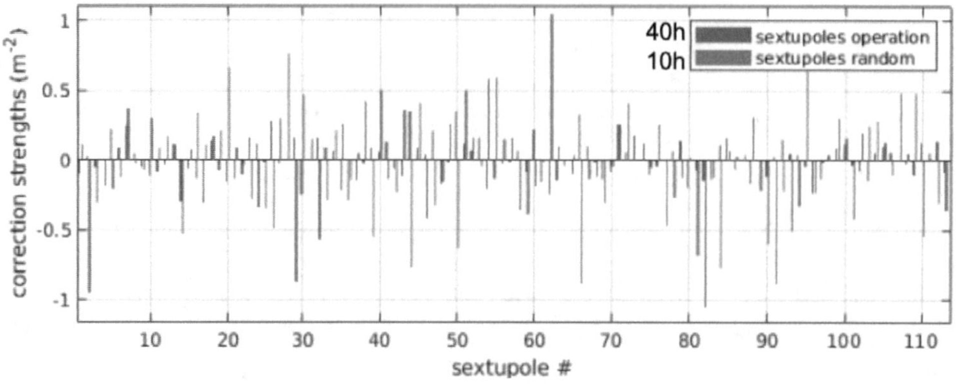

Figure 5. Correction strengths of the first 115 normal sextupoles empirically optimized for operation and randomly set to reduce the lifetime.

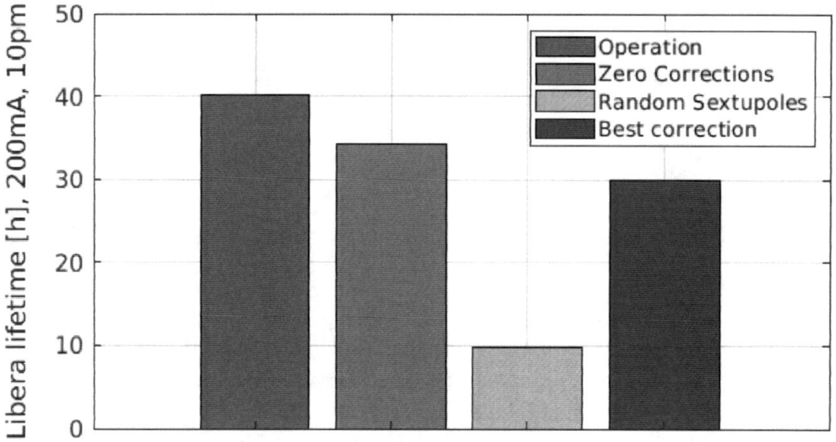

Figure 6. Beam lifetime for the operation setting (red), with a sextupole correction to zero (red), with random sextupole corrections (yellow) and for the best NOECO correction (purple).

part of the optics model. The same exercise was done with a SD defocussing sextupole magnet. In this case, the magnet was not identified individually, however the lifetime was partially restored as for the SF sextupole case. A further test was performed introducing in the SR random sextupoles correction values (192 sextupoles) such as to obtain a lifetime reduction. The values were chosen empirically. For the peculiar configuration during measurements the Touschek Lifetime was reduced from 40 h to 10 h by sextupole correction strengths extracted from a Gaussian distribution with standard deviation of $0.4\,\mathrm{m}^{-2}$ ($\sim 3\%$ of the main sextupole field). These are about ten times larger than the values corrected in the initial simulations. The NOECO correction was then applied and a sextupole correction setting was defined. The random setting and the computed correction are presented in figure 5.

There is no correlation among the two settings, however the lifetime after correction is almost as good as in the case with all sextupole corrections set to zero, as can be seen in figure 6.

Referring to figure 6 the lifetime for the empirically optimized operation sextupole setting is also shown. The NOECO correction is again able to recover a good fraction of the lost lifetime due to the random sextupole settings, but it is unable to reach the performances obtained by simply setting all sextupoles to their design gradients, thus not applying any correction at all.

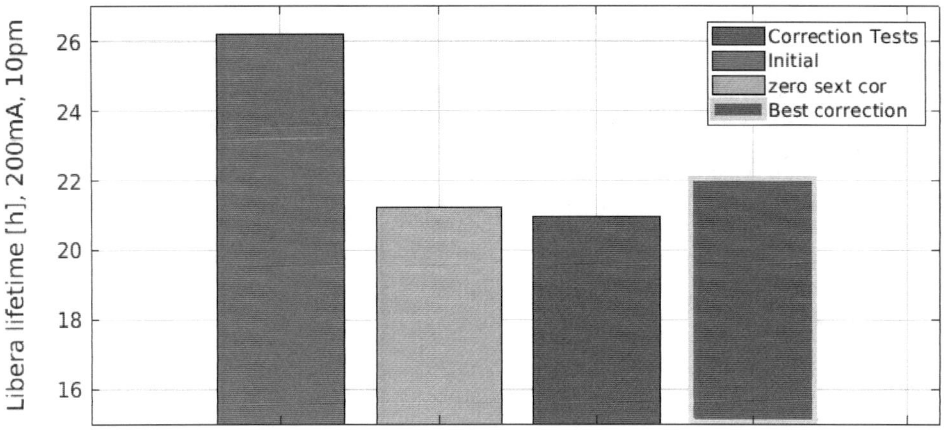

Figure 7. Beam lifetime for the operation setting (red), with a sextupole correction to zero (yellow), with NOECO correction (blue). The best set of correction parameters is shown by a green edge.

The sextupole corrections (obtained by empiric optimization) presently set in the SR are about 1% of the main sextupole fields, three times larger then the values used for simulations.

A final test was performed, starting from zero correction strengths, to seek for an equivalent or improved operation setting defined by NOECO. The zero correction strengths settings has a purely periodic pattern, except for the 12 sextupoles adjacent to the injection. The individual calibration factors of the sextupoles are included in such setting. Also, all sextupoles are combined function with steerers and skew quadrupoles, with a calibration curve taking in account also the main sextupole main coil in order to provide the correct strengths for all multipoles. The sextupole currents are thus not periodic, with variations up to more than 3% in each family. In case the NOECO correction could find a better solution than the one obtained by empiric optimization with specialized knobs, then a large amount of machine dedicated time could be spared, avoiding empirical optimizations during storage ring restarts activities. The test was in fact a failure. As shown in figure 7 the NOECO correction could not produce any solution with lifetimes similar to the ones used for operation. It could however, in some of the cases tested, mildly improve the zero sextupoles correction strengths setting.

5. Conclusions

The NOECO correction tested for the EBS SR has shown a great potential in simulations and in spotting localized sextupolar errors. Nevertheless it was not possible for the moment to obtain a better sextupole setting compared to the operational one. Future machine dedicated time will be devoted to further study the NOECO correction. In particular those will take place after a better non linear model of the lattice optics will be available. Slow ORM measurements (without FOFB) with larger energy deviation at low currents (to avoid safety interlocks) will be investigated. To obtain the best possible measurement, refined and updated ORM fit procedures will be investigated, in particular looking at the BPM and correctors scale and rotation factors and applying a more restrictive selection on the BPMs used for correction, if needed. Individual BPM weights will be measured and included as shown in the original work in [1]. A NOECO correction including also skew sextupole components will be considered. Finally, more iterations of the NOECO correction will be applied in a sequence until convergence of the computed corrections.

References

[1] Olsson D, Andersson A and Sjöström M 2020 *Phys. Rev. Accel. Beams* **23** 102803
[2] Carmignani N *et al.* 2022 *Proc. 13th Int. Particle Accelerator Conf. (IPAC'22)* (Bangkok, Thailand) pp 2552–55
[3] MATLAB 2020b *version 7.10.0 (R2010a)* (Natick, Massachusetts: The MathWorks Inc.)
[4] Nash B *et al.* 2015 *Proc. 6th Int. Particle Accelerator Conf. (IPAC'15)* (Richmond, Virginia, USA) pp 113–16
[5] Liuzzo S, Carmignani N, Andrea F, Perron T, Scheidt K, Taurel E, Torino L and White S 2019 *J. Phys. Conf. Ser.* **1350** 012022
[6] Liuzzo S *et al.* 2015 *6th International Particle Accelerator Conference* p TUPWA014
[7] Liuzzo S M *et al.* 2022 *Proc. 13th Int. Particle Accelerator Conf. (IPAC'22)* (Bangkok, Thailand) pp 1401–04

Cooling challenges in a NEG-coated vacuum chamber of a light source

S Talebi Motlagh[1,2], A Danaeifard[1], J Rahighi[1], F Saeidi[1] and F Zamani[2]

[1] Iranian Light Source Facility Institute for Research in Fundamental Sciences (IPM), Tehran, Iran
[2] University of Kashan, Kashan, Iran

E-mail: s.talebimotlagh@ipm.ir

Abstract. In a Light Source, unused synchrotron radiation is being distributed along the walls of the chambers. Due to the small conductance of the chambers, vacuum pumping will be based on the distributed concept, and then non-evaporable getter (NEG) coating is extensively used. The vacuum chambers are made of copper alloys tube, and cooling circuits are welded to the chamber to remove the heat load from the radiation generated. Filler metal creates a brazed joint between the water-cooling pipe and the vacuum chamber body. The thermal conductivity of the fillers is less than the vacuum chamber body. On the other hand, the velocity of the water in the cooling pipe is a critical parameter in thermal calculations that must be taken into account. So, in this paper, we study and investigate the effects of the filler metal and the cooling water velocity on cooling the NEG-coated chambers.

1. Introduction

The ILSF storage ring lattice is based on 20 five-bent achromats; Each achromat contains three-unit cells and two matching cells. The unit cells have a 3.9° bending magnet, while the matching cells deflect the beam 3.15° [1]. The radiation of these five dipole magnets is uniform in the horizontal direction (plane of the storage ring), while in the vertical direction, it follows a narrow Gaussian profile.

The irradiate power and the radiation power density on the vacuum surface are simulated by Synrad+ software [2]. It is assumed that the chamber body will absorb all the radiation, and no reflection would happen. Hence, the facets' sticking factor has been considered equal to 1. The cooling of this chamber and its challenge will be studied in the present work.

The commonly used thermal absorber design criteria are [3]:

- The maximum cooling wall temperature $T_{Max}^{C\,W}$ should be lower than water boiling temperature $T_{boiling}$ at the pressure of the water in the cooling tubes
- The maximum temperature of the chamber T_{Max}^{Ch} must be significantly lower than the melting point of the copper and the brazing temperature.
- The maximum temperature rise in the chamber should be less than 300°C for Glidcop and 150°C for oxygen-free high thermal conductivity (OFHC) copper, which are also used at APS [4].

2. Theory

The total radiative power of all the bending magnets in a ring is determined by the electron's energy E [GeV], the bending magnet's field B [T], and the electron beam current I_b[mA], as the following equation [5]:

$$P_{total}[\text{kW}]=26.6E^3[\text{GeV}]I_b[\text{A}]B[\text{T}] \tag{1}$$

Correspondingly, the power density on the beam axis can be obtained as follows:

$$\frac{dP}{d\Omega}\left[\frac{\text{W}}{\text{mrad}^2}\right]|_{\psi=0}=5.44E^4[\text{GeV}]I_b[\text{A}]B[\text{T}] \tag{2}$$

Since the magnitude of the magnetic field in the dipoles of the ILSF storage ring is about B=0.567 T according to equation (2), at a current of 400 mA and an energy of 3 GeV, the total output power due to synchrotron radiation will be 162.95 kW. It means that for 100 dipole magnets in the ring, each bend chamber wall receives 162.95 kW radiation power.

A maximum value of 8.35 W/mm^2 for the power density has been obtained analytically.

Most of the power will be distributed along the water-cooled vacuum chambers. So, a cooling circuit channel is welded to each chamber to remove the heat load from the radiation generated as seen in figure 1. In this research, a simple model of a curved vacuum chamber in a dipole without photon extraction is studied to investigate the thermal effects, as seen in figure 2. Although the electrons travel parallel with the chamber in a curved path but the radiant photons go in a straight line. So, significant radiation collisions will be happened along the second half of the chamber. Therefore, the initial part of the chamber will have an ambient temperature, while the final part will be warm due to the radiation, as discussed here.

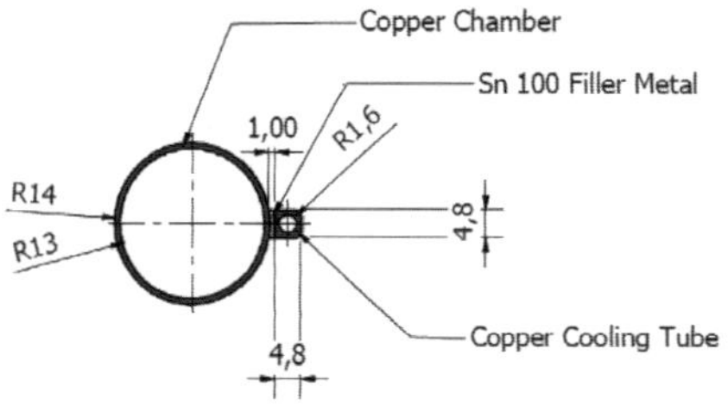

Figure 1. The ILSF chamber cross-section view.

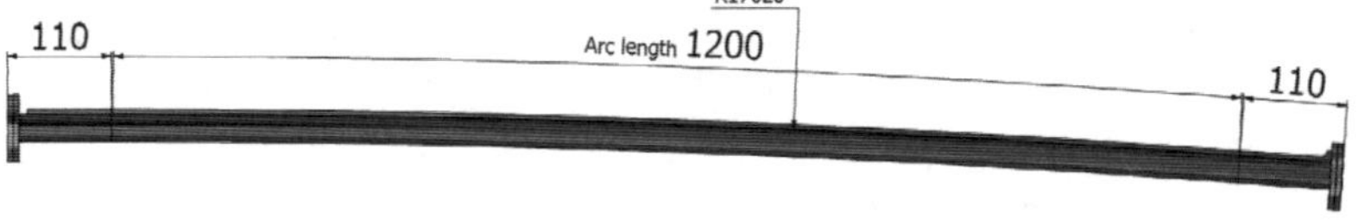

Figure 2. Simple model of the vacuum chamber in a dipole without photon extraction.

The vacuum chambers of the ILSF storage ring will be fabricated using an oxygen-free copper alloy. The small amount of silver in this alloy helps the chamber increase its resistance to softening.

The velocity of water in the cooling pipes should be less than 3 m/s to keep the flow-induced vibrations within acceptable levels and larger than 1 m/s to avoid the accumulation of air in the circuits.

A filler metal (such as SN100C) will be used to create a low-temperature brazing joint between the water-cooling pipe and the vacuum chamber body. The thermal conductivity of this alloy is less than the conductivity of copper. The thermal conductivity coefficient of OFS copper is 388 W/m°K while the thermal conductivity coefficient of SN100C is 64 W/m°K. A thickness of 1 mm against brazed filler metal has been considered in the thermal simulations, which is a strict assumption. Nevertheless, it will be sure that it will not encounter any problems in practice. The effect of this filler conductance has been studied in this work.

3. Simulations

3.1. Computational fluid dynamics

Computational Fluid Dynamics (CFD) model is based on the fundamental equations governing fluid dynamics: mass, momentum, and energy conservation. CFD helps anticipate fluid flow behavior based on mathematical models using software tools. This method is widely used and accepted as a proper engineering tool in the industry. The CFD simulation process involves several different steps involved in fluid flow analysis.

3.2. Finite element analysis (FEA)

FEA is a famous method to solve numerically differential equations moving up in engineering and mathematical modeling. The FEA subdivides an extensive system into smaller, simpler parts called finite elements to solve a problem. The thermal analysis of the presented chamber has been performed based on FEA. The ANSYS software [6] is used to simulate the thermal distribution of the chamber body. The result of the thermal analysis will be used for stress analysis in the next step.

The heat generated on the vacuum chamber's inner surfaces is transferred to the water that passes through the cooling pipe by the convection process. Several quasi-experimental equations are available to calculate the convective heat transfer coefficient from the pipe wall to the water. Since the heat is transferred from the hotter body to the water pipe walls, the heat absorbed is approximately equal to the radiation heat.

In FEA, cooling is entirely defined by the convective heat transfer coefficient and water temperature.

CFD modeling has not been commonly used in the past for absorber cooling calculations in the synchrotrons. In Ref [7] authors refer to a heat transfer film coefficient value that varies from 10 to 20 $W/m^2 °K$ with no or little precisions concerning the cooling channel dimensions and the flow rate.

For our case, it is decided to study a simple model of a cooling channel with a water flow rate (1-3 m/s), to extract the maximum and average heat transfer coefficient in the cooling tube wall. Also, the maximum and average temperatures of the chamber body are interesting to compare with FEA results. All CFD simulations were made using the k-ε turbulence model [8], and then it is compared with the results of FEA.

Heat transfer coefficient h will be calculated by ANSYS from the equation $q''=h\left(T^{C\,W}-T^{\infty}\right)$ where q'' is the heat transfer rate, $T^{C\,W}$ is the cooling wall temperature and T^{∞} is the temperature of the reference. The cooling calculations using CFD have been done with three different flows. In this method, it is assumed that the inlet water temperature is 25 degrees. Heat flux due to synchrotron radiation colliding with the inner wall of the vacuum chamber was calculated using Synrad+ software. The cooling tube was hypothesized to be brazed into the vacuum chamber with a filler material with about 1 mm thickness. The maximum and average temperature of the vacuum chamber was simulated using Fluent software [6]. The maximum wall temperature of the cooling pipe is too critical since it should be less than the boiling temperature of the cooling water.

The heat transfer coefficient between the wall of the cooling tube and water was calculated by the software. Calculated convection coefficients at different locations of the cooling tube wall have different values because the wall temperature of the cooling pipe can be different in dissimilar locations, figures 3 and 4. The maximum value of the convection coefficient of the cooling pipe wall and the average of this quantity for the whole wall are obtained that all listed in table 1.

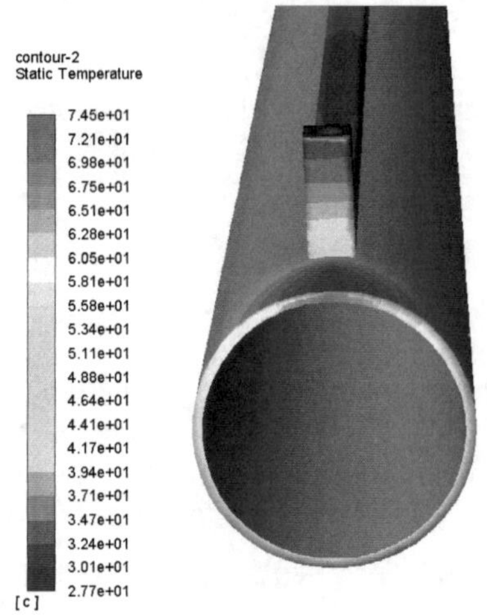

Figure 3. CFD simulations of temperature distribution with 3 m/s flow rate.

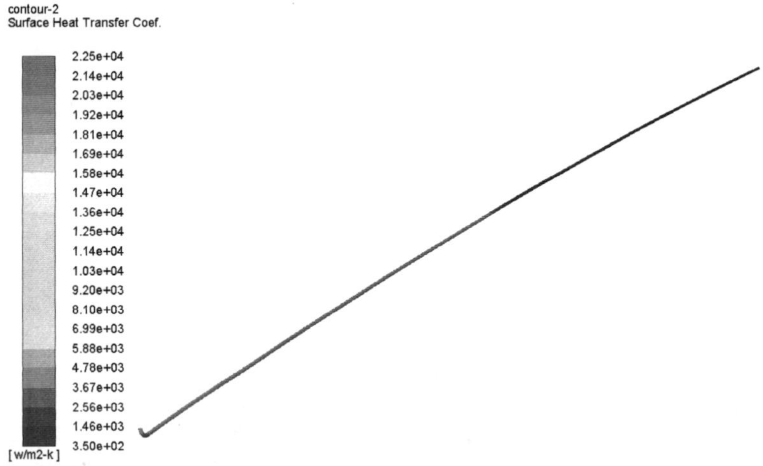

Figure 4. CFD simulations of heat transfer coefficient between the wall of the cooling tube and water with 3 m/s flow rate.

Table 1. CFD results for different water flow rates.

Flow rate(m/s)	$h_{Ave}(w/m^2{}^\circ c)$	$h_{Max}(w/m^2{}^\circ c)$	T^{Ch}_{Ave}(°C)	T^{Ch}_{Max}(°C)	$T^{C\,W}_{Max}$(°C)
3	5800	22500	44	74.5	50
2	4080	17800	51	78.5	60
1	2257	11800	71	98	87

In the next step, with the heat transfer coefficient obtained with CFD, the cooling calculations of the chamber with the FEA modeling were performed, figures 5 and 6. Cooling calculations were done with the maximum value of the heat transfer coefficient and its mean value, which are shown in table 2. In this method, a single convection coefficient should be considered for the entire wall of the cooling pipe chamber. As with the CFD method, the maximum mean temperature of the chamber is simulated.

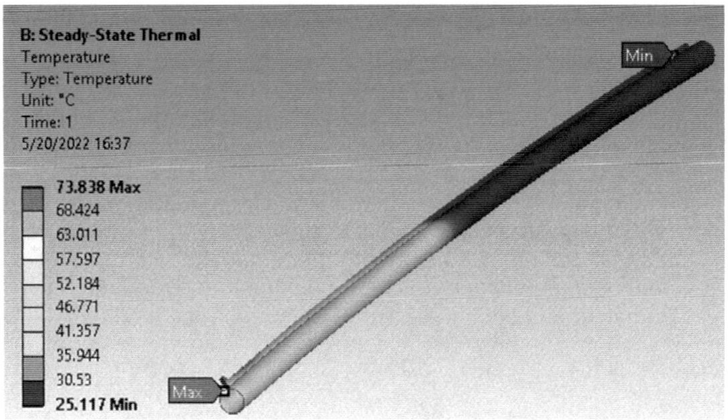

Figure 5. FEA simulations of temperature distribution with h=22500 W/m^2°C.

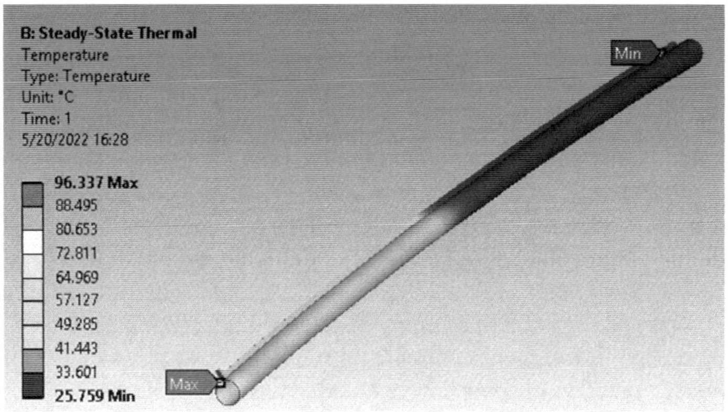

Figure 6. FEA simulations of temperature distribution with h=5800 W/m^2°C.

Table 2. FEA simulation with the heat transfer coefficient results obtained from CFD.

Flow rate(m/s)	h(W/m^2°C)	T^{Ch}_{Ave}(°C)	T^{Ch}_{Max}(°C)	$T^{C\,W}_{Max}$(°C)
3	22500	34	74	37
3	5800	46	96	61
2	17800	35	76	39
2	4080	54	109	75
1	11800	38	81	44
1	2257	73	143	113

Assuming that the material used as a filler in welding has the same heat transfer coefficient as the copper, the chamber cooling simulations have been done again. Comparing the results with the situation in which the actual conductivity of the filler was considered can help us understand this effect. For our simple case, considering the filler thermal conductivity, FEA thermal analysis shows that the maximum temperature of this chamber is about 8 degrees higher than a similar chamber in which the filler's thermal conduction is assumed to be equal to the copper. For the average chamber temperature, this difference is about 2°C; the results are shown in table 3.

Table 3. Simulation results, assuming the thermal conductivity of filler material as the same as the thermal conductivity of copper.

Flow rate(m/s)	$h(w/m^2{}^\circ c)$	T^{Ch}_{Ave}(°C)	T^{Ch}_{Max}(°C)
3	22500	32	66
3	5800	44	89
2	17800	33	68
2	4080	52	101
1	11800	36	73
1	2257	71	136

4. Conclusion

To estimate the amount of heat transfer coefficient in vacuum chamber cooling simulations, cooling simulations with different flow rates were done by FEA. After obtaining the heat convection coefficient in these simulations, it was seen that the maximum amount of heat convection coefficient of the cooling pipe and water could be used in finite element calculations. In the process of welding the cooling pipe to the vacuum chamber, in practice in some places the distance between them may be more than usual. Such a gap will be filled by the filler. The filler at the welding place increases the temperature of the chamber locally. Considering the maximum temperature of 150°C for the chamber and a filler with a thickness of 1 mm increases the maximum temperature of the chamber by 8°C, which is not negligible in the calculations and may lead to further problems. So, it is recommended to consider the maximum temperature of 140°C on the design level for the enclosures.

References

[1] Ahmadi E, Rahighi J, Jazayeri SM, Ghasem H and Jafarzadeh M 2016 *Proc. 7th Int. Particle Accelerator Conf. (IPAC'16)* (Busan, Korea) pp 2858-60
[2] Kersevan R and Ady M 2019 *Proc. 10th Int. Partile Accelerator Conf. (IPAC'19)* (Melbourne, Australia) pp 19-24
[3] Zhang L, Biasci JC and Plan B 2002 *Proc. 2nd Int. Work. on Mechanical Engineering Design of Synchrotron Radiation Equipment and Instrumentation (MEDSI02)* (Argonne, Illinois, USA) pp 5-6
[4] Sharma S, Rotela E and Barcikowski A 2000 High heat-load absorbers for the APS storage ring. Tech. rep. Argonne National Laboratory
[5] Clarke J A 2004 *The science and technology of undulators and wigglers* (OUP Oxford)
[6] Fluent software `https://www.ansys.com`
[7] Sharma S 2014 *Mechanical Engineering Design of Synchrotron Radiation Equipment and Instrumentation Conf. (MEDSI 2014)* (Melbourne, Australia)
[8] 2013 *Fluent AN. ANSYS Fluent tutorial guide, version 15.0* (Pennsylvania, USA)

Sensitivity of EEHG simulations to dynamic beam parameters

D Samoilenko[1], W Hillert[1], F Pannek[1], S Ackermann[2], E Ferrari[2], N Mirian[2], P Niknejadi[2], G Paraskaki[2], L Schaper[2], F Curbis[3], M Pop[3] and S Werin[3]

[1] University of Hamburg, 20148 Hamburg, Germany

[2] Deutsches Elektronen-Synchrotron DESY, 22607 Hamburg, Germany

[3] MAX IV Laboratory, 22363 Lund, Sweden

E-mail: dmitrii.samoilenko@desy.de

Abstract. Currently, the Free electron laser user facility FLASH at DESY is undergoing a significant upgrade involving the complete transformation of one of its beamlines to allow external seeding. With the Echo-Enabled Harmonic Generation (EEHG) seeding method, we aim for the generation of fully coherent XUV and soft X-ray pulses at wavelengths down to 4 nm. The generated FEL radiation is sensitive to various electron beam properties, e.g., its energy profile imprinted either deliberately or by collective effects such as Coherent Synchrotron Radiation (CSR). In dedicated particle tracking simulations, one usually makes certain assumptions concerning the beam properties and the collective effects to simplify implementation and analysis. Here, we estimate the influence of some of the common assumptions made in EEHG simulations on the properties of the output FEL radiation, using the example of FLASH and its proposed seeding beamline. We conclude that the inherent properties of the FLASH1 beam, namely the negatively chirped energy profile, has dominant effect on the spectral intensity profile of the radiators output compare to that of the CSR induced chirp.

1. Introduction

Echo-Enabled Harmonic Generation (EEHG) [1] is an external seeding technique for Free Electron Lasers (FEL). In comparison with the classical Self Amplified Spontaneous Emission (SASE) scheme, seeding techniques offer temporally coherent, narrow-bandwidth FEL radiation with much better shot-to-shot stability [2]. In comparison with other seeding schemes, EEHG provides higher conversion efficiency at high harmonics of the seed laser wavelength and more robustness with respect to the initial beam quality [3]. FLASH2020+ [4] is a major upgrade of the existing FLASH (Free electron LASer in Hamburg) facility, which includes the reconstruction of the FLASH1 beamline to allow external seeding. The EEHG option in FLASH1 beamline will be used to generate soft X-ray radiation with wavelengths down to 4 nm. The realization of EEHG is more challenging at shorter wavelengths because of the precise phase space transformations in the dedicated seeding section of the beamline at higher harmonics. The essential components of the seeding section are the two modulators and the two magnetic chicanes, as depicted in figure 1. Each component induces specific longitudinal energy correlations, which have to be carefully chosen and transported through the beamline. Any deviations from the design beam parameters at each point of the beamline could have a detrimental effect on the performance of EEHG-based FEL. Therefore, it is crucial to investigate this section's dynamic beam parameters in detail. Particle tracking simulations are a widely used tool for investigating beam dynamics, which can help anticipate detrimental effects and indicate ways of mitigating them. At the same time, the accuracy of the simulations in each particular case is restricted by the underlying approximations and assumptions. Some of these assumptions are related to the functionality of the simulation code. For example, in EEHG simulations performed with `Genesis 1.3` [5] the effect of Coherent Synchrotron Radiation (CSR) in the chicanes is not taken into account, even though it can have a noticeable effect on the electron beam parameters [6]. Other assumptions are made deliberately by the user to make the implementation or interpretation of the simulations more straightforward. For example, while tracking the electron beam through the EEHG seeding section with `elegant` [7], one can neglect the initial electron beam energy chirp. The effect of the initial electron beam energy modulations on EEHG itself is well studied elsewhere (see, e.g., [8]), but an interplay between the initial chirp and the modulations induced by the CSR might be possible. In this work, we use the example of the future EEHG beamline at FLASH to see how the assumptions mentioned above can change the properties of the output FEL radiation.

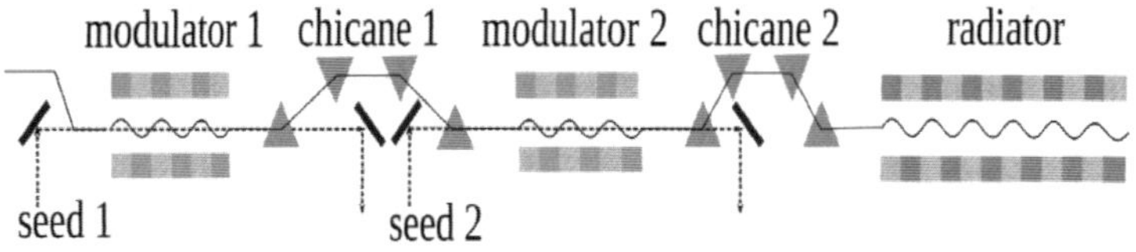

Figure 1. Schematic of the simulated setup.

2. Methods

The simulations are performed in two steps. First, we start at the entrance of the first EEHG modulator, where the electron distribution is generated by `elegant` according to the beam parameters specified in table 1. Two ideal matched initial electron beam distributions are considered: in one, the energy chirp is 0 MeV/ps (no chirp), and in the other −15 MeV/ps, which is the expected value for 4 nm working point in FLASH1. All other beam parameters are the same. The beam is then tracked in `elegant` through the seeding section (shown in figure 1) up to the entrance of the radiator section. The parameters of the seeding section are given in the table 1, where $A_{1,2}$ is the energy modulation in terms of the initial energy spread in the first and the second modulator respectively and $R_{56}^{(1,2)}$ is the dispersion strength of the first and the second chicane respectively. In this work, we neglect the CSR in the second chicane as well as in both modulators, because it is expected to be much less pronounced than in the strong first chicane. The CSR in the first chicane is calculated by the built-in method [9] implemented in `elegant`'s `CSRCSBEND` element. The method uses 1D-model to calculate longitudinal CSR-induced energy kicks along the dipoles of the chicane and applies them to the electron bunch. The model calculates the CSR in free space, ignoring the effect of the resistive chamber walls of the chicane. The effect of shielding and wakefields from the chamber walls will be considered in future work. Further information about the element and the application of the model can be found in `elegant`'s manual [10].

The particle distribution at the entrance of the first radiator, produced by `elegant`, is then converted into the format of `GENESIS` particle distribution and imported into the simulation for the radiator section. The most important parameters of the radiator are also given in table 1.

Table 1. Simulation parameters.

Initial beam parameters	
Central energy	1350 MeV
Slice energy spread	150 keV
Bunch length rms	96 µm
Peak current	500 A
Normalized emittance	0.6 mmmrad
Seeding section parameters	
Seed lasers wavelength	300 nm
A_1	3.10
$R_{56}^{(1)}$	7.05 mm
A_2	5.18
$R_{56}^{(2)}$	81.25 µm
Radiator section parameters	
Target wavelength	4 nm
Length	2.508 m × 11
Undulator period	3.3 cm

3. Results and discussion

Figure 2(a) shows the energy profile of the initial electron beam with (dashed blue line) and without (solid blue line) the initial −15 Mev/ps energy chirp. The plot also shows the current profile (solid black line) for the reference. Figure 2(b) shows the energy profile of the electron beam at the exit of the first chicane in the cases with (orange lines) and without (blue lines) the CSR effect. From the plot, we see that the energy profile of the electron beam changes in the presence of the CSR especially in the region of the high current. Based on the results of [6] we expect an effect on the spectrum of the output FEL radiation. One can also see from the plot that the amplitude of the CSR-induced energy modulation is much smaller than the initial chirp.

Figure 3 shows the spectrum of the FEL output in the vicinity of the target wavelength when power saturation in the radiator is reached. From the plot, we immediately see that the quality of the spectrum has deteriorated because of the CSR-induced energy modulation. Despite the complex shape of the spectrum, we see indications that the maximum radiation power is shifted due to the CSR (see table 2). We also estimate the RMS bandwidths of the spectra, which are given in table 3. From this consideration, we conclude that the RMS bandwidth of the FEL radiation changes by one order of magnitude for the unchirped electron beam due to the CSR.

Table 2. Resonant wavelength shift.

Initial chirp	**no CSR** (10^{-4}**nm**)	**CSR** (10^{-4}**nm**)
0 MeV/ps	-	2.66 (0.007%)
−15 MeV/ps	3.73 (0.009%)	5.31 (0.013%)

Finally, we repeat the simulations for the case when the initial electron beam has a linear chirp of −15 MeV/ps. Figure 4 shows the spectral shapes with (solid orange line) and without (solid blue line) the CSR for the initially chirped electron beam. The plot also includes the spectrum

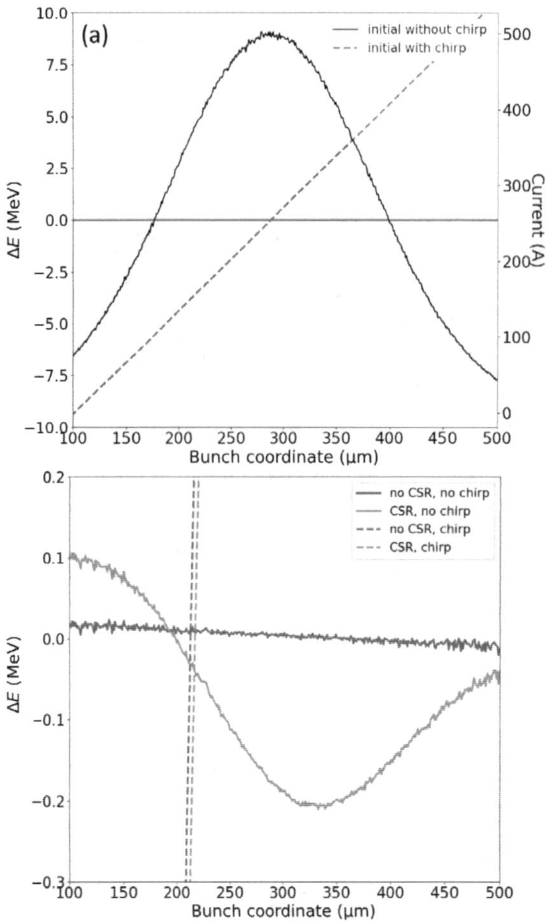

Figure 2. (a) Energy profile of the initial electron beam with the current profile (solid black line); (b) Energy profile of the electron beam at the exit of chicane 1. The head of the bunch is on the left.

Table 3. RMS bandwidths of the output FEL spectra.

Initial chirp	**no CSR** (10^{-4}**nm**)	**CSR** (10^{-4}**nm**)
0 MeV/ps	0.17	1.31
−15 MeV/ps	0.81	1.11

for the CSR-affected unchirped beam (shaded orange) from figure 3 for the reference. One can see that the multiple peak structure is not observed in the spectrum with the chirp present. This result is in agreement with the theory from [6], that the multiple peaks in the spectrum originate from the regions of the bunch, which have different local chirp values. When a relatively strong initial linear chirp dominates the electron beam energy profile, this mechanism of spectral broadening is expected to have less importance than the spectral broadening due to the initial electron beam energy chirp described in [8]. From the bandwidth values given in table 3 we see that the relative spectral broadening due to the CSR is only about 30% for the initially chirped electron beam. The wavelength shift due to the CSR is still observed; however, it is comparable

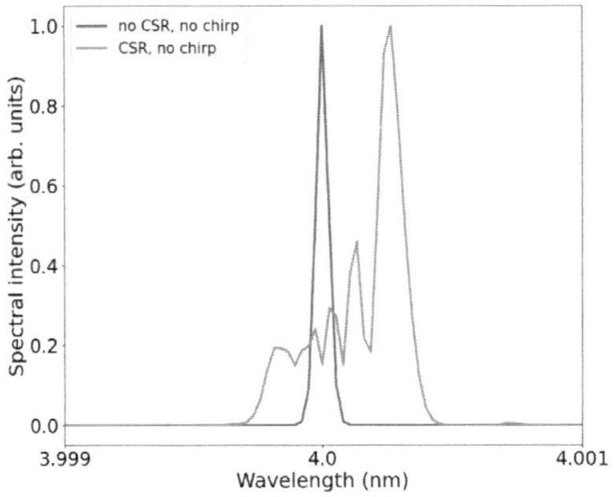

Figure 3. Spectral shape of the EEHG FEL pulse in the saturation mode in the case of no initial electron beam energy chirp.

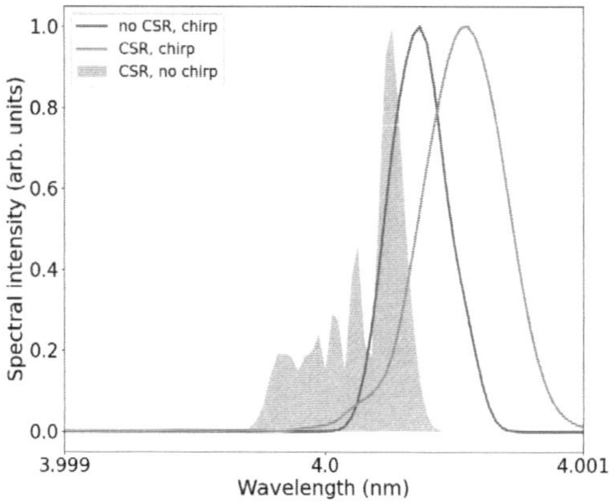

Figure 4. Spectral shape of the EEHG FEL pulse in the saturation mode in the case of 15 MeV/ps initial electron beam energy chirp.

to the bandwidth in this case.

4. Conclusion

We have investigated how the dynamic energy profile of the electron beam in the EEHG seeding section can affect the output FEL radiation spectrum. We considered the energy profile of the electron beam at the exit of the first chicane with and without CSR (in free space) effects in the chicane. In addition, we simulated the spectral shape of the FEL radiation by the beam in both cases. From comparing the two cases, we have concluded that taking CSR into account can significantly affect the EEHG-based FEL radiation spectrum. The effect of CSR on the spectrum in the presence of a linear energy chirp in the electron beam was investigated as well. We concluded that the effect of the CSR-induced energy modulation in the presence of the chirp is significantly less critical. Future work will extend this work to include internal beam scattering and the effects of chamber walls.

Acknowledgments

The authors gratefully acknowledge the Gauss Centre for Supercomputing e.V. (www.gauss-centre.eu) for funding this project by providing computing time through the John von Neumann Institute for Computing (NIC) on the GCS Supercomputer JUWELS [11] at Jülich Supercomputing Centre (JSC). The work is supported by BMBF within "05K2019-STAR" project. The authors also acknowledge the funding through the Lund-Hamburg "IFELD" collaboration.

References

[1] Stupakov G 2009 *Phys. Rev. Lett.* **102** 074801
[2] Hemsing E *et al* 2016 *Nat. Photonics* **10** 512–515
[3] Rebernik Ribič P *et al* 2019 *Nat. Photonics* **13** 555–561
[4] Schaper L *et al* 2021 *Appl. Sci.* **11** 9729
[5] Reiche S 1999 *Nucl. Instr.* **429** 243–248
[6] Samoilenko D, Hillert W, Pannek F, Niknejadi P, Paraskaki G, Curbis F, Pop M and Werin S *Proc. 12th Int. Particle Accelerator Conf. (IPAC'21)* (Campinas, Brazil) pp 1622–24

[7] Borland M *Proc. 6th International Computational Accelerator Physics Conference (ICAP 2000)* (Darmstadt, Germany) LS-287
[8] Hemsing E, Garcia B, Huang Z, Raubenheimer T and Xiang D 2017 *Phys. Rev. Accel. Beams* **20** 060702
[9] Borland M 2001 *Phys. Rev. Spec. Top. Accel Beams* **4** 070701
[10] Borland M User's Manual for elegant `https://ops.aps.anl.gov/manuals/elegant_latest/elegant.html`
[11] Alvarez D 2021 *Journal of large-scale research facilities* **7** A183

Emittance feedback for the Diamond-II storage ring using resonant excitation

Shaun Preston, Teresia Olsson and Beni Singh
Diamond Light Source, Harwell Campus, Didcot OX11 0DE, UK
E-mail: shaun.preston@diamond.ac.uk

Abstract. In the Diamond Light Source storage ring, the vertical emittance is kept at 8 pm rad during operation to maintain the source brightness for the users. This is achieved by a feedback which modifies the skew quadrupole strengths, but has disadvantages such as the introduction of betatron coupling and vertical dispersion. For the proposed Diamond-II upgrade, the storage ring will have a much smaller horizontal emittance, meaning a significantly larger coupling would be required to reach the target vertical emittance, negatively affecting the off-axis injection process. To solve this problem, a feedback using the transverse multibunch feedback striplines to drive the beam at a synchrotron sideband is planned. By driving the beam resonantly in this way, the emittance can be increased without modification of the optics. This paper describes simulations of the effects of linear and non-linear optics on the excitation as well as the impact of the machine impedance for the Diamond-II storage ring.

1. Introduction

In synchrotron light storage rings, the emittance of the beam must be controlled during operations to stabilise the source brightness and lifetime for users as conditions vary due to long-term drifts and insertion device (ID) gap movement. At the existing Diamond storage ring, the vertical emittance is kept at 8 pm rad by a feedback which modifies the skew quadrupole strengths [1], but this introduces betatron coupling and vertical dispersion. An upgrade of the ring is planned [2] to significantly reduce the horizontal emittance, meaning the new Diamond-II storage ring will require a larger coupling to reach the same vertical emittance, affecting the off-axis injection. Studies at BESSY II and MAX IV [3–6] have shown that the emittance can be blown up by driving the beam at a synchrotron sideband. The purpose of those studies was to increase the emittance of a single bunch to provide light for timing users while operating with a multi-bunch fill pattern [7], but at Diamond the same method is planned to be used for an emittance feedback acting on all bunches without affecting the optics. The emittance will be measured using the existing pinhole cameras and then the feedback will adjust the gain of an excitation from the multi-bunch feedback striplines, to keep the emittance at the target value. The vertical emittance will first be corrected to a few pm rad using LOCO [8] and then increased to 8 pm rad with the emittance feedback. This paper presents simulations of the effect of linear and non-linear optics as well as impedance on the optimal excitation frequency for the Diamond-II storage ring.

2. Simulation setup

Simulations were performed in Elegant [9, 10] including broadband impedance given by the sum of the resistive-wall and geometric contributions [11]. A bunch of 1,000 particles was tracked to equilibrium using a one-turn map including radiation damping, quantum excitation and non-linear optics terms (second and third order chromaticity, and amplitude-dependent tune shifts). The bunch was excited vertically using a zero-length kicker with a sinusoidal kick angle given by

$$\theta = A\cos(2\pi ft) \tag{1}$$

where A is the maximum kick angle, f the excitation frequency and t the arrival time at the kicker.

3. Sideband distortion

Figure 1 shows the beam oscillation and size as a function of excitation frequency. The symmetry of the synchrotron sidebands is broken by the impedance and non-linear terms. The impedance widens and shifts them closer to the tune while the non-linear terms narrow the lower sidebands.

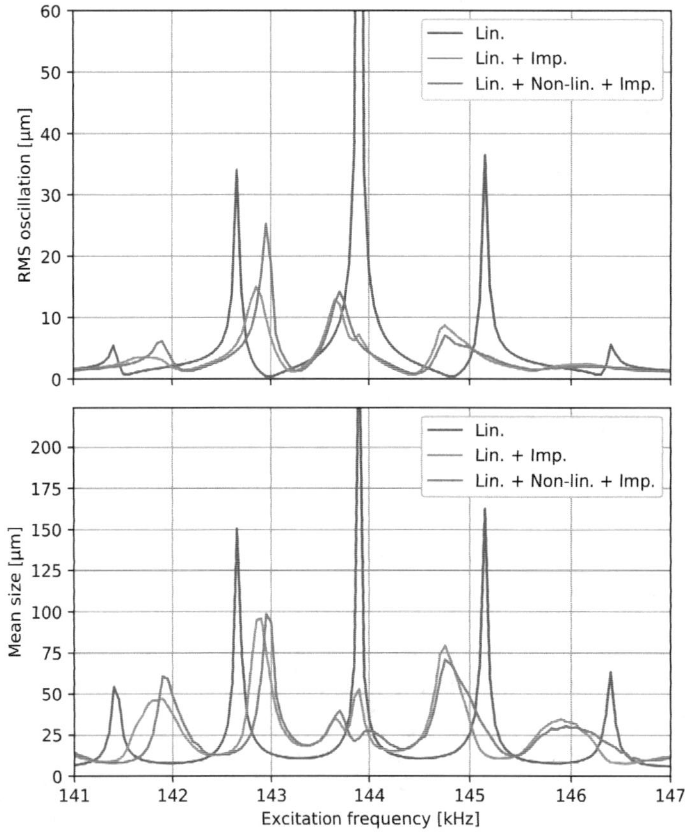

Figure 1. Beam oscillation and size as a function of excitation frequency due to linear chromaticity, non-linear terms and impedance at nominal bunch current 0.32 mA and vertical chromaticity 2.33 for a kick angle of 50 nrad.

4. Impedance effects

4.1. Transverse

Figure 2 shows how the beam oscillation and size vary with current due to transverse impedance. The effect comes from the vertical impedance as the horizontal contribution has no impact. At the upper synchrotron sidebands, the beam oscillation and size decrease with current while the opposite is true at the lower sidebands. It is also seen that the tune peak bifurcates and becomes weaker with current, for both the beam oscillation and size. Typically, the first synchrotron sidebands produce the greatest increase in size to oscillation compared to the second sidebands, but with increasing current, the second lower sideband also produces a large increase in beam size and maintains a comparatively small oscillation, making it a potential candidate for excitation.

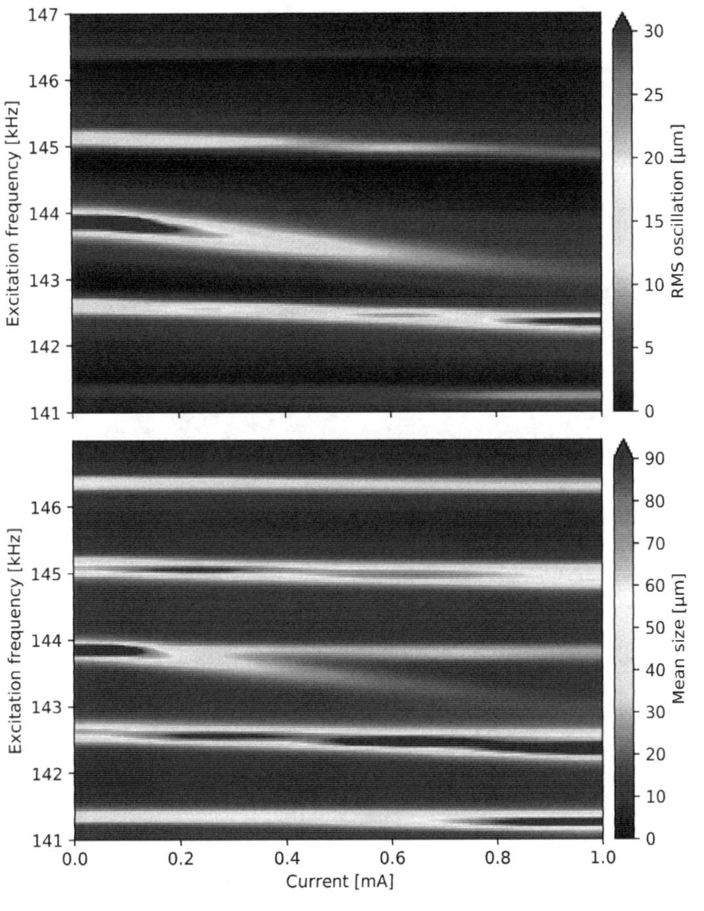

Figure 2. Beam oscillation (top) and size (bottom) as a function of current due to transverse impedance at vertical chromaticity 2.33 for a kick angle of 50 nrad. Non-linear terms and longitudinal impedance are not included.

4.2. Longitudinal

Figure 3 shows how the beam oscillation and size vary with current due to longitudinal impedance. The upper and lower synchrotron sidebands demonstrate similar behaviour and shift closer to the vertical betatron tune, while becoming broader as the bunch current is increased.

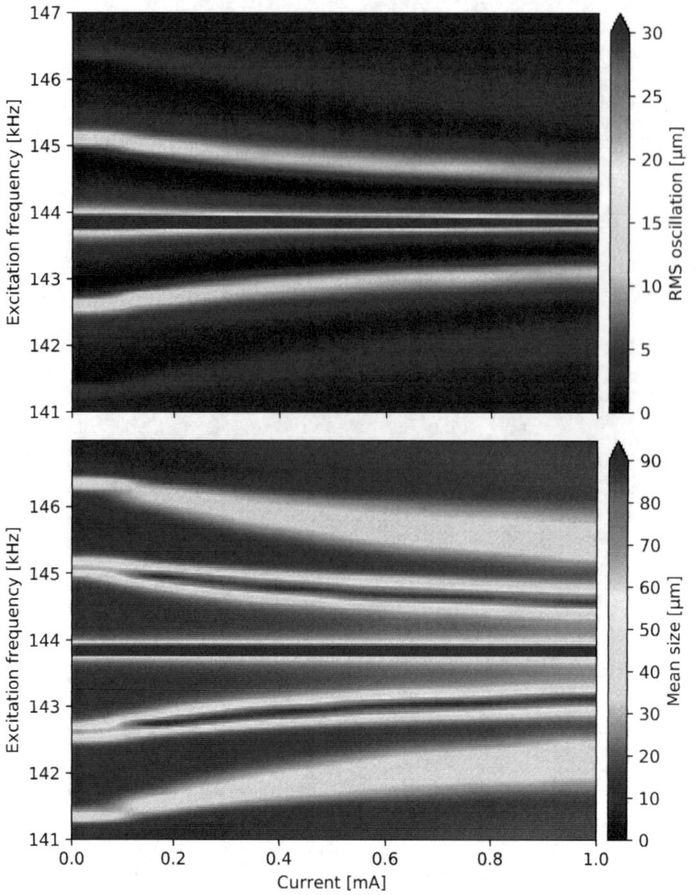

Figure 3. Beam oscillation (top) and size (bottom) as a function of current due to longitudinal impedance at vertical chromaticity 2.33 for a kick angle of 50 nrad. Non-linear terms and transverse impedance are not included.

4.3. Transverse + longitudinal

Figure 4 shows the ratio of the beam size to the oscillation including both transverse and longitudinal impedance to highlight how the optimal excitation frequency changes with current. For low current, the first lower synchrotron sideband produces the largest ratio, but at nominal bunch current (0.32 mA) the first upper sideband becomes better. The figure also shows that the sidebands are sensitive to changes in linear chromaticity. The first upper synchrotron sideband is the best candidate at chromaticity 1, becoming broader at high currents, but with increasing chromaticity and current there is little difference in exciting between upper and lower sidebands. In general, higher chromaticity produces larger ratios.

5. Emittance growth

As the emittance of the beam changes due to ID gap changes and other external factors, the feedback will have to adjust the kick angle given by the stripline kicker to maintain the target emittance. It is therefore of interest to understand the relationship between kick angle and emittance increase. If the emittance increase at the chosen excitation frequency is small, it will have to be compensated for with a larger kick, increasing the oscillation of the beam. The

Journal of Physics: Conference Series **2420** (2023) 012025 doi:10.1088/1742-6596/2420/1/012025

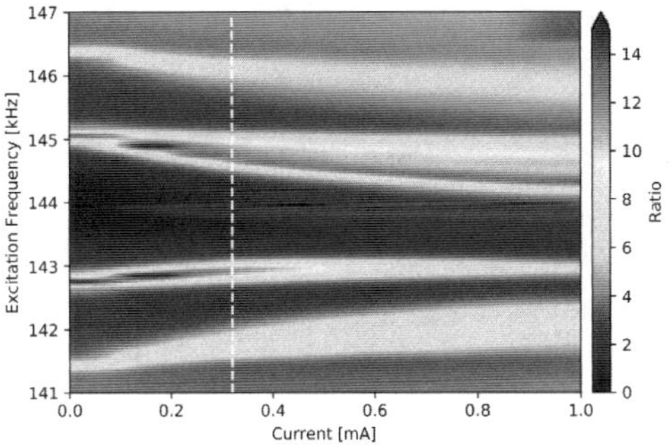

(a) Vertical chromaticity 1.

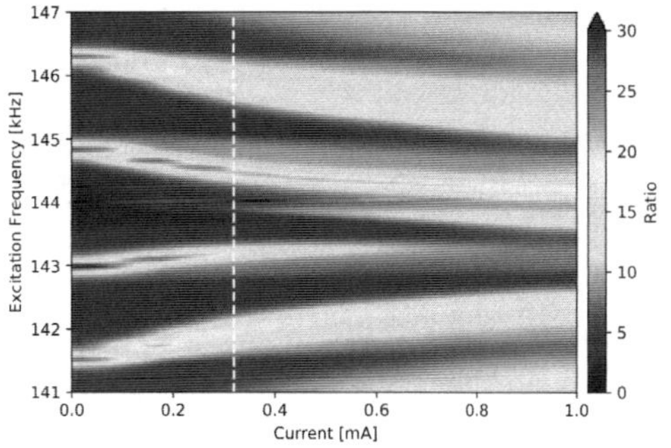

(b) Vertical chromaticity 2.33.

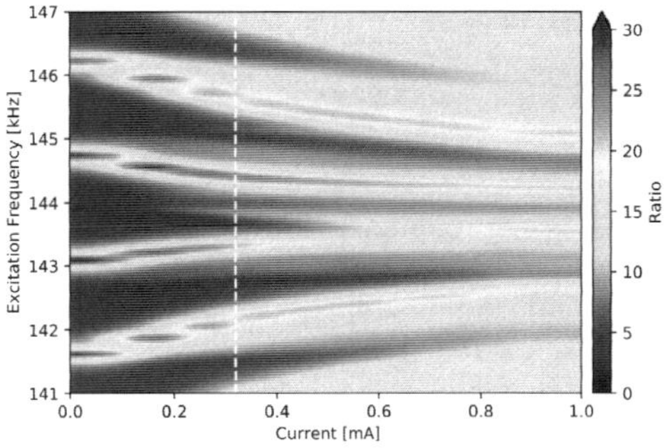

(c) Vertical chromaticity 3.

Figure 4. Ratio of the mean beam size to RMS oscillation as a function of current due to transverse and longitudinal impedance at different vertical chromaticity for a kick angle of 50 nrad. A dashed white line shows the nominal bunch current (0.32 mA). Non-linear terms are not included.

maximum kick angle used in the simulations was far below the technical limitations of the multi-bunch feedback system which implies that compromising between the ratio and size is irrelevant for Diamond-II, since a large blow up of the emittance can be achieved with a very small kick angle. In figure 5, it can be seen that the emittance grows quadratically for small kick angles whereas the beam size grows linearly. It is also shown that the impedance has a significant effect on the growth rate. As the kick angle is increased to hundreds of nrad, amplitude-dependent tune shifts start to dominate, which causes the growth of the emittance to plateau.

6. Conclusions and future work

The studies show that the upper first synchrotron sideband is the optimal excitation frequency because it gives the best ratio between beam size and oscillation. The studies also show that both the non-linear optics terms and impedance cause a broadening of the sidebands, which suggests that the beam can be excited at a large range of frequencies to achieve emittance increase. The impedance however has a much larger effect on the optimal frequency than the non-linear terms for the nominal current and target emittance of the Diamond-II storage ring. Studies are ongoing of the effect of a harmonic cavity since the ring is planned to operate with a passive superconducting third order harmonic cavity. Benchmarking between simulations and measurements is also planned at the existing Diamond storage ring to understand how well the simulations predict the real behaviour with current. Of special interest is how the magnitude of

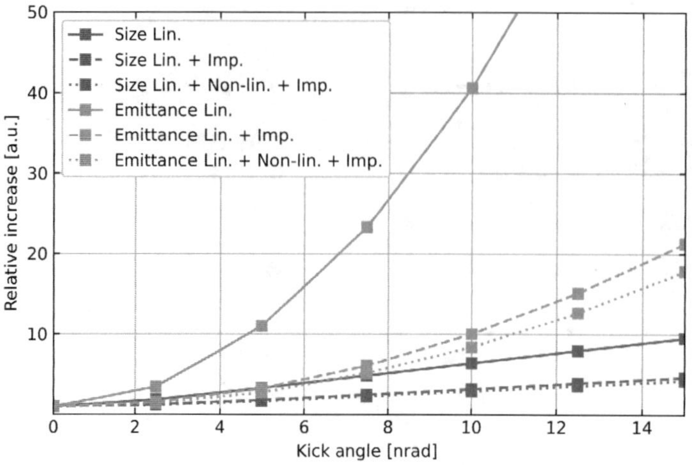

Figure 5. Relative increase in beam size (σ/σ_0) and emittance (ϵ/ϵ_0) at the first upper synchrotron sideband as a function of kick angle due to linear chromaticity, non-linear terms and impedance at the nominal bunch current 0.32 mA and vertical chromaticity 2.33.

the kick angle in simulation compares to the required kick angle in the machine to achieve the same emittance increase. It is also of interest to study if the multi-bunch feedback can help to mitigate the beam oscillation driven by the excitation and thus improve the ratio between the beam size and oscillation.

Acknowledgments
The authors wish to thank Alun Morgan for explaining the principle of the planned emittance feedback and Jonas Kallestrup for useful discussions and suggestions as well as input on this paper.

References
[1] Martin I P S, Abbott M G, Apollonio M, Hickin D and Bartolini R 2013 *Proc. 4th Int. Particle Accelerator Conf. (IPAC'13)* (Shanghai, China) pp 249–51
[2] Diamond-II Tech. rep. Diamond Light Source `https://www.diamond.ac.uk/Home/About/Vision/Diamond-II.html`
[3] Hwang J G, Koopmans M, Ries M, Schälicke A and Müller R 2019 *Nucl. Instrum. Methods Phys. Res., Sect.* A **940** 387–92
[4] Müller R, Birke T, Falkenstern F, Holldack K, Jankowiak A, Ries M, Schälicke A *et al.* 2017 *Proc. 8th Int. Particle Accelerator Conf. (IPAC'17)* (Copenhagen, Denmark) pp 2574–77
[5] Olsson T, Andersson Å *et al.* 2017 *Proc. 8th Int. Particle Accelerator Conf. (IPAC'17)* (Copenhagen, Denmark) pp 2750–52
[6] Olsson T, Andersson A, Olsson D K *et al.* 2018 *Proc. 9th Int. Particle Accelerator Conf. (IPAC'17)* (Vancouver, BC, Canada) pp 4300–03
[7] Holldack K, Ovsyannikov R, Kuske P, Müller R, Schälicke A, Scheer M, Gorgoi M, Kühn D, Leitner T, Svensson S *et al.* 2014 *Nature Communications* **5** 1–7
[8] Safranek J 1997 *Nucl. Instrum. Methods Phys. Res., Sect.* A **388** 27–36
[9] Borland M 2000 Elegant: A flexible sdds-compliant code for accelerator simulation
[10] Borland M and Wang Y 2006 Pelegant: A parallel accelerator simulation code for electron generation and tracking
[11] Fielder R, C Chao H C and W Wang S 2022 *Proc. 13th Int. Particle Accelerator Conf. (IPAC'22)* (Bangkok, Thailand) pp 2257–60

Virtual commissioning of the European XFEL for advanced user experiments at photon energies beyond 25 keV using low-emittance electron beams

Ye Chen, Frank Brinker, Winfried Decking, Matthias Scholz, Lutz Winkelmann and Zihan Zhu[1]

Deutsches Elektronen-Synchrotron DESY, Notkestr. 85, 22607 Hamburg, Germany
[1] on leave from SINAP, Shanghai, China

E-mail: ye.lining.chen@desy.de

Abstract. Growing interests in ultra-hard X-rays are pushing forward the frontier of commissioning the European X-ray Free-Electron Laser (XFEL) for routine operation towards the sub-ångström regime, where a photon energy of 25 keV (0.5 Å) and above is desired. Such X-rays allow for larger penetration depths and enable the investigation of materials in highly absorbing environments. Delivering the requested X-rays to user experiments is of crucial importance for the XFEL development. Unique capabilities of the European XFEL are formed by combining a high energy linac and the long variable-gap undulator systems for generating intense X-rays at 25 keV and pushing the limit even further to 30 keV. However, the FEL performance relies on achievable electron bunch qualities. Low-emittance electron bunch production, and the associated start-to-end modelling of beam physics thus becomes a prerequisite to dig into the XFEL potentials. Here, we present the obtained simulation results from a virtual commissioning of the XFEL for the user experiments at 25 keV and beyond, including the optimized electron bunch qualities and corresponding FEL lasing performance. Experimental results at 30 keV from the first test run are presented.

1. Introduction

Linear accelerator based self-amplification spontaneous emission (SASE) free-electron lasers (FEL) produce extremely short, brilliant and coherent X-ray pulses [1–7]. This made it possible using the X-rays to probe distances at the atomic scale and explore the dynamics of atomic and molecular process on their natural length and time scale. The XFEL has greatly set forward the frontiers of resolving the structure of matters with X-rays over the last decade.

The XFELs capable of operating at short wavelengths of 0.10 nm to 0.01 nm, that is, at high photon energies of about 12.40 keV to 123.98 keV, can provide the so-called hard X-rays in the sub-ångström regime. These hard X-rays are of significant importance to temporally and spatially-resolved analysis and reconstruction of materials with various degrees of crystallinity. Harder X-rays above 12.40 keV are more beneficial to characterizing semi and noncrystalline materials, observing structural in-situ phase transitions, acquiring structural information, reconstructing material spatial distributions, etc.

At the European X-ray Free-Electron Laser (EuXFEL) [2], hard X-ray pulses have been delivered for routine user experiments since 2017. At a nominal electron beam energy of 14 GeV, stable, high-intensity SASE performance is achieved at photon energies up to 14 keV in routine user runs [8–10], while in test runs, good lasing signals towards 25 keV have been achieved with only limited tuning time. The actual lasing capability of the European XFEL in an even deeper sub-ångström emission regime is not yet demonstrated, although relevant theoretical studies have been carried out in [11, 12] under specific working modes and conditions.

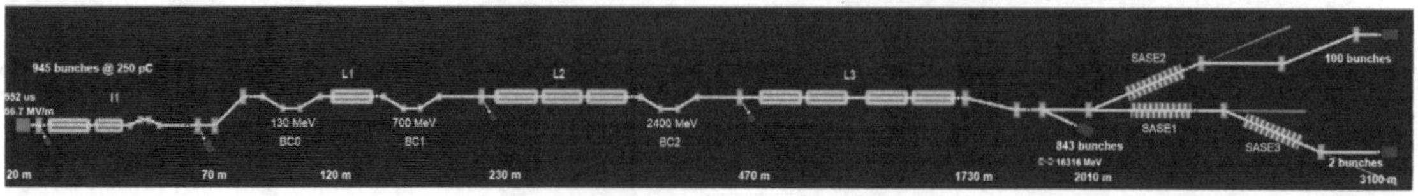

Figure 1. Schematic layout of the European X-ray Free-Electron Laser. I1: injector section; Li: the i-th linear accelerator; BCi: the i-th bunch compressor. SASEi: the i-th undulator beamlines in which SASE1 and SASE2 are hard X-ray lines while SASE3 is a soft X-ray line.

Table 1. Machine operation parameters.

Parameter	Value	Unit
Bunch charge	250	pC
Bunch shaping aperture [0]	1.0	mm
Cathode laser pulse shape	Gauss	n/a
Cathode laser pulse length[1]	3.0	ps
Cathode accelerating gradient	56.7	MV/m
Beam energy at BC0[2]	130	MeV
Beam energy at BC1	700	MeV
Beam energy at BC2	2400	MeV
Beam energy downstream L3[3]	16300	MeV
R56 at BC0	-50	mm
R56 at BC1	-50	mm
R56 at BC2	-30	mm
Undulator period[4]	4	cm
Undulator length	175	m

[0] in diameter
[1] root mean square value
[2] BCi stands for the i-th bunch compression stage, see figure 1.
[3] L3 stands for the main linac, see figure1.
[4] for the undulators in SASE1 and SASE2 beamlines, see figure 1.

2. Simulation

Here we present the obtained results from a virtual commissioning of the EuXFEL by start-to-end beam physics simulations and SASE optimization at high photon energies of 25 keV and above. Table 1 shows typical machine operation parameters which are employed correspondingly in the follow-up numerical studies. The codes OCELOT [13] and GENESIS [14] are used. The EuXFEL photoinjector [15] is optimized in the simulations taking into account the collective effects as studied in detail in [16]. The close-to cathode beam dynamics is carefully considered using the 3D approach as reported in [17].

Figure 2 shows the slice emittance of an optimized electron bunch at the injector exit (i.e. in section I1 as shown in figure 1). This is done by scanning the phase of the gun and the solenoid strength in the gun section. Thermal emittance of the photocathode is considered as the lower limit of an overall optimized bunch emittance [18]. At the European XFEL, cesium-telluride cathodes are used, the thermal emittance of which may intrinsically depend on the manufacture recipe. Here our optimization results are shown for three different thermal emittance values of the cathode, 0.85 μm/mm (blue), 0.99 μm/mm (black) and 1.11 μm/mm (red). The case with a thermal emittance of 0.99 μm/mm is selected based on previous studies (as reported in Ref. [19]) for further tracking simulations.

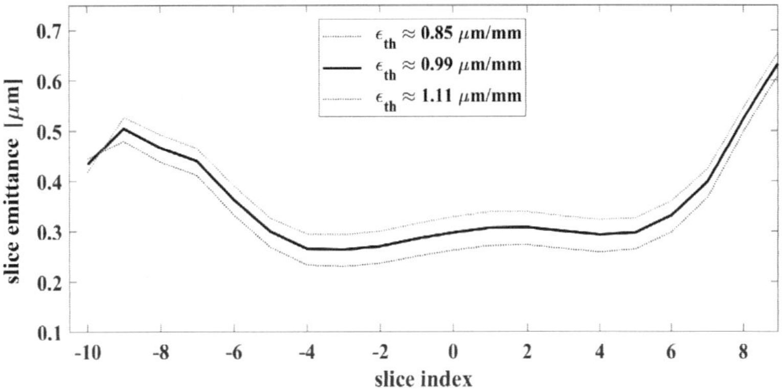

Figure 2. Optimized slice emittance at the injector exit considering different thermal emittance values of the cesium-telluride cathodes as tested at PITZ [19].

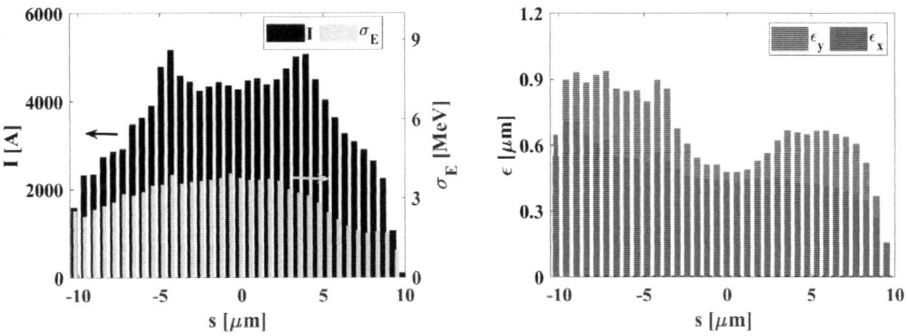

Figure 3. Bunch qualities in front of the undulator beamline SASE1. Left: peak current (left axis) and slice energy spread (right axis); Right: slice horizontal (green) and vertical (red) emittance.

Figure 3 presents the optimized bunch qualities at the undulator beamline (i.e. SASE1 in figure 1) entrance via start-to-end beam physics simulations. RF parameters (amplitude and phase) of the accelerators (i.e. I1, L1, L2 in figure 1) are optimized compressing the bunch to 4.5 kA peak current after the second compression stage (i.e. BC2 in figure 1) according to the available accelerating voltages of the facility. The beam energy set-points and momentum compaction factors at individual bunch compression locations are given in table 1. Collective effects through the whole accelerator beamline such as wakefields, space-charge and coherent synchrotron radiation are all included. As shown, the left plot illustrates an obtained bunch peak current of about 4.5 kA (left axis) and a slice energy spread of about 3.5 MeV (right axis). The right plot gives the bunch horizontal (green bars) and vertical (red bars) slice emittance.

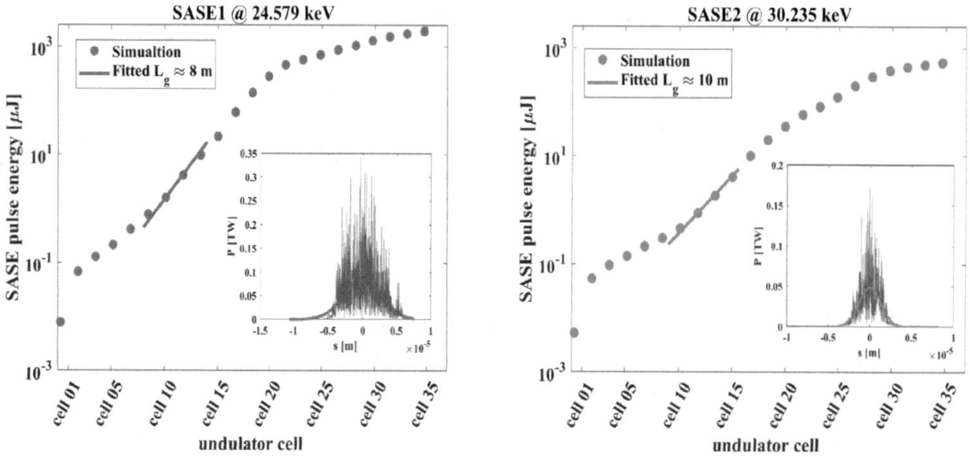

Figure 4. Simulated SASE intensities at 24.58 keV (left) and 30.24 keV (right) for beamlines SASE1 and SASE2, respectively.

Using the obtained bunch distribution as shown in figure 3, SASE optimization is carried out. As shown in figure 4, an optimized SASE intensity of about 1.8 mJ and 0.5 mJ are shown at 24.579 keV for SASE1 (left) and 30.235 keV for SASE2 (right), respectively. The corresponding fitted gain lengths are about 8 m and 10 m. The figure insets show the power distributions by the end of the undulator beamlines. In these simulations, both linear and quadratic tapers are optimized. An average beta function of 32 m is applied.

3. First test run up to 30 keV

First test run of the EuXFEL aiming for high photon energies above 25 keV was carried out last year on the request of the on-site users at DESY. Due to a temporary technical limit in the transport capability of the photon beam line, a target photon energy of 24 keV was preliminarily determined for SASE1 and 30 keV was requested for SASE2.

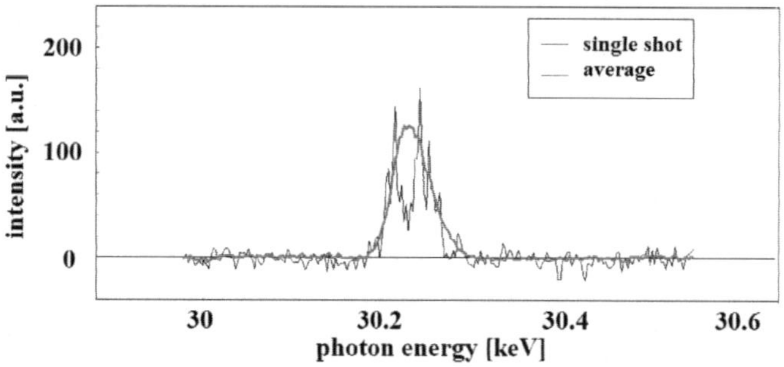

Figure 5. SASE spectrum detected by HIREX [20] for SASE2 with a central photon energy of 30.235 keV.

After short-term SASE tuning, in both SASE1 and SASE2 undulator lines the electron bunch lased at the aimed photon energies. The HIgh REsolution hard X-ray single-shot (HIREX) spectrometer [20] was used to verify the achieved photon wavelengths to be 24.579 keV and 30.235 keV at SASE1 and SASE2, respectively. The SASE spectrum at 30.235 keV is shown in

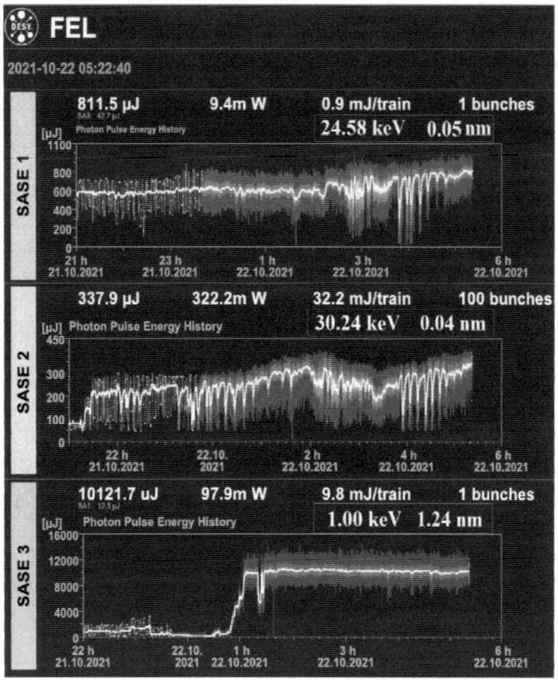

Figure 6. First test run of the EuXFEL delivering hard X-rays of about 0.8 mJ and 0.3 mJ at photon energies of 24 keV and 30 keV for SASE beamlines 1 and 2, respectively, with a simultaneous SASE pulse delivery of about 10 mJ at 1 keV for the SASE beamline 3.

figure 5. It should be noted, that for the first time, the FEL facility has ever been operated at such high photon energies. It is also worth mentioning that fairly good lasing intensities have been achieved in both beamlines at these photon wavelengths in the sub-ångström working regime.

Figure 6 shows the overall FEL performance panel corresponding to the test run in October 2021. As shown, about 0.8 mJ SASE intensity is achieved at 24.58 keV (0.5 Å) at SASE1 and about 0.3 mJ is obtained at 30.24 keV (0.4 Å) at SASE2. Additionally, the SASE3 beamline is simultaneously lasing undisturbed at 1 keV with an intensity of 10 mJ [9] during SASE tuning for the other two undulator lines.

4. Conclusion

Virtual commissioning of the European XFEL is performed based on start-to-end electron beam physics simulations and SASE performance studies at high photon energies up to 30 keV in the sub-ångström working regime. The optimized simulation results have shown millijoule-level lasing performance. First short-term test run of the EuXFEL in October 2021 has demonstrated SASE intensities of about 0.8 mJ and 0.3 mJ at 24.58 keV and 30.24 keV, respectively, which, to the best of our knowledge, are the highest demonstrated SASE intensities at these wavelengths of the hard X-rays that are delivered to the users. More detailed characterization of the whole accelerator and undulator beam lines is planned. Comprehensive numerical and experimental studies will follow to reduce the existing discrepancies between measurement and simulation.

Acknowledgments

The authors acknowledge support from Deutsches Elektronen-Synchrotron DESY (Hamburg, Germany), a member of the Helmholtz Association HGF and European XFEL GmbH (Schenefeld, Germany). The authors thank the operation teams at DESY and the European XFEL for their experimental support.

References

[1] Dohlus M and Rossbach J 2013 *Landolt-Boernstein New Sewries: Synchrotron Radiation and Free-Electron Lasers* vol 21C (Berlin/Heidelberg, Germany: Springer)
[2] Decking W *et al.* 2020 *Nat. Photonics* **14** 391
[3] Prat E *et al.* 2020 *Nat. Photonics* **14** 748
[4] Emma P *et al.* 2010 *Nat. Photonics* **4** 641
[5] Ishikawa T *et al.* 2012 *Nat. Photonics* **6** 540
[6] Kang H *et al. Nat. Photonics* **11** 708
[7] Zhao Z T *et al* 2017 *Proc. 38th Int. Free Electron Laser Conf. (FEL'19)* (Santa Fe, NM USA) pp 182–84
[8] Nölle D 2019 *Proc. 39th Int. Free Electron Laser Conf. (FEL'19)* (Hamburg, Germany) pp 766–71
[9] Fröhlich L 2019 *Proc. 39th Int. Free Electron Laser Conf. (FEL'19)* (Hamburg, Germany) pp 335–38
[10] Scholz M 2019 *Proc. 39th Int. Free Electron Laser Conf. (FEL'19)* (Hamburg, Germany)
[11] Chen Y, Brinker F, Decking W, Scholz M and Winkelmann L 2021 *Applied Sciences* **11** 10768
[12] Bazyl D, Chen Y, Dohlus M and Limberg T 2021 CW Operation of the European XFEL: SC-Gun Injector Optimization, S2E Calculations and SASE Performance *Preprint* ArXiv:2111.01756
[13] OCELOT `https://github.com/ocelot-collab`
[14] GENESIS `http://genesis.web.psi.ch/index.html`
[15] Brinker F 2016 *Proc. IPAC'16 (Busan)* p 1044
[16] Zagorodnov I, Tomin S, Chen Y and Brinker F 2021 *Nucl. Instrum. Methods Phys. Res., Sect.* A **995** 165111
[17] Chen Y, Zagorodnov I and Dohlus M 2020 *Phys. Rev. Accel. Beams* **23** 044201
[18] Chen Y *et al.* 2020 *AIP Advances* **10** 035017
[19] Huang P W *et al* 2019 *Proc. 39th Int. Free Electron Laser Conf. (FEL'19)* (Hamburg, Germany) pp 473–76
[20] Kujala N *et al.* 2020 *Rev. Sci. Instrum.* **91** (2020)

Twin delayed deep deterministic policy gradient for free-electron laser online optimization

M Cai[1,2], Z H Zhu[1,2], K Q Zhang[3], C Feng[1,2,3], L J Tu[1,2], D Gu[3] and Z T Zhao[1,2,3]

[1]Shanghai Institute of Applied Physics, Chinese Academy of Sciences, 201800 Shanghai, China
[2]University of Chinese Academy of Sciences, 100049 Beijing, China
[3]Shanghai Advanced Research Institute, Chinese Academy of Sciences, 201204 Shanghai, China

E-mail: zhangkaiqing@zjlab.org.cn, fengchao@zjlab.org.cn

Abstract. X-ray free-electron lasers (FEL) have contributed to many frontier applications of nanoscale science which benefit from its extraordinary properties. During FEL commissioning, the beam status optimization especially orbits correction is particularly significant for FEL amplification. For example, the deviation between beam orbit and the magnetic center of undulator can affect the interaction between the electron beam and the FEL pulse. Usually, FEL commissioning requires a lot of effort for multi-dimensional parameters optimization in a time-varying system. Therefore, advanced algorithms are needed to facilitate the commissioning procedure. In this paper, we propose an online method to optimize the FEL power and transverse coherence by using a twin delayed deep deterministic policy gradient (TD3) algorithm. The algorithm exhibits more stable learning convergence and improves learning performance because the overestimation bias of policy gradient methods is suppressed.

1. Introduction

X-ray free-electron lasers (FEL) open new chapters to various frontiers of scientific applications in biology, chemistry, and material science for its abilities to generate femtosecond and nanoscale pulses with gigawatt peak power and tunable wavelength down to less than 0.1 nm [1–3]. In recent years, several FEL facilities worldwide are constructed and operated successfully, which indicates a new era of X-ray science. To ensure the stable operation of FEL facilities, a robust and collimated beam orbit is generally required to achieve precise overlaps between the electron beam and radiation. The deviation between the beam orbit and the center of magnetic elements can induce a significant decrease in the peak power and transverse coherence of FEL radiation [4].

For a traditional FEL commissioning, beam orbit alignment can be roughly achieved by adjusting the current of correctors. However, the precise beam orbit alignment is usually difficult to be implemented and the beam orbit can change with the variation of beam status. Moreover, the effect of beam orbit alignment will also rely on the beam orbit stability. Therefore, manual beam orbit alignment and optimization require a lot of effort in the accelerators which is a time-varying dynamics system of multi-dimensional parameters [5, 6]. In recent years, deep reinforcement learning (DRL) methods have been adopted in the commissioning and optimization of FEL facilities since they can work at different operating points and do not require labeled datasets compared to supervised learning methods [7, 8].

This work proposes an online optimization method in FEL based on a model-free off-policy actor-critic algorithm tailored to Markov decision processes. The modified DRL method can achieve good convergence and reduce overestimation by improving policy gradient approaches. Due to practical restrictions such as radiation safety and the potential for damage to the hardware of facility from erroneous online state, optimization with previous FEL simulation is usually the preferred solution. In this paper, we assess the algorithm in a simulated FEL environment firstly, and the simulation results of two methods are compared.

2. Twin delayed deep deterministic policy gradient

As a subfield of machine learning, reinforcement learning (RL) has advantages in solving control tasks that conform to Markov decision processes [9]. In tasks with sufficient nonlinearity, complexity and time-varying systems like the FEL tuning process, RL is a more appropriate consideration than traditional control methods.

Reinforcement learning aims to find the optimal policy π_φ, with parameters φ, by maximizing the discounted sum of rewards $R_t = \sum_{i=t}^{T} \gamma^{i-t} r(s_i, a_i)$, where γ is a discount factor. The action-value function describes the expected return after taking an action in state s_t following policy π:

$$Q^{\pi}(s_t, a_t) = E_{s_i \sim E, a_i \sim \pi}[R_t | s_t, a_t]$$

The function can be estimated recursively through Bellman equation:

$$Q^{\pi_\varphi}(s_t, a_t) = E_{s_{t+1} \sim E}[r(s_t, a_t) + \gamma Q^{\pi_\varphi}(s_{t+1}, \pi_\varphi(s_{t+1}))]$$

A neural network function approximator parameterized by θ^Q can be trained by minimizing the loss:

$$L(\theta^Q) = E_{s_t \sim E, a_t \sim \pi_{\varphi_i}}[(Q(s_t, a_t | \theta^Q) - y_t)^2]$$

where $y_t = r(s_t, a_t) + \gamma Q(s_{t+1}, \pi_\varphi(s_{t+1}) | \theta^Q)$.

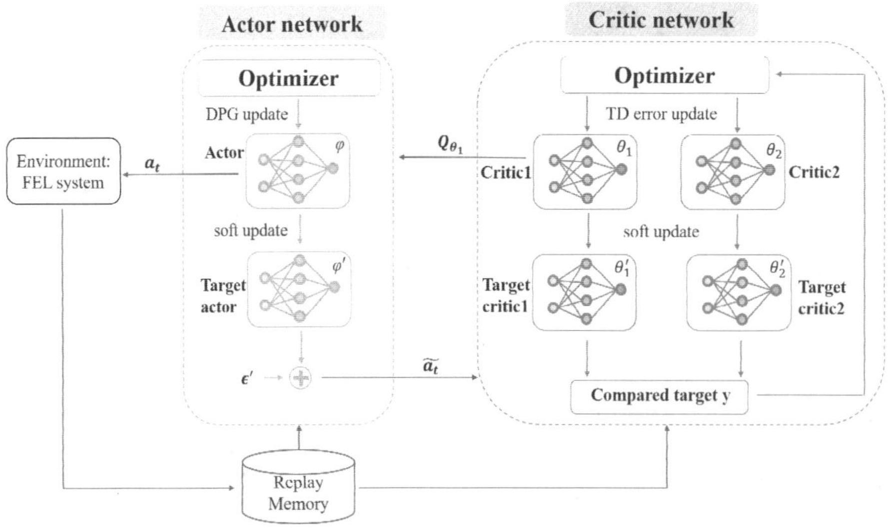

Figure 1. Structure of twin delayed deep deterministic policy gradient.

Deep Deterministic Policy Gradient (DDPG) is a model-free Q-learning method in continuous action space [10] by combining Deep Q-network [11] and Deterministic Policy Gradient [12].

However, the performance of the Q-learning algorithm is known to be influenced by the systematic overestimation of values because of the output prediction noise [13]. If overestimation bias accumulates, policy updates will be negatively affected during the FEL online optimization with relatively large orbit jitter. As a potential solution, Twin Delayed Deep Deterministic Policy Gradient (TD3) [14] solves the problem by using several techniques on the DDPG. Figure 1 describes the structure of TD3.

The first technique is Clipped Double-Q Learning. In TD3, additional bias can be reduced by learning two Q-functions and selecting the smaller Q value:

$$y^{target} = r + \gamma \min_{i=1,2} Q_{\theta_{i'}}(s', \pi_{\varphi_1}(s'))$$

Unlike the actor-critic network, the target update is presented by two critic networks. The value target in Clipped Double Q-learning contributes no additional overestimation as compared to the standard Q-learning target.

Less frequent policy updates in TD3 allow the value network to become more reliable and error-free, resulting in a lower variance value estimate and a better policy network. The parameters are updated according to $\theta' \leftarrow \tau\theta + (1-\tau)\theta'$, which maintains a small temporal difference error and slows down the updating process.

Adding noise to the target policy and averaging over mini-batches can reduce the impact of inaccuracies induced by function approximation error:

$$y = r + \gamma Q_{\theta'}(s', \pi_{\varphi'}(s') + \epsilon)$$

$$\epsilon \sim clip(\mathcal{N}(0, \sigma), -c, c)$$

Moreover, a regularization approach can be used in the training stage to smooth the target policy and solve the overfitting issue of value estimate in the deterministic policy.

3. FEL simulation

Simulations with GENESIS 1.3 [15] are carried out using the typical parameters of the Shanghai Soft X-Ray Free Electron Laser User Facility (SXFEL-UF), as given in table 1, to test RL methods at different setpoints. In this simulation, the methods of TD3 and DDPG are used to correct the orbit between the undulators and achieve the SASE FEL optimization. According to the simulated result with an ideal orbit, the radiation can get saturation with 6 undulators as shown in figure 2, thus 6 correctors are used to correct the orbit and ensure a better FEL amplification. Before optimization, we have added horizontal and vertical offsets of about 0.12-0.15 mm at the very beginning of the undulator because the entrance parameters of the actual facility are not always ideal. With these offsets, the laser power dropped from 1.16 GW to around 0.05-0.18 GW. Misalignments between undulators are not considered during the training.

Figure 2. Layout of the undulator system. Six undulator segments are used for SASE amplification to saturate at the typical wavelength.

In the optimization process, actions performed by the agent are the magnetic strengths of six correctors in the horizontal and vertical directions. The electron beam trajectory is a group of

Table 1. Main parameters of the simulation.

Parameter	Value
Beam average energy	1.5 GeV
Peak current	800 A
Energy spread	0.014%
Average beam radius (RMS)	50 μm
Undulator length	126 × 2.35 cm
FEL wavelength	3.72 nm

position coordinates of the electron beam along the undulator line, and it is also the state of the environment. To maximize the FEL power, we set the reward function as $I/I_0 - 1$, where I is the current power of each step and I_0 is the initial power before optimization with orbit offsets.

4. Results and discussion

In the FEL simulation environment, we utilize TD3 and DDPG with the parameters listed in table 2 to conduct beam tuning trails with different random seeds. The learning rate of actor network is 0.0001 and critic network is 0.0003. Actor and critic networks have two layers, with 256 and 512 nodes respectively. Each trail runs for about 2000 steps. The learning results with TD3 and DDPG agents are shown in figure 3. According to figure 3, the reward function converges as the learning process of TD3 agent accumulates, and the FEL power gets saturation finally. Whereas, the DDPG agent cannot be able to get the stabilization condition in limited episodes.

Table 2. Network parameter settings of DDPG and TD3 algorithm.

Parameter	Value
Actor learning rate	0.0001
Critic learning rate	0.0003
Deep neural network size	256 × 512
Batch size	128
Optimizer	Adam
Discount factor	0.99

Figure 4 shows the initial FEL gain curve and optimized FEL gain curves with these two methods in a tuning task. In our optimization, a special case with an initial offset of 150 μm at both horizontal and vertical positions is adopted to simulate the actual FEL facility with a nonideal entrance orbit. For this case, there will be an offset between the optimal orbit and the magnetic center of undulators as well. According to figure 4, TD3 algorithm optimizes the output power from the initial 0.078 GW to 0.813 GW approximately, which is significantly higher than the DDPG algorithm. In figure 5, the true value is estimated using the average discounted sum of rewards over randomly initialized 10 steps following the current policy. Compared with TD3, there exists an apparent overestimation bias that occurs on the DDPG learning procedure. Rapidly rising values but limited laser power gains during DDPG optimization indicate inaccurate action evaluation and problematic strategy updates.

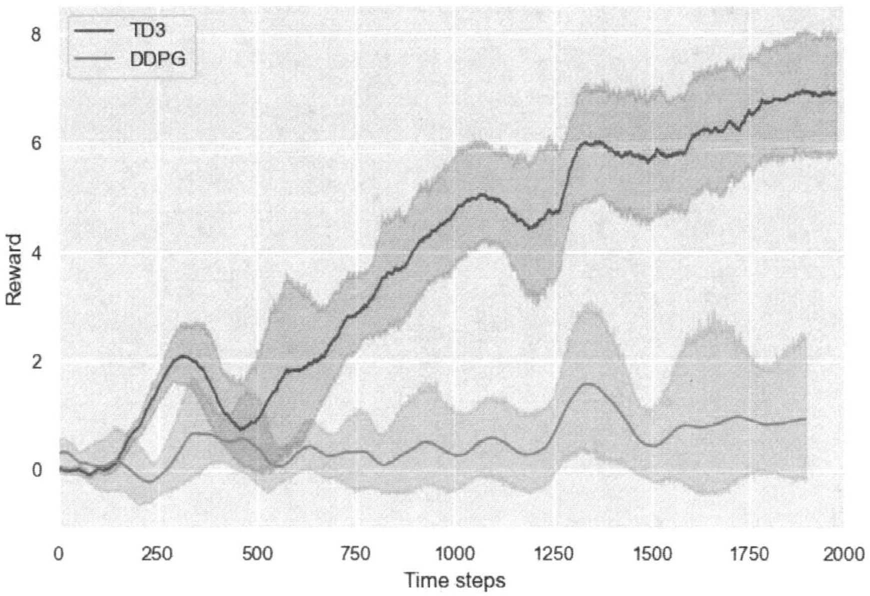

Figure 3. Learning curves comparison between TD3 and DDPG agents for the FEL tuning tasks. The shaded area represents the range of confidence intervals when aggregating 10 training curves with randomly initialized network parameters by an estimator.

Figure 6 shows the initial beam orbit and optimized beam orbit with DDPG and TD3 method. For an actual FEL amplification, a relatively straight orbit can ensure better interaction between the electron beam and radiation. Therefore, we try to find a straight and optimized orbit to obtain a better FEL amplification in our simulation, as shown in figure 6. According to figure 4, we can also find that the final peak power with TD3 optimization is larger than that with no optimization and DDPG optimization.

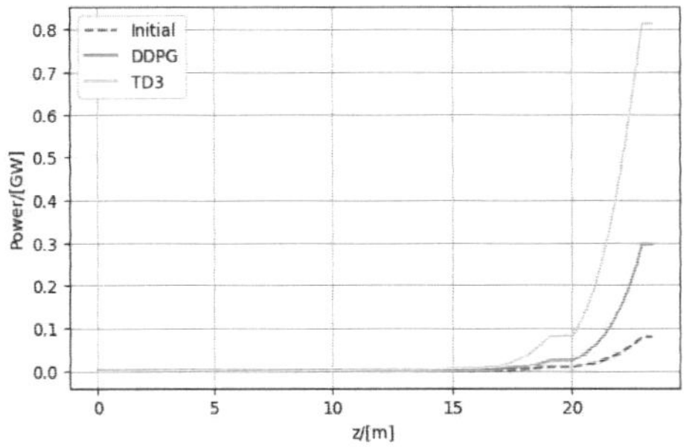

Figure 4. The initial FEL gain curve and optimized FEL gain curves with DDPG and TD3 optimization.

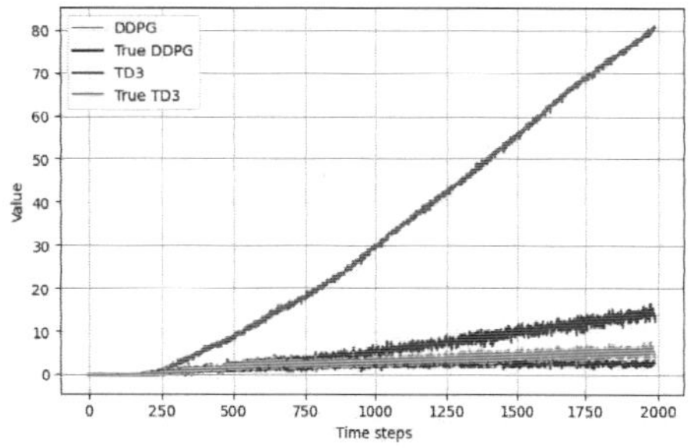

Figure 5. Estimated value and true value of DDPG and TD3 in FEL optimization.

5. Conclusion

We have tried to optimize the electron beam trajectory in undulators by using policy gradient methods to simultaneously control multiple corrector magnets. According to the preliminary

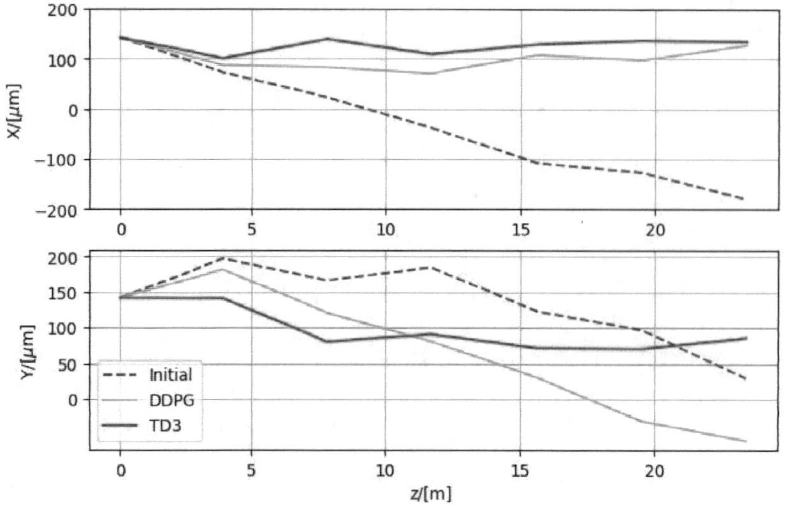

Figure 6. The initial beam orbit and optimized beam orbit with DDPG and TD3 methods.

simulation results, TD3 can considerably enhance the learning speed and performance of DDPG because it addresses the overestimation bias issue. The proposed method can perform tasks that involve management of more magnetic parameters and more complex requirements by simply modifying network structure since no prior physical knowledge and datasets are required. This method is being tested in the SXFEL facility, and it may be applied to the FEL optimization for the whole undulators and varying wavelengths in the future.

Acknowledgments

The authors would like to thank Zhen Wang and Zheng Qi for fruitful discussions in free-electron laser physics and simulations. This work is supported by the National Natural Science Foundation of China (12105347, 12275340) and Shanghai Science and Technology Committee Rising-Star Program (20QA1410100).

References

[1] Chapman H N 2019 *Annu. Rev. Biochem.* **88** 35-58
[2] Coffee R N, Cryan J P, Duris J, Helml W, Li S, and Marinelli A 2019 *Philos. Trans. R. Soc. London* A **377** 20180386
[3] Marangos J P 2017 *Philos. Trans. R. Soc. London* A **377** 20170481
[4] Zeng L, Feng C, Gu D, Li J, and Zhao Z 2018 *Nucl. Instrum. Methods Phys. Res., Sect.* A **905** 104
[5] Duris J *et al* 2020 *Phys. Rev. Lett.* **124** 124801
[6] Scheinker A, Bohler D, Tomin S, Kammering R, Zagorodnov I, Schlarb H, Scholz M, Beutner B, and Decking W 2019 *Phys. Rev. Accel. Beams* **22** 082802
[7] O'Shea F H, Bruchon N, and Gaio G 2020 *Phys. Rev. Accel. Beams* **23** 122802
[8] Bruchon N, Fenu G, Gaio G, Lonza M, O'Shea F H, Pellegrino F A, and Salvato E 2020 *Electronics* **9** 781
[9] Puterman M L 2014 *Markov Decision Processes: Discrete Stochastic Dynamic Programming* (New York: Wiley-Interscience)
[10] Lillicrap T P, Hunt J J, Pritzel A, Heess N, Erez T, Tassa Y, Silver D and Wierstra D 2015 Continuous control with deep reinforcement learning *Preprint* ArXiv:1509.02971
[11] Mnih V, Kavukcuoglu K, Silver D, Graves A, Antonoglou I, Wierstra D and Riedmiller M 2013 Playing Atari with Deep Reinforcement Learning *Preprint* ArXiv:1312.5602
[12] Silver D, Lever G, Heess N, Degris T, Wierstra D, and Riedmiller M 2014 *Proc. 31st Int. Conf. on Machine Learning* (Beijing: PMLR) pp 387–95

[13] Thrun S, Schwartz A 1993 *Proc. of the 1993 Connectionist Models Summer School Hillsdale* vol 6 (New York: Lawrence Erlbaum) pp 1–9
[14] Fujimoto S, Hoof H, Meger D 2018 *Proc. 35th Int. Conf. on Machine Learning* (Stockholmsmässan: PMLR) pp 1587–96
[15] Reiche S 1999 *Nucl. Instrum. Methods Phys. Res., Sect.* A **429** 243

Journal of Physics: Conference Series **2420** (2023) 012028 doi:10.1088/1742-6596/2420/1/012028

Start-to-end simulations for bunch compressor and THz SASE FEL at PITZ

A Lueangaramwong, X Li, P Boonpornprasert, M Krasilnikov and F Stephan

Deutsches Elektronen-Synchrotron, 15738 Zeuthen, Germany

E-mail: anusorn.lueangaramwong@desy.de

Abstract. The magnetic bunch compressor was designed as part of a THz accelerator source being developed at the Photo Injector Test facility at DESY in Zeuthen (PITZ) as a prototype for pump-probe experiments at the European XFEL. As an electron bunch is compressed to achieve higher bunch currents for the THz source, the beam dynamics in the bunch compressor was studied by numerical simulations. A start-to-end simulation optimizer including coherent synchrotron radiation (CSR) effects has been developed by combining the use of ASTRA, OCELOT, and GENESIS to support the design of the THz source prototype. In this paper, we present simulation results to explore the possibility of improving the performance of the THz FEL at PITZ by using the developed bunch compressor.

1. Introduction

An accelerator-based THz source prototype for pump-probe experiments at the European XFEL is in the final stage of installation at the Photo Injector Test Facility at DESY in Zeuthen (PITZ). The electron beam from the RF gun is further accelerated to a beam momentum of ~17 MeV/c by the CDS booster cavity. In order to achieve a THz Self-amplified spontaneous emission (SASE) free electron laser (FEL) with pulse energy in the mJ-range, highly space-charge-dominated electron beams such as 4 nC beams with a bunch current of about 160-200 A are set as a nominal setting [1–3] with a challenge in beam transport and matching to the undulator. Alternatively, relatively-lower-charge beams such as 1.5-2.5 nC beams with an average bunch current near 200 A are considered as an option when using a bunch compressor [4]. In this option, the lower-charge beams may also provide experimental benefits such as less challenges in beam transport and matching, etc.

The bunch compressor using a magnetic chicane has been foreseen and assigned as a part of the prototype for various applications such as the SASE FEL, seeded FEL, superradiant radiation, etc [4, 5]. The chicane consists of four rectangular dipole magnets with identical strength and length. Due to limited available space in the PITZ tunnel, this chicane has a vertical bending plane with an angle of 19 degrees in order to make a vertical clearance above the original PITZ beamline components. These dipole magnets are obsoleted from the HERA beamline [6]. While the magnetic field of the dipole magnet is yet to be measured, the effective length of each dipole of 0.327 m is estimated from the magnetic field simulation via the program CST Studio Suite [7]. Therefore, R_{56} of this chicane is 0.215 m.

We study the bunch compressor option at PITZ by simulating the beam dynamics and SASE FEL. A start-to-end simulation optimizer has been developed by combining the use of the programs ASTRA [8], OCELOT [9], and GENESIS 1.3 [10] to support our test experiments on the THz source prototype. Note that coherent synchrotron radiation (CSR) effects influencing the performance of the bunch compression was discussed in [4]. In this paper, we report simulation results such as optimized FEL pulse energy with booster phases as a beam-energy-chirp tuning knob, while discussing possible beam transports using available quadrupole magnets.

2. Start-to-end beam dynamic simulation

The start-to-end beam dynamic simulation [4] was previously developed with a combination of the programs ASTRA [8], OCELOT [9], and IMPACT-t [11]. ASTRA is firstly used to track the electron from the cathode to the booster, and then OCELOT and IMPACT-t are set up to investigate the bunch compression performance including CSR effects. While the CSR simulation with OCELOT is benchmarked with IMPACT-t, IMPACT-t is replaced by OCELOT for faster optimization in the investigation of beam transport and FEL performance. The FEL performance parameters such as saturation length and pulse energy are estimated by both the program GENESIS and a semi-analytical calculation of M. Xie's solutions in the framework of the one-dimensional model [12, 13] implemented to OCELOT evaluation functions. In this semi-analytical calculation, when the electron bunch is much longer than the FEL cooperation length ($\sim$400 µm), the electron bunch is sliced and the SASE power growth of each slice is estimated.

3. Simulation of beam transport and matching in chicane

Since several quadrupole magnets prior to the chicane at PITZ are installed [14], it is possible to transport the beam with constraints. In this study, the first constraint is described by a following mathematical condition at the middle z-position of the chicane with the Courant–Snyder (CS) parameters,

$$\alpha_x = \alpha_y = 0. \tag{1}$$

This condition is applied to all cases in a scan of bunch charge and booster phase. Note that other constraints can also be considered in a future study to support beam transport to the undulator.

In order to obtain solutions of equation (1), the simplex algorithm [15] is combined with the use of OCELOT including the space charge and the CSR effects. First, OCELOT generates a Gaussian distribution at the selected position in front of the chicane – a reference screen position – with assumed CS parameters and beam emittance obtained from ASTRA simulation and tracks the beam to the middle z-position of the chicane in each iteration. The simplex algorithm is used to find the CS parameters that satisfy the condition defined by equation (1). Solutions of the simplex algorithm depend on initial guesses of CS parameters, since equation (1) does not define a unique beam transport.

The second constraint dictates the beam transport and matching after the chicane. Due to limited available space in the PITZ tunnel [14, 16], only one quadrupole magnet is installed at 0.45 m downstream to the chicane exit. To transport the beam with a distance of 3.8 m from the chicane exit to a quadrupole triplet, installed for matching the beam into the undulator, solutions with relatively low beam size and divergence in one of the transverse axes at the chicane exit is preferred. Thus, the horizontal root-mean-squared (rms) beam size σ_x at the chicane exit must be smaller than the vertical one σ_y.

Once such a solution is obtained, the simplex algorithm is again combined with OCELOT to instead tune strengths of several quadrupole magnets prior to the chicane in order to achieve the earlier-solved CS parameters. Here the beam distribution simulated by Astra is taken as input. Figure 1 shows the transport of a model beam of 1.5 nC along the chicane, where a beam

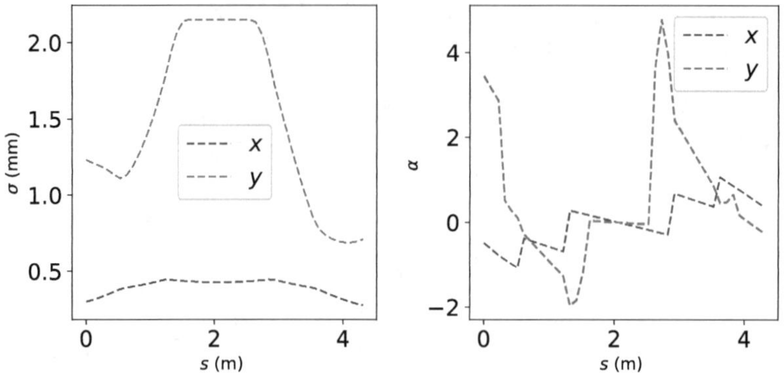

Figure 1. Evolution of root-mean-squared transverse beam size (left) and Courant–Snyder parameter α (right) of a 17 MeV/c beam with a bunch charge of 1.5 nC along a beam path s throughout the chicane ending at 0.453 m downstream to the final dipole magnet, where $s = 0$ indicates the location with a longitudinal distance of 0.238 m prior to the first dipole magnet.

waist ($\alpha_x, \alpha_y \sim 0$) is achieved in both horizontal and vertical planes in the middle z-position of the chicane. Meanwhile, the beam is small and converging in the horizontal plane but is large and diverging in the vertical plane. This guarantees the further transport of the beam using the single quadrupole that follows the chicane.

4. SASE FEL simulations

In this section, we firstly discuss the semi-analytical calculation and its results for LCLS-I undulator parameters. The semi-analytical calculation is performed by solving the M. Xie's solutions with the compressed beam parameters [12, 13] in order to obtain an optimized CS parameter β_x giving maximum FEL pulse energy. Then the optimized β_x becomes a primary goal of tuning quadrupole magnet strengths prior to the undulator by the simplex algorithm with beam tracking to the undulator using OCELOT including the space charge effect. Furthermore, several constraints such as CS parameters $\alpha_x = \alpha_y = 0$, rms transverse beam size along the tracking $\sigma_x, \sigma_y < 4$ mm are implemented to the goal of the simplex algorithm.

In this study, we scan the bunch charge, booster phase, and the (ASTRA) initial Gaussian laser pulse length to find the maximum FEL pulse energy, see details of the parameters in [4]. Due to the space charge effect, β_x from the tracking with the calculated quadrupole magnet strengths is larger than the optimized β_x and FEL pulse energy in the case of 1.75 nC bunch charge drops from 950 to 276 µJ. Note that the initial laser pulse length is represented in full width at half maximum (FWHM).

Figure 2 shows the results of the semi-analytical calculation with a resonant wavelength of 100 $\pm$2 µm, as the saturation length is under 4.6 m. They preliminarily indicate that the maximum FEL pulse energy is obtained from the initial laser pulse length of 8 ps in FWHM with optimized booster phases for each bunch charge. Note that all optimized booster phases correspond to under-compression in the chicane, see figure 6 in [4] showing the full compression phases. Therefore, we further investigate FEL simulations with GENESIS using the initial pulse length of 8 ps FWHM and the expected bunch charges in the same scan method.

GENESIS is also used to perform FEL simulations with a 3.4-meter-long LCLS-I undulator (currently installed at PITZ for proof-of-principle experiments). In the first step using GENESIS, a simulated transverse beam profile after the compression is represented by CS parameters. Thus, CS parameters are optimized (by scanning) for maximum FEL pulse energy simulated by GENESIS, while the original transverse emittance is maintained. For simplicity, the CS

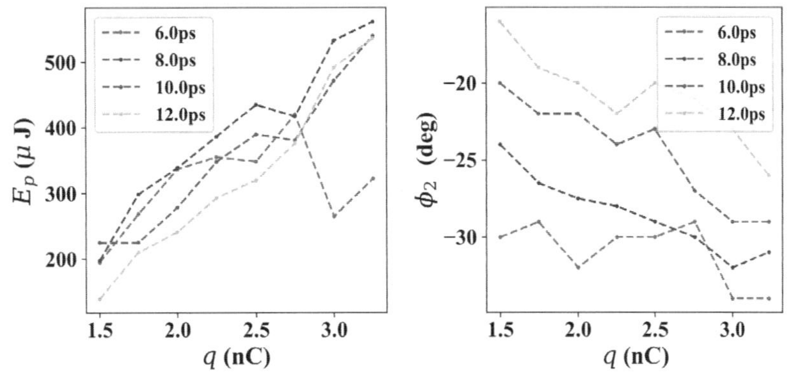

Figure 2. Optimized FEL pulse energy via the semi-analytical calculations and corresponding booster phase with respect to MMMG phase as a function of a bunch charge with initial laser pulse length of 6, 8, 10, and 12 ps (FWHM).

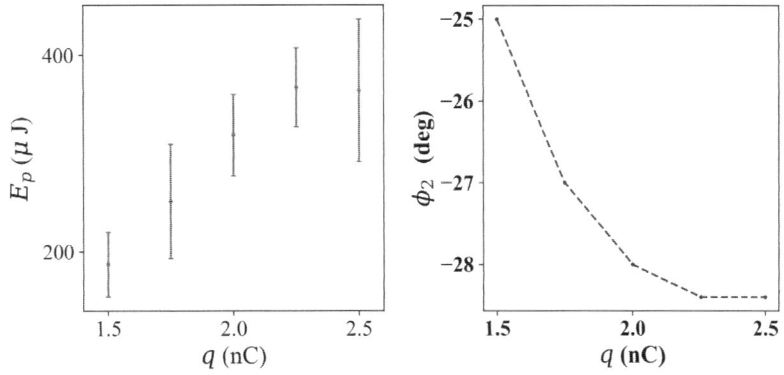

Figure 3. Optimized FEL pulse energy via GENESIS (left) and corresponding booster phase (right) as a function of bunch charge with initial laser pulse length of 8 ps (FWHM).

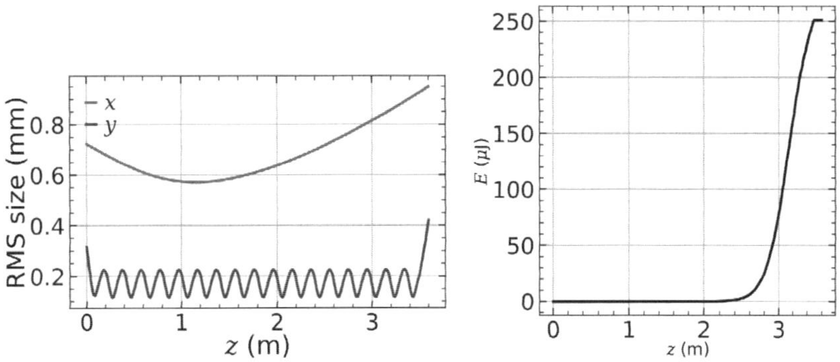

Figure 4. Evolution of the rms transverse beam size (left) and FEL pulse energy (right) of 17-MeV beam with a bunch charge of 1.75 nC along the longitudinal axis throughout LCLS-I undulator via GENESIS.

parameters $\beta_x = 2.605, \beta_y = 0.231, \alpha_x = 0.695, \alpha_y = 2.174$ are obtained from the scan using a 2.5 nC beam with booster phase ϕ_2 of -32 degree with respect to maximum mean momentum gain phase (MMMG phase) and are set for all study cases using bunch charge of 1.5-2.5 nC and booster phases between -24 and -32 degree.

Figure 3 shows GENESIS results for the beam with initial laser pulse length of 8 ps FWHM. They show reasonable agreement with the earlier semi-analytical calculations, whereas the GENESIS simulation yields the relatively smaller FEL pulse energy. For instance, the FEL pulse energy of ~250 µJ is obtained for the 17-MeV beam, which is slightly lower than the semi-analytical result. As a result of the CS parameter scan, figure 4 shows that rms transverse beam size of the 17-MeV beam with a bunch charge of 1.75 nC is expected to be smaller than 1 mm, while the FEL pulse energy does not reach the saturation.

5. Summary

A start-to-end beam dynamic simulation tool is developed by a combination of the several codes to optimize the PITZ bunch compressor option. A FEL pulse energy of ~250 µJ is obtained from a 1.75 nC beam (smaller than the nominal bunch charge) with optimized booster phases via GENESIS for the setup of the proof-of-principle experiments.

Acknowledgments

This work was supported by the European XFEL research and development program. We would like to thank T. Weilbach, F. Mueller, and A. Oppelt for technical support at PITZ, and thank T. Limberg, I. Zagorodnov, and M. Dohlus for their expert advice on the bunch compressor study.

References

[1] Boonpornprasert P *et al.* 2014 *Proc. 36th Int. Free Electron Laser Conf. (FEL'14)* (Basel, Switzerland) pp 153–58
[2] Boonpornprasert P 2019 *Investigations on the capabilities of THz production at the PITZ facility* Ph.D. thesis Universität Hamburg Germany
[3] Krasilnikov M *et al.* Developments on accelerator based THz source at PITZ *under preparation*
[4] Lueangaramwong A *et al.* 2021 *Proc. 12th Int. Particle Accelerator Conf. (IPAC'21)* (Campinas, Brazil) pp 3285–88
[5] Shaker H *et al.* 2019 *Proc. 39th Int. Free Electron Laser Conf. (FEL'19)* (Hamburg, Germany) pp 45–47
[6] Voss G A 1988 *Nucl. Phys. B Proc. Suppl.* **3** 525
[7] CST Studio Suite `https://www.cst.com`
[8] Flöttmann K 2017 ASTRA: a space charge algorithm version 3.2, User's Manual
[9] Agapov I *et al.* 2014 *Nucl. Instrum. Methods Phys. Res., Sect.* A **768** 151
[10] Reiche S 2018 Genesis1.3 code, Version 2 `http://genesis.web.psi.ch/`
[11] Qiang J, Lidia S, Ryne R D and Limborg-Deprey C 2006 *Phys. Rev. Spec. Top. Accel Beams* **9** 044204
[12] Xie M *Nucl. Instrum. Methods Phys. Res., Sect.* A **445** 59
[13] Saldin E L, Schneidmiller E A and Yurkov M V 2010 *New J. Phys.* **12** 035010
[14] Weilbach T *et al.* 2021 *Proc. 12th Int. Particle Accelerator Conf. (IPAC'21)* (Campinas, Brazil) pp 1528–31
[15] Nelder J A and Mead R 1965 *Comput. J.* **7** 308
[16] Weilbach T *et al.* 2022 *Proc. 13th Int. Particle Accelerator Conf. (IPAC'22)* (Bangkok, Thailand) pp 1033–36

Journal of Physics: Conference Series **2420** (2023) 012029 doi:10.1088/1742-6596/2420/1/012029

Reconstruction and beam-transport study of the cERL dump line for high-power IR-FEL operation

N Nakamura, Y Tanimoto, N Higashi, K Harada, M Shimada, T Uchiyama, T Nogami, A Ueda, S Nagahashi, T Obina, R Takai, H Sagehashi, K Nigorikawa, O Tanaka, R Kato and H Sakai

High Energy Accelerator Research Organization (KEK), Tsukuba, Japan.

E-mail : norio.nakamura@kek.jp

Abstract. Infrared (IR) SASE-FEL emission was successfully generated for macro pulses of about 1 μs with the maximum repetition rate of 5 Hz at the cERL. In the future, high-power FEL operation will be planned to increase electron bunches drastically with energy recovery. Therefore, the dump line was redesigned and reconstructed to increase the momentum acceptance and to improve the magnet system for avoiding serious beam loss in the high power FEL operation. Furthermore, the first beam-transport study was performed by transporting the beam directly from the injector to the beam dump through the reconstructed dump line. In this paper, we present the reconstructed dump line and the beam-transport study. The new dump line can be a model for dump lines of high-power ERL-FELs.

1. Introduction

A significant FEL pulse energy was generated at the cERL IR-FEL [1] in Burst mode where a macro pulse of about 1 μs is repeated at the maximum frequency of 5 Hz. In the next step, high-power FEL operation in CW mode should be carried out with energy recovery by increasing electron bunches drastically. However, momentum spread of the electron beam increases due to the FEL emission and the space charge effects and may cause serious beam loss by exceeding the momentum acceptance of the cERL downstream of the FEL. Therefore, we reconstructed the dump line in Autumn 2020 to greatly increase the momentum acceptance and to improve the magnet system for more flexible beam control. Then we carried out the beam-transport study of the new dump line in March 2021 by injecting the beam directly from the injector without passing the recirculation loop.

2. Momentum acceptance

Beam loss must be efficiently suppressed in the high-power FEL operation at the cERL [2] in terms of radiation safety and machine protection, because the beam current is much increased in CW mode as compared that in Burst mode. The electron beam can have large momentum spread due to the FEL emission and the longitudinal space charge effects and cause significant beam loss at the dispersion sections by exceeding the apertures of the beam ducts. To avoid it, the momentum acceptance downstream of the FEL must be sufficiently large compared to the momentum spread of the electron beam. The main dispersion sections downstream of the FEL are the 2nd arc and the dump line as shown in figure 1. In the FEL operation, we have so far used Burst mode, where the electron beam is dumped without energy recovery at the movable dump before the dump line. In the high-power FEL operation, the high-current beam is decelerated in the main linac for energy recovery and transported to the beam dump through the dump line.

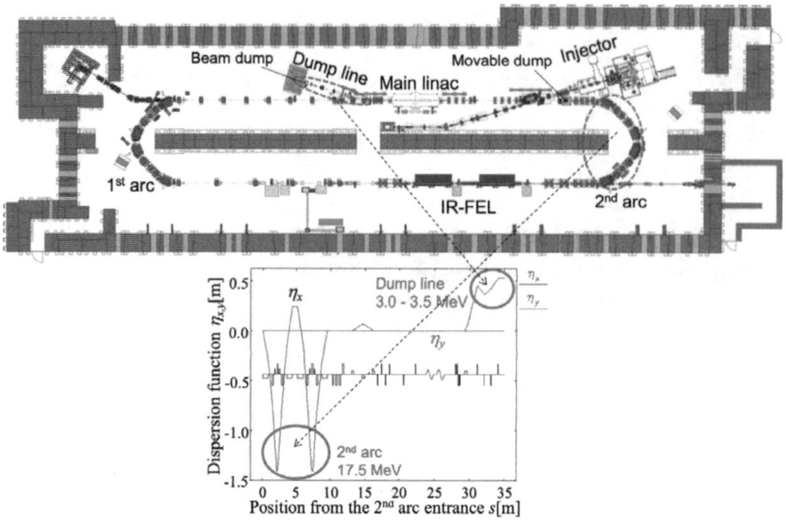

Figure 1. Layout of the cERL including the two dispersive sections, the 2nd arc and dump line, downstream of the IR-FEL with the dispersion functions.

The momentum acceptance is defined with the maximum dispersion function $\eta_{x,\,max}$ and the beam-duct aperture A_x in the dispersion section by the following equation.

$$\left(\frac{\Delta p}{p}\right)_{MA} = \frac{A_x}{\eta_{x,max}} \tag{1}$$

In equation (1), the effect of the betatron beam size is neglected for simplicity.

Table 1. Parameters of the 2nd arc and the dump line before and after reconstruction.

2nd arc	
Beam energy	17.5 MeV
Aperture	35 mm
Maximum dispersion	1.4 m
Momentum acceptance	2.50 %
Dump line before reconstruction	
Beam energy	3.53 MeV
Aperture	42.45 mm
Maximum dispersion	0.523 m
Momentum acceptance	8.12 % (1.62 % [a])
Dump line after reconstruction	
Beam energy	3.53 MeV
Aperture	48.70 mm
Maximum dispersion	0.341 m
Momentum acceptance	14.28 % (2.86 % [b])

[a,b] Normalized by the beam momentum at the 2nd arc.

In the cERL, two beams with different momenta pass for energy recovery between the exit of the injector and the entrance of the dump line. In the injection and dump chicanes, the lower-momentum beam is deflected by 16 degrees for injection and dump and the higher-momentum beam makes closed bump orbits. The ratio of the lower and higher momenta must be less than 1/5 in order that the higher-momentum beam does not hit the beam ducts in the injection and dump chicanes. Therefore, the beam energy at the dump line must be 3.5 MeV or lower when the beam energy in the recirculation loop including the FEL and the 2nd arc is set to 17.5 MeV. Table 1 shows the parameters of the two dispersion sections downstream of the IR-FEL, the 2nd arc and the dump line before and after the reconstruction. The values within parentheses in the table are the momentum acceptances of the dump line normalized by the momentum at the 2nd arc for comparing the momentum acceptances of the different momenta. The normalized momentum acceptance before the reconstruction is significantly smaller than that of the 2nd arc. This means that beam loss at the dump line can be the largest in the cERL. Therefore, we decided to redesign and reconstruct the dump line for the purpose of improving the normalized momentum acceptance up to that of the 2nd arc at least. The parameter values for the dump line after the reconstruction are also shown in table 1 and described in detail in the next section.

3. Reconstruction of dump line

The layout of the dump line before and after the reconstruction is shown in figure 2.

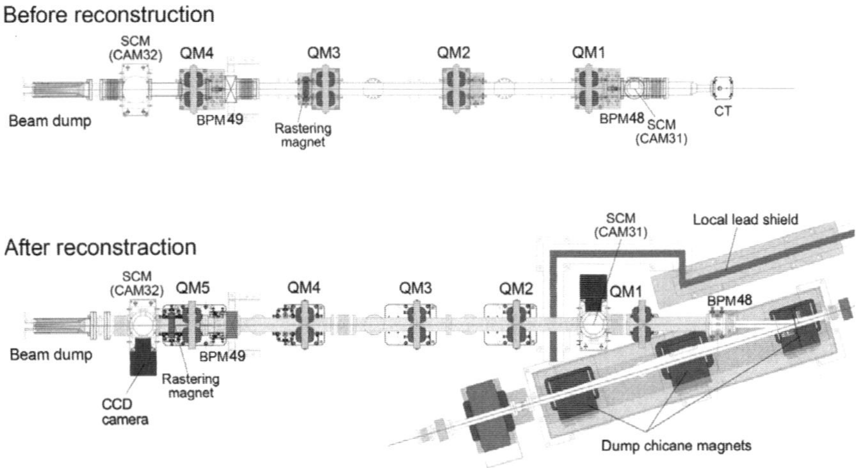

Figure 2. Layout of the cERL dump line before (upper) and after (lower) reconstruction. The beam line with the dump chicane, the local lead shield and the CCD cameras for the screen monitors (SCMs) are the same before and after the reconstruction and omitted in the upper figure.

In the reconstruction, the aperture of the circular beam duct downstream of the first quadrupole magnet QM1 in the dump line was increased from 42.45 to 48.70 mm in inner radius. The beam duct is made of stainless steel with 2.1-mm thickness. The maximum aperture was limited by the bore radius of the quadrupole magnets (51 mm) in the dump line. Furthermore, the quadrupole magnet QM1 was moved as upstream as possible (40 cm upstream) and as a result the maximum dispersion function was reduced from 0.523 to 0.341 m as shown in table 1. The normalized momentum acceptance of the dump line was increased from 1.62 to 2.86 %, which is larger than that of the 2nd arc.

The number of the quadrupole magnets in the dump line was increased from four to five to control the beam more flexibly. QM1, QM3 and QM5 have sub-coils for tuning the horizontal and vertical beam orbits. The rastering magnet [3], which scans the beam position horizontally and vertically for reducing the thermal load on the beam dump, was also moved from just after QM3 to just after QM5 so that the beam position and size at the beam dump should not depend on field strength of the quadrupole magnet. The layout of the rastering magnet and the beam dump is shown in figure 3. The distance between them is 697 mm. The rastering image at the beam dump is estimated by using the beam position movement at the screen monitor CAM32, which is placed at 216 mm from the rastering magnet. The screen monitors CAM31 and CAM32 with the half aperture of 50 mm were reused in the new layout. The beam position monitor BPM48 with the aperture of 42.5 mm was reused and BPM49 was renewed to increase the half aperture to 50 mm. The current transformer (CT) was removed.

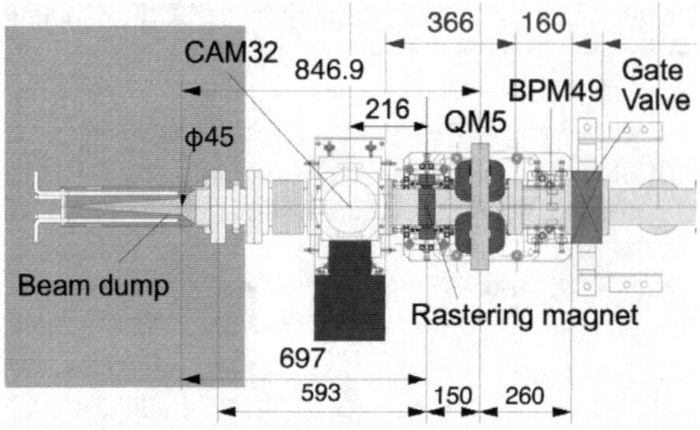

Figure 3. Layout of the rastering magnet and the beam dump.

4. Beam-transport study

The first beam-transport study of the reconstructed dump line was carried out on March 18 in 2021. The beam line used for the study and the direction of the beam transport are shown in figure 4.

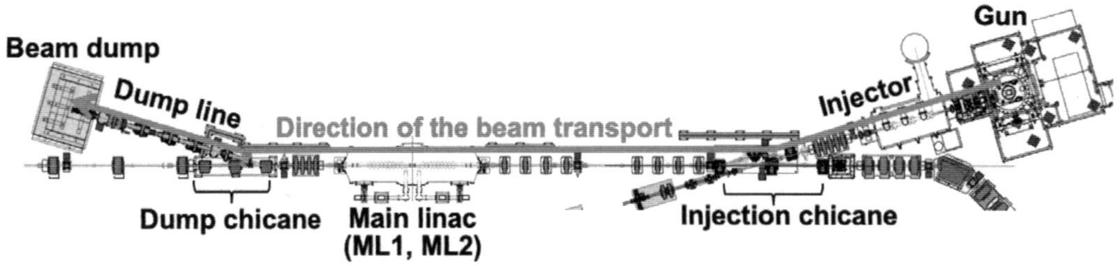

Figure 4. Layout of the used beam line and the beam direction in the beam-transport study.

In the study, the electron beam with the bunch charge of 60 pC was operated in Burst mode, which was the same as that of the IR-FEL operation. The injection energy in front of the main linac with two superconducting cavities, ML1 and ML2, is 5 MeV and the used parameters from the electron gun to the entrance of ML2 were the same as those of the FEL operation carried out in March 2021. The beam energy is 10 MeV between ML1 and ML2. The beam was decelerated by using the ML2 cavity and transported to the dump line in the beam-transport study, though it was accelerated to about 17.5 MeV in the FEL operation.

The 10-MeV beam between the ML1 and ML2 cavities was decelerated to 3.5 MeV in the ML2 cavity. First the RF amplitude and phase of the ML2 cavity were adjusted to decelerate the beam on crest to 3.5 MeV. Figure 5a shows the beam profiles measured by CAM31 and CAM32. The measured horizontal and vertical beam sizes in rms were 3.6 and 2.5 mm at CAM 31 and 3.4 and 1.1 mm at CAM32. Next the RF phase of the ML2 cavity was shifted by 30 degrees with keeping the beam energy at 3.5 MeV by the RF amplitude. In this off-crest deceleration by the ML2 cavity, the momentum spread became very large compared to that of the on-crest deceleration. This experimentally simulated an electron beam with a large momentum spread caused by the IR-FEL emission and/or the space charge effects. The horizontal beam size greatly increased to about 12 mm at CAM31 ($\eta_x \sim 0.28$ m) as shown in figure 5b. However, the beam sizes could be well controlled within a few mm at CAM32 by the new magnet system in the dump line.

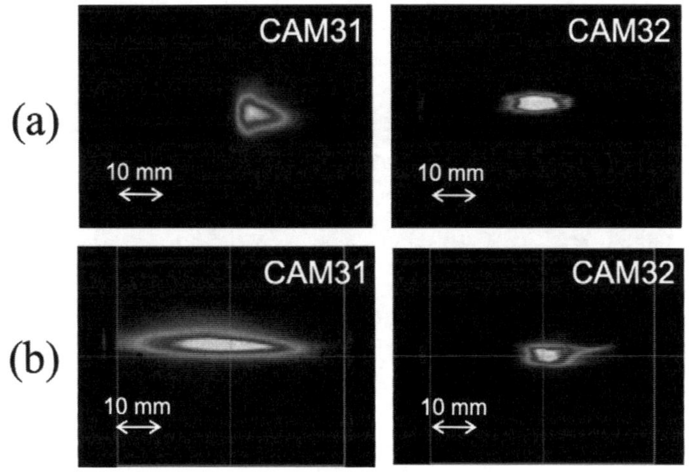

Figure 5. Profiles of the 3.5-MeV beam observed at CAM31 and CAM32 for the (a) on-crest and (b) off-crest deceleration of the ML2 cavity.

The beam was transported to the beam dump while the beam position was transversely being scanned by the rastering magnet. Figures 6a and 6b show traces of the beam positions measured at CAM32 with the rastering magnet on for on-crest and off-crest deceleration of the ML2 cavity. The rastering circle size in this study was set to 6.6 – 6.8 mm at CAM32 in diameter and hence 21 – 22 mm at the entrance of the beam dump. The rastering frequency was set to 10 Hz, at which the beam position filled the rastering circle twice and returned to the initial position.

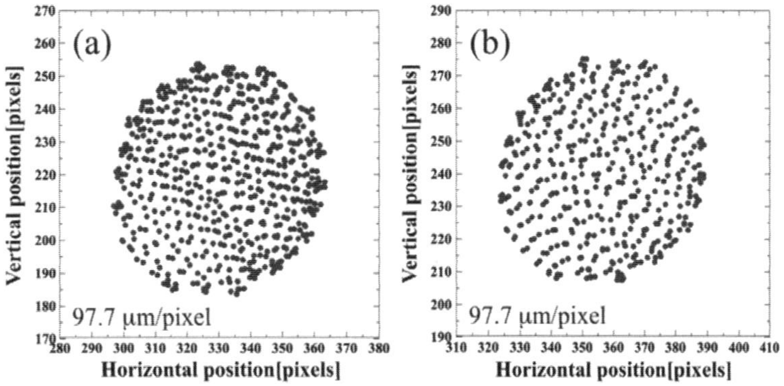

Figure 6. Beam position movement by the rastering magnet at CAM32 for the (a) on-crest and (b) off-crest deceleration of the ML2 cavity.

Figure 7 shows the betatron and dispersion functions, the momentum spread and beam sizes in rms from the ML2 cavity to the beam dump for the off-crest deceleration of the ML2 cavity. The optics design and beam simulation were performed by using the simulation code Elegant [4] with the initial distribution at the center of the main linac calculated by GPT [5]. The longitudinal space charge (LSC) effects and coherent synchrotron radiation (CSR) effects were taken into consideration in the Elegant simulation. The momentum spread reaches to about 3.4 % and the maximum horizontal beam size exceeds 11 mm. As shown in figure 7d, the new magnet system of the dump line can well adjust both horizontal and vertical beam sizes to about 2 mm at the beam dump, which is suitable for the rastering system. The simulation result is roughly consistent with the measurement result of the beam-transport study.

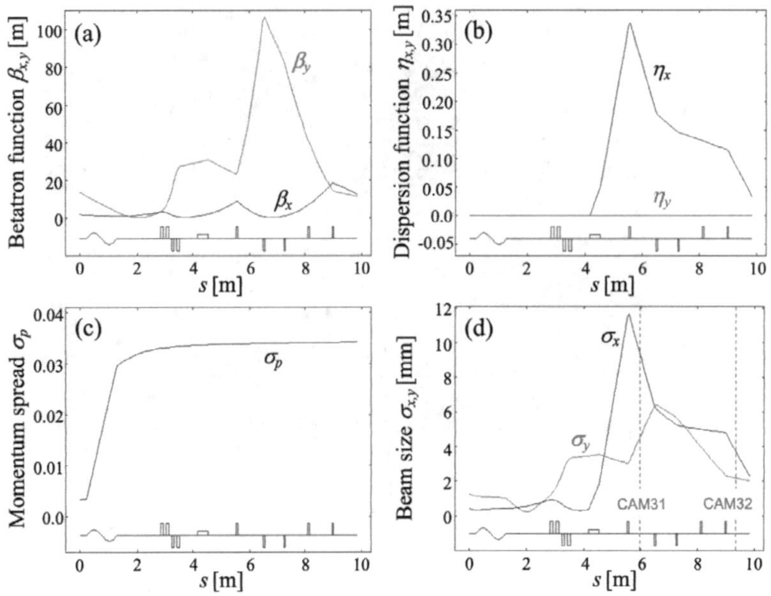

Figure 7. (a) Betatron and (b) dispersion functions, (c) momentum spread and (d) beam sizes along the dump line from the ML2 cavity to the beam dump calculated for the off-crest deceleration.

5. Summary

We summarize our presentation on the reconstructed dump line and the beam-transport study below.

- The largest beam loss had a possibility of occurring at the dump line in the high-power IR-FEL operation at the cERL, because the dump line had the minimum normalized momentum acceptance.
- The dump line was reconstructed in Autumn 2020 to improve the momentum acceptance by 76 % and to enhance flexibility of the magnet system in beam control.
- The first beam-transport study was done in March 2021 and as a result the beam was successfully transported with rastering on to the beam dump through the reconstructed dump line.
- The new cERL dump line is expected to well work in the high-power FEL operation and to serve as a model for dump lines of high-power ERL-FELs.

Acknowledgments

We would like to thank all the members of the cERL IR-FEL team. This work is supported by a NEDO project "Development of advanced laser processing with intelligence based on high-brightness and high-efficiency laser technologies."

References

[1] Honda Y *et al.* 2021 *Rev. Sci. Instrum.* **92** 113101
[2] Akemoto M *et al.* 2018 *Nucl. Instrum. Methods Phys. Res., Sect.* A **877** 197
[3] Harada K, Nakamura N, Obina T, Shimada M, Miyajima T, Honda T, Nogami T, Tanimoto Y, Honda T and Takai R 2016 *Proc. 13th Annual Meeting of Particle Accelerator Society of Japan (PASJ2016)* (Chiba Japan) pp 585–88
[4] Borland M 2000 elegant: A Flexible SDDS-Compliant Code for Accelerator Simulation Tech. rep. LS-287 Advanced Photon Source
[5] Tanaka O A, Higashi N and Miyajima T 2021 *Proc. 12th Int. Particle Accelerator Conf. (IPAC'21)* (Campinas, Brazil) pp 4531–34

Tungsten Electron Emitter (TE^2) with direct heated cathode by plasma stream

K I Thoma[1,2], T Dönges[1], M Droba[1], O Meusel[1], H Podlech[1] and K Schulte-Urlichs[1,2]

[1]Institute for Applied Physics, Goethe-University Frankfurt, Max-von-Laue Straße 1, 60438 Frankfurt, GER

[2]GSI Helmholtzzentrum für Schwerionenforschung GmbH, Planckstraße 1, 64291 Darmstadt, GER

E-mail: thoma@iap.uni-frankfurt.de

Abstract. At Goethe-University, a novel concept of heating metallic cathodes is currently under investigation. In the scope of the ARIES collaboration WP16, an RF-modulated electron gun was developed and manufactured for application in electron lenses for space charge compensation. The goal of this project is to increase the intensity of primary beams, especially in low energy booster synchrotrons like the SIS18 and SIS100 at GSI/FAIR or the SPS at CERN. The gun was designed to produce electron currents of 10 A at extraction voltages of 30 kV. The tungsten electron emitter ($\mathbf{TE^2}$) and the grid electrode were designed and manufactured to be integrated in the extractor of the original volume type ion source. Significant effort was put into a robust and flexible design with highly reliable key components. The cathode is heated by a plasma stream generated in the plasma chamber of the source. Different heating options of the cathode are currently being studied. This contribution presents the working principles of the electron gun and first measurements results of cathode heating.

1. Introduction

Common electron sources use oxide cathodes or photo cathodes for electron beam production [1]. Some exotic cathode materials are currently used [2]. All of these technologies are less robust against vacuum contamination, vacuum discharges (Townsend discharge) and secondary particle impact. Therefore, it was decided to use a pure tungsten cathode for the production of the intense electron beam with an emission density of $J = 0.453\,\mathrm{Acm^{-2}}$. The heating of the cathode is challenging, as a temperature range of 2200-2800 °C is needed for an adequate electron emission following Richardson's law. Inspired by the concept of hybrid ion sources [3], the direct heating of the cathode by the use of a plasma stream, generated by an arc discharge, was used. For the proof of concept, an existing gas discharge ion source was modified and the plasma electrode was replaced by a tungsten cathode. The flexibility of the plasma generator enables the guiding of the arc discharge to the cathode and the variation of arc power and power density distribution on the cathode. The resulting high cathode temperatures need an effective cooling of cathode and grid flange. On the other hand thermal conductivity determines the temperature distribution on the extraction surface. A careful layout of stainless steal thermal spacers led to heat balance and resulted in the desired temperature distribution. The spacers were mounted to connect the water cooled cathode flange and the tungsten cathode.

Three options to create a time structure on the extracted electron beam were discussed. The first one is a pulsed extraction voltage. This option is not favourable because of beam mismatch and the high power load. A pulsed arc discharge, leading to a pulsed heating of the cathode, is only suitable for duty cycles of a few percent with low pulse frequencies due to the thermalisation time of the cathode. Therefore, a grid modulated beam extraction is used. A high voltage modulation can be applied to synchronise the time structure of the electron beam with the ion bunch evolution during a synchrotron cycle. A careful layout protects the grid from heat load and provides a cut-off frequency of about 600 MHz. The design of the electron source presented in this paper was chosen to be the prototype of the IRME-gun [4] [5]. Additionally, a research project was started for the investigation of electron emission of metal surfaces in the near of phase transitions or plasma chemistry.

2. Design of the electron gun and test stand

The design of the electron gun TE^2 is based on an existing volume type ion source. All parts are optimized regarding water cooling, high voltage break downs, vacuum conditions and outside isolation [6] [7].

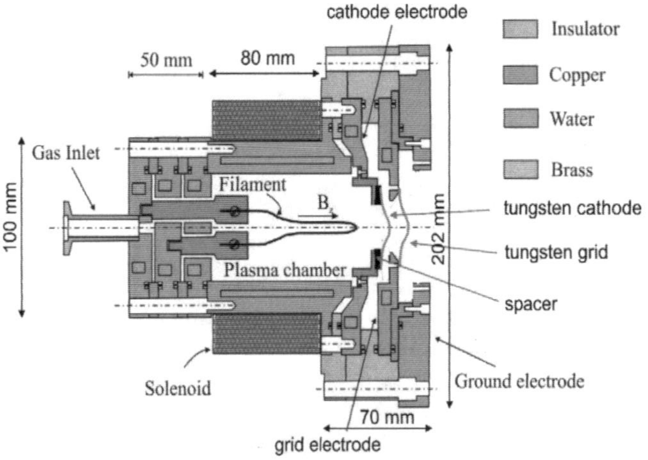

Figure 1. Cross sectional view of TE^2.

The ion source design was transformed into an RF modulated electron gun by replacing the plasma electrode by a copper cathode flange with tungsten cathode. The screening electrode of the former accel-decel extraction system was replaced by a grid flange with tungsten grid. The ground electrode was shaped for an optimal beam matching.

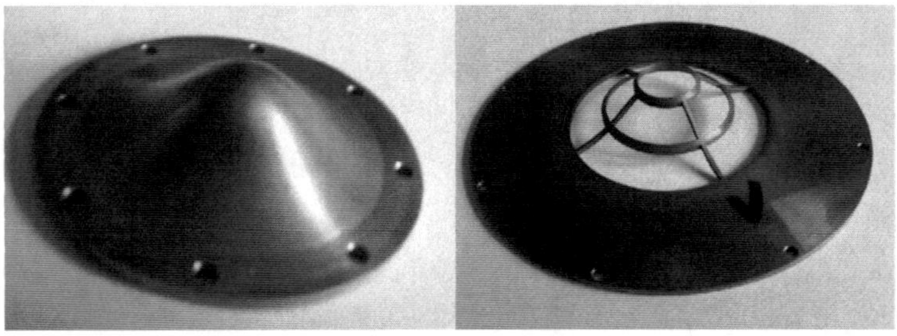

Figure 2. Left: robust tungsten cathode with a purity factor of 99.98 %, Right: grid for beam modulation.

In a first step, cathode and grid are shaped to gain a Gaussian electron density distribution, but all parts are mounted as flexible as possible (see figure 2). Different shapes of cathode and grid can be investigated in the future as well as hollow or elliptical cathodes. Therefore, TE^2 is well suited as a test device for innovative electron beam extraction schemes.

Journal of Physics: Conference Series **2420** (2023) 012030 doi:10.1088/1742-6596/2420/1/012030

The plasma generator is electrically insulated against the cathode flange, and enables different modes for the arc discharge as shown in figure 3. On the left hand side both, plasma generator and cathode are biased with the same arc voltage leading to an arc discharge with a centered plasma stream in the middle of the plasma chamber. A floating of the cathode shifts the plasma stream in the direction of the cathode (middle), whereas a floating of the plasma chamber shifts the plasma stream in the opposite direction. The direction of the plasma stream, as well as the arc discharge power, can be manipulated by applying a small potential difference between the plasma generator and cathode. Additionally, the total heating power can be adjusted with discharge current and working gas pressure. The hole electron source is embedded in an axial magnetic field of a solenoid to produce intense magnetised electron beams. The influence of the magnetic field on the plasma stream has to be studied in detail.

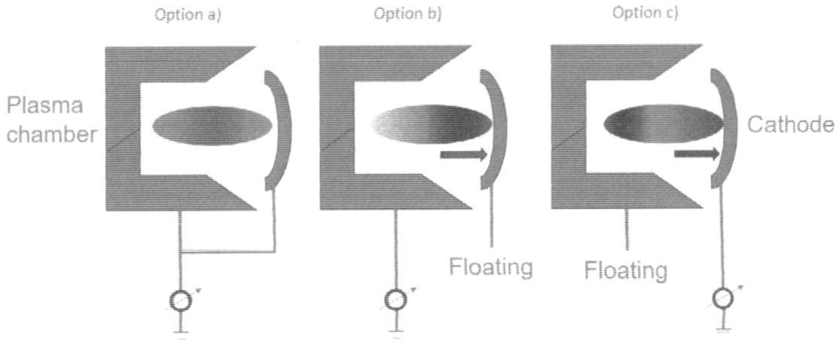

Figure 3. Three different modes to connect plasma chamber and cathode for manipulation of the arc discharge.

TE^2 was integrated in a test bench to proof the concept of cathode heating, to evaluate the performance with respect to thermal stress and electrical insulation. GSI/FAIR provided two solenoids with a maximum field strength of 0.6 T to the experiments. The first solenoid houses the electron source, and the second one is connected to the diagnostic chamber. The latter is equipped with a faraday cup, a ccd-camera for optical inspection of the cathode and grid as well as a pyrometer for the measurement of surface temperature distribution. TE^2 is connected with a high voltage terminal, which can provide an electric power up to 20 kW and a beam loading 4 kW. Therefore in the future, the second solenoid will house the beam dump with power recovering system.

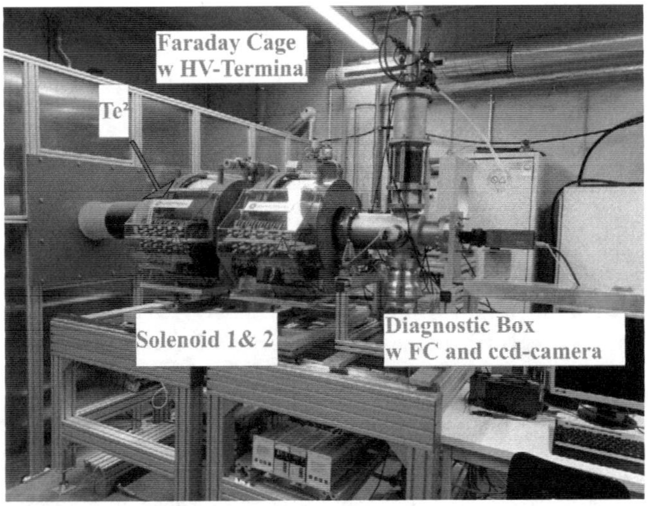

Figure 4. HV-Terminal with connection to the beamlines. TE^2 is inside the first of the two solenoids. The diagnostic chamber is equipped with a Faraday-cup, a camera and pyrometer.

3. Plasma ignition

TE^2 uses a hot filament driven gas discharge for plasma production. The filament is a Tungsten wire with 1mm in diameter which produces a small amount of electrons to fulfil Paschen's law. Therefore it has to be heated with a power of 780 W. First experiments using air as working gas showed a degradation of the cathode-plasma interaction surface, by layers of tungsten carbide and tungsten blue, which indicate temperatures between 700 and 1400 °C at the surface. In the next step, Argon was used, but the degradation of the filament by sputtering reduced the life time of the filament. Finally, Helium is used as a inert gas with a moderate sputter coefficient and the gas pressure is 0.01 hPa. No degradation of the surfaces, neither at the plasma chamber non at the cathode was observed. The aging of the filament is estimated by the control system by measuring the filament resistance continuously and it predicts the time still left till the next maintenance. After replacing the filament with a new one, a baking routine is necessary to prevent microscopic cracks, which in turn reduces the life time of the filament. In future, it is planned to investigate an RF driven gas discharge for plasma stream production to overcome the disadvantages of the use of a hot filament.

4. Cathode heat-up

The heat-up of the cathode was observed by the use of a ccd-camera. Figure 5 shows two images

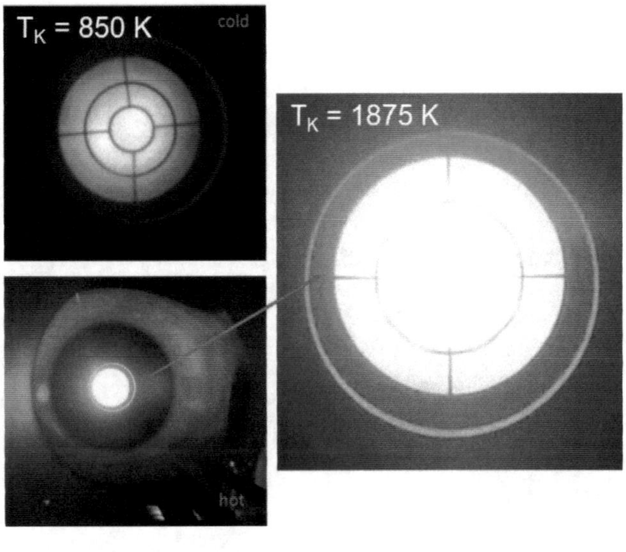

Figure 5. An example of cathode heat-up measurement and proof of principle, measured temperature given at emissivity factor $\varepsilon = 1$.

of the cathode at different temperatures. The image of the "cold" one at 850 K was used to evaluate the adjustment of the plasma stream in the center of the cathode area to provide a symmetric electron density distribution. The given temperature was measured by the use of the pyrometer in the center of the cathode. The change of the color across the cathode indicates the temperature gradient. This will be verified in a next step by the use of a pyrometer with an spatial resolution of $1\,mm^2$. The grid is visible clearly in front of the cathode. After increasing the arc discharge power, the temperature of the cathode increases up to 1875 K. Due to the high luminescence, it was not possible to observe a temperature gradient as well as the heat-up of the grid.

In a next step, different modes for the gas discharges were investigated. The mode using the floated cathode was found to be most efficient. The cathode temperature for this mode is plotted as a function of arc discharge power in figure 6 and shows a nearly linear behaviour. This facilitates the adjustment of cathode temperature, and therefore the electron beam current by the control system.

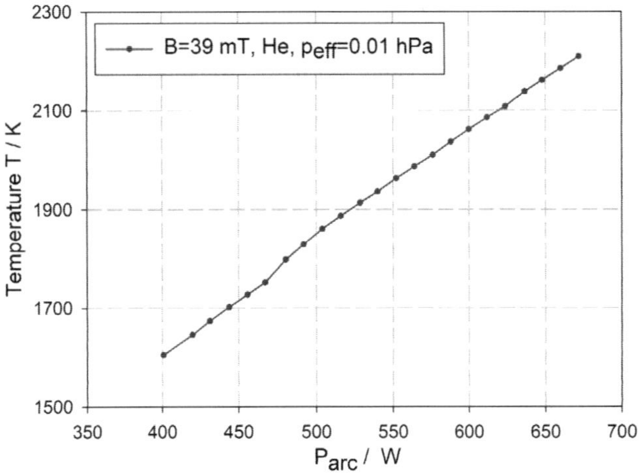

Figure 6. Power of the arc discharge as a function of the temperature measured with pyrometer at a magnetic field of 39 mT and Helium gas with a residual gas pressure of 0.01 hPa.

Direct heating using a plasma stream may suffer under fluctuation in the arc power and plasma density. The control system of TE^2 stabilizes the arc power to provide constant surface temperatures and therefore constant electron currents. Figure 7 shows a histogram of the arc power recorded for low median arc power of 280 W for test of the control loop. A variation of $\Delta P = \pm 1.5\,\%$ was observed over a time of ten minutes and the control frequency was chosen to be 10 Hz. Experiments emphasizing long term stability and reliability are planned, and will show the impact on the electron current fluctuation of the extracted electron beam.

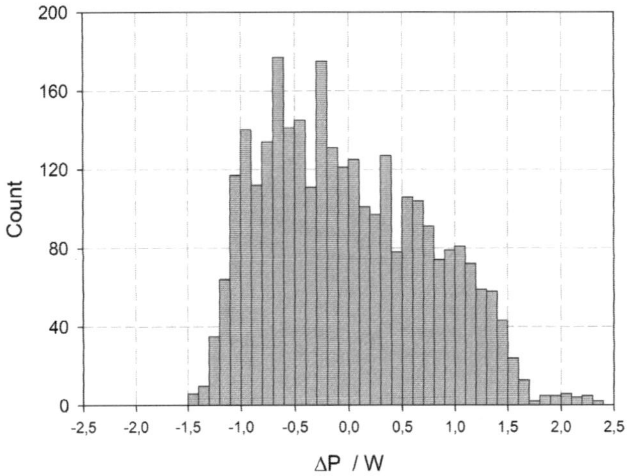

Figure 7. Arc power stabilisation by the TE^2 control system.

5. Conclusion

TE^2 was designed to use a new concept for direct heating of robust metal cathodes. Experiments show that it is possible to reach temperatures suitable to extract high intensity electron beams. The flexibility of the technical layout enables a variation of different cathode and grid shapes, which will be important for the development of e-lenses for space charge compensation or hollow e-lenses for beam collimation. Furthermore, TE^2 enables the research on electron emission in the near of phase transitions or plasma chemistry.

Acknowledgements
This project has received funding from European Union's Horizon 2020 Research and Innovation programme under Grant Agreement No 730871. We express our thanks to ARIES collaboration (workpackage 16: Electron lenses for synchrotrons). We also want to express our thanks to GSI/FAIR for the technical support and fruitful discussions.

References
[1] Joseph A Eichmeier M K T 2008 *Vacuum Electronics* (Springer)
[2] Pahl H, Bidault N, Khatri G, Pikin A and Wenander F J C 2022 *Phys. Rev. Accel. Beams* **25**(1) 013402
[3] Keller R, Luft P, Regis M, Wallig J, Monroy M, Ratti A, Syversrud D, Welton R and Anderson D 2005 *AIP Conf. Proc.* **773** 104–6
[4] Schulte-Urlichs K *et al.* 2022 Tests of electron gun at test stand completed (aries milestone report D16.3) Tech. rep. CERN
[5] Schulte-Urlichs K *et al.* 2022 *Proc. 13th Int. Particle Accelerator Conf. (IPAC'22)* (Bangkok, Thailand) pp 2027–30
[6] Joshi N, Droba M, Meusel O and Ratzinger U 2009 *Nucl. Instrum. Methods Phys. Res., Sect.* A **606** 310–3
[7] Schulte-Urlichs K *et al. Proc. 12th Int. Particle Accelerator Conf. (IPAC'21)* (Campinas, Brazil) pp 1880–3

Journal of Physics: Conference Series **2420** (2023) 012031 doi:10.1088/1742-6596/2420/1/012031

Studies of a Ka-band high power klystron amplifier at INFN-LNF

M Behtouei[1], B Spataro [1], L Faillace[1], A Leggieri [2], F Di Paolo [2], F Marrese[2], L Valletti[2], S Fantauzzi[2], G Torrisi[3], M Carillo[4], F Bosco[4], A Mostacci[1,4], L Palumbo[4,5] and M Migliorati[4,5]

[1]INFN-LNF, Via E. Fermi 40, Frascati, Italy
[2] Università degli Studi di Roma "Tor Vergata", Via del Politecnico, 1-00133-Roma, Italia
[3] Istituto Nazionale di Fisica Nucleare (INFN/LNS) Lab. Nazionali del Sud, Catania, Italy
[4]Department of SBAI, Sapienza University of Rome, Piazzale Aldo Moro 5, Rome, Italy
[5]Istituto Nazionale di Fisica Nucleare (INFN-Roma1), Rome, Italy

E-mail: Mostafa.Behtouei@lnf.infn.it

Abstract. In the framework of the Compact Light XLS project [1], a Ka-band linearizer with electric field ranging from 100 to 150 MV/m is requested [2, 3, 4]. In order to feed this structure, a proper Ka-band high power klystron amplifier with a high efficiency is needed. This paper reports a possible solution for a klystron amplifier operating on the TM_{010} mode at 36 GHz, the third harmonic of the 12 GHz linac frequency, with an efficiency of 44% and 10.6 MW radiofrequency output power. We discuss also here the high-power DC gun with the related magnetic focusing system, the RF beam dynamics and finally the multiphysics analysis of a high- power microwave window for a Ka-band klystron providing 16 MW of peak power.

1. Introduction

Recent developments in accelerator science require higher power at higher frequencies. At the state of the art no stand-alone power sources can respond to these requirements, with the exception of systems composed by large combination of several power sources and pulse compression devices like SLEDs and corrugated waveguide structures.

In this paper we develop the design of a Ka-band klystron amplifier with an efficiency of 44% providing 10.6 MW radiofrequency output power. The paper is organized as follows: section 1 reports the electron gun injector design and magnetostatic simulation performed with CST Particle Studio [5] , section 2 discusses the interaction structure which receives a beam current of 50 A and 480 kV with a beam radius of 1 mm confined in a drift tube of 1.2 mm radius. For the RF beam dynamics, the software KLyC [6] has been used. Much effort has been focused to obtain a sufficiently short structure suitable to accommodate a considerably narrow focusing magnet and to obtain the perveance needed for the required beam power output ensuring the maximum efficiency. In the section 3 we discusses the multiphysics analysis of a high- power microwave window for a Ka-band klystron providing 16 MW of peak power.

2. Electron gun injector design and magnetostatic simulation

The beam trajectory and the electric field equipotential lines are shown in figures 1a and 1b. The cathode-anodes geometry was optimized to adjust the electric field equipotential lines to

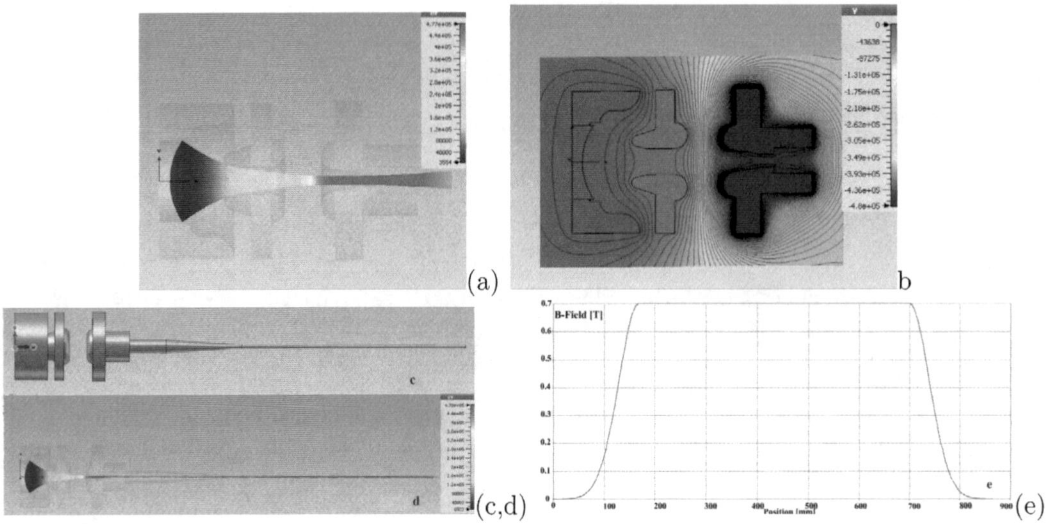

Figure 1. a) The dual anodes electron gun design from CST for a beam current of 50 A and the cathode-anode voltage of 480 kV. a) Beam trajectory b) equipotential lines c) 3D model of the gun d) beam trajectory along the propagation direction e) the axial magnetic field distribution.

obtain a beam current extraction of 50A. Figures 1c, d and 1e show the 3D model of the gun and of the beam pipe, the beam trajectory along the propagation direction and the axial magnetic field distribution, respectively.

The dual anode structure with an intermediate electrode allows us to overcome the strong space- charge force on the cathode and improve the transverse focusing [7], thus getting a low perveance. As a result, with this configuration, we are able to reduce the magnetic field value down to the normal conducting regime to confine the beam propagation in a pipe of 1.2 mm. This solution can be compared with our previous solution [8] using a magnetic focusing system working in superconducting regime. The magnetic field needed to compress the beam is reduced from 3.2 T to 0.7 T with this type of electron gun. The micro-perveance of this configuration is 0.15 $AV^{-3/2}$ which is a factor 2 less than the previous design [8]. The advantage of designing a low perveance electron gun is to get a weaker space charge, and, consequently, a stronger bunching. The maximum electric field on the focusing electrode is about 210 kV/cm almost the same of the previous configuration with standard anode. The design parameters of the gun with the focusing magnetic field along the beam axis for the beam current of 50 A are listed in table 1.

3. Interaction structure design

The numerical electromagnetic modeling has been set-up and performed based on the 1.5D large signal code KlyC [6] by adding the desired gain, bunch and output cavities. The layout of the interaction structure is shown in figure 2a.

Our proposed structure is composed by 9 cavities where the first three cavities and the fifth one operate in the first harmonic of the input signal while the fourth one operates on the second harmonic and the last four cavities are a system of coupled cavities operating at the third harmonic of the drive frequency (f=12 GHz).

The cavity 4 is designed following the principle of the core oscillation method (COM) that is based on the non-monotonic bunching technique, where, along the length of the klystron, electrons at the periphery of the bunch gradually approach the bunch center. Simultaneously,

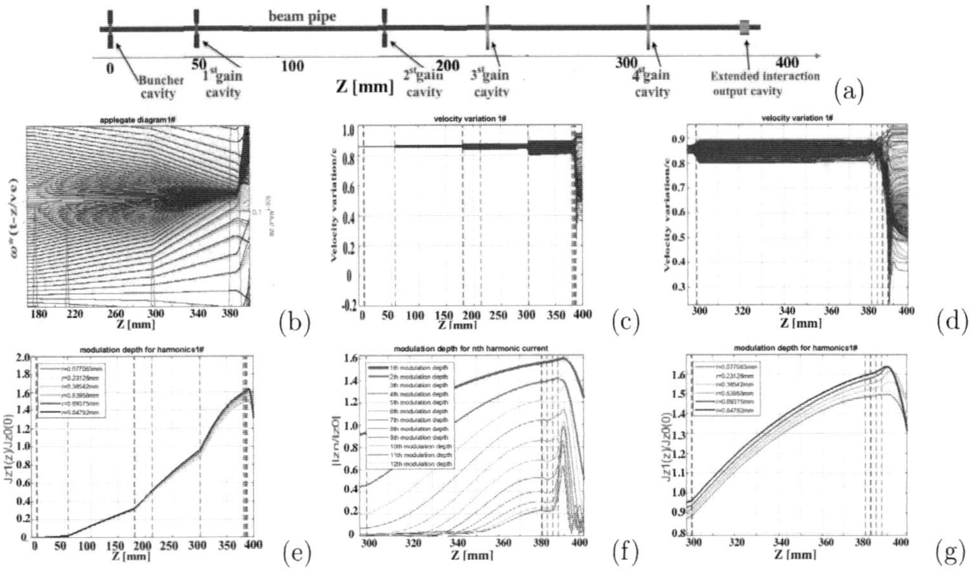

Figure 2. a) Layout of the interaction structure for the klystron with 50 A beam, b) the phase grouping, c and d) velocity variation of klystron with 50A, e) modulation depth of the 1° harmonic current at different beam substrates f) modulation currents normalized to the DC current, J_{z0} as function of the longitudinal coordinate z and g) modulation depth of the 1st harmonic current.

Table 1. Design parameters of the gun with the focusing magnetic field along the beam axis.

Design parameters	
Beam power [MW]	24
Beam voltage [kV]	480
Beam current [A]	50
Micro-perveance [I/V$^{3/2}$]	0.15
Cathode diameter [mm]	45.5
Pulse duration, min-max range [µs]	0.1- 0.2
Repetition rate [Hz]	10
Minimum beam radius in magnetic system [mm]	0.82
Nominal radius [mm]	0.925
Max EF on focusing electrode [kV/cm]	210
Electrostatic compression ratio	106:1
Beam compression ratio	697:1
Emission cathode current density [A/cm^2]	2.67
Transverse Emittance of the beam [mrad-cm]	0.63π

the core of the bunch experiences an oscillation in its phase, due to the space charge that causes it to expand, and the momentum delivered by successive bunching cavities that causes it to contract. The cavity RF fields have a weak effect on the periphery electrons (phase of $\pm\pi$ with respect to the core); as a result, COM klystrons require a substantial increase in their interaction length to capture all electrons in the bunch (a full saturated bunch) [9].

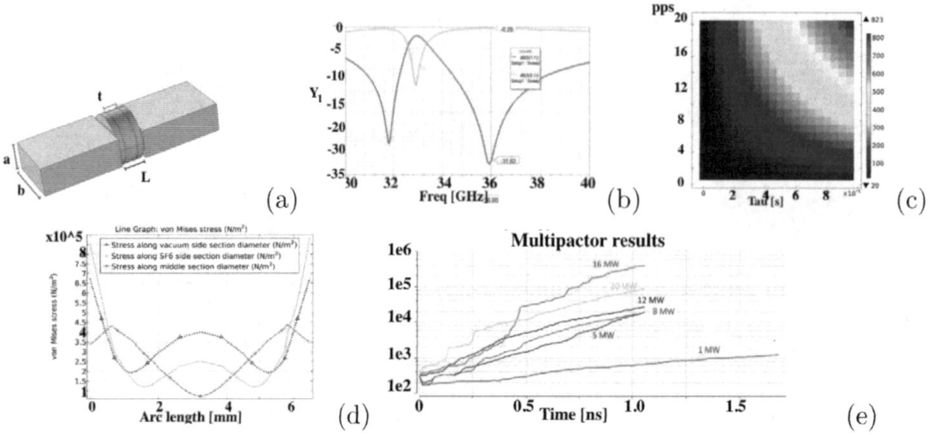

Figure 3. a) Pill-box window, b) scattering parameters of the pill-box window are plotted respect to the first propagating mode, c) steady state average Temperature (K) chart of the window, d) Von Misses Stress on the outer surfaces diameters and e) multipactor results.

The output circuit is an extended interaction output cavity that integrates four gaps. Extended interaction cavities can distribute the necessary retarding RF voltage at an output cavity over several gaps, thereby reducing the RF gradient and the danger of RF breakdown [9].

The phase grouping is shown in figure 2b. The electron beam has a considerably reduced energy, down to a velocity of -0.09 c along the z direction to achieve a high efficiency. The velocity variations are shown in figures 2c and 2d. In order to avoid reflected electrons, a minimal velocity, below -0.1c, has been considered in the output cavities.

The modulation depth of the 1st harmonic current at different beam substrates is shown in figure 2e. In this figure, J_{z1} is the 1st harmonic current normalized to the DC current, J_{z0}. The 1st harmonic content has a specific distribution on different substrates of the beam. It has been observed that, with respect to reentrant cavities, in pill-box cavities the gap electric fields for different radial positions exhibit reduced intensity spread. In order to exploit this feature and to have a reduced gain, needed for the last stage, the last gain cavity is a low R/Q pill-box cavity.

From figure 2f, it can be observed that the superior harmonic content grows while the signal is sufficiently compressed; this operation is demanded mostly to the last gain cavity. By observing the current distribution, the contribution to the superior harmonic content is mostly given by the inner part of the beam (the inner beam currents have the same behavior of harmonic currents), while the outer one behaves like the 1° harmonic. Hence, the inner substrate of the beam should be richer in superior harmonic content and the outer beam substrate poorer (see figure 2g).

The output power of 10.6 MW and a total efficiency (considering losses) of 44% are obtained with an excellent minimal velocity around 0.2c that ensures no electron reflections and a maximum electric field less than about 134 kV/mm.The electric and geometric parameters of the proposed klystron are listed in table 2.

4. Design of a window for the Ka-band klystron

We presented a complete multiphysics design approach to design a window for the Ka-band klystron providing a peak power of 16MW [10] with an operating safety margin compared to 10.6 MW klystron output power. The proposed window is a pill-box type shown in figure 3a. The outer waveguides are WR28 type, while the inner one is a circular section where the ceramic window is brazed. After the optimization of the electromagnetic performances, we analyzed the effect of RF heating effect and the stress of the pressure on the window.

Table 2. Electric and geometric parameters of the proposed klystron.

CN*	HN**	F_0 [MHz]	R/Q [Ω]	M	Q_e	Q_{in}	z [mm]
1	1	12000	111	0.9597	30	4957	0.000
2	1	12038	111	0.9597	100000	4957	56.969
3	1	12250	111	0.9597	100000	4957	180.115
4	2	23790	27	0.8868	100000	1956	212.859
5	1	12095	33	0.9567	100000	2772	300.219
6	3	35255	117	0.6717	100000	4059	380.000
7	3	35255	117	0.6717	100000	4059	382.500
8	3	35255	117	0.6717	100000	4059	385.000
9	3	35255	117	0.6717	47	4059	387.500

* Cavity number.
** Harmonic number.

The scattering parameters of the pill-box window are shown in figure 3b. In figiure 3c we demonstrated that the average temperatures reached by the window depend on pulse per second (pps) and pulse width (τ) with 16 MW of peak power, with operational mode TE_{10}-TE_{11}. Figure 3d shows the stress along the diameters of the vacuum-side, air-side surfaces and of the middle section of the cylindrical window. Multipactor problems have also been analyzed (see figure 3e) and mitigated, since this phenomenon is often cause of window failures.

The design reported in this work shows that the microwave window stimulated by a pulse of 200 ns and 16 MW of peak power does not show a significant increase in temperature and it is possible to avoid the harmful phenomenon of multipactor by TiN Coating of the microwave window.

5. Conclusion

In this paper, we presented a design of a low perveance dual anode electron gun for a high efficiency klystron in order to feed a phase space linearizer operating on the third harmonic of the main linac frequency. The simulation has been carried out with CST Particle Studio. The RF beam dynamics has been discussed for this configuration. Finally a multiphysics design approach has been considered for a window of a Ka-Band klystron providing 16 MW of peak power.

Acknowledgements

This work was partially supported by the Compact Light XLS Project, funded by the European Union's Horizon 2020 research and innovation program under grant agreement No. 777431 and was also partially funded by ARYA and MICRON projects of the Vth Committee of the INFN.

References

[1] D'Auria G, Cross A, Zhang L and Nix L 2021 Conceptual Design Report of the CompactLight X-ray FEL
[2] Behtouei M, Faillace L, Ferrario M, Spataro B and Variola A 2020 *J. Phys.: Conf. Series* **1596** 012021
[3] Spataro B, Behtouei M, Faillace L, Variola A, Dolgashev V, Rosenzweig J, Torrisi G and Migliorati M 2021 *Nucl. Instrum. Methods Phys. Res., Sect.* A **1013** 165643
[4] Behtouei M, Faillace L, Spataro B, Variola A and Migliorati M 2020 *Nucl. Instrum. Methods Phys. Res., Sect.* A **984** 164653
[5] 2020 CST Studio Suite solvers, electromagnetic simulation and compatibility, cad
[6] Cai J and Syratchev I 2019 *IEEE Trans. Plasma Sci.* **47** 1734–41
[7] Shang F, Shang L and Li J 2016 The measurement system of the electron gun with double-anode structure *Proc. 7th Int. Particle Accelerator Conf. (IPAC'16)* (Busan, Korea) pp 3954–6

[8] Behtouei M, Spataro B, Di Paolo F and Leggieri A 2021 *Vacuum* **191** 110377
[9] Caryotakis G 2004 High Power Klystrons: Theory and Practice at the Stanford Linear Accelerator CenterPart I Tech. rep. SLAC National Accelerator Laboratory, Menlo Park, CA, USA
[10] Marrese F, Valletti L, Fantauzzi S, Leggieri A, Behtouei M, Spataro B and Di Paolo F 2022 *JOMe* **21** 157–70

Photocathode performance characterisation of ultra-thin MgO films on polycrystalline copper

C Benjamin[1,2], H M Churn[2,3], T J Rehaag[1], L B Jones[2,3], G R Bell[1] and T C Q Noakes[2,3]

[1] University of Warwick, Coventry, United Kingdom
[2] ASTeC, STFC Daresbury Laboratory, United Kingdom
[3] Cockcroft Institute of Accelerator Science, Daresbury, United Kingdom

E-mail: christopher.benjamin@warwick.ac.uk

Abstract. The performance expected from the next generation of electron accelerators is driving research into photocathode technology as this fundamentally limits the achievable beam quality. The performance characteristics of a photocathode are most notably: normalised emittance, brightness and energy spread. Ultra–thin oxide films on metal substrates have been shown to lower the work function (WF) of the surface, enhancing commonly utilised metal photocathodes, potentially improving lifetime and performance characteristics.

We present the characterisation of two MgO/Cu photocathodes grown at Daresbury. The surface properties such as: surface roughness, elemental composition and WF, have been studied using atomic force microscopy (AFM), X-ray photoelectron spectroscopy (XPS) and ultraviolet photoelectron spectroscopy (UPS). The photoemissive properties have been characterised with quantum efficiency (QE) measurements at 266 nm. Additionally, we measure the Transverse Energy Distribution Curves (TEDC) for these photocathodes under illumination at various wavelengths using ASTeC's Transverse Energy Spread Spectrometer (TESS) and extract the Mean Transverse Energy (MTE).

1. Introduction

The development and improvement of the next generation electron sources require photocathodes with a high quantum efficiency (QE), low emittance, fast response times and robustness. The photoemissive properties of photocathodes are governed by its surface characteristics. The surface roughness and the work function (WF) strongly influence the quantum efficiency and intrinsic emittance. Metal photocathodes are predominantly used due to their high durability and fast response time. However, their relatively high WF leads to a low QE and the requirement for UV drive lasers [1].

The mean transverse energy (MTE) is a photoemissive property which is directly dependent on the excess energy in the photoemission process, where the excess energy is the difference between the photon energy and the WF of the sample. The MTE can be approximated using the model derived by Dowell and Schmerge [2]:

$$MTE = \frac{1}{3}\left(h\omega - \phi\right) \tag{1}$$

where $h\omega$ is the incident photon energy and ϕ is the WF of the photocathode, which was measured using ultraviolet photolectron spectroscopy (UPS) (see table 1).

One possible method of improving the performance of photocathodes is by the use of surface preparation or treatment, lowering the surface WF and increasing the QE. Surface dielectric films on metals have been shown to produce a surface with a lower WF [3] and previous works [4, 5] on MgO thin films on Ag(100) have demonstrated this reduction in WF and the potential for QE enhancement. Therefore, MgO films have the potential to improve the QE, while simultaneously increasing the robustness of the sample due to the chemical stability of MgO forming an overlayer to protect from residual gases in the system.

2. Experimental details

2.1. Sample preparation

Two 6 mm diameter polycrystalline copper cathodes, with a surface roughness of $R_a < 30$nm, (labeled DaCB–7 and DaCB–14) were supplied by Surface Preparation Laboratory (SPL). Cathodes were cleaned *in situ* using cycles of Ar^+ bombardment (2 keV beam energy for 20 mins) and annealing at 500 °C, with surface contamination being assessed using X-ray photoelectron spectroscopy (XPS). Once clean, an ultra-thin MgO film layer was deposited by thermal evaporation of Mg in a Chell K-Cell Miniature Knudsen Evaporation Cell in an O_2 partial pressure of 5×10^{-7} mbar. Sample preparation, QE, XPS and ultraviolet photoelectron spectroscopy (UPS) were all conducted within the same UHV system with a base pressure of 3×10^{-9} mbar.

2.2. Surface characterisation

XPS spectra were obtained using an non-monochromated Al K_α X-ray source (1486.7 eV) and a Thermo Alpha 110 analyser. The analyser transmission function was determined experimentally using the technique described by Ruffiuex et al. [6], and the WF of the analyser was measured using the Fermi edge of Ag. Core region spectra were acquired with a pass energy of 20 eV. Analysis was conducted using CasaXPS [7].

The surface roughness of DaCB–7 was measured using atomic force microscopy (AFM) conducted at the University of Warwick using a Bruker dimension icon AFM after all other work was completed. AFM scans over 10 μm × 10 μm and smaller were repeated over two locations on the surface. The photocathode was transported under ultra high vacuum conditions, exposed to maximum pressure of 5×10^{-10} mbar during transport. This reduced surface contamination and formation of hydroxides and carbonates.

QE measurements used a Crylas FQSS Q4 266 nm, 1 kHz pulsed laser source coupled with a ×3 beam expander and 2 mm circular aperture, and then a 2.0 OD reflective filter yielding an optical power of 19.2 μW illumination on the sample. A HV extraction electrode was placed close to the sample and a lock-in amplifier was used to measure the total yield photocurrent, thus ensuring any DC current from external source is not sampled and noise is reduced.

2.3. Energy spread measurement

Samples were transferred to the transverse energy spread spectrometer (TESS) under ultra high vacuum (UHV) conditions ($< 10^{-10}$ mbar). TESS captures the photoemission footprint of a photocathode when illuminated by various wavelengths of of light [8]. The mean transverse energy (MTE) is then extracted from its radial distribution [9].

3. Experimental results

Both polycrystalline copper substrates underwent the same surface preparation and *in situ* cleaning. Sample preparation was continued to the point at which XPS showed no surface contamination from adventitious carbon or oxidation. Following this, the QE of each substrate was measured to be 1×10^{-6} and 8×10^{-7} for DaCB–7 and DaCB–14 respectively.

Table 1. QE and WF of Substrate and Thin Film Sample, the error for the WF is ± 0.1 eV.

Sample	Substrate WF (eV)	QE	MgO Film WF (eV)	QE
DaCB–7	4.9	1×10^{-6}	3.7	7.0×10^{-5}
DaCB–14	5.0	8×10^{-7}	3.2	4.5×10^{-5}

Previously reported copper QE values are significantly higher [1]. The WF of both samples (table 1) were measured and fall within reference Cu values [10], albeit on the high end. The photon energy at 266 nm is 4.66 eV, lower than the measured WF of both Cu substrates. Consequently, a low QE is expected due to the Fermi distribution at room temperature.

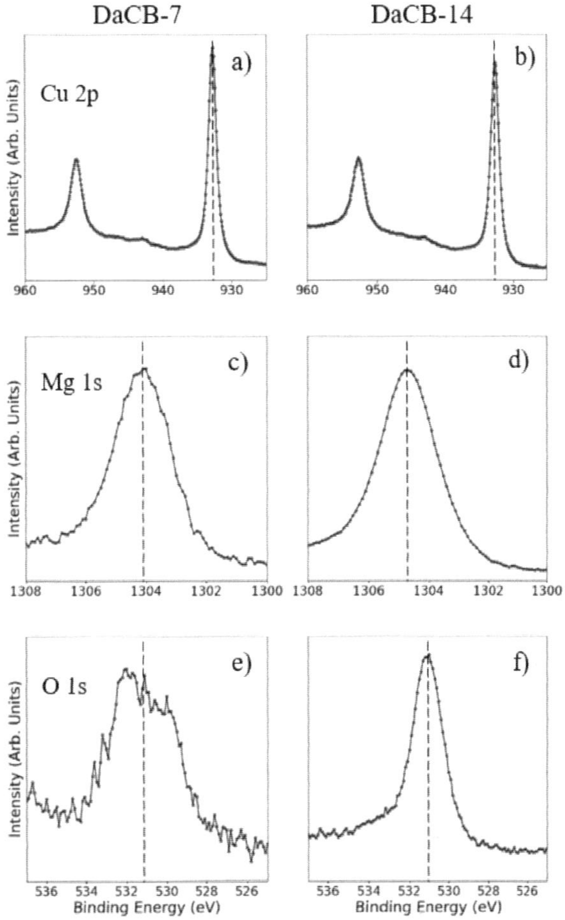

Figure 1. XPS core region scans of DaCB–7 (a, c and d) and DaCB–14 (b, d and f) following ultra-thin film deposition. XPS spectra were acquired for the core regions; Cu 2p (a, b), Mg 1s (c, d) and O 1s (e, f), showing a ultra-thin film MgO layer on the Cu substrate.

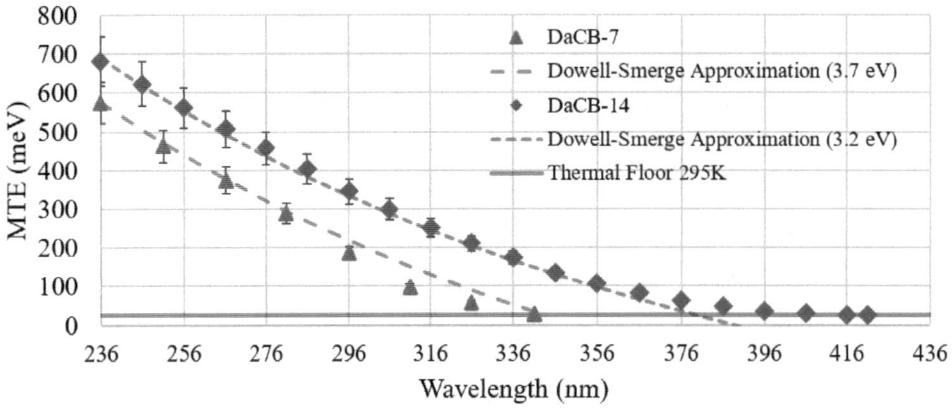

Figure 2. MTE measurements for DaCB–7 and DaCB–14 photocathodes. Dashed lines show the MTE predicted by equation (1). The MTE thermal floor is defined by k_bT at room temperature is shown by the solid red line.

Figure 1 shows the post-deposition surface elemental composition of both photocathodes, DaCB–7 and DaCB–14. In regards to DaCB–7, the Cu $2p_{3/2}$ (figure 1a) has a peak position of 932.6 eV and a FWHM of 1.3 eV, suggesting a clean copper surface free from any significant oxidation from the O_2 exposure during deposition. The lineshape of the O 1s (figure 1e) suggests two components contributing to the core region spectrum; MgO and Cu_2O. The MgO film is stoichiometric as there is no observable peak at the binding energy (BE) 1303 eV in the Mg 1s spectrum, figure 1c. Using the method outlined by Cumpson (2000) [11], we estimate the MgO film thickness to be 0.24 nm ± 0.02 nm.

Similar features are also observed in the DaCB–14 XPS spectra. The copper substrate has remained relatively oxide-free following deposition and no observable Mg metallic component is present in the Mg 1s, figure 1d. A slight asymmetry in the O 1s shown in figure 1f once again suggesting a trace amount of Cu_2O, although relatively this is significantly less prominent than for the previous sample. We estimate the film thickness to be thicker at 0.71 nm ± 0.07 nm.

Figure 2 shows the measured MTE for both DaCB–7 and DaCB–14 photocathodes under different illumination wavelengths. All data points have an experimental error of ± 10 %. Overall, the data shows the direct dependence of the MTE on the excess energy as predicted by Dowell [2]. The plotted lines represent the MTE approximation, given in equation (1).

At 266 nm the MTE was 374 meV and 507 meV for DaCB–7 and DaCB–14 respectively. Previous work on copper photocathodes reported values of 137.2 meV and 162.6 meV for two different Cu polycrystalline samples [12].

At threshold emission both samples achieved a MTE value of 28 meV and 26 meV, commensurate with the minimum energy determined by the temperature of the system (k_bT), 25 meV at room temperature. The two photocathodes varied greatly at the photoemission threshold with a difference of 80 nm, reflecting the measured difference in WF.

Figure 3a shows surface topography over a 10 μm × 10 μm AFM scan. The distinctive step terrace array in the surface topography indicates film growth. We measured a RMS roughness of 4.1 nm over the entire surface, less than the quoted surface roughness at the time of the substrate manufacture by SPL. Figures 3b and 3c, show the forward and backward lateral force measurements (LFM). Faint lines can be observed at the steps terraces, as well as a distinct 'snakelike' contrast.

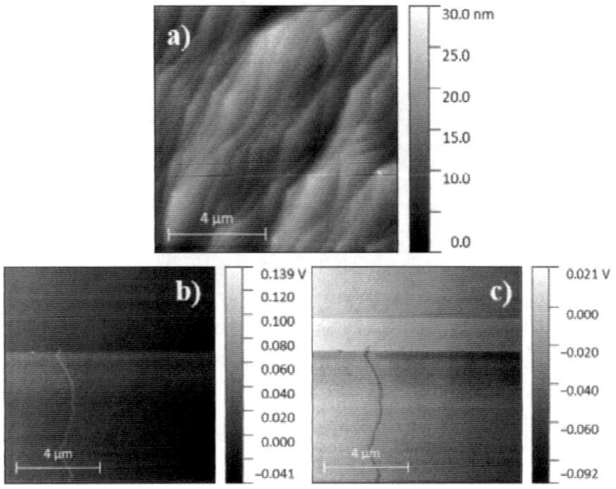

Figure 3. $10\,\mu\text{m} \times 10\,\mu\text{m}$ AFM and LFM scans of DaCB–7 photocathode: a) AFM height showing step terrace features; b) LFM image of forward scan; c) LFM image of backward scan showing minimal friction contrast.

4. Discussion

Relative to the Cu substrates, the deposition of the ultra-thin MgO film delivers a substantial improvement in QE at 266 nm. This can be related to the reduction in the surface WF by -1.2 eV and -1.8 eV for DaCB–7 and DaCB–14 respectively. The MTE agrees strongly with the MTE approximation discussed above [2] showing a direct dependence on the excess energy defined by equation (1) in the photoemitted electrons.

Variations between the photocathodes in their MTE and WFs can be attributed to the thickness and quality of the film. The thinner film, DaCB–7, has a smaller relative shift in the work function when compared to the thicker DaCB–14. With reference to figure 1, the XPS suggests Cu_2O surface contamination on DaCB–7 which would greatly affect the observed surface WF. Conversely, DaCB–7 exhibits a factor of 70 increase in QE relative to the bare substrate surface, compared to only 56 for DaCB–14. This difference in QE may be ascribed to the increased thickness of the MgO film on DaCB–14 attenuating the photoemitted electron yield.

As discussed above, at their respective threshold energies the photocathodes both approached the thermal limit of 25 meV. Based on the surface roughness measured for DaCB–7 of 4.1 nm RMS, we conclude that the surface roughness is not a contributing factor for the MTE in DaCB–7. The same conclusion may also be drawn for DaCB–14 due to the comparative sample preparation and surface treatment.

5. Conclusion

In this paper we show the potential, predicted by Chang [4], of ultra-thin MgO film metal photocathodes. The photocathodes presented exhibit a WF reduction of more than 1 eV relative to the substrate and an increase in QE of more than an order of magnitude. Furthermore, their MTE precisely follows that predicted by Dowell and Schmerge [1]. This significant and measurable improvement in photocathode performance and the potential for enhanced robustness and operational lifetime in an accelerator environment, highlights the use of MgO films as a candidate for future photocathode applications.

References

[1] Dowell D, Bazarov I, Dunham B, Harkay K, Hernandez-Garcia C, Legg R, Padmore H, Rao T, Smedley J and Wan W 2010 *Nucl. Instrum. Methods Phys. Res., Sect.* A **622** 685–97
[2] Dowell D H and Schmerge J F 2009 *Phys. Rev. Spec. Top. Accel Beams* **12** 074201
[3] Prada S, Martinez U and Pacchioni G *Phys. Rev.* B **78** 235423
[4] Chang V, Noakes T C Q and Harrison N M *Phys. Rev.* B **97** 155436
[5] Jaouen T, Jézéquel G, Delhaye G, Lépine B, Turban P and Schieffer P *Appl. Phys. Lett.* **97** 232104
[6] Ruffieux P, Schwaller P, Gröning O, Schlapbach L, Gröning P, Herd Q C, Funnemann D and Westermann J *Rev. Sci. Instrum.* **71** 3634
[7] Fairley N, Fernandez V, Richard-Plouet M, Guillot-Deudon C, Walton J, Smith E, Flahaut D, Greiner M, Biesinger M, Tougaard S, Morgan D and Baltrusaitis J *Appl. Surf. Sci. Adv.* **5** 100112
[8] Jones L B, Cash R, Fell B, Middleman K, Militsyn B, Noakes T, Gorshkov D, Scheibler H and Terekhov A 2013 *Proc. 35th Int. Free-Electron Laser Conf. (FEL2013)* (New York, USA) pp 290–3
[9] Jones L B, Scheibler H E, Gorshkov D V, Terekhov A S, Militsyn B L and Noakes T C Q *J. Appl. Phys.* **121** 225703
[10] Kawano H *Prog. Surf. Sci.* **83** 1–165
[11] Cumpson P J *Surf. Interface Anal.* **29** 403–6
[12] Jones L, Beaver T, Militsyn B, Mistry S, Noakes T and Valizadeh R 2017 *Proc. 8th Int. Particle Accelerator Conf. (IPAC'17)* (Copenhagen, Denmark) pp 1580–3

Material normal energy distribution for field emission analyses from monocrystalline surfaces

J I Mann[1], Y Li[1], J K Nangoi[2], T Arias[2] and J B Rosenzweig[1]

[1] Department of Physics and Astronomy, UCLA, Los Angeles, CA 90024, USA
[2] Department of Physics, Cornell University, Ithaca, NY 14850, USA

E-mail: jomann@physics.ucla.edu

Abstract. Electron field emission is a complicated phenomenon which is sensitive not only to the particular material under illumination but also to the specific crystalline orientation of the surface. Summarizing the ability for a crystal to emit in a particular direction would be of great use when searching for good field emitters. In this paper we propose a material normal energy distribution which describes the ability of the bound electrons to tunnel under an intense electric field. This framework breaks a computationally expensive 3-D system down to a source distribution representation applicable for more efficient 1-D models. We use the Fowler-Nordheim framework to study the yield and MTE (mean transverse energy) from sources including gold, copper, and tungsten in both monocrystalline and polycrystalline forms. We find an increase in effective work function for field emission in the (111) direction for gold and copper associated with the Bragg plane intersections of the Fermi surface.

1. Introduction

Electrons are commonly sourced through field emission. Nanotips, for instance, are used in a static field emission setup for electron microscopy [1]. Field emission is also the culprit of dark current [2] and may lead to breakdown [3] in RF cavities. Strong laser field emission is a growing topic of interest where a nanostructured cathode is used to enhance laser fields to field-emitting regimes [4].

Field emission is typically analyzed in 1-D using the Fowler-Nordheim (FN) framework [5] where each electron impinging on the surface has a probability of tunneling and emitting. This probability, for a transversely isotropic surface potential, is invariable on the transverse momentum and is therefore only dependent on the normal energy, summarized by the vacuum motive energy $M(x) = U(x) - E_\mathrm{n}$ with U the vacuum potential and E_n the energy directed normal to the surface. The motive energy is integrated within the classically forbidden region to determine the tunneling probability. It is thusly important to understand the distribution of source current within the material with respect to the normal energy when finding the field emission current.

The FN model for field emission is typically considered using an isotropic free electron gas (FEG) scheme. In this case, there is a constant current density in energy space, permitting simple analytical results in the limits of low or high field strength [5]. However, we do not expect all crystalline surfaces to behave like a FEG in field emission. We thus propose a material normal energy distribution (MNED) which embodies the electron density or normal current density as a function of the energy directed towards the surface E_n.

The MNED for the FEG at zero temperature is found by integrating over slices of the Fermi sphere,

$$\frac{dj}{dT_{\mathrm{n}}} = \frac{E_{\mathrm{f}} - T_{\mathrm{n}}}{2\pi^2} = -\frac{1}{2\pi^2}E_{\mathrm{n}} \tag{1}$$

where $T_{\mathrm{n}} \in [0, E_{\mathrm{f}}]$ is the normal-directed kinetic energy, E_{f} is the Fermi energy of the material, and E_{n} is the normal energy relative to the Fermi level. Equations (1) and (2) are in Hartree atomic units (a.u.) with $\hbar = m_{\mathrm{e}} = -q_{\mathrm{e}} = 1$.

The MTE (mean transverse energy) as a function of the normal energy is found to be,

$$\mathrm{MTE}(T_{\mathrm{n}}) = \frac{1}{2}(E_{\mathrm{f}} - T_{\mathrm{n}}) = -\frac{1}{2}E_{\mathrm{n}} \tag{2}$$

and it is conserved in our model between bulk and vacuum due to transverse momentum conservation.

2. Methods

We use density-functional theory (DFT) to determine the MNED of real materials through two approaches: slab and bulk calculations. The former is used as a direct, but coarse, calculation of the MNED so that we may verify our bulk calculation results, which attain a much higher resolution by employing some basic assumptions for an indirect approach. This verification is done only for Au (100) and (111).

We use the code JDFTx [6] for all DFT calculations. We use SG15 norm-conserving pseudopotentials [7] for all ions and the Perdew-Burke-Ernzerhof GGA exchange-correlation (XC) functional [8].

2.1. Slab system analysis

The Au (100) and (111) slab systems consist of 15 layers where the outermost 6 layers' positions are optimized. The vacuum separation between the periodic slabs' outermost layers is 46.26 a.u. in length, about 6 lattice constants, before optimization. This slab system allows us to directly calculate the wavefunction decay rate in vacuum, κ, which may then be related to the normal energy (relative to the Fermi level) by $E_{\mathrm{n}} = -\frac{1}{2}\kappa^2 + W$ for bound electronic states. To find a single value for κ for each eigenstate we average the modulus of the wavefunction in the transverse dimensions. We fit an exponential to the wavefunction starting where the potential is within 0.8 eV of vacuum level and ending before the wavefunction's magnitude falls below the noise floor.

2.2. Bulk system analysis

The bulk calculations use a single unit cell with one atom. This is significantly cheaper computationally than the slab system and allows us to use a single DFT calculation for any surface orientation. We find that allowing the lattice to relax results in a roughly unchanged lattice constant and that the band structures produced match previous calculations [9, 10], indicating a good pseudopotential and XC functional choice for gold, copper, and tungsten.

For a given state with crystal momentum $\mathbf{k}$ we stipulate that the total energy E (relative to the Fermi level) and the transverse crystal momentum $\mathbf{k}_{\|}$ are conserved between bulk and vacuum, and we can therefore determine the decay rate of the wavefunction in vacuum simply with $\kappa = \sqrt{2(W - E) + k_{\|}^2}$. This then leads to the normal energy within the material $E_{\mathrm{n}} = T_{\mathrm{n}} - E_{\mathrm{f}} = E - \frac{1}{2}k_{\|}^2$.

The transverse momentum conservation is true up to a reciprocal lattice vector – the inclusion of larger momenta outside the first Brillouin zone (BZ) would, in general, lead to larger transverse kinetic energies and lower normal energies. However, states near the Fermi level, which contribute most to field emission, interact with the lattice weakly and therefore occupy these higher transverse modes less.

The normal current contribution of each state is taken to be $dj = \frac{2}{(2\pi)^3} v_{gn} d^3\mathbf{k}$, with $v_{gn} = \frac{\partial E}{\partial k_\perp}$ the group velocity projected along the surface normal and the factor of 2 accounting for spin degeneracy.

We find the distribution of current with respect to the normal energy by Monte-Carlo integration over the first BZ, using the Wannier interpolation method [11] to find the energy and group velocity. Samples with a negative normal group velocity are discarded as they would not contribute to the tunneling current. When finding the polycrystalline averaged MNED we use a random unit vector as the surface normal for each sample. The MTE is found by taking $\frac{1}{2}k_\parallel^2$ to be the transverse energy of each sample and an average, weighted by the samples' normal current contributions, is done within each normal energy bin. We use about 50 million samples within the first BZ.

3. Results

The Au MNED density is shown in figure 1. The surface and bulk calculations roughly agree with each other in shape, particularly considering that the uncertainty in normal energy for the

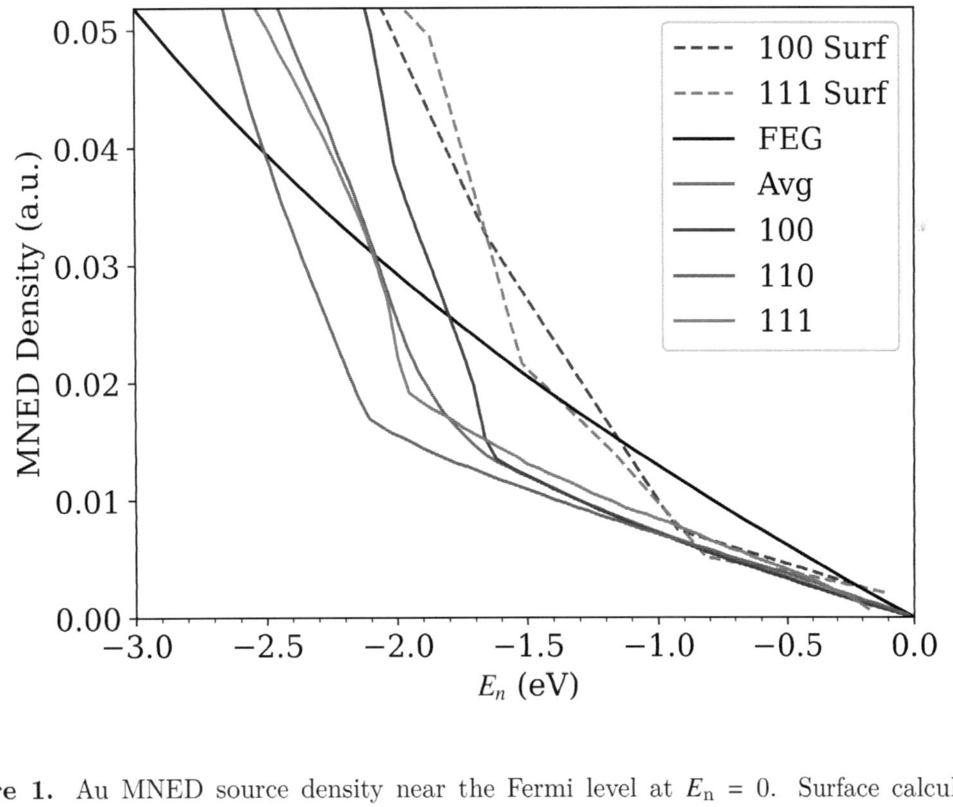

Figure 1. Au MNED source density near the Fermi level at $E_\mathrm{n} = 0$. Surface calculations (dashed) for the (100) (blue) and (111) (red) surfaces agree well with their bulk calculation (solid) counterparts, considering their 0.8 eV uncertainty in E_n. The polycrystalline average (green) and (110) surface (magenta) for from the bulk calculations are also included. All cases exhibit FEG-like qualities close to the Fermi level, albeit with a lower density.

surface calculations is 0.8 eV due to the potential's non-instantaneous transition to vacuum and our fitting scheme. The Au MNED current matches the FEG result surprisingly well (much like Cu in figure 2) while the MNED density does not. By this observation the normal group velocity, or the factor between current and density, cannot follow a simple FEG calculation. We thus have no method of calculating the MNED current for a slab system.

The Cu MNED current and MTE are shown in figure 2. The Cu (100), (110), and averaged surfaces agree well with the FEG predictions near the Fermi level. The Cu and Au (111) MNED currents strongly deviate from the FEG predictions close to the Fermi level. The maximum normal energy attained is less than the Fermi level and the minimum MTE is found to be non-zero due to the Bragg plane intersection of the Fermi surface for these materials in this direction, as we will discuss in the Discussion section.

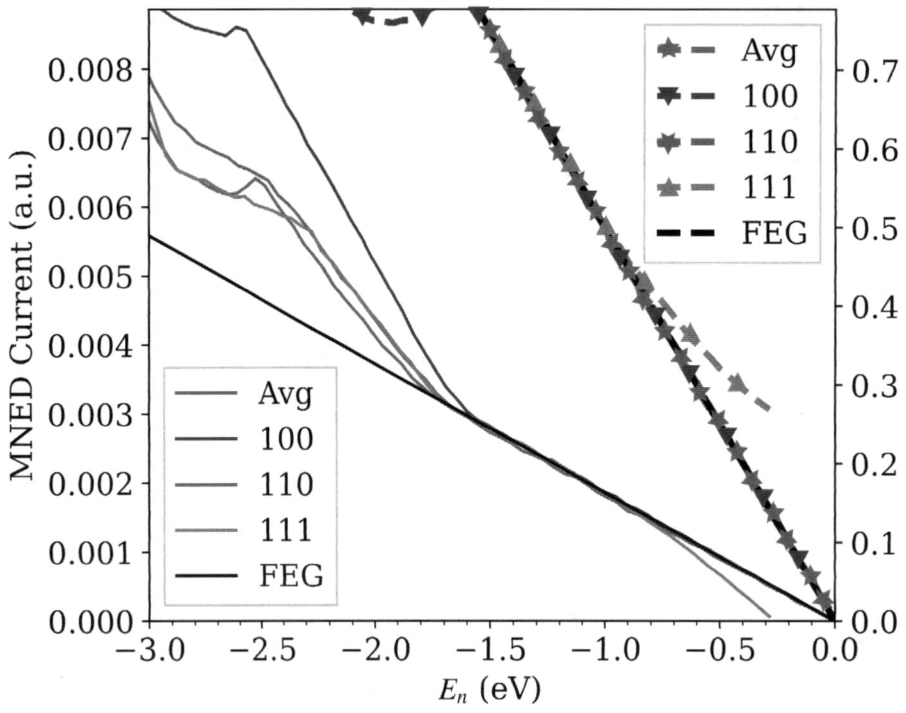

Figure 2. Cu MNED source current (solid) and MTE (dashed) near the Fermi level at $E_n = 0$. The polycrystalline (green), (100) (blue), and (110) (magenta) curves agree surprisingly well with the FEG (black) near the Fermi level. The (111) (red) result notably terminates 0.27 eV below the Fermi level and attains a minimum MTE of 270 meV due to the Fermi surface Bragg plane intersection in this direction.

The W MNED current and MTE are shown in figure 3. Tungsten behaves less like a FEG than gold and copper [12] and so we expect these results to deviate further, which is observed. Regardless, the distributions are roughly linear near the Fermi level and may be approximated by a modified FEG with larger currents, i.e. by multiplying equations (1) and (2) by a constant.

We find these modified FEG constants for the Au, Cu, and W results by fitting a line between the modified Fermi level (at the highest occupied normal energy) and where the linear trend seems to stop at lower normal energies. This data, as well as the increase in work function

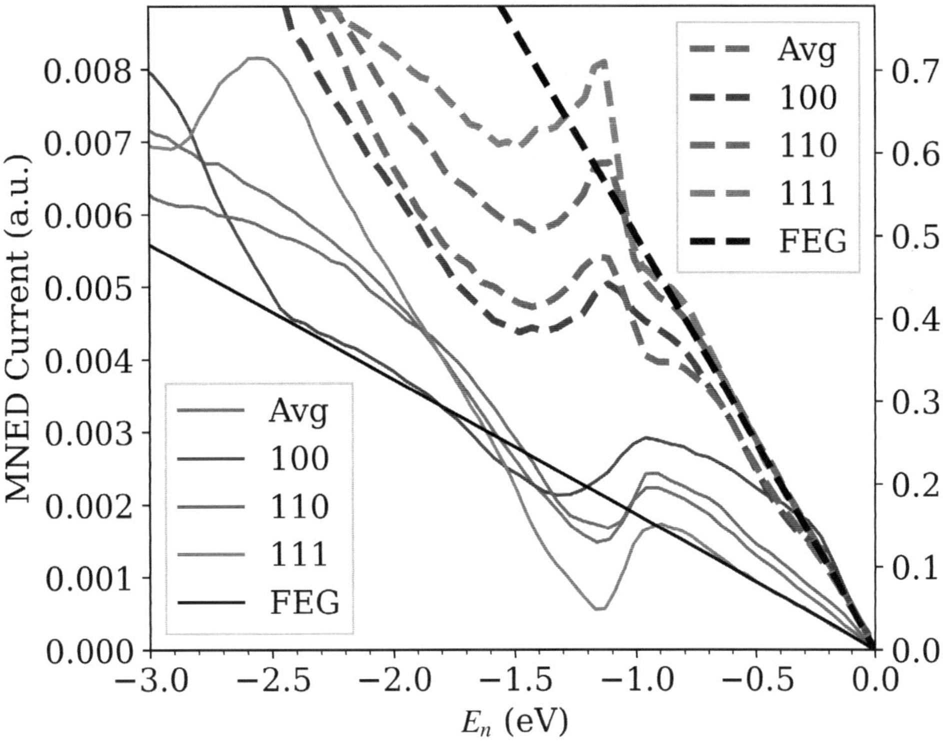

Figure 3. W MNED source current (solid) and MTE (dashed) near the Fermi level at $E_n = 0$. The polycrystalline (green), (100) (blue), and (110) (magenta) curves behave similarly to the FEG (black). The (111) (red) result agrees most with the FEG.

Table 1. Observed MNED current and MTE slopes relative to the slopes for the FEG in equations (1) and (2) respectively, with effective work function increase and minimum MTE contribution for each studied material surface. The fitting is done between the (modified) Fermi level and where the linear trend roughly ceases. "(Avg)" refers to the directionally averaged calculation for polycrystalline materials. For surfaces Au(100), Au(110), Au(Avg), Cu(100), Cu(110), and W(111) no significant deviation from the FEG was found. The W(100) and W(110) surfaces are only fit out to $E_n = -0.13$ eV.

Surface	**MNED Slope**	**MTE Slope**	ΔW (eV)	**Minimum MTE** (meV)
Au(111)	1.25	0.73	0.16	170
Cu(Avg)	0.98	1	-	-
Cu(111)	1.48	0.54	0.27	270
W (Avg)	1.42	1	-	-
W (100)	3.58	1.06	-	-
W (110)	1.92	0.97	-	-

(due to the modified normal energy Fermi level) and minimum MTE for Au and Cu (111), are summarized in table 1 for the cases where a deviation from the FEG was observed.

4. Discussion

The field emission work function anisotropy for Au and Cu (111) surfaces is perhaps the most notable result here. The Fermi surfaces of these metals intersect the (111) Bragg planes [12], which may be understood through the weak binding model. This intersection forms a disc where the total energy is less than the Fermi level everywhere except for the edge where the transverse momentum is non-zero. This indicates that the normal energy is also less than the Fermi level. This feature combined with our model results in an increased effective work function and MTE at and around the (111) surface normal.

Work function anisotropy is a well-studied phenomenon with various methods of measurement, oftentimes with different approaches resulting in different values [13]. It is understood to originate primarily due to the surface disturbance, and not any particular bulk-originating anisotropy [14]. Here we have found a possible bulk-originating work function increase on the order of 100's meV specifically for field emission. This effect compounds with the photoemission work function for the given surface and is comparable to the surface-originating anisotropies.

We otherwise find that, for orientations where there is no Bragg plane intersection, these materials behave akin to a FEG, albeit with more or less total current or MTE near the Fermi level. Of particular note is the large source current for the W(100) surface where we expect about 3.6 times the current we would otherwise get from a pure FEG calculation at low fields. This is quite substantial as, for instance, the choice of potential shape for a Fowler-Nordheim model results in a prefactor which may vary in magnitude by factors on the order of 2 [15], making this observed increase in current impactful for such studies.

5. Conclusion

We found that gold, copper, and tungsten generally abide well by the FEG model for applications in electron field emission. Some orientations of tungsten have a higher electron supply than calculated by a purely FEG calculation, namely the W(100) surface which has nearly 3.6 times the surface current. An orientation-averaged calculation in the spirit of polycrystalline studies also shows good FEG-like behavior as expected. We also find that, for Au and Cu (111) surfaces, the work function and intrinsic MTE are elevated due to the Fermi surfaces' Bragg plane intersections. These results will have a significant impact on the currents calculated using a Fowler-Nordheim model, illuminating the importance of studying even a metallic cathode's electronic distribution.

Future work includes surface calculations of tungsten as well as improved vacuum fitting methods and an increase in sample number by Wannier interpolation. Additionally, the bulk method may be bolstered by some analytical work which may reveal a simpler method of calculating the MNED and MTE slope near the Fermi level. Other common field-emitting materials including semi-conductors are also under consideration.

Acknowledgments

This research is supported by the Center for Bright Beams, U.S. National Science Foundation grant PHY-1549132.

References

[1] Crewe A V, Isaacson M and Johnson D 1969 *Rev. Sci. Instrum.* **40** 241–6
[2] Xiang R, Arnold A, Kamps T, Lu P, Michel P, Murcek P, Vennekate H, Staats G and Teichert J 2014 *Phys. Rev. Spec. Top. Accel Beams* **17** 043401
[3] Keser A C, Antonsen T M, Nusinovich G S, Kashyn D G and Jensen K L 2013 *Phys. Rev. Spec. Top. Accel Beams* **16** 092001

[4] Mann J I, Arias T A, Lawler G E, Nangoi J K and Rosenzweig J B 2021 Simulations of Nanoblade-Enhanced Laser-Induced Cathode Emissions and Analyses of Yield, MTE, and Brightness *Proc. 12th Int. Particle Accelerator Conf. (IPAC'21)* (Campinas, Brazil) pp 2957–60
[5] Fowler R H and Nordheim L 1928 *Proc. R. Soc. London, Ser.* A **119** 173–81
[6] Sundararaman R, Letchworth-Weaver K, Schwarz K, Gunceler D, Ozhabes Y and Arias T A 2017 *SoftwareX* **6** 278–84
[7] Hamann D R 2013 *Phys. Rev.* B **88** 085117
[8] Perdew J P, Burke K and Ernzerhof M 1996 *Phys. Rev. Lett.* **77** 3865–8
[9] Prandini G, Marrazzo A, Castelli I, Mounet N and Marzari N 2018 *npj Comput. Mater.* **4**
[10] Lejaeghere K *et al* 2016 *Science* **351** 3000
[11] Marzari N, Mostofi A A, Yates J R, Souza I and Vanderbilt D 2012 *Rev. Mod. Phys.* **84** 1419–75
[12] Gaĭdukov Y P 1970 *Soviet Physics Uspekhi* **13** 194–203
[13] Kawano H 2022 *Prog. Surf. Sci.* **97** 100583
[14] Smoluchowski R 1941 *Phys. Rev.* **60** 661–74
[15] Forbes R G 2008 *J. Appl. Phys.* **103** 114911

Journal of Physics: Conference Series **2420** (2023) 012034 doi:10.1088/1742-6596/2420/1/012034

Towards high-repetition rate petawatt laser experiments with cryogenic jets using a mechanical chopper system

M Rehwald[1,2], S Assenbaum[1,2], C Bernert[1,2], C B Curry[3,4], M Gauthier[3], S H Glenzer[3], S Göde[5], C Schoenwaelder[3,6], U Schramm[1,2], F Treffert[3,7] and K Zeil[1]

[1] Helmholtz-Zentrum Dresden - Rossendorf, Institute of Radiation Physics, Bautzner Landstr. 400, 01328 Dresden, Germany
[2] Technische Universität Dresden, 01062 Dresden, Germany
[3] High Energy Density Science Division, SLAC National Accelerator Laboratory, Menlo Park, California 94025, USA
[4] University of Alberta, Edmonton, Alberta T6G 1H9, Canada
[5] European XFEL GmbH, Holzkoppel 4, 22869 Schenefeld, Germany
[6] Friedrich-Alexander Universität Erlangen-Nürnberg, 91054 Erlangen, Germany
[7] Technische Universität Darmstadt, 64289 Darmstadt, Germany

E-mail: `m.rehwald@hzdr.de`

Abstract. Laser-plasma based ion accelerators require suitable high-repetition rate target systems that enable systematic studies at controlled plasma conditions and application-relevant particle flux. Self-refreshing, micrometer-sized cryogenic jets have proven to be an ideal target platform. Yet, operation of such systems in the harsh environmental conditions of high power laser induced plasma experiments have turned out to be challenging. Here we report on recent experiments deploying a cryogenic hydrogen jet as a source of pure proton beams generated with the PW-class ultrashort pulse laser DRACO. Damage to the jet target system during application of full energy laser shots was prevented by implementation of a mechanical chopper system interrupting the direct line of sight between the laser plasma interaction zone and the jet source.

1. Introduction

Laser-driven particle sources have been studied extensively over the last two decades [1]. The interest originates from unique beam properties that are useful for a number of applications ranging from ultrafast electromagnetic field probes [2, 3] and high flux neutron converter for material radiography [4] through isochoric heating of warm dense matter [5] and inertial confinement fusion [6] to injection sources for conventional accelerator structures [7] and medical applications [8–10]. Ongoing research in this field aims for the realization of a sufficiently high repetition rate which is needed to achieve application-relevant particle yields. Therefore target systems are required that are suitable for the challenges arising from the high laser shot rates. This includes target insertion and alignment as well as mitigation of debris produced by the evaporation of the target material during the shot potentially causing coating and damage of sensitive optical components. Recently developed renewable cryogenic jets allow for debris-free operation and rapid injection of fresh targets [11, 12]. No debris is produced as the material is gaseous at ambient temperatures and removed by the vacuum pumps.

The target system has to withstand the harsh environment of the high-power laser plasma interaction. When the laser pulse hits the target, energy is transferred to target electrons which can be accelerated to relativistic energies. This results in broad emission of radiation as the electrons recirculate inside the target and further interact with the laser light. The most energetic electrons are emitted from the target, building up charge separation fields, that lead to the acceleration of ions. The charge imbalance in the interaction region drives return currents in the target assembly, and the radiation pressure from the high intensity laser launches spherical shock waves which have a component along the jet axis. These combined effects of energy dissipation into vacuum (radiation, charged particles) or within the bulk of the target may not only cause problems with electronic systems (motors, diagnostics etc..) but can also damage the jet target system itself (the nozzle aperture in particular).

At laser powers in the 100 TW range, cryogenic hydrogen jet targets were successfully implemented and operated in a number of experiments [13–15]. Damage was mainly prevented by focusing the laser at a sufficient distance from the target source. This comes at the cost of increased shot-to-shot fluctuations due to the pointing jitter of the flowing jet. Extending the scope of applications of laser-driven beams demands higher particle energies that can only be realized with petawatt-class laser systems. This intuitively worsens the potential for damage. Besides further increasing the distance between target source and interaction point which would substantially decrease the hit rate, replacing the damaged part after every shot was the second option until now. Clearly, both options are incompatible with systematic studies with high repetition rates and therefore more advanced methods are needed.

In this paper, we report on implementation and demonstration of two new methods to shield the target system from damage caused by the high-power laser plasma interaction. First, the solid jet between the interaction point and the target system is interrupted by the help of a cutting laser pulse synchronized to main driver laser pulse. This prevents the energy transfer in the bulk of the target. The second approach is the use of mechanical cutting with a rotating blade completely blocking the line-of-sight between interaction zone and nozzle. Such chopper systems are an established technology in atomic and molecular gas jets as velocity filters. The increased protection requires temporal synchronization of the blade position. The mechanical cutting technique enabled systematic studies with PW-class laser systems and the jet target. This allows the proton acceleration performance to be evaluated at different experimental settings (e.g. different temporal laser pulse shapes) which benefits from the virtually unlimited number of shots possible with the self-replenishing jet target.

2. Experimental setup

The experimental setup for laser proton acceleration is illustrated in figure 1. The Titanium:Sapphire based laser system DRACO [16] is used to deliver linearly polarized pulses with up to 23 J energy on target and a duration of 30 fs yielding peak intensities of about $6.8 \cdot 10^{21}\,\mathrm{W/cm^2}$ at a focal spot size of 2.6 µm (FWHM). A solid hydrogen jet target [11, 12] was produced by injecting pure liquid hydrogen at a temperature of about 18 K through an aperture of 5 µm diameter into vacuum where evaporative cooling caused solidification. With a velocity of around 100 m/s, the jet is continuously refreshing. Time-resolved high resolution shadowgraphy images were recorded employing an off-harmonic optical probe laser system (515 nm wavelength) [17] to characterize and align the hydrogen jet. Laser-accelerated protons were recorded by a Thomson parabola spectrometer (TPS), implemented in laser forward direction. The TPS was equipped with a microchannel plate (MCP) containing a phosphor screen that was imaged onto a camera for on-line readout.

In the harsh environment of the high-power laser plasma interaction, radiation and charged

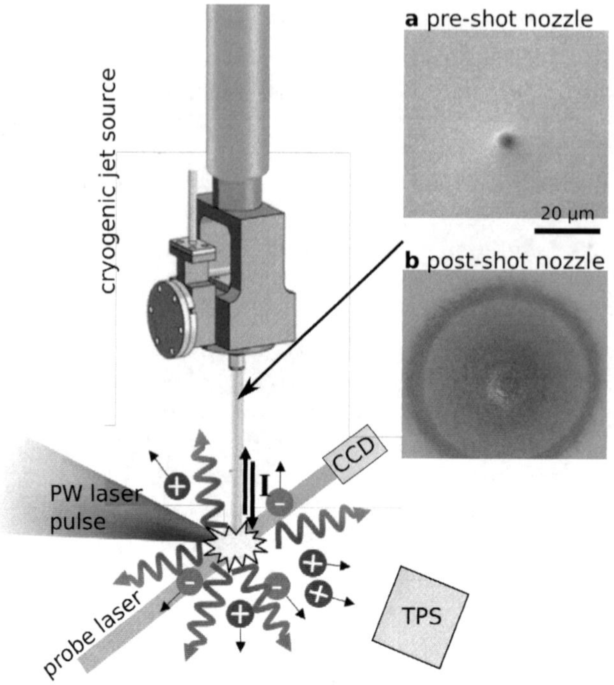

Figure 1. Schematic of the experimental setup: DRACO PW laser pulses (red) are focused onto a cylindrical cryogenic hydrogen jet target (light blue). A synchronized optical probe laser pulse with 515 nm wavelength (green) was used as a backlighter for shadowgraphy imaging. A Thomson parabola spectrometer (TPS) was positioned along the laser propagation direction for proton energy spectra measurement. Strong charged particles and electromagnetic radiation emission as well as laser-driven currents and shocks from the laser-plasma interaction impact the 5 μm nozzle aperture (a) and destroy it in a single laser shot (b).

particle emission as well as the energy transferred along the jet axis can damage the target system (see figure 1). The micrometer-sized aperture nozzle is hereby the most fragile part and by that prone to being damaged, as demonstrated by the microscopic images of a nozzle before and after a single laser shot without protection measures in figure 1a and b, respectively. The laminar flow of the liquid through the nozzle gets disturbed or even completely blocked and the jet operation stops. As such, only a very limited number of shots would be achievable when using this setup without additional protection. To mitigate the damaging effect, two concepts for shielding the aperture were implemented and tested. While laser cutting only disrupts the solid connection between the interaction point and the sensitive nozzle, mechanical cutting completely blocks the direct line-of-sight.

2.1. Laser cutting

A cutting laser beam is introduced into the experimental setup (figure 2a) to study if interruption of the cryogenic jet can prevent damaging of the nozzle. The cutting pulse was generated from a portion of the high-intensity beam and focused to a spot size of 50 μm. Irradiation with pulse energies E_{cut} of at least 0.3 mJ is sufficient for ionizing the solid hydrogen as the peak intensity of the laser pulse is estimated to be in the order of 10^{14} to 10^{15} W/cm^2 (depending on E_{cut}) and thus larger than the barrier-suppression ionization threshold for hydrogen atoms

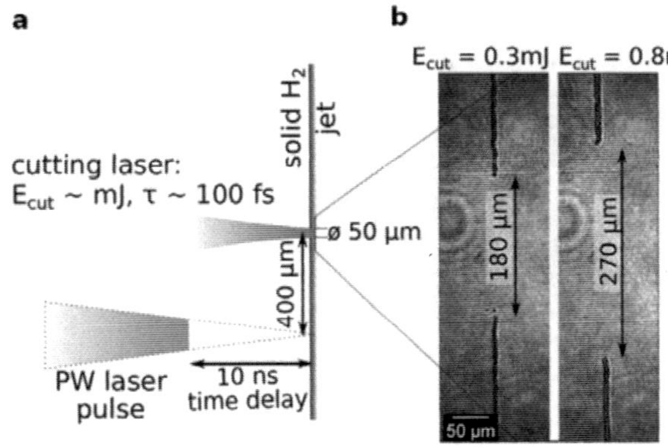

Figure 2. a) Arrangement of the cutting laser beam featuring an energy E_{cut} on the mJ-level, a pulse duration $\tau \approx 100\,\mathrm{fs}$ and a focal spot size of 50 μm. The cutting beam hits the jet target 400 μm above the focus position of the high-intensity laser with a time offset of several tens of nanoseconds. Optical probing images (b) indicate a few 100 μm wide gap in the solid jet 10 ns after the irradiation with the cutting pulse.

($1.37 \times 10^{14}\,\mathrm{W/cm^2}$) [18]. After a duration of 10 ns the solid hydrogen is completely vaporized on 180 μm length (see left image in figure 2b) and a gap in the jet is formed which represents the introduced interruption. This duration is sufficiently short with respect to the delay of several tens of nanoseconds between cutting and high-intensity laser pulse. Intuitively, an increase in pulse energy to $E_{\mathrm{cut}} = 0.8\,\mathrm{mJ}$ (right image in figure 2b) leads to a larger gap size as the ionization threshold intensity is exceeded within a larger area. It is important to note that the laser cutting does not result in any observable modifications of the solid jet in the region where the interaction of the high-intensity laser takes place with the target at about 0.4 mm below the cutting region. However, the application of laser cutting during high-power shots resulted in little improvement only as the jet did not stop immediately but became unstable due to accumulating damage to the aperture. The energy emitted by radiation and charged particles (traveling across the vacuum gap) is therefore assumed to be the main cause of nozzle damage and needs to be shielded mechanically. As a positive outcome for future studies, laser cutting enables precise and stable generation of gaps in the target, jet snippets which may be of particular interest for investigating mass limited targets [19].

2.2. Mechanical chopper

A second approach to mitigate the aperture damage is to utilize a mechanical cutting device, referred to as *mechanical chopper*, that intercepts the energy transfer along the jet and blocks the line-of-sight. Crucial for the error-free operation is a precise synchronization of the blade position in time and space with respect to the laser pulse arrival. The upper limit for the temporal synchronization is defined by the duration that the jet needs to travel the distance between the nozzle and the interaction point. For a typical distance of 10 mm and a flow velocity of $\approx 100\,\mathrm{m/s}$, a precision better than 100 μs is required. A more accurate synchronization, of course, allows for a smaller distance to the nozzle and thus improved target stability.

For the implementation in the experiment, a fast spinning blade is used, which is inserted between the aperture and the interaction point and which is driven by a direct current electric motor. The blade is made of few millimeters long, diamond-shaped metal pins (1 mm in size) on the outer edges of a center section made of plastic to isolate the motor from return currents. The rotation of the electric motor is synchronized to a 10 Hz laser trigger signal by a phase-

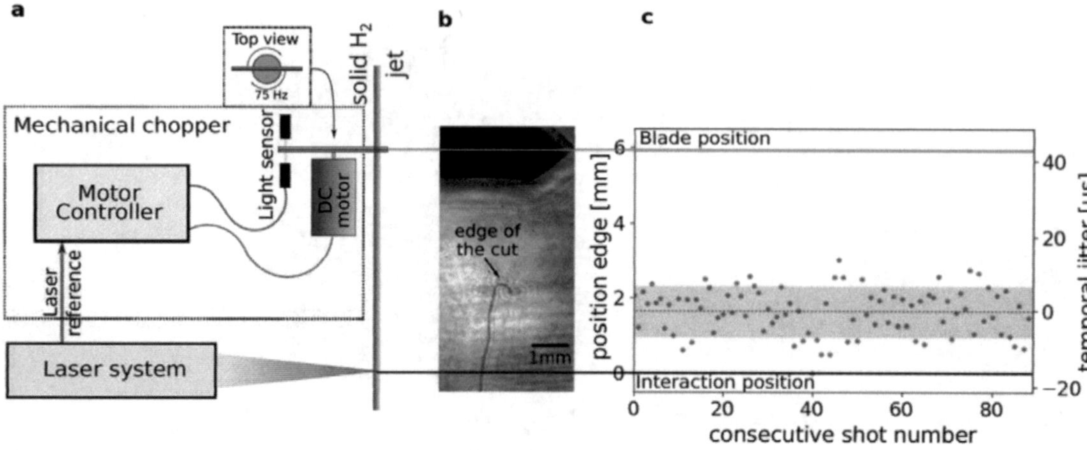

Figure 3. a) Schematic drawing of the components of the mechanical chopper (DC motor with a rotating blade, a light barrier, a motor controller and a laser reference signal). b) A shadowgraphy image generated by back-illumination with the optical probe beam, that is synchronized to the high-intensity laser, shows the chopper blade (black bar in the top of the image), the cryogenic jet in the bottom of the image and the cutting edge. The chopper blade is located 4 mm below the nozzle. c) Characterization of the stability of the mechanical chopper. Red and black horizontal lines illustrate the position of the blade as well as the interaction point. Blue dots indicate the position of the jet edge obtained from probe images for 90 consecutive images at 1 Hz. The blue shaded area illustrates the standard deviation interval.

locked loop motor speed controller (based on *Thorlabs MC2000B-EC*) as illustrated in figure 3a. The blade rotates with a much higher frequency, typically 75 Hz, to increase the relative phase stability. The actual blade rotation is further detected by a light sensor consisting of a laser diode and a photodiode. The light sensor provides feedback to the controller allowing to adjust the rotation speed with respect to the trigger signal.

The on-shot position of the blade is measured and adjusted using a low magnification imaging system using the probe laser to achieve the cutting synchronized to the irradiation with the high-intensity laser pulse. The recorded image in figure 3b illustrates the situation where the cut edge is positioned about 2 mm above the interaction point. Momentum transfer from the blade causes a small curl at the edge of the jet resulting in a deflection of the target within the first 0.5 mm below the cutting point. However, the jet stability is not affected at greater distances. A study of the temporal stability of the mechanical cutting is shown in figure 3c. The blue data points indicate the position of the edge of the cut (see probe image in figure 3b) for 90 consecutive images at 1 Hz. The stability of the edge position can be characterized by a standard deviation of 0.6 mm while 1.6 mm marks the highest offset from the average.

The implementation of the mechanical chopper in the laser-plasma experiment showed that damage to the aperture was successfully prevented by cutting the jet and blocking the line-of-sight. This enabled systematic studies of laser proton acceleration at the petawatt level.

3. Laser proton acceleration results

During a measurement campaign of several days, more than 1000 high-intensity laser shots were focused onto the cryogenic jet without damaging the aperture demonstrating the success of the chopper concept. Broadband proton beams were accelerated with large divergence from the plasma similar to what was published in [13] for lower laser pulse energies. Cut-off energies of

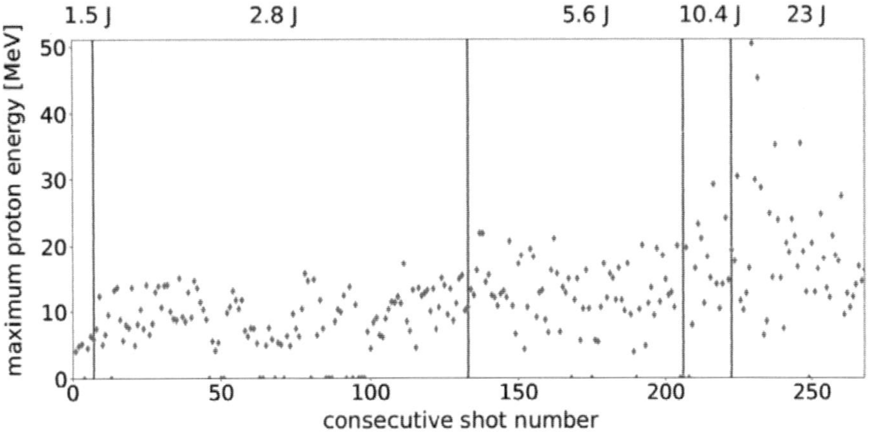

Figure 4. Maximum proton energies measured in laser forward direction for a total of 268 consecutive shots of a single experiment day using different pulse energies on target (displayed at the top). The very same nozzle was used without observing any noticeable performance degradation during the following days of the experiment.

proton energy spectra measured along the laser main axis being characteristic for the accelerator performance are given in figure 4 for one day of operation. Consecutive shots applied at different laser pulse energies on target are shown. Starting at low values, the laser pulse energy was slowly increased while monitoring for potential damage to the target system. No degradation could be detected with respect to the maximum proton energies as shown by the consistent trend of the data points in figure 4. Protons with a kinetic energy above 4 MeV, the lower detection threshold of the TPS, were detected in at least 90% of the shots. The large shot-to-shot fluctuation is caused by the non-perfect spatial overlap of the laser focus spot and the target that can be independently monitored using the optical shadowgraphy images. While best performance is obtained for perfect hits, larger lateral distance to the jet axis is in general yielding lower maximum proton energies.

4. Conclusion and outlook

We demonstrated laser proton acceleration using an ultrashort pulse PW-class laser systems with a self-refreshing, debris-free cryogenic hydrogen jet target. Recording of a large number of shots was enabled through the implementation of a chopper device that effectively protects the target system from damage due to radiation generated in the laser-plasma interaction. It allowed for systematic studies of proton and ion acceleration regimes in a variety of experimental configurations (to be published elsewhere [20]). It should be noted that such chopper concepts could also be of interest for other target systems where small sensitive nozzles are applied to produce solid or liquid targets for high-power laser experiments [15, 21–23].

In order to further increase the shot-to-shot stability in the experiment, an improved chopper assembly is under development utilizing e.g. a high precision position feedback. The new system guarantee a strongly decreased timing jitter of the rotating blade in the order of few microseconds and will therefore allow to operate the laser plasma interaction even closer to the nozzle. Besides the improved synchronization, the goal is to design a compact assembly to minimize the footprint around the target interaction point.

As an interesting alternative beyond the rotating device, compact piezo driven chopper systems are under consideration. The smaller form factor of piezo designs and much simpler synchronization to the laser pulse arrival time may allow for more flexible on-demand operation schemes.

Acknowledgments
The work of C.B.C., M.G., S.H.G., F.T. and C.S was supported by the U.S. Department of Energy, Office of Science, Fusion Energy Science under FWP 100182. C.B.C. acknowledges partial support from the Natural Sciences and Engineering Research Council of Canada (NSERC). F.T. acknowledges support from the National Nuclear 657 Security Administration (NNSA).

References

[1] Albert F *et al.* 2021 *New J. Phys.* **23** 031101
[2] Borghesi M *et al.* 2002 *Phys. Plasmas* **9** 2214–20
[3] Quinn K *et al.* 2009 *Phys. Rev. Lett.* **102** 3–6
[4] Roth M *et al.* 2013 *Phys. Rev. Lett.* **110** 44802
[5] Patel P K, Mackinnon A J, Key M H, Cowan T E, Foord M E, Allen M, Price D F, Ruhl H, Springer P T and Stephens R 2003 *Phys. Rev. Lett.* **91** 125004
[6] Fernández J C, Honrubia J, Albright B J, Flippo K A, Gautier D C, Hegelich B M, Schmitt M J, Temporal M and Yin L 2009 *Nucl. Fusion* **49** 065004
[7] Busold S *et al.* 2014 *Instrum. Methods Phys. Res., Sect.* A **740** 94–98
[8] Bulanov S V, Esirkepov T Z, Khoroshkov V S, Kuznetsov A V and Pegoraro F 2002 *Phys. Lett. A* **299** 240–47
[9] Kroll F *et al.* 2022 *Nat. Phys.* **18** 316–22
[10] Yogo A *et al.* 2009 *Appl. Phys. Lett.* **94** 181502
[11] Kim J B, Göde S and Glenzer S H 2016 *Rev. Sci. Instrum.* **87** 11E328
[12] Curry C B, Schoenwaelder C, Goede S, Kim J B, Rehwald M, Treffert F, Zeil K, Glenzer S H and Gauthier M 2020 *J. Vis. Exp.* e61130
[13] Obst L *et al.* 2017 *Sci. Rep.* **7** 10248
[14] Gauthier M *et al.* 2017 *Appl. Phys. Lett.* **111** 114102
[15] Polz J *et al.* 2019 *Sci. Rep.* **9** 16534
[16] Ziegler T *et al.* 2021 *Sci. Rep.* **11** 7338
[17] Bernert C *et al.* 2022 *Sci. Rep.* **12** 1–11
[18] Protopapas M, Keitel C H and Knight P L 1997 *Rep. Prog. Phys.* **60** 389–486
[19] Buffechoux S *et al.* 2010 *Phys. Rev. Lett.* **105** 015005
[20] Rehwald M *et al. Nat. Phys.* submitted
[21] Koralek J D *et al.* 2018 *Nat. Commun.* **9** 1353
[22] George K M *et al.* 2019 *High Power Laser Sci. Eng.* **7** e50
[23] Garcia S, Chatain D and Perin J P 2014 *Laser Part. Beams* **32** 569–75

Proposal for a compact neutron generator based on a negative deuterium ion beam

Kouichi Jimbo[1], Toshiyuki Shirai[1], Ka-Ngo Leung[2] and Karl van Bibber[2]

[1] National Institutes for Quantum Science and Technology, Accelerator and Medical Physics, Chiba, 263-8555, Japan
[2] Department of Nuclear Engineering, University of California Berkeley, Berkeley CA 94720, USA

E-mail: kouichi.jimbo@qst.go.jp

Abstract. Interest in high intensity generators of neutrons for basic and applied science has been growing, and thus the demand for an economical neutron generator has been growing. A major driver for the development of high intensity neutron generators are studies of neutron disturbance in integrated circuits, for which a compact generator that can be easily accommodated in an ordinary size lab would be highly desirable. We have investigated possible designs for neutron generators based on the D-D fusion reaction, which produce direction dependent mono-energetic neutrons with carry-off energy larger than 2.45 MeV. Specifically, we find a negative deuterium ion beam most attractive for this application, and plan to construct such a system with a negative deuterium ion beam of 200 keV energy and 100 mA current as a prototype of this concept.

1. Introduction

The National Institute for Quantum Science and Technology (QST, Chiba) hosted a long-standing program of neutron irradiation of biological specimens under pathogen-free background conditions, the Neutron exposure Accelerator System for Biological Effect Experiment (NASBEE). Neutrons were produced by 4 MeV positive hydrogen and deuterium ion beams striking on a beryllium target through the ^{9}Be(p,n) and ^{9}Be(d,n) reactions [1]. Unfortunately, due to the large size of the neutron generating system, and its high operational cost, its operation had to be terminated.

Similarly, there is an unmet need of widely available neutrons for radiation effect studies on integrated circuits. Such studies are imperative, particularly for electronics in space and high-altitude flight, as cosmic rays collide with nuclei in the atmosphere, producing approximately ten neutrons per cm^2 per hour. These cosmic rays create secondary neutrons of 1 MeV or higher [2]. These neutrons undergo scattering in integrated circuits and cause soft errors in them, also known as single-event upsets. Soft errors are a phenomenon by which one or more bits within the data on the semiconductor device have their values reversed. A soft error does not damage the semiconductor device itself. Given the large and growing number of high-altitude aircrafts and satellites being deployed, there is a great need for chip testing and certification capacity, particularly with fast neutrons, but below 3 MeV. This need also points to the necessity of new types of neutron generators [3].

In Japan, accelerator driven neutron generators based on the $^{9}Be(p,n)$ reaction are commonly used; these involve the acceleration of protons to ~10 MeV directed onto a beryllium target. There are currently many neutron generators of this kind are available [4]. However, the requirement of an ion source and an accelerator makes this design very large and very expensive. Furthermore, the target area becomes highly activated due to the high neutron flux emitted at all angles, not just those neutrons impinging on the sample of interest. The maintenance expenses likewise are not small.

The two examples above, from the biomedical and semiconductor sectors, are indicative of the general demand for commercial neutron applications, and which are increasing over all energy ranges. Thus, we would like to develop a compact neutron generator with its power supply system, that could be affordably acquired, and occupy an ordinary size laboratory room of a university or a private company.

2. D-D reaction based neutron generators

Neutron generators based on the D-D fusion reaction, and isotopically producing neutrons of carry off energies larger 2.45 MeV (zero deuterium ion incident kinetic energy) have a long and successful history. The mono-energic energy-angle distribution changes only as a function of incident deuterium ion energy. One design of a deuterium ion source-driven neutron generator, based on a self-loading target [5], has been developed and utilized for a broad range of basic and applied science interests (geochronology, medical isotope studies, nuclear data for reactor design, etc.) at the Nuclear Engineering Department of University of California Berkeley.

The near mono-energetic neutrons with an energy of approximately 2.8 MeV at zero degrees with respect to the deuterium ion beam direction was obtained. The basic concept for this particular device, the High Flux Neutron Generator (HFNG) owes to K.-N. Leung, with elaborations and upgrades being carried out largely by students and postdocs [6,7].

In the HFNG, positive deuterium ions are accelerated up to 125 keV after extraction from the high atomic ratio RF-driven plasma ion source. The symmetric planar water-cooled (cathode) target is positioned in between two opposing plasma ion sources, 10 cm away on each side (although typically only one is used). The target has a thin slot machined edgewise in it, allowing a cassette of small samples to be inserted within 8 mm of where the beam strikes the target, i.e. the production point of the neutrons [6,7]. As the target is at negative high voltage, secondary emission electrons from the target will accelerate back to the ion source, impeding the beam irradiation and damaging the ion source, unless these secondary electrons are suppressed.

In the case of the HFNG, suppression of secondary electrons is accomplished by a specially designed shroud around the target, held at a more negative voltage, e.g. if the target is at −100 kV, the shroud would generally be at −102.5 kV; this solution worked very well. To produce a large number of neutrons and produce a flat profile at the sample cassette, multi-aperture electrodes are used. An electrode with only a single aperture is used, when one wants to measure a cross section as a function of energy over the available range 2.2-2.8 MeV. This can be done accurately and all at once, with a sample holder that positions different samples at several angles with respect to the plasma beam axis, taking advantage of the energy-angle correlation of the emitted neutrons.

The target is made of a thin layer either welded or explosion-bonded onto the water-cooled copper plate. Titanium is a very favourable target both due to its good thermal properties, but most importantly due to its very high (and reversible) affinity for hydrogen, which allows the surface of the target to be fully loaded within half an hour, at which time the HFNG reaches maximum neutron output. Under continuous bombardment of deuterium ions, the titanium eventually erodes, requiring a fresh target to be swapped in; this feature of the source is only a minor inconvenience and not a fundamental problem, solved by having two or three targets whereby one is undergoing a replacement of its titanium layer while the irradiation continues with a fresh target. A maximum of 10^{10} n/sec was obtained continuously under best conditions, resulting in a flux on the surface of small samples in the slot well of $10^{(8\text{-}9)}$ n/cm^2-sec.

3. Neutron generator using a negative deuterium ion beam

In principle, secondary emission electrons do not arise when negative ions impinge on a positive biased target. If negative ions are employed, one does not need a shroud with finite aperture to control the secondary emission electrons and therefore one can achieve an even larger ion extraction area with a multi-aperture cathode. In this case, the neutron output is simply limited to the negative deuterium ion beam current. Another positive feature is that there is no molecular negative deuterium ion. A single aperture (~1-mm-diameter) negative deuterium ion source is already in use at the Sandia National Laboratory neutron generator [8]. However, a negative deuterium ion source with multi-aperture electrodes has not yet been utilized for neutron generators even though it is being used in neutral beam injection systems (NBIs) for a future fusion reactor.

A large area negative deuterium ion source with filament cathodes has been developed at QST, Naka as NBIs for JT 60, a large plasma confinement tokamak for nuclear fusion experiments. Twenty-five years ago, a 10-A negative deuterium ion beam had been accelerated up to 410 keV and 2.3×10^{12} n/sec neutrons were detected near the water-cooled cylindrical copper beam dump [9]. At that time those neutrons had been considered to be an obstacle to the development of a NBI. This is a demonstration that 10^{12} n/sec can be produced with a 10 A 400 keV negative deuterium ion beam striking on a copper target.

A RF deuterium multi-cusp ion source (without filaments) was developed at Lawrence Berkeley National Laboratory (LBNL), in which RF antenna is placed outside the ion source chamber [10]. Its maintenance under the influence of radiation is very simple since the antenna is located outside the vacuum. A long service interval for the ion source is also possible. If we replace the filaments with the RF antenna in the QST negative ion source, we can construct a new neutron generator of which the generator maintenance is much easier. This new neutron generator combines the RF negative deuterium multi-cusp ion source and the copper target and will be very compact.

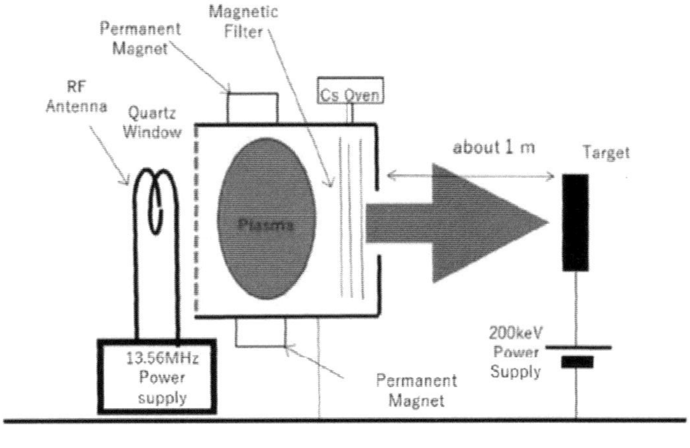

Figure 1. A conceptual image of the compact negative deuterium ion neutron generator.

4. Search for the axion at the NICA

We would like to construct a negative deuterium ion beam system of 100 mA 200 keV as a possible compact neutron generator, using a water-cooled copper target. The heat conductivity of copper is very high, and the copper target is effectively cooled and maintain neutron production for a long time without a target replacement. The deuterium ion source is situated at ground potential while the copper target is at a high positive potential. A vacuum system with gas handling apparatus and a special water circulation system to supply the target with cooling water will be employed. A 13.56 MHz RF power generator will be used to operate the multi-cusp deuterium ion source. A dc high voltage supply is needed to accelerate negative deuterium ions up to 200 keV, see figure 1.

5. Conclusion

We plan to construct a prototype compact neutron generator which consists of a negative deuterium ion beam system with a copper target; this will position us to demonstrate an effective D-D fusion mono-energetic neutron generator, in which neutron energies depend on the direction.

References

[1] Ogawara R *et al.* 2020 *Nucl. Instrum. Methods Phys. Res., Sect.* A **949**
[2] Kobayashi K 2016 *J. Particle Accelerator Society of Japan* **13** 245
[3] Iwashita H *et al.* 2020 *IEEE Trans. Nucl. Sci.* **67** 2363
[4] Hirota K 2018 *J. Particle Accelerator Society of Japan* **15** 69
[5] JKim J 1977 *Nucl. Instrum. Methods* **145** 9
[6] Ayllon Unzueta M *et al.* 2018 *Nucl. Instrum. Methods Phys. Res., Sect.* A **903** 193
[7] Waltz C *et al.* 2017 *Appl. Radiat. Isot.* **125** 124
[8] Antolak A J *et al.* 2016 *Nucl. Instrum. Methods Phys. Res., Sect.* A **806** 30
[9] Kuriyama M *et al.* 1995 *Proc.16th Int. Symp. on Fusion Engineering* **1** pp 491-496
[10] Hahto S K *et al.* 2004 *Rev. Sci. Instrum.* **75** 355

Performance of automated synchrotron lattice optimisation using genetic algorithm

X Zhang[1] **and S L Sheehy**[1,2]

[1] The University of Melbourne, Melbourne, Australia

[2] Australian Nuclear Science and Technology Organisation (ANSTO), Sydney, Australia

E-mail: xuanhaoz@student.unimelb.edu.au

Abstract. Rapid advances in superconducting magnets and related accelerator technology opens many unexplored possibilities for future synchrotron designs. We present an efficient method to probe the feasible parameter space of synchrotron lattice configurations. Using this method, we can converge on a suite of optimal solutions with multiple optimisation objectives. It is a general method that can be adapted to other lattice design problems with different constraints or optimisation objectives. In this method, we tackle the lattice design problem using a multi-objective genetic algorithm. The problem is encoded by representing the components of each lattice as columns of a matrix. This new method is an improvement over the neural network based approach in terms of computational resources. We evaluate the performance and limitations of this new method with benchmark results.

1. Introduction

The high level goal of this work is to assess whether the 'art' of lattice design can be economically achieved by a computer without assuming existing standard lattice configurations known to accelerator physicists. This paper is a continuation of the effort to develop an automated lattice optimisation algorithm. In the previous implementation of this algorithm [1], a neural network based approach was used to generate new lattice structures. There were two main problems with the neural network based approach. Firstly, new lattices were generated piece-wise based on a feedback loop controlled by tracking the motion of a test particle. This required each successful neural network to have *a priori* encoded values to produce the exact feedback values in order to create a desirable lattice. This is possible in principle, however this system is an overly complicated mapping between the optimisation quantities and the actual machine lattice structure. Secondly, the neural networks are very sensitive to small perturbations of the node values and resulted in drastically different lattice structures. Therefore it becomes computationally expensive to optimise the neural network values for target lattice structures with a large number of elements. The work for this paper tackles both of these problems by introducing a different method for generating lattices. In addition, the newest version of the genetic algorithm was implemented to improve the performance when optimising higher dimensional objectives.

2. Lattice matrix encoding

One major challenge with applying any kind of optimisation algorithm is how to represent the problem and parse it to the algorithm, this is referred to as the encoding of the problem. A matrix encoding method was used in this work, such that any lattice structure with n elements and at most p attributes per element can be represented by a $p \times n$ matrix. For example, a simple FODO cell with separate function dipole magnets is represented by an 8×4 matrix as shown in figure 1. Each column explicitly represents a different element of the lattice including empty drift spaces (E). The rows encapsulate all the attributes needed to describe every element in the lattice, where *k0* and *k1* are the dipole and quadrupole coefficients as used by MADX [2].

Element	QF	E	B	E	QD	E	B	E
k0	0	0	1.07	0	0	0	1.07	0
k1	0.23	0	0	0	-0.23	0	0	0
Length	0.3	0.43	0.87	0.43	0.3	0.43	0.87	0.43

Figure 1. Example matrix encoding of a FODO cell with separate function dipoles.

It should be noted that this matrix encoding can include any number of attributes by adding more rows, such as higher order coefficient terms of non-linear magnets. Also, it does not matter to the optimisation algorithm where the actual starting point of the lattice is defined. This property is particularly useful when optimising for a periodic structure such as a ring.

As part of the optimisation algorithm, the aim is to probe as much of the available parameter space as possible. In other words: to explore lattice configurations that we might not think would be stable based on our experience. This calls for a method to randomly sample from the entire parameter space that is bounded by a few basic design requirements such as the total bending angle or betatron tune. Note that there is no constraint on the number of magnetic elements or the total length of arc. Using the matrix encoding method, it is possible to initialise a collection of random lattice structures by building the lattices one element at a time. This means, starting each lattice matrix from a randomly chosen element (drift, dipole, quadrupole, or combined function dipole) and randomly choosing the attribute values necessary for that element. This process is repeated by adding new columns to the lattice matrices until the total bending angle of the lattice exceeds the requirement.

3. Genetic algorithm

The principle for the multi-objective genetic algorithm (MOGA) applied in this work is based on Darwinian evolution. The fittest members of a population are most likely to survive or pass on their traits to future generations as the population evolves. The algorithm used in this work is the constrained and adaptive version of the reference-point-based many-objective evolutionary algorithm (NSGA-III) [3, 4]. Compared to algorithms applied in previous works [1], NSGA-III has improved performance at preserving population diversity for problems with higher dimensional objectives ($N > 3$) as well as reduced computation resource needs. Specifically, NSGA-III allows the user to define a set of reference points within the objective space and change the selection pressure based on the population distribution around each reference point.

A brief outline of the custom implementation in context of this work is provided below, specifics of the NSGA-III are not reproduced here.

1 Build a population of N randomly initialised lattice matrices based on the required total bending angle.

2 Evaluate the one-turn matrix, objective values, and penalty values for each new member of the population.

3 Sort the population into non-domination fronts using non-dominated sorting.

4 Initialise new generation of R randomly initialised matrices and $N - R$ offsprings from the previous generation using genetic operators.

5 Evaluate the one-turn matrix, objective values, and penalty values for each new member.

6 Rank the combined population of $2N$ members and associate each member with the closest reference point.

7 Select N members starting from the first non-domination front. For the last front to be included, select members that would maximise the diversity of associated reference points.

8 Adaptively change the reference point positions to reduce over clustering around any one point.

9 Start new generation and repeat steps 4-8.

3.1. Non-dominated sorting and genetic operators

NSGA-III and previous versions of this algorithm all rely on the idea of non-domination levels, which is a relative measure of the constraint violation and objective satisfaction of the members of a population. As a result of the lattice matrix encoding and the random initialisation process, a large portion of the initial population corresponds to unstable lattice structures. This creates a problem of how to quantify the degree of divergence between unstable lattices. One naive solution is using the trace value of the first order one-turn transfer matrix $\mathrm{Tr}(M_1)$. As our results in the next section demonstrate, $\mathrm{Tr}(M_1)$ is an effective measure of divergence. Our algorithm is able to reach stable lattice structures by favouring unstable lattices with smaller $\mathrm{Tr}(M_1)$ values.

In addition to the original NSGA-III constrained-dominate definition, we added the following. A member p is defined to constrained-dominate another member q if any of the following is true.

- If p is stable in at least one transverse plane and q is unstable in both planes.
- If both are unstable in one plane, but $\mathrm{Tr}(M_1)$ of p in the unstable plane is smaller then that of q.
- If both are unstable in both planes, but the sum of $\mathrm{Tr}(M_1)$ of p in both planes is smaller then that of q.

The *crossover* and *mutation* operators are used to create new lattice matrices from existing members for each generation of the algorithm. The crossover operator selects pairs of matrices from the parent population and creates two new matrices by swapping columns of the parent matrices at the same crossover point. A binary tournament selection process [5] is used to select parent pairs from a randomly selected subset of the population, the non-domination rank is used as the comparison criterion and the size of the subset correlates to the selection pressure. New matrices created by the crossover operator are neither guaranteed to result in better lattice structures compared to their parents nor even stable lattices, however this does not seem to have significant impact on the convergence of the algorithm and can be improved in the future by applying a repair operator [6]. One limitation of the crossover operator is that it does not allow for subtraction of elements from existing matrices, therefore the number of magnetic elements in new matrices can only increase in number. An additional lattice simplification step could be added to the algorithm later if desired.

The mutation operator only acts on a single lattice matrix at a time, it randomly changes the attribute of a randomly selected position on the matrix within the bounds of allowed values for that position. Attributes on the same column are also changed as needed to reflect the mutated attribute. For example, if the *Element* attribute was mutated from QF to CF on the column [QF, $k0 = 0$, $k1 = 0.75$, $l = 0.3$], then a new non-zero $k0$ attribute would also be generated for that column.

4. Results

We report the results of a benchmark study carried out to check the performance of the algorithm. For this study, the optimisation objectives were the following.

- f_1: Minimise $\max(\beta_x)$.
- f_2: Minimise $\max(\beta_y)$.
- f_3: Minimise number of segments.
- f_4: Minimise number of magnetic elements.

While the constraints were set to mimic a long closed dispersion arc.

- Total bending angle, $B_{\mathrm{tot}} = 180°$.
- Horizontal tune, $Q1 = 0.836$.
- $\mathrm{Initial}(\beta, \alpha, D, Dp) = \mathrm{Final}(\beta, \alpha, D, Dp)$.

Each constraint violation is normalised using the sigma function $(|x|/(1+|x|))$ and the total penalty value is calculated by summing the normalised constraint violations. A maximum penalty tolerance of 0.2 was used to select feasible members from the population. Note the vertical tune was left unconstrained to allow an extra degree of freedom such that the algorithm can populate the Pareto front in the objective space, rather than confined to a single solution that would minimise the constraint violation for a specific vertical tune. A population of 288 random members were initialised and allowed to evolve for 1000 generations. Initially, the objective space was evenly divided into 10 partitions along each objective and a set of 286 reference points were generated to cover the 4D simplex formed by the partitions [7]. The population size was chosen to maximise the efficiency of the reference point association procedure following [4].

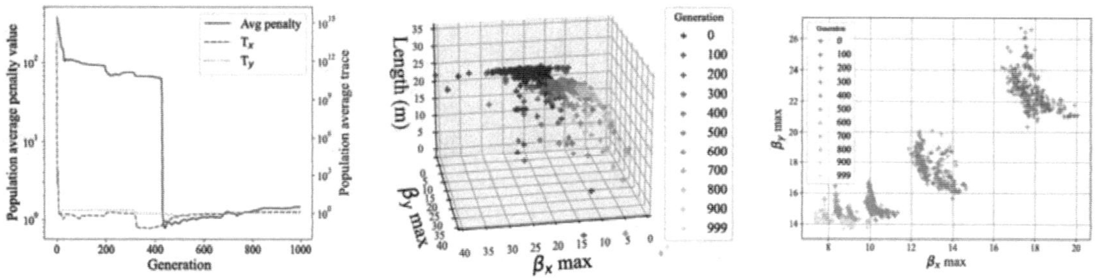

Figure 2. (left) The solid line shows the population average penalty value, the dashed and dotted lines are the population average horizontal and vertical $\mathrm{Tr}(M_1)$ values. (mid) Feasible members projected onto 3D objective space, and (right) 2D objective space. The colours from blue to yellow denotes the evolution from generation 1 to 1000.

The algorithm was able to converge to stable lattice structures within 20 generations using the $\mathrm{Tr}(M_1)$ ranking method as shown on figure 2(left). Interestingly, the population average penalty value on figure 2(left) sharply decreases around generation 500, then slowly increases towards generation 1000. The sharp drop can be correlated to a sudden success in matching the total bending angle constraint. While the slow increase after generation 500 is the algorithm attempting to minimise the objectives while exploiting the tolerance range set for the penalty value. Figures 2(mid) and 2(right) shows the progression of the population getting closer to the Pareto optimum front over the generations. The distribution on the (β_x max, β_y max) plane matches our expectation of a trade-off relationship. One interesting example of the best performing member is shown in figure 3. Although not feasible to build, it is interesting that

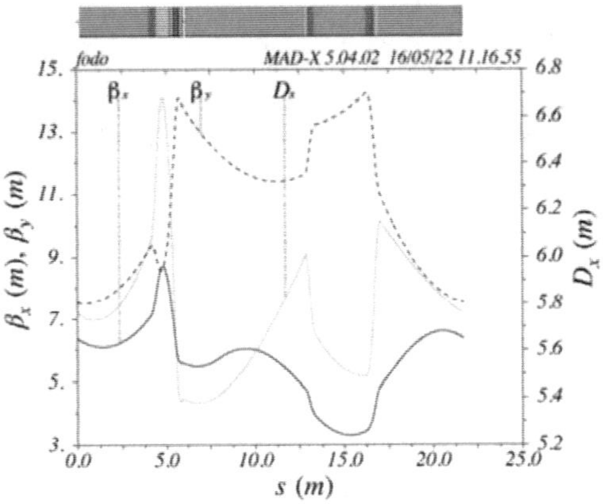

Figure 3. Lattice functions of a member in the first non-dominated front, representing one Pareto trade-off point in the objective space. Green segments are pure dipoles, yellow and blue segments are horizontally focusing and defocusing combined function dipoles respectively.

the algorithm has found a design with long dipole segments and a few closely packed almost alternating focusing segments in between.

The algorithm presented in this work is able to efficiently converge to stable lattice structures and populate the Pareto front of the supplied objectives. However, this feature is also the limitation of the algorithm, since it will attempt to 'cheat' the objectives by slightly tweaking the attributes of a feasible member in order to fill out the objective space. This is in contrast to the original intent of searching for completely different lattice structures to fill out the objective space. As a result, members in the first non-domination front all feature a similar structure as shown in figure 3.

Having prior knowledge of lattice design and accelerators more generally, there are now a number of other improvements to be made to the basic algorithm, these are the following. Initialise new lattice matrices conditionally to better represent physical requirements such as leaving extra drift space between each new magnetic element. Introduce repair operator after combination and mutation operations to improve the likelihood of creating new stable structures. Optimise the selection pressure to preserve diversity of candidate structures while maintaining a reasonable convergence rate towards stable candidates. Seeding the algorithm using several stable lattice structures or combining the converged stable candidate pool from several different random initialisations to improve diversity of structures.

5. Conclusion

We presented an improved method to generate lattice structures using a lattice matrix encoding method. Our implementation of NSGA-III is capable of finding a Pareto front for a set of given constraints, however it is limited by the lack of a penalty function that measures the diversity of lattice layouts. We plan to address this limitation in future works and apply the algorithm to a full ring structure with slow extraction constraints.

References

[1] X Zhang, S L Sheehy and E Benedetto 2021 *Proc. 12th Int. Particle Accelerator Conf. (IPAC'21) (Campinas, Brazil)* p 616-9
[2] CERN - Accelerator Beam Physics Group 2020 *MAD - Methodical accelerator design*
[3] K Deb and H Jain 2014 *IEEE Trans. Evol. Comput.* **18** 577-601
[4] H Jain and K Deb 2017 *IEEE Trans. Evol. Comput.* **18** 602-22
[5] B Miller and D Goldberg 1995 *Complex Syst.* **9**
[6] M Ehrgott 2005 *Multicriteria Optimization* (Berlin: Springer)
[7] I Das and J E Dennis 1998 *SIAM J. Optim.* **8** 631-57

High current heavy ion beam investigations at GSI-UNILAC

H Vormann[1], W Barth[1,2,3], M Miski-Oglu[1,2], U Scheeler[1], M Vossberg[1] and S Yaramyshev[1]

[1] GSI Helmholtzzentrum fuer Schwerionenforschung GmbH, Darmstadt, Germany
[2] HIM Helmholtz Institute Mainz, Mainz, Germany
[3] JGU Johannes-Gutenberg-University Mainz, Mainz, Germany

E-mail: h.vormann@gsi.de

Abstract. The GSI Universal Linear Accelerator UNILAC and the synchrotron SIS18 will serve as injector for the upcoming FAIR-facility. The UNILAC-High Current Injector will be improved and modernized until FAIR is commissioned and the Alvarez post stripper accelerator is replaced. The reference heavy ion for future FAIR-operation is uranium, with highest intensity requirements. To re-establish uranium beam operation and to improve high current beam operation, different subjects have been explored in dedicated machine investigation campaigns. After a beam line modification in 2017 the RFQ-performance had deteriorated significantly; new rods have been installed and the RF-working point has been redefined. Also the Superlens-performance had become unsatisfactory; improved with a modified RF-coupler. With a pulsed hydrogen gas stripper target the uranium beam stripping efficiency could be increased by 65%. Various work has already been carried out to establish this stripper device in routine operation. With medium heavy ion beams a very high beam brilliance at the end of transfer line to SIS18 was achieved. Results of the measurement campaigns and the UNILAC upgrade activities will be presented.

1. Introduction

Before the year 2021, high current uranium beam machine experiments were conducted in October 2015 and July 2016 for the last time, only at the GSI-High Current Injector (HSI) and the gas stripper section. At this time only three of the five Alvarez DTL post stripper tanks were available, due to work on the post stripper RF-amplifier systems of the UNILAC (figure 1). The achievable high current beam brilliance at injection into the heavy ion synchrotron SIS18 was estimated only by using front-to-end high-current measurements with a proton beam performed in 2014 [1-14].

During the long shutdown in the whole year 2017 there was no beam operation at all. After this shutdown, the beam times 2018 and 2019 had to be re-organized, as the HSI-RFQ performance had decreased strongly and required a time-consuming repair. In 2020 uranium beam operation was skipped due to the Corona crisis (figure 2).

The measurement campaigns in 2021 and 2022 have been conducted with high intensity heavy ion beams (uranium, bismuth, tantalum, xenon) and with medium and light ion beams (Argon, p^+), to fully characterize the high intensity beam properties of the UNILAC.

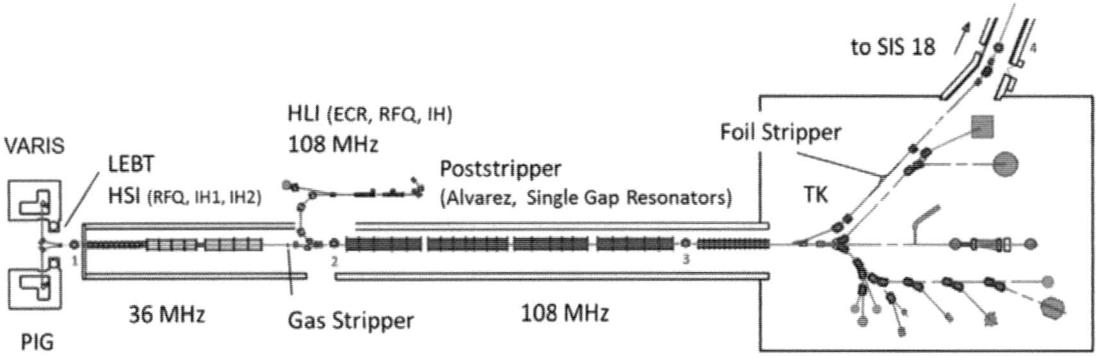

Figure 1. Overview of the GSI Universal Linear Accelerator UNILAC [4].

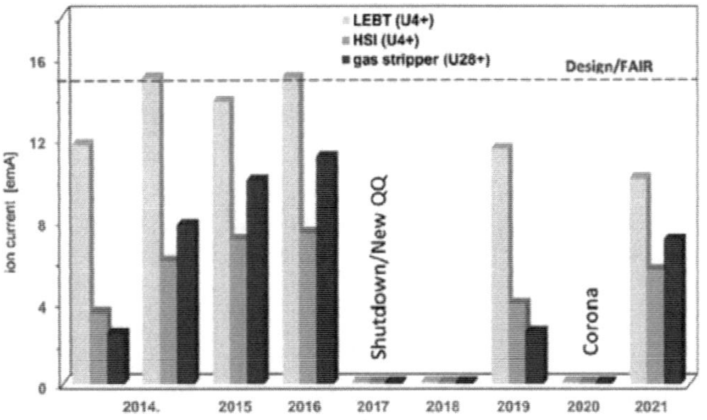

Figure 2. Uranium beam intensity at the UNILAC HSI [4].

2. HSI RFQ optimisation

The RF performance of the HSI-RFQ had decreased strongly after a beam line modification during the shutdown 2017, when the RFQ had been kept under atmosphere conditions for almost one year. During the recommissioning campaign in 2018 only 70% of the nominal RF-voltage could be reached. Apparently the copper surface conditions were significantly degraded, due to many years of operation and additional humidity influence. Therefore new electrodes (rods) have been produced and installed (2018-2019). After successful recommissioning with light ions and also with U^{5+}, the working point of the HSI-RFQ has been re-defined: With a medium heavy ion beam (Ar^{2+} and Ar^{1+}), applying different RF-voltages for acceleration, the transmission through the HSI has been scanned in a wide range from voltages far below the working point to high voltages well above. The result was a surprisingly long plateau of almost full transmission and sufficient beam performance. The working point was defined at an RF voltage closely above the plateau beginning. This was confirmed by measurements with uranium, the FAIR design ion (A/Z = 59.5), and also with ions of even higher mass-over-charge ratio: $^{181}Ta^{3+}$ (A/Z = 60.3) and $^{124}Xe^{2+}$ (A/Z = 62). In figure 3 the RF voltage normalized to A/Z is displayed for the different cases.

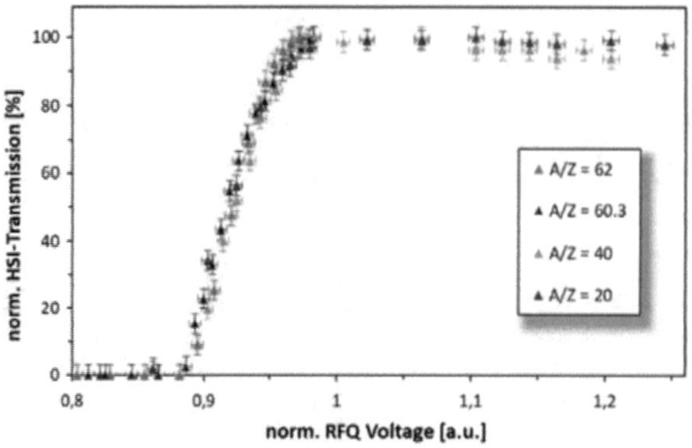

Figure 3. HSI-RFQ working point re-defined [4].

3. Stripper development

Investigations with a pulsed hydrogen gas target in the HSI gas stripper section (1.4 MeV/u) in 2015 and 2016 showed that the stripping efficiency for the desired U^{28+} fraction is by 65% higher than with the nitrogen gas target. Technical and safety issues do not yet allow a routine operation of a hydrogen gas target so far. Only short test periods (3 days each) have been performed with the pulsed hydrogen gas stripper since then. After an upgrade of the stripper gas cell the optimal H2 target thickness of up to 14 mueg/cm^2 (for stripping to charge state 28+) was available in 2021, confirming an absolute stripping efficiency of 21%, while the efficiency for the standard N2 gas stripping is only 13%. Technical developments have been pushed in parallel, using a dedicated test stand (vacuum operated, but without ion beam). While for the initial tests the used pulsed valves had been specified for gasoline fuels, the valves currently used are designed for gaseous media and let expect improved durability [10]. In order to provide safety and reliability GSI typical multi user operation of up to 15 different users, a dedicated on call service of the gas stripper experts is established. In the recent beam test campaign (2022) an advanced control rack was used, allowing reliable parallel operation of five virtual accelerators.

For the injection into SIS18 the transversal beam emittance must be small, in particular the horizontal emittance. For pulsed gas H2-stripper operation, the vertical beam emittance for high intensity uranium beam (7.0 emA, U^{28+}) is increased, while the horizontal emittance is decreased, confirming former emittance measurements with $^{209}Bi^{26+}$ at an H2-target. The horizontal U^{28+} beam brilliance (figure 4) at 1.4 MeV/u scales inverse with the pulse current. It could be shown that the horizontal beam brilliance inside the beam core is increased by a factor of three by applying the H2-stripper target instead of the N2-target.

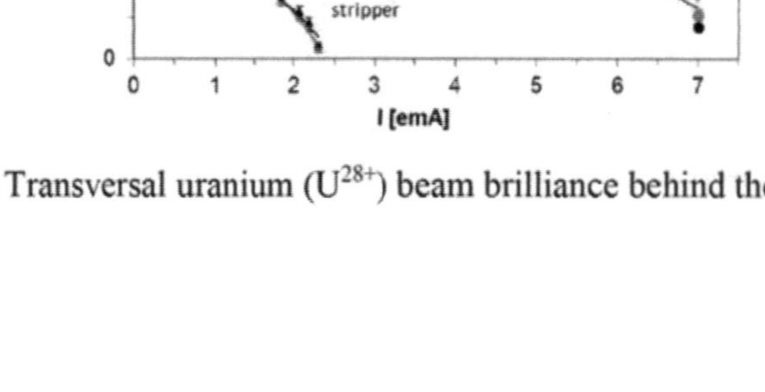

Figure 4. Transversal uranium (U^{28+}) beam brilliance behind the gas stripper.

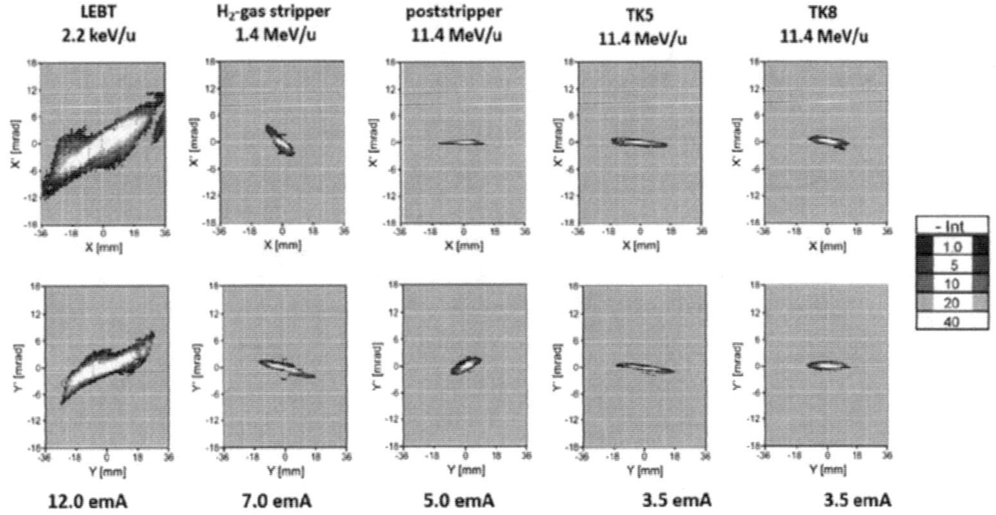

Figure 5. Transversal $U^{4+/28+}$ emittance shape along the UNILAC.

In order to achieve beam energies of up to 1 GeV/u in the SIS18, higher charge states of the ion beam can be provided by foil stripping at full UNILAC-energy (11.4 MeV/u) in the transfer channel to SIS18. Recently a high intensity uranium beam (5 emA U^{28+}) has been used to investigate beam emittance blow up due to straggling effects at the carbon foil targets. To minimize beam spot enlargement, the foil thickness was reduced from 600 to 400 mueg/cm^2 at a remaining stripping yield in the desired charge state (73+). However, the energy loss could thereby be reduced while in particular the horizontal emittance growth is significantly lower (30%).

4. Front-to-End measurements

The horizontal beam emittance is a critical value to estimate the high-current capability of a synchrotron injector. The behaviour of the UNILAC was investigated in 2019, 2020 and 2022 with high intensity Bismuth and Uranium beams.

The measured transversal emittances and the beam pulse currents for uranium, shown in figure 5, were measured at LEBT-section (U^{4+}), behind the H2-gas stripper (U^{28+}), behind poststripper and in the middle (TK5) and at the end (TK8) of the transfer channel. For bismuth beam, an increase of the vertical emittance and a decrease of the horizontal emittance occurred behind gas stripper and stays different along the complete UNILAC. At the transfer channel end a three times smaller horizontal emittance was measured compared to the vertical plane. For injection into SIS18, a very small horizontal emittance of 0.42 mm mrad (4*rms, 90%, normalized) was measured. From the LEBT to the end of the transfer channel (TK8), no net emittance growth has been observed in the horizontal direction, whereas the vertical emittance increases by a factor of 5. (Note: Particle losses of > 40% occurred along post stripper and transfer channel, may distort the emittance growth balance).

Comparable emittance measurements were performed with uranium beam as shown in figure 6. The effect of asymmetric emittance growth and reduction behaviour was less pronounced compared to the bismuth beam emittance (4*rms, norm.). At least, also for uranium beams the horizontal emittance growth is lower when using the hydrogen stripper, appr. 35% compared to the vertical emittance.

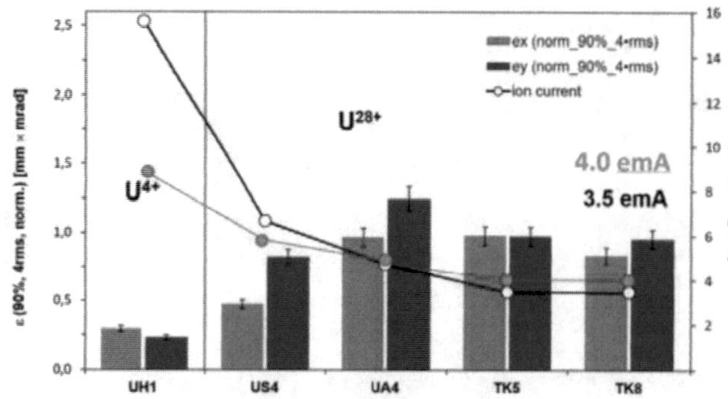

Figure 6. Transversal $U^{4+/28+}$emittance along UNILAC [15].

The recent machine experiments (April 2022) confirmed the results from 2019 and 2020 applying hydrogen gas stripping. The high current emittances measured at the end of the transfer channel (TK8) are displayed in figure 7. The measured horizontal emittance was sufficiently small: 0.46 mm mrad (4*rms, 90%, normalized) for the U^{28+} beam, and 0.42 mm mrad for the $U73^{+}$ beam, fitting well into the acceptance of SIS18 of 0.75 mm mrad. The emittance blow up for the stripped U^{73+} beam does not result in measurable larger emittance size.

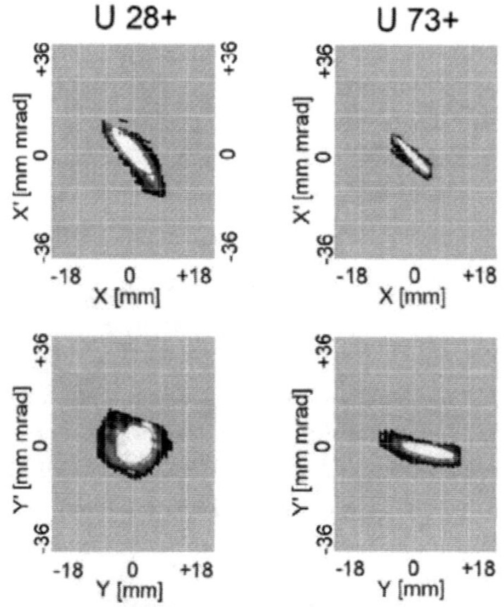

Figure 7. Transversal emittance at TK8, applying pulsed hydrogen gas stripper, 4.0 emA U^{28+}, 1.0 emA U^{73+}(2022).

Of significant relevance were the recent investigations of uranium beam matching from the UNILAC to the SIS18 injection. The fast current transformer analysis, displayed in figure 8, shows an RF capture efficiency inside SIS18 of more than 90% as a result of the optimization of the longitudinal beam shape applying the UNILAC rebuncher cavities behind the poststripper and in the transfer channel.

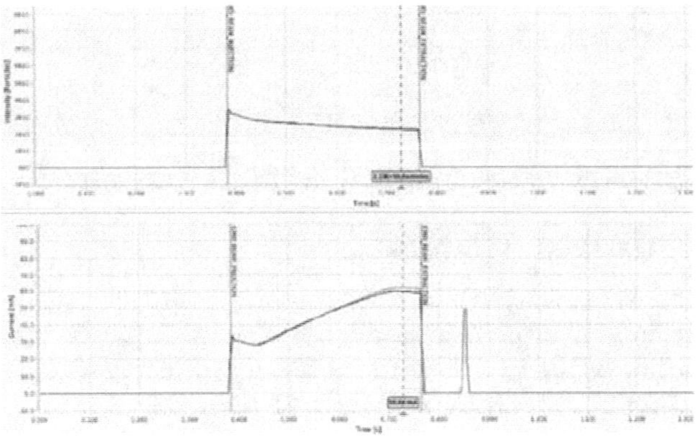

Figure 8. RF capture of the UNILAC beam in the SIS18.

5. Summary

Among several GSI linac projects [16-23], one of the most important ones is an upgrade of UNILAC for FAIR. After an exchange of the HSI-RFQ rods and the re-definition of its working point, the full spectrum of ions can be accelerated again, after a five years interruption.

With the beam brilliance of the high current uranium beam is now sufficient to fill the SIS18 up to 30% of the space charge limit. This significant improvement in beam brilliance was achieved by using the pulsed hydrogen gas stripper. With heavy ion beams (Bi, U) very low horizontal emittances along the UNILAC up to the SIS injection have been achieved, as well as sufficient longitudinal matching to the SIS18 applying dedicated rebuncher settings.

Acknowledgements
The authors would like to thank all colleagues of the involved GSI departments for the support.

References

[1] Barth W *et al.* 2017 *Phys. Rev. Accel. Beams* **20** 050101
[2] Barth W *et al.* 2007 *Nucl. Instrum. Methods Phys. Res., Sect.* A **577** 211-4
[3] Barth W *et al.* 2020 *J. Instrum.* **15** T12012
[4] Barth W *et al.* 2022 *Phys. Rev. Acc. Beams* **25** 040101
[5] Yaramyshev S *et al.* 2006 *Nucl. Instrum. Meth-ods Phys. Res., Sect.* A **558** 90-94
[6] Barth W *et al.* 2015 *Phys. Rev. Accel. Beams* **18** 050102
[7] Adonin A and Hollinger R 2014 *Rev. Sci. Instrum.* **85** 02A727
[8] Yaramyshev S *et al.* 2015 *Phys. Rev. Spec. Top. Accel Beams* **18** 050103
[9] Barth W *et al.* 2015 *Phys. Rev. Accel. Beams* **18** 040101
[10] Gerhard P *et al.* 2018 *Proc. 29th Linear Accel. Conf. (LINAC'18)* (Beijing, China) pp 982-7
[11] Scharrer P *et al.* 2015 *J. Radioanal. Nucl. Chem.* **305** 837-42
[12] Scharrer P *et al.* 2017 *Nucl. Instrum. Methods Phys. Res., Sect.* A **863** 20-25
[13] Adonin A *et al.* 2016 *Rev. Sci. Instrum.* **87** 02B709
[14] Appel S *et al.* 2017 *Nucl. Instrum. Methods Phys. Res., Sect.* A **852** 73
[15] Barth W 2022 presentation in the GSI Machine meeting
[16] Barth W *et al.* 2018 *Phys. Rev. Accel. Beams* **21** 020102
[17] Lauber S *et al.* 2020 *Phys. Rev. Accel. Beams* **23** 114201
[18] Herfurth F *et al.* 2015 *Physica Scripta* **2015** 014065
[19] Schwarz M *et al.* 2019 *Nucl. Instrum. Meth. Phys. Res. Sect.* A **951** 163044
[20] Busold S *et al.* 2014 *Nucl. In-strum. Meth. Phys. Res. Sect.* A **740** 94–98
[21] Lauber S *et al.* 2021 *Rev. Sci. Instrum.* **92** 113306
[22] Barth W *et al.* 2017 *EPJ Web Conf.* **138** 01026
[23] Ratzinger U *et al.* 2005 *AIP Conf. Proc.* **773** 249–53

Journal of Physics: Conference Series **2420** (2023) 012038 doi:10.1088/1742-6596/2420/1/012038

Design and beam dynamics study of disk-loaded structure for muon LINAC

K Sumi[1], T Iijima[1], K Inami[1], Y Sue[1], M Yotsuzuka[1], H Ego[2], M Otani[2], N Saito[2], T Mibe[2], M Yoshida[2], Y Kondo[3], K Moriya[3], Y Takeuchi[4], Y Nakazawa[5] and H Yasuda[6]

[1] Nagoya University, Nagoya, Aichi, Japan
[2] High Energy Accelerator Research Organization (KEK), Tsukuba, Ibaraki, Japan
[3] Japan Atomic Energy Agency (JAEA), Tokai, Naka, Ibaraki, Japan
[4] Kyushu University, Fukuoka, Japan
[5] Ibaraki University, Mito, Ibaraki, Japan
[6] University of Tokyo, Tokyo, Japan

E-mail: `ksumi@hepl.phys.nagoya-u.ac.jp`

Abstract. The disk-loaded structures (DLS) in the muon LINAC are under development for the J-PARC muon g−2/EDM experiment. Four DLSs with an accelerating gradient of 20 MV/m take charge of muon acceleration from 40 MeV to 212 MeV, which corresponds to 70% to 94% of the speed of light. The quasi-constant gradient type TM01-$2\pi/3$ mode DLSs with gradually varying disk spacing was designed and it was confirmed that the cumulative phase slip due to the mismatch between muon and phase velocity can be suppressed to less than 2 degrees at the frequency of 2592 MHz. In addition, the optimum synchronous phase and the lattice were investigated to satisfy the requirements of the total emittance less than 1.5π mm mrad and the momentum spread less than 0.1% in RMS.

1. Introduction

The world experimental average value of the muon anomalous magnetic moment, g−2, deviates from the standard model prediction by 4.2 times the standard deviation [1]. This discrepancy may be a sign of new physics beyond the standard model. The muon electric dipole moment, EDM, which is unobserved, could also be enhanced if new physics exists.

The experiment in J-PARC plans to validate the discrepancy of muon g−2 and to search muon EDM with novel techniques: low emittance muon beam, compact storage magnet, no electrostatic focusing, and three-dimensional spiral injection [2]. The low emittance muon beam is generated by thermalized muonium ionization followed by their acceleration by the muon LINAC. The requirements for beam quality are low transverse normalized emittance of approximately 1.5π mm mrad and momentum of 300 MeV/c with a small spread of less than 0.1% in root-mean-square (RMS).

The muon linac consists of four kinds of structures: radio frequency quadrupole (RFQ), inter-digital H-mode drift-tube linac (IH-DTL), disk and washer coupled cavity linacs (DAW-CCL), and disk-loaded traveling wave structures (DLS), matching the varying muon velocity for the short-time — sufficiently shorter than the muon lifetime of 2.2 µs — acceleration to suppress decay loss. The DLS section consists of four DLSs approximately 2 m in length and takes charge of acceleration from 40 MeV to 212 MeV in kinetic energy, corresponding velocity range of 70-94% of the speed of light. The transport line includes two quadrupole magnets between each DLS. The requirement for the DLS is a high accelerating gradient of 20 MV/m.

2. Structure design

Since DLS is an accelerating structure that has been in use for relativistic electron acceleration and has proven to have a sufficiently high gradient, we apply it for high-velocity muon acceleration. The essential feature of DLS for muon acceleration, different from DLS for electron acceleration, is that disk spacing (cell length) varies proportionally to muon velocity [3]. Since assuming a constant gradient is convenient for the velocity calculation of muon through the DLS section, a quasi-constant gradient type with linear tapering iris apertures is adopted.

An operating frequency is one of the most important parameters in an accelerating structure design. Therefore, we consider two different operating frequencies and summarize the parameters of two types of DLS calculated using SUPERFISH [4] in table 1. At first, the L-band DLS was designed with an operating frequency of 1296 MHz, the same frequency as DAW-CCL, to ensure longitudinal acceptance. However, this structure required excessive RF power compared with L-band klystron power of 40 MW to obtain an accelerating gradient of 20 MV/m [5]. Therefore, the S-band DLS with twice the operating frequency is designed for higher efficiency and shows its accelerating gradient is approximately 60% higher than the L-band DLS with the same power. A narrower iris aperture results in a higher accelerating gradient, but the current iris aperture of each DLS is determined so that group velocity is faster than 1%c to ensure sufficient coupling strength for field stability [5].

Table 1. Parameters of the first cell in the DLS section of two different frequencies.

Parameters	L-band[5]	S-band
Structure type	disk-loaded traveling wave quasi-constant gradient type	
Resonant mode	TM01-$2\pi/3$	
Frequency [MHz]	1296	2592
Muon velocity ($\beta = v/c$)	0.695	
Cell length (D) [mm]	53.698	26.885
Iris aperture ($2a$) [mm]	43.370	25.875
Cylinder diameter ($2b$) [mm]	181.147	92.070
Disk thickness [mm]	5.000	
Quality factor (Q)	17116	11289
Shunt impedance (Z) [MΩ/m]	29.15	31.16
Group velocity (v_g/c) [%]	1.22	1.52
Field attenuation factor (α) [/m]	0.0648	0.163
Cell accelerating gradient (E_{acc}) [MV/m/MW$^{1/2}$]	1.94	3.13

For the above consideration, we select S-band and design four DLSs with the synchronous phase (ϕ_s) of −13 degrees, determined based on the discussion in the next section. The designed and calculated cell parameters are plotted in figure 1. Some parameter symbols are as in table 1 and others are the kinetic energy (W), the RF power (P), the peak surface electric field (E_{surf}), and the modified Poynting vector (S_c) [6]. W is calculated assuming the constant energy gain per unit length:

$$\Delta W = e\overline{E_{\text{acc}}}\cos\phi_s, \tag{1}$$

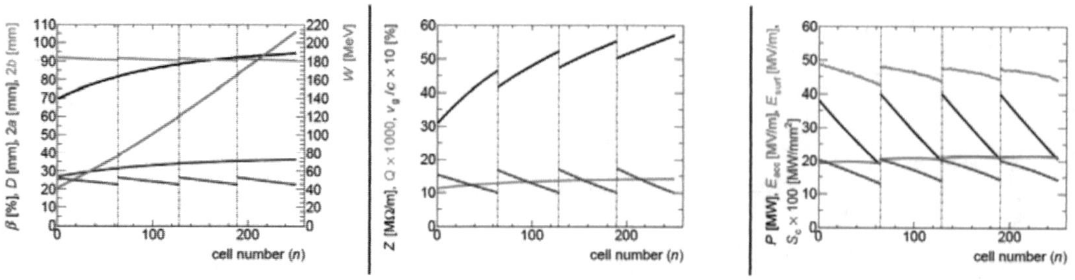

Figure 1. The S-band DLS cell parameters; first DLS (n=1-64), second DLS (n=65-128), third DLS (n=129-189) and fourth DLS (n=190-250).

where $\overline{E_{\mathrm{acc}}}$ is the average of E_{acc} for each DLS: 19.6, 20.8, 21.3, and 21.4 MV/m. β is naturally obtained by W. Since we choose $2\pi/3$ mode, the cell length is determined as:

$$D = \beta\lambda/3, \tag{2}$$

where $\lambda = 115.661$ mm is the wavelength of 2592 MHz RF. $2a$ is optimized by iterative calculation with the different taper angle until E_{acc} at the first and last cells become the same. After optimization, the distribution of E_{acc} is arched and the difference of E_{acc}, defined as $(\max(E_{\mathrm{acc}}) - \min(E_{\mathrm{acc}}))/\overline{E_{\mathrm{acc}}}$, is approximately 2% for all DLSs. $2b$ is adjusted to have a resonant frequency of 2592 MHz for each cell. Z and Q depend on D, and v_g depends on $2a$. The RF power of DLS1 is 38 MW which is tuned to obtain a 300 MeV/c muon beam at the end of the DLS section. The power dissipation is calculated considering a 3% deterioration of Q. E_{surf} and S_c are evaluated as the criteria for accelerating field stability in the situation of 40 (or 38) MW flat-top RF input at a pulse length of 1 µs. Since the Kilpatrick limit [7] at 2592 MHz is 43.8 MV/m, the bravery factor is within the range from 0.97 to 1.11. S_c is sufficiently below the safety threshold of 5 MW/mm^2 [6] at a pulse length of 200 ns.

3. Beam dynamics simulation

The longitudinal acceptance depends on ϕ_s. An approximate separatrix at the entrance of the DLS section is shown in figure 2, where assuming $\overline{E_{\mathrm{acc}}} = 20$ MV/m, $\phi_s = -13$ degrees, and no velocity change. The contour plot in figure 2 shows the simulated beam distribution after acceleration through DAW-CCL. The ratio of the loss particles, out of the separatrix, are estimated to be approximately 0.6% of the total. Although a larger acceptance is desirable considering the possibility of the emittance growth due to the beam mismatch upstream, a larger $|\phi_s|$ results in a smaller energy gain and a larger transverse RF defocusing force.

An average RF defocusing force per cell is described by the approximate formula

$$F_x(x) = -\frac{\pi}{\beta\gamma^2}\frac{x}{\lambda}eE_{\mathrm{acc}}\sin\phi_s = -m_\mu c^2\beta^2\gamma K_1 x, \tag{3}$$

where x is the horizontal distance from the beam axis, m_μ is the muon mass, γ is the relativistic factor, and K_1 is the quadrupole focusing strength. The lattice function of the first DLS and its downstream transport line is defined as shown in figure 3. The field strength of each quadrupole magnet is optimized considering the Twiss-beta function. We conduct the parameter scan by the horizontal/vertical phase advance ($\sigma_{\mathrm{x/y}}$) and find the optimum lattice whose maximum Twiss-beta in the first DLS is minimum under the condition that the phase advance is less than 80 degrees considering the possibility of beam instability. The gradient of each quadrupole magnet is approximately 20 T/m. The Twiss-beta functions of the optimized lattice under periodic boundary conditions are plotted in figure 3.

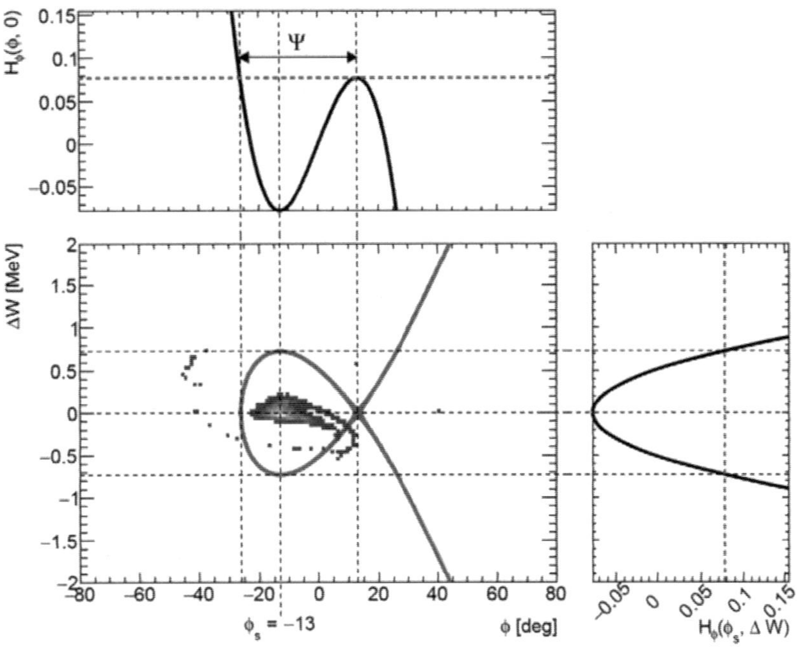

Figure 2. The longitudinal beam distribution including the separatrix and the Hamiltonian H_ϕ.

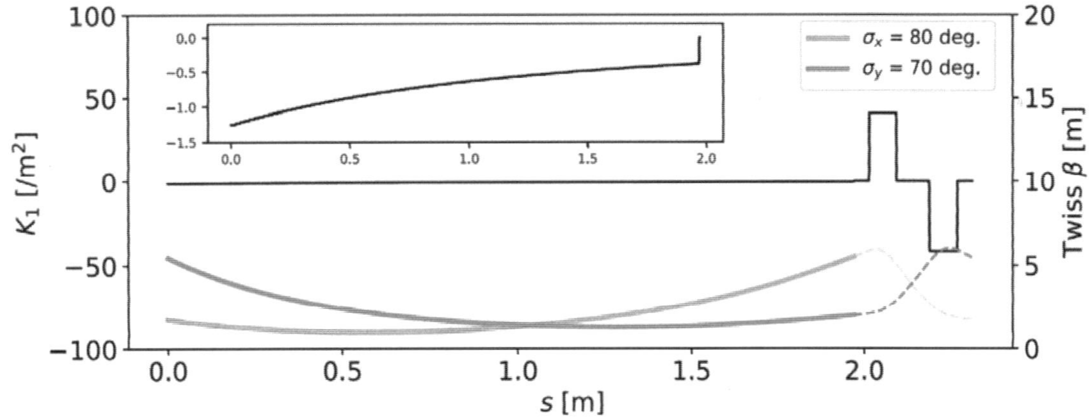

Figure 3. The lattice function of the first DLS section and two quadrupole magnets and the Twiss-beta function under periodic boundary conditions.

The particle tracking simulation is conducted using GPT [8]. As a first step, the simulation with a reference muon is performed to validate the phase slip. Due to the quasi-constant gradient, the difference between actual and ideal energy gain causes a phase slip. The large phase slip can result in large beam loss because of narrow longitudinal acceptance. The cell phase slip is defined as $\Delta\phi = \varphi_{\text{beam}} - \varphi_{\text{RF}}$, where φ_{beam} corresponds to the time beam takes to fly per cell, and $\varphi_{\text{RF}} = 120$ degrees is ideal RF phase advance per cell. $\Delta\phi$ and the cumulative phase slip are evaluated by ideal muon tracking as shown in figure 4. In the actual machine, it is estimated

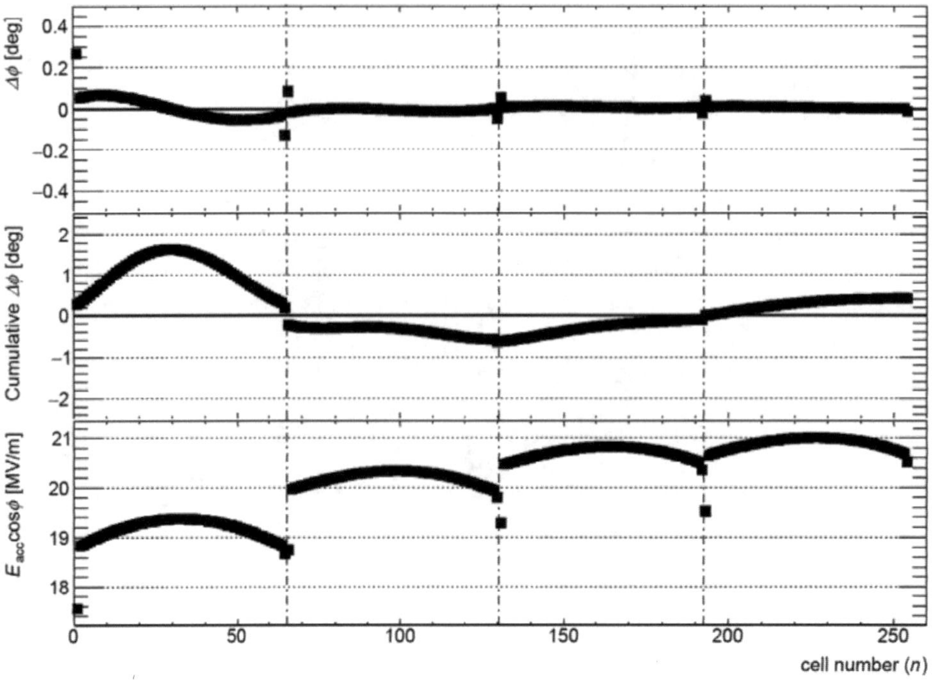

Figure 4. The cell phase slip (top), the cumulative phase slip (middle), and the average electric field which affects the ideal muon.

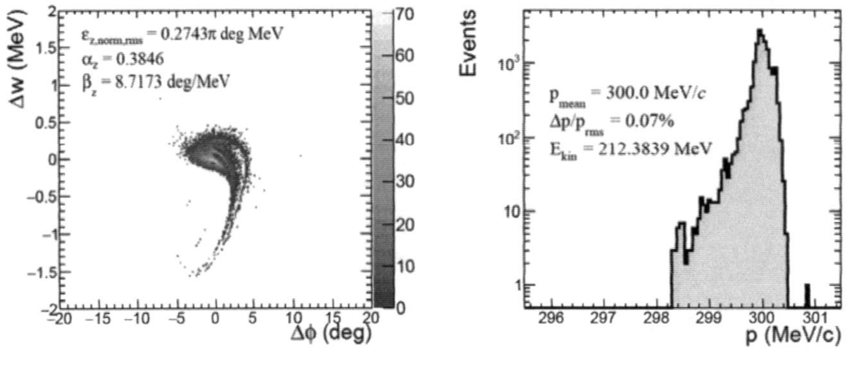

Figure 5. The longitudinal beam distributions at the exit of the DLS section.

that the phase slip of a few degrees to φ_{RF} occurs due to frequency error caused by machining error. Therefore, compared to the phase slip due to machining error, the $\Delta\phi$ of up to 1.5 degrees is sufficiently suppressed.

Then, the particle tracking simulation is conducted using realistic beam distribution estimated from the upstream accelerator simulation. The accelerated beam distributions are shown in figure 5. The momentum spread is 0.07% in RMS and is sufficiently small. Figure 6 shows the beam envelope and the transverse emittance through the DLS section. The envelope is enough smaller than DLS apertures and the normalized RMS emittance does not grow during acceleration thanks to beam matching at the entrance of the DLS section using the initial condition obtained in figure 3. The transmission rate is 99.4%, and all beam loss is due to longitudinal acceptance.

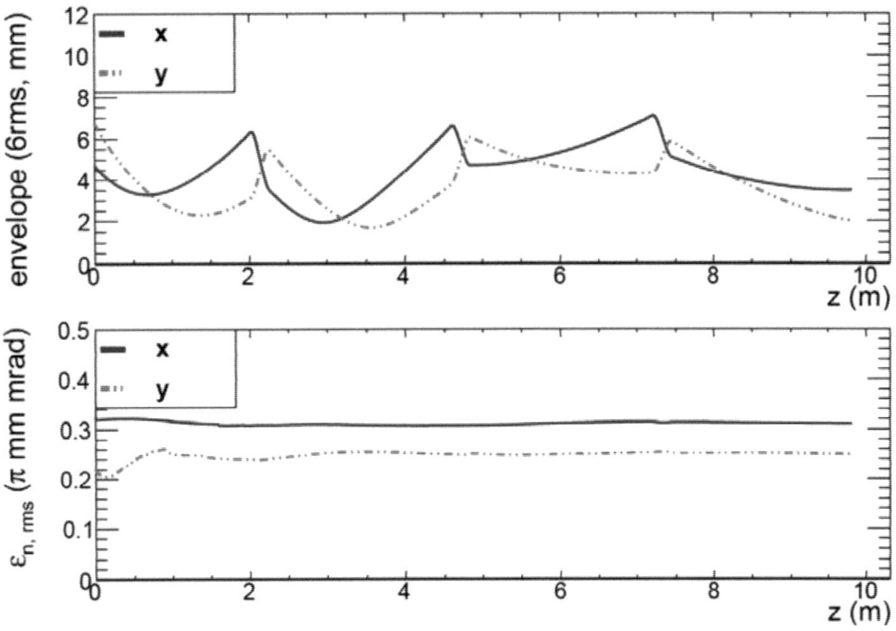

Figure 6. The beam envelope of six RMS (top) and the transverse normalized RMS emittance through the DLS section.

4. Summary and prospects

We discussed the operating frequency, the synchronous phase, and the lattice of the DLS section for muon acceleration. The designed S-band DLS is simulated to accelerate the muon beam with sufficient accelerating gradient without large phase slip and to satisfy the requirements for momentum spread and emittance. We are now designing the coupler cell and will fabricate the first and last regular cells and two coupler cells to evaluate actual parameters and to demonstrate the tuning method.

Acknowledgments

This work is supported by JSPS KAKENHI Grant Numbers JP18H03707, JP18H05226, JP20H05625, 21K18630, 21H05088, 22H00141, 22J20870, JST FOREST Program (Grant Number JPMJFR212O), and the natural science grant of the Mitsubishi Foundation. This paper is based on results obtained from a project commissioned by the New Energy and Industrial Technology Development Organization (NEDO).

References

[1] Abi B *et al.* 2021 *Phys. Rev. Lett.* **126** 141801
[2] Abe M *et al.* 2019 *Prog. Theor. Exp. Phys.* **2019** 053c02
[3] Kondo Y *et al.* 2017 *J. Phys.: Conf. Ser.* **874** 012054
[4] Billen J H and Young L M 1996 *Poisson Superfish* (LA-UR-96-1834)
[5] Sumi K *et al.* 2021 *Proc. 18th Annu. Meet. PASJ (PASJ2021)* (QST-Takasaki Online, Japan) pp 133-7
[6] Grudiev A *et al.* 2009 *Phys. Rev. Spec. Top. Accel. Beams* **12** 102001
[7] Kilpatrick W D 1953 Tech. rep. University of California Radiation Laboratory (UCRL-2321)
[8] Pulser Physics General Particle Tracer `http://www.pulsar.nl/gpt/`

The Zgoubidoo python framework for ray-tracing simulations with Zgoubi: applications to fixed-field accelerators

M Vanwelde[1], C Hernalsteens[1,2], R Tesse[1], E Gnacadja[1], E Ramoisiaux[1], N Pauly[1]

[1] Service de Métrologie Nucléaire, Université libre de Bruxelles, Brussels, Belgium
[2] CERN, 1211 Geneva 23, Switzerland

E-mail: marion.vanwelde@ulb.be

Abstract. The study of beam dynamics in accelerators featuring main magnets with complex geometries such as Fixed Field Accelerators (FFAs) requires simulation codes allowing step-by-step particle tracking in complex magnetic fields, such as the Zgoubi ray-tracing code. To facilitate the use of Zgoubi and to allow readily processing the resulting tracking data, we developed a modern Python 3 interface, Zgoubidoo, using Zgoubi in the backend. In this work, the key features of Zgoubidoo are illustrated by detailing the main steps to obtain a non-scaling FFA accelerator from a scaling design. The results obtained are in excellent agreement with prior results, including the tune computation and orbit shifts. These results are enhanced by Zgoubidoo beam dynamics analysis and visualization tools, including the placement of lattice elements in a global coordinate system and the computation of linear step-by-step optics. The validation of Zgoubidoo on conventional scaling and non-scaling FFA designs paves the way for future uses in innovative FFA design studies.

1. Introduction

The magnetic field of FFAs is constant in time, which allows rapidly accelerating different kinds of beams, in particular high-intensity beams (high repetition rate), in a wide energy range (variable energy beam). Their magnetic structure also enables strong focusing, either focusing using alternating fields or focusing using spiral magnet edges. These key points allow FFAs to be accelerators of choice for some applications: FFAs have already been considered for neutron spallation, muon accelerators, and medical applications [1–5]. There are two main variants of FFAs: scaling FFAs and non-scaling FFAs. Scaling FFAs, which were the first to be studied, allow having constant tunes with energy by imposing a strong condition on the magnetic field, while non-scaling FFAs relax this condition to have easier to build magnets and to reduce the orbit shift at the expense of a non-zero tune variation. The complex geometry and magnetic fields and the absence of a true reference trajectory due to its energy dependence make simulation codes featuring step-by-step tracking the tools of choice. The Zgoubi ray-tracing code [6–8] allows tracking particles in any magnetic and electric field, particularly in highly non-linear magnetic fields, which is the case for FFAs. In addition, Zgoubi realistically models the magnet fringe fields and allows the user to implement complex and arbitrary misalignment of the magnetic elements, and shape of the magnet edges. It can be used for the optics study, helpful for the accelerator design, but also for the final study of the lattice, using measured field maps with realistic fringe fields and possible magnetic imperfections. Because of these advantages, Zgoubi has been extensively used to study FFA lattices [5, 9]. To facilitate the use of Zgoubi and allow data analyzes and simplified visualizations, we developed a modern Python 3 interface using Zgoubi in the backend: Zgoubidoo [10]. This library provides a fully featured and easy-to-use interface to Zgoubi. It can be used both to generate Zgoubi input files and run the simulations in parallel, as well as to perform more advanced tracking results analyzes, including linear and non-linear beam dynamics studies. Additionally, Zgoubidoo also has modules ensuring the correct placement of the lattice elements in space and the plotting of the beamline and tracking results, including particle trajectories in the global reference frame. Zgoubidoo thus has all the necessary tools to study all kinds of accelerators, especially FFAs. This work illustrates the main features of Zgoubidoo on well-known scaling and non-scaling FFA designs [11, 12]. First, the main steps to model and simulate a scaling lattice with a large field index [11] are presented. Then, the magnet simplification steps allowing to transform this scaling design into a non-scaling one following the steps presented in [12] are detailed. The complete analysis of conventional FFA lattices with Zgoubidoo, from the closed orbit search to the linear tune computation, validates its different modules and paves the way for further studies of FFA designs.

2. Model definition of the scaling lattice

The first lattice studied with Zgoubidoo is scaling and consists of 12 identical triplet cells, with radial focusing (sector-alternating field) and a nominal energy of 250 MeV. The working point of this lattice is chosen in the second stability region of Hill's equations, with a phase advance per cell larger than 180°, which implies choosing a large field index. It allows reducing the orbit shift while remaining a scaling lattice and therefore having energy-independent tunes [11]. Zgoubidoo allows studying this lattice: closed orbit search, particle tracking, linear and non-linear beam dynamics, and parametric study to set the working point. Zgoubidoo's class structure makes it straightforward to create the lattice with a few efficient commands: it is possible to build a Zgoubi input file with realistic default values for the element parameters in Zgoubidoo. As an example, figure 1 shows the simple creation of a cell consisting of three FFA magnets.

For FFA components, Zgoubi allows having up to 5 FFA magnets in the same element (keyword 'FFAG' [13]). In the Zgoubi input file, it involves duplicating blocks of lines to have the required number of magnets and changing the desired parameters in the right place, which is error-prone. Zgoubidoo allows to readily define FFA elements by giving the magnet parameters in the form of lists. After running the simulations on the previously created beamline, Zgoubidoo allows the automatic tracking result collection and the plotting of the retrieved tracks along with a 2D representation of the beamline. The Zgoubidoo plotting module represents the FFA elements with their magnets (and magnet shape) and their entire integration zone. The user can thus verify the correct magnet placement in the FFA element and the correct shape for the magnet edges. Figure 2 shows one of the cells of the considered scaling lattice.

In addition to this plotting module, Zgoubidoo has a powerful survey module that associates a global reference to the ring and thus places all the elements in space. Contrary to Zgoubi, which only uses local frames, Zgoubidoo makes it possible to know the global geometry of the ring, allowing better visualization of the lattice. Figure 3 shows the entire accelerator ring, along with the closed orbits for different energies. We can observe the horizontal orbit shifts and the geometric similarity of the orbits of the particles at different energies, which characterizes a scaling machine. Additionally, this figure shows the difference in orbit shift between working points set in the two stability regions: the orbit shift for the working point in the second stability region is equal to 0.17 m, which is significantly reduced compared to the orbit shift in the first stability region (~ 0.83 m). These results correspond perfectly to prior results [11].

```
kin = zgoubidoo.Kinematics(250 * units.MeV)
beam_for_tracks = Objet2('BUNCH', BORO=kin.brho)
beam_for_tracks.add(np.array(closed_orbit))
FDF_cell = FFAG(f"FDF_cell",
    N= 3, RM = r0, AT = cell_angle,
    ACN=[10.20, 15.0045, 19.80741, 0, 0] * units.degree,
    BZ0=[B0F, B0D, B0F, 0.0, 0.0] * units.kilogauss,
    K=[k, k, k, 0.0, 0.0],
    OMEGA_E=[1.2, 1.2, 1.2, 0, 0] * units.degree,
    OMEGA_S=[-1.2, -1.2, -1.2, 0.0, 0.0] * units.degree,
)
zi = zgoubidoo.Input(name ="FDF_cell",
    line = [beam_for_tracks, FDF_cell])
```

Figure 1. Code example that builds the lattice cell consisting of three FFA scaling magnets and the reference particle with a few Zgoubidoo commands. The magnet parameters are given in the form of lists, allowing to readily manage Zgoubi keywords with several magnets in their integration zone.

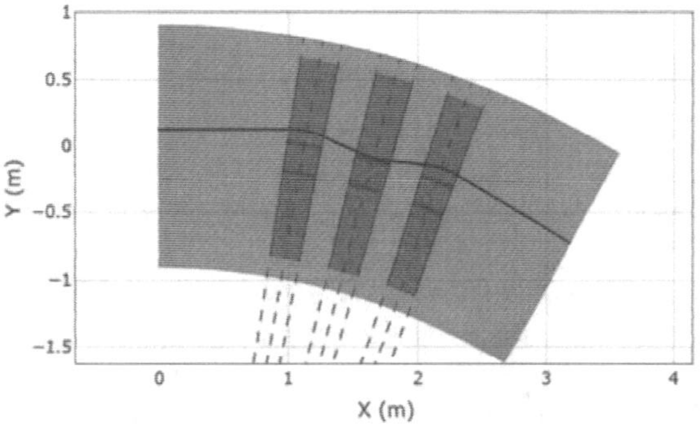

Figure 2. One cell of the model scaling FFA lattice. Zgoubidoo's plotting module allows a 2D representation of the cell. The FFA magnets are shown in dark blue, while the FFA element integration zone is shown in light blue. The dotted lines indicate the straight magnet faces and the mean radius of the FFA element.

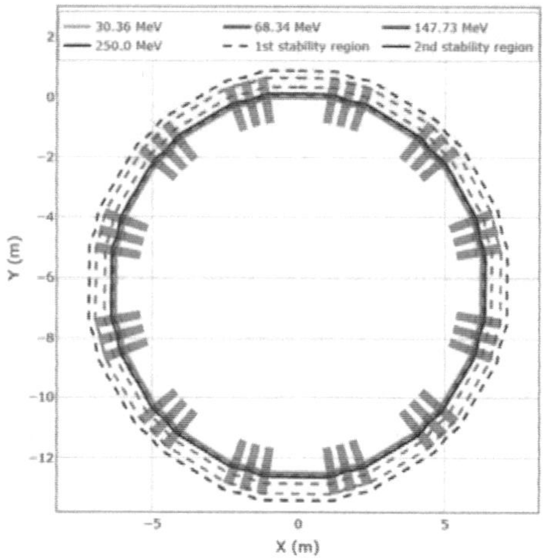

Figure 3. Representation of the entire ring. Zgoubidoo's survey module correctly places the elements and the particle trajectories in a global reference frame.

3. From a scaling to a non-scaling design

It is possible to transform the design of the scaling lattice to obtain a non-scaling lattice with simplified and more compact magnets. We have reproduced with Zgoubidoo different magnet simplification steps allowing such a design transformation [12]: the first step consists in expanding the scaling FFA non-linear field in terms of multipoles and truncating this series at the lowest orders (up to n=5 for a dodecapole); The second step transforms the sector-magnets into rectangular-shaped magnets; The last simplification consists in the alignment of the three magnets parallel to each other with different transverse offsets between the magnets. Figure 4 shows the three simplifications on the baseline cell.

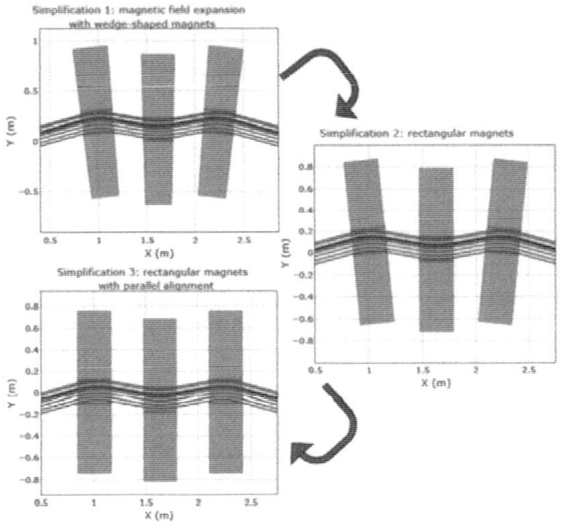

Figure 4. Non-scaling cell evolution with the different magnet simplifications presented in Ref. [12]. The cell consists of three magnets whose field is a truncated multipolar expansion. The magnets are first sector then rectangular, either facing the ring center or aligned with each other.

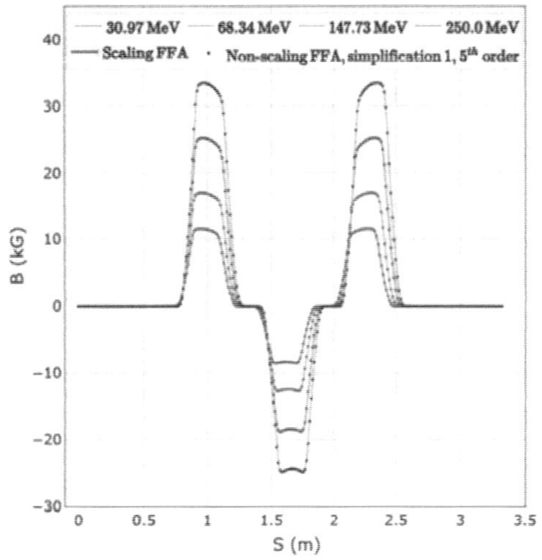

Figure 5. Magnetic field along the particle trajectory at different energies in a cell composed of scaling FFA elements and a cell with magnets whose magnetic field is truncated at the 5th order.

To adapt the lattice for each simplification, in addition to changing the magnet shape and type, it is necessary to fit the magnetic field and radius of each element to have the same integrated magnetic field as in the scaling FFA element. Zgoubidoo simplifies the use of the complex Zgoubi FIT command, especially by managing the numbering associated with the parameters of each magnet. Zgoubidoo also allows to readily retrieve the results of a FIT command by storing them in easily usable data structures in Python. By looking for the appropriate magnet parameters (nominal magnetic fields and offsets), it is possible to have an integrated field equivalent to that obtained in a scaling FFA. Figure 5 shows, for the first simplification, the magnetic field seen by particles at different energies, either in a scaling FFA element or in a magnet whose magnetic field is truncated at order 5. By correctly adapting the magnet parameters, the magnetic field is very similar in both cases. The magnets being rectangular for the other two simplifications, it is no longer possible to have closed orbits comparable to those in the scaling FFA. The magnetic field seen by the particles will thus be different, but the integral of the magnetic field in the successive magnets will always remain close to the scaling value.

Finally, Zgoubidoo allows the computation of linear optics at each integration step. Tracking results from Zgoubi are stored in data frames for data analysis: the user no longer has to read text files but can do everything using Python 3 and associated libraries (pandas, NumPy). From the tracking results in the Frenet-Serret reference frame, Zgoubidoo can compute the transfer matrices at each integration step, as well as the Twiss parameters. As an example, figure 6 shows the β-functions for two energies and a truncation of the multipole expansion at the 3rd order or the 5th order. We observe that the β-functions are more similar for a higher-order truncation because a particle with an offset will see the influence of higher-order terms in the magnetic field.

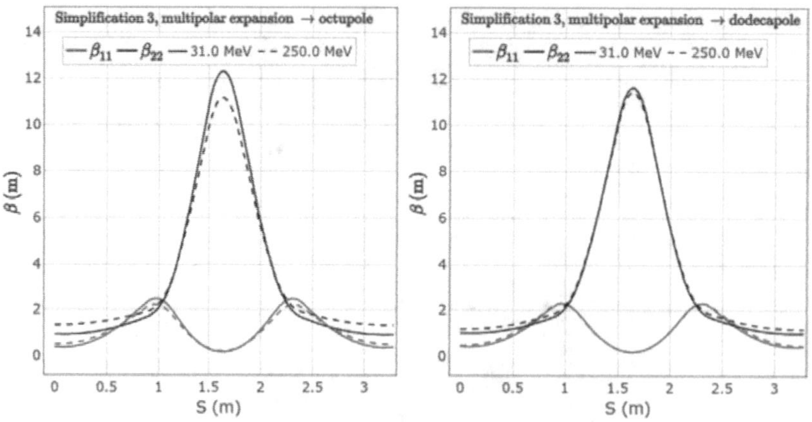

Figure 6. Horizontal and vertical β-functions (respectively β_{11} and β_{22}) for different energies obtained for the third simplification and a multipolar expansion truncated at the 3rd or 5th order.

Finally, it is also possible to compute the cell and ring tunes for different energies and for all the simplifications. Figure 7 shows the excellent agreement between the tune evolution obtained with Zgoubidoo and the original results obtained in Ref [12].

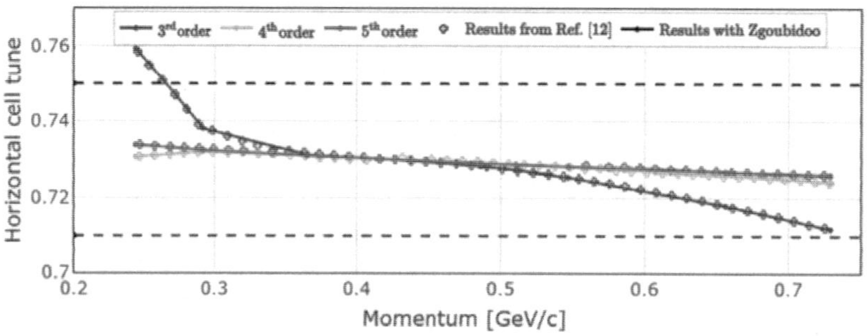

Figure 7. Horizontal tune evolution with momentum, for the last magnet simplification step and different truncation orders for the field multipolar expansion. There is an excellent agreement between original results using Zgoubi [12] and results obtained with Zgoubidoo.

4. Conclusion

We used Zgoubidoo to study FFA lattices and illustrate the steps to transform a scaling FFA design into a non-scaling one. It has been shown that Zgoubidoo significantly enhances the use of Zgoubi, from the preparation of parametrized input files to the processing of tracking data, while offering additional advanced analysis and visualization. Zgoubidoo provides a set of numerical tracking methods and tools allowing the beam dynamics analysis, from the search for closed orbits to the computation of the linear and non-linear beam dynamics. The study of conventional FFA lattices with Zgoubidoo provides results in excellent agreement with prior results, including the correct linear optics computation, which shows that Zgoubidoo can be used extensively for further studies of FFAs accelerators.

Acknowledgments
Marion Vanwelde is a Research Fellow of the Fonds de la Recherche Scientifique – FNRS.

References

[1] Machida S 2021 *Proc. 3rd J-PARC Symposium (J-PARC2019)* (Tsukuba, Japan) p 011005
[2] Sheehy S L *Proc. Cyclotrons'16* (Zurich, Switzerland) pp 374–9
[3] Méot F 2019 RACCAM: An example of spiral sector scaling FFA technology Tech. Rep. BNL–211536-2019-NEWS, 1507116
[4] Antoine S *et al.* 2009 *Nucl. Instrum. Methods Phys. Res., Sect.* A **602** 293–305
[5] Peach K J *et al.* 2013 *Phys. Rev. Spec. Top. Accel Beams* **16** 030101
[6] Méot F 1999 *Nucl. Instrum. Methods Phys. Res., Sect.* A **427** 353–6
[7] Méot F and Berg J 2020 Zgoubi `http://sourceforge.net/projects/zgoubi/`
[8] Méot F 2014 *Nucl. Instrum. Methods Phys. Res., Sect.* A **767** 112–25
[9] Fourrier J, Martinache F, Méot F and Pasternak J 2008 *Nucl. Instrum. Methods Phys. Res., Sect.* A **589** 133–42
[10] Hernalsteens C, Tesse R and Vanwelde M 2022 Zgoubidoo `https://ulb-metronu.github.io/zgoubidoo/`
[11] Machida S 2009 *Phys. Rev. Lett.* **103** 164801
[12] Sheehy S, Peach K, Witte H, Kelliher D and Machida S 2010 *Phys. Rev. Spec. Top. Accel Beams* **13** 040101
[13] Méot F Zgoubi users' guide Tech. Rep. BNL-98726-2012-IR

Recent results of beam loss mitigation and extremely low beam loss operation of J-PARC RCS

P K Saha[1], K Okabe[1], T Nakanoya[1], Y Shobuda[1], H Harada[1], F Tamura[1], H Okita[1], M Yoshimoto[1] and H Hotchi[2]

[1] J-PARC Center, JAEA and KEK Joint Project, Tokai-mura, Naka-gun, Ibaraki-ken, Japan
[2] KEK, Tokai, Ibaraki-ken, Japan

E-mail: saha.pranab@j-parc.jp

Abstract. In the 3-GeV RCS (Rapid Cycling Synchrotron) at J-PARC (Japan Proton Accelerator Research Complex), multi-turn H^- charge-exchange injection is performed by using a thin stripped foil. The residual radiation at the injection area caused by the uncontrolled beam loss occurred by foil scattering of the circulating beam is a serious issue for regular maintenance works. In addition, the beam loss at the collimator section and its downstream caused for a large emittance beam also should be reduced, especially at high intensity operation. For that purpose we have minimized injection beam size and implemented a smaller size stripper foil. The circulating beam hitting rate is reduced by using a smaller foil, while an optimized vertical angle of the smaller injection beam for vertical transverse painting also gave a reduction of the circulating beam emittances and resulted a further significant beam loss mitigation at the collimator section and its downstream. As a result, the residual radiation after user operation at 700 kW beam power was also measured to be significantly reduced.

1. Introduction

The 3-GeV RCS at J-PARC delivers high intensity proton beam to both MLF (Materials and Life Science Experimental Facility) and the MR (30-GeV Main Ring Synchrotron) [1]. The injection beam energy is 400 MeV, which is accelerated to 3 GeV at repetition rate of 25 Hz and simultaneously delivered to the MLF and MR. The designed beam power is 1 MW (8.33×10^{13} protons/pulse), while at the latest it is 800 kW to the MLF and nearly 800 kW equivalent beam power to the MR. As more than 90 % of the beam is delivered to the MLF, a beam loss reduction in the RCS for operation to the MLF is highly important.

Due to multi-turn (307 turns in 0.5 ms) charge-exchange injection of H^- performed by using a stripper foil, the foil scattering beam losses of the circulating beam is the dominant uncontrolled beam loss and high residual radiation sources at injection area. Figure 1 shows a layout of the RCS injection area and a schematic view of the transverse painting (TP) process performed at injection. The TP at injection is adopted to produce a wider beam profile required by MLF to minimize beam density on the neutron production target as well as to reduce circulating beam hitting rate on the foil by large betatron oscillation occurred by varying horizontal closed orbit with 4 horizontal painting magnets (PBH1-4) and varying vertical angle of the injection beam by using two vertical painting magnets (PBV1,2) placed at the injection beam transport (BT) [2,3]. The average foil hits of the circulating beam at a painting area of 200π mm mrad for the MLF can be kept to only 7, but similar to other facilities the residual radiation at the injection area caused by the foil scattering beam losses is very high even at a lower beam power [4–7]. The foil thickness is 333 µg/cm^2 to achieve a stripping efficiency of 99.7 %, while the un-stripped beams are further stripped to protons by secondary foils and disposed at the injection beam dump (I-Dump) [8]. Due to nonlinear nature of the space charge (SC) effect, the beam loss at high-intensity also increases non-linearly, especially at the collimator section and affects the downstream 1st arc section. The beam loss reduction at the collimator section is also highly important for stable operation at 1 MW beam power in a near future. In this research we have succeeded to reduce both uncontrolled beam losses at the injection area caused by the foil scattering of the circulating beam as well as the beam losses caused by the beam halos at the collimator section and its downstream.

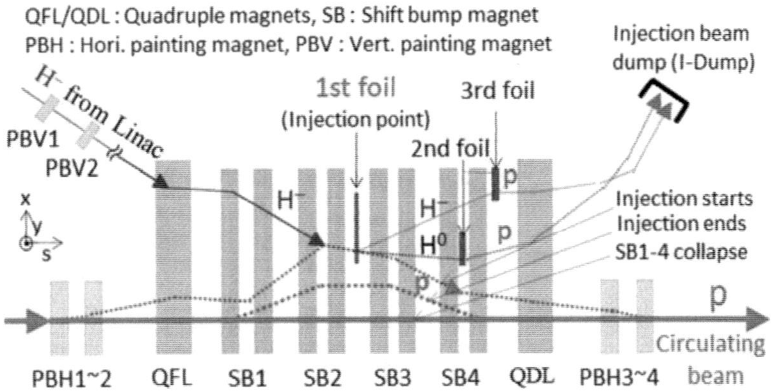

Figure 1. Layout of the RCS injection area and a schematic view of the TP process at injection. The PBH1-4 and PBV1-2 are used for horizontal and vertical painting, respectively.

2. Present approach and numerical simulation results

Figure 2 shows a schematic view of our present approach. We minimized vertical size of the injection beam at the 1st stripper foil by manipulating its vertical twiss parameter (β_y) to reduce vertical size of the foil. The foil hitting rate of the circulating beam and the corresponding beam losses caused by the foil scattering can be thus reduced. A smaller vertical size of the injection beam allows us to reduce both sides of the vertical foil instead of a single side if done for the horizontal size, as the foil is horizontally mounted. Figure 3 shows the measured vertical profiles of the injection beam at the 1st foil location for an original β_y of 8 m and minimized one of 2.4 m depicted by the black and red lines, respectively. The original width of 2 mm (σ) was reduced to nearly half of 1.1 mm to cover well by a vertical foil size of 14 mm. There was no missing H$^-$ measured at the I-Dump. It is worth mentioning that the existing magnet configuration and the aperture of the BT gives a β_y minimization of around 2 m.

Figure 4 shows a schematic demonstration of vertical TP done by changing vertical angle (y') of the injection beam by using VPB1 and VPB2. The edge of the injection beam determines the maximum painting area, where a y' of -3.4 mrad gives a painting area of 200π mm mrad for the MLF. As emittance of the injection beam is unchanged, then the beam ellipse for a smaller β_y is changed as shown in the right figure. The angle of the injection beam can be thus minimized to keep the painting area unchanged. Such an optimization of the vertical painting improves vertical beam distribution by minimizing the number of large amplitude particles as compared to that with a bigger β_y of the injection beam. While due to SC effect the maximum beam emittance goes beyond 200π mm mrad at high intensity, the beam loss caused by the large amplitude particles at the collimator section can also be reduced with a smaller injection β_y.

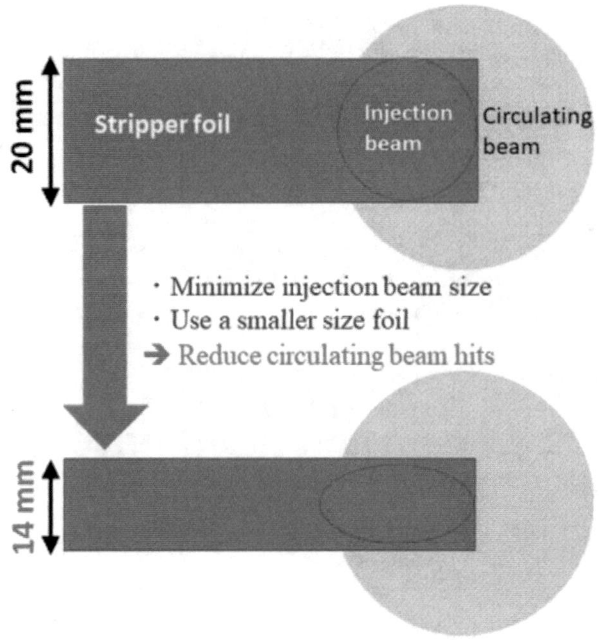

Figure 2. Present approach to reduce circulating beam hits on the foil by minimizing vertical size of the injection beam at the foil and using a smaller size vertical foil.

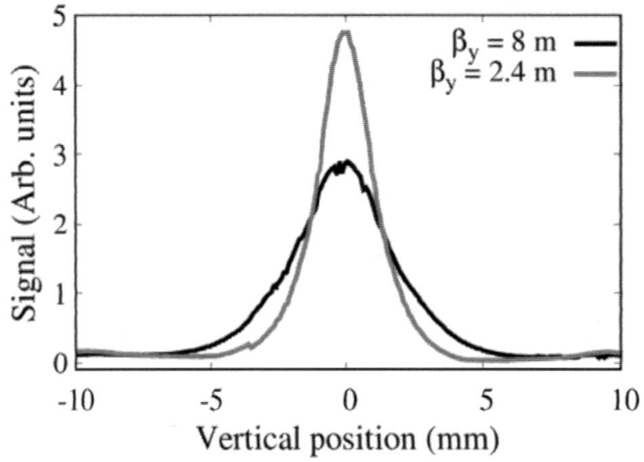

Figure 3. Measured vertical profiles of the injection beam at the foil. The profile width (σ) by manipulating β_y was reduced to 1.1 mm from that of its original 2 mm.

Figure 5 shows numerical simulation results of beam survival for β_y of 8 m (black) and 2.4 m (red). The simulation was done for a beam power of 700 kW by taking into account measured twiss parameters of the injection beam including foil scattering of the beam passing through the foil and all realistic machine parameters. The vertical foil size for a bigger and smaller β_y was 20 mm and 14 mm, respectively. A significant beam survival can be improved with a smaller β_y by reducing the total beam loss as much as 45 %. In this simulation the average of foil hits of each injected particle for a smaller β_y and a smaller foil was also estimated to be nearly 30 % reduced as compared to that for a bigger β_y with a bigger foil. A reduction of the foil hits would

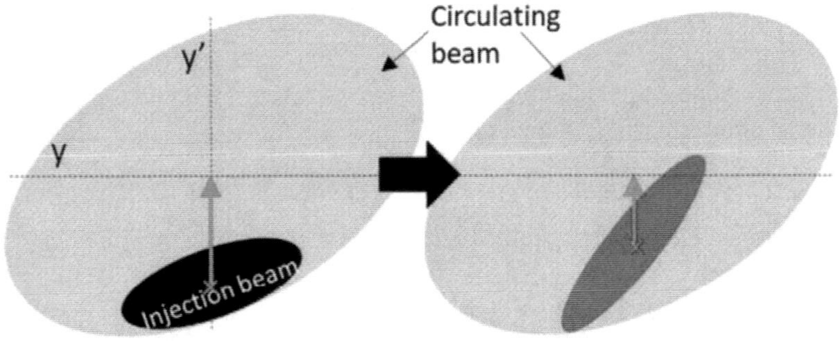

Figure 4. Schematic view of vertical TP done by varying y' of the injection beam. The y' can be minimized for a smaller β_y while keeping a same painting area to reduce large amplitude particles in the circulating phase space.

thus reduce the uncontrolled beam loss and the corresponding residual radiation at the injection area caused by the large angle foil scattering of the beam.

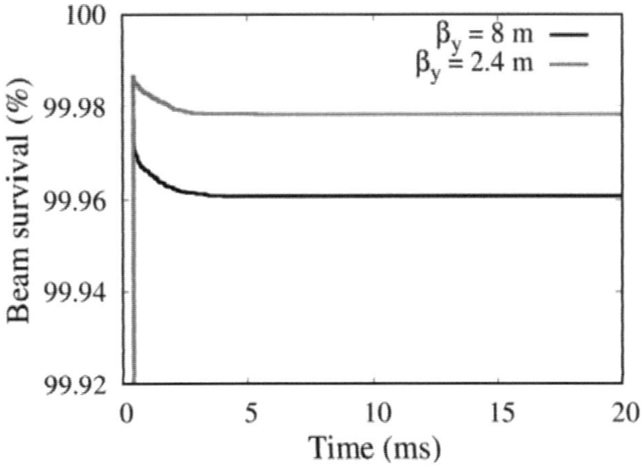

Figure 5. Simulation results of beam survival improvement at 700 kW by minimizing injection β_y from 8 m to 2.4 m and reducing the foil size 20 mm to 14 mm, respectively.

3. Experimental results

Similar to the simulations, we also performed experimental studies for beam loss and foil hit measurements by using two β_y of the injection beam. Figure 6 shows measured signals of the beam loss monitors (BLM) placed at the collimator and the 1st arc sections. Each BLM signal is integrated for the whole cycle of 20 ms (injection to extraction). The measurement was done an equivalent beam power of 700 kW. The beam loss with a minimized β_y of 2.4 m (red) was measured to be 42 % reduced in average as compared to that with an original β_y of 8 m (black), where more than 50 % reductions have also been obtained at several points. It is worth mentioning that the absolute beam losses at the 1st arc section is comparatively much lower than the collimator section. The high voltage of the BLM devices are set higher to measure even a lower beam loss signal. The measurement result is quite consistent with simulation result as shown for a beam survival improvement (figure 5).

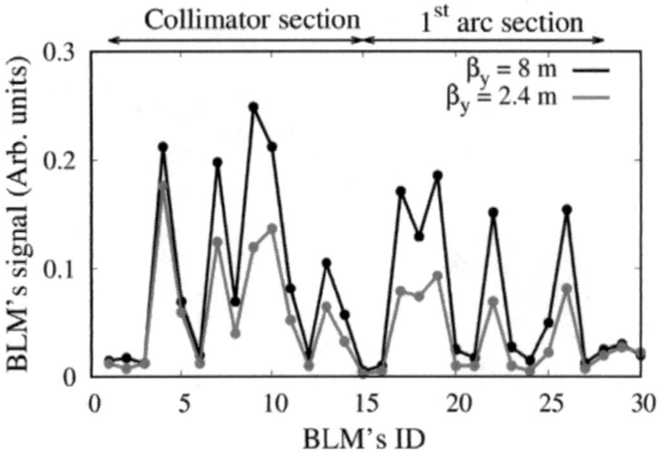

Figure 6. Measured beam loss as a function of BLM's ID at a beam power of 700 kW. A beam loss reduction of 42 % (in average) is obtained by minimizing the injection β_y.

We have also measured a reduction of circulating beam hitting rate on the foil for a smaller β_y and a smaller foil at 1 MW beam power. The measurement was done by using a plastic scintillator counter type BLM placed 90° above the foil in the horizontal direction. Secondary particles such as, γ rays were measured, generated from the lost primary protons at the nearby beam pipe due to large angle scattering at the foil. The measurement result is shown in figure 7. A reduction of the signal intensity for a smaller β_y (red) as compared to that for a bigger β_y reflects a reduction of the foil hits, which was obtained to be 30 % from a ratio of the integrated signals, and was also consistent with an expected reduction of 27 %. A similar reduction of the residual radiation at the injection area can be thus expected.

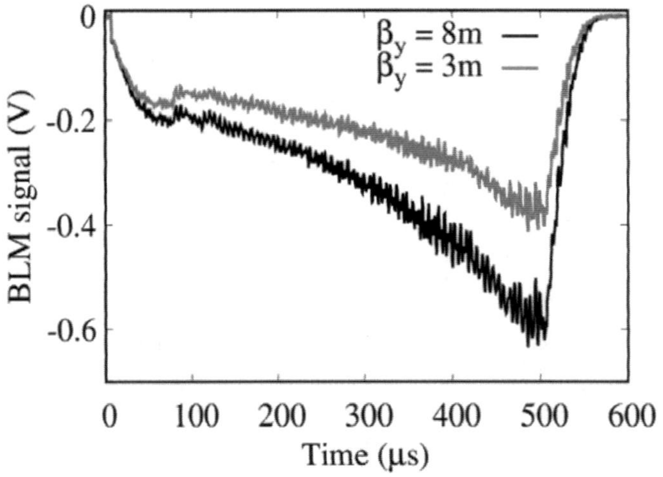

Figure 7. Measured signals of secondary particles generated from the lost primary protons caused to large angle foil scattering during injection period. A foil hit reduction of 30 % was achieved with a smaller β_y.

We have implemented a smaller injection β_y and a smaller vertical size foil of 14 mm for RCS operation at 700 kW beam power. Figure 8 shows a comparison of the measured residual

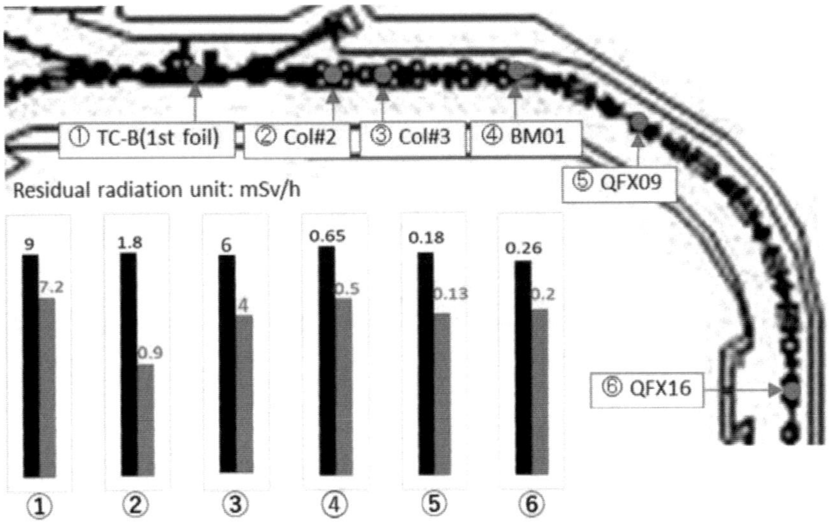

Figure 8. Comparison of residual radiation (on contact) at the injection, collimator and 1st arc sections with a bigger (black) and smaller (red) β_y measured after 1 month operation at 700 kW beam power. The residual radiation by implementing a smaller β_y was significantly reduced.

radiation (on contact) at the injection, collimator and 1st arc sections after operation with a bigger (black) and smaller (red) injection β_y. For each case, the operation was performed for 1 month and measurement was done after 4 hours cooling from beam stop. Similar to the measured beam loss, the residual radiation was also successfully reduced significantly by implementing a minimized injection β_y. It is worth mentioning that a reduction of the residual radiation outside the collimator section, such as injection and 1st arc section have significant importance due frequent access on these areas for regular maintenance works. At present RCS beam power to the MLF is increased to more than 800 kW, and there is no issues with using a smaller size foil. A smaller size foil with a smaller injection β_y will also be tested at 1 MW beam power at the end of June 2022.

4. Summary

We have achieved a significant reduction of the beam loss at J-PARC RCS by minimizing β_y of the injection beam and reducing vertical size of the stripper foil. The β_y was minimized to 2.4 m from 8 m and the corresponding vertical size (in σ) was reduced to 1.1 mm from 2 mm so as the vertical foil size to 14 mm from 20 mm, respectively. The average foil hits of each circulating proton was measured to be 30 % reduced and was consistent with an estimated value of 27 %. The uncontrolled beam loss at the injection area caused by foil scattering of the circulating beam as well as the beam losses at the collimator section and its downstream at a beam power of 700 kW was obtained to be 42 % reduced in average, which was also consistent to that expected in the numerical simulation. As a result, the residual radiation after 1 month operation at 700 kW beam power with a smaller β_y was measured to be significantly reduced as compared to that with a bigger β_y. A smaller β_y and a smaller size foil have been successfully implemented for the present RCS operation at 800 kW and will also be tested at 1 MW beam power operation in June 2022.

References

[1] High-intensity Proton Accelerator Project Team 2003 Accelerator technical design report for J-PARC Tech. rep. KEK report 2002-13 and JAERI-Tech 2003-044

[2] P.K. Saha *et al.* 2009 *Phys. Rev. Spec. Top. Accel Beams* **12** 040403

[3] Hotchi H *et al.* 2012 *Phys. Rev. Spec. Top. Accel Beams* **15** 040402

[4] Holmes J A and Plum M A 2012 *Proc. 52th ICFA Advanced Beam Dynamics Workshop (HB2012)* (Beijing, China) pp 254–58

[5] Huang M Y, Wang N, Wang S and Zu S Y, 2012 *Proc. 52th ICFA Advanced Beam Dynamics Workshop (HB2012)* (Beijing, China) pp 78–80

[6] Goddard B, Aiba M, Bracco C, Carli C, Meddahi M and Weterings W, 2010 *Proc. 1st Int. Particle Accelerator Conf. (IPAC'10)* (Kyoto, Japan) pp 3951–53

[7] Kato S, Yamamoto K, Yoshimoto M, Harada H and Kinsho M 2013 *Phys. Rev. Spec. Top. Accel Beams* **16** 071003

[8] Saha P K *et al.* 2020 *Phys. Rev. Accel. Beams* **23** 082801

Gap in pagination due to unavailable paper.

Pages 253-258

Investigation, simulation and first measurements of a 2 m long electron column trapped in a Gabor-Lens device

K I Thoma, M Droba and O Meusel

Institute for Applied Physics, Goethe-University Frankfurt, Max-von-Laue Straße 1, 60438 Frankfurt, Germany

E-mail: thoma@iap.uni-frankfurt.de

Abstract. Various Gabor-Lenses (GL) were investigated at Goethe University. Confinements of sufficient electron densities ($n_e \sim 1 \cdot 10^{15}\,m^3$) were reached without any external source of electrons. Focusing of ion beams by low energy was demonstrated, long term stability and reproducibility were approved. Main differences compared to experiments and investigations of the pure non-neutral in Penning-Malmberg traps [1] are higher residual gas pressure and therefore higher collision rates, higher bulk temperatures, self-sustaining electron production process, much higher evaporation cooling rate. GL2000 is a new 2 m long device and was mainly designed for focusing of ion beams in energy ranges up to GeV but also for investigation of non-neutral plasma parameters. The confined electron column is much longer compared to previous constructed Lenses. This makes ion and hadron beam focussing much more efficient, in addition new physical phenomena can be expected and investigated. Simulation results of steady- and thermal equilibrium states with various external parameters and first measurements will be presented. The first operational tests show that it is possible to confine a two-meter long electron column.

1. Introduction

The Gabor-Lens (GL) [2], is an optical device used for focusing or defocusing of charged particle beams by a trapped electron column. Recently, an overview over a wide range of space charge lenses was given by Alexey Goncharov [3]. In a Gabor-Lens the radial confinement is provided by an axial magnetic field produced by coils. The longitudinal confinement is given by a potential well generated by a system of electrodes in axial direction. The electron cloud is created due to rest gas ionization inside the lens, it does not need any external source for the electron production, like other traps for example the Penning trap. The confined electron cloud fulfills plasma criteria for collective behaviour in many cases (Debye screening, number of particles in Debye-sphere, plasma frequency) it can be described as a non-neutral plasma [4]. Over recent years various GLs were systematically investigated at Goethe University. Confinement of sufficient electron density ($n_e \sim 1 \cdot 10^{15}\,m^{-3}$) was reached without any external source of electrons. Focusing of light and heavy ion beams by low energy was demonstrated, long term stability and reproducibility were approved. Thick aperture lenses were built and successfully tested [5]. Diagnostic methods based on optical detection, X-ray detection, RF-detection, rest gas monitor and spectrometer were developed and used for characterization of different non-neutral plasma states [6]. Experiments with GL as a focusing device and as an electron target for low energy ion beams were carried out to examine interactions between beam ions and electron cloud [7]. The confinement of an electron column in a toroidal configuration was also tested [8]. The radius r to length l ratio was subsequently changed from device to device. The focal length, chromatic aberrations of the lens as well as the charge exchange and charge recombination rates need to be investigated in the future.

The electron temperatures ranged typically from 10 eV to 1000 eV in these experiments, Debye lengths were calculated therefore between 1 mm and 10 cm with a typical cloud radius between 1-10 cm dependent on used external parameter settings (anode voltage, magnetic field, residual gas pressure).

2. Motivation

With the new device GL2000, built in 2019, an aspect ratio of r/l = 0.0375 is realized. To our knowledge, this is the lowest aspect ratio with longest overall on-axis distance of about two meters designed to date. The focal length f of a Gabor-Lens is inversely proportional to its length [4].

$$\frac{1}{f} = k^2 \cdot L = \frac{q n_e}{4\varepsilon_0} \frac{L}{W_b}, \tag{1}$$

where k is the focusing strength, L the length of the device and W_b in eV is the energy of the beam.

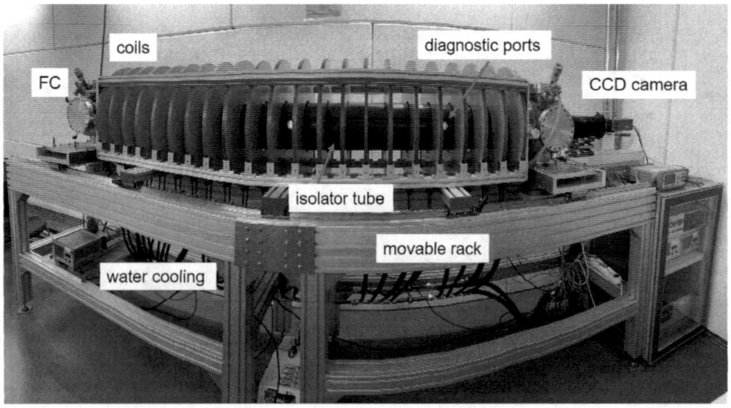

Figure 1. Setup of the experiment GL2000 in a concrete shielding.

For a given electron density $n_e \sim 1 \cdot 10^{15}\,m^{-3}$ GL2000 can be used for the focusing of ion beams with higher kinetic energy W_b compared to previous experiments. Further the focusing of relativistic (W_b in GeV range) highly charged ion beams can be studied and arrays of GL2000 applied as a final focus for fixed target experiments as discussed in NA61/Shine Collaboration at CERN [9]. It also enables new possibilities in research with very long confined static electron clouds.

Static charged particle columns are comparable with ion or electron beams passing through long accelerator or transport chains (e.g. LINAC, Synchrotrons) for long times. Inspired by Paul-Trap-Simulator of the Princton Plasma Physics Laboratory [10] it is planned to investigate the non-neutral plasma confined in GL2000 on the premise that it is a charged particle beam propagating in a periodic transport channel of several kilometers length in an experiment of only two meters. Basic physical properties of the charged particle ensembles can be studied while changing external conditions continuously.

In the GL2000 experiment, it will be possible to excite plasma oscillations, for example through a pulsed magnetic field or by using RF-waves. Due to the length of two meters, it is possible to excite and measure low frequency waves while minimizing the boundary effects.

3. Experimental design

GL2000 consists of a two-meter-long anode tube, which leads to a very small r/l ratio with an inner diameter of 164.3 mm. The anode tube is made of stainless steel and equipped with six diagnostic ports for in situ diagnostics. The magnetic field is generated by 22 water-cooled copper coils in a pancake configuration (figure 1). These are arranged equidistantly over a length of two meters. Small position deviations of about 1 cm only exist in the area of the diagnostic ports. The coil excitation current can be set to a maximum of 500 A, resulting in a maximum on-axis field of $B_z = 0.5\,T$.

The individual pancake connections allow the generation of various magnetic field configurations, such as homogeneous or alternating with existing gradients. Furthermore it is possible to pulse one or more pancakes or to use them as a detector for current fluctuations induced by the confined non-neutral plasma. In a standard electric field configuration, the closing copper electrodes are connected to ground and a positive high voltage is applied to the anode tube. The anode in this setup has been designed for a potential Φ_A up to 30 kV. The ground electrodes are electrically insulated against all flanges, so that investigations in the so-called Penning-mode are possible. In the Penning-mode, in contrast to previous configurations, the anode is at ground potential while the electrodes are charged negatively. The GL2000 experiment will be equipped with various diagnostic instruments that can be installed in the setup or are developed for this experiment.

4. Simulation of electrons inside the GL

First numerical simulations have been performed with the self-consistent code GABORM to evaluate physical phenomena and to compare the numerical model with the experimental results [11]. By the use of an additional numerical code more information about the interaction of the electrons with RGA and RGI or the production rate with respect to the rise time can be studied. Another interesting aspect is the cooling by inelastic electron-atom collisions as a function of the RGA-density or RGA-density gradient. Therefore a numerical algorithm was developed for a simulation of a single electron collision with RGA using Bethe-Bloch formula. It delivers the number of impacts N, the electron velocity v, the collision frequency t, and the mean free path λ for different GL settings. Impact ionization and total cross sections σ are approximated by given parameters [12]. This simple model shows that at the beginning of the confinement only electrons created at the edge of the potential well contribute to the ionization process. The electron collides with atoms with the velocity v at the distance of the mean free path length λ. In doing so it releases energy at the level of the binding energy, in this example, of helium $E_B = 24.59\,eV$. The electron continues to collide with residual gas atoms until it has lost all its kinetic energy.

Using the electron energy and residual gas density as input parameters, the algorithm calculates the electron movement from the first to the last impact. The code helps to estimate the electron production rate and to identify the energy transfer from anode to electron to RGA as a function of confinement time. Later it will be implemented in a Monte-Carlo-Code for detailed simulation of the production, interaction and loss processes in Gabor-Lenses. Figure 2 shows the dependence of the total ionization cross-section for helium as a function of the confinement time of a single electron.

The comparison of production and loss rates will also deliver information about the minimum field configuration in combination with the residual gas pressure which is needed to provide the confinement inside a GL. For a given residual gas pressure of $1 \cdot 10^{-5}$ hPa a minimum anode voltage of $\Phi_{A,min} = 6\,kV$ and a minimum magnetic field strength of $B_{z,min} = 11\,mT$ is needed to provide a stable confinement in GL2000, for example.

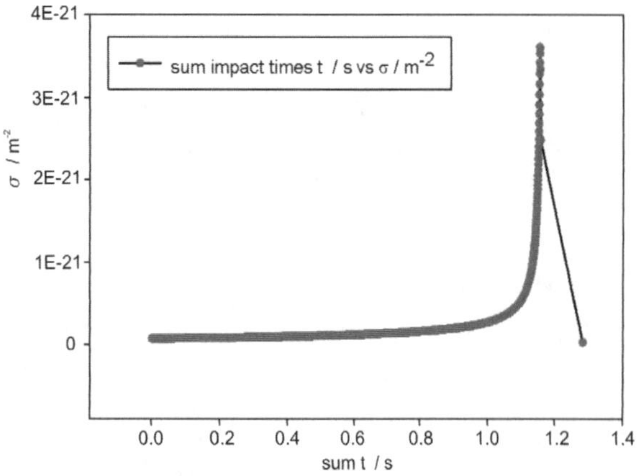

Figure 2. Neutral helium total ionization cross-section as a function of the summation of impact time. In this example the electron energy is set on 30 keV at t=0 s and the density of the residual gas atoms is $1 \cdot 10^{17}\,\mathrm{m}^{-3}$. The electron production rate is $953.125\,\mathrm{s}^{-1}$.

5. First measurements

After construction and assembly, GL2000 was installed in a dedicated concrete shielding because X-ray radiation occurs due to the application of high anode voltage. The wave length of the radiation depends on the longitudinal confinement, with respect to the energy of the confined electrons. The intensity of the radiation is inversely proportional to the magnetic field [13]. This means the higher magnetic field prevents electron losses at the inner surfaces of the lens. On the other hand the electron losses can be used for a scrubbing procedure to remove impurities from the surface. Therefore the so called "conditioning procedure" of GL2000 was started in DC-mode with $\Phi_{\mathrm{A,min}} = 6\,\mathrm{kV}$ to prevent sparks and unwanted discharges. The magnetic field was then reduced from 18 mT to 9.1 mT, as shown in figure 3 to increase the radial electron losses. The modified operation function [7] is a superposition of the maximum densities

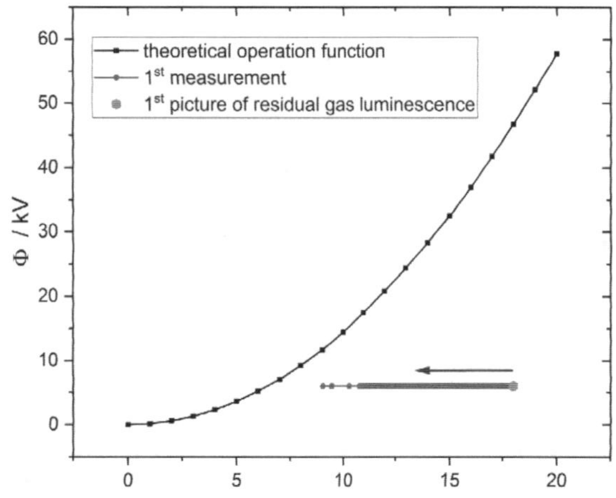

Figure 3. Operation function and first conditioning procedure of GL2000. The arrow shows the direction of the measurements.

$\mathrm{n_{e,max,rad}} = \mathrm{n_{e,max,long}}$ and results in:

$$\Phi_A = \frac{eR_P^2(1 + 2ln(\frac{R_A}{R_P}))B_z^2}{8m_e}, \qquad (2)$$

where $\mathrm{R_P}$ is the radius of the electron cloud and $\mathrm{R_A}$ is the radius of the anode tube.

Thus equation (2) gives the specific operation function (OF) for a Gabor-Lens depending on its geometry. For GL2000 the OF is shown in figure 3. In the first run of GL2000 argon gas was used as residual gas at a pressure of $1 \cdot 10^{-5}$ hPa. The magnetic field was adjusted at 18 mT. After applying an anode voltage the CCD camera observed immediately a residual gas fluorescence induced by the confined electrons, see figure 4. Assuming that the excited state of the atoms has a short lifetime and the argon gas is distributed homogeneously, the luminescence represents the electron density distribution within the GL. In addition to the excitation, electron impact ionization occurs. While the electrons are trapped in the GL, the argon ions are expelled out of the volume in longitudinal direction. The properties of ion stream current and momentum distribution provide a global information about the electron cloud [14]. It also effects to a load on the HV power supply for the anode. During commissioning, the ion current was measured by the Faraday cup. The residual gas pressure rises immediately after applying a new magnetic field strength. The data were measured after the desorption had subsided and the pressure was again constantly close to the originally adjusted value. This means that the data is not determined by the increased residual gas pressure during electron scrubbing.

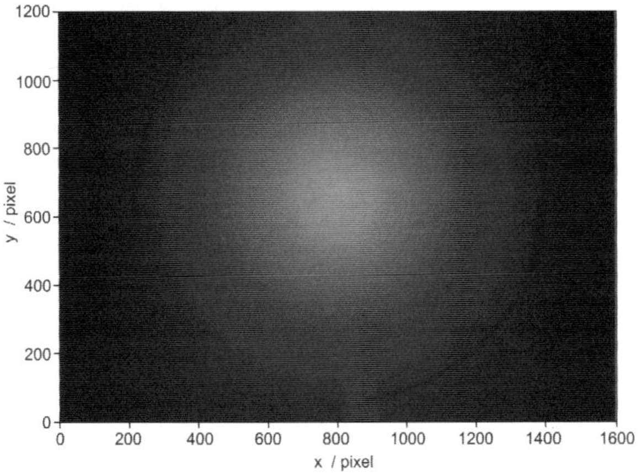

Figure 4. Residual-gas luminescence of argon during electron cloud confinement with GL settings: $\Phi_A = 6$ kV, $B_z = 18$ mT.

The loss current at the anode tube as well as the Faraday cup current decreases for a decreasing B_z.The residual gas luminescence also becomes less intense when the magnetic field is reduced. The probability of ionization and induced fluorescence is a function of the confined electron density, when the average kinetic energy of the ensemble is assumed to be constant. Then load power, ion current and luminescence should correlate with respect to the electron density. The result indicates that the electron density is decreasing. The experimental data show fluctuations in the Faraday cup as well as in the loss current signals in the interval of B_z from 11 mT to 14 mT. These fluctuations also coincided with the visual observations of the residual gas luminescence, which at times became very faint and then increased in intensity. Below 11 mT no confinement was observed.

6. Conclusion

A low aspect ratio Gabor-Lens was designed and constructed for focusing of high energy ion beams. Phase one of the project was the functional test and conditioning of GL2000 and the equipped diagnostic tools. In comparison with other GLs an increased minimum potential and magnetic field for a stable electron confinement at a given pressure was observed. This was assumed due to the low aspect ratio of GL2000 and therefore higher radial electron loss rate.

The first measurement results give an indication of an imbalance between loss and production rates of the electrons confined in GL2000.

In a next step it is planned to measure the electron density as a function of the confinement strength. Experiments for the estimation of radial electron losses are in preparation.

References

[1] Malmberg J H and deGrassie J S 1975 *Phys. Rev. Lett.* **35** 577–80
[2] GABOR D 1947 *Nature* **160** 89–90
[3] Goncharov A 2013 *Rev. Sci. Instrum.* **84** 021101
[4] Meusel O 2006 *Fokussierung und Transport von Ionenstrahlen mit Raumladungslinsen* Ph.D. thesis Goethe University Frankfurt a. M. Germany
[5] Schulte K *et al. Proc. 4th Int. Particle Accelerator Conf. (IPAC'13)* (Shanghai, China) pp 3806–8
[6] Schulte K, Droba M, Meusel O and Ratzinger U *Proc. 10th European Workshop on Beam Diagnostics and Instrumentation for Particle Accelerators (DIPAC'11)* (Hamburg, Germany) pp 266–8
[7] Schulte K 2014 *Studies on the focusing performance of a Gabor lens depending on nonneutral plasma properties* Ph.D. thesis Goethe University Frankfurt a. M. Germany
[8] Thoma K I 2018 *Erste Charakterisierung eines nichtneutralen Plasmas bei toroidalem Magnetfeldeinschluss* Master's thesis Goethe University Frankfurt a. M. Germany
[9] Gazdzicki M 2019 Light ion beams for na61/shine beyond LS3
[10] Gilson E P, Davidson R C, Efthimion P C, Richard M and Hong Q 2003 *Laser Part. Beams* **21** 549–52
[11] Pozimski J and Meusel O 2005 *Rev. Sci. Instrum.* **76** 063308
[12] Kim Y K and Rudd M E 1994 *Phys. Rev.* A **50** 3954–3967
[13] Klaproth S 2017 *Development of a Control System for Space Charge Lenses based on Experimental Data* Master's thesis Goethe University Frankfurt a. M. Germany
[14] Klaproth S *et al. Proc. 8th Int. Particle Accelerator Conf. (IPAC'17)* (Copenhagen, Denmark) pp 166–9

Journal of Physics: Conference Series **2420** (2023) 012043 doi:10.1088/1742-6596/2420/1/012043

Status of the development of the electron lens for space charge compensation at GSI

K Schulte-Urlichs[1,2], M Droba[2], T Dönges[2], O Meusel[2], D Ondreka[1] H Podlech[2], P Spiller[1] and K I Thoma[1,2]

[1]GSI Helmholtzzentrum für Schwerionenforschung GmbH, Planckstraße 1, 64291 Darmstadt, Germany

[2]Institute for Applied Physics, Goethe-University Frankfurt, Max-von-Laue Straße 1, 60438 Frankfurt, Germany

E-mail: k.schulte-urlichs@gsi.de

Abstract. At GSI a prototype electron lens for space charge (SC) compensation is currently being designed and main components as the RF-modulated electron gun are already under commissioning. The goal of this project is the (partial) compensation of SC forces within the ion beam by an overlapping electron beam. This may help to increase the intensity of primary beams, especially in the FAIR facility and potentially all large synchrotrons operated at the SC limit. For an effective SC compensation, the generated electron beam needs to follow the transverse and longitudinal beam profile of the ion bunch structure. The requirements are maximum currents of 10 A and grid modulation to cover a broad frequency range from 400 kHz to 1 MHz. The RF-modulated electron gun was designed and manufactured in the scope of the ARIES collaboration and is currently being tested at the E-Lens Lab of Goethe-University Frankfurt. A dedicated test bench was built for commissioning of the major e-lens components and diagnostics. In this contribution the overall set-up will be presented putting special emphasis on the beam dynamics and collector design as well as simulation results of the electron gun.

1. Introduction

The prototype electron lens for space charge compensation is being designed for the integration into the SIS18 and potentially the SIS100 synchrotron at GSI/FAIR. However, the SIS18 will serve as testbed to demonstrate the space charge compensation scheme and the electron lens with an interaction length of 3.36 m will be installed in one of the available drift sections [1]. The SIS18 ion beam has an elliptical cross section over the length of the interaction region and has to be transversally embedded into a homogeneous electron beam while the pulse needs to be exactly matched to the bunch form in order to avoid over-compensation of the bunch head and tail as presented in [2]. During acceleration of the ion beam the SC tune spread reaches its maximum. Therefore, the prototype electron lens shall support operation within the ramp leading to high requirements on frequency and bandwidth of the e-gun modulator.

A modulated electron gun was designed based on the parameters of the SIS18 and constructed in the scope of the ARIES collaboration WP16 [3]. In the course of this project, a test stand at Goethe-University Frankfurt was build to commission the gun. But it will also serve as test bench for further e-lens components as well as the diagnostics. In the following the status of test stand and the electron lens development with its underlying components will be presented.

2. Electron lens design and beam dynamic studies

The space charge compensation (SCC) electron lens for SIS18 is currently designed at GSI and a 3D model of the actual, integrated layout is presented in figure 1. The layout of the magnetic

Figure 1. Preliminary 3D model of the SCC electron lens.

system comprises two identical solenoids in the gun branch, two solenoids in the collector branch, two toroids, a main solenoid as well as two correction dipoles, all covered in an iron housing. It has been designed for a maximum longitudinal field of 0.6 T. As a result of the vertical field of the toroids two corrector dipoles are required to compensate the deflection of the ion beam. Due to lack of space in the synchrotron, the dipoles are integrated into the toroids but are positioned not to affect the electron trajectories.

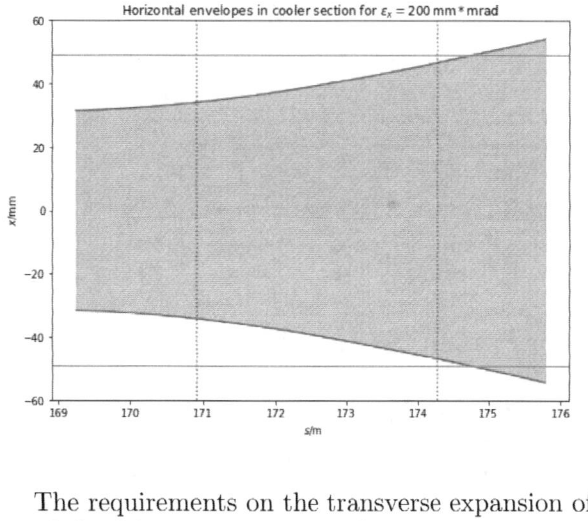

Figure 2. Calculated ion beam envelopes in the interaction section of the SCC electron lens. The red lines mark the interaction region, while the yellow lines represent the required horizontal size of the electron beam.

The requirements on the transverse expansion of the electron beam in the interaction region are defined by an acceptance of ϵ_x=200 mm·mrad. Within the length of the interaction section l=3.36 m the envelope of the ion beam (depicted in figure 2) opens up from 35 mm to 46 mm in x and 22 mm in y.

The ion bunch has to be transversally embedded into the electron beam to avoid resonances as a result of non-linearities by the beams' transverse field. However, the cathode size is limited with respect to homogeneous heating, for this reason it seems favorable not to expand the cathode to the required ion beam radius but rather expand the electron beam in the main solenoid in the interaction section. According to the magnetic expansion factor [4]

$$k = \frac{r_{gun}}{r_{main}} = \sqrt{\frac{B_{main}}{B_{gun}}} \tag{1}$$

a beam expansion of 49.5 mm is reached by lowering the magnetic field in the main solenoid to 0.3 T. For the correct set-up of the compensation scheme, it has to be taken into account that electrons experience a drift motion inside the toroidal fields and an ExB-rotation under the influence of the beams' space-charge especially in the main solenoid.

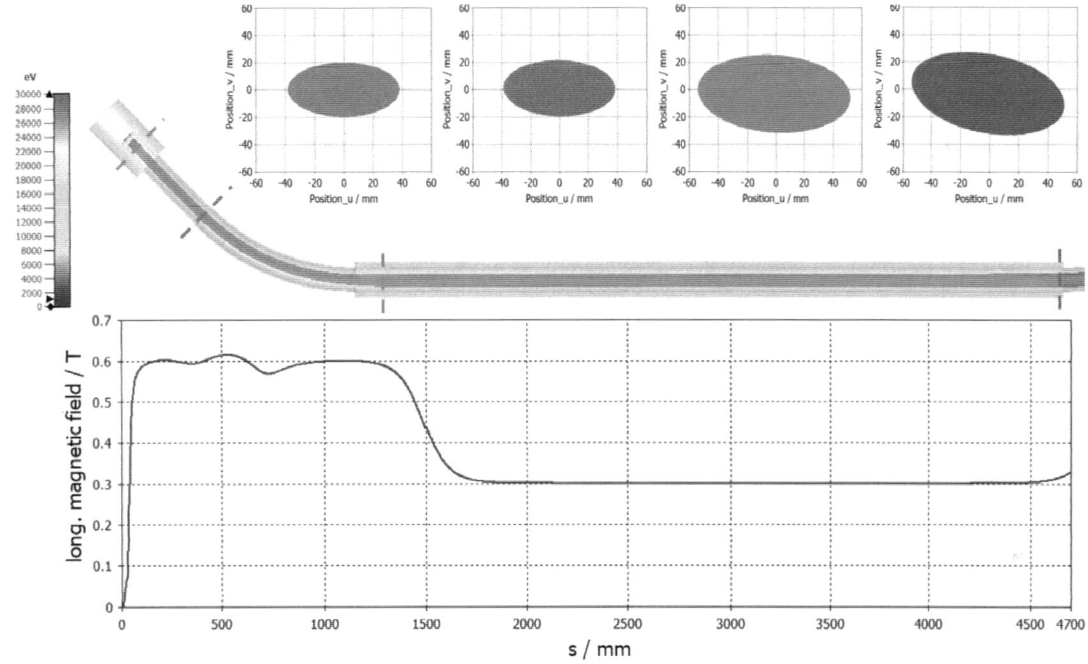

Figure 3. Electron beam dynamics along the electron lens. The color code indicates the beam distribution along the beam path (top). Calculated longitudinal magnetic field along beam path (bottom).

Figure 3 presents the beam transport simulation calculated using CST-PS [5] of an elliptic beam distribution with a=35 mm, b=20 mm, I=23.8 A and U_a=30 kV tracked through the electron lens device considering space-charge. This beam distribution was calculated for a gun perveance of 4.8 μP but without modulation grid. The colour code indicates the location of the beam distributions along the beam path: in blue is the generated electron beam distribution behind the gun, in red is the beam distribution after passing the first straight section, in magenta is the beam distribution at the entrance, and in green is the beam distribution at the exit of the main solenoid. The magnetic field was calculated using CST EM Studio and was imported into the beam transport simulation in order to save calculation time.

3. Test stand

At Goethe-University Frankfurt a dedicated laboratory (E-Lens Lab) was provided to construct a test stand for investigation of electron beam dynamics, diagnostics and commissioning of major electron lens components. It was equipped with a high voltage platform to power the electron gun and recycle the beam in an energy recovery system that is planned to be installed.

Figure 4 shows a picture of the test stand. It will be successively used in different stages according to the commissioning requirements. In the first stage, pyrometric measurements to characterize the heating process of the cathode are performed. For this reason, instead of a beam dump a pyrometer is installed after the second solenoid. In the second stage, a Faraday cup (FDC) will be installed in the diagnostic box and first low current beam extraction experiments as well as beam modulation will be performed. In the third stage the electron gun will be powered to 18 kV (5 A beam current) and the beam will be dumped in a high-performance FDC designed for 27 kW beam power deposition. For these three design stages accompanying beam transport simulations have been done (see [6] for more details). In the final stage the gun can be powered to its design parameters and the collector is installed in the second solenoid.

Figure 4. Test stand for the main electron lens components and diagnostics at Goethe-University Frankfurt.

The testing of the electron gun designed within the scope of ARIES (IRME-gun) is currently performed at the test stand (see figure 4) at Goethe-University Frankfurt and first results of the heat-up experiments look very promising. More details on the gun design can be found in [7].

4. Electron gun and collector

4.1. Electron gun

The IRME-gun makes use of a novel method for cathode heating. The cathode as well as the modulation grid are made of tungsten due to its robustness, maintenance and easy handling. But it requires temperatures about 2500°C in order to generate sufficient emission currents. The necessary heating is provided by the generation of an arc discharge in a plasma chamber located right behind of the cathode.

The cathode of the gun under commissioning has a round Gaussian profile and was developed to study the influence of non-linear transverse current density distribution on the space charge compensation scheme. Nonetheless, an electrode design for generation of an elliptic homogeneous electron beam profile is required for the SCC lens and different concepts for creation of an elliptic beam as well as different grid designs are currently being investigated.

A straightforward option is, of course, the usage of an elliptic cathode. Another option is beam shaping by using quadrupole coils integrated into the gun solenoid. By choosing the latter option, a variation of the beams' aspect ratio is possible during operation. For both options an electrode system was designed and optimized to have a perveance of 4.8 μP. Besides this, different grid designs with beamlet and target structure are being studied with respect to beam control and modulation (see figure 5). In the current electrode design the grid has a positive voltage of 800 V compared to cathode in order to extract the required beam current. This leads to a power deposition of almost 4 kW and secondary electron generation between cathode and grid. The influence of the secondary electrons will be further investigated numerically as well as experimentally at the test stand. The extracted current for the beamlet grid is I_{ex}=11.7 A and I_{ex}=12 A for the target-style grid.

Figure 5 shows as an example a symmetric gun design with cathode radius r_c=26.5 mm and an integrated quadrupole to vary the aspect ratio of the electron beam.

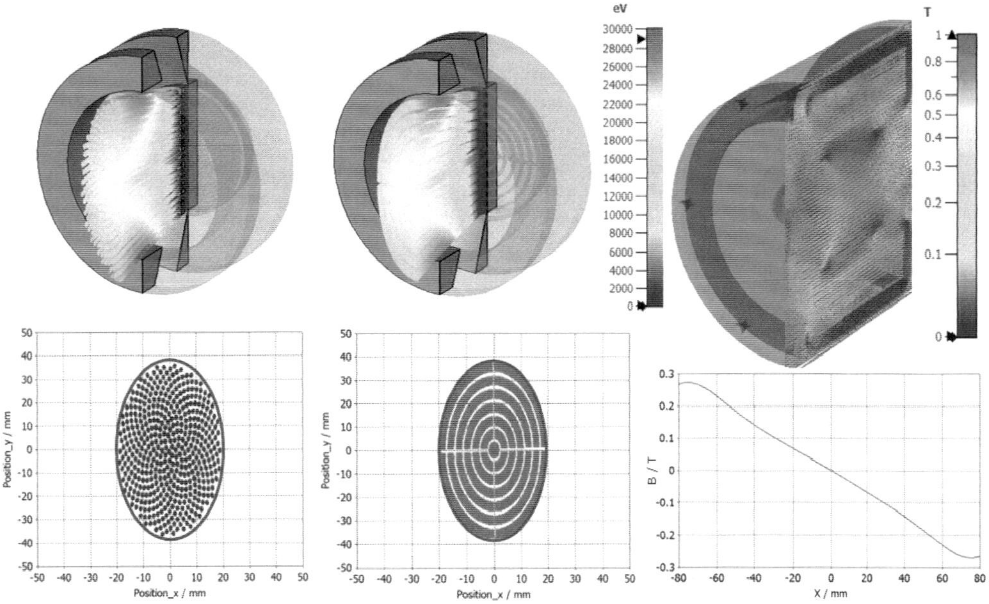

Figure 5. Simulated beam extraction for a beamlet-style grid (left), a target-style grid (middle) and magnetic system including a quadrupole for beam shaping with corresponding B_x (right).

Another aspect of the studies is also the evaluation of the beam dynamics of the extracted beam in the interaction region, which has either a target or a beamlet structure depending to the grid. The beam dynamic studies of the different beamlet structures will be performed in the next step.

4.2. Collector design

The maximum power of the extracted electron beam is almost 300 kW for modulation frequencies between 400 kHz and 1 MHz. For this reason, the beam needs to be decelerated before being dumped in the collector. The challenge in the collector design is to reduce the deposited beam power to a minimum but still fully dump the beam while depressing secondary electrons that are generated at the collector wall. In the actual collector set-up the beam is decelerated, while being expanded by the fringe field of the collector solenoid and finally dumped on the collector walls. While the electron beam is decelerated the beam potential can grow in the order of a few kV and has to be considered in the choice of the collector potential. Regarding the secondary electron emission (SEE), simulations using CST particle studio based on the Furman model [8] indicate that especially reflected electrons have to be mitigated. Different collector designs were studied in order to reduce the SEE. But a final mitigation strategy and for this reason a final collector set-up is still under investigation.

Acknowledgments
This work was supported by the European Union's Horizon 2020 Research and Innovation Program under Grant Agreement No 73087. We would like to express our gratitude to our collaborators of the ARIES workpackage 16 from CERN and Riga Technical University.

References
[1] Artikova S, Boine-Frankenheim O, Meusel O, Oeftiger A, Ondreka D, Schulte-Urlichs K and Spiller P 2021 *J. Instrum.* **16** P03044
[2] Boine-Frankenheim O and Stem W 2018 *Nucl. Instrum. Methods Phys. Res., Sect.* A **896** pp 122–28
[3] Vretenar M 2017 *ARIES* `https://aries.web.cern.ch`
[4] Stancari G *et al.* 2021 *J. Instrum.* **16** P05002
[5] CST Studio Suite 2022 `https://www.cst.com`
[6] Meusel O and Schulte-Urlichs K 2021 Test bench of IRME-gun at Goethe-University Frankfurt Tech. rep. *ARIES* ARIES-MS57
[7] Dönges T, Droba M, Meusel O, Podlech H, Schulte-Urlichs K and Thoma K I 2022 *Proc. 13th Int. Particle Accelerator Conf. (IPAC'22)* (Bangkok, Thailand) pp 667–670
[8] Furman M A and Pivi M T F 2002 *Phys. Rev. Spec. Top. Accel Beams* **5** 124404

Simulating quasi-integrable optics with space charge in the IBEX Paul trap

J A D Flowerdew[1], **D J Kelliher**[2], **S Machida**[2] **and S L Sheehy**[1,3,4]

[1] University of Oxford, Oxford, UK
[2] STFC Rutherford Appleton Laboratory, Oxford, UK
[3] University of Melbourne, Parkville, Victoria, Australia
[4] Australian Nuclear Science and Tehcnology Organisation (ANSTO), Sydney, Australia

E-mail: jake.flowerdew@physics.ox.ac.uk

Abstract. The intensity frontier has called for new initiatives in hadron accelerator design in order to accommodate space charge dominated beams. Octupoles are often used to damp beam instabilities caused by space charge, however the insertion of octupole magnets leads to a nonintegrable lattice which reduces the area of stable particle motion. One proposed solution is Quasi-Integrable optics (QIO), where the octupoles are inserted between sections of a specific lattice insertion called a T-insert. An octupole with a strength that scales as $1/\beta^3(s)$ is applied in the drift region, where the horizontal and vertical beta functions are equal, to create a time independent octupole field. This leads to a lattice with a time-independent Hamiltonian which is robust to small perturbations. IBEX is a Paul trap which allows the transverse dynamics of a collection of trapped particles to be studied, mimicking the propagation through multiple quadrupole lattice periods, whilst remaining stationary in the laboratory frame. In order to test QIO at the IBEX experiment, it has recently undergone an upgrade to allow for the creation of octupole fields. We present our design of the IBEX experiment upgrade along with simulation results of our proposed experiment to test QIO with space charge.

1. Introduction

Space charge forces are the result of Couloumb interactions between charged particles in a beam. As higher intensity accelerators are being designed and built, resonances and instabilities caused by space charge can become the limiting factor in the intensities these machines can reach. Building test accelerators to study high intensity beams can be costly and, when carrying out beam loss studies, accelerator components can be activated and damaged. Therefore, simulations are a vital part of the accelerator design process, however they are computationally intensive when reproducing space charge forces over long timescales (tens of thousands of turns) and can never be a replacement for experimental verification.

These challenges led to the design and construction of linear Paul traps to investigate transverse beam dynamics more efficiently at Hiroshima University, Japan [1], Princeton University, US [2] and, most recently, the Intense Beams Experiment (IBEX), at the Rutherford Appleton Laboratories (RAL), UK [3]. IBEX is a table-top sized experiment that can replicate the transverse betatron motion in alternating gradient accelerators in a dispersion- and chromaticity-free environment.

This paper aims to simulate a lattice proposed by the theory of Nonlinear Integrable Optics (NIO) to damp coherent resonances created by space charge.

2. The IBEX Paul trap and upgrade

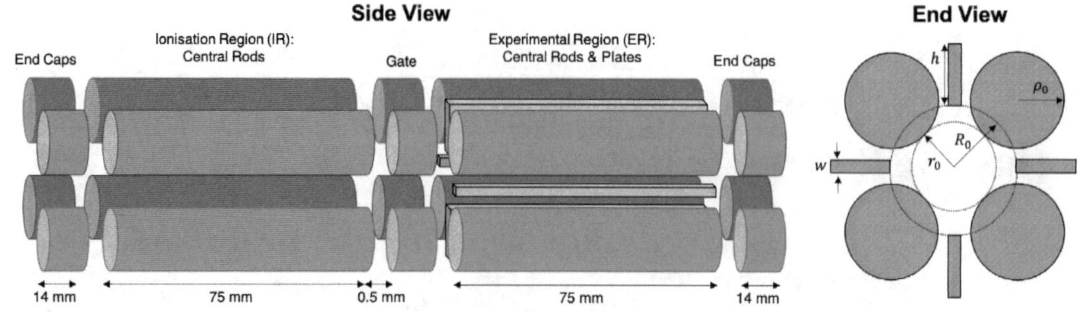

Figure 1: Schematic of the IBEX trap nonlinear upgrade. Opposing RF voltages are applied to the red and blue outlined rods for transverse confinement of ions. A DC voltage is applied to the end caps and gate electrodes to provide longitudinal trapping. In IBEX $r_0 = 5\,\mathrm{mm}$ and $\rho_0 = 5.75\,\mathrm{mm}$. Four additional plates between the rods are present in the nonlinear trap at an inscribed radius $R_0 = 8.5\,\mathrm{mm}$ to enable the creation of octupole fields.

The IBEX trap was recently upgraded to allow for the excitation of sextupole and octupole fields. The trap consists of two sections, an ionisation region (IR) and an experimental region (ER) as seen in figure 1. The design was adapted from the Hiroshima group's nonlinear trap [4]. Argon gas is introduced into the trap vessel and is ionised in the IR with an electron gun. Typically, a sinusoidal RF voltage is applied to the central rods with a maximum peak-to-peak of 300 V and frequency of 1 MHz. Voltages of the same form but opposite polarity are applied to the blue and red outlined rods in figure 1 to provide transverse confinement of the ions. Longitudinal confinement of the ions is achieved by applying a DC offset to the end caps and gate electrodes. Once the argon gas is ionised, the DC voltage on the gate is dropped to allow ions to pass into the experimental region.

Adjusting the peak voltage applied to the central rods is analogous to changing the quadrupole strength in an accelerator, which in turn changes the betatron tune in both the horizontal and vertical planes. Ions can be stored for around 1 s in IBEX, corresponding to 10^6 RF periods. Octupole and sextupole fields can be created by applying voltages to the plate electrodes in the ER. The DC voltage on the end caps is then dropped and the ions are directed onto a Micro-Channel Plate (MCP) detector. The number of ions stored in the trap is controlled by adjusting the length of time that the electron gun is on. This allows for a wide range of intensities to be studied within the trap. Due to the low energy of the ions ($< 1\,\mathrm{eV}$), high intensity beam loss studies can be carried out in the trap without damaging or activating components.

3. Nonlinear integrable optics

Current synchrotrons and linacs utilize a system of alternating focusing and defocusing quadrupole magnets that are used to confine a beam of charged particles. The use of linear magnets leads to a Hamiltonian of the system that can be made time-independent and hence integrable. However, in reality these linear lattices are susceptible to perturbations such as magnet errors and space charge forces, and so components such as sextupoles and octupoles are often used to apply higher order corrections. In general, the addition of these nonlinear components creates a non-integrable system, which limits the available area of phase-space where the particle motion is non-chaotic. Therefore, it is desirable to design a lattice which includes nonlinear components in such a way that it remains integrable. If an integrable, nonlinear lattice exists, then the system will always be close to an integrable solution, even when effects such as magnet misalignment and space charge are included.

The theory of NIO [5] proposed an integrable, nonlinear lattice consisting of a linear T-insert and a drift region for a nonlinear magnet insert. The three conditions of the T-insert lattice are (1) $n\pi$ (where n is an integer) phase advance over the linear section to provide quasi-periodic motion through the drift region. (2) Equal beta functions in the drift region ($\beta_x = \beta_y$). (3) For Quasi-Integrable optics, the octupole strength will scale with $1/\beta^3(s)$.

The Integrable Optics Test Accelerator (IOTA), Fermilab [6] will test the fully integrable solution, which requires a complex elliptical potential in the drift region. However, the elliptical potential is a challenge to create experimentally. Therefore, IBEX will first test the quasi-integrable case, which uses an octupole potential of the form

$$V(x, y, s) = \frac{\kappa}{\beta^3(s)} \left(\frac{x^4}{4} + \frac{y^4}{4} - \frac{3x^2y^2}{2} \right), \tag{1}$$

where κ is a constant and $\kappa/\beta^3(s)$ defines the strength of the octupole field. Assuming equal beta functions in the horizontal and vertical direction ($\beta_x(s) = \beta_y(s) = \beta(s)$), the octupole potential creates the time-independent Hamiltonian

$$H_N = \frac{p_{xN}^2 + p_{yN}^2}{2} + \frac{x_N^2 + y_N^2}{2} + U(x_N, y_N) \tag{2}$$

where $U(x_N, y_N) = \beta(s)V(x, y, s)$ and the coordinate transform, $z_N = z/\sqrt{\beta(s)}$ is used. The time-independent Hamiltonian becomes an integrable of motion, creating a quasi-integrable lattice which should be robust to small perturbations. In this work, a two cell T-insert lattice close to the 1/4 integer resonance was created. The lattice was tested in the presence of space charge and the benefits of applying a quasi-integrable octupole were studied.

4. Simulation results

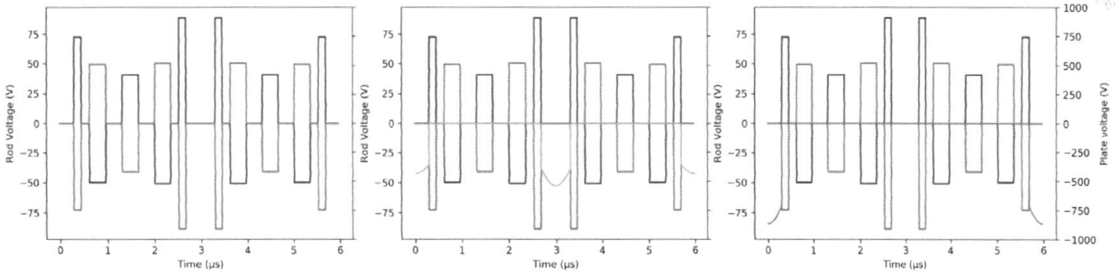

Figure 2: Voltage waveform, applied to the rods in IBEX (red and blue) to create the two-cell T-insert super-period. **Left:** No octupoles applied in the drift region. **Center:** Octupole pulse with $1/\beta^3$ strength scaling (orange) applied in the drift region to create Quasi-Integrable lattice. **Right:** Octupole pulse with $1/\beta^3$ strength scaling (green) with twice the strength but applied only once per super-cell.

The particle-in-cell (PIC) code VSim 11.0 [7] was used to create the T-insert lattice in a 2D simulation of the IBEX trap. Figure 2 plots the voltage waveform applied to the rod electrodes in IBEX to create the two cell T-insert lattice. The lattice was designed to have a π phase advance through the T-insert region and equal horizontal and vertical beta functions within the drift region. Two distinct cells with mirror symmetry were designed to create a super-period with tunes of $Q_x = 1.276$ and $Q_y = 1.277$, close to the $1/4$ integer resonance. On the left of figure 2 is the T-insert lattice with no octupole applied in the drift region. In the center of figure 2 is the Quasi-Integrable (QI) lattice where the octupole is turned on in the drift region and varies as $1/\beta^3(s)$. On the right of figure 2, a comparison lattice is used where the octupole is turned off every other cell. The octupole strength is twice as large in this lattice to make the integrated octupole strength the same as the QI lattice, however this lattice is non-integrable.

A 2D Gaussian distribution of 25,000 Ar^+ macro particles was tracked in a VSim 11 model of IBEX for 1000 T-insert lattice super-periods (2000 cell periods). The charge ratio of macro particles to physical particles was 1:1000. The distribution of 2.5×10^7 physical particles created an incoherent space charge tune shift of $\Delta Q_{\text{rms}} = -0.036$. This tune shift was chosen to excite coherent oscillations which satisfied the following resonant condition [8]

$$Q_0 + C_2 \Delta Q_{\text{rms}} = \frac{1}{2}\left(\frac{5}{2}\right) \tag{3}$$

where $C_2 = 3/4$ [9].

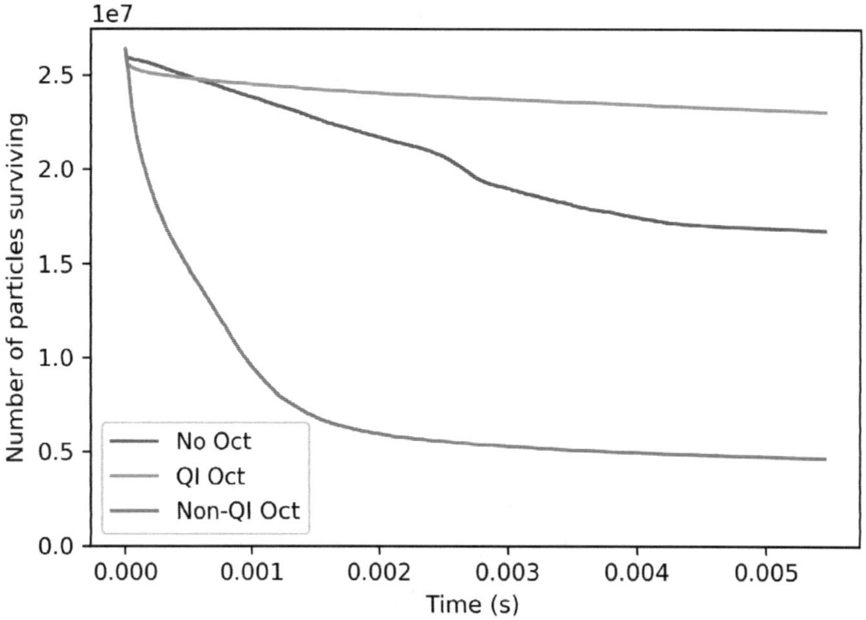

Figure 3: Rate of particle loss over 1000 super-periods of the T-insert lattice. (Blue) No octupoles applied in the drift region. (Orange) Octupoles turned on in the drift region which meet the quasi-integrability condition. (Green) Octupoles turned on with twice the strength but applied only once per super-period.

The number of particles surviving over 1000 T-insert super-periods is plotted against time in figure 3. The linear T-insert lattice with no octupoles applied is plotted in blue and can be seen to lose $36.5 \pm 0.3\,\%$. When the octupoles are turned on and they meet the quasi-integrability condition (orange), the particle loss is reduced to $12.6 \pm 0.2\,\%$. If we compare the QI lattice (figure 2, center) to a nonlinear lattice which breaks the integrability condition (figure 2, right) we get $82.4 \pm 0.2\,\%$ particle loss, which is plotted in green in figure 3.

The mechanisms behind the particle loss can be observed from the phase space plotted in figure 4. Figure 4a is the (x,x') phase space of a matched Gaussian distribution of 25,000 macro particles with no space charge (ratio of macro particles to physical particles was 1:1 where space charge is negligible). The phase space is plotted after 500 super-periods of the linear lattice (figure 2, left) and resembles the initial Gaussian distribution it started with.

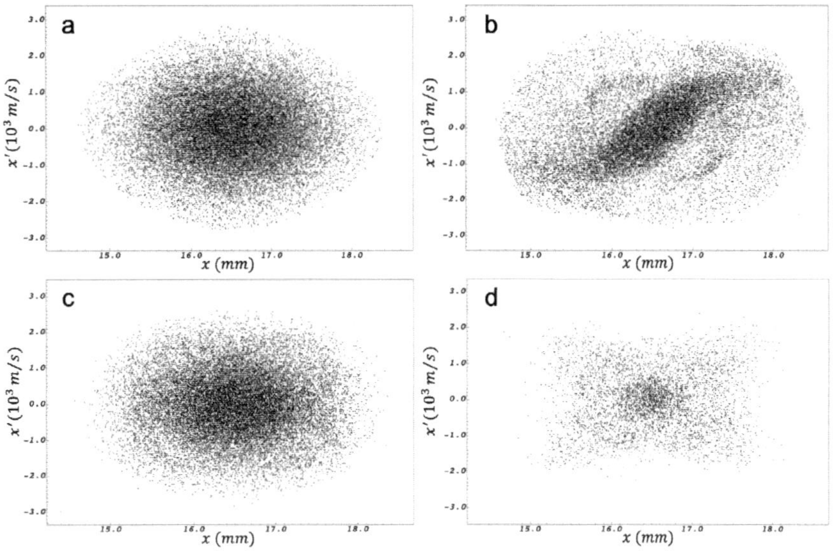

Figure 4: (x,x') phase space plotted after 500 super-periods of the T-insert lattice. a) Octupoles off in the drift region (no space charge). b) Octupoles off in the drift region. c) Octupoles on and QI conditions met. d) Octupoles on but QI conditions broken.

In figure 4b the phase space is plotted after 500 super-periods through the linear lattice but with space charge. Due to the resonant condition in equation (3) being met, the 2nd order coherent resonance is excited. This can be seen from the two arms forming in the phase space of figure 4b.

Figure 4c compares the phase space of the lattice when the octupoles are turned on and the quasi-integrability condition is met. The coherent resonance caused by the space charge tune shift has been damped by the octupole field and the distribution is returned to what resembles a Gaussian distribution. It is comparable to the phase space in figure 4a when no space charge was present.

In order to show the benefits of Quasi-Integrable Optics (QIO) over the use of ordinary octupoles to dampen this coherent resonance, we also compared a lattice which broke the integrability condition (figure 2, right). Figure 4d shows the phase space when just one octupole of twice the strength is turned on per two cell super-period. The four arms that appear in the phase space show that the 4th order incoherent resonance is being excited and is responsible for the significant particle loss seen in figure 3.

5. Conclusion

These simulation results show that a coherent resonance can be damped with octupoles when arranged in a lattice prescribed by QIO. The results also show that deviating from the conditions of quasi-integrability can result in the octupole driving it's own set of 4th order resonances. The IBEX Paul trap has undergone an upgrade to allow for the creation of octupole fields within the trap with the aim to confirm these simulations with detailed experiments.

References

[1] Okamoto H, Wada Y and Takai R 2002 *Nucl. Instrum. Methods Phys. Res., Sect.* A **485** pp 244–54

[2] Davidson R C, Efthimion P C, Majeski R, Qin H and Shvets G 2001 *Nucl. Instrum. Methods Phys. Res., Sect.* A **464** pp 502–11
[3] Sheehy S L, Carr E J, Martin L K, Budzik K, Kelliher D J, Machida S and Prior C R 2017 *J. Phys. Conf. Ser.* **874** pp 4–10
[4] Fukushima K and Okamoto H 2015 *Plasma and Fusion Research* **10** 1401081
[5] Danilov V and Nagaitsev S 2010 *Phys. Rev. Spec. Top. Accel Beams* **13** pp 1–10
[6] Antipov S *et al.* 2017 *J. Instrum.* **12** T03002
[7] VSim 11.0. Tech-X. `https://txcorp.com/vsim/`.
[8] Okamoto H and Yokoya K 2002 *Nucl. Instrum. Methods Phys. Res., Sect.* A **482** pp 51–64
[9] Sacherer F J 1968 Transverse Space-Charge Effects in Circular Accelerators. Thesis D.Phil. University of California Lawrence Radiation Laboratory Berkley California

Horizontal beam response at extraction conditions at the Heidelberg Ion-beam Therapy Centre

E C Cortés García, E Feldmeier and Th Haberer

Heidelberg Ion-Beam Therapy Centre (HIT), Department of Radiation Oncology, Heidelberg University Hospital, Heidelberg, Germany

E-mail: Cristopher.CortesGarcia@med.uni-heidelberg.de

Abstract. The Heidelberg Ion-Beam Therapy Centre's synchrotron makes use of the sextupole driven RF-KO method near the third-order resonance in order to slowly extract the beam that is delivered to the patients. The horizontal beam response of a coasting beam was studied experimentally and with simulations at extraction conditions in order to deduce regions of interest for an optimal excitation signal spectrum. Two narrow frequency regions were found where the beam reacts coherently. With these information an RF signal was proposed for the resonant slow extraction.

1. Introduction

The Heidelberg Ion-Beam Therapy Centre is a state-of-the-art facility capable of delivering four ion species of proton, ^{4}He, ^{12}C and ^{16}O and is equipped with the first heavy ion gantry worldwide. After the beam has reached it's final nominal energy in the synchrotron, it is slowly extracted by means of the sextupole driven RF-KO method [1,2]. The horizontal tune is brought near a third-order resonance and this is actively fed further by resonance sextupoles. Under these conditions the beam's horizontal emittance is slowly blowed up with an external EM field. To ensure that the the extracted particle count remains constant, several techniques were implemented [3].

In this contribution the experimental data of the horizontal beam response of a coasting beam under extraction conditions and a simulation of it are presented. For the design of excitation RF signals for the slow extraction the typical approach was based on direct phenomenological studies (measurements and simulation) of the spill [4–7]. The study of the BTF at extraction conditions can aid to the design of the spectrum of the RF signal, since it can directly unveil the areas in frequency-space where the beam reacts coherently.

2. BTF measurement

In this section the BTF measurement of a coasting carbon-ion beam $^{12}C^{6+}$ with kinetic energy $E_{\text{kin}} = 124.25\,\text{MeV/u}$ is presented. The experimental setup for the measurement of the Beam Transfer Function (BTF) is shown in figure 1. Relevant parameters of the linear ion-beam optics can be found in table 1. The tune values were infered from the BTF signal measured at extraction flattop with sextupole magnets off. The chromaticity of the machine was measured by introducing an offset on the end frequency of the accelerating cavity to produce a momentum offset and determining the tune shift through the BTF measurement with sextupole magnets off. The value presented in table 1 is meant to be the specific chromaticity of the machine, namely $\xi = \frac{\Delta Q}{Q} / \frac{\Delta p}{p}$.

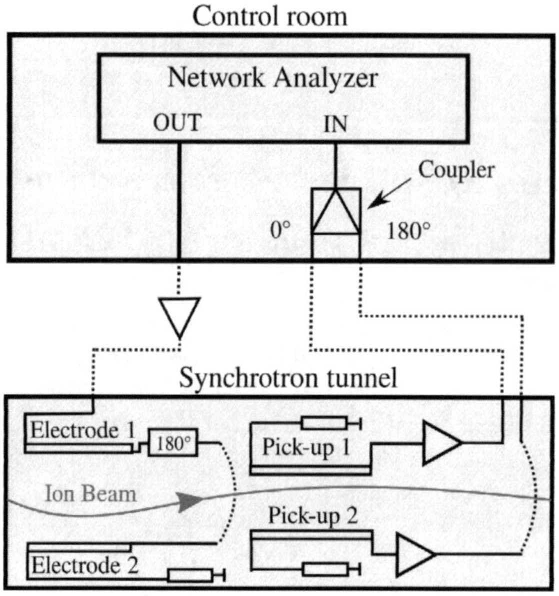

Figure 1. BTF experimental setup. The network analyzer (NA) generates an excitation signal and scans over a frequency span of interest. The response of the beam is then measured with two pick-up electrodes.

The slip factor was determined from the experimental value of the synchrotron frequency f_S. The momentum-spread was infered from the width of the Schottky pick-up signal of the second harmonic of the revolution frequency.

Table 1. Ion optical and beam parameters of the HIT synchrotron. The momentum spread is at FWHM.

Parameter	Symbol	Nom. value	Meas. value
Hor. tune	Q_x	1.67895	1.6795(1)
Ver. tune	Q_y	1.755	1.720(6)
Slip-factor	$\|\eta\|$	0.4766	0.44(2)
Hor. chroma.	ξ_x	-0.66	-0.72(6)
Mom. spread	σ_δ	-	$1.2{\cdot}10^{-3}$

After the beam has arrived to the nominal flattop energy, it is debunched and the tune is brought near the third order the resonance. Simultaneously the sextupoles are excited to the nominal strength for extraction, this corresponds for the resonance sextupoles to a strength of $k' = 0.9\,\mathrm{m}^{-3}$. This process lasts approx. 500 ms to ensure adiabaticity and that the particle distribution converges to a stationary distribution. In contrast to an ion-beam optics diagnostic run, the BTF at extraction conditions poses an additional challenge. Since the beam is prepared to be extracted, the dynamical aperture is considerably reduced, hence any external excitation leads easily to extraction (or beam loss) and thus spoils the signal. In order to recover a reasonable signal, the power of the external stimulus was dimmed such that particle loss during the measurement could be avoided.

The sweep time of the investigated frequency span around the betatron resonance was set to ten seconds and equal for all shots. The investigated frequency span $f \in [18.0479, 18.0979]$MHz was scanned in 701 frequency steps. The read-out of the stored current was used to monitor and confirm that no beam loss was present during the measurement. The measured signal is illustrated in figure 2.

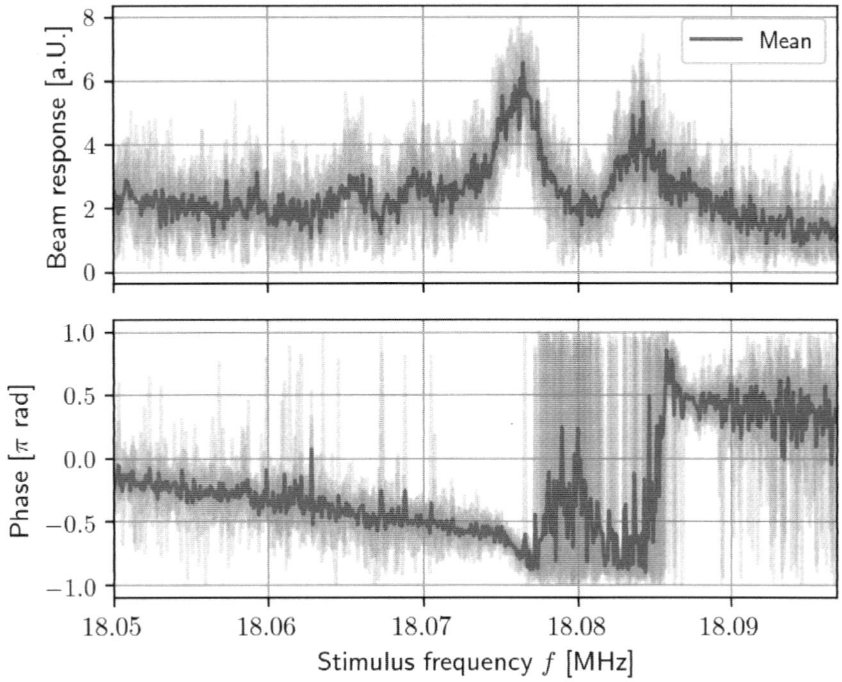

Figure 2. Measured horizontal beam response at extraction conditions. Orange background: Overlapp of eight different shots. Blue solid curve: Mean value of the eight recorded shots.

The beam response (see figure 2 upper panel) shows clearly two regions where the beam reacts coherently to external periodic excitations. This result is in complete agreement with previous studies [4]. Since the measurement had to be performed with a dimmed excitation signal to avoid beam loss, the recorded signal was at background noise level. Note that the beam response is in linear scale. By averaging over many shots the background noise contribution can be suppresed, as it is illustrated by the blue solid curve in figure 2. The phase (see figure 2 lower panel) appears to show a jump near the position of the peak nearest to the third order resonance $f_{\text{res}} = 18.1$ MHz. The apparent phase jumps in the frequency region from $f \in [18.075, 18.085]$ MHz come from the equivalence of an additive degree of freedom of 2π; the signal jumps from $-\pi$ to π and viceversa.

In the next section the simulation of the BTF measurement will be presented.

3. BTF simulation

The BTF measurement was simulated under extraction conditions, where a bunch of particles was tracked turn-by-turn while externally excited by a harmonic stimulus with a frequency (tune) range near the betatron resonance. To start, an initial particle distribution, whose transversal probability density function reads

$$\rho(\vec{x}) = \frac{1}{\sqrt{2\pi\varepsilon}} \exp\left(\vec{x}^T \Sigma \vec{x}\right), \qquad \Sigma = \varepsilon \begin{pmatrix} \beta & -\alpha \\ -\alpha & \gamma \end{pmatrix} \tag{1}$$

Table 2. Parameters for the simulation of the BTF measurement.

Parameter	Value
N particles	10^4
Turns(recorded)	2600(325)
Max. kick	2.5 µrad/turn
Tune span	[1.6679, 1.6841]
Tune steps	200
$\varepsilon_{x,y}$ (rms)	2 mm mrad

where α, β, γ are the corresponding ion-beam optical functions from the Courant-Snyder parametrization and the emittance ε for each transversal plane, was tracked for 10^4 turns under extraction conditions (with nominal sextupole fields for extraction) in order to recover a stationary distribution. This stationary distribution was excited externally over a tune region of interest (see table 2) over 2600 turns for each sampled tune step. Relevant parameters of the simulation can be found in table 2. The simulation was performed with the MAD-X tracking module [8]. The momentum-spread was the same as in table 1.

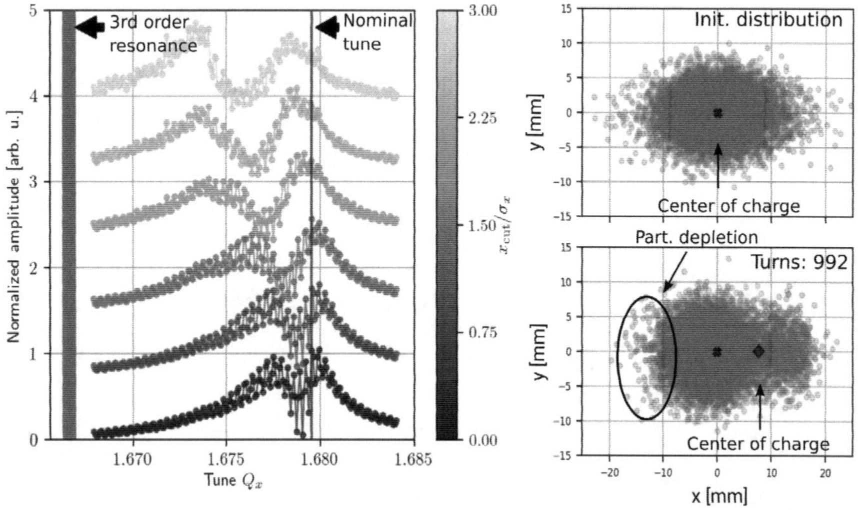

Figure 3. Horizontal beam response as function of the cutoff parameter.

For the analysis of the simulated data the considerations described in [9, 10] were implemented. In principle, the region corresponding to the core maintains the same particle density such that for a coasting beam the core delivers a simple DC signal to the pick-up electrodes. In contrast, the particle density outside of the core (which form the so called lunettes [9]) shows a more significant change over time due to the external harmonic excitation, thus the AC signal measured at the pick-up electrodes is mostly given by this density modulation. On top of this, the stripline Schottky pick-up electrodes' non-linear sensitivity enhance the contribution of particles with higher amplitudes to the measured signal.

To account for this phenomenon, the particles inside a region of $x_{\rm cut} \in [-2.1\sigma_x, 2.1\sigma_x]$ were masked-out and the remaining particles were considered for the calculation of the center of charge. σ_x refers to the horizontal spatial standard deviation of the initial particle distribution (with no external harmonic excitation). The dependency of the simulated signal as function of the parameter x_{cut} is illustrated in figure 3. Without this consideration the simulated BTF signal peaks more prominently at the nominal tune value (zero amplitude betatron frequency) and the second peak is considerably overshadowed and shifted away from the resonance. Note that the splitting behaviour of the simulated signal is always present. There is no trivial way of determining from *a-priori* considerations the cutoff region.

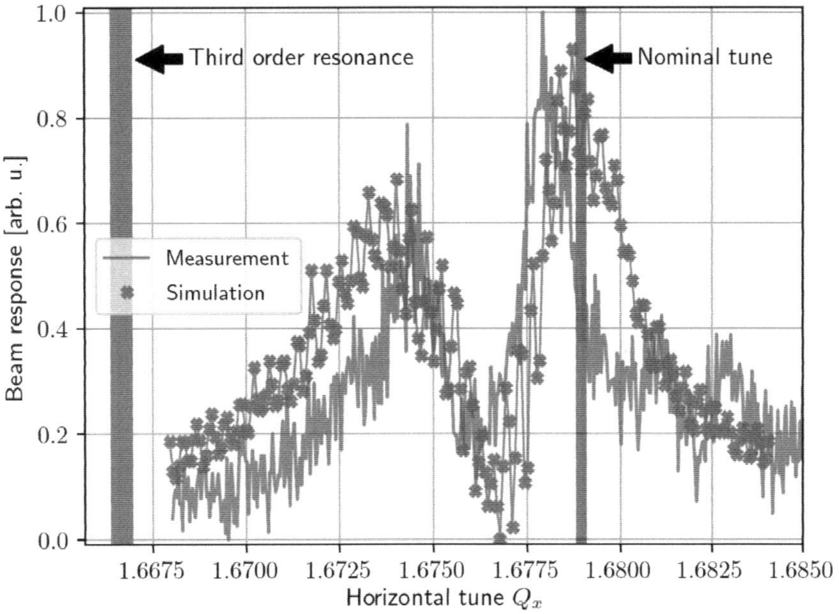

Figure 4. Horizontal beam response.

The result of the simulation and a comparison to the measured signal are illustrated in figure 4. From the BTF two regions of interest can be recognized. The tune regions appear to be sharply localized such that narrow bands would suffice for driving the beam's coherent response efficiently. Other tune regions appear to have a weaker influence on the beam response at extraction conditions at HIT. Therefore an excitation signal for the emittance blow-up scheme has to carry mainly frequencies in its spectrum that cover these two regions of interest.

4. Summary and conclusion

In this contribution the measurement of the BTF of a coasting carbon ion beam $E_{\mathrm{kin}} = 124.25\,\mathrm{MeV/u}$ and relevant ion beam optics and beam parameters were presented. The signal showed two frequency regions at which the beam reacts coherently. The BTF measurement was simulated under extraction conditions. Considering the Hereward effect [9] together with the non-linearity of the strip-line pick-up sensitivity, the simulation and the measurement agree well with each other. With this information an excitation RF signal for the slow RF KO extraction can be designed, such that the frequency regions, where the beam reacts coherently strong, are covered by the signal's spectrum.

References

[1] Hiramoto K and Nishi M 1992 *Nucl. Instrum. Methods Phys. Res., Sect.* A **322** 154–160
[2] Tomizawa M *et al.* 1993 *Nucl. Instrum. Methods Phys. Res., Sect.* A **326** 399–406

[3] Schoemers C, Feldmeier E, Naumann J, Panse R, Peters A and Haberer T 2015 *Nucl. Instrum. Methods Phys. Res., Sect.* A **795** 92–99
[4] Noda K, Furukawa T, Shibuya S, Muramatsu M, Uesugi T, Kanazawa M, Torikoshi M, Takada E and Yamada S 2002 *Nucl. Instrum. Methods Phys. Res., Sect.* A **492** 241–252
[5] Noda K *et al.* 1996 *Nucl. Instrum. Methods Phys. Res., Sect.* A **374** 269–277
[6] Krantz C *et al.* 2018 *Proc. 9th Int. Particle Accelerator Conf. (IPAC'18)* (Vancouver, BC, Canada) pp 1084–86
[7] Sorge S, Forck P and Singh R 2018 *J. Phys. Conf. Ser.* **1067** 052003
[8] Deniau L, Grote H, Roy G and Schmidt F 2021 *The MAD-X Program (Methodical Accelerator Design) Version 5.07.00 User's Reference Manual* (CERN)
[9] Hereward H G 1969 Landau damping by non-linearity Tech. rep. CERN Geneva CERN-MPS-DL-69-11
[10] Hereward H G 1965 The elementary theory of Landau damping Tech. rep. CERN Geneva CERN Yellow Reports: Monographs CERN-1965-020

Excitation of the σ_{l_l} = 90° resonance by the cavity rf accelerating fields

J -M Lagniel

Grand Accélérateur National d'Ions Lourds, 14000 Caen, France

E-mail: lagniel@ganil.fr

Abstract. In rf linacs the longitudinal focalization is done by si-nusoidal forces and at high accelerating fields the zero-current longitudinal phase advance per longitudinal fo-cusing period σ_{0l_l} can be high. The nonlinear compo-nents of the sinusoidal rf field (sextupolar, octupolar and higher order components) can then excite parametric resonances, including the 4th-order resonance (σ_{l_l} = 90°) when σ_{0l_l} is higher than 90°, inducing strong longitudinal emittance growths and acceptance reductions. As pointed out in previous papers, the longitudinal beam dynamics is therefore complex, even when the space-charge forces are ignored. The parametric resonance excitation by the rf field is analyzed before discussing the additional effect of the space-charge forces, in particular to explain why the zero-current longitudinal phase advance per trans-verse focusing period σ_{0l_t} is not a relevant parameter. Examples are given in the SPIRAL2 linac case.

1. Parametric resonances, basic definitions

As done in [1] and [2], the choice done here is to study the parametric resonances in the longitudinal plane independently of the study of the radial longitudinal coupling resonances in order to avoid unnecessary complications and confusions. As usual (e.g. [3] chapter X, section IV), the resonance zoology is defined considering the linear equation of motion of individual particles perturbed by linear (n = 1) or nonlinear (n > 1) periodic perturbing forces

$$\frac{d^2\delta\varphi}{dz^2} + \sigma_l^2\,\delta\varphi + a_{n,k}\begin{matrix}[\delta\varphi_0\sin(\sigma_l z)]^n \\ [\delta\varphi_0\cos(\sigma_l z)]^n\end{matrix}\cos(k\sigma_p z) = 0 \qquad (1)$$

with σ_l the phase advance per unit of length of the unperturbed particle motion, σ_p and k the phase advance per unit of length and the harmonic number of the perturbing force respectively. Considering that a resonance condition is fulfilled when a component (n, k) of the perturbing force has a "frequency" equal to the unperturbed particle oscillation "frequency" σ_l [3], a resonance occurs when

$$(i+1)\,\sigma_l = k\,\sigma_p$$

with i even integer from 0 to n for n even and i odd integer from 1 to n for n odd.

Using $\sigma_{l\_l} = \sigma_l L_l$ the longitudinal phase advance per period of the longitudinal focusing system (gaps or cavities), $\sigma_{l\_p} = \sigma_l L_p$ the longitudinal phase advance per period of the perturbation and $\sigma_p = 2\pi/L_p$ with L_p the period of the perturbation, the resonance condition is

$$\sigma_{l\_p} = \left(L_p/L_l\right) \sigma_{l\_l} = \frac{2\pi k}{i+1} \tag{2}$$

$(i+1)$ is the order of the resonance and k is the harmonic of the perturbation.

For example, particles with $\sigma_{l\_l} = 90°$ can be excited by the $a_{n=3,\,k=1}$ component of the perturbing force when $L_p = L_l$ (fourth order resonance), or can be excited by the $a_{n=1,\,k=1}$, $a_{n=3,\,k=1}$ and $a_{n=3,\,k=2}$ components of the perturbing force when $L_p = 2L_l$ (half integer and fourth order resonances).

2. Excitation by the cavity rf field

Following [1] and [2], the non-linear components of the rf accelerating field must be considered as main perturbing forces in the parametric resonance studies. This can be understood starting from the equation of motion of the phase oscillations around the synchronous particle obtained in smooth approximation ([3], chapter XVI section II-D).

$$\frac{d^2\delta\varphi}{dz^2} + K_{dp}\frac{d\delta\varphi}{dz} + \left[\frac{2\pi q\, E_0\, T_{\beta_s}}{m_0 c^2 \lambda\, \beta_s^3 \gamma_s^3}\right]\left[\cos(\varphi_s + \delta\varphi) - \cos\varphi_s\right] = 0 \tag{3}$$

The effect of the damping term K_{dp} which has a great importance on the longitudinal dynamics at high acceleration rates [1, 2] is not discussed here to focus the analysis on the rf field nonlinear focusing force which can be expressed by its Taylor series

$$\begin{aligned}[\cos(\varphi_s + \delta\varphi) - \cos\varphi_s] = & - [\sin\varphi_s]\,\delta\varphi && \text{“quadrupole” (linear focusing)}\\ & - [\cos\varphi_s/2!]\,\delta\varphi^2 && \text{“sextupole”}\\ & - [\sin\varphi_s/3!]\,\delta\varphi^3 && \text{“octupole”}\\ & + [\cos\varphi_s/4!]\,\delta\varphi^4 && \text{“decapole”} + \ldots\end{aligned} \tag{4}$$

Equation (3) is obtained using the smooth approximation, then averaging the effect of the rf field over the longitudinal period L_l. The effect of the gap or cavity periodicity must be analyzed including the harmonics of the rf field $E_z(z) = E_0 + \sum E_k \cos(k\, 2\pi z/L_l)$ in the longitudinal focusing force of equation (3), then to each term of the Taylor series of (4). The longitudinal focusing of the rf field can then be seen as a nonlinear focusing equation (4) and with $k = 0$ in (1), plus $\delta\varphi^n \cos(k\, 2\pi z/L_l)$ perturbing terms which induce resonances when the resonance conditions are fulfilled.

This was demonstrated in [1] and [2] using theoretical field distributions, figure 1 shows that, as expected, this is also true in the case of the SPIRAL2 superconducting linac cavity field [4].

The different plots in figure 1 are Poincaré maps done with the SPIRAL2 low beta cavities (L_l = 1.19 m), working at different fields and tuned in buncher mode (no damping) for deuterons at the RFQ output energy (0.732 MeV/A).

With kE = E/E_{max} and E_{max} = 6.5 MV/m the nominal field, for $\sigma_{0l\_l} = 53.9°$ at kE = 0.06. Figure 1 shows that, as expected, there is 8 resonance islands at $\sigma_{l\_l} = 45°$ and a beginning of acceptance reduction due to the high-order resonance overlap in the separatrix area. As expected, the 6th order parametric resonance $\sigma_{l\_l} = 60°$ is present at $\sigma_{0l\_l} = 70.3°$ (kE = 0.10), with a significant acceptance reduction due to higher-order resonance overlap in the separatrix area. As expected, the fourth-order resonance is present at $\sigma_{0l\_l} = 93.6°$ (kE = 0.17) and $\sigma_{0l\_l} = 102.6°$ (kE = 0.20), with the 4 resonance islands present in the stable region around the synchronous particle in the first case and in the chaotic area in the second case. The drastic longitudinal acceptance reduction due to even more high-order resonance overlap in the separatrix region above $\sigma_{0l\_l} = 90°$ is also clearly visible in both cases.

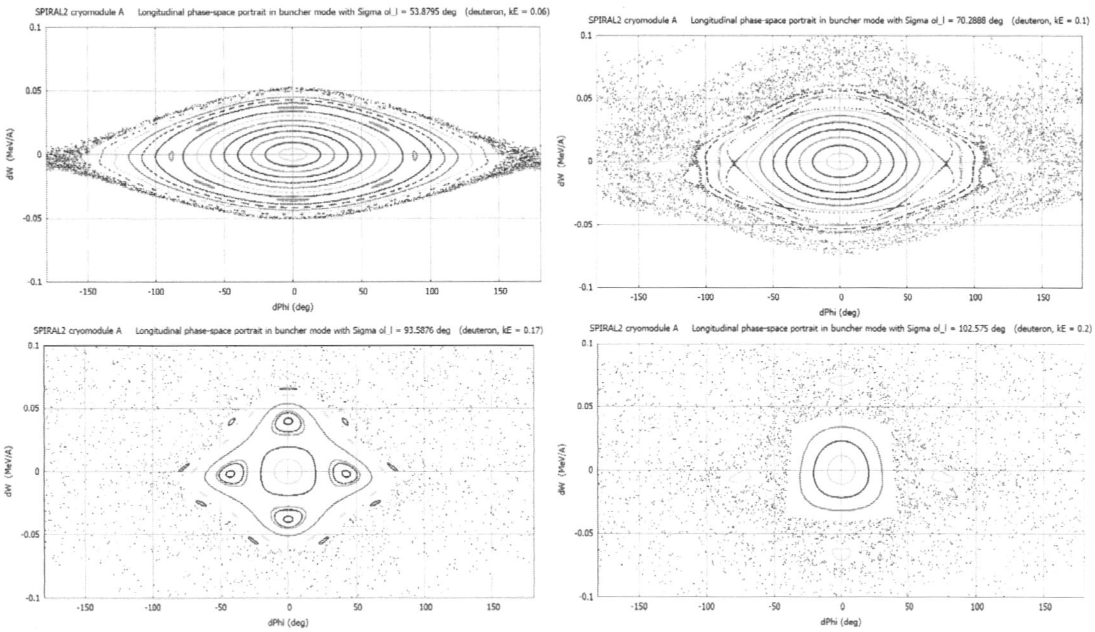

Figure 1. Longitudinal phase-space portrait for SPIRAL2 low beta cavities tuned in buncher mode (φ_s = -90°) at the RFQ output energy (0.732 MeV/A). Top left: σ_{0l_l} = 53.9° (kE = 0.06), top right σ_{0l_l} = 70.3° (kE = 0.10), bottom left σ_{0l_l} = 93.6° (kE = 0.17), bottom right σ_{0l_l} = 102.6° (kE = 0.20).

As foreseen by equation (4) with cos $\varphi_s = 0$, there is only even order resonances present in figure 1. Not shown, but in the $\sigma_{0l\_l} > 90°$ cases the chaotic region extends at high energies, particles injected out of the reduced acceptance will then form "high-energy tails" (up to $\delta W >$ 1.2 MeV/A above the RFQ output energy of 0.732 MeV/A).

One can finally point out the strong constraint imposed by the $\sigma_{0l\_l} < 90°$ rule that can impose the use of cavity fields much lower than the nominal ones in SC linacs, e.g. a limitation at less than 10% of the nominal field for the nominal tuning in the SPIRAL2 linac low energy part.

All of this pushes to better study the effect of the damping reducing the effect of the resonances at high accelerating rates [1, 2]. Considering that at the linac design stage, the longitudinal acceptance could be sufficiently preserved to leave the possibility to use higher accelerating fields.

3. Excitation by the space-charge field

With no loss of generality, the model for the space charge forces assuming the bunch to be a three-dimensional ellipsoid of uniform density introduced by Lapostolle [5] can be used. Following Gluckstern [6], using this model the longitudinal space charge force dependency (via the Poisson equation) on the bunch dimensions is usually (0.8 < ξ <5 range in [6]) given by

$$\mu_l \propto \frac{1}{\sqrt{a_x\, a_y}\, b^2} \tag{5}$$

with a_x and a_y the bunch radial sizes and b the bunch longitudinal size.

For matched beams, the longitudinal envelope oscillation has the period L_l of the longitudinal focusing system (gaps or cavities); all the harmonics at the longitudinal focusing period L_l are then always present in the $b^2(z)$ contribution to the space charge "perturbing force", equation (5).

In linacs the horizontal and vertical emittances are usually equals and the phase advances in the horizontal and vertical planes are traditionally chosen equals. In the case of matched beams in a solenoid or quadrupole doublet focusing (SPIRAL2 case), all the harmonics at the $L_t = L_l$ transverse period are also always present in the $a_x a_y$ contribution to the space charge "perturbing force".

In the case of matched beams in an alternating focusing with quadrupoles (FODO or FOFODODO) and choosing z = 0 such that the lattice has an even symmetry, the bunch radial dimensions can be expressed by their Fourier series

$$a_x(z) = a_0 + \sum_{k=1}^{\infty} a_k \cos\left(k\frac{2\pi z}{L_t}\right)$$
$$a_y(z) = a_x\left(z + \frac{L_t}{2}\right) = a_0 + \sum_{k=1}^{\infty} (-1)^k\, a_k \cos\left(k\frac{2\pi z}{L_t}\right) \tag{6}$$

The product $a_x\, a_y$ inducing the variations of the longitudinal space charge force with respect to the radial bunch size variations (5) is then of the form

$$a_x a_y = \left[a_0^2 + \sum_{k=1}^{\infty} (-1)^k \frac{a_k^2}{2}\right] + \sum_{k=1}^{\infty}\sum_{l=1}^{\infty} A_1(k,l) \cos\left(2k\frac{2\pi z}{L_t}\right)$$
$$+ \sum_{k=1}^{\infty}\sum_{l=1}^{\infty} A_2(k,l) \cos\left(2(k+l)\frac{2\pi z}{L_t}\right) \tag{7}$$

There is therefore no "space-charge excitation" in the longitudinal plane at a period equal to the radial focusing period, the first component ($k = 1$) is at $L_t/2 = L_l$ in the FODO case and at $L_t/2 = 2\, L_l$ in the FOFODODO case.

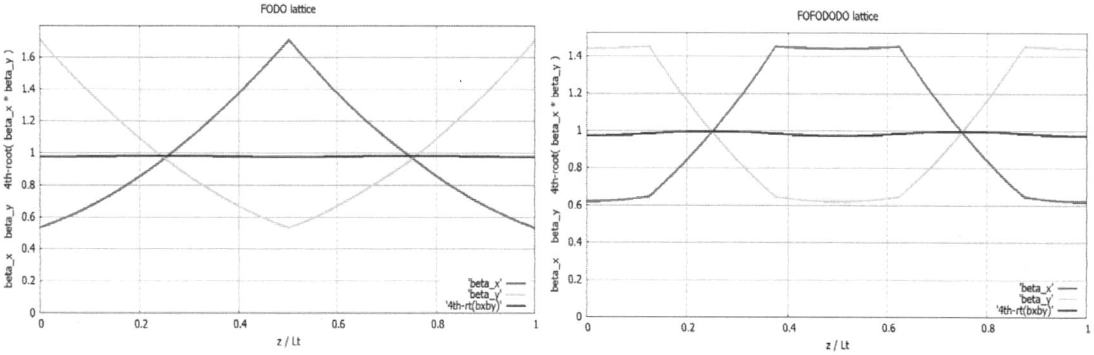

Figure 2. Evolution of the Courant-Snyder beta functions (red and green) and of the contribution of the radial envelope oscillations in the longitudinal space-charge perturbing force (blue) for matched beams in FODO (left) and FOFODODO (right) lattices with thin lenses tuned such that $\sigma_{t\_t} \sim 63°$.

Figure 2 also shows that the oscillatory part of the radial envelope contribution to the longitudinal space-charge force (blue curve, $\sqrt{a_x\, a_y}$ term in equation (5)) is very low, then has a small contribution compared to the one of the longitudinal envelope oscillations (b^2 term in equation (5)). One can then also conclude that in the case of the retained assumptions, the radial envelope oscillations have week effects on the excitation of parametric resonances in the longitudinal plane in both FODO and FOFODODO cases.

4. Summary

Many publications systematically discuss the excitation of the fourth-order parametric resonance σ_l= 90° by the space-charge forces. After [1] and [2], it is again pointed out here that the first source of excitation of this resonance (as well as the other parametric resonances in the longitudinal plane) is the rf field of the cavities.

Concerning the space-charge effects, we have shown that in the linac case (equal tune and emittance in the horizontal and vertical planes) there is no excitation in the longitudinal plane at a period equal to the radial focusing period in the FODO and FOFODODO cases, and that the radial envelope oscillations have week effects in both cases.

The periodicity of the main longitudinal motion perturbations (rf field and space-charge) is the longitudinal lattice. As pointed out several times, it is a mistake to consider the σ_{l_t} longitudinal phase advances per transverse focusing period speaking of parametric resonances in the longitudinal plane.

For linac tunings at high acceleration rates, the strong effect of the longitudinal phase oscillation damping [1, 2] would need to be taken into account.

Acknowledgements

Many thanks to the GANIL colleagues for their support and to M. Eshraqi (ESS) for his encouragements and judicious suggestions for the improvement of this paper.

References

[1] Lagniel J -M 2014 *Proc. 27th Int. Linear Accelerator Conf. (LINAC14)* (Genève, Switzerland) pp 572-4
[2] Lagniel J -M 2021 French Physical Society meeting (Roscoff, France)
[3] Bruck H 1966 *Accélérateurs circulaire de Particules* (Institut National des Sciences et Techniques Nucléaires) Eng. trans 1972 *Circular Particle Accelerators* Tech. rep. Los Alamos Scientific Laboratory (LA-TR-72-10 Rev.)
[4] Orduz A 2022 *Proc. 13th Int. Particle Accelerator Conf. (IPAC'22)* (Bangkok, Thailand) pp 70-73
[5] Lapostolle P 1965 Effets de la charge d'espace dans un accélérateur linéaire à protons Tech. rep. CERN AR/Int SG/65-15
[6] Gluckstern R L 1970 *Space charge effects Linear Accelerators* ed P.M. Lapostolle P M and Septier A L (North Holland Publishing Company) section C.1.3c

Study of transverse resonance island buckets at CESR

S T Wang and V Khachatryan
CLASSE, Cornell University, Ithaca, NY, 14853, USA
E-mail: sw565@cornell.edu

Abstract. A 6-GeV lattice with the horizontal tune near a 3rd-order resonance line at $3\nu_x$=2 is designed for studying the transverse resonance island buckets (TRIBs) at the Cornell Electron Storage Ring (CESR). The distribution of 76 sextupoles powered individually is optimized to maximize the dynamic aperture and achieve the desired amplitude-dependent tune shift α_{xx} and the resonant driving term h_{30000}, which are necessary conditions to form stable island buckets. The particle tracking simulations are developed to check and confirm the formation of TRIBs at different tunes with clearing kicks in this TRIBs lattice. Finally, the lattice is loaded in CESR and the TRIBs are successfully observed when the horizontal fractional tune is adjusted to 0.665, close to the 3rd-order resonance line. Bunch-by-bunch feedback is also explored to clear the particles in the main bucket and the island buckets, respectively.

1. Introduction

Nonlinear resonances are critical topics in the beam dynamics of accelerator physics, which have been studied theoretically and experimentally for many years [1]. Normally a storage ring operates at tunes far away from resonant lines to avoid the negative impact such as instability, poor lifetime, and increased emittance. However, by taking advantage of the resonance nature, one application of the ring operating at the 3rd-order resonance line is to extract particles in multi-turns at CESRN-PS and ELSA [2]. Recently, MLS and BESSY-II have demonstrated the stable two-orbit operation by utilizing the transverse resonance island buckets (TRIBs) [3, 4]. The second orbit for the beam in the island decreased the x-ray pulse frequency by 3 or 4 times which expands the possibilities for timing experiments [5].

TRIBs form in the vicinity of a resonance line, which can be achieved by adjusting the horizontal tune along with the families of sextupoles (harmonic and normal) to tune the amplitude-dependent tune shift (ADTS) to stabilize the beam [3]. This approach is effective but empirical tuning of the sextupoles may be required. With this method, TRIBs have been successfully observed at MLS, BESSY-II, and MAX-IV [6]. To understand how TRIBs can form at CESR, a relatively high-energy (6-GeV) storage ring, we choose a different approach by designing a lattice with the horizontal tune near the 3rd integer line and a new sextupole distribution, at which TRIBs could form easily by adjusting the horizontal tune only.

In this paper, we discuss the criteria of optimizing the sextupoles for the TRIBs. A particle tracking simulation is developed to demonstrate the formation of TRIBs at different tunes. The lattice is then loaded into CESR and the TRIBs are successfully observed while adjusting horizontal tune to 0.665. Clearing kicks using feedback kicker is explored to clear the particles in the core or island buckets.

Table 1. CESR parameters.

Beam Energy (GeV)	E_0	6.0
Circumference (m)	L	768.438
Transverse Damping time (ms)	$\tau_{\mathrm{x,y}}$	12.0, 14.6
Longitudinal Damping time (ms)	τ_{z}	8.2
Synchrotron tune	Q_{s}	0.027
Horizontal tune	Q_{x}	16.556
Vertical Tune	Q_{y}	12.636
Horizontal Emittance (nm·rad)	ϵ_{x}	~ 28
Energy spread	σ_{p}	8.2×10^{-4}

2. Lattice design

Cornell Electron Storage Ring (CESR) is a 6-GeV accelerator located on the Cornell University campus. Since 2008, CESR serves as a dedicated light source for x-ray users, namely Cornell High Energy Synchrotron Source (CHESS). In 2018, one sextant of the ring was upgraded with double bend acromat to reduce the emittance and accommodate more compact undulators [7]. The main accelerator parameters are listed in table 1. CESR magnets including 113 quadrupoles, 12 dipole quadrupoles, and 76 sextupoles are all individually powered, which provides great flexibility for lattice design and complex nonlinear dynamics studies. Details of CESR lattice are described in [7].

As shown in table 1, the nominal horizontal fractional tune (Q_{x}) during CHESS operation is 0.556, far from the 3rd-order line $3\nu_{\mathrm{x}}=2$ (0.667). It is very unlikely that TRIBs would appear by only adjusting the tune near 0.667. Thus, we first optimize quadrupoles to create a linear lattice with the design tunes at (16.643, 12.579) closer to the 3rd-order line while preserving most optics parameters as the normal lattice. Then 76 sextupoles are optimized to meet the conditions of forming TRIBs as well as maximizing the dynamic aperture (DA).

The formula of the stable fixed points (SFP) have been derived in [1] as shown in equation (1),

$$J_{SFP}^{1/2} = \frac{3G_{3,0,l}}{4\alpha_{xx}}\left(1 \pm \sqrt{1 - \frac{16\alpha_{xx}\delta}{9G_{3,0,l}^2}}\right), \tag{1}$$

where J_{SFP} is the particle's action at the SFP, $G_{3,0,\mathrm{l}}$ is the resonance strength at the 3rd-order resonance $3\nu_{\mathrm{x}}=l$, α_{xx} is the detune coefficient of ADTS, and $\delta=Q_{\mathrm{x}} - \frac{l}{3}$. As equation (1) implies, the formation of TRIBs depends on three variables $G_{3,0,\mathrm{l}}$, α_{xx}, and Q_{x}. Thus, adjusting the tune (Q_{x}) and the ADTS (α_{xx}) will help the formation of TRIBs.

For sextupole optimization, appropriate target values of $G_{3,0,\mathrm{l}}$ and α_{xx} are needed. Suppose we approach the 3rd-order line from below and observe TRIBs at $Q_{\mathrm{x}}=0.654$ ($\delta=-0.013$) with $J_{\mathrm{SFP}}=1.0 \times 10^{-5}$ m·rad, which is within the range that our instrument [8] can measure the TRIBs. Since $\delta < 0$, $\alpha_{\mathrm{xx}} > 0$ will satisfy the bifurcation condition $\frac{16\alpha_{\mathrm{xx}}\delta}{9G_{3,0,1}^2} \leq 1$ more easily. Using equation (1) with the above requirements, we obtain one set of values $G_{3,0,1}=0.1$ and $\alpha_{\mathrm{xx}}=1347$ for the sextupole optimization. In the optimization requirement, resonant driving terms except h_{30000} and h_{22000}, which are directly related to $G_{3,0,1}$ and α_{xx}, respectively, are minimized to maximize the DA. The horizontal and vertical chromaticities are fixed at 1.

3. Simulation

Once the linear lattice with desired sextupoles are optimized, simulations are implemented to check the lattice properties. All the simulation programs discussed below are based on BMAD code library [9]. First, 20 particles with different action are tracked through the TRIBs

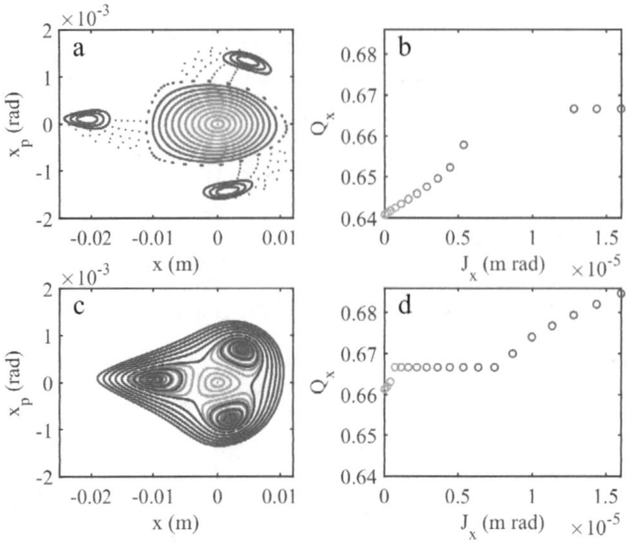

Figure 1. Particles' xx_p plots at two lattice tunes (a) Q_x=0.643 and (c) 0.661. The particle's Q_x as a function of its action in (b) and (d) are calculated from the TBT data plotted in (a) and (c), respectively.

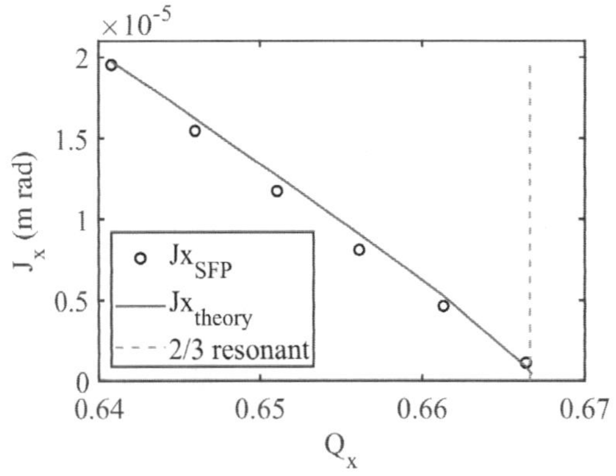

Figure 2. The particle action at the SFP as a function of lattice tune.

lattice at different Q_x for 1000 turns while RF, radiation damping and excitation are turned off. Figures 1(a) and (c) plot the xx_p of all particles at Q_x=0.643 and 0.661, respectively. Three stable islands are clearly visible at both tunes in the phase space. From FFT analysis of the turn-by-turn (TBT) x coordinates, the particles' Q_x are calculated and shown as a function of their actions in figures 1(b) and (d). Both plots show that Q_x increases as the particle's action increases, indicating a positive ADTS coefficient ($\alpha_{xx} > 0$). When the particles are in the stable islands, their tunes are exactly 2/3. Several particles are lost while tracking through the lattice with Q_x=0.643 so that their tunes are absent in figure 1(b).

From the xx_p phase space plots in figure 1, the action of SFP at different Q_x is extracted and plotted in figure 2. The extracted Jx_SFP decreases as the tune approaches the 3rd-order resonance line, indicating the island separation in the horizontal plane shrinks. The simulated Jx_SFP agrees very well with the calculated Jx_theory using equation (1).

Frequency map analysis is performed to check the DA of the TRIBs lattice at different tunes.

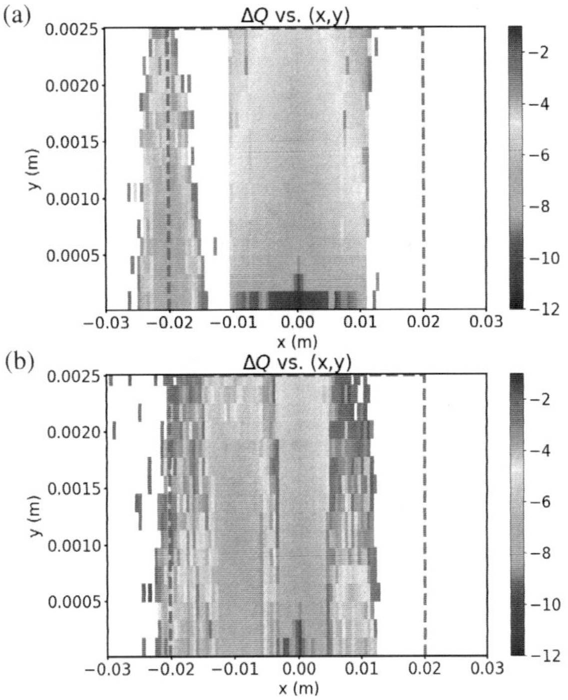

Figure 3. Dynamic aperture at two different horizontal tunes (a) Q_x=0.643 and (b) 0.661. The red dash lines indicate the physical apertures at element 1.

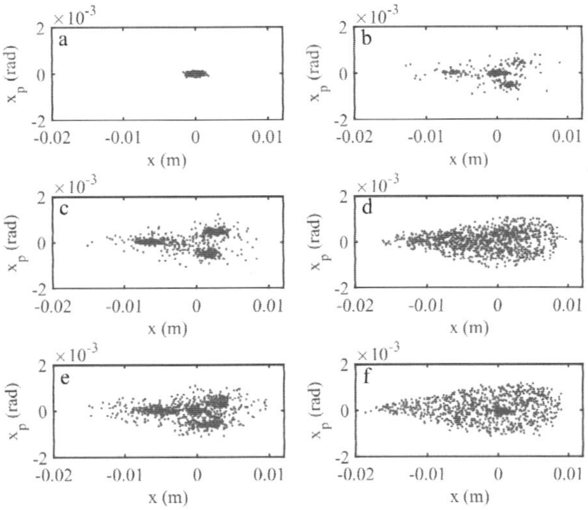

Figure 4. The 1000-particles distribution in the xx_p phase space at Q_x=0.6649 (259.4 kHz) at (a) turn 1 and (b) turn 40000, when applying a sinusoidal kick at f=259.4 kHz with the kick amplitude of (c) 0.5 μrad and (d) 5 μrad at turn 40000, and at f=261.35 kHz with the kick amplitude of (e) 0.5 μrad and (f) 5 μrad at turn 40000.

Figures 3(a) and (b) show the DA at Q_x=0.643 and 0.661, respectively. It is interesting that both the island and main buckets are visible in these plots. The horizontal DA of the island bucket is much less than the main buckets at Q_x=0.643 while comparable to the main bucket at Q_x=0.661, which is consistent with the phase space plots in figure 1.

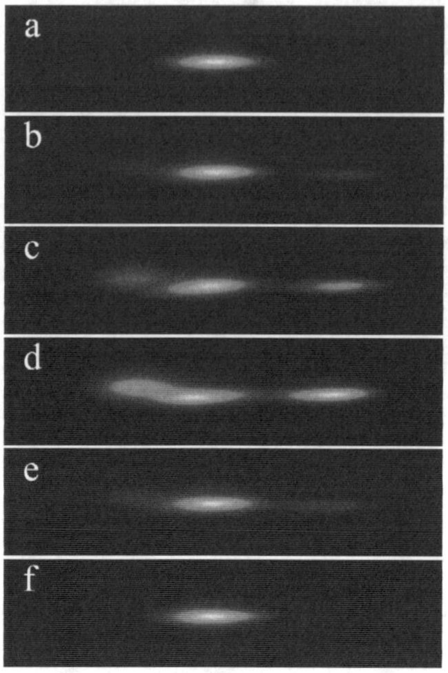

Figure 5. Beam profiles at Q_x= (a) 0.6613 (258.0 kHz), (b) 0.6631 (258.7 kHz), (c) 0.6648 (259.35 kHz), and (f) 0.6700 (261.4 kHz). (d) and (e) are recorded at the same tune 0.6638 (259.35 kHz) but with applying a sinusoidal kick at f=259.35 kHz and 261.35 kHz, respectively.

More realistic tracking simulation including radiation damping and excitation is implemented to check the TRIBs formation in the TRIBs lattice. Starting with an initial distribution with design emittance ϵ_x=30 nm·rad (figure 4(a)), 1000 particles are tracked through the lattice at different horizontal tunes. Since the horizontal radiation damping time in CESR is 12 ms (~5000 turns), 4×10^4 turns is used for tracking to cover ~8 damping times. When $Q_x < 0.6613$ (258 kHz), the particles remain in the similar distribution through 4×10^4 turns as seen in figure 4(a). When $Q_x > 0.6639$ (259 kHz), the particles diffuse from the core to the island buckets. Figure 4(b) shows the particles occupy three islands as well as the core after tracking 4×10^4 turns at Q_x=0.6649 (259.4 kHz).

Starting with the distribution in figure 4(b) and applying a sinusoidal kick at f=259.4 kHz same as the core tune, most particles are driven from the core to island buckets (figure 4(c)) after tracking 4×10^4 turns. However when the sinusoidal kick amplitude is greater than 3 µrad, the particles are not clustered in the islands but distributed in the xx_p space evenly (figure 4(d)).

When starting tracking with the initial distribution as in figure 4(c) with the sinusoidal kick at f=261.35 kHz near the island tune (0.667), the particles are converged back to the core bucket but not fully cleared from the islands buckets after 4×10^4 turns (figure 4(e)). When increasing the kick amplitude to 5 µrad, the particles are cleared from island buckets but most of them are evenly distributed in the phase space (figure 4(f)).

4. Experiment

After confirming the formation of TRIBs in simulation, we loaded the TRIBs lattice into CESR. Once the optics correction were made (phase and orbit), we adjusted Q_x from 0.6537 to 0.667 while viewing and recording the beam profile from synchrotron radiation using vBSM [8]. The beam profile images from a 1-mA single positron bunch observed at different tunes are shown in figure 5. When $Q_x < 0.663$ (258.7 kHz), the beam stays in the core (figure 5(a)). When Q_x is near 0.663 (258.7 kHz), the TRIBs start to form (figure 5(b)) while most particles are still in the core bucket. When Q_x=0.6648 (259.35 kHz), more particles diffuse to the island buckets (figure 5(c)). A bunch-by-bunch feedback was then used to apply a sinusoidal kick (~0.1 μrad) with the same frequency as the core tune 259.35 kHz to the beam. This clearing kick drives the particles from the core to the island buckets (figure 5(d)). If the sinusoidal kick's frequency was set to 261.35 kHz, this clearing kick drives the particles from the islands to the core (figure 5(e)). But there are still some particles remaining in the island buckets. When Q_x is above the 3rd-order resonant line 0.667 (260.1 kHz), all the particles diffuse back to the core island (figure 5(f)). The experimental observation agrees well with the simulation results. The behavior of the clearing kick is consistent with the results reported in [3].

5. Conclusion

We studied the conditions of TRIBs formation near a 3rd-order resonance line and designed a special lattice with optimized sextupole distribution to meet the requirement. Tracking simulation is then developed to confirm the formation of TRIBs. Finally, the lattice was loaded into CESR and the TRIBs were observed at Q_x=0.6648 (259.35 kHz). In addition, the sinusoidal kicks at 259.35 kHz or 261.3 kHz help to clear the particles in the core or island buckets, respectively. In conclusion, we have demonstrated a new approach to design and observe TRIBs in a high-energy storage ring, which provides guidance for future systematic TRIBs study.

Acknowledgments

We thank David Rubin and Robert Meller for valuable discussion, Joel Brock and Ernest Fontes for supporting the project. The work was supported by National Science Foundation under award number PHYS-1757811 and DMR-1829070.

References

[1] Lee S Y 1999 *Accelerator Physics* (USA: World Scientific)
[2] Cappi R *et al* 2004 *Phys. Rev. Spec. Top. Accel. Beams* **7** 024001
[3] Ries M *et al* 2015 *Proc. 6th Int. Particle Accelerator Conf. (IPAC'15)* (Richmond, VA, USA) pp 138-40
[4] Goslawski P *et al* 2019 *Proc. 10th Int. Particle Accelerator Conf. (IPAC'19)* (Melbourne, Australia) pp 3419-22
[5] Holldack K *et al* 2020 *Commun. Phys.* **3** 61
[6] Tavares P F *et al* 2019 *Proc. 10th Int. Particle Accelerator Conf. (IPAC'19)* (Melbourne, Australia) pp 1185-90
[7] Shanks J *et al* 2019 *Phys. Rev. Spec. Top. Accel. Beams* **22** 021602
[8] Wang S T *et al* 2013 *Nucl. Instrum. Methods Phys. Res., Sect.* A **703** 80
[9] Sagan D 2006 *Nucl. Instrum. Methods Phys. Res., Sect.* A **558** 356

Journal of Physics: Conference Series **2420** (2023) 012048 doi:10.1088/1742-6596/2420/1/012048

Fast orbit response matrix measurement via sine-wave excitation of correctors at Sirius

M M S Velloso, F H de Sá and M B Alves

Brazilian Synchrotron Laboratory (LNLS), Campinas, Brazil

E-mail: matheus.velloso@lnls.br

Abstract. Sirius is the new 4th generation storage ring based synchrotron light source built and operated by the Brazilian Synchrotron Light Laboratory (LNLS). In this work, we report on the implementation at Sirius of a fast method for orbit response matrix (ORM) measurement which is based on sine-wave parallel excitation of orbit corrector magnets' strength. This "AC method" has reduced the ORM measurement time from $\sim$ 25 minutes to 2.5–3 minutes and displayed increased precision if compared to the standard serial measurement procedure. When used as input to the Linear Optics from Closed Orbits (LOCO) correction algorithm, the AC ORM yielded similar optics corrections with less aggressive quadrupoles strength changes.

1. Introduction

1.1. Orbit response matrix and Sirius' setup

At Sirius, 160 *beam position monitors* (BPMs) read horizontal and vertical displacements of the electron beam. The BPMs data is arranged in a 320-component vector $\mathbf{u} = (x_1, x_2, \ldots, x_{160}, y_1, y_2, \ldots, y_{160})^\intercal$. A $\Delta\theta_j$ kick from the j-th corrector magnet (CM) causes an orbit distortion which is measured by the i-th BPM as the combination

$$\Delta u_i = \sum_{j=1}^{n} M_{ij} \Delta\theta_j. \tag{1}$$

M_{ij} are the entries of the orbit response matrix (ORM), which relates the orbit change due to CMs strength variations. At Sirius, $n = n_x + n_y = 280$ is the total number of CMs, with $n_x = 120$ and $n_y = 160$ being the number of horizontal (CHs) and vertical correctors (CVs), respectively. In matrix notation, the orbit distortion reads

$$\begin{bmatrix} \Delta\mathbf{x} \\ \Delta\mathbf{y} \end{bmatrix} = \begin{bmatrix} M_{xx} & M_{xy} \\ M_{yx} & M_{yy} \end{bmatrix} \begin{bmatrix} \Delta\boldsymbol{\theta}_{\text{CHs}} \\ \Delta\boldsymbol{\theta}_{\text{CVs}} \end{bmatrix}, \tag{2}$$

which highlights the diagonal blocks M_{xx} and M_{yy}, and off-diagonal blocks M_{xy} and M_{yx} of the ORM.

The ORM is essential to orbit correction, where we wish to minimize $\chi^2 = |\mathbf{u} - \Delta\mathbf{u}|^2 = |\mathbf{u} - M\Delta\boldsymbol{\theta}|^2$, $\Delta\boldsymbol{\theta}$ being the vector with entries $\Delta\theta_j$. The matrix also encodes information about the storage ring linear optics and is the input to the model-based correction algorithm LOCO [1, 2].

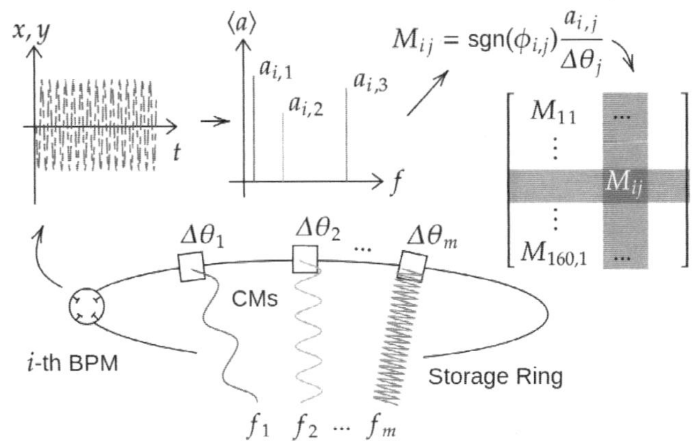

Figure 1. Parallel AC measurement of ORM: beam is excited by different CMs, each one at a different frequency. Spectral signature in beam motion reveals the amplitudes $a_{i,j}$, induced by the j-th CM to the beam as read by the i-th BPM.

1.2. Fast measurement procedure: The "AC Method"

If we perform kicks to the beam using only one CM, say, the j-th CM, equation (1) reduces to $\Delta u_i = M_{ij}\Delta\theta_j$, giving $M_{ij} = \Delta u_i/\Delta\theta_j$. Therefore, by serially kicking the beam corrector by corrector and measuring the corresponding orbit distortions we can reconstruct the ORM column by column. This is the traditional procedure for measuring the ORM. At Sirius, it usually takes about 25 up to 30 minutes to be completed.

The alternative method we report here is based on the parallel, alternating excitation of the beam. This "AC method" was first implemented at the Diamond Storage Ring [3] and later at ALBA [4] and NSLS-II [5], where it proved to be a faster and reliable ORM measurement method. The general idea is to sinusoidally drive the beam by CMs at different frequencies so the harmonic signature in the BPMs readings holds information about several CMs' excitation at the same acquisition.

In the i-th BPM time series, we fit the beam motion to harmonic components at the CMs frequencies by solving the linear problem

$$\begin{bmatrix} \cos(2\pi f_1 t_1) & \sin(2\pi f_1 t_1) & \dots \\ \cos(2\pi f_1 t_2) & \sin(2\pi f_1 t_2) & \dots \\ \vdots & \vdots & \\ \cos(2\pi f_1 t_n) & \sin(2\pi f_1 t_n) & \dots \end{bmatrix} \begin{bmatrix} b_{i1} \\ c_{i1} \\ \vdots \\ b_{im} \\ c_{im} \end{bmatrix} = \begin{bmatrix} u_i(t_1) \\ u_i(t_2) \\ \vdots \\ u_i(t_n) \end{bmatrix}, \tag{3}$$

where the cosines and sines columns at frequencies f_j repeat up to frequency f_m. We solve for the Fourier components $b_{i,j}$ and $c_{i,j}$ by least-squares, thus extracting the amplitudes $a_{i,j} = \sqrt{b_{i,j}^2 + c_{i,j}^2}$ and phases $\phi_{i,j} = \mathrm{atan2}(b_{i,j}, c_{i,j})$ from the beam motion imprinted by the CM oscillating at frequency f_j. The expected orbit distortions are $\Delta u_i(t_n) = \sum_j a_{i,j}\sin(2\pi f_j t_n + \phi_{i,j})$, so the ORM entries read

$$M_{ij} = \mathrm{sgn}(\phi_{i,j})\frac{a_{i,j}}{\Delta\theta_j}, \tag{4}$$

where $\phi_{i,j} \in (-\pi, \pi]$ and $\mathrm{sgn}(\cdot)$ is the sign function. In summary, we condense the measurement process to the frequency domain, as illustrated by figure 1.

2. Beam response

In order to choose adequate driving frequencies, we sought to characterize the beam frequency response compared to the BPMs noise baseline by constructing a signal-to-noise ratio vs.

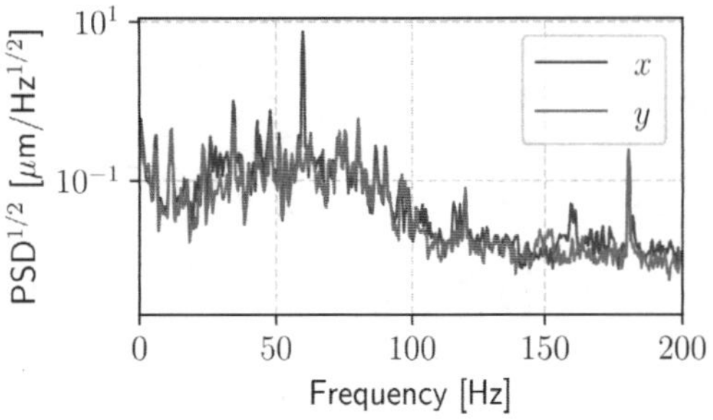

Figure 2. An example of PSD estimated for BPMs' readings.

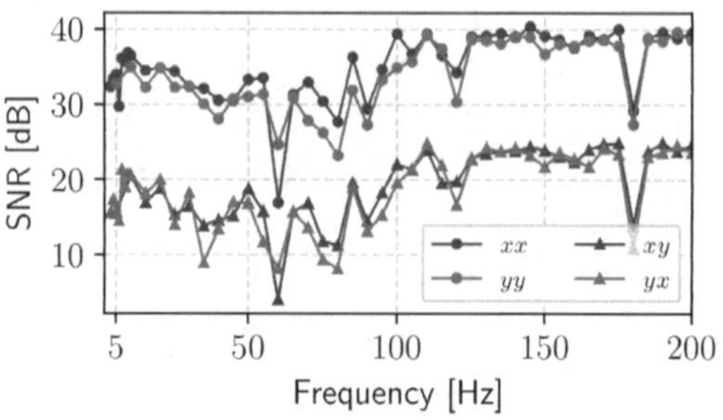

Figure 3. SNR for orbit distortions due to kicks on the same axis (xx, yy) and on different axes (xy, yx).

frequency. We implemented scripts to set one CM to operate with alternating sine-wave excitations while other CMs remain static. One CH drove the beam at fixed strength of $\Delta\theta = 5$ µrad with frequencies ranging from 1 to 200 Hz, one frequency at a time, with steps of 5 Hz. Subsequent orbit distortions were captured by all the BPMs, whose acquisition trigger was synced with the CMs' trigger. We repeated this procedure for one CV. The BPMs were set to read 5500 points at $\sim$ 1 kHz sampling rate after receiving the trigger event, i.e. they collected positions for about 5.5 s. In the time series for each BPM, we fitted the recorded beam displacements to harmonic components and solved for the amplitude $a_{i,j}$ of the i-th BPM orbit readings excited by the CM at the f_j frequency.

To determine noise at BPMs readings we acquired $T_{\text{acq}} = 9.9$ s of CM excitation-free orbits, at BPMs sampling rate of $\sim$ 1 kHz and estimated the Power Spectral Density (PSD) of the BPMs readings (figure 2). Noise at frequency f_j on the i-th BPM readings was evaluated as the square-root of the signal variance $\sigma_{i,j} = \sqrt{\text{PSD}(f_j) \times \delta f}$, with $\delta f = T_{\text{acq}}^{-1}$ being the frequency resolution. The average ratio between amplitude and noise at a given frequency, $\langle a_{ij}/\sigma_{i,j} \rangle_i$, was adopted as the signal-to-noise ratio (SNR) in units of dB (the last i subscript indicates the average over BPMs). The xx (yy) line in figure 3 represents the SNR for x (y) orbit distortions due to CH (CV) excitations, while the xy (yx) line indicates the SNR for x (y) orbit distortions due to CV (CH) excitations The constructed SNRs discourage the use of frequencies multiples of 60 Hz.

3. Fast ORM measurements

We configured parallel alternating excitation of CMs and acquired data of multi-frequency excitation in order to reconstruct the ORM by the AC method. We excited the beam by driving the CMs of each of the 20 sectors present in the Sirius storage ring. At each sector, the 6 CHs drove frequencies $f_x = 3, 7, 13, 19, 29, 37$ Hz, while the 8 CVs drove frequencies $f_y = 5, 11, 17, 23, 31, 41, 47, 59$ Hz. We chose prime numbers to avoid problems with harmonics which might arise from non-linearities in the beam response function.

The start of BPM acquisition and the activation of CMs were synced by the same triggered event from the timing system, and a delay of 25 ms was set for the CMs to start driving the beam. Since the chosen frequencies are integers in units of Hz, all the CMs performed an integer number of oscillations for 4 s at 5 µrad while the BPMs recorded positions for $\sim$ 4.1 s. This was done to be sure we knew exactly when the beam excitation started and ended in the time series. The fitting algorithm was restricted to fit data within the time window in which CMs were actually driving the beam, and the integer number of oscillations guarantee the orthogonality of the data vectors at the frequencies which we used to fit the data. Measurements with these

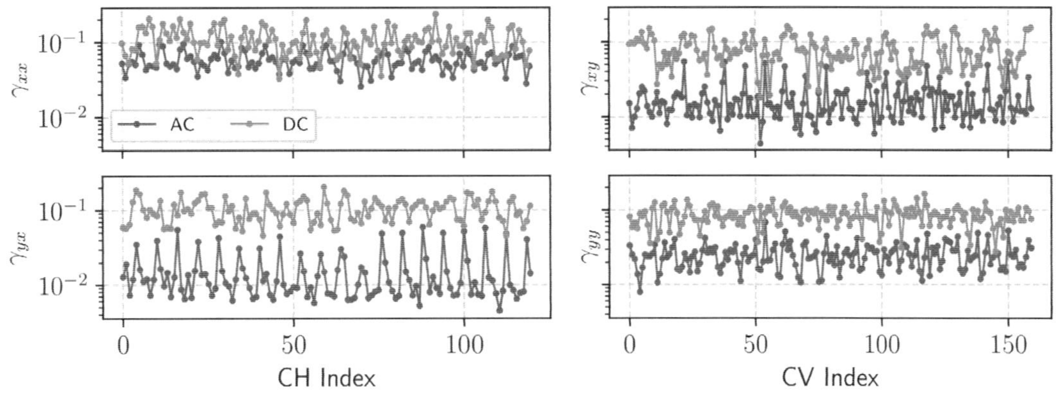

Figure 4. Mean column deviation for four consecutive AC and DC ORMs measurements: matrices are measured and the variance of the entries are averaged for each one of its blocks. The square-root of the average is taken, resulting in a quantity γ_j condensing a column's deviation. The AC methods presents lower deviations across measurements.

settings took around 2.5–3 min to be completed.

To evaluate the resemblance between AC and DC ORMs we performed the following analysis: the cosine $\cos\theta_j$ between the j-th column vectors $\mathbf{v}_{\mathrm{AC},j}$, $\mathbf{v}_{\mathrm{DC},j}$ of the AC and DC ORMs is an estimator of these vectors' signature correlation: the resemblance between the columns signature is higher when this estimator is closer to 1. In our measurements, the correlation residue $|1-\cos\theta_j|$ between the AC and DC ORMs columns is on average $\sim 3\%$ for off-diagonal blocks (M_{xy}, M_{yx}) and $\sim 0.03\%$ for diagonal blocks (M_{xx}, M_{yy}), indicating a good agreement between the matrices signature.

3.1. *The method's precision*

For comparing consecutive ORM measurements we defined a variance matrix

$$\sigma_{ij}^2 = \frac{1}{N-1}\sum_{k=1}^{N}(M_{ij}^k - \langle M\rangle_{ij})^2 \tag{5}$$

with entries indicating the variance between the ij entries of the measured M^k matrices with respect to the average matrix $\langle M\rangle$. For each corrector, i.e. for each ORM column, we considered the average $\left\langle\sigma_{ij}^2\right\rangle_i$, which condenses the mean variance for a given corrector's column in the ORM. Four consecutive AC and DC measurements were carried and the variance matrix was calculated for each block $uv \in \{xx, xy, yx, yy\}$ of the ORMs. We also defined a γ mean deviation vector, with entries $\gamma_j = \sqrt{\left\langle\sigma_{ij}^2\right\rangle_i}$. Figure 4 shows γ_{uv}, the γ vector for the uv ORM blocks. The higher precision of the AC method is evident from the lower values of the γ vectors.

3.2. *Performance at linear optics correction*

The LOCO algorithm fed with an AC ORM provided optics corrections similar to those achieved previously at Sirius with DC-measured ORMs, as figure 5 shows. The "AC LOCO" was able to deliver similar optics with subtler changes. In the DC ORM LOCO, the STD of percentual changes in quadrupoles' trim coils integrated gradient is 0.67%, while it is 0.42% for the AC LOCO.

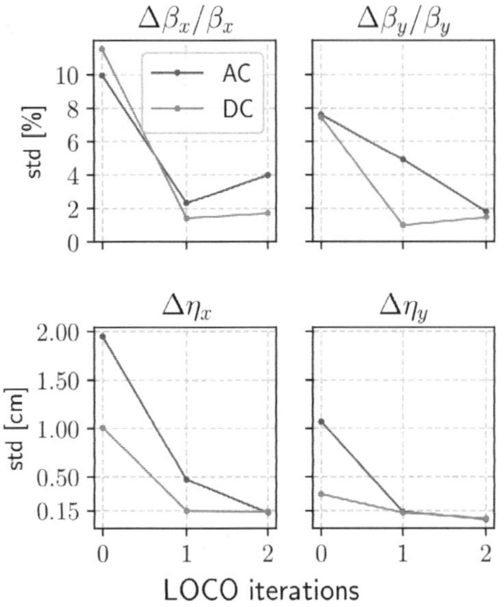

Figure 5. Comparison between LOCO fitted models: standard deviation (STD) of horizontal and vertical beta-beatings and dispersion function errors for the AC and DC ORMs as LOCO iterates.

4. Conclusions

The "AC ORM" method sped up ORM measurements at Sirius by almost ten times, while also displaying increased precision and delivering similar optics correction when used as input to the LOCO algorithm. Our measurement process still needs further improvements so it can be used for orbit correction. A proper characterization of the beam transfer function would elucidate the calibration of amplitudes needed to determine the ORM entries with correct scale factors and reveal additional phases introduced by the vacuum chamber and magnets. This will allow a more accurate determination of the beam motion phases and the use of higher frequencies to drive the beam, e.g., in the range $[120 - 180]$ Hz, for which the SNR is well-behaved.

References

[1] Safranek J 1997 *Nucl. Instr. Meth.* **388** 27–36
[2] Alves, M B 2021 *Proc. 12th Int. Particle Accelerator Conf. (IPAC'21)* (Campinas, Brazil) pp 825–8
[3] Martin I P S, Abbott M G, Bartolini R, Furseman M J and Rehm G 2014 *Proc. 5th Int. Particle Accelerator Conf. (IPAC'14)* (Dresden, Germany) pp 1763–5
[4] Martí Z *et al.* 2017 *Proc. 8th Int. Particle Accelerator Conf. (IPAC'17)* (Copenhagen, Denmark) pp 365–7
[5] Yang X, Smaluk V, Yu L H, Tian Y and Ha K 2017 *Phys. Rev. Accel. Beams* **20** 054001

Reverse bend option for a 6 GeV storage ring lattice

L Hoummi, N Carmignani, L R Carver, F Cianciosi, S M Liuzzo, T Perron and S White

ESRF, Grenoble, France

E-mail: lina.hoummi@esrf.fr

Abstract. Several high-energy synchrotron facilities adopted the Hybrid Multi-Bend Achromat scheme (HMBA) developed for the ESRF-EBS [1]. The considered lattice has been developed for a generic 6 GeV storage ring (SR) of 1100 m circumference [2]. It includes a short bending (SB) magnet at the center of the cell, and achieves a ~70 pm rad equilibrium horizontal emittance. The optics of such SR are modified introducing reverse bending magnets [3, 4] to further reduce the natural horizontal emittance to 53 pm rad. The impact of such modification on dynamic aperture and lifetime is assessed and optimized.

1. Introduction

The present work has been developed for a generic high-energy storage ring of 6 GeV and 1100 m circumference [2]. The lattice considered is an adaptation of the ESRF-EBS H7BA lattice [5] with 40 cells, achieving ~70 pm rad horizontal emittance. Figure 1 displays the optics of a standard cell of the presented ring.

Further reduction of the horizontal emittance is achievable with a fixed lattice layout, by controlling the dispersion function at the entrance of the dipole magnets, to approach the Theoretical Minimum Emittance (TME) conditions [6]. Reverse bends (RB), which are mechanically displaced quadrupoles with regards to the machine axis, generate a negative dipolar field of resulting angle [4]:

$$\theta^{\mathrm{RB}} = -k_1 l \Delta x \tag{1}$$

with k_1 the strength of the quadrupole, l its length and Δx the horizontal displacement. Such quadrupoles help modulate the dispersion function and derivative to approach the TME in the dipoles [3]. Previous attempts were conducted on a similar lattice [7].

2. Angular correction

All focusing quadrupoles in the dispersive area were used as reverse bends to optimise the emittance reduction. Their location is displayed in figure 1. The introduction of negative angles perturbs both the geometry and the dispersion in the standard cell, reducing the total angle per cell. The correction of the total angle is pursued using two methods:

- **Transposed method** The total reverse bending angle per cell $\theta^{\mathrm{RB}}_{\mathrm{tot}}$ is defined as the sum of all negative bending angles, expressed per RB in equation (1). It is compensated by shifting the standard dipole angles to recover the geometry of the cell. The angular distribution

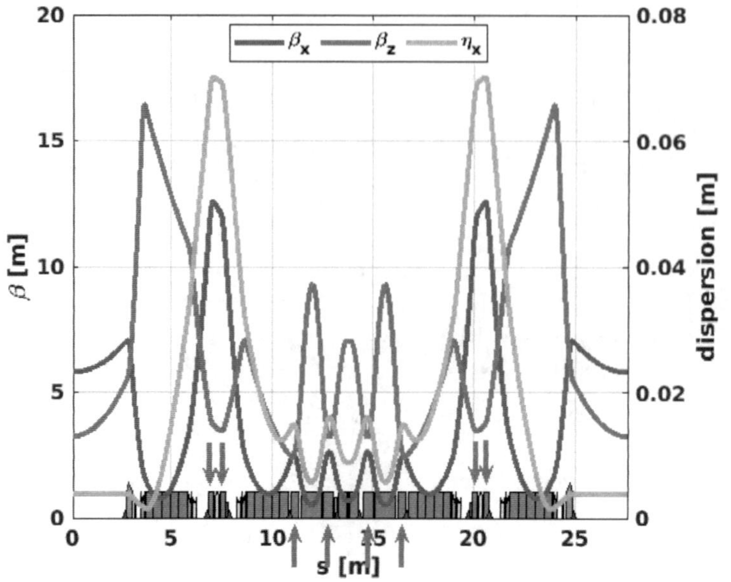

Figure 1. Location of the reverse bends (red arrows) in the lattice. All dipoles are represented in blue.

along the cell simply becomes $\theta^{\mathrm{RB}} = -\frac{|\theta^{\mathrm{RB}}_{\mathrm{tot}}|}{N_{\mathrm{RB}}}$ for RB and $\theta_{\mathrm{dip}} = \theta_0 + \frac{|\theta_{\mathrm{tot}}|}{N_{\mathrm{dip}}}$ for the standard bending magnets, where N_{RB}, N_{dip} are respectively the number of RB and dipoles in the cell, and θ_0 the bending angle of the dipoles in the nominal lattice.

- **Proportional method** The total angle is restored proportionally over the whole cell, applying the following operation on all bending magnets: $\theta_{\mathrm{new}} = \frac{2\pi\theta_{\mathrm{old}}}{N_{\mathrm{cells}}*a}$, where a is the cell angle before correction, and N_{cells} the number of cells of the machine. This method conserves the perturbation of the dispersion of the reverse bends, for which we expect a stronger emittance reduction.

Both methods were tested and compared in Accelerator Toolbox (A.T.) [8], for different total reverse bending angles $\theta^{\mathrm{RB}}_{\mathrm{tot}}$. To ease the comparison, $\theta^{\mathrm{RB}}_{\mathrm{tot}}$ was equally distributed among the reverse bends. As the dispersion is perturbed by the insertion of the reverse bends and by the correction of the total angle, the lattice optics are rematched at each step in $\theta^{\mathrm{RB}}_{\mathrm{tot}}$. Figure 2 superimposes the variation of the emittance and momentum compaction factor with $\theta^{\mathrm{RB}}_{\mathrm{tot}}$ for both methods considered. Table 1 compares the main characteristics of the lattices without and with reverse bends at the minimum emittance generated with the two methods detailed above. The smallest natural horizontal emittance (59 pm rad) is achieved with the proportional method, with similar total reverse bending angle and energy spread, and lower radiation losses.

3. Optimisation of the reverse bending angle distribution

To find the lowest horizontal natural emittance, the reverse bends are freely varied, conserving the proportional method for angular correction. Four knobs, respecting the cell axial symmetry, are used to seek for the minimum horizontal emittance. The inhomogeneous angular distribution frees the total reverse bending angle $\theta^{\mathrm{RB}}_{\mathrm{tot}}$, which optimum was previously limited to the parabolic evolution of the $\mathcal{H}$-function with the dispersion functions. The total reverse bending angle is varied by ± 0.01 degrees between consecutive iterations of a minimization process.

Each iteration randomly generates five angular distributions in the reverse bends. For each of

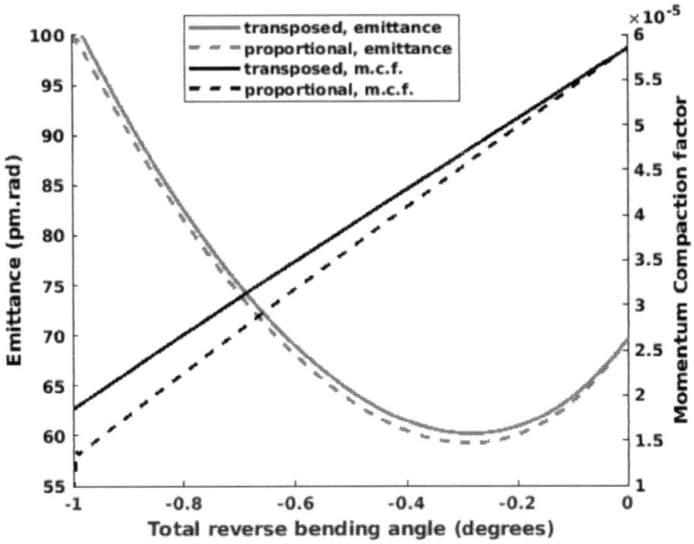

Figure 2. Evolution of the emittance and the momentum compaction factor (m.c.f.) with the total reverse bending angle, for the two methods of angle restoration.

Table 1. Comparison of main standard cell parameters for the two lattices with reverse bends of minimum natural horizontal emittance obtained with two methods of angular correction.

Parameter	Nominal	Method	
		Transposed	Proport.
θ_{tot}^{RB}(degrees)	0	-0.27	-0.29
ϵ_x(pm.rad)	70.0	66.9	59.2
σ_E	9×10^{-4}	10^{-3}	10^{-3}
m.c.f.	6×10^{-5}	5×10^{-5}	4.5×10^{-5}
J_x	1.54970	1.91472	1.93536
Radiation loss	50 keV	56 keV	54 keV

these distributions the total angle is corrected with the proportional method and the optics are matched. For a given θ_{tot}^{RB}, the distribution of minimum emittance is chosen. This ring is then reintroduced as the input of the next scan. The scan controls the horizontal damping partition number for sake of longitudinal stability.

The variation of both the horizontal natural emittance and the momentum compaction factor during the conducted scans are displayed in figure 3. The natural horizontal emittance first follows a parabolic variation with θ_{tot}^{RB} and then the inhomogeneous distribution of the reverse angles allows the increase of θ_{tot}^{RB}, further reducing the emittance. The optimisation is stopped at -0.5 degrees after convergence.

4. Lattice option with 53.1 pm rad emittance

After selection of the optimum standard cell option using the previous scan, high-β_x injection cells are matched to the ESRF-EBS ones [5]. The standard cells reverse bends were included in the injection cells, to close the ring geometry. The circumference of the ring was increased to set

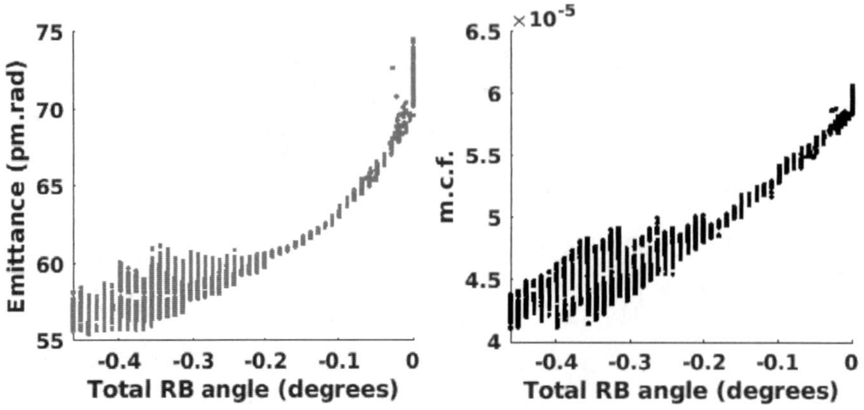

Figure 3. Variation of (left) the natural horizontal emittance and (right) the momentum compaction factor during the cycle of the algorithm. The large number of points per angle originates from the random reverse-angle distribution and the two-sided evolution of the total reverse bending angle $\theta_{\rm tot}^{\rm RB}$.

the harmonic number to 1296, for a RF frequency of 352.2 MHz [9]. Control of the dispersion at the middle of the standard cell allowed a total reduction in horizontal emittance to about 53 pm rad for the ring with reverse bends. Main parameters are listed for the lattices with and without reverse bends in table 2.

Table 2. Comparison of main parameters for the nominal lattice and its option with reverse bends.

Parameter	**Nominal**	**RB option**
Circumference	1100 m	1103 m
$(Q_{\rm x}, Q_{\rm y})$	(95.21, 33.34)	(95.21, 34.34)
$\epsilon_{\rm x}$(pm.rad)	70.0	53.1
$\theta_{\rm tot}^{\rm RB}$ (degrees) per cell	0	-0.35
$\sigma_{\rm E}$	8.6×10^{-4}	1.0×10^{-3}
m.c.f.	6×10^{-5}	5×10^{-5}
Energy loss	2.0 MV	2.2 MV
Chromaticity	(7,6)	(1,5)
RF Frequency	352.0 MHz	352.2 MHz
Harmonic number	1294	1296
Nat. bunch length	2.9 mm	3.4 mm

Table 3 lists the characteristics of the reverse bending magnets and their corresponding horizontal shift. The main contribution in the emittance reduction comes from the central part of the lattice (between 11 and 17 m in figure 1), with a maximum 2.1 mm shift required. To compare, the maximum transverse displacement values $\Delta x_{\rm max}$ in the ESRF-EBS are determined by analysis of the full cell assembly drawings and are reduced by about 1 mm considering the presence of kapton foils under the dispersion bump and the vacuum chambers, and specific vacuum chambers shapes and positioning of heating wires for the central reverse bends [10]. The displacement of high-gradient quadrupoles exceeds the maximum available. Vacuum chambers may need specific design adjustments compared to those of EBS.

Table 3. Horizontal shifts Δx of the implemented reverse bends.

Name	QF4[AE]	QF4[BD]	QF6	QF8
Length (m)	0.21	0.21	0.39	0.48
k_2 (m^{-2})	2.6	2.2	4.52	4.42
Δx (mm)	0.18	0.29	2.1	1.9
Δx_{max} (mm)	2.4-3.4	2.4-3.4	1.2-2.2	1.2-2.2

5. Dynamic aperture and lifetime

Optimisation of the sextupoles started with a scan on the ring chromaticity to optimise both the Touschek lifetime (TL) and the transverse dynamic aperture. Particles were tracked for 2000 turns and the momentum acceptance was computed over four standard cells for computing time considerations. The Touschek lifetime was calculated assuming the same uniform filling pattern as ESRF-EBS, 200 mA in 1296 bunches [11]. Figure 4 gathers the resulting dynamic aperture area and the Touschek lifetime for the scanned chromaticities. The best compromise found between Touschek lifetime at high current and on-momentum dynamic aperture at the injection point was found for the chromaticity (1,5).

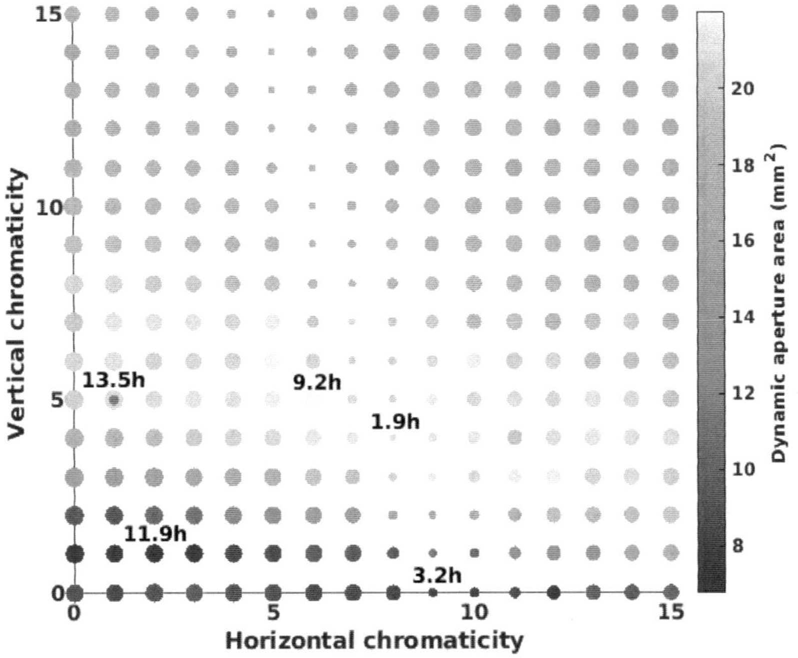

Figure 4. Touschek lifetime (dot size) and dynamic aperture area (color) variation with chromaticity for the H7BA ring with reverse bends, for $Z_N = 0.67\,\Omega$ and $\epsilon_V = 10\,\text{pm rad}$. The maximum TL and DA were obtained for a chromaticity (1,5) (red dot).

Figure 5 compares the dynamic aperture (DA) of the nominal lattice and its option with reverse bends of chromaticity (1,5) with high-beta injection cells to increase the dynamic aperture at injection. Particles are tracked starting at the injection point for 5000 turns. The standard cell with reverse bends presents a reduction of its dynamic aperture which remains comparable to the nominal one. The high-β_x injection cells improve the horizontal transverse acceptance at injection for the lattice without reverse bends as expected, but not for the lattice with RB. Their impact on the lifetime is also detrimental on the Touschek lifetime: when the 40-fold symmetry ring with reverse bends presents a 32 hours lifetime for 200 mA in 1296 bunches, the insertion of the injection cells with reverse bends reduce it to 13.5 hours. Tuning of the injection is therefore mandatory to enhance the DA at injection point, with less impact on the Touschek lifetime.

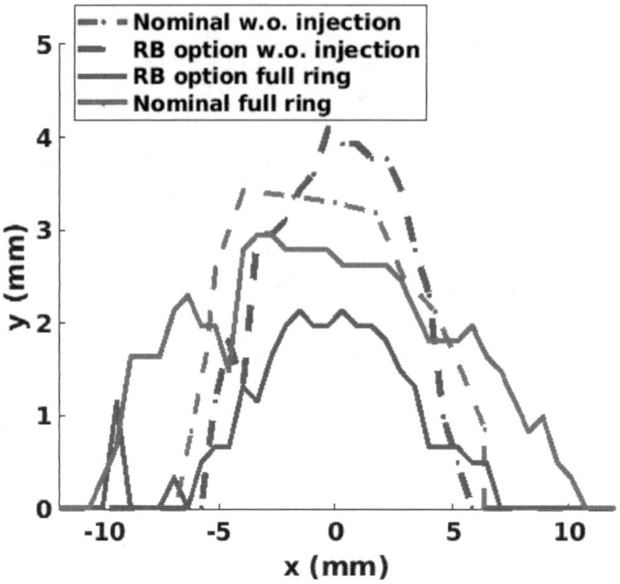

Figure 5. Dynamic aperture of the standard cell (dots) and ring with injection cells (plain) for the lattices with (black)and without (red) reverse bends.

6. Conclusion

A H7BA lattice for a 6 GeV 1100 m storage ring was presented, with a 70 pm rad natural horizontal emittance. Further reduction of its emittance can be achieved with reverse bending magnets with no additional constraint to the optics. The angle correction procedure was discussed. The proportional correction was selected for it achieved lower emittance and lower radiation losses.

Further optimisation of the reverse bending angle distribution was achieved in A.T., by tuning four RB knobs. A minimum emittance of 53 pm rad was achieved with this algorithm while conserving horizontal shifts at the limit of feasibility for each magnet. The full ring includes high-β_x injection cells like for ESRF-EBS. Nonetheless, the dynamic aperture was not increased at the injection point and a reduction of about 20 hours in Touschek lifetime was calculated. Further studies will focus on upgrading the injection cells in terms of optics and optimising the injection sextupoles to restore the high dynamic aperture and reduce their detrimental effect on the beam lifetime.

References

[1] Farvacque L *et al.* 2013 *Proc. 7th Int. Particle Accelerator Conf. (IPAC'13)* (Shanghai, China) pp 79–81
[2] Liuzzo S M *et al.* 2021 *Proc. 12th Int. Particle Accelerator Conf. (IPAC'21)* (Campinas, Brazil) pp 1466–9
[3] Streun A *et al.* 2014 *Nucl. Instrum. Methods Phys. Res., Sect.* A **137** 148-54
[4] Riemann B and Streun A 2019 *Phys. Rev. Accel. Beams* **22** 021601
[5] ESRF Orange book `https://www.esrf.fr/about/upgrade`
[6] Wiedemann H 2007 *Particle Accelerator Physics* 3rd ed (Berlin, Germany: Springer) p 305

[7] Dyubkov V S, Kulevoy T and Tsyplakov E D 2021 *Proc. 27th Russian Particle Accelerator Conf. (RuPAC'21)* (Alushta, Russia) pp 280–2
[8] 8Nash B *et al.* 2015 *Proc. 6th Int. Particle Accelerator Conf. (IPAC'15)* (Richmond, VA, USA) pp 113–6
[9] Serriere V *et al.* 2011 *Proc. 2nd Int. Particle Accelerator Conf. (IPAC'11)* (San Sebastian, Spain) pp 68–70
[10] ESRF 2018 EBS Technical design report Tech. Rep. ESRF
[11] Revol J -L *et al.* 2021 *Proc. 12th Int. Particle Accelerator Conf. (IPAC'21)* (Campinas, Brazil) pp 3929–32

Generalisation and longitudinal extension of the Genetic Lattice Construction (GLC) algorithm

S Reimann[1,2], M Droba[1], O Meusel[1] and H J Podlech[1,3]

[1] IAP, Goethe University, Frankfurt am Main, Germany
[2] GSI Helmholtz Centre, Darmstadt, Germany
[3] HFHF, Helmholtz Forschungsakademie Hessen für FAIR, Germany

E-mail: `s.reimann@gsi.de`

Abstract. The GLC algorithm [1] allows the construction of efficient transfer lines with defined imaging properties using a minimum number of quadrupole elements. This work describes a generalisation of this algorithm to make it applicable to the use of arbitrary beam optical elements. This includes an extension to longitudinal phase space.

1. Introduction

The design of a matching section of a particle accelerator can be formulated as an optimisation problem that can be solved algorithmically. The goal of such an optimisation is to achieve an optimal fit of a given particle distribution to an arbitrary acceptance, with minimum energy and material requirements (i.e., with the lowest possible field strengths and the fewest possible elements). The formulation of the individual problem can be reduced to the problem of minimising a single-valued fitness function over a multidimensional cube. This requires finding a suitable, unambiguous parameterisation to encode both the transfer line geometry and the field strengths used (phenotype) by a list of numbers (genotype).

2. Parametrisation

Given a general unspecific transport line of length L, in which at least one section of length L_{opt} can be equipped with new elements. If $L = L_{\text{opt}}$, then there are no constraints on the position of new elements. Furthermore, there is a set of μ freely positionable beam manipulating components $\{C_1, \cdots, C_\mu\}$. These can also be meta-elements consisting of a fixed combination of components. In addition, there are those ν components whose positions are fixed from an optimisation point of view, which means that they are located outside L_{opt}. In sum, this gives $N = \mu + \nu$ components. Each of these components $C_n \in \{C_1, \cdots, C_{\mu+\nu}\}$ has a certain number of λ_n free parameters $x_{n,1} \cdots x_{n,\lambda_n}$. These parameters can be e.g. quadrupole strengths, accelerating gap voltages, but also the lengths of the respective elements. The total number of free parameters is then

$$\eta_p = \sum_{n=1}^{N} \lambda_n \, . \tag{1}$$

A position must now be defined for each free component from $\{C_1, \cdots, C_\mu\}$. The possibility of arbitrary permutations of the components should be implicitly included. For this purpose, a relative position $s_n \in [0,1]$ is defined for each component. The actual position l_n results then from

$$l_n = s_n \cdot \left(L_{\text{opt}} - \sum_{n=1}^{\mu} L_n \right), \tag{2}$$

where L_n is the length of the component C_n. Due to free positioning, the number of free parameters increases by μ to $\eta = \eta_p + \mu$.

Every possible realisation of a transport path that satisfies the given boundary conditions can be assigned to a point in an η-dimensional subspace of $\mathbb{R}^\eta$. The parameters are normalised to the range $[-1,1]$ with respect to the individual device limits. So this corresponds to a point

$$\boldsymbol{x} = (\tilde{x}_1 \ldots \tilde{x}_{\eta_p}\, \tilde{s}_1 \ldots \tilde{s}_\mu) \tag{3}$$

in the η hypercube. Here $\tilde{x}_n$ and $\tilde{s}_n$ are respectively the values of x_n and s_n normalised to the interval $[-1,1]$. I want to emphasise that each design solution for such a transfer line corresponds to a point in this space. There is a distinct assignment. Each of these points can now be assigned a value of a fitness function

$$F(\boldsymbol{x}) = 1 - T(\boldsymbol{x}) + \epsilon(\boldsymbol{x}) \tag{4}$$

that describes the performance of the associated transfer line with respect to beam transmission T. Its minimum F_{min} corresponds to the largest transmission T_{max}. The term $\epsilon(\boldsymbol{x})$ represents the parameterisation of further optimisation goals, such as lowest possible field strengths or minimum component dimensions, and is defined as a sub-norm in the simplest case, e.g.

$$\epsilon(\boldsymbol{x}) = ||(\tilde{x}_1, \cdots, \tilde{x}_{\eta_p})|| = \sqrt{\tilde{x}_1^2 + \cdots + \tilde{x}_{\eta_p}^2}\,. \tag{5}$$

The minimum of the fitness function encodes the optimal design solution for any given problem.

2.1. Example: a single quadrupole

A drift section of length $L = 20\,\text{m}$ is given at the end of which an aperture limitation is attached. This has a diameter of 4 cm horizontally and 10 cm vertically. An asymmetric, divergent particle distribution is chosen such that most of the beam is lost on the aperture limitation. For a quadrupole of length $L_q = 1\,\text{m}$, the goal is to find a position $l = s \cdot (L - L_q)$ within this distance and a gradient (kL_q) at which the transmission becomes maximum (figure 1).

According to equations (2) and (3) the parameters

$$x_1 = (kL_q) \quad \text{and} \quad s_1 = \frac{l}{(L - L_q)} \tag{6}$$

are used for the general optimisation. The problem has $\eta = 2$ degrees of freedom and each solution here corresponds to one point

$$\boldsymbol{x} = \begin{pmatrix} \tilde{x}_1 \\ \tilde{s}_1 \end{pmatrix} \tag{7}$$

in the associated unit square. Since it is a 2-dimensional problem, the function $F(\boldsymbol{x}) = F(\tilde{x}_1, \tilde{s}_1)$ can be visualized over the entire domain of definition (figure 2). F has only one minimum in this case, therefore a simple gradient method will also lead to success here. In general, however, this is not the case, because due to the quadrupole symmetry, already from the use of a second quadrupole further secondary minima will occur and it is a priori not clear which one corresponds to the global minimum.

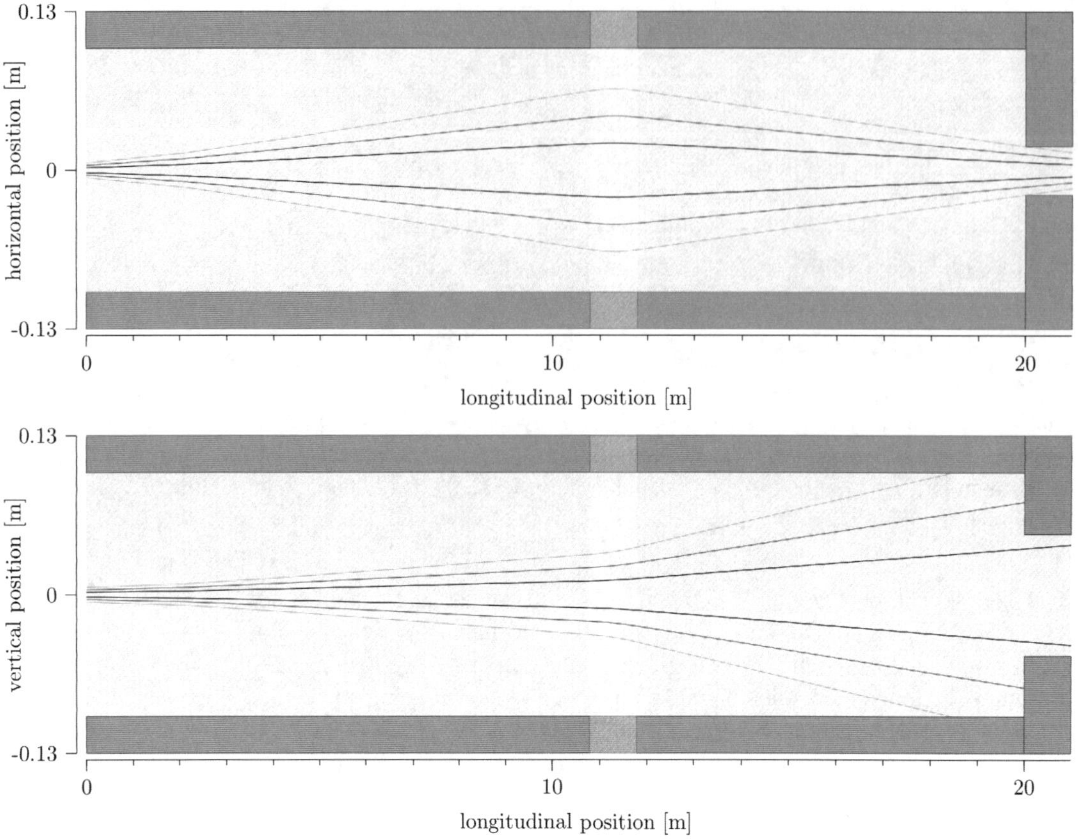

Figure 1. Representation of the solution with the maximum achievable transmission of $T = 74.5\%$ in this test case [2, page 82]. The black lines represent 1σ of the distribution, the dark grey 2σ and the light gray 3σ.

To determine the value of the fitness function at a certain point, a particle tracking simulation must be performed. The number of particles or the consideration of space charge and higher order effects are determined by the interested scientist. For this purpose, for example, the common matrix-based codes [3–7] can be used. However, depending on the specification of the problem and the desired accuracy, the number of necessary simulations grows rapidly. In particular, it grows exponentially with the number of dimensions of the problem space and soon exceeds all technically feasible possibilities to complete the simulation in a reasonable time.

3. Optimisation techniques

In cases of large multidimensional search spaces, natural-analog and metaheuristic optimisation methods have proven to be very useful, which was used also for many accelerator related optimisation studies [8–13]. For GLC, the following methods were investigated and compared (table 1):

- genetic algorithms [14]
- particle swarm optimisation [15]
- simulated annealing [16]
- BOBYQA [17]

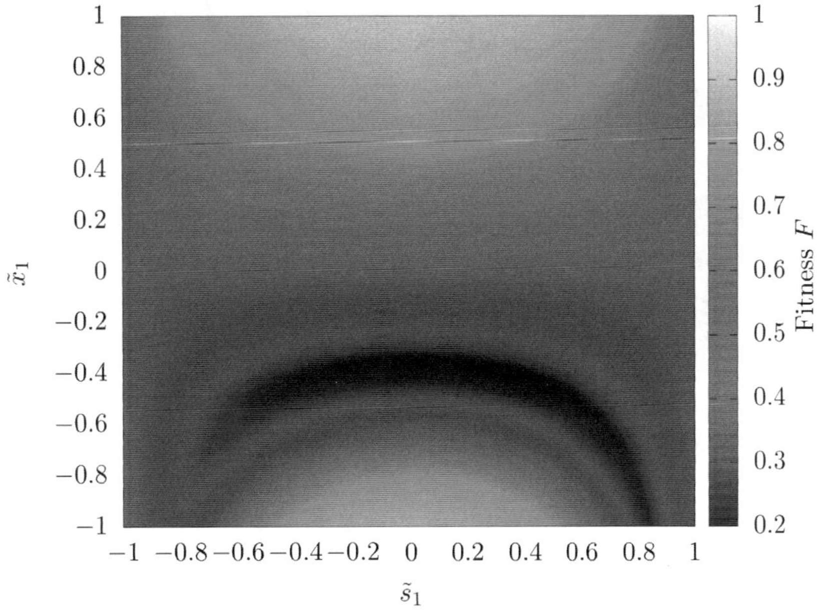

Figure 2. The result of the parameter scan [2, page 81]. Shown is the value of the fitness function depending on the normalised parameters $\tilde{s}_1$ and $\tilde{x}_1$. The minimum is located at $\tilde{x}_1 = -0.35$, $\tilde{s}_1 = 0.135$.

Table 1 shows a comparison of the optimisation methods in terms of the number of particle tracking simulations needed to find the optimal solution. Similar results are also obtained for other problems. The BOBYQA algorithm usually converges 2 orders of magnitude faster than all other methods. Unfortunately, it becomes increasingly unreliable in finding the optimum for more complex problems and in fractured parameter space topologies. For most examples, the genetic algorithm proved to be the best compromise between speed and probability of success. The topic of machine learning is currently very dynamic and it can be expected that new

Table 1. Comparison of a set of optimisation methods with respect to their performance. All heuristic algorithms are reliably able to find the optimum and differ only in the number of necessary optimisation cycles and therefore in the number of necessary tracking simulations. For the comparison, no special optimisation of the algorithms' hyper-parameters was performed.

optimisation method	**# tracking simulations**
gradient descent	63 ± 12
scan $\Delta = 0.005$	160000 ± 0
genetic algorithm	1790 ± 1048
particle swarm	7500 ± 6100
simulated annealing	8600 ± 5800
BOBYQA	26 ± 6

algorithms and methods will bring further improvements. Since the GLC method is generally independent of the underlying optimisation algorithm, the algorithm can easily be replaced to directly benefit from future developments. A collection of further examples, as well as the application of GLC in 2 accelerator projects (first, the construction of an RFQ matching section in the planned HBS accelerator [18] and second, the equipping of the H4 beam line at CERN north area with Gabor lenses [19]) can be found in [2].

4. Longitudinal extension

The extension of the algorithm to the longitudinal phase space is implicitly included in the parameterisation described above. The motion of the particles must then, necessarily be described via their path in the 6 dimensional phase space. In addition to the transverse location and momentum coordinates (x, x', y, y') at orbit position s, the longitudinal distance to the central particle and the momentum deviation $(z, \Delta p/p)$ or, equivalently phase and energy deviation $(\Delta\phi, \Delta W)$ or (z, z') must be additionally considered.

$$z = \phi(\beta\lambda) \ , \quad z' = \frac{\Delta W}{\beta^2\gamma^3 mc^2} \ , \quad \frac{\Delta p}{p} = \frac{\Delta W}{\beta^2\gamma mc^2} \tag{8}$$

For all beam optical elements, the corresponding 6x6 transfer matrices must then be used in the particle tracking codes. In this way, accelerating gaps and elements based on them, such as buncher cavities, single gap resonators, but also complete RF accelerators can be used as elements. The latter can be defined, for example, as compound meta components. The free parameters are then usually the position, the integral accelerating voltage and the phase of the synchronous particle. In this way, complex matching sections for 6-dimensional acceptance targets of following accelerator sections can be designed automatically.

4.1. Example: a single gap

Given is a matching section of length $L = 10\,\mathrm{m}$ in which an accelerating gap is to be positioned to match a beam to a given target acceptance ellipsoid. The main constraint is the phase acceptance of $\pm 30°$. All other limiting quantities were chosen to be no real constraint. Calculations were performed with the Accelerator Construction Set [2] (3dKV space charge, no compensation, 1 mA protons, $\frac{\Delta p}{p} = 0.001$, gaussian beam $\sigma_\phi = 25°$). There are 3 degrees of freedom in this example: the position of the gap mid s_{gap}, the gap voltage U_{gap} and the phase ϕ_{gap}. Thus each solution corresponds to one point

$$\boldsymbol{x} = \begin{pmatrix} \tilde{U}_{\text{gap}} \\ \tilde{\phi}_{\text{gap}} \\ \tilde{s}_{\text{gap}} \end{pmatrix} . \tag{9}$$

After an average of 40 generations (corresponding to 4000 simulation runs), the GLC solver converged. The result is shown in figure 3.

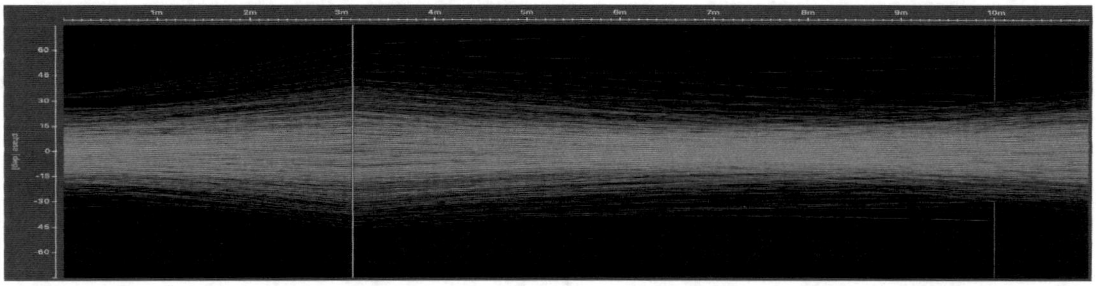

Figure 3. Solution with the maximum achievable transmission of $T = 91\%$ in this test case. The gap position is 3.12 m.

Scanning voltage and phase with the gap position held fixed gives a projection of the fitness function onto the U-ϕ-plane (figure 4).

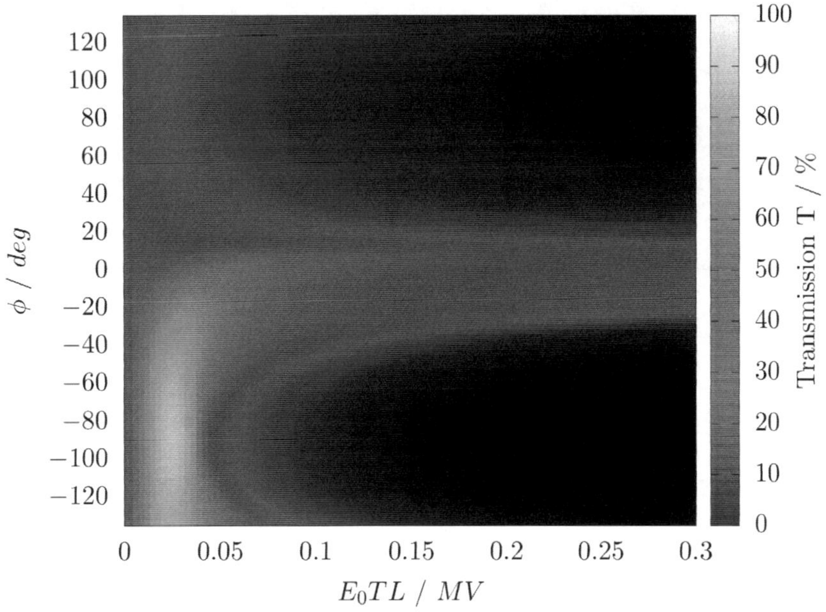

Figure 4. The result of the parameter scan with fixed gap position. Shown is the value of the transmission depending on the parameters ϕ and $U = E_0TL$. The optimum is located at $U = 25.3\,\mathrm{kV}$, $\phi = -92°$.

5. Error studies

In all cases where positioning errors were induced already during the phenotype construction within the optimisation process, there was evidence [2] that genetic lattice construction favours solutions that are robust to positioning errors. In this way, error tolerance could be implemented early in the design phase and the need for dedicated error studies and subsequent iterations of the design process could be eliminated.

6. Conclusion

The GLC algorithm is applicable to a large class of design problems and, together with precise particle tracking codes, GLC provides an excellent tool for the ion-optical design of future linear accelerators or for the optimisation of beam transport lines and matching sections [2].

References

[1] Reimann S, Droba M, Meusel O and Podlech H 2019 *J. Phys. Conf. Ser.* **1350** 012110

[2] Reimann S 2021 *Investigations of the automatic design and optimisation of beam transport lines in particle accelerators with dedicated imaging properties* Ph.D. thesis Goethe University, Frankfurt am Main, Germany

[3] Brown K L, Carey D C, Iselin F C and Rothacker F 1973 *TRANSPORT: a computer program for designing charged-particle beam transport systems; 2nd ed.* CERN Yellow Reports: Monographs (Geneva: CERN)

[4] Grote H and Schmidt F 2003 *Proc. of the 2003 Particle Accelerator Conf. (PAC2003)* (Portland, Oregon USA) pp 3497–9

[5] Franczak B 1984 *MIRKO — an interactive program for beam lines and synchrotrons* (Berlin: Springer Berlin Heidelberg) p 170–5

[6] Uriot D and Pichoff N 2020 *TraceWin* (Paris, France: CEA Saclay)
[7] Bryant P J 2000 *Proc. 7th European Particle Accelerator Conf. (EPAC'00)* (Vienna, Austria) pp 1357–9
[8] Gavaldà X, Keil J and Wanzenberg R 2018 *Proc. 9th Int. Particle Accelerator Conference (IPAC'18)* (Vancouver, Canada) pp 3015–7
[9] Appel S and Reimann S 2019 *J. Phys. Conf. Ser.* **1350** 012104
[10] Tang Z X and Pei Y J 2014 *Proc. 5th Int. Particle Accelerator Conf. (IPAC'14)* (Dresden, Germany) pp 3875–7
[11] Reimann S 2022 *Proc. 13th Int. Particle Accelerator Conf. (IPAC'22)* (Bangkok, Thailand) pp 922–4
[12] Velotti F M, Barnes M J, Carlier E, Dutheil Y, Fraser M, Goddard B, Magnin N, Ramjiawan R L, Renner E, Trappen P V and Waagaard E 2022 *Proc. 13th Int. Particle Accelerator Conf. (IPAC'22)* (Bangkok, Thailand) pp 961–4
[13] Zhang X and Sheehy S 2022 *Proc. 13th Int. Particle Accelerator Conf. (IPAC'22)* (Bangkok, Thailand) pp 38–41
[14] Goldberg D 1989 *Genetic algorithms in search, optimization, and machine learning* (Reading, Mass: Addison-Wesley Publishing Company)
[15] Eberhart R and Kennedy J 1995 *Proc. 6th Int Symp. on Micro Machine and Human Science (MHS'95)* (Nagoya, Japan) pp 39–43
[16] Kirkpatrick S, Gelatt C D and Vecchi M P 1983 *Science* **220** 671–80
[17] Powell M 2009 The bobyqa algorithm for bound constrained optimization without derivatives Tech. Rep. Department of Applied Mathematics and Theoretical Physics Cambridge University
[18] Brückel T, Gutberlet T E, Baggemann J, Böhm S, Doege P, Fenske J, Feygenson M, Glavic A, Holderer O, Jaksch S, Jentschel S *et al.* 2020 *Conceptual Design Report Jülich High Brilliance Neutron Source (HBS)* vol 8 (Jülich, Germany: FZ Jülich)
[19] Sherjan A, Droba M, Meusel O, Reimann S and Thoma K I 2022 *Proc. 13th Int. Particle Accelerator Conf. (IPAC'22)* (Bangkok, Thailand) pp 663–6

Parallelization of Radia magnetostatics code

A Banerjee[1], O Chubar[2], G Le Bec[3], J Chavanne[3], B Nash[4], C Hall[4] and J Edelen[4]

[1]Stony Brook University, Stony Brook, NY 11794, USA
[2]Brookhaven National Laboratory, Upton, NY 11973, USA
[3]ESRF, 38000 Grenoble, France
[4]RadiaSoft LLC, Boulder, CO 80301, USA

E-mail: anushka.banerjee@stonybrook.edu

Abstract. Radia 3D magnetostatics code has been used for the design of insertion devices for light sources over more than two decades. The code uses the magnetization integral approach that is efficient for solving permanent magnet and hybrid magnet structures. The initial version of the Radia code was sequential, its core written in C++ and interface in the Mathematica language. This paper describes a new Python interfaced parallel version of Radia and its applications. The parallelization of the code was implemented on C++ level, where the semi-analytical calculations of interaction matrix elements and resultant magnetic fields were parallelized using the Message Passing Interface. The parallel performance results are encouraging, particularly for magnetic field calculation post relaxation where a ~600 speedup with respect to sequential execution was obtained. The new parallel Radia version facilitates designs of insertion devices and lattice magnets for novel particle accelerators.

1. Introduction

The Radia code has been continuously improved for nearly two decades since its inception and development in the Insertion Devices laboratory of the European Synchrotron Radiation Facility (ESRF). From establishing boundary integral method-based calculations as an alternative and better approach than the conventional finite element, for e.g., FLUX3D [1], Radia has been widely used to compute fields for undulators. Over the last few years, Radia has not only provided better time performance and precision results when compared to FEM codes [2], it provides excellent agreement between the calculated and measured field values corresponding to relatively simple structures like quadrupole and sextupole [3], or complicated structures like the iron-dominated electromagnetic structure: SOLEIL Undulator HU256 [4].

Following the concepts of object-oriented programming, the core part of Radia is written in C++ and was initially interfaced to run serially on Wolfram Mathematica and Igor Pro only. Currently, the new, open-source version of Radia is already available in Python interface [5], with ongoing developments to make it executable on web-based graphical interface like Jupyter Notebook [6]. Radia's Python interface has been used for magnetic "cross talk" computations of the recent ESRF-Extremely Brilliant Source upgrade, where a large number of sequential calculations with different input parameters were performed in parallel at the ESRF cluster. The calculated values had good agreement with the measured values (relative errors in the 10^{-4} range) [7].

2. Methods used for parallelization

With an increased development in the field of light sources, the importance of magnetostatics code has increased by manifolds compared to what it was a few decades earlier. Despite Radia's stellar performance when compared to conventional FEM based code, the target computations involving 3D magnetic simulations of the insertion devices and accelerator magnets are quite complicated and CPU-intensive, with further increasing complexities. With an aim to further improve Radia's overall performance, parallelization tasks were undertaken.

MPI is a communication protocol and is used as the industry standard for the message passing model where a certain application comprises of a set of tasks which are assigned their own local memory whose location can be in the same machine or across several machines. Data exchange to conduct the operation by tasks is established by sending and receiving messages [8]. MPI provides programmers the flexibility to use it as a low-level approach with a detailed control on the flow of data, or as a high-level programming approach with parallel libraries designed to provide optimized performance without going into the depths of the MPI algorithm [9].

Solving any 3D magnetostatics problems in Radia comprises of 3 subsequent steps: calculation of elements of a large (often tens GBs memory size) matrix called Interaction Matrix and describing magnetic interaction between 'active' sub-volumes of a magnet geometry, created by segmentation (1), performing a relaxation procedure on it to determine values of the magnetization vector in all the "active" sub-volumes (2), and computing magnetic field and/or field integrals or other characteristics of magnetic fields, created by the sub-volumes with magnetization or current density (3). The two sections– the generation of the interaction matrix (1) and the calculation of magnetic field values after relaxation (3) are 'embarrassingly parallel' algorithms and hence have been parallelized using MPI at the C++ level.

Interaction matrix is a dense matrix in Radia, that may occupy a large memory, depending on geometry and its segmentation. Relaxation of this matrix may require a large number of iterations, in particular when solving complicated iron-dominated geometries [2, 4]. At each of these iterations, a relatively small number of multiplications needs to be done to take into account (or update) magnetization vector in each sub-volume (created by segmentation). This makes application of MPI not very efficient for parallelizing this algorithm. Instead shared memory-based parallelization, implemented using native multithreading capabilities introduced in C++11 standard library has been considered for parallelizing the specific section of the Radia code. The structure used for the tests is discussed in detail in the following section titled 'Test Magnetic Structure', followed by the outcomes detailed in 'Results of Benchmarking'.

3. Test magnetic structure

To evaluate the parallel performance of the new version of the Radia 3D magnetostatics code, we used a 3D model of a 14 mm-period, hybrid, in-vacuum, Cryo-cooled Permanent Magnet Undulator (CPMU) at 4 mm magnetic gap (see figure 1). This model comprises of an upper and a lower array with each of these arrays including an alternating sequence of permanent magnets (PrFeB with 1.67 T remnant magnetization) and soft iron poles (Vanadium Permendur with $\sim$2.29 T magnetization at saturation). The model geometry is generated by mirroring with respect to the three orthogonal symmetry planes passing through the magnetic centre. The geometry (specifically the poles) has been extensively subdivided to achieve solving accuracy better than 0.001 with respect to the peak magnetic field. 22680 independent sub-volumes are produced because of implementing subdivision before applying the symmetries, consequently yielding a dense interaction matrix of size $\sim$2 GB. The final calculated on-axis vertical field after solving the geometry for the magnetization in the sub-volumes is shown in figure 2.

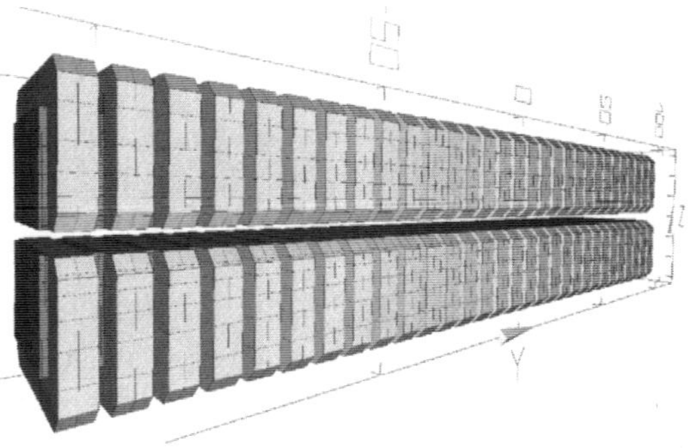

Figure 1. Radia 3D model of the hybrid, in-vacuum, CPMU used for our parallel performance tests.

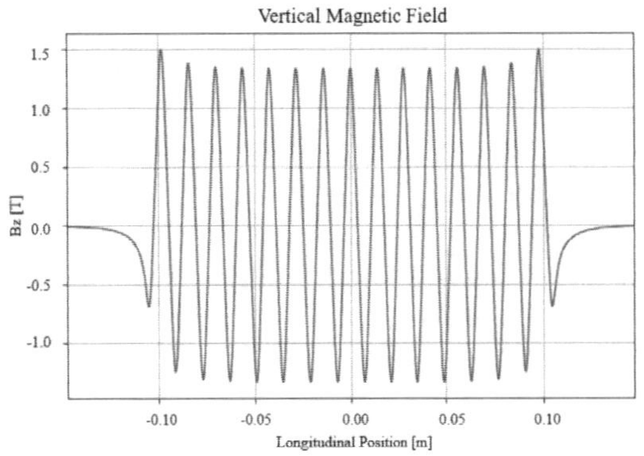

Figure 2. Calculated on-axis vertical magnetic field along the CPMU structure using Radia.

4. Results of benchmarking

Tests to evaluate parallel performance on many server nodes were conducted at the National Energy Research Scientific Computing Centre (NERSC), specifically the NERSC supercomputer Cori. Cori is a Cray XC40 with a theoretical peak performance of 30 petaflops. Execution times for each section were obtained at the end of each test, iteratively performed starting from 1 process (i.e., serial execution) and going up to 1024 processes and then averaged over 3 runs for each process number, thereby estimating the performance characteristics.

The parallel version of the Radia code performed well for both the cases – computation of elements of the interaction matrix and computation of magnetic field post relaxation of this matrix. For both the instances, a systematic decline in the required execution times was observed with increasing number of processes, see figure 3, that can be interpreted as the corresponding computational speedup achieved because of parallelizing the Radia code, leading to a maximum speedup value of $\sim$200 for the setting up of interaction matrices, and approximately $\sim$600 for computation of field values post relaxation, see figure 4.

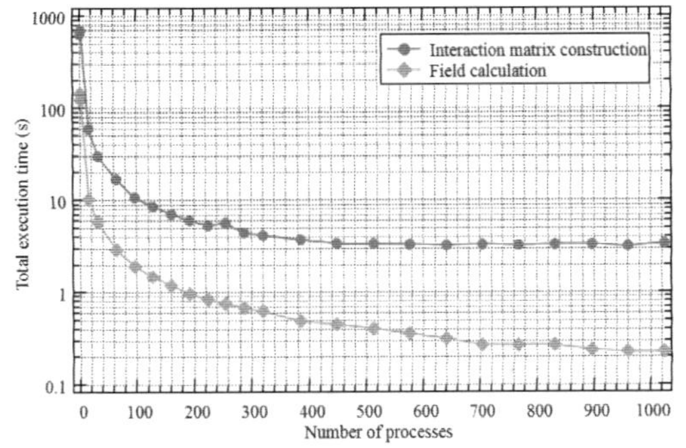

Figure 3. Mean time to set up the interaction matrix and calculate magnetic field, both as a function of the number of MPI processes.

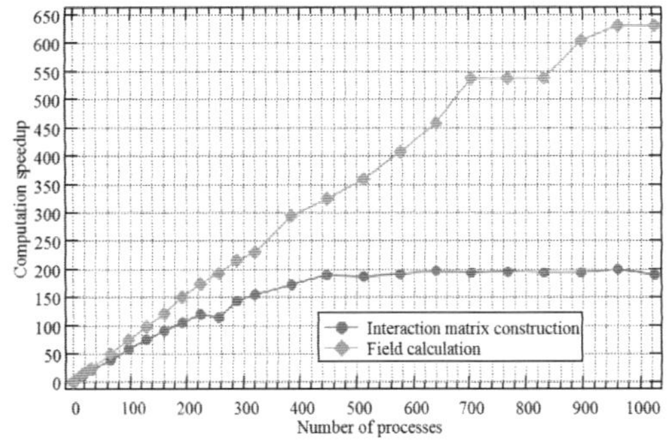

Figure 4. Total speedup in time values to set up interaction matrix and calculate field vs. number of MPI processes.

The performance corresponding to computation of elements of the interaction matrix demonstrates significant speed gain only until number of parallel processes is lesser than 400, whereas no such parallel performance saturation is observed for the function of field calculation after relaxation. Thus, profiling of the parallel performance was implemented at the C++ level, where time stamps were introduced at the beginning and end of specific functions in the code, to evaluate their corresponding execution times. Functions specifically related to communication and exchange of data between processes were profiled, namely 'MPI_Send' and 'MPI_Recv'. Primarily, 'MPI_Send' is executed by workers after they have computed the data packets to be sent to the master, and 'MPI_Recv' is performed by the master to receive the data packets sent by each worker. The plots demonstrating the total time required for the execution of 'MPI_Send' by each worker and 'MPI_Recv' by the master are given in figure 5 for setting up of interaction matrix and figure 6 for calculation of field post relaxation.

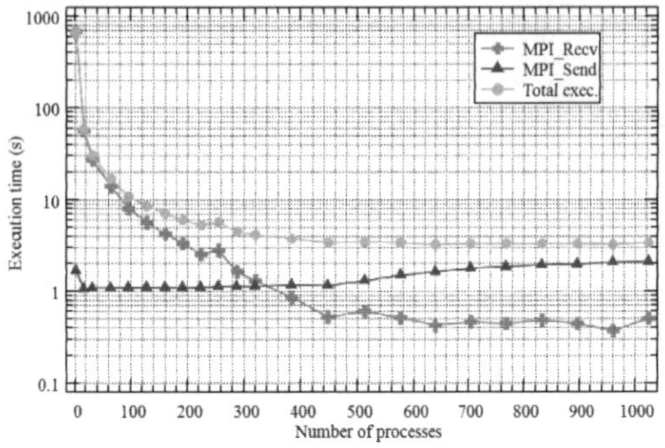

Figure 5. Total time to execute 'MPI_Send' by each worker, 'MPI_Recv' by master, and complete execution of interaction matrix generation vs. number of MPI processes.

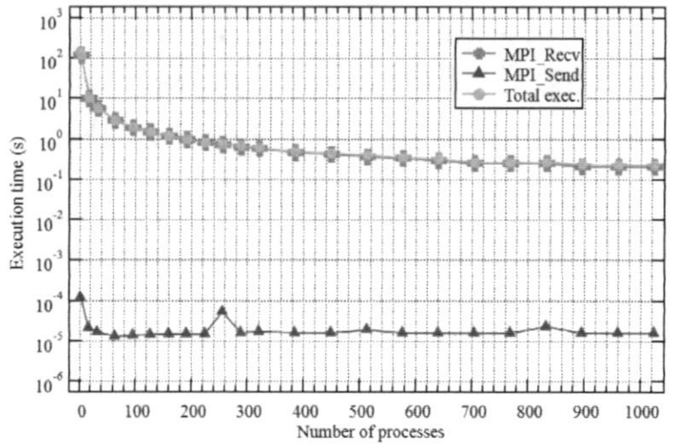

Figure 6. Total time to execute 'MPI_Send' by each worker, 'MPI_Recv' by master, and complete execution of field calculation vs. number of MPI processes.

For interaction matrix construction (figure 5), it can be observed that at some point near the saturation, the time spent by a worker in 'MPI_Send' exceeds the total time spent by the master in 'MPI_Recv', such a crossover is absent in case of the field calculation post relaxation, where the time spent by a worker in 'MPI_Send' is always much smaller than the time spent by the master in 'MPI_Recv', as observed in figure 6. The role of the master is to receive data after the calculation by workers. Thus, initially (for lesser number of workers), for interaction matrix generation, master time is mostly spent waiting for workers to finish the computation. But with increasing number of workers, the computation time decreases (i.e., each worker executes lesser number of calculations) and eventually master is unable to receive quickly enough all the computed data, thus delaying the workers who have already finished computing and instead get queued to send the data. This saturation is absent in the case of calculation of the field after the relaxation, when each worker performs large number of summing up operations to account for contributions from all sub-volumes of the 3D geometry to the field value at a given point in space. This is confirmed by the fact that in our tests, the time spent by a worker on calculation of one element of the interaction matrix was only 6.6 µs, while the calculation of the magnetic field after the relaxation by a worker at one observation point took a much longer time, 22.5 ms. These observations explain the earlier saturation of the overall speedup curve in the case of the interaction matrix generation, as compared to the field calculation after the relaxation.

Scalability of shared memory parallelization when compared to MPI, is weaker, but the resource sharing is better optimized. Preliminary tests to estimate the parallel performance of the relaxation of interaction matrix yielded a speedup of approximately 10 times compared to sequential execution of the code. The estimated overall speedup for the test geometry is ~41 when using 400 processes, the extent of parallelization dependent on the number of field calculations performed. Further algorithm-specific implementations are required to reach higher levels of parallelism.

5. Conclusion
The upgraded version of parallel Radia is now available for the Python interface, and it offers considerable speedups in different types of 3D magnetostatics computations [5]. Efforts are ongoing to improve its parallel performance in order to efficiently meet the computation requirements of the next generation light sources.

Acknowledgements
This work was supported by the US DOE BES SBIR grant no. DE-SC0018556.

References
[1] Chubar O, Elleaume P and Chavanne J 1998 *J. Synchrotron Radiat.* **5** 481-84
[2] Elleaume P, Chubar O and Chavanne J 1997 *Proc. 17th Particle Accelerator Conf. (PAC'97)* (Vancouver, B.C., Canada) pp 3509-11
[3] Andersson A, Chubar O and Lindgren L J 1998 *Proc. 6th European Particle Accelerator Conf. (EPAC'98)* (Stockholm, Sweden) p 1207
[4] Chubar O, Benabderrahmane C, Chavanne J, Elleaume P, Marcouillé O and Marteau F 2004 *Proc. 9th European Particle Accelerator Conf. (EPAC'04)* (Lucerne, Switzerland) pp 1675-7
[5] Radia `https://github.com/ochubar/Radia`
[6] Hall C *et al.* 2022 *The 14th Int. Conf. on Synchrotron Radiation Instrumentation (SRI2021)* (Humburg, Germany) unpublished
[7] Le Bec G, Chavanne J, Liuzzo S and White S 2021 *Phys. Rev. Accel. Beams* **24** 072401
[8] Kang S J, Lee S Y and Lee K M 2015 *Adv. Multimedia* **2015**
[9] Gropp W D and Lusk E 2003 *European Parallel Virtual Machine/Message Passing Interface Users' Group Meeting* 27

Gap in pagination due to unavailable paper.

Pages 318-324

Beamline design and optimisation for high intensity muon beams at PSI

Eremey Valetov (for the HIMB project)

Paul Scherrer Institut, CH-5232 Villigen PSI, Switzerland

E-mail: `eremey.valetov@psi.ch`

Abstract. The High Intensity Muon Beams (HIMB) project at the Paul Scherrer Institute (PSI) will provide muon intensities of the order of 10^{10} muons/s for particle physics and material science experiments, two orders of magnitude higher than the state of the art, which is currently available also at PSI. In particle transport simulations for the HIMB, we use *G4beamline* with measured $^{+}$ cross-sections and with variance reduction. We also use the codes *COSY INFINITY*, *TRANSPORT*, and *TURTLE* for some studies. We perform asynchronous Bayesian optimisation of the beamlines on a computing cluster using *G4beamline* and the optimisation package *DeepHyper*. We performed numerous studies for the design of the HIMB, and we produced various results, including the muon transmission, beam phase space, polarisation, and momentum spectrum.

1. Introduction

Next-generation muon experiments at Paul Scherrer Institute (PSI), such as Mu3e phase II, require muons to be delivered at unprecedented rates. The Mu3e experiment [1] searches for indications of Beyond-Standard-Model physics by attempting to detect the neutrinoless decay of a muon into three electrons, which would be practically unobservable (at the 10^{-55} level) in the Standard Model of particle physics as a charged lepton flavour violation (cLFV) [2].

A next-generation $\to$ e (MEG) experiment [3], searching for the highly suppressed cLFV decay of a muon into an electron and a photon, would also benefit from the substantially higher muon rates. Other intensity frontier particle physics experiments that would rely on an increase in the available muon rates are envisaged [3].

In addition to particle physics, novel concepts for performing characterisations of samples using the muon spin rotation method (SR) such as pixel-based detectors or microbeams also require an increase in available muon rate [3].

With the presently available rate of $\sim 10^8$ muons/s, to achieve the sensitivity of 10^{-16}, phase II of the Mu3e experiment would have to run for more than 13 years. The High Intensity Muon Beams (HIMB) project [4] seeks to deliver muons at $\sim 10^{10}$ muons/s at a proton current of 2.4 mA, making such sensitivities feasible [3].

To provide a high-intensity muon beam at 10^{10} muons/s into the experimental areas, the HIMB project will use a new graphite target "TgH" with a slanted target design, which increases the surface muon rate; high-acceptance capture solenoids close to the target; and solenoid focusing instead of quadrupole focusing in the beamlines MUH2 and MUH3 for a higher transmission rate. A partial layout of the HIMB is shown in figure 1. The MUH2 beamline has

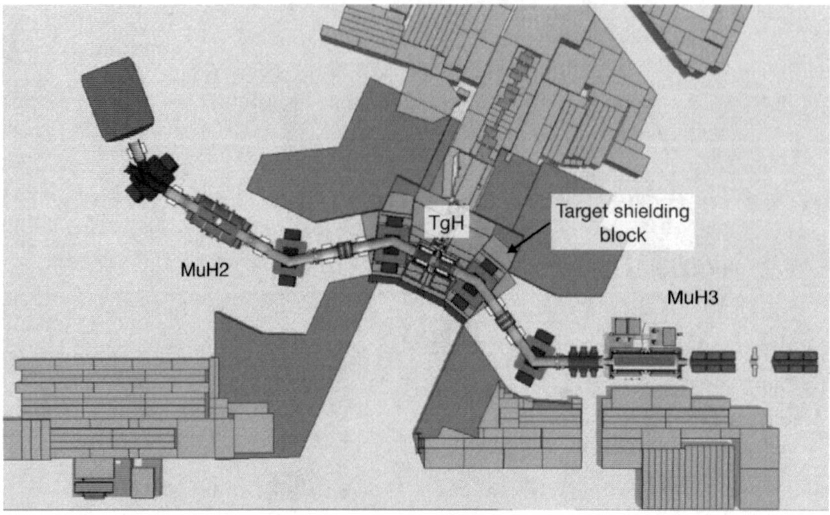

Figure 1. Layout of the HIMB target station and beamlines. The MUH3 beamline is not fully shown; it extends about 38 m from target TgH to the final focus of its branch MUH3.3 and also has a branch MUH3.2.

only solenoid focusing, while the MUH3 beamline has solenoid focusing in the first two straight sections and conventional quadrupole focusing further downstream. The HIMB will also benefit from a newly designed layout that uses lower bending angles and large-aperture dipoles.

We performed optimisations, studies, modelling, and simulations for the target station and the muon beamlines of the HIMB project.

The scope of this paper is the author's collaborative contributions to the HIMB project. For further information about the project, please see the IMPACT conceptual design report [4]. The IMPACT project comprises the HIMB and the TATTOOS projects.

2. Particle transport

For simulations of particle transport and the production of surface muons at the target station, we mainly use *G4beamline* [5], based on a custom build of *Geant4* [6] with PSI's own measured $^{+}$ cross-sections [7] and a splitting factor for $^{+}$ production and decay for variance reduction. The measured cross-sections were found to be more precise than the default *Geant4* cross-sections, which deviated from the experimental data by a factor of up to ten (see Ref. [7]). The new target station was simulated with an equivalent of 10^{11} protons, and the produced muons were then used as the initial beam in particle transport simulations of the beamlines delivering the beam to the MUH2 and MUH3 experimental areas and the beamlines' final foci.

The first HIMB *G4beamline* models were composed of *G4beamline*'s built-in particle optical elements and objects, which required lower computing resources in optimisation than using field maps. We added field maps of particle optical elements as they became available.

2.1. Target station TgH

We performed optimisations of target TgH design and operational parameters using *G4beamline* by simulating a proton beam impinging on the target and recording the muon counts at the upstream and downstream ends of the capture solenoid in the MUH2 or MUH3 beamline, and in some cases, further downstream.

In one of the optimisations, we studied the muon flux as a function of the angles between each of the muon beamlines and the proton beamline in the horizontal plane. We also considered the

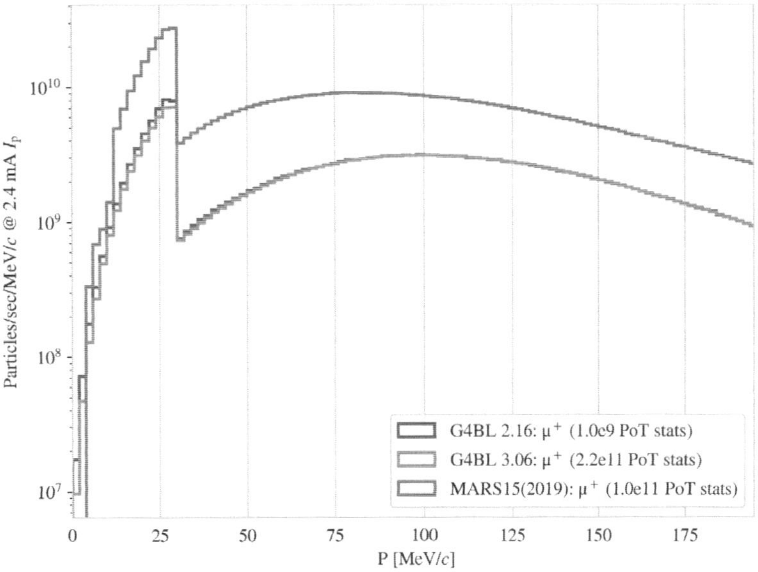

Figure 2. Muon production by target TgH, recorded 150 mm downstream from the target by a virtual detector of radius 250 mm. A comparison between *G4beamline* 3.06, *G4beamline* 2.16 (both with measured $^+$ production cross-sections but with *Geant4*'s built-in $^+$ decay cross-sections), and *MARS* [8, 9] is made. For a comparison of *Geant4*'s and *MARS*'s double-differential $^{\pm}$ production cross-sections with experimental data, where built-in cross-sections were used for *both* codes, see the HARP collaboration's paper Ref. [10].

counter-clockwise target rotation angles of 0, 5, and 10 degrees relative to the proton beamline. Without rotation, the target's surface would be perpendicular to the proton beam, because it is a surface muon production target. Our optimisation study confirmed the previous results that a slanted target provides a 30–60% [7] and, in one case, a ∼100% higher surface muon flux than a non-slanted target. Beamline orientations at other than 90 degrees relative to the proton beamline did not show significantly higher muon fluxes. The selected configuration uses muon beamlines placed perpendicularly to the proton beam, a 10 degree target rotation relative to the proton beamline, a 3.47 mm target thickness (providing a 20 mm proton beam path through the target), and a 100 mm target length.

We also performed optimisation studies for the aperture of a conical shield between TgH and the capture solenoid; for the use of co-directional versus contra-directional currents in the MUH2 and MUH3 capture solenoids, which are on the opposite sides of TgH; and for a non-zero incoming vertical angle of the proton beam to compensate for the magnetic field of the solenoids at the target.

We performed a particle production study for $^+$ and $^+$ at TgH using *G4beamline* and *MARS*, and we recorded the particle counts at virtual detectors placed at distances 150mm (see figure 2) and 250 mm from the target.

2.2. Transport beamlines MUH2 and MUH3

The layouts of the MUH2 and MUH3 beamlines are based on maximising the muon transmission to the experiments while having at least two dipole magnets in each beamline for radiation shielding purposes. Considering the somewhat flexible constraints of the floor plans of the experimental halls, we optimised the layout of the MUH2 and MUH3 beamlines using the asynchronous Bayesian optimiser of the tool DeepHyper [11] with computing resources provided by Argonne Leadership Computing Facility, with some additional optimisation studies using grid searches. A transmission of 1.34×10^{10} muons/s in the MUH2 beamline into the experimental area was achieved. In the process of optimisations, we noticed that, as expected, the transmission tended to increase as the bending angles of the dipoles decreased.

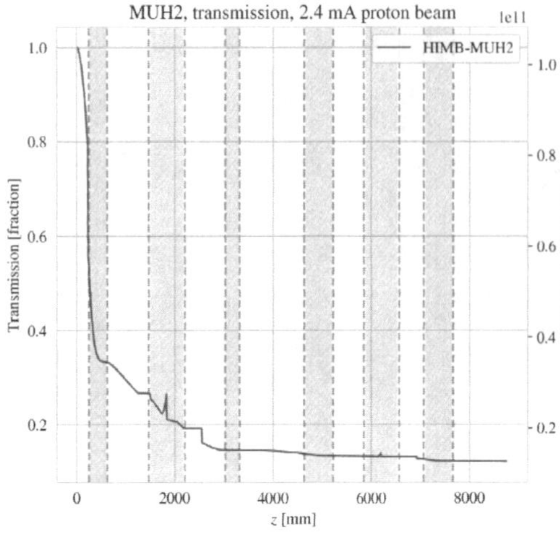

Figure 3. Muon transmission in the MUH2 beamline (preliminary model), plotted versus the longitudinal position. Dipoles and solenoids are denoted by green and red vertical columns, respectively. The optimisation of the currents was only for transmission in this case. The vertical spikes at the dipole centers are a *G4beamline* artifact due to a change in the centerline direction.

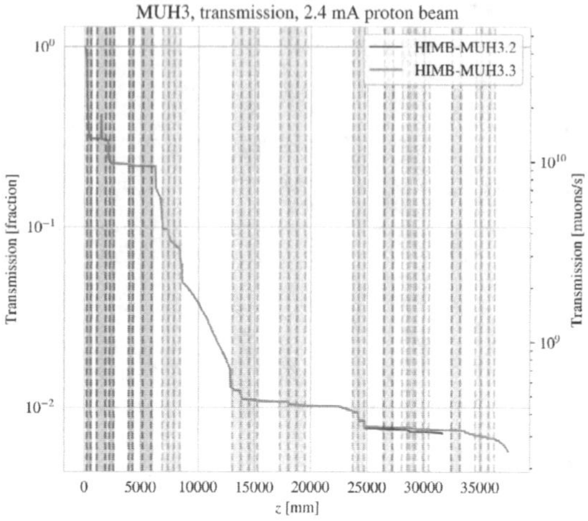

Figure 4. Muon transmission *within the momentum bite* of 25.38 to 29.79 MeV/c to the ends of the MUH3.2 (blue) and MUH3.3 branches (orange) of the MUH3 beamline (preliminary model), plotted versus the longitudinal position. Dipoles, solenoids, and quadrupoles are denoted by green, red, and yellow vertical columns, respectively. The optimisation of the currents was only for transmission in this case.

We performed numerous optimisations of design parameters (e.g., particle optical element sequences, drift lengths, offsets, and rotations) and operational parameters (e.g., currents or fields), and studies for the design of the MUH2 and MUH3 beamlines. Optimised transmissions for MUH2 and MUH3 beamline models are shown in figures 3 and 4, respectively. Two examples of design studies are given in the remainder of this subsection.

We studied the possibility of using a thin aluminium or titanium degrader foil in conjunction with a downstream dipole to suppress the positron contamination, placed at an intermediate focus to reduce multiple scattering. However, our *G4beamline* simulations have shown that when these inserts are thick enough to reduce the positron contamination substantially, the muon flux is also substantially impacted. Thus, it was decided to keep the separator in the design of the MUH2 beamline for positron removal.

We studied the option of an additional branch of the MUH3 beamline for the parasitic $^{-}$ beam, starting with a bidirectional bend at the first dipole "ASH31" of the MUH3 beamline, in the opposite directions for the $^{-}$ and $^{+}$ beams. This beamline would extend for two straight sections after the dipole ASH31, use only quadrupole focusing, and provide up to $\sim 10^7$ muons/s at a proton current of 2.4 mA.

2.3. MUH2 and MUH3 final foci

For final focus optimisation, we calculate the transfer map of the final focus using the transfer map based code *COSY INFINITY* [12], with the quadrupole strengths and other parameters as differential-algebraic system knobs. Then focusing is achieved by minimising the respective transfer map elements, which can be done analytically for low-order expansions in terms of the system knobs. We use *COSY INFINITY*'s built-in Levenberg–Marquardt "LMDIF" and simulated annealing optimisers.

Because the beamlines are mainly modelled using *G4beamline*, we developed a matching *COSY INFINITY* model of the *G4beamline* model of the MUH3.2 branch's final focus. The agreement between the *G4beamline* and *COSY INFINITY* models was with a difference of $\sigma(\Delta x) = 1.7\,\mathrm{mm}$ and $\sigma(\Delta y) = 3.4\,\mathrm{mm}$ (the latter being larger because of quite long distribution tails) for a beam passing through the MUH3.2 final focus, including a quadrupole doublet, a dipole magnet "ASS31", and a quadrupole triplet, without collimation at the entrance of the dipole. To put the differences seen between *COSY INFINITY* and *G4beamline* into context, the goal for the beam spot at the final focus is $\sigma_{x,y} \lesssim 20\,\mathrm{mm}$. The differences of 1.7 and 3.4 mm stated above are thus acceptable.

A new iteration of final focus optimisation studies is ongoing, with a field map of the dipole calculated using *Radia*.

3. Field computations

The MUH3 beamline will use a horizontal steering magnet SSL32 and a septum magnet ASS31 to switch the beam delivery between the MUH3.2 and the MUH3.3 branches, with a "shared" mode where the delivery is to both. Using the boundary integral method code *Radia* [13], we computed the field maps of SSL32 and ASS31. We computed the electric field of the spin rotator SpinRot1, which is in the third straight section of the MUH3 beamline, using the boundary element method (BEM) in *COMSOL* [14]. The BEM method tends to be more accurate than the finite element method, which is also available in *COMSOL*, for such calculations.

4. Conclusion

We performed particle transport and beam optics optimisations and studies for the design of the novel HIMB beamlines at PSI, which will deliver 10^{10} muons/s to next-generation intensity frontier experiments. For that, we developed *G4beamline* models of the MUH2 and MUH3 beamlines, as well as *COSY INFINITY* and *TRANSPORT* models of parts of the MUH3 beamline[1]. We calculated the magnetic field of the dipoles SSL32 and ASS31 using the boundary integral method code *Radia* and the electric field of the spin rotator SpinRot1 using the BEM solver in *COMSOL*.

Acknowledgments

We are thankful to Andreas Knecht, Peter-Raymond Kettle, and Hubertus Luetkens for productive and interesting discussions. This project has received funding from the European Union's Horizon 2020 research and innovation programme under the Marie Skłodowska-Curie grant agreement No. 884104. This research used resources of the Argonne Leadership Computing Facility, which is a DOE Office of Science User Facility supported under Contract DE-AC02-06CH11357.

References

[1] Arndt K *et al.* (Mu3e Collaboration) 2021 *Nucl. Instrum. Methods Phys. Res., Sect.* A **1014** 165679
[2] de Gouvêa A and Vogel P 2013 *Prog. Part. Nucl. Phys.* **71** 75–92

[1] The code *TURTLE* is also used in the HIMB project.

[3] Aiba M *et al.* 2021 Science Case for the new High-Intensity Muon Beams HIMB at PSI (*Preprint* hep-ex/2111.05788)
[4] Barth L *et al.* 2022 IMPACT conceptual design report Tech. rep. Paul Scherrer Institut Switzerland
[5] Muons, Inc 2019 G4beamline http://www.muonsinternal.com/muons3/G4beamline
[6] Agostinelli S *et al.* 2003 *Nucl. Instrum. Methods Phys. Res., Sect.* A **506** 250–303
[7] Berg F and other 2016 *Phys. Rev. Accel. Beams* **19** 024701
[8] Mokhov N, Aarnio P, Eidelman Y, Gudima K, Konobeev A, Pronskikh V, Rakhno I, Striganov S and Tropin I 2014 *Prog. Nucl. Sci. Tech.* **4** 496–501
[9] Mokhov N V and James C C 2017 The MARS code system user's guide version 15(2016) Tech. rep. Fermi National Accelerator Laboratory
[10] Apollonio M *et al.* (HARP Collaboration) 2009 *Phys. Rev.* C **80**(3) 035208
[11] Balaprakash P, Salim M, Uram T D, Vishwanath V and Wild S M 2018 *Proc. IEEE 25th Int. Conf. on High Performance Computing (HiPC 2018)* (Bengaluru, India) p 42–51
[12] Makino K and Berz M 2006 *Nucl. Instrum. Methods Phys. Res., Sect.* A **558** 346–50
[13] ESRF 2016 Radia https://www.esrf.fr/Accelerators/Groups/InsertionDevices/Software/Radia
[14] COMSOL AB 2021 COMSOL https://www.comsol.com

Modeling a Nb_3Sn cryounit in GPT at UITF

S Pokharel[1,2], A S Hofler[2] and G A Krafft[1,2]

[1] Center for Accelerator Science, Old Dominion University, Norfolk, Virginia, 23529, USA
[2] Thomas Jefferson National Accelerator Facility, Newport News, Virginia, 23606, USA

E-mail: spokh003@odu.edu

Abstract. Nb_3Sn is a prospective material for future superconducting radio frequency (SRF) accelerator cavities. Compared to conventional niobium, the material can achieve higher quality factors, higher temperature operation, and potentially higher accelerating gradients ($E_{acc} \approx 96$ MV/m). In this work, we performed modeling of the Upgraded Injector Test Facility (UITF) at Jefferson Lab utilizing newly constructed Nb_3Sn cavities. We studied the effects of the buncher cavity and varied the gun voltage from 200-500 keV. We have calibrated and optimized the SRF cavity gradients and phases for the Nb_3Sn five-cell cavities' energy gains with the framework of the General Particle Tracer (GPT). Our calculations show the beam goes cleanly through the unit. There is full energy gain out of the second SRF cavity but not from the first SRF cavity due to non-relativistic phase shifts.

1. Introduction

The desire to reduce construction and operating costs of future SRF accelerators motivates the search for higher-performing alternative materials. Nb_3Sn is a very promising alternative material for SRF accelerator cavities. Nb_3Sn possesses a high critical temperature ($T_c \approx 18.3$ K) and superheating field ($H_{sh} \approx 425$mT) [1, 2] giving it the potential for higher intrinsic quality factor (Q_0) than niobium (Nb), 4.2 K operations, and accelerating gradients of ≈ 96 MV/m. Nb_3Sn has a critical temperature about twice that of Nb, allowing it to achieve a high $Q_0 > 10^{10}$ at approximately two times higher operating temperatures than Nb. Changing the operating temperature from 2.0K, typical for Nb, to 4.3 K for Nb_3Sn while maintaining Q_0 in the 10^{10} to 10^{11} range would reduce energy consumption and thus cryogenic operating costs by as much as an order of magnitude, and would substantially decrease infrastructure costs for the cryogenic plant. Nb_3Sn cavities have a very high quality factor even above 4K. They can be cooled with cryocoolers for cavity cooling, which significantly cuts capital and installation costs, enabling compact and potentially even mobile applications. Many studies and tests have been done at Thomas Jefferson National Laboratory (JLab) [3, 4, 5] for Nb_3Sn and we performed the numerical simulation study of two 5-cell cavities coated with Nb_3Sn through the accelerator layout at UITF for the first time.

Based on the UITF beamline layout with a new SRF booster that eliminates the need for a warm capture section, this paper describes the transport of the electron beam through the elements of a 12 m beamline which consist of the photocathode electron gun, solenoids and magnets, buncher cavity, and accelerating superconducting radio frequency (SRF) cavities. With the General Particle Tracer (GPT) [6] simulations, we will describe the designed beam size, the initial electron beam distribution, calibration, and optimization of the Nb_3Sn cryounit

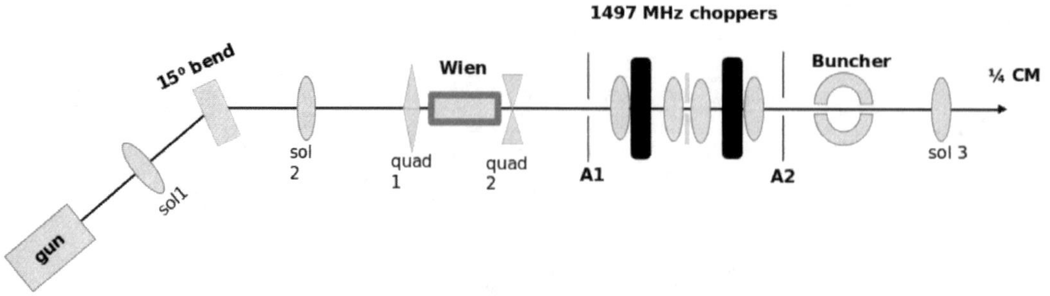

Figure 1. Cathode to the quarter cryomodule (QCM) layout of the Upgraded Injector Test Facility (UITF) at JLab.

using different gun voltages 200 keV, 350 keV, and 500 keV. Also, we will describe the beam properties that would be expected without a capture section.

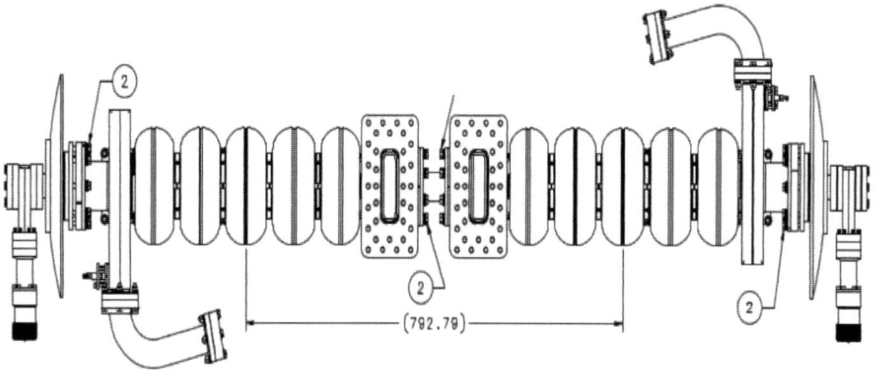

Figure 2. Nb_3Sn pair cryomodule with each of five cells separated by 792.79 mm between their centers [7].

2. Beamline setup and simulation details

Figure 1 is the schematic representation of the keV beamline at the Upgraded Injector Test Facility (UITF) at JLab. The beam dynamics simulation was performed by General Particle Tracer (GPT) [6]. For the simulation, we have used a straight beamline; 15° dipole, RF choppers, beam diagnostics, Wien apertures, etc. are omitted. The laser pulses are getting absorbed by the photocathode inside the electron gun cavity causing the gun to emit the electrons. The electrons are then accelerated to 200 keV initially by the electric field in the DC gun. The electron beam is transported through the solenoids, quadrupoles, and buncher cavity of 750 MHz before reaching the SRF cavities of 1497 MHz. There are two cavities inside the SRF cryomodule, both of them are 5-cell Nb_3Sn cavities separated by 792.79 mm between their centers as shown in figure 2. Again, we performed simulations with a different DC gun electric fields to accelerate the beam to 300 keV and 500 keV.

3. Initial distribution

For the particle distribution at the cathode in simulation, the beam is assumed to have a Gaussian distribution in t, x, y, p_x, and p_y following the profile of the laser. The transverse beam size is 213 μm, the laser pulse length is 21.3 ps, and the transverse emittance is 0.061 mm mrad. We have used 1D profiles for the DC gun and SRF cavities, a 2.5D field map for RF buncher, and 2D and 3D field maps for the solenoids. The beam current used is 1 nA. Thus, the bunch charge is calculated by 1 nA with 750 MHz CW (continuous wave) mode frequency, which is 1.333×10^{-18}C. The applied macro particle number is 2500.

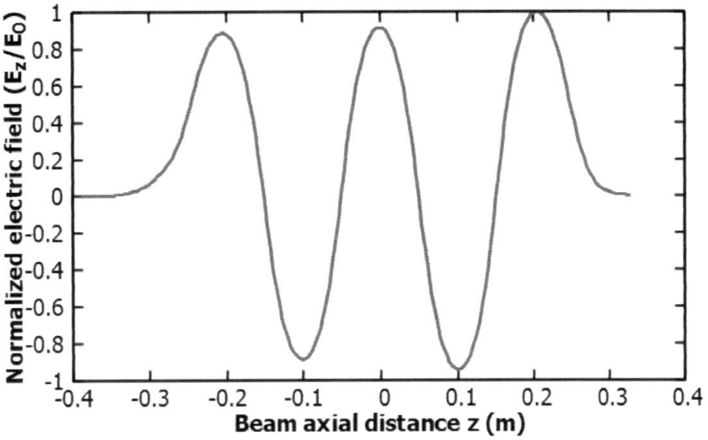

Figure 3. 1D field map used for Nb_3Sn cryounit in the computer simulation.

4. Nb_3Sn unit acceleration

We incorporated the buncher cavity into the study and looked at various SRF cavity gradients/phases. The 1D field map for the cryounit in the simulation is shown in figure 3.

Using the field map for the cryounit, we first calibrated the accelerating gradient using the second cavity where the beam is relativistic. Once calibrated, we have optimized the crest phases for each cavity at different accelerating gradients with only the first cavity on. The reason is that the different acceleration rates lead to different phase shifts. Then, the actual energy gains were achieved after cresting had been completed. From the computer simulations, we found that

$$\frac{\mathrm{E_0}}{\mathrm{E_{acc}}} = 2.1086 \tag{1}$$

for the given field map. Where $\mathrm{E_0}$ is the peak value of the electric field, and $\mathrm{E_{acc}}$ is the accelerating gradient.

5. Results and discussion

After cresting each of the cavities, the actual energy gains achieved were calculated. The energy gains vs. accelerating gradients are plotted for each of the cavities for different gun voltages 200 kV, 350 kV, and 500 kV. They are shown in figures 4 and 5. The energy gain of an arbitrary particle with charge e traveling through a SRF gap [8, 9] is:

$$\Delta W = e\int_{-\infty}^{\infty} \mathrm{E(0,z)} \cos\left(\frac{\omega \mathrm{z}}{\beta \mathrm{c}} + \phi_{\mathrm{s}}\right) \mathrm{dz} = e\mathrm{V_0 T}\cos\phi_{\mathrm{s}} \tag{2}$$

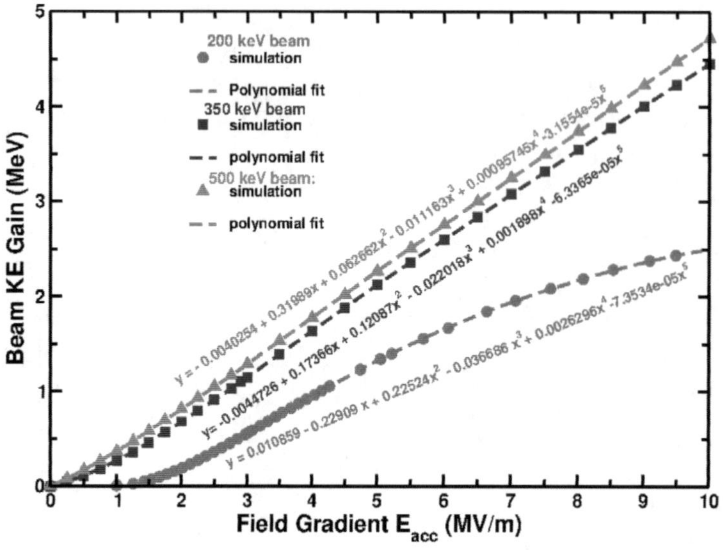

Figure 4. Beam kinetic energy (KE) gain for the first 5-cell cvaity with second five-cell cavity off for different gun voltages 200 kV, 350 kV, and 500 kV.

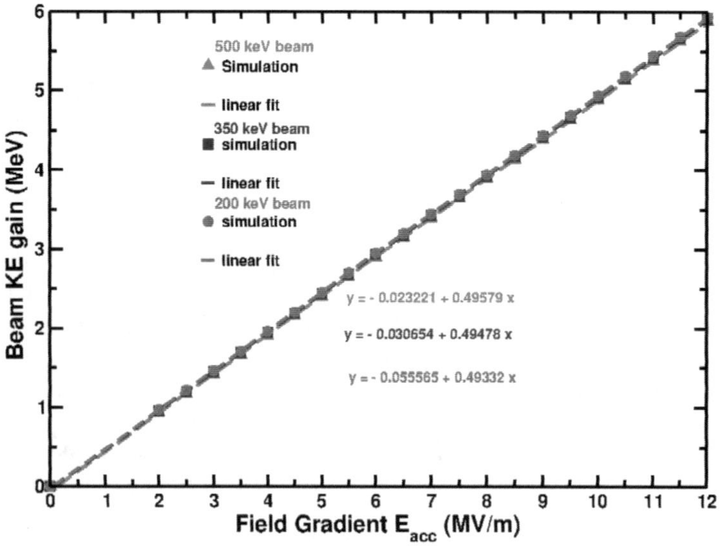

Figure 5. Beam kinetic energy (KE) gain for the second 5-cell cavity (first cavity at 3.6 MV/m) for different gun voltages 200 kV, 350 kV, and 500 kV.

where $V_0 = \int_{-\infty}^{\infty} E(0, z)\ dz = E_0 L$ is an axial RF voltage, z is the beam axial distance, L is the effective length of the RF cavity and is given by $L = \beta\lambda_{RF}/2$, and $E_0 T = E_{acc}$, ϕ_s is the synchronous phase of the particle. The transit time factor T is a measure of the reduction of energy gain caused by the sinosoidal time variation of the filed in the SRF gap and is given by

$$T = \frac{\int_{-\infty}^{\infty} E(0, z) \cos\left(\frac{\omega z}{\beta c}\right)\ dz}{\int_{-\infty}^{\infty} E(0, z)\ dz} - \tan\phi_s \frac{\int_{-\infty}^{\infty} E(0, z) \sin\left(\frac{\omega z}{\beta c}\right)\ dz}{\int_{-\infty}^{\infty} E(0, z)\ dz} \tag{3}$$

where ω is the RF angular frequency, λ_{RF} is the RF wavelength, β the normalized velocity, and c is the speed of light in vaccum.

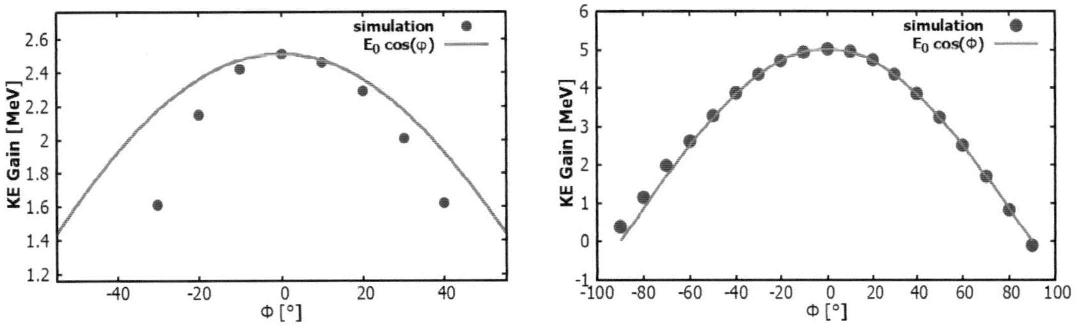

Figure 6. The energy gains vs. phase of-crest (Φ) for the first (left) and second cavity (right) for the accelerating gradient of 10.10 MV/m for 200 keV beam.

For the first 5-cell cavity as shown in figure 4, the energy gain obtained from the simulation is not linear with the accelerating gradient inside the cavity. The non-linear relation for the beam (200 keV, 350 keV, and 500 keV) is best fitted by the fifth order of the polynomials. However, the best non-linearity is obtained for the 200 keV beam than the 350 keV, and 500 keV cases.

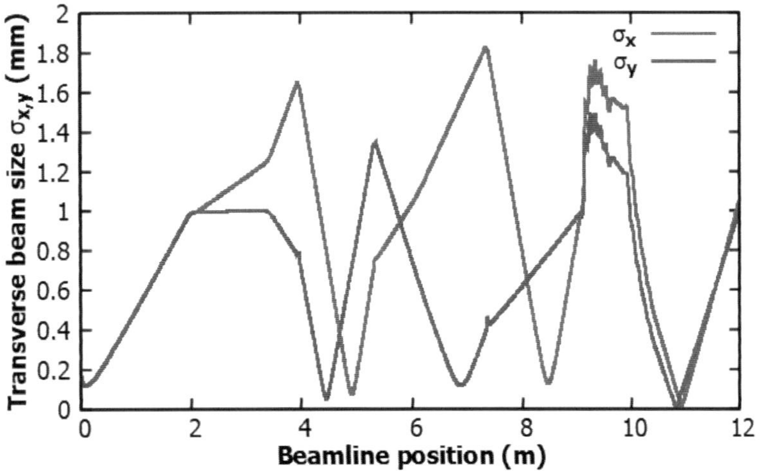

Figure 7. Transverse beam sizes (σ_x and σ_y) for the 200 keV beam.

There are several issues to explain the "non-linearity". First, the beam is not relativistic when it enters the cavity as the electrons traversing the cavity are not velocity-of-light and there is phase shifts and incomplete accelerations. The second isssue is the transit time factor. And lastly the effective length of the 5-cell cavity changes for the non-relativistic beam. The cavity length specified L = $\beta\lambda_{RF}/2$ works poorly for low β cavities. So the energy gain is not linear with the field inside the cavity (E_{acc}). For the different settings, the deviation of the energy gain from a linear model in a two-gap RF cavity for a non-relativistic beam was calculated by A. Shemyakin [10]. We observed exactly the same non-linearity for the first 5-cell cavity.

For the second 5-cell cavity, the beam is relativistic and the energy gain obtained from the simulation is linear with the accelerating gradients inside the cavity. The linearity for the beam (200 keV, 350 keV, and 500 keV) are best approximated by the linear fit as shown in figure 5.

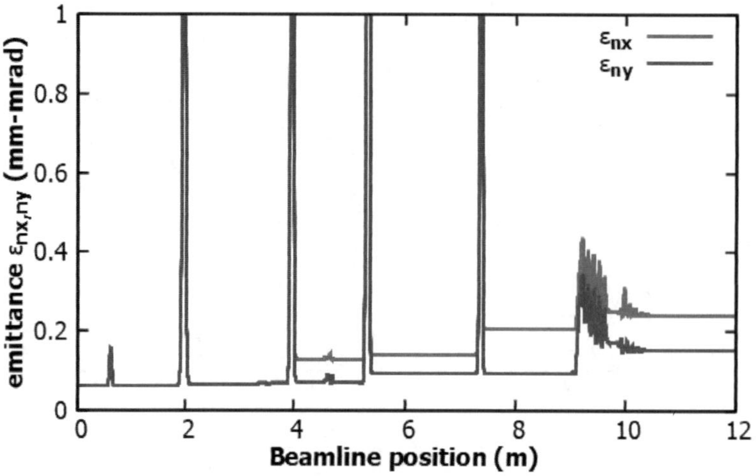

Figure 8. Normalized transverse emittances (ϵ_{nx} and ϵ_{ny}) for the 200 keV beam.

Again, setting the electric field gradient for both of the 5-cell cavities $E_{acc} = 10.10$ MV/m, we plotted energy gain vs. the phase off-crest for them for a 200 keV beam which is shown in figure 6. From the figure, it is clearly seen that there is an incomplete acceleration from the first 5-cell cavity. But the full energy gain is achieved from the second 5-cell cavity as the beam entering the cavity is relativistic. Also, the gain is the cosine function of the phase relative to the crest.

Also, we observed the beam characteristics downstream after the cryounit for the 200 keV beam setting the field gradient (E_{acc}) for the first cavity and second cavity to 3 MV/m and 5 MV/m, respectively. The beam has a bunch length of 1.10 ps, average kinetic energy of 6.86 MeV, and energy spread σ_E of 14.95 keV. The beam envelope (transverse beam sizes and normalized transverse emittances) along the beamline are shown figures 7 and 8 respectively.

6. Conclusion

We have calibrated and optimized the Nb_3Sn cryounit at UITF at Jefferson Lab using GPT modeling. For all the studied beam cases (200 keV, 350 keV, and 500keV), the beam goes cleanly through the unit. The beam is not fully relativistic for the first 5-cell cavity; there is no full energy gain out from the cavity. The energy gain is not linear with the field inside the cavity whereas the beam is relativistic for the second 5-cell cavity, there is almost full energy gain out from the cavity. The energy gain is linear with field inside the cavity. Also, we found that using the velocity-of-light structure for the non-relativistic beam is suitable for the desired beam requirement after the quarter cryo-module at Upgraded Injector Test Facility (UITF). Jefferson Lab intends to install such Nb_3Sn cavities in UITF and accomplish the world's first beam acceleration by Nb_3Sn.

Acknowledgments

This work was supported by the United States Department of Energy, Office of Science, Office of Nuclear Physics under contract DE-AC05-06OR23177.

References

[1] Godeke A 2006 *Supercond. Sci. Technol.* **19** R68
[2] Cetelani G and Sethna J P 2008 *Phys. Rev.* B **78** 224509

[3] Pudasaini U, Eremeev G, Ciovati G, Kelley M J, Parajuli I, Reece C E and Sayeed Md N 2018 *Proc. 9th Int. Particle Accelerator Conf. (IPAC'18)* (Vancouver, Canada) pp 1798–1803
[4] Ciovati G, Cheng G, Pudasaini U and Rimmer R A 2020 *Supercond. Sci. Technol.* **33** 07LT01
[5] Eremeev G, Clemens W, Macha K, Reece, C E, Valente-Feliciano A M, Williams S, Pudasaini U and Kelley M, 2020 *Rev. Sci. Instrum.* **91** 073911
[6] General Particle Tracer (GPT) code-Pulsar Physics `http://www.pulsar.nl/gpt/`
[7] Macha K (Mar 2021) JL0041172–C75 same as Nb3Sn pair.pdf *private communication*
[8] Wangler T P 2008 *RF Linear Accelerators* (New York: Wiley-VCH) 2nd ed.
[9] Lee S Y 2004 *Accelerator Physics* (Singapore: World Scientific)
[10] Shemyakin A 2021 Energy Gain in a Two-gap RF Cavity *Preprint arXiv:2107.13041*

Review of alignment and stability tolerances for advanced light sources

A Khan, S Sharma and V Smaluk
BNL, NSLS-II, Upton, NY, 11973-5000, USA
E-mail: akhan1@bnl.gov

Abstract. Alignment and mechanical-stability specifications are essential to the performance of low-emittance storage rings. Beam dynamics simulations are usually performed to establish these specifications. However, the simulation procedures and the input parameters related to magnet positions are not well established which leads to differences in the final specifications. In this paper we discuss important parameters of the mechanical/structural systems of the storage ring that impact on the alignment and stability specification. We reviewed the alignment and stability specifications used at modern light sources across the world that will help to propose an efficient model for a low-emittance upgrade of NSLS-II.

1. Introduction

Achieving a good level of performance in the low-emittance light sources places demanding requirements on the field quality and alignment precision of the magnets. Identifying and rectifying the major sources of machine errors is a major task during the commissioning of a new accelerator, and efforts to reduce errors and improve machine performance are frequently continued throughout the facility's lifetime. Modeling the sensitivity to various errors is also an important part of the design process, which includes studies of diagnostics to identify the sources of errors and the correction systems to compensate for them. Thus, to ensure any facility will perform as expected by reducing uncertainty, advancing design in a cost-effective way requires finding the proper alignment and stability tolerances. This is achieved by considering the work of others, and developing the suitable models for accommodating results.

There are two types of errors, static and dynamic, concerning lattice magnets. Static errors, such as misalignment and field errors, are time independent or change slowly over time. These affect dynamic aperture, dispersion, beta beating, lifetime and chromaticity. We can reduce the impact of such errors by deploying a diagnostic and corrective system. Dynamic errors, such as jitter in the power supply, floor motion, girder vibrations, and temperature variations, are ones that change with time on a scale of milliseconds to hours, as discussed in the section on noise sources. These errors affect a variety of beam properties, including orbit stability, and if we can monitor them in real time, the correction can be implemented in real time as well. Detailed beam dynamics simulations give the specification of upper limits on these errors in a machine.

It is also necessary to think through the specification of tolerances in order to avoid unnecessary costs. In this report, we discuss the alignment and stability tolerances of different light sources.

Journal of Physics: Conference Series **2420** (2023) 012055 doi:10.1088/1742-6596/2420/1/012055

2. Overview of alignment tolerances

The misalignment can be caused by the occurrence of some mounting errors and other errors in the manufacturing and assembly process of the magnets. Considering the parameters change (for convenience from the computational viewpoint) and the physical meaning of the parameter deviations, the random nature of errors is assumed [1].

There are three most important factors from beam dynamics point of view to decide the alignment tolerances for a light source: *beta beating, beam orbit and dynamic aperture.* The general steps for such simulations are as outlined in Refs. [2, 3]:

- Apply misalignment errors and beam-offset to the lattice model
- Perform trajectory correction until the beam reaches one turn
- Perform global trajectory correction until sufficient multi-turn transmission is achieved
- Perform global trajectory correction including RF cavities until closed orbit is found

Table 1. Specifications for magnets and girders alignment in modern light sources around the world. Each element is misaligned randomly in a Gaussian distribution with given RMS amplitude σ. Where NA^{+} corresponds to Not Applicable and NA^{*} corresponds to Not Available.

Machine	**E** (GeV)	**C** (m)	**Magnet-to-magnet** (µm)	**Cut-off** ($\pm\sigma$)	**Girder-to-girder** (µm)	**Cut-off** ($\pm\sigma$)
ALS-U	2	196.5	10	2	50	2
APS-U	6	1103.608	30	2	100	1
Diamond-II	3.5	560.574	25	2	150	2
ESRF-EBS	6	843.97	60	2.5	NA^{+}	NA^{+}
NSLS-II	3	792	30	1	100	1
SIRIUS	3	518.4	40	1	80	1
MAX-IV	3	528	20	1	NA^{*}	NA^{*}
SOLEIL-II	2.75	353.74	30	2	50	2

Table 1 summarizes the magnet-to-magnet and girder-to-girder alignment specifications in advanced light sources around the world, including ALS-U [4], APS-U [5], Diamond-II [6], ESRF-EBS [7], NSLS-II [8], SIRIUS [9], MAX-IV [10] and SOLEIL-II [11].

Alignment specifications at APS-U are specified for two different stages: (1) the alignment requirements that have to be achieved as a zeroth-order machine alignment prior to the beam commissioning, and (2) the requirements that have to be ensured during standard machine operation [12].

The ESRF-EBS facility described that girder-to-girder tolerances are not of particular significance to the facility with loose magnet-to-magnet tolerance. From a beam dynamics perspective, as long as magnet-to-magnet tolerances (single magnet rms position errors) are met at every location in the machine, including between magnets standing on adjacent girders, the desired dynamic aperture and lifetime are achievable. It worked effectively for ESRF-EBS, well beyond expectations/simulations [13].

At SOLEIL-II, magnet tolerances are tight 30 µm and much tighter for neighboring girders (50 µm in H and 30 µm in V). The major issue is that machine lattice is so compact that considerable offset in sextupole magnets would have a significant impact on the lattice performance [14].

Journal of Physics: Conference Series **2420** (2023) 012055 doi:10.1088/1742-6596/2420/1/012055

3. Overview of stability tolerances

One of the most critical parameters for synchrotron light source users is beam stability. It encompasses beam position, angle, beamsize, emittance, beam energy, and energy spread stability. The general stability requirements for orbit initially of 10% of the beam sizes is changing with the advancement of beamline experiments [15]. Table 2 specifies the typical beam stability requirements for moderately demanding synchrotron radiation experiments, which helps to define the stability tolerances for a particular facility [16].

Table 2. Typical beam stability requirements for moderately demanding synchrotron radiation experiments [16].

Measurement parameter	Stability Requirement
Intensity variation $\Delta I/I$	$\ll 1\%$ of normalized I
Beam position and angle	$<$ 2-5% of beam σ and σ'
Energy resolution	$< 10^{-4}$
Timing stability	$<$ 10% bunch length
Data acquisition rate	$10^{-3} - 10^{5}$Hz

A variety of sources can compromise the system's stability once the accelerator support system has been installed and the components aligned over a large range of time scales, ranging from milliseconds to years, with disturbance amplitude decreasing with increasing frequency. These sources are characterised in different time-scale as following [15]:

3.1. Noise sources

- Short term (t < 1 hour): Ground vibration induced by human activities, mechanical devices like compressors and cranes or external sources like road traffic potentially attenuated by concrete slabs, amplified by girder resonances and spatial frequency dependent orbit responses, cooling water circuits, power supply noise, electrical stray fields, booster operation, etc.
- Medium term (t < week): Movement of the vacuum chamber (or even magnets) due to changes of the synchrotron radiation induced heat load especially in decaying beam operation, water cooling, tunnel and hall temperature variations, day/night variations, gravitational sun/moon earth tide cycle.
- Long term (t > 1 week): Ground settlement and seasonal effects (temperature, rainfall) resulting in alignment changes of accelerator components, including girders and magnets. Disturbances on this time scale are not a problem for users, since experiments are regularly realigned.

The frequency content of motion due to vibrations is given by the Power Spectral Density (PSD) which depends on the geology of the site and cultural noise amplified by the girders and supports. The square root of integrated PSD gives the RMS beam motion induced by vibration. We only discuss the magnet-girder assembly specifications and thermal specifications in this report.

3.2. Magnet-girder assembly specifications

For a better mechanical stability, it is required that all structures (stands and mechanics from the floor to optical component) must have high natural frequency for eigenmodes that influence the resolution, spot sizes, etc. A standard vibration diagnostic of mechanical structures starts with an operational response measurement at various sites. The amplification factor, peak frequencies, natural frequencies, and the outline of the structure's mode form are all analyzed initially, followed by the vibration amplitude at a few important spots. If the level of vibration is lower than the tolerance, the structure is stable. The lowest natural frequency of magnet-girder assembly of different light sources is given in table 3.

Table 3. Lowest natural frequency of the magnet-girder assembly of different light sources [9, 11, 17–19].

Machine	Frequency (Hz)
APS-U	50
ESRF-EBS	55
MAX-IV	50
NSLS-II	30
Diamond-II	50
HEPS	40
SPring-8-II	110
SIRIUS	120
SOLEIL-II	97

There are two main groups with different beam height facilities throughout the world: one with 1.2 m beam height, such as NSLS-II, ESRF, SOLEIL, SPring-8, and the other with 1.4 m beam height, such as APS-U, ALS-U, DIAMOND, and so on. From the standpoint of vibration and stability, as well as tunnel cost, a short beam height is preferable. However, the availability of insertion devices, their manufacturability, and the worker's working environment are frequently factors in determining this height.

Table 4. Uncorrelated Vibration Tolerances [6, 8, 10, 20]. Where H and V stands for horizontal and vertical.

Machine	f (Hz)	Magnet (nm)	Girder (nm)
APS-U	1-100	10 (H/V)	20 (H/V)
MAX-IV	> 5	20-30 (V)	
NSLS-II	> 4	25 (H), 150 (V)	70 (H), 600 (V)
Diamond-II	1-100	14.7 (H), 21.7 (V)	

Table 4 shows the differences in the vibration tolerances across different facilities. One explanation for this difference could be that, while vibration tolerances are important for providing engineers with simple design constraints, they are based on broad generalizations about the character of mechanical system motion, which could be inaccurate. For example, physicists model magnet-to-magnet vibration within a girder as uncorrelated motion, but in reality, this motion is mostly coupled, and so the amplification factor will differ. It is difficult to compare the beam dynamics of different magnet grouping arrangements because the specifications are based on a specific arrangement.

3.3. Thermal specifications

Thermal excursions are typically caused by day–night temperature variations or by heat loads from thermal system. The thermal system of an accelerator facility consists of heat sources (cables, RF, absorbers, electromagnets etc), the air ventilation and the water cooling system, high thermal inertia components (girders, slabs, concrete walls), and external disturbances including experimental hall and groundwater temperature. Long term stability affected by thermal drift is a crucial task, especially in the first phase of a facility. It might take a long time before the site gets into equilibrium. There is even a risk of periodic drift with seasons.

The thermal specifications for different facilities are given as following:

- **APS-U:** APS-U standards to keep tunnel air temperature stable within ±0.1 °C. This specification has been demonstrated at multiple locations on a one-week time scale and is meant to be maintained throughout the storage-ring tunnel and over weeks/months, rather than only in a few areas over a week.
- **ALS-U:** During steady-state operation, the temperature specification is ±0.05 °C. A study was conducted to examine historical building performance based on data collected from thermal sensors over time. The average temperatures in the storage-ring tunnel ranged from 22.4 – 25.2 °C across the various zones. The majority of the variances are due to differing operational modes.
- **SOLEIL-II:** The extension of the air-conditioning systems to the injector complex, the new booster ring, and special care for the electronics of the storage ring component, whose electronics boards are housed in thermo-regulated cabinets, all necessitate temperature regulation of better than ±0.05 °C.
- **ESRF-EBS:** The ESRF machine had two major thermal issues: first, air-conditioning requires a temperature ramp up to 2 °C along the sector, and second, the storage ring required a four-day warm-up period to attain stable orbit. The ESRF-EBS specifications are outlined to ±0.1 °C by adding the tunnel-sector supplementary cooling systems [21].
- **NSLS-II:** To ensure acceptable thermal stability of the storage ring magnets, process water and tunnel air temperature fluctuations are maintained within ±0.1 °C with 1 hour time scale. The viscoelastic damping pads are used to reduce the thermal bending of the girder by more than one order of magnitude [22].

4. Concluding remarks

For a low-emittance synchrotron, design optimization is a trade-off between performance, environmental conditions, and manufacturing capabilities. Alignment and mechanical-stability specifications of the storage rings of several light sources were reviewed in detail. The differences in the specifications are related to optimization parameters in beam dynamics simulations as well as to input parameters representing mechanical and structural systems of the storage ring. Simulation procedures for the alignment specifications are relatively mature and yield similar results. Stability specifications (both vibrational and thermal) have significant differences, in part because they are not yet firmly grounded in beam dynamics simulations. This study will help us to set up an efficient design model for simulations of the low-emittance NSLS-II upgrade using the Accelerator Toolbox [23].

Journal of Physics: Conference Series **2420** (2023) 012055 doi:10.1088/1742-6596/2420/1/012055

Acknowledgement
The authors wish to thank Hellert Thorsten (ALS-U), Simone Liuzzo (ESRF-EBS) and Vadim Sajaev (APS-U) for presenting their work at NSLSII and fruitful discussions.

References
[1] Abdushukurov D 2015 *Proc. Int. Conf. on Stability and Control Processes, in Memory of V I Zubov (SCP 2015)* (St. Petersburg, Russia) pp 136–7
[2] Hellert T, Amstutz P, Steier C and Venturini M 2019 *Phys. Rev. Accel. Beams* **22** 100702
[3] Sajaev V 2019 *Phys. Rev. Accel. Beams* **22** 040102
[4] Allézy A *et al.* 2018 ALS-U conceptual design report Tech. rep. LBNL, Berkeley, CA, USA
[5] Fornek T E 2019 Advanced Photon Source Upgrade Project Final Design Report Tech. Rep. APSU-2.01-RPT-003 Argonne National Lab. (ANL), Argonne, IL, USA
[6] Abraham C *et al.* 2019 Diamond-II Conceptual Design Report Tech. rep. Diamond Light Source, OX, UK
[7] Farvacque L, Bec G L, Biasci J C, Bouteille J, Jacob J and et al 2018 EBS Storage Ring Technical Report Tech. rep. European Synchrotron Radiation Facility (ESRF)
[8] Dierker S 2007 National Synchrotron Light Source II Preliminary Design Report Tech. Rep. BNL-94744-2007 Brookhaven National Lab., Upton, NY, USA
[9] Rodrigues A, Arroyo F, Bagnato O, Citadini J, Farias R, Franco J, Leao R J, Liu L, Marques S, Neuenschwander R *et al.* 2018 *Proc. 9th Int. Particle Accelerator Conf. (IPAC'18)* (Vancouver, Canada) pp 2886–9
[10] Magnuson M, Fahlman M, Uhrberg R, Johansson L *et al.* 2006 MAX IV Conceptual Design Report (CDR) Tech. rep. MAX IV Laboratory, Sweden
[11] Nadji A *Proc. 12th Int. Particle Accelerator Conf. (IPAC'21)* (Campinas, Brazil) pp 463–5
[12] Redaelli S 2009 Dynamic alignment in particle accelerators Tech. rep. CERN
[13] Liuzzo S 2021 Private Communication
[14] Nadolski L S and Sharma S 2021 Private Communication
[15] Hettel R O 2002 *Rev. Sci. Instrum.* **73** 1396–1401
[16] 2018 BES Light Sources Beam Stability Workshop https://www.aps.anl.gov/BES-Light-Sources-Beam-Stability-Workshop/
[17] Liu Z, Nudell J, Preissner C A, Collins J T and Cease H *Proc. 9th Mechanical Engineering Design of Synchrotron Radiation Equipment and Instrumentation Conf. (MEDSI'16)* (Barcelona, Spain) pp 254–6
[18] Jiao Y, Xu G, Cui X H, Duan Z, Guo Y Y, He P, Ji D H, Li J Y, Li X Y, Meng C *et al.* 2018 *J. Synchrotron Radiat.* **25** 1611–18
[19] Tanaka H, Ishikawa T, Goto S, Takano S, Watanabe T and Yabashi M *Proc. 7th Int. Particle Accelerator Conf. (IPAC'16)* (Busan, Korea) pp 2867–70
[20] Nudell J, Liu Z, Sajaev V and Preissner C *Proc. 10th Mechanical Engineering Design of Synchrotron Radiation Equipment and Instrumentation (MEDSI'18)* (Paris, France) pp 95–98
[21] Tampigny B, Dabin Y, Thomas F, Bouteille J, Favacque L, Marchial T, Favier F, Roux P, Buisson J, Raimondi P *et al. Proc. 9th Mechanical Engineering Design of Synchrotron Radiation Equipment and Instrumentation Conf. (MEDSI'16)* (Barcelona, Spain) pp 262–6
[22] Sharma S, Doom L, Jain A, Joshi P, Lincoln F, Ravindranath V *et al. Proc. 24th Particle Accelerator Conf. (PAC'11)* (New York, USA) pp 2082–6
[23] Terebilo A *Proc. 19th Particle Accelerator Conf. (PAC'01)* (Chicago, USA) pp 3203–5

Measurements of radiation fields from a ceramic break

Yoshihiro Shobuda, Takeshi Toyama, Masahiro Yoshimoto and Shuichiro Hatakeyama

J-PARC Center, JAEA/KEK, Ibaraki, JAPAN

E-mail: yoshihiro.shobuda@j-parc.jp

Abstract. Ceramic breaks are used in synchrotrons for many purposes. For example, they are inserted between the Multi-Wire Profile Monitor (MWPM) on the injection line at the Rapid Cycling Synchrotron (RCS) in J-PARC to completely prevent the wall currents accompanying beams from affecting the MWPM. On the other hand, from the viewpoint of suppressing beam impedances and the radiation fields from the ceramic breaks, it would be preferable that the inner surface of the ceramic break is coated with Titanium Nitride (TiN), or covered over capacitors. In this report, we measure the radiation fields from the ceramic break with and without capacitors as well as the beam profile and investigate the effect of the ceramic breaks on the measurements.

1. Introduction

The RCS in J-PARC [1] has been realizing the high-intensity beams [2–4] by accumulating H^- injection beams from the LINAC. During the injection painting process [5], the H^- beams are transformed into the proton beams after being hit on the foil at the injection point. When two bunched beams, each containing 4.15×10^{13} particles per bunch, are accelerated from 400 MeV to 3 GeV at a repetition rate of 25 Hz, a 1 MW beam can be performed at the RCS.

Precise adjustment of the injection beams on the phase space area during the painting process is important to realize the high intensity beams with low beam loss rate. Hence, seven MWPMs [6] are installed, where MWPM1 is on the injection line to the RCS, MWPM2-5 are on the merging area in the RCS, and MWPM6-7 are on the dump line to measure the unstripped H^- beams. Based on the design concept of RCS, MWPM1 is sandwiched by two ceramic breaks to completely prevent the wall currents with H^- beams from interfering with the measurements with MWPM1 (see figure 1).

On the other hand, suppressing beam coupling impedances [7, 8] is a critical issue to accomplish high intensity beams. Since the impedance of ceramic breaks is closely related to the radiations from the ceramic break [9–13], covering capacitors on the ceramic breaks is an effective way to suppress both the impedance and radiations from the ceramic breaks. Therefore, it is important to directly measure the radiations from several types of ceramic break, and investigate the effects of ceramic breaks on the monitors.

The MWPM1 and the ceramic breaks next to it are good tools because the residual dose around there is kept below at most ten micro-sievert per hour even after 4 hours of 830 kW beam operation stop of the RCS. In this report, we measure the radiations from 15 mm long ceramic

Journal of Physics: Conference Series **2420** (2023) 012056 doi:10.1088/1742-6596/2420/1/012056

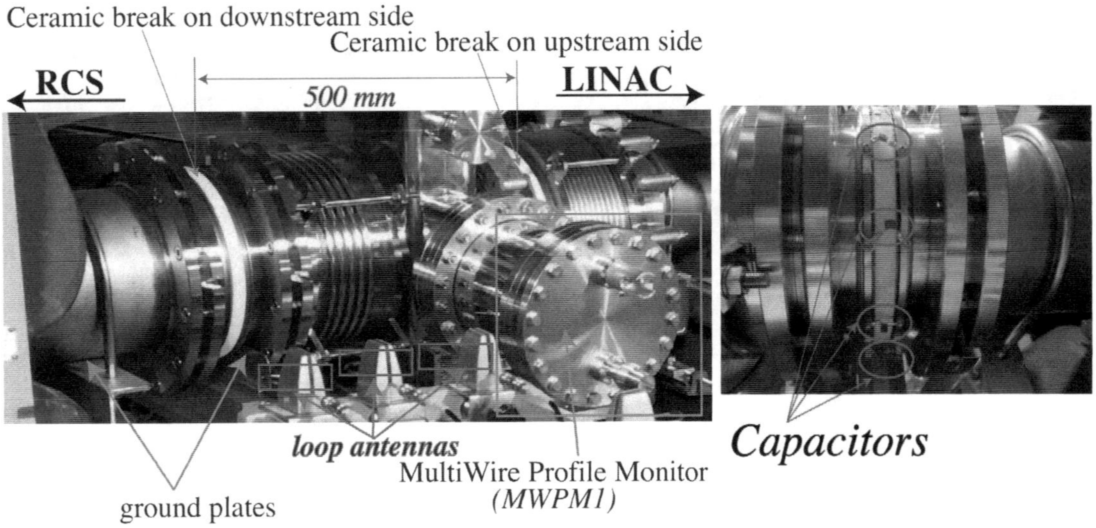

Figure 1. MWPM1 and the ceramic breaks without (left) and with capacitors (right).

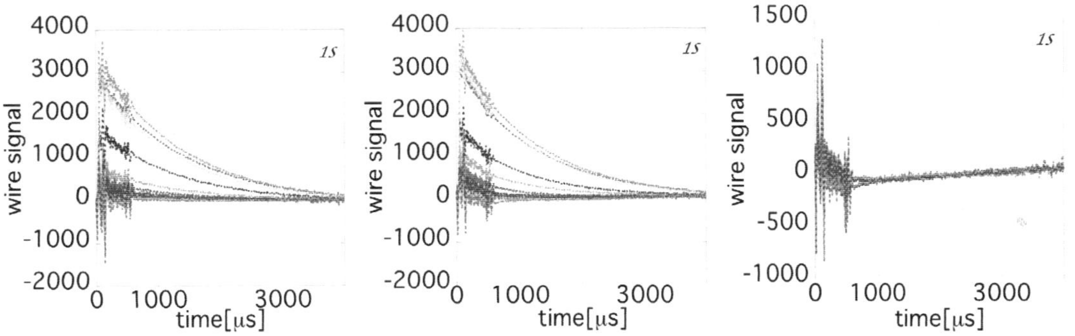

Figure 2. Signals if the ceramic break is covered by capacitors (left), or by Al with (center) and without beams (right).

break with 101 mm inner and 111 mm outer radii beside MWPM1 by which the transverse beam profile is measured under several conditions, after test-measuring the magnetic field in the electromagnetic anechoic chamber in JAXA [14].

2. Measurements of the transverse beam profile

MWPM1 consisting of 27 wires measures the transverse beam profile in u and v axes, which are tilted 45° from the horizontal and vertical axes. Each wire moves 0.1 mm/s filling the intervals of respective wires, while 100 LINAC beams pass there with a 1 Hz repetition rate [6]. The integration of respective wire signals over 4 ms after being excited by a hit of the beam provides the beam density at the wire positions, so that a smooth bunch distribution is obtained after 100 LINAC beam shots.

Here, let us cover the ceramic break on the downstream side by twelve 1 μF capacitors [15] with 2 MHz resonant frequency or by Aluminum (Al) sheet to investigate the effects on the beam

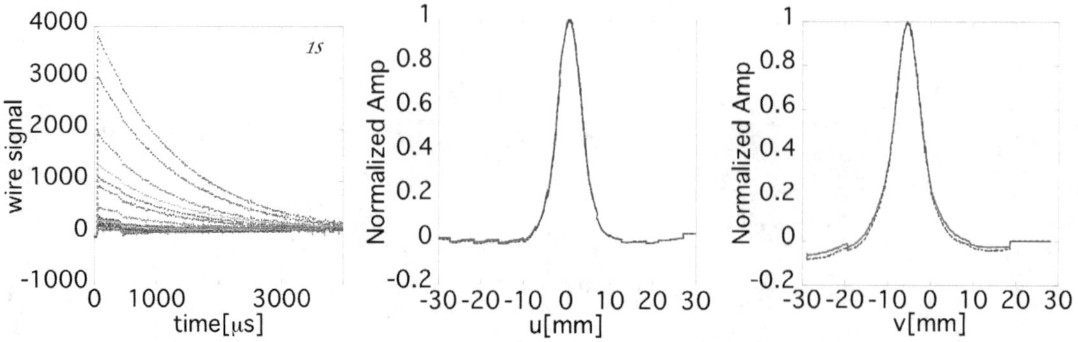

Figure 3. Wire signals (left) and beam profiles (center/right).

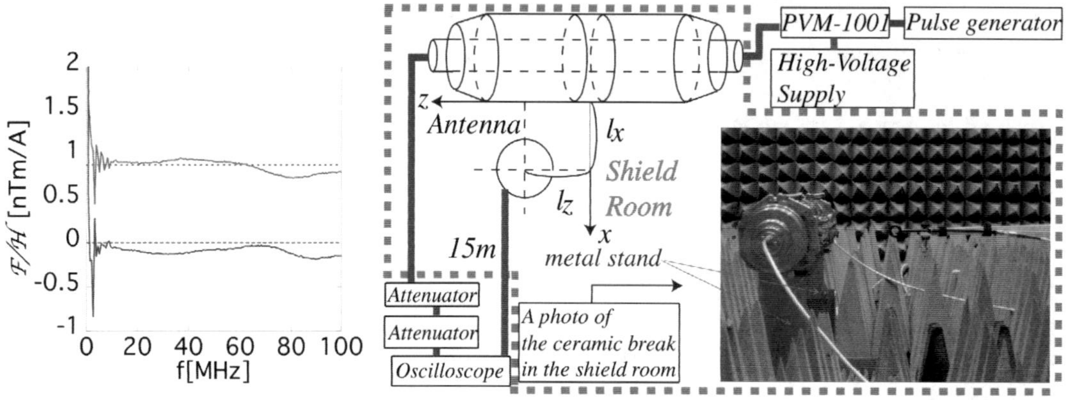

Figure 4. The frequency dependence of measured magnetic flux (left), and the measurement setup including a photo of the ceramic break in the shield room in JAXA (right).

profiles. The measured 27 wire signals are shown in figure 2. We can identify the noise from 0 to 0.7 ms in both cases (left/center). Though the capacitors are expected to suppress the low-frequency components of currents, high-frequency components contribute to the noise. This is a systematic error by the trapezoid bump magnets [16] at the injection area on the RCS rather than the wall currents with beams because it is excited even in the case without beams (see the right). As a result, it can be eliminated in principle to produce the smooth beam profiles.

Finally, we have identified that the noise is excited through the ground plates in figure 1. After the plates are removed, the error diminishes on the signals even when the downstream and upstream ceramics are covered over capacitors and Al, respectively, as shown in the left panel of figure 3, and no integration gate dependence is found in the produced distribution in the center/right figures, where the red and blue lines are obtained by setting the integration gates over 0-4 ms and 0.7-4 ms on the signals, respectively. The identification of the noise path enables the capacitors to perfectly suppress the adverse effect on the monitors from the nearby magnets.

3. Measurements of radiations

Now, let us investigate whether the capacitors can suppress the radiation around the ceramic break beside MWPM1, after measuring the magnetic fields from a 10 mm long ceramic break

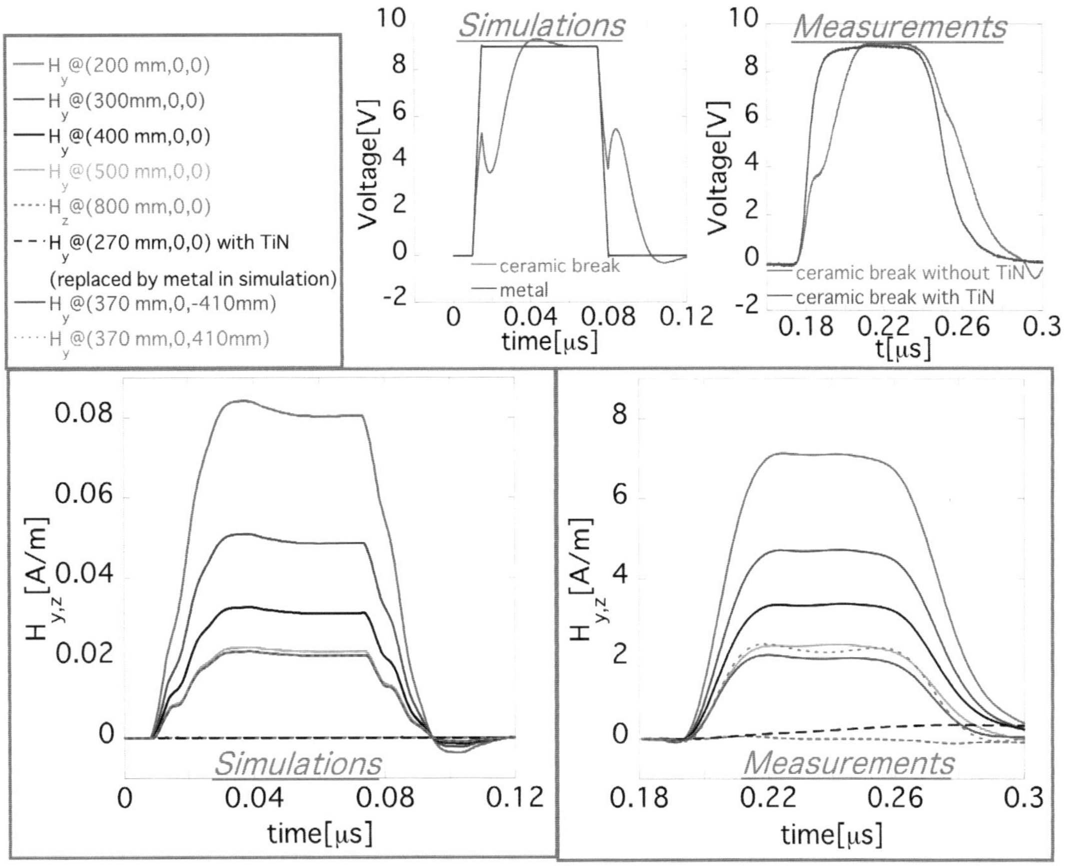

Figure 5. Simulations and measurements.

with and without TiN coating in the electromagnetic anechoic chamber in JAXA by using a 30 mm diameter loop antenna [17] whose upper resonant frequency is 1.5 GHz. The left figure of figure 4 shows the frequency dependence of measured magnetic flux divided by the magnetic field at the center of the loop. The red and blue lines show the real and imaginary parts, while the dashed lines show the calculated theoretical results [18].

The right figure of figure 4 shows a schematic of the measurement setup and a photo of the ceramic break in the shield room in JAXA, where the ceramic breaks are supported by the metal stand ensuring the ground path for the wall current [11]. The ceramic breaks were sandwiched between two stainless-steel chambers. The inner conductor was placed in the center of chambers, making a coaxial structure with 50 Ω characteristic impedance. The upstream side was connected to the pulse generator [19], which was amplified up to 1 kV by the pulse voltage module (PVM-1001 [20]). We generated a pulse with a rise time of 5 ns and a fall time of 5 ns with a flat top time of 60 ns. The port on the downstream side and the antenna were connected to an oscilloscope [21] via 40 dB attenuator [22], and 15 m coaxial cable [23], respectively. The position of the antenna is specified by $(l_x, 0, l_z)$ in the right figure. By rotating the antenna, we measured the magnetic flux in the longitudinal and vertical directions.

Figure 5 shows the simulations by CST [24] and 300 times averaged measurements. The upper-center/right and the other panels show the signals at the downstream port and magnetic

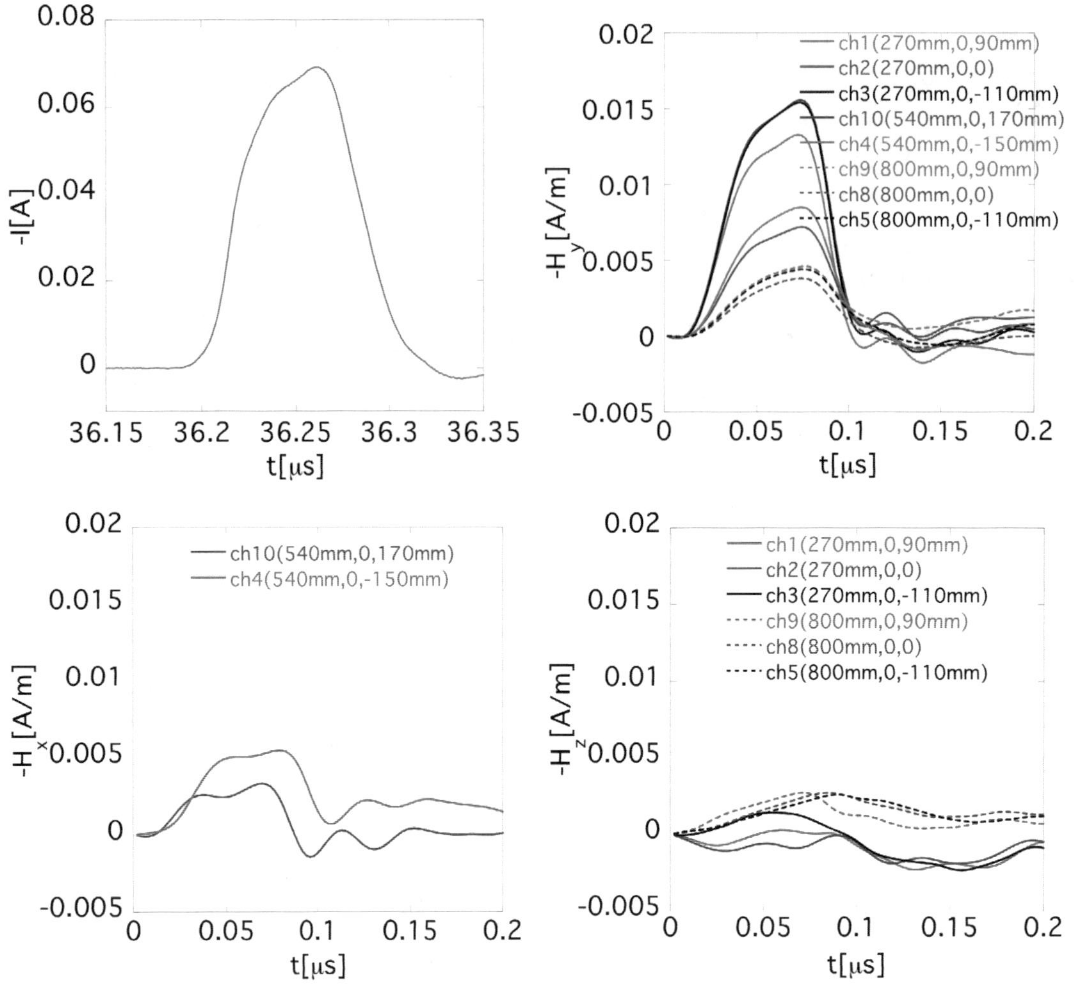

Figure 6. The observed wall current (upper-left) and magnetic fields (the others) from the ceramic breaks, which are averaged over 256 times.

fields at $(l_x, 0, l_z)$, respectively. The simulations for metal are calculated after the ceramic break is replaced with a perfectly conductive (PEC) pipe. The measured magnetic field is obtained after integrating the measured induced voltage on the antenna over time under the assumption that the magnetic flux is constant in the loop. The upper-center/right panels illustrate the modulated pulses monitored at the downstream port due to the impedance mismatch, which is significant for the ceramic break without TiN coating. The input pulse of simulation is 100 times lower than that of measurement, making both results of radiations different by 2 orders, sustaining both output pulses in the same order. The simulations and measurements are in agreement except in the case where the antenna is positioned at (200 mm, 0, 0), which means that the uniformity of the magnetic flux on the antenna is significantly violated in the loop. Hence, the antennas are placed at (270 mm, 0, 0) even in the nearest position at the following measurements in the RCS. The measured purple solid and brown dashed lines are almost identical and only the vertical component H_y is excited under this idealistic environment, as suggested by the simulation and theory [13]. Moreover, the measurements demonstrate that the magnetic

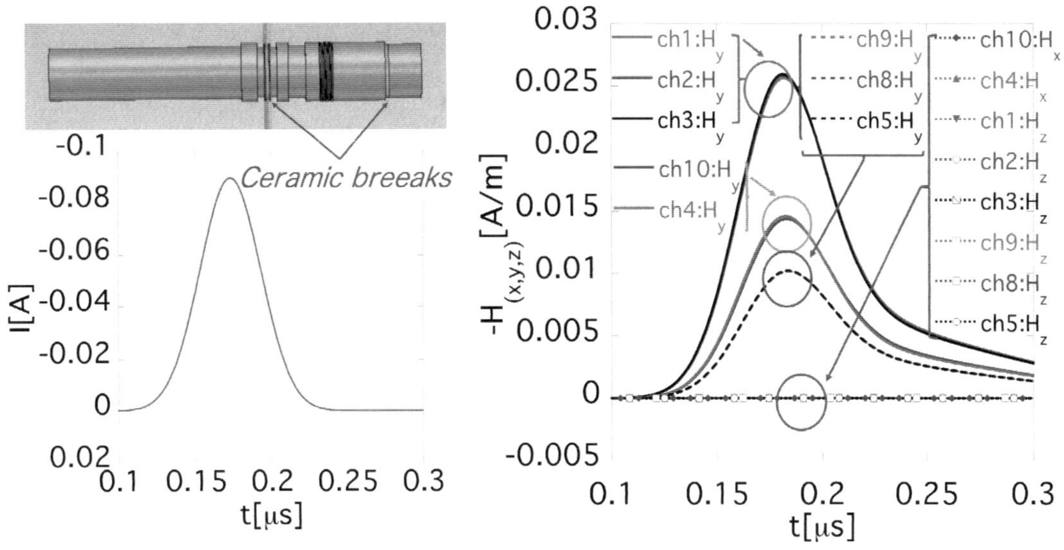

Figure 7. The simulation setup (upper-left), the input beam (lower-left), and the simulated magnetic fields (right).

field is significantly shielded by TiN coating with 10 nm thickness as expected [9–12].

Now, let us measure the radiation from the ceramic break when H^- beam with $\beta = 0.7$, 90 ns width, and -4.45 nC passes through it in the RCS. The longitudinal beam shape is measured by a wall current monitor to be confirmed. The positions of eight antennas are identified by [**ch1** (270 mm, 0, 90 mm), **ch2** (270 mm, 0, 0 mm), **ch3** (270 mm, 0, -110 mm)], [**ch10** (540 mm, 0, 170 mm), **ch4** (540 mm, 0, -150 mm)], and [**ch9** (800 mm, 0, 90 mm), **ch8** (800 mm, 0, 0 mm), **ch5** (800 mm, 0, -110 mm)], where (**ch1**, **ch3**), (**ch10**, **ch4**), and (**ch9**, **ch5**) are almost symmetric for $z = 0$ plane.

Figure 6 and figure 7 show the measurements and simulations without capacitors neither TiN coating. Compared to the test-stand case, the situation is too complicated to perform accurate simulations from viewpoints of modeling the real environment. Then, the simulation setup is simplified as in the left panel of figure 7. Nonetheless of the simplification, the results feature the measurements, because H_y partially reflects the longitudinal bunch shape, is almost symmetric for the $z = 0$ plane, and systematically decreases as l_x increases in both the simulations and measurements. The shape distortion may be found by more precisely observing the tail part of H_y. Though the simulation suggests the radiations should not be emitted in the longitudinal and horizontal directions, they are produced in the measurements owing to the objects in J-PARC. One drawback of this simplified simulation is that the results are 1.5 times larger than the measurements, which may be improved by modifying the input pulse shape.

Now, let us simulate two cases, where the capacitors are attached to the downstream only, or PEC is covered over the upstream only. As shown in figure 8, the simulations suggest the radiations from both ceramics can be comparable at **ch1**, although the distances from the downstream and the upstream to **ch1** are 285 mm and 650 mm, respectively. The corresponding measurements at **ch1** reveal the contribution from upstream is not negligible. Hence, let us compare the results of **ch1** with **ch4**, where the distances from the 'upstream' ceramic breaks to **ch1** and **ch4** are both 650 mm, when the capacitors are attached only to the 'downstream' ceramic break because theoretical results predict that the magnetic fields are enhanced near

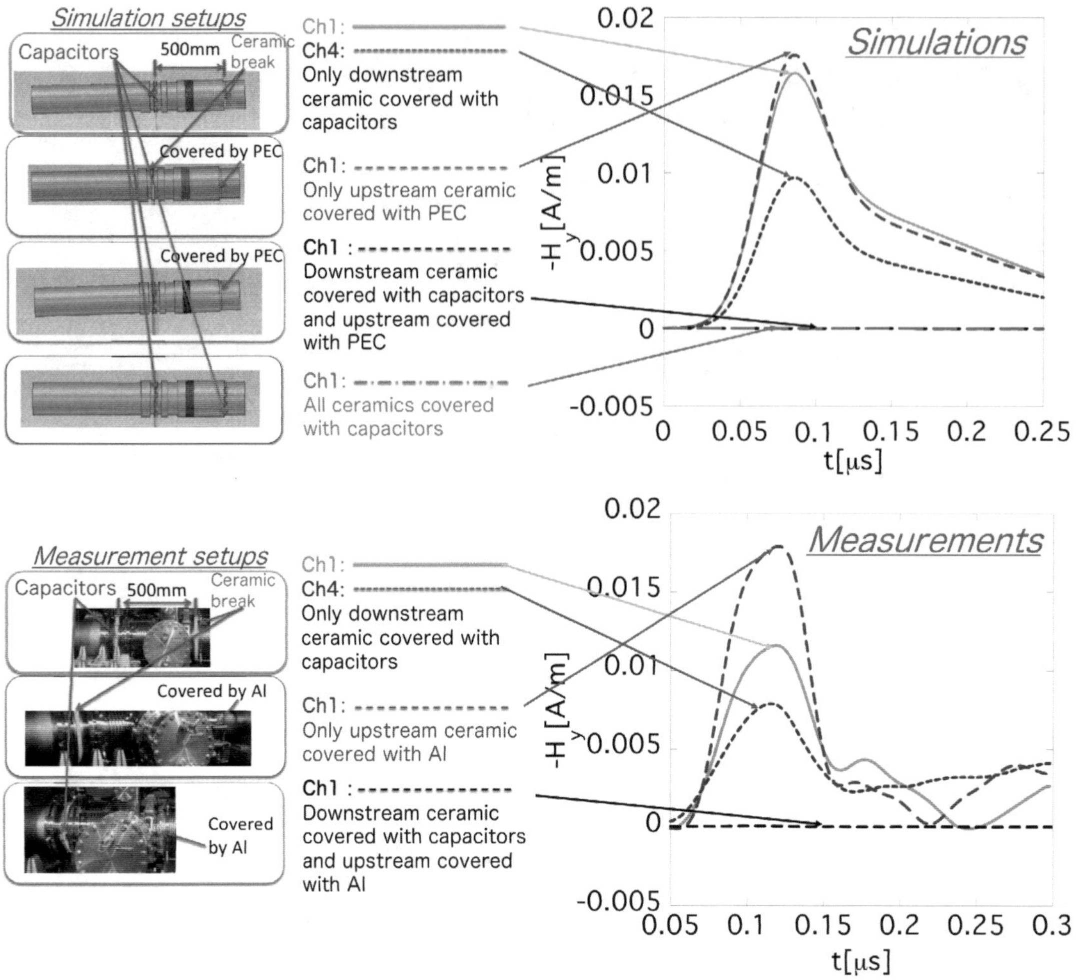

Figure 8. The simulation and measurement setups (left), simulated and measured magnetic fields (right) at **ch1** (270 mm, 0, 90 mm) and **ch4** (540 mm, 0, -150 mm).

the chamber wall [13]. Both the simulations and measurements demonstrate the enhancement, nonetheless of the existence of MWPM1.

To cure the situation, let us cover the upstream ceramic break with Al sheet (or capacitors) as well as the downstream one with capacitors, as shown in figure 8. The simulations and measurements in figure 8 demonstrate the remarkable suppression of the radiations. Practically, the ceramic break with 10 nm thin TiN coating can be more beneficial because the induced voltage on the ceramic break can provide the precise bunch shapes [11], suppressing the radiations.

4. Summary

Since the chamber walls significantly enhance the magnetic field from the ceramic breaks, simplified simulations are effective to evaluate it when the ceramic breaks are close to monitors. The noise stemming from the ground should be carefully investigated when we cure the ceramic breaks.

Acknowledgments
This work was supported by KAKENHI (17K05124).

References
[1] Japan Accelerator Research Complex `http://j-parc.jp/index-e.html`
[2] Hotchi H *et al.* 2017 *Phys. Rev. Accel. Beams* **20** 060402
[3] Hotchi H *et al.* 2017 *Proc. 8th Int. Particle Accelerator Conf. (IPAC'17)* (Copenhagen Denmark) pp 2470–73
[4] Hotchi H *et al.* 2017 *J. Phys. Conf. Ser.* **874** 012059
[5] Saha P K, Shobuda Y, Hotchi H, Hayashi N, Takayanagi T, Harada H and Irie Y 2009 *Phys. Rev. Accel. Beams* **12** 040403
[6] Hatakeyama S and Yoshimoto M 2013 *Proc. 10th Annual Meeting of Particle Accelerator Society of Japan* (Nagoya, Japan) pp 1082–86
[7] Chao A W 1993 *Physics of Collective Beam Instabilities in High Energy Accelerators* (New York: Wiley)
[8] Zotter B W and Kheifets S 1998 *Impedances and wakes in high-energy particle accelerators* (Singapore: World Scientific)
[9] Shobuda Y, Chin Y H and Takata K 2009 *Phys. Rev. Accel. Beams* **12** 094401
[10] Shobuda Y, Chin Y H and Takata K 2014 *Phys. Rev. Accel. Beams* **17** 091001
[11] Shobuda and Toyama T 2020 *Phys. Rev. Accel. Beams* **23** 092801
[12] Shobuda Y 2022 *Prog. Theor. Exp. Phys.* **2022** 053G01
[13] Shobuda Y, Chin Y H and Takata K 2007 *Phys. Rev. Accel. Beams* **10** 044403
[14] Japan Aerospace Exploration Agency `https://global.jaxa.jp/`
[15] Murata `https://www.murata.com/en-us/products/productdetail?partno=RCER72E105K5B1H03B`
[16] Shobuda Y, Irie Y and Igarashi S 2009 *Phys. Rev. Accel. Beams* **12** 032401
[17] ETS·LINDGREN, 7405 E & H Near Field Probe Set
[18] Smith G S 1993 Loop antennas *Antenna Engineering Handbook 3rd edition* ed R C Johnson and H Jasik (New York: McGraw-Hill) chapter 5
[19] Tektronix, KEITHLEY 3390 50MHz Arbitrary Waveform Generator
[20] BNC DEI High Voltage, PVM-1001-P +950V 10ns OEM Pulse Generator Module
[21] Teledyne Lecroy, WavePro 715Zi-A
[22] Tamagawa Electronics Co. Ltd, UFA-100NPJ-20
[23] Huber+Suhner, SUCOFLEX 106
[24] CST Studio Suite `http://www.cst.com`

Investigation of RF heating for the multipole injection kicker installed at SOLEIL

A Gamelin[1], P Alexandre[1], R Ben El Fekih[1], J Da Silva Castro[1], M El Ajjouri[1], A Letresor[1], L S Nadolski[1], R Ollier[1], S Thoraud[1], M Sacko[2] and S Taurines[2]

[1]Synchrotron SOLEIL, Gif-sur-Yvette, France
[2]Avantis Concept, Saint-Cere, France

E-mail: alexis.gamelin@synchrotron-soleil.fr

Abstract. During the commissioning of the new Multipole Injection Kicker (MIK) pulsed magnet at SOLEIL synchrotron, an anomalously high heating of the MIK chamber and flanges was found. To better manage the heat load, fans directed toward the MIK were added to improve the air-cooling flow. This allowed the nominal current to be reached in all operation modes while keeping reasonable temperatures on the MIK. Post-installation investigations subsequently showed that the initial estimate of the maximal heat load was in agreement with the measured temperature in several operation modes both with and without the additional fans. In this article, we present the complete study, starting from the impedance calculation to thermal simulations, and comparison with the measured data with beam.

1. Multipole injection kicker

Non-linear kicker magnets, such as the Multipole Injection Kicker (MIK), are key components for compact injection scheme of some 4th generation light sources [1, 2]. Non-linear kicker magnets are pulsed magnets characterized by an off axis field region where the injected beam is located and a field free region on the stored beam path. They are developing since their first use at KEK-PF [3] and BESSY II [4] as they allow for different types of on-axis and off-axis injections types in rings with small dynamic aperture [2]. The MIK provides a combined quadrupolar and octupolar pulsed field which allows for a near-transparent injection compared to the usual bump-based scheme using several kickers [1].

The MIK, designed and built by SOLEIL in collaboration with MAX IV, is composed of a racetrack chamber made of monocrystaline sapphire where grooves are precisely machined to hold copper conductors in the direction parallel to beam propagation, see figure 1. The inside of the chamber is coated with titanium to shield the beam from the sapphire body, allowing the image current to flow over the titanium layer and reducing the thermal load. But the coating thickness can not be too large as it induces a disturbance of the magnetic field and generates a pulsed quadrupolar component via eddy currents [5]. A compromise must therefore be found on the coating thickness to get a magnet with good performance with a contained thermal load.

A first prototype of the MIK was installed at MAX IV 3 GeV ring in 2017 where heating due to the 0.8 µm thin titanium coating was found to be an issue [1]. The problem was solved in 2019 when an improved chamber with a thicker coating of 3.5 µm was installed. The MIK is now routinely used for day to day operation at MAX IV with excellent performance [1].

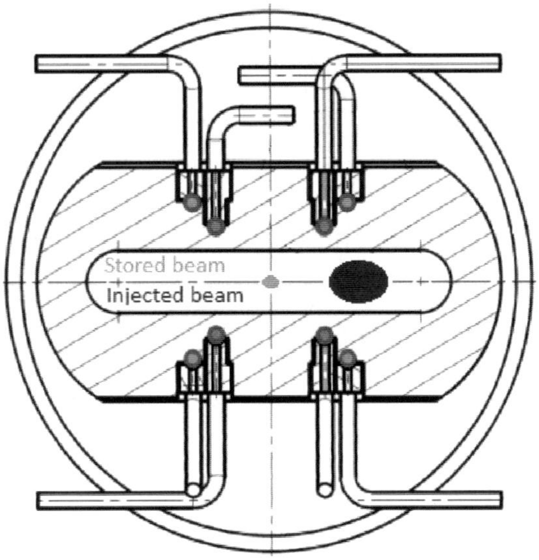

Figure 1. Cross section of the MIK showing the position of the copper conductors (red and green dots), the injected and stored beam.

Another MIK device was installed in the 2.75 GeV storage ring of SOLEIL in January 2021 and is being commissioned [6–8]. As the SOLEIL MIK also presents anomalously high temperatures, a post-installation investigation was initiated to review the RF heating estimate, the choice of the coating thickness and understand the observed high temperatures. This article describes the findings from this investigation and presents the different components: impedance calculations, temperature measurements with beam and thermal simulations.

2. Impedance calculations

The impedance calculations shown here for the MIK chamber and flanges, the two elements where significant heating is observed, were performed prior to installation and are based on the MIK mechanical design that is described in detail in [9].

2.1. MIK chamber

The MIK is composed of a single 33 cm long racetrack chamber with an inner aperture of 46.8 mm in horizontal and 7.8 mm in vertical plane. Its geometric (longitudinal) impedance is thus negligible compared to the resistive wall one produced by the titanium coating deposited on the sapphire body.

The coating thickness was measured at 3.5 µm using control pellets used during the deposit. The DC resistance of the coating was also measured at 3.43 Ω using the two probe method. From this measurement, assuming that the coating is homogeneous, its resistivity can be estimated between $2.83 \times 10^{-6}\,\Omega\,\mathrm{m}$ and $3.73 \times 10^{-6}\,\Omega\,\mathrm{m}$ depending whether the deposition is considered to be present only on the horizontal sections or on the whole chamber.

The measured resistivity value of the coating is quite different from the usual titanium resistivity in its metallic form: $4.27 \times 10^{-7}\,\Omega\,\mathrm{m}$. The reason for this difference could be related to the sputtering method used to realize the titanium deposition, leading to an amorphous structure of the titanium, more resistive than the usual metallic form.

The resistive wall impedance of the MIK chamber can then be computed using the multi-layered flat chamber analytic model implemented in the `ImpedanceWake2D` code [10, 11] or,

Journal of Physics: Conference Series **2420** (2023) 012057 doi:10.1088/1742-6596/2420/1/012057

neglecting the effect of the high relative permittivity ϵ_r of the sapphire body, using equation (5) of [12]. Input values for the `ImpedanceWake2D` code are shown in table 1.

Table 1. Input values for `ImpedanceWake2D` code for the "amorphous titanium" MIK chamber model.

Layer	Thickness	DC resistivity	ϵ_r
Vacuum	3.9 mm	∞	1
Titanium	3.5 µm	$3 \times 10^{-6}\,\Omega\,\mathrm{m}$	1
Sapphire	11.1 mm	$1 \times 10^{12}\,\Omega\,\mathrm{m}$	10 ($\approx$ 9.3/11.5)
Vacuum	∞	∞	1

The real part of the longitudinal impedance obtained using both methods is shown in figure 2 for both an "amorphous titanium" layer using the measured resistivity and a metallic layer. The good agreement between both methods shows that the 3.5 µm titanium coating effectively shields the beam from the sapphire high relative permittivity and poor conductivity. In uniform filling at 500 mA (416 bunches), with the measured bunch length of 19.8 ps, the expected thermal load of each model yields 112 W for the "amorphous titanium" impedance and 17 W for the metallic titanium one.

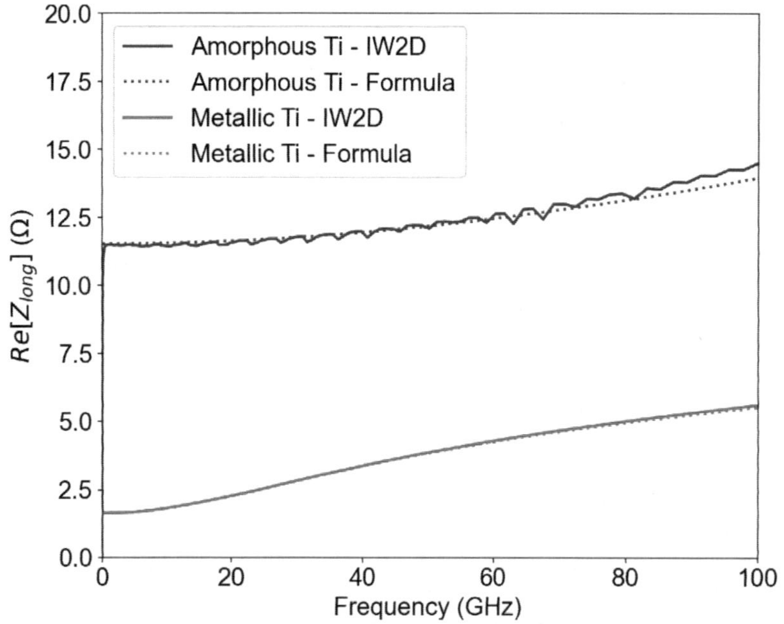

Figure 2. Real part of the longitudinal impedance of the MIK chamber.

At SOLEIL, the highest thermal stress comes from the 8 bunch filling pattern where a total current of 110 mA is stored. For this mode, the expected thermal load was about 156 W for a maximum power density of 3.6 W mm^{-2} in the center of the MIK chamber. This power density was estimated to be low compared to the one generated by synchrotron radiation in near by areas so a thermal simulation was deemed unnecessary at the time.

2.2. MIK flanges

The junction between the MIK chamber and its flanges is ensured by a brazed sapphire/copper interface, then a brazed copper/stainless steel interface and finally a stainless steel/flange weld [9]. The flanges are made of 316L stainless steel and have the same racetrack cross section as the MIK chamber. The flanges also integrate small bellows shielded from the beam by RF fingers. The flange vertical aperture increases locally from the 8 mm of the MIK to 9 mm at the RF fingers location to decrease again to 8 mm at the upstream/downstream connection with taper-absorber elements. The small cavity created by this aperture variation leads to two trapped modes at 16.7 GHz and 19.2 GHz with a corresponding amplitude of 247 Ω and 40 Ω in the longitudinal impedance spectrum.

For the uniform filling mode at 500 mA and for full coincidence with beam spectrum line, the power deposited by the first mode was estimated to be around 2.8 W for a maximum power density of 0.5 W mm^{-2}. The resistive wall heating for the same operation mode was evaluated around 2.0 W for a maximum power density of 0.4 W mm^{-2}.These flanges were designed to accommodate the MIK aperture to the cross-section of the SOLEIL straight section. Considering the lack of timely fallback solutions, it was decided to keep this design despite the trapped mode, as the expected heat load was low and there was no danger of coupled bunch instabilities.

3. Beam measurements

After the MIK installation, significant heating was measured at the downstream extremity of the MIK chamber where the temperature reached 130 °C for a beam current of 300 mA. To contain the temperature rise and allow the nominal current to be reached in all operation modes, the following steps had to be taken: the MIK outer frame was removed, two heatsinks were installed on the MIK chamber body and three fans directed toward the MIK were used to improve the air-cooling flow.

As the heat load was not linearly proportional to the stored current, the synchrotron radiation was quickly ruled out as a potential contributor. Temperature sensors installed on the MIK and thermal camera measurements made right after a beam dump then shown a asymmetry between the upstream and downstream regions of the MIK. In addition to data taken in the different user operation modes, additional measurements were taken with the fans off during a dedicated machine development shift. The two observed cases were selected to mimic the usual operation modes parameters, in particular bunch length, but with an overall lower beam current in order not to overheat the MIK. Measured data is compared to simulated data in table 2.

4. Thermal simulations

The thermal simulation was performed following the methodology described in [13, 14]. Firstly, 3D maps of the dissipated power were generated at different frequencies (DC for the resistive wall, mode frequencies for the geometric impedance) using `CST` wakefield solver [15]. Then these maps

Table 2. Comparison between the simulated and measured temperatures of the MIK chamber.

Current	Filling mode	Fans	Simulated	Upstream	Downstream
500 mA	Uniform (416 bunches)	On	61 °C	52 °C to 69 °C	86 °C to 125 °C
500 mA	Uniform (416 bunches)	Off	335 °C		
110 mA	8 bunches	On	72 °C	64 °C to 74 °C	113 °C to 164 °C
110 mA	8 bunches	Off	425 °C		
100 mA	1/4 (104 bunches)	Off	72 °C	69 °C to 72 °C	85 °C to 86 °C
25 mA	2 bunches	Off	107 °C	103 °C to 112 °C	136 °C to 138 °C

were re-scaled using the deposited power computed at that frequency using the beam properties and impedance as input. After this procedure, the different power maps corresponding to each operation modes can be used as input in a thermomechanical simulation software like `ANSYS` [16] and mapped to the 3D geometry.

For the thermal simulation, convection was imposed on all MIK surfaces exposed to air with an heat transfer coefficient of $5\,\mathrm{W\,m^{-2}\,K^{-1}}$ when the fans where off (free convection) and of $46.2\,\mathrm{W\,m^{-2}\,K^{-1}}$ when they were switched on (forced convection). The value for the heat transfer coefficient with fans was computed by taking into account airflow rate of the fans and the MIK chamber geometry.

The simulated temperatures on the MIK chamber are compared to those measured in the upstream and downstream regions in table 2. The range of measured values correspond to 4 measurement points in the upstream region (2 of them providing too low values were discarded as outliers, maybe because of a weak thermal contact with the MIK chamber) and 5 measurement points in the downstream region. The simulated value reported corresponds to the maximum value but the simulated temperature distribution is very homogeneous (a few °C of difference at maximum) as the heat is due to the resistive wall.

When the fans are turned off, the temperature spread along the different measurement points is much more limited than when the fans are switched on as the fans air flow is not uniform on the MIK body. There is a good match between the simulated temperatures and the ones measured in the upstream region for all cases. It seems to indicate that the impedance model, using measured resistivity, and the methodology is correct. But the measured temperatures in the downstream region are consistently higher than in the upstream one and therefore do not match with the simulation.

Two possible hypotheses were advanced to explain the abnormal heating observed in the downstream region: a local degradation of the titanium coating or a partial loss of the RF contact on the neighbouring flange.

The flanges were simulated with the same method but in this case the temperature rise is very moderate because the total deposited power is lower than 5 W for the worst case. The extreme case of no RF contact has also been simulated but gives much too high temperatures due to the huge deposited power (1.6 kW for the worst case). The measured temperatures on the downstream flange are always lower than the those measured on the downstream MIK region, e.g. 75 °C for the 100 mA case and 115 °C for the 25 mA case. So it seems that the most probable explanation is a local degradation of the coating.

When the MIK and its flanges are simulated together with perfect thermal contact, the simulated temperature on the MIK decreases a bit for the case without fans, from 10 °C to 20 °C, compared to table 2 as some of the heat is transferred in the flanges. This transfer is expected and can explain the high temperatures at the downstream flange if the coating is degraded in that very region. The temperature drop on the MIK chamber compared to table 2 could either indicate that the coating resistivity is underestimated or more likely that the perfect thermal contact hypothesis is too coarse and that a refined model using an experimental thermal contact conductance should be used [17]. The thermal simulation performed with `ANSYS` was also benchmarked against `CST` Conjugate Heat Transfer (CHT) solver and gave similar results.

5. Conclusion

This investigation showed that the initial thermal load estimate using the measured resistivity was correct but that its consequence were underestimated. The imbalance between the upstream and downstream region still need to be better understood and a future visual inspection of the MIK during a machine shutdown could help in this respect.

Future MIK devices could be improved on this aspect by using thermal simulation during the design phase to optimize the heat transfer and integrate cooling fans in the base design. The heat amount could also be reduced by a slight increase of the coating thickness, but at the cost of increased field perturbations.

References

[1] Alexandre P, Ben El Fekih R, Letrésor A, Thoraud S, Da Silva Castro J, Bouvet F, Breunlin J, Andersson A and Fernandes Tavares P 2021 *Nucl. Instrum. Methods Phys. Res., Sect.* A **986** 164739
[2] Tordeux M A *et al.* 2021 *Proc. 12th Int. Particle Accelerator Conf. (IPAC'21)* (Campinas, Brazil) pp 796–98
[3] Takaki H, Nakamura N, Kobayashi Y, Harada K, Miyajima T, Ueda A, Nagahashi S, Shimada M, Obina T and Honda T 2010 *Phys. Rev. Spec. Top. Accel Beams* **13** 020705
[4] Dressler O, Atkinson T, Dirsat M, Kuske P and Rast H 2011 *Proc. 2nd Int. Particle Accelerator Conf. (IPAC'11)* (San Sebastian, Spain) pp 3394–96
[5] Kallestrup J *et al.* 2019 *Proc. 10th Int. Particle Accelerator Conf. (IPAC'19)* (Melbourne, Australia) pp 1547–50
[6] Ollier R, Alexandre P, Ben El Fekih R and Nadolski L S 2021 *Proc. 12th Int. Particle Accelerator Conf. (IPAC'21)* (Campinas, Brazil) pp 3525–28
[7] Nadolski L S *et al.* 2021 *Proc. 12th Int. Particle Accelerator Conf. (IPAC'21)* (Campinas, Brazil) pp 3945–48
[8] Ollier R *et al.* 2022 *Proc. 13th Int. Particle Accelerator Conf. (IPAC'22)* (Bangkok, Thailand) pp 2675–78
[9] Da Silva Castro J, Alexandre P, Ben El Fekih R and Thoraud S 2018 *Proc. 10th Mechanical Engineering Design of Synchrotron Radiation Equipment and Instrumentation (MEDSI'18)* (Paris, France) pp 48–49
[10] Mounet N and Métral E 2010 *Proc. 46th ICFA Adv. Beam Dynamics Workshop on High-Intensity and High-Brightness Hadron Beams (HB 2010* (Morschach, Switzerland) pp 353–57
[11] Mounet N ImpedanceWake2D
[12] Migliorati M, Belli E and Zobov M 2018 *Phys. Rev. Accel. Beams* **21** 041001
[13] Teofili L, Migliorati M, Carbajo Perez D, Giordano F, Lamas Garcia I and Mazzacano G 2018 *Proc. 9th Int. Particle Accelerator Conf. (IPAC'18)* (Vancouver, Canada) pp 3452–55
[14] Teofili L 2019 *A Thermomechanical and Electromagnetic Approach For The Design Of High-Intensity Accelerator Components* Ph.D. thesis Roma La Sapienza
[15] CST studio suite 3D EM simulation and analysis software `www.cst.com`
[16] ANSYS `www.ansys.com`
[17] Vega Cid L, Abánades A, Barnes M, Mostchmann F, Vlachodimitropoulos V and Weterings W 2019 *Nucl. Instrum. Methods Phys. Res., Sect.* A **916** 296–305

Space charge analysis for low energy photoinjector

M Carillo[1,4], M Behtouei[2], F Bosco[1,3,4], L Faillace[1,2], L Ficcadenti[4],A Fukasaway[3] L Giuliano[1,4], A Mostacci[1,4], M Migliorati[1,4], B Spataro[2], J Rosenzweig[3] and L Palumbo[1,4]

[1]Sapienza University of Rome, 00161 Rome, Italy
[2]INFN-LNF, 00044 Frascati, Italy
[3]UCLA, Los Angeles, 90095 California, USA
[4] INFN-Sez.Roma1, 00161 Roma, Italy

E-mail: `martina.carillo@uniroma1.it`

Abstract. Beam dynamics studies are performed in the context of a C-Band hybrid photo-injector project developed by a collaboration between UCLA/Sapienza/INFN-LNF/RadiaBeam [1, 2].These studies aim to explain beam behaviour through the beam-slice evolution, using analytical and numerical approaches. An understanding of the emittance oscillations is obtained starting from the slice analysis, which allows correlation of the position of the emittance minima with the slope of the slices in the transverse phase space (TPS). At the end, a significant reduction in the normalized emittance is obtained by varying the transverse shape of the beam while assuming a longitudinal Gaussian distribution. Indeed, the emittance growth due to nonlinear space-charge fields has been found to occur immediately after moment of the beam emission from the cathode, giving insight into the optimum laser profile needed for minimizing the emittance.

1. Introduction

In the new generation of high-brightness beams [3] a fundamental role is played by the achievement of emittance compensation [4, 5]. This scenario includes the design of the Hybrid C-band photo-injector and the in-depth study of the dynamics of its output beam [6]. To do this, different approaches are exploited, such as beam slice analysis and the analytical study of the phenomena under consideration, including space charge fields.

2. Beam slice analysis

Slice analysis has a key impact to understand beam dynamics and in the emittance compensation process. Starting from the definition of the beam rms emittance it is possible to separate four different components due to slice splitting [7]. In this analysis we will neglect the emittance terms due to the linear and non-linear contributions of the transverse offsets between the slices themselves because they give a negligible contribution to the total projected emittance [8].

The transverse rms emittance is defined starting from the second order moments according to the equation [9]:

$$\varepsilon_{rms}^2 = \sigma_x^2\sigma_{p_x}^2 - \sigma_{x,px}^2. \tag{1}$$

Let us take a beam of N particles, divided into S slices, each one populated by a variable number of particles M_i. σ^2 moments can be written as function of sum of the slice second order moments:

$$\sigma_x^2 = \langle x^2 \rangle = \sum_{i=1}^{S}\sum_{j=1}^{M_i} \frac{x_j^2}{N} = \sum_{i=1}^{S} \frac{M_i}{N}\langle x^2 \rangle_i = \sum_{i=1}^{S} \frac{M_i}{N}\sigma_{x,i}^2 \tag{2}$$

where $\langle\rangle$ defines an ensemble average, and in the same way:

$$\sigma_{p_x}^2 \equiv \sum_{i=1}^{S} \frac{M_i}{N}\sigma_{p_x,i}^2 \ , \ \sigma_{xp_x} \equiv \sum_{i=1}^{S} \frac{M_i}{N}\sigma_{xp_x,i} \tag{3}$$

By replacing equation (2), and (3) in to equation (1) the rms emittance becomes:

$$\varepsilon_{rms}^2 = \sum_{i=j}\left(\sigma_{x,i}^2\sigma_{p_x,i}^2 - \sigma_{xp_x,i}^2\right) + \sum_{i\neq j}\frac{M_iM_j}{N^2}\left(\sigma_{x,i}^2\sigma_{p_x \cdot j}^2 - \sigma_{xp_x,i}\sigma_{xp_x,j}\right) \tag{4}$$

where the first sum in the left-hand side is the emittance definition of $i-$th slice $\varepsilon_{rms_i}^2$, and the second one defines the correlated slices emittance expression, ε_{corr}^2. Since our goal is to optimize the emittance compensation process, a study on the double emittance oscillation at the gun exit [10] is performed. We assume, as initial conditions, the ones at the exit from the photoinjector and we study the evolution in a long drift.

Every single slice satisfies the envelope equation of a bunch in a free space, evolving under the effect of space charge force and emittance pressure, and each slice has different initial conditions $(\sigma_{0_i}, \sigma'_{0_i})$ but also different charge denisty λ_i and energy γ_i. The second order moment in equation (4) can be written as function in terms of of the envelope function σ and its derivative, with respect to the longitudinal position z, σ':$\sigma_x = \sigma$, $\sigma_{xp_x} = \sigma\gamma\sigma'$ and $\sigma_{p_x} = \gamma\sigma'$. Assuming for simplicity, that each slice is populated by the same number of particles $M = S/N$, the correlated emittance term can be written as:

$$\varepsilon_{corr}^2 = \frac{1}{2S^2}\sum_{i\neq j}(\sigma_i\gamma_j\sigma'_j - \sigma_j\gamma_i\sigma'_i)^2. \tag{5}$$

2.1. Floettmann model

In Floettmann emittance compensation analysis [11] using the slice model, it was demonstrated that to understand the emittance growth [12], it is well to know the orientation of the rms distribution in phase space, in addition to the slice size. For thin slice, the slope m of the distribution can be fitted by a straight line in the phase space: $m = \gamma\langle xx'\rangle/\langle x^2\rangle$, where x and x' represent the phase space coordinates, and $\langle x\rangle = \langle x'\rangle = 0$ is assumed. It is possible to derive the derivative of the beam envelope as function of the slope parameter m. In fact by definition:

$$\sigma' = \frac{1}{2\sigma_x}\frac{\partial}{\partial z}\langle x^2\rangle = \frac{\langle xx'\rangle}{\langle x^2\rangle^{1/2}}, \tag{6}$$

so that,

$$m = \frac{\gamma\langle xx'\rangle}{\langle x^2\rangle} = \frac{\gamma\sigma'}{\sigma}. \tag{7}$$

Thus the spread of m of all slices shrinks: the sizes of the slices and their speed of rotation evolve differently due to the difference in the space charge forces which they are subject to. Taking into account only two slices, the condition for obtaining a minimum of emittance is:

$$\Delta m = 0. \tag{8}$$

In Floettmann the analysis takes into account two beam slices, respectively with high and low perveance, and he proved equation (7), studying that the evolution of the slope m of the other slices with the intermediate perveance must be included in the interval of the two examined slices. In the two-slice case, the evolution equation is a second order equation, and hence the condition in equation (8) is reached in two positions. For this reason, two minima are observed in the emittance oscillation.

2.2. Multi-slice model

The goal is to explain why in a real physical beam, where the number of slices is greater than two, only two minima of emittance are still observed. For this reason the Floettmann analysis has been extended to a multi-slice model.

The relation in equation (7) is exploited by substituting it in the correlated emittance equation of the slices (equation (5)):

$$\varepsilon_{corr}^2 = \frac{1}{2S^2}\sum_{i\neq j}(\sigma_i m_j \sigma_j - \sigma_j m_i \sigma_i)^2 = \frac{1}{2S^2}\sum_{i\neq j}\sigma_i^2\sigma_j^2(m_i - m_j)^2. \tag{9}$$

The validity of this model is confirmed for two slice. As can been see from the multi-slice equation (9), it is not so easy to identify a condition on m in the multi-slice case. In fact, the squared slope difference in the phase space for each pair of beam slices is weighted by the product of their slice sizes. To show that, we applied this analysis to the case of the C-band hybrid photoinjector. The beam is split into 10 slices as described in [13], and evolves in a one meter drift long.

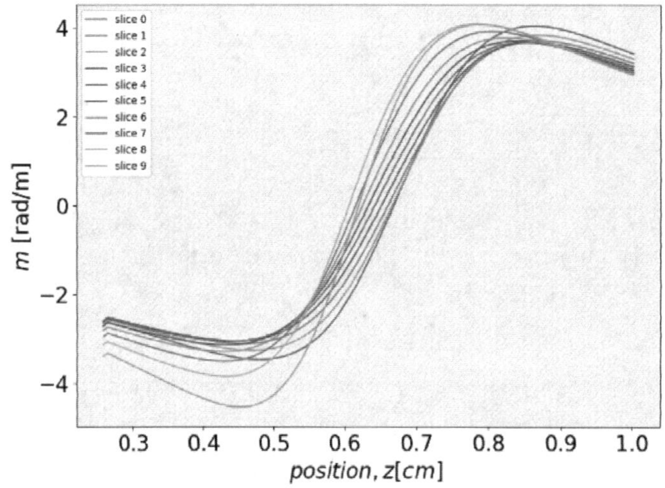

Figure 1. Evolution of the TPS slope m (equation (7)) as a function of the position in the drift. Each color represents one of the 10 slices the bunch has been divided into, where slice number 0 and 9 represents respectively the tail and the head slice.

Starting from the knowledge of the distributions in the TPS, the evolution of m for each slice along the drift is obtained using equation (7): the slices exiting from the photoinjector have a negative m value as they are still affected by the focusing effect of the solenoid. After reaching the minimum, the space charge forces dominate the focusing effects and the slices rotate in the opposite direction until they reach a positive defocusing slope.

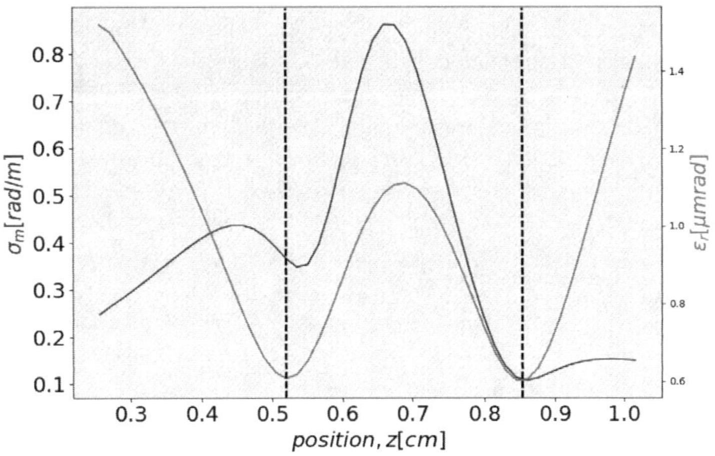

Figure 2. The slope spread σ_m of the TPS (blue curve) as a function of the drift position, in comparison with the rms emittance oscillation (red curve). The two vertical black dotted lines represent the exact position of the two emittance minima.

Whereas in the two-slices Floettman model the emittance minimum condition is reached when the slices are exactly overlapping (equation (8)), in the multi-slice model the minimum emittance conditions should occur when the difference of all the slopes of the slices m_i is minimal. For this reason the trend of the spread of m of the 10 beam slices along the drift has been studied and compared, in figure 2, with the emittance oscillation. What figure 2 shows is that, in the multi-slice case, the positions of the emittance minima do not coincide precisely with the positions for which the spread is the smallest one, but they remain in a close range. The reason for this behavior is implicit in equation (9). In fact it is not possible to extract a direct dependence on the spread σ_m, since the differences $|m_i - m_j|$ are weighted by the dimensions of the slices themselves, which change along the drift.

In any case, observing the slices evolution in the TPS, it is possible to obtain information on why only two minima of emittance occur even in the case with many slices. The slices leaving the photoinjector have a negative slope in the TPS (figure 1), which indicates their focusing behavior. Since they rotate with different speeds, they can reach a condition of maximum alignment with each other already in the focusing half-plane, and therefore identify the first emittance minimum. When, due to compression, the effect of space charge forces becomes dominant with respect to the effect of the focusing kick, the slices change direction of rotation in the phase space and assume a positive slope value. Also in this half-plane they will reach a second alignment condition, which determines the identification of the second emittance minimum. We could expect that continuing to have different speeds, the slices can re-align other times and therefore identify other positions of minimum emittance. What actually happens is that the slices now have a de-focusing behavior, and therefore their size tends to increase significantly. This means that even if they were re-aligned among themselves, the weight of the contribution of the size product in equation (9) would be so significant that the emittance value would still be high, and therefore not significant for the emittance compensation goal.

3. Space charge at the cathode

Beam shaping at the photocathode injection has been investigated with two different transverse laser distributions assumed with the aim of emittance compensation [2]. The first form is a simple uniform transverse distribution in r ("flat-top"), while the second one has the shape of a 1σ Cut-Gaussian. Indeed, using a uniform transverse laser illumination, the optimized rms normalized emittance has been found to be near 0.75 $mm - mrad$, while when one uses the truncated Gaussian laser profile on the photocathode, this normalized emittance decreases dramatically, to 0.46 $mm - mrad$. Slice analysis was used to explain the reason for this.

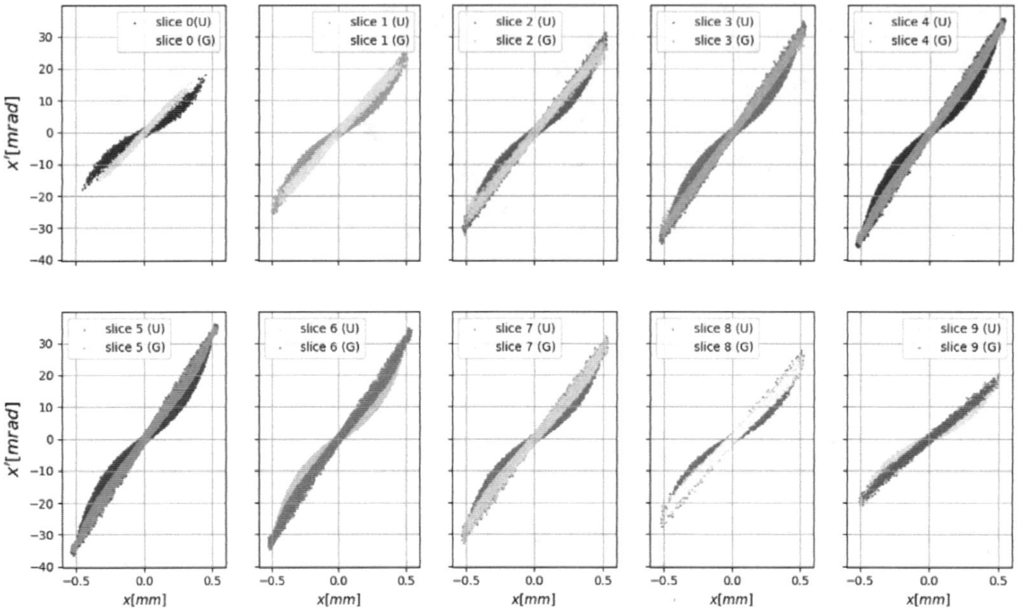

Figure 3. Slices comparison in the TPS for the beam just generated by the cathode. The two different colors in each plot represent the Cut-Gaussian and Flat-top distributions respectively, for each of the 10 slices.

Both beams are divided into 10 slices of equal length, and the TPS is analyzed. As shown in figure 3, it is evident that, by overlapping the same beam slices, the beam with the Cut-Gaussian distribution has a more linear trend than that of the Uniform case. This effect is visible already near of the photo-cathode, where the two distributions are generated. The reason is in the space charge forces expression [14].

Starting from the scalar potential solution of the not-homogeneous wave equation and using the Green function method [15], it is possible to obtain the radial space charge field expression as a function of the radial ($R(r)$) and longitudinal ($\lambda(z)$) distributions, for a fixed longitudinal position $\hat{z}$:

$$E_r^{(sc)}(r, \hat{z}) = -Q \int \frac{dk}{2\pi} e^{ik\hat{z}} \tilde{\lambda}(k) \frac{\partial \mathcal{S}(r,k)}{\partial r}, \tag{10}$$

where $\mathcal{S}$ represents the surface integral:

$$\mathcal{S}(r,k) = 2\pi \int_0^a \tilde{G}(r, r', k) R(r') r' dr', \tag{11}$$

and k is the wave vector in the Fourier space.

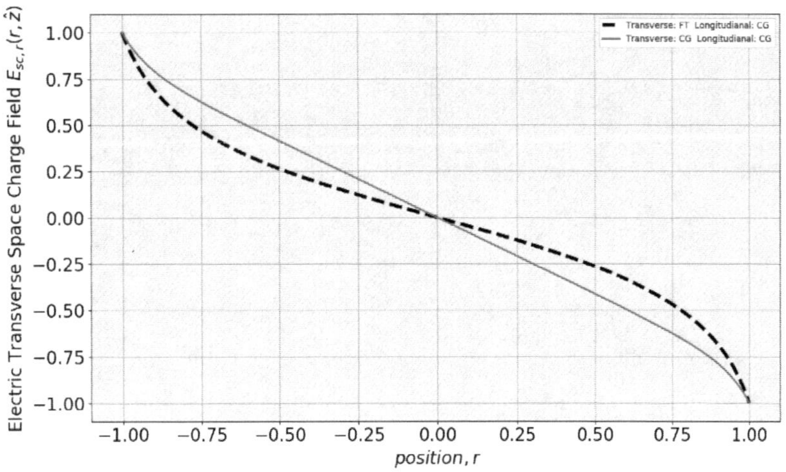

Figure 4. Normalized radial space charge field as a function of the normalized radial position. The field is compared with the same Gaussian longitudinal distribution while the transverse distributions varies from 1σ Cut-Gaussian (red curve) to Flat-Top (black dashed curve).

Plotting equation (10) for the two distributions, the behaviour in figure 4 is obtained. This shows that in the Cut-Gaussian case the space charge forces are more linear than in the Flat-Top case and, therefore, the contribution to the emittance is lower since the blow-out regime is reached earlier.

4. Conclusion and future goals

Several criteria for emittance compensation were analyzed using the slice analysis. In particular, the Floettmann model for the emittance oscillation has been extended to the multi-slice case, allowing the understanding of the presence of only two emittance minima, even in the multi-slice case.

Furthermore, a study on the dependence of the emittance on the laser distribution on the photocathode surface was performed, leading to derive the generic expression of the space charge fields for newly generated beams (equation (10)). The latter paved the way for a more complete study of the distributions used in the field of high-brightness accelerators, in order to reduce the beam emittance.

Acknowledgments

This work is supported by DARPA GRIT under contract no. 20204571 and partially by INFN National committee V through the ARYA project.

References

[1] Faillace L *et al.* 2021 *Proc. 12th Int. Particle Accelerator Conf. (IPAC'21)* (Campinas, Brazil) pp 2714–17
[2] Faillace L *et al.* 2022 *Phys. Rev. Accel. Beams* **25** 063401
[3] Rosenzweig J B *et al.* 2019 *Phys. Rev. Accel. Beams* **22** 023403
[4] Ferrario M *et al.* 2010 *Phys. Rev. Lett.* **104** 054801
[5] Rosenzweig J 2003 *Fundamentals of beam physics* (Oxford: Oxford University Press)
[6] Bosco F *et al.* 2021 *Proc. 12th Int. Particle Accelerator Conf. (IPAC'21)* (Campinas, Brazil) pp 3185–88
[7] Mitchell C 2015 A General Slice Moment Decomposition of RMS Beam Emittance *Preprint* ArXiv:1509.04765
[8] Niemczyk R 2021 *Subpicosecond-resolved emittance measurements of high-brightness electron beams with space charge effects at PITZ* Ph.D. thesis University of Hamburg

[9] Serafini L and Rosenzweig J B 1997 *Phys. Rev.* E **55** 7565
[10] Ferrario M *et al.* 2007 *Phys. Rev. Lett.* **99** 234801
[11] Floettmann K 2017 *Phys. Rev. Accel. Beams* **20** 013401
[12] Migliorati M *et al.* 2013 *Phys. Rev. Spec. Top. Accel Beams* **16** 011302
[13] Carillo M *et al.* 2021 *Proc. 12th Int. Particle Accelerator Conf. (IPAC'21)* (Campinas, Brazil) pp 3240–43
[14] Ferrario M, Migliorati M, and Palumbo L 2016 Space charge effects *Preprint* ArXiv:1601.05214
[15] Jackson J D 1998, *Classical Electrodynamics* (New York City:John Wiley and Sons)

Modeling and mitigation of long-range wakefields for advanced linear colliders

F Bosco[1,2,3], O Camacho[3], M Carillo[1,2], E Chiadroni[1,4], L Faillace[4], A Fukasawa[3], A Giribono[4], L Giuliano[1,2], N Majernik[3], A Mostacci[1,2], L Palumbo[1,2], J Rosenzweig[3], B Spataro[4], C Vaccarezza[4] and M Migliorati[1,2]

[1]Department of Basic and Applied Science for Engineering, La Sapienza University, Via A. Scarpa 16, 00185 Rome RM, Italy

[2]INFN-Sez. Roma1, Piazzale A. Moro 5, 00185 Rome RM, Italy

[3]Department of Physics and Astronomy, University of California Los Angeles, 405 Hilgard Ave., Los Angeles, CA 90095, United States of America

[4]Laboratori Nazionali di Frascati, INFN, Via E. Fermi, 00044 Frascati RM, Italy

E-mail: fabio.bosco@uniroma1.it

Abstract. The luminosity requirements of TeV-class linear colliders demand use of intense charged beams at high repetition rates. Such features imply multi-bunch operation with long current trains accelerated over the km length scale. Consequently, particle beams are exposed to the mutual parasitic interaction due to the long-range wakefields excited by the leading bunches in the accelerating structures. Such perturbations to the motion induce transverse oscillations of the bunches, potentially leading to instabilities such as transverse beam break-up. Here we present a dedicated tracking code that studies the effects of long-range transverse wakefield interaction among different bunches in linear accelerators. Being described by means of an efficient matrix formalism, such effects can be included while preserving short computational times. As a reference case, we use our code to investigate the performance of a state-of-the-art linear collider currently under design and, in addition, we discuss possible mitigation techniques based on frequency detuning and damping.

1. Introduction

Advanced experiments in the field of high energy particle physics rely on high luminosity electron/positron linear colliders working at the TeV scale. In the last decades, several conceptual facilities of such type have been investigated exploring techniques ranging from X-band ($\sim 12\,$GHz) rf-linacs to superconductivity and plasma acceleration [1–3]. A recent design proposed by SLAC [4] exploits advanced rf concepts to achieve high gradient performances relying on the mature and consolidated experience with C-band (5.712 GHz) technology [5, 6]. The design project, known as C^3 or the "Cool Copper Collider", allows to work with $\sim$ 120 MeV/m accelerating gradients by means of state-of-the-art techniques such as distributed coupling [7] and cryogenic cooling of the accelerating structures [8, 9].

In all the above cases, the collective interaction among charged bunches constitutes a major concern since it can cause unstable motion due to *beam break-up* (BBU) effects [10, 11]. In particular, particles traveling off-axis in the accelerating structures excite parasitic dipole fields in the form of higher order modes (HOMs) supported by the structures themselves which deflect the trajectories of the trailing charges. In this paper we introduce a dedicated tracking code studying the long-range wakefield interaction caused by self-induced dipole modes in linacs and we exploit such a tool to investigate the C^3 machine. The code describes the bunches within the rf-pulse as a sequence of rigid macro-particles with no internal structure performing transverse oscillations in presence of an external focusing optics and acceleration. The combination of such fields in an alternating gradient FODO lattice with length L_c, average betatron function β_x and phase advance $\mu_x = L_c/\beta_x$ provides the following transfer map [12]

$$\begin{pmatrix} x \\ x' \end{pmatrix} \mapsto \sqrt{\frac{\gamma}{\gamma + \gamma' L_c}} \begin{pmatrix} \cos\mu_x & \beta_x \sin\mu_x \\ -\frac{1}{\beta_x}\sin\mu_x & \cos\mu_x \end{pmatrix} \begin{pmatrix} x \\ x' \end{pmatrix} \tag{1}$$

which is responsible for the adiabatic damping of the transverse oscillations and characterizes the motion in absence of collective effects.

2. The process of beam break-up

The interaction between two particles with time separation τ induced by a resonant mode with angular frequency ω_0, shunt impedance $R_\perp$ and quality factor Q is described by the dipole wake-function per unit length

$$w_\perp(\tau) = \frac{\omega_0 R_\perp}{Q} e^{-\alpha\tau} \sin(\omega_n \tau) \tag{2}$$

with $\omega_n = \omega_0\sqrt{1-(2Q)^{-2}}$ and $\alpha = \omega_0/2Q$ [13]. Deflected charges moving off-axis cause further excitation of the resonant mode intensifying the parasitic interaction. A quantitative analysis for such a process has been introduced by Mosnier [14] whose approach describes the bunch train as a sequence of point-like macro-particles accelerated inside a machine with uniform betatron function where the mutual interaction is due to a single dipole HOM which exhibits the same parameters in each cell of the linac. Such a model shows that the motion becomes highly unstable if the ratio of the HOM frequency and the bunch repetition rate is close to an integer so that the beam-mode interaction fulfills a resonant condition. Mitigation is possible by breaking the coherent interplay through a spread in frequency of the dipole modes [15, 16]. Indeed, it can be shown that frequency spread reduces the HOM quality factor with a consequent shortening of the corresponding decay time.

3. The matrix formalism for wakefields

In this section we describe the approach we use in our tracking code for the inclusion of collective effects. The interaction of charged particle beams with a resonant mode can be described in terms of a matrix formalism which allows to account for both the longitudinal and transverse wakefield effects [17, 18]. Deflecting modes admit a lumped element shunt-circuit representation whose voltage accounts for the transverse kick applied to the particles: $c\Delta p_x = qq_0x_0w_\perp(\tau) \doteq qV_\perp(\tau)$, where the subscript zero refers to source particle quantities. Such a voltage is induced by the passing bunches which perturb the resonator and evolves according to the following matrix equation

$$\begin{pmatrix} V(t+\tau) \\ \dot{V}(t+\tau) \end{pmatrix} = e^{-\alpha\tau} \begin{pmatrix} \cos\omega_n\tau + \frac{\alpha}{\omega_n}\sin\omega_n\tau & \frac{1}{\omega_n}\sin\omega_n\tau \\ -\frac{\omega_0^2}{\omega_n}\sin\omega_n\tau & \cos\omega_n\tau - \frac{\alpha}{\omega_n}\sin\omega_n\tau \end{pmatrix} \begin{pmatrix} V(t) \\ \dot{V}(t) \end{pmatrix} \tag{3}$$

As new bunches enter the cavity, the system is continuously perturbed and, therefore, the matrix-transformation alone does not fully describe the evolution. The correct operator has the form $\mathcal{T}(\cdot) = \mathbf{p} + M(\cdot)$ where the argument is multiplied by the matrix M in equation (3) and the additive vector $\mathbf{p}$ accounts for the perturbation. The latter is given by equation (2) and its derivative in the limit $\tau \to 0$: reminding that $V(\tau) \propto w_\perp(\tau)$ one has $\mathbf{p} = q\omega_n \frac{\omega_0 R_\perp}{Q} \Delta x \begin{pmatrix} 0 \\ 1 \end{pmatrix}$ where Δx is the transverse displacement of the bunch from the axis which is responsible for the excitation of dipole modes. In Ref. [19] the details of such a formalism were discussed showing that it reduces the number of operations needed to evaluate the deflecting kick introduced by the wakefields.

3.1. Benchmark test for the matrix formalism

In the remaining of this section we show an example validating the matrix formalism in presence of long-range wakefields. In particular, we consider a fictitious linac described in [14] by Mosnier in order to demonstrate his analytical model for beam break-up effects. In the approximation of moderate BBU in a single betatron wavelength, Mosnier's analysis allows to find the amplitude of the transverse oscillation for each bunch while it provides an asymptotic solution for strong BBU. Table 1 shows the main parameters characterizing the bunch train, the linear accelerator and the resonant mode responsible for the BBU effects.

Table 1. Parameter list for the comparison with Mosnier's theory.

Parameter	**Value**
Number of bunches	100
Bunch population	$2 \cdot 10^9$
Bunch separation	40 ns
Injection energy	10 GeV
Accelerating gradient	100 MeV/m
Avg betatron function	10 m
Linac length	4900 m
f_0 (HOM)	15.7 GHz
Q (HOM)	100
$R_\perp/Q$ (HOM)	$0.45\,\mathrm{M\Omega/m^2}$

In figure 1, a comparison between Mosnier's analytic approach and our tracking code is shown. In this example each bunch is injected with an offset of $50\,\mu$m and the amplitude of the transverse oscillation at the end of the linac is shown together with the maximum envelope provided by Mosnier's theory.

4. The C^3 linac case

The working point of the C^3 linac is described in [4]. A total charge of 75 nC is delivered within a 250 ns long macro-pulse (1428 rf-bucket) which accommodates 75 electron bunches (1 nC each) with a separation of 19 rf periods ($\sim$ 3.3 ns). The electron beams are injected at 10 GeV and accelerated up to a final energy of $\sim$ 1 TeV with an average gradient of 117 MeV/m (*i.e.*, overall length $\gtrsim$ 12 km) while being focused by a FODO lattice with $\langle\beta_x\rangle = 4$ m.

A study of the higher order modes ringing in the unit accelerating cell was performed with the electromagnetic software CST Studio Suite [20]. A set of 20 modes (excluding degeneracy) was found in the frequency range 5 GHz $< f <$ 20 GHz nine of which possess dipole field components. In the following we investigate the wakefield effects induced by the excitation of such modes and how they affect the beam trajectories.

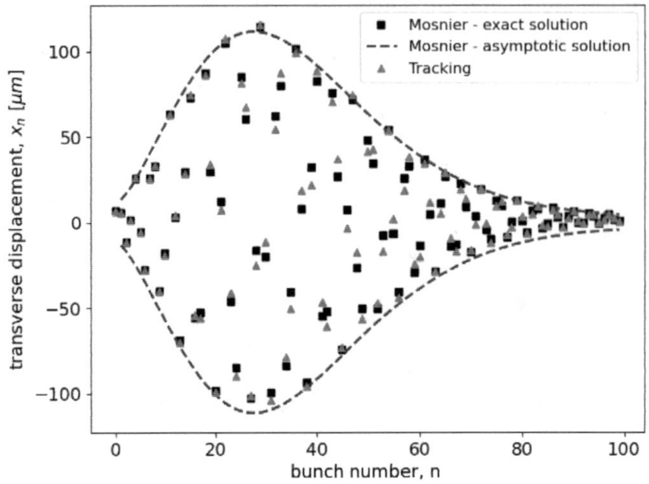

Figure 1. Final amplitude of the transverse oscillations in presence of BBU effects. Comparison with Mosnier's analytical approach.

4.1. Alignment errors
In the example of figure 2 we discuss the effects of injection and alignment errors by comparing the amplitude of the transverse oscillations for each bunch at the end of the linac. The first two curves (blue and red) assume that the accelerating sections are perfectly aligned but all the bunches are injected with a transverse offset $\Delta x = 50$ and $100\,\mu$m respectively. For the last two curves (green and magenta) the injection occurs on-axis but the linac sections are affected by gaussian random offsets with standard deviation $\sigma_{\Delta x} = 50$ and $100\,\mu$m. In order to mitigate statistical fluctuations, the final amplitudes in the latter case are obtained by averaging 50 runs of the tracking code. It can be noticed that in case of random misalignments the dipole modes are continuously excited regardless of the adiabatic damping process and, thus, we observe larger oscillations compared to the case of a simple injection error.

4.2. Detuning techniques
As the previous example has shown, mitigation techniques aimed to keep the amplitude of the transverse oscillations under control are usually necessary. A possible approach is to introduce a frequency spread of the dipole modes. Indeed, slight geometric variations of the unit-cell allow to shift the resonant frequencies of the HOMs without affecting the fundamental mode. Such fluctuations cause each HOM to exhibit a different resonant frequency throughout the subsequent linac sections breaking the coherent interaction with the pulsed beam. In order to apply this technique, a set of seven cavities has been designed from small variations of the original geometry. As an example, figure 3 shows the frequency variations for the TM_{110}-like mode in the unit cell.

Such information can be exploited to introduce a section-by-section frequency spread in our tracking code as we show in figure 4. Here the case of random alignment errors with $\sigma_{\Delta x} = 100\,\mu$m is considered again with the introduction of either uniform or gaussian spread of the HOM frequencies. Breaking the interaction pattern cancels the BBU-effects completely as the displacement of each bunch at the interaction point is close to the amplitude of the first beam.

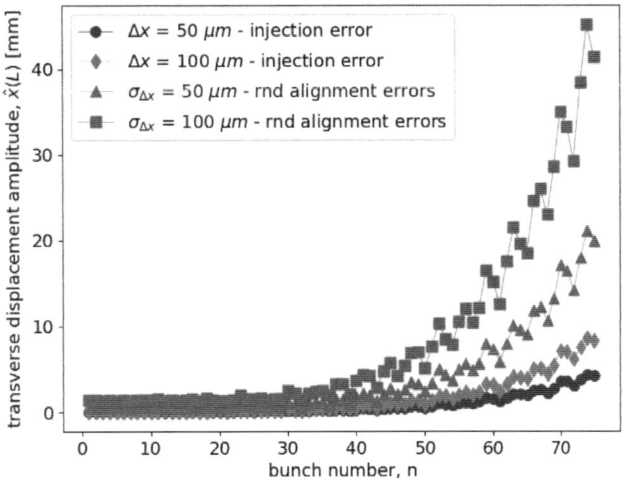

Figure 2. Amplitude of the transverse oscillation for each bunch in the rf-pulse at the interaction point. Both the cases of injection errors and randomly misaligned linac sections are investigated.

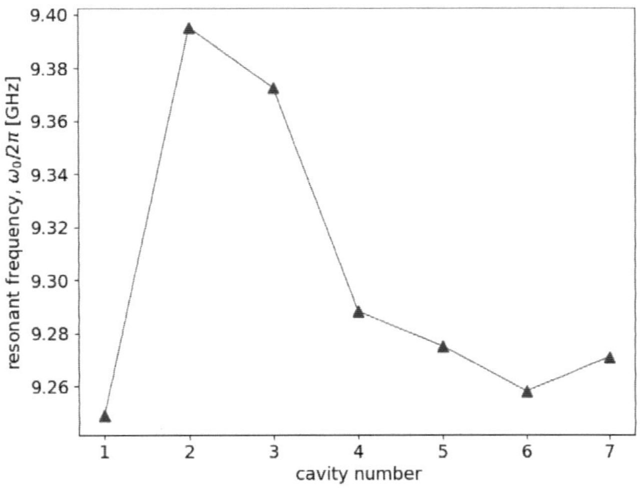

Figure 3. Spread of the resonant frequency of the TM_{110}-like mode for seven geometric variations of the unit-cell cavity.

5. Conclusion

In this paper we introduced a simple and fast tool which allows to investigate BBU effects in linear accelerators operating in multi-bunch mode. In particular, we have studied the performance of a state-of-the-art collider for which the main issues arising from injection and alignment errors were discussed. In addition, mitigation techniques based on a frequency spread of the higher order modes are investigated in order to compensate the beam break-up effects.

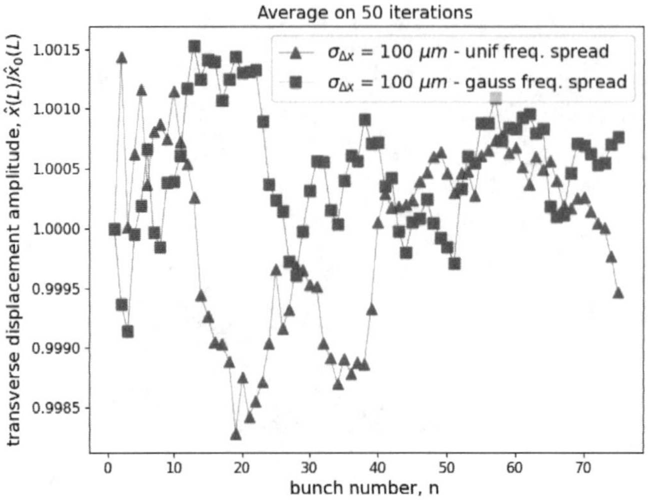

Figure 4. Frequency spread in presence of randomly misaligned sections ($\sigma_{\Delta x} = 100\,\mu$m). The cases of uniform and gaussian spread of the HOMs are investigated and displacements are normalized to the amplitude of the first bunch.

Acknowledgments
This work is supported by DARPA under Contract N.HR001120C0072, by DOE Contract DE-SC0009914 and DE-SC0020409, by the National Science Foundation Grant N.PHY-1549132 and by INFN through the project ARYA.

References

[1] Aicheler M, Burrows P N, Catalan N, Corsini R, Draper M, Osborne J *et al.* 2018 The compact linear collider (CLIC) – project implementation plan Tech. rep. CERN Geneva arXiv:1903.08655, CERN-2018-010-M
[2] Bambade P, Barklow T, Behnke T, Berggren M *et al.* 2019 The international linear collider: a global project *Preprint* ArXiv:1903.01629
[3] Seryi A, Hogan M, Pei S, Raubenheimer T, Tenenbaum P, Katsouleas T *et al.* 2009 A concept of plasma wake field acceleration linear collider (PWFA-LC) Tech. rep. SLAC Stanford SLAC-PUB-13766
[4] Bane K, Barklow T, Breidenbach M *et al.* 2018 An advanced NCRF linac concept for a high energy e^+e^- linear collider *Preprint* ArXiv:1807.10195
[5] Adriani O, Albergo S, Alesini D, Anania M, Angal-Kalinin D, Antici P *et al.* 2014 Technical design report EuroGammaS proposal for the ELI-NP gamma beam system *Preprint* ArXiv:1407.3669
[6] Pedrozzi M *et al.* 2010 SwissFEL injector conceptual design report Tech. rep. PSI Bericht
[7] Tantawi S, Nasr M, Li Z, Limborg C, and Borchard P 2020 *Phys. Rev. Accel. Beams* **23** 092001
[8] Cahill A D, Rosenzweig J B, Dolgashev V A, Li Z, Tantawi S G and Weathersby S 2018 *Phys. Rev. Accel. Beams* **21** 061301
[9] Cahill A D, Rosenzweig J B, Dolgashev V A, Tantawi S G and Weathersby S 2018 *Phys. Rev. Accel. Beams* **21** 102002
[10] Panofsky W K H and Bander M 1968 *Rev. Sci. Instrum.* **39**
[11] Chao A W 1993 *Physics of Collective Beam Instabilities in High Energy Accelerators* (New York, USA: Wiley)
[12] Reiser M 1994 *Theory and Design of Charged Particle Beams* (New York, USA: Wiley)
[13] Palumbo L, Vaccaro V G and Zobov M 1995 Wake fields and impedance *CAS Accelerator School* pp 331–90
[14] Mosnier A 1993 Instabilities in Linacs *CAS Accelerator School*
[15] Bane K L F and Gluckstern R L 1993 *Particle Accelerators* **42** pp 123–69
[16] Thompson K A, Adolphsen C and Bane K L F 1993 *Proc. 15th Particle Accelerator Conf. (PAC'93)* (Washington, DC USA) pp 3342–44

[17] Migliorati M and Palumbo L 2015 *Phys. Rev. Spec. Top. Accel Beams* **18** 031001
[18] Migliorati M, Aumon S, Koukovini-Platia E, Huschauer A, Métral E, Sterbini G and Wang N 2018 *Phys. Rev. Spec. Top. Accel Beams* **21** 120101
[19] Bosco F, Camacho O, Carillo M, Chiadroni E, Faillace L, Fukasawa A *et al.* 2022 A fast tracking code for evaluating collective effects in linear accelerators *Preprint* ArXiv:2208.06466
[20] CST Studio Suite `www.cst.com`

Strong-strong simulations of coherent beam-beam effects in the EIC

J Qiang[1], Y Hao[2], Y Luo[3], C Montag[3], D Xu[3] and F Willeke[3]

[1]Lawrence Berkeley National Laboratory, Berkeley, CA 94720, USA
[2]Michigan State University, East Lansing, MI 48824, USA
[3]Brookhaven National Laboratory, Upton, NY 11973, USA

Email: jqiang@lbl.gov

Abstract. The high luminosity electron ion collider (EIC) will provide great opportunities in nuclear physics study and is under active design. The coherent effects due to the beam-beam interaction of two colliding beams can cause beam size blow-up and degrade the luminosity in the EIC. In this paper, we report on the study of coherent beam-beam effects in the EIC design using self-consistent strong-strong simulations. These simulations show the coherent dipole and quadrupole mode instabilities in the tune working point scan and bunch intensity scan.

1. Introduction
The electron-ion collider (EIC) as a gluon microscope has been approved by the Department of Energy as the next major scientific facility that probes the detailed physics inside the nucleus with deep inelastic scattering using polarized high energy electron [1]. The EIC consists of two colliding rings, a hadron ring of 41-275 GeV and an electron storage ring of 5-18 GeV. The nominal design goal is to attain a peak luminosity of $\sim 10^{34}$/cm2/s. The coherent instability driven by beam-beam interactions could cause beam size blow-up and degrade the peak luminosity [2-5]. Such an instability depends on the choice of transverse tune working points and beam bunch intensities as seen in the following strong-strong simulations.

2. Computation tool
The computational tool used in this study is a self-consistent strong-strong beam-beam code, BeamBeam3D [6,7]. The BeamBeam3D is a parallel three-dimensional particle-in-cell code to model beam-beam effects in high-energy ring colliders. This code includes a self-consistent calculation of the electromagnetic forces (beam-beam forces) from two colliding beams (i.e. strong-strong modeling), a linear and nonlinear high-order transfer map model for beam transport between collision points, a stochastic map to treat radiation damping, quantum excitation, a single map to account for chromaticity effects, a feedback model, an impedance model, and a Bremsstrahlung model. Here, the beam-beam forces can be from head-on collision, offset collision, and crossing angle collision. These forces are calculated by solving the Poisson equation using a shifted integrated Green function method, which can be computed very efficiently using an FFT-based algorithm on a uniform grid. For the crossing angle collision, the particles are transformed from the lab frame into a boosted Lorentz frame following the procedure described by Hirata [8] and by Leunissen et al. [9], where the beam-beam forces are calculated in the same way as the head-on collision. After the collision the particles are transformed back into the laboratory frame. The BeamBeam3D code can handle multiple bunches from each beam collision at

multiple interaction points (IPs). The parallel implementation is done using a particle-field decomposition method to achieve a good load balance.

3. Simulation results

The nominal design parameters used in the simulations are given in the following table.

Table 1. EIC design parameters.

	electron	proton
Energy (GeV)	10	275
Tune	(0.08,0.06)	(0.228,0.21)
Particles (10^{11})	1.72	0.69
Emittance (nm)	(20,1.1)	(11.3,1.0)
(cm)	(55,5.6)	(80,7.2)
Beam-beam para.	(0.088,0.1)	(0.01,0.012)
Chromaticity	(1,1)	(1,1)
Damp time (k turn)	(4,4,2)	
Crab freq. (MHz)	394	197

These design parameters were chosen to produce ~10^{34}/cm2/s peak luminosity for the collision of a 10 GeV electron bunch and a 275 GeV proton bunch with 25 mrad crossing angle [10]. Pairs of crab cavities are used on both sides of the collision point to correct the crossing angle for both colliding beams. The nominal beam-beam parameters for the electron beam are (0.088, 0.1), and (0.01, 0.012) for the proton beam. The large beam-beam parameters result in strong coherent beam-beam effects. Such effects can cause instability due to the collective dynamic interactions between the electron beam and the proton beam.

In the EIC design, the transverse tune working points have to be carefully selected. Figure 1 shows the proton beam and electron beam horizontal centroid evolution with different choice of electron beam horizontal and vertical tunes. In this scan, the proton beam tune working point is fixed. It is seen that for the electron horizontal tune between 0.1 and 0.14, both electron and proton beam centroid become unstable.

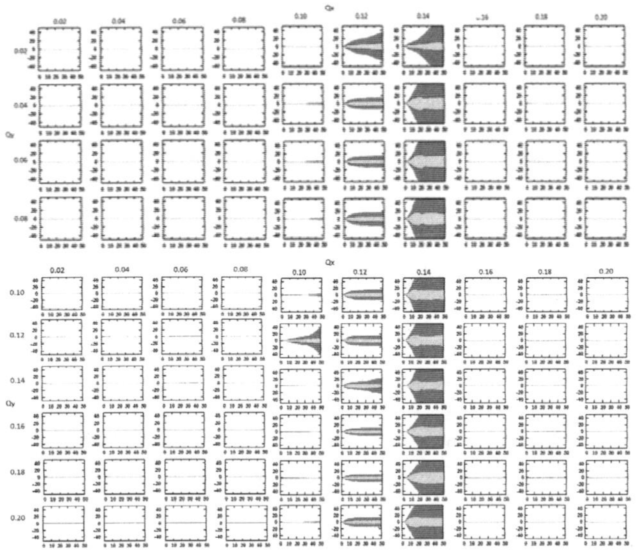

Figure 1. Proton beam (red) and electron beam (green) horizontal centroid evolution with electron beam horizontal tune (Qx) and vertical tune (Qy) scan.

To study the mechanism underlying these instabilities, we selected one electron tune working point (0.1,0.1). Figure 2 shows spectra of electron beam horizontal centroid evolution during three time windows: first 4000 turns, after 2000 turns, another 3000 turns. It is seen that spectrum amplitude of a single mode around 0.24 grows through three windows while the other mode amplitudes stay about the same. This suggests that this single becomes unstable and results in beam centroid blow-up.

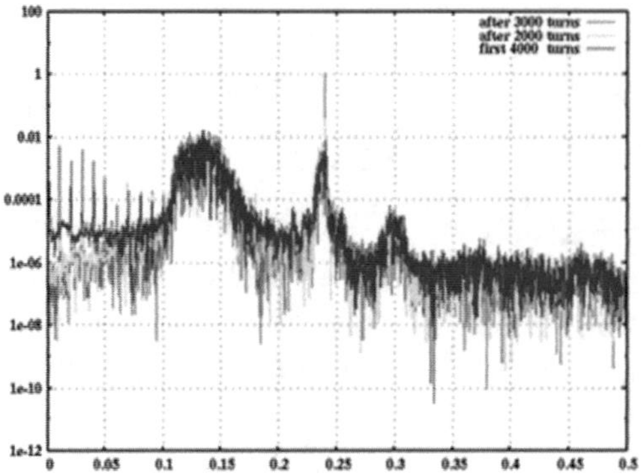

Figure 2. Spectra of electron centroid evolution during three periods of evolution.

Figure 3 shows the electron beam and proton beam horizontal centroid spectra with six electron horizontal tunes and one fixed vertical tune. It is seen that as the electron horizontal tune changes, the coherent electron mode frequency moves, while the coherent mode frequency from the proton beam stays the same. The electron beam becomes unstable only for the horizontal tunes between 0.1 and 0.14. This suggests a coupling resonance between the coherent electron mode and the coherent proton mode.

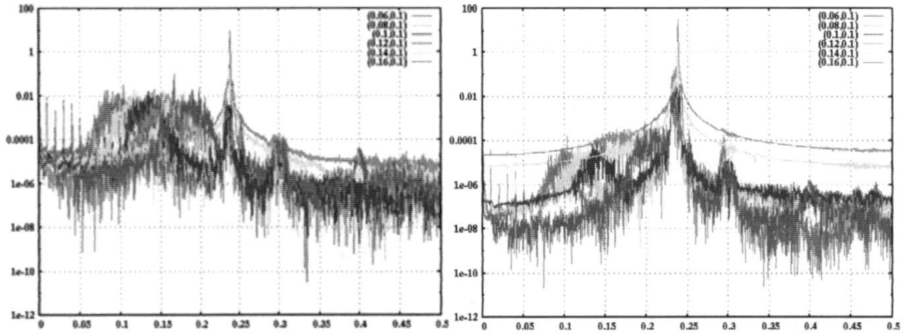

Figure 3. Electron (top) and proton (bottom) beam horizontal centroid spectra with different electron horizontal tunes

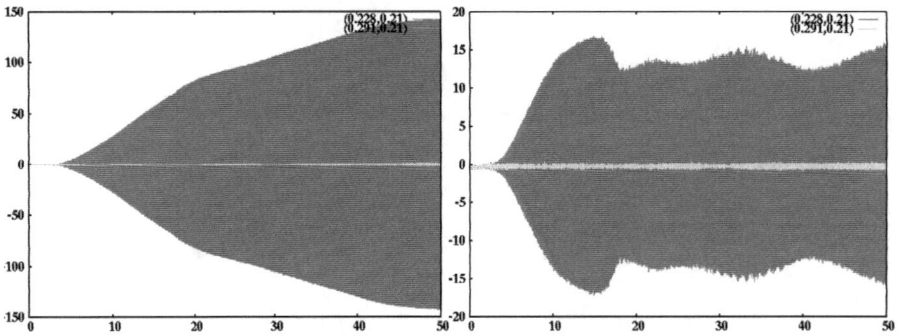

Figure 4. Proton beam (top) and electron beam (bottom) horizontal centroid evolution with two proton beam horizontal tunes.

The above coherent dipole mode instability can be avoided by moving the proton beam working point. Figure 4 shows the proton beam and electron beam horizontal centroid evolution with two proton beam horizontal tunes. It is seen that by moving the proton beam tune away from the resonance, both beams become stable.

Besides scanning the electron beam transverse working points, we also scanned the proton beam transverse tunes. Figure 5 shows the peak luminosity as a function of proton beam horizontal and vertical tunes. It is seen that the good luminosity stays below the diagonal line of the tune space. Near the diagonal line, coherent coupling resonance occurs and causes the degradation of luminosity.

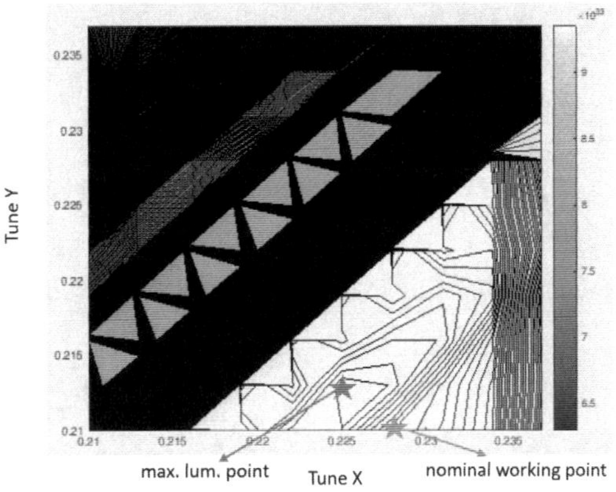

Figure 5. Peak luminosity as a function of proton beam horizontal and vertical tunes.

Coherent instability also occurs during the beam intensity scan. Figure 6 shows electron beam and proton beam RMS size evolution with several proton beam intensities. Here, 1 means the nominal proton beam intensity, 0.8 means 0.8 times the nominal intensity. It is seen that when the proton beam lowers to 0.4 times the nominal proton intensity, both the electron beam and the proton beam become unstable.

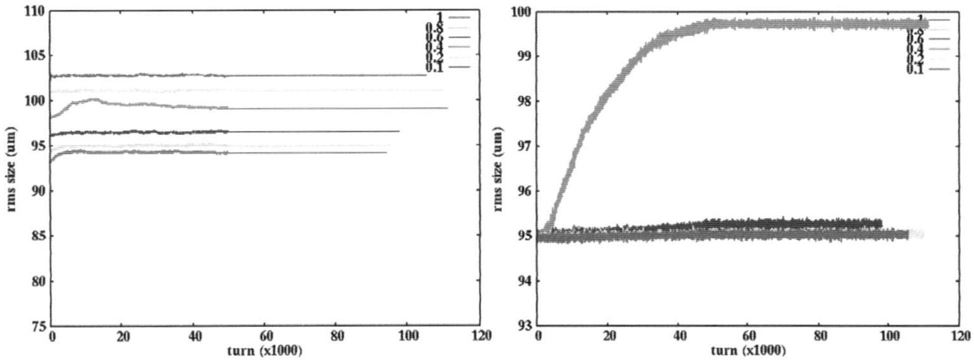

Figure 6. Electron beam (top) and proton beam (bottom) horizontal RMS size evolutions with several proton beam bunch intensities.

To understand the mechanism of this instability, we calculated the tune footprints of the electron beam in figure 7. It is seen that as the proton beam intensity decreases, the beam-beam forces become weaker, and the electron beam tune footprint become smaller. There is no particular lower order resonance for the x0.4 proton intensity. This suggests that instability should not be a dipole mode instability.

Figure 8 shows the spectra of electron and proton beam horizontal RMS size evolution through these proton beam intensity scan. It appears that for x0.4 proton intensity the electron beam quadrupole mode hits the 5th order resonance with too small tune spread to stabilize the beam. Further lower the proton intensity moves away from the resonance, while further increase the proton intensity results in a larger tune spread and stabilize the instability through the Landau damping.

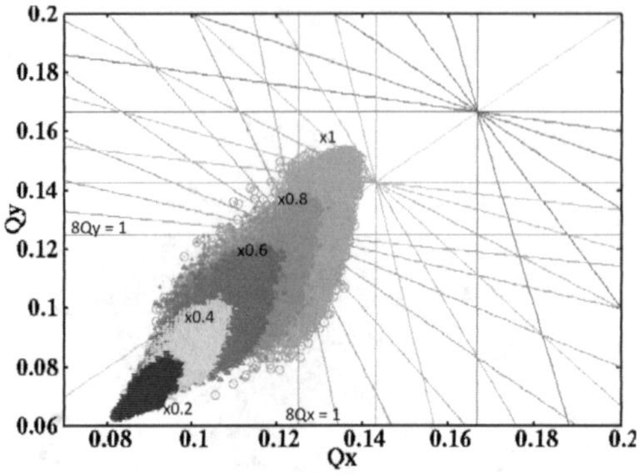

Figure 7. Electron beam tune footprint with several proton beam bunch intensities.

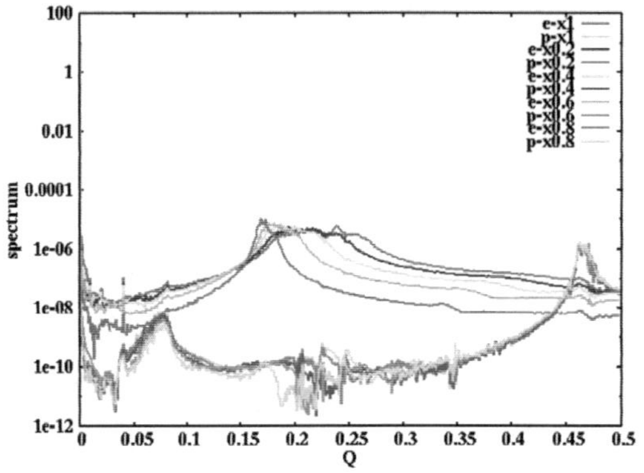

Figure 8. Spectra of electron beam and proton beam horizontal RMS size evolution with several proton beam bunch intensities.

We also scanned electron beam intensity as shown in figure 9. It is seen that proton beam emittance growth rate increases quickly after the electron intensity attains 3.5 x10^{11}. This is due to the large proton beam-beam parameter (~0.019) across the 4th order coherent resonance and results in large emittance growth.

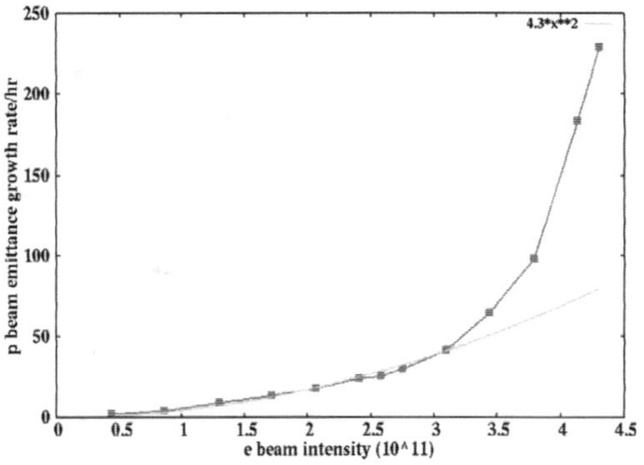

Figure 9. Proton beam emittance growth rate as a function electron beam bunch intensity.

4. Summary

In this study, using the self-consistent strong-strong beam-beam simulations, we observed coherent instabilities of both dipole and quadrupole modes in the transverse tune working point scan and the beam intensity scan. These instabilities can be avoided by appropriately choosing working points and beam intensity parameters.

Acknowledgments

Authors wishing to acknowledge discussions with Dr. M. blaskiewicz in this study. This work was supported by the Director of the Office of Science of the US Department of Energy under Contract no. DEAC02-05CH11231and used computational resources at National Energy Research Scientific Computing Center.

References

[1] Wang J B ed. 2021 Electron-Ion-Collider at Brookhaven National Laboratory Conceptual Design Report
[2] Chao A W and Ruth R D 1985 *Particle Accelerators* **16** 201
[3] Hao Y and Ptitsyn V 2010 *Phys. Rev. Spec. Top. Accel Beams* **13** 071003
[4] White S, Buffat X, Mounet N and Pieloni T 2014 *Phys. Rev. Accel. Beams* **17** 041002
[5] Ohmi K, Kuroo N, Oide K, Zhou D and Zimmermann F 2017 *Phys. Rev. Lett.* **119** 134801
[6] Qiang J, Furman M A and Ryne R D 2004 *J. Comp. Phys.* **198** 278
[7] Qiang J, Furman M and Ryne R 2002 *Phys. Rev. Spec. Top. Accel Beams* **5** 104402
[8] Hirata K 1995 *Phys. Rev. Lett.* **74** 2228
[9] Leunissen L H A, Schmidt F and Ripken G 2000 *Phys. Rev. Spec. Top. Accel Beams* **3** 124002
[10] Luo Y *et al.* 2021 *Proc. 12th Int. Particle Accelerator Conf. (IPAC'21)* (Campinas, Brazil) pp 3808–11

The report of machine studies related to the vertical beam size blow-up in SuperKEKB LER

S Terui[1], T Ishibashi[1], K Ohmi[1], Y Funakoshi[1], H Fukuma[1], R Ueki[1], Y Ohnishi[1], M Tobiyama[1] and T Nakamura[2]

[1] High Energy Accelerator Research Organization (KEK), Tsukuba, Japan
[2] High Energy Accelerator Research Organization (KEK) J-PARC Center, Tokai, Japan

E-mail: sterui@mail.kek.jp

Abstract. In the low energy ring (LER) for positrons in the SuperKEKB, a vertical beam size blow-up was observed when the bunch current was approximately 1 mA. If a beam size blow-up occurs, the design luminosity cannot be achieved. Therefore, beam size blow-ups must be pre-vented. According to calculations, the bunch current threshold of the transverse mode coupling instability (TMCI) is 2 mA or more, and the observed value is 50% or smaller. Ordinary TMCI cannot explain this vertical beam size blow-up. This paper shows that the cause of the vertical beam size blow-up can be determined by analyzing factors such as beam oscillation. The study results showed that the vertical beam size blow-up in the LER was caused by a -1 mode instability.

1. Introduction

SuperKEKB is a high-luminosity electron-positron collider with asymmetric energies of 7 GeV (electron, high energy ring: HER) and 4 GeV (positron, low energy ring: LER) [1]. SuperKEKB adopts a nanobeam scheme in which the vertical beta function (β_y^*) at the collision point is squeezed to the limit. The design value of β_y^* in the LER is 0.27 mm [2]. The minimum value of β_y^*= 0.8 mm was achieved in 2022.

In the 2020c run (October 19, 2020–December 18, 2020), a vertical beam size blow-up was observed when the bunch current was increased. The vertical beam size blow-up subsided when the vertical collimator gap was widened, even under the same bunch current conditions [3]. This observation established that the instability causing the vertical beam size blow-up is related to the wake generated by the vertical collimator. The impedance of the collimator was calculated using GdfidL [4], assuming a bunch length of 6 mm. The following measurements were performed at β_y^*=1 mm in both the LER and HER, under the condition that the sum of the vertical beta function at the collimator location multiplied by the kick factor of the vertical collimator in the LER is approximately 31 × 10^{15} V/C, with fewer bunches (30 - 393) than the number of bunches under normal physics experiments (about 1600) to reduce the effect of multi-bunch instability and the synchrotron tune (ν_s) in the LER is 0.0227.

2. Vertical beam size blow-up in the LER during single beam condition

The beam sizes were measured using an X-ray beam size monitor [5]. Figure 1 shows the bunch current on the horizontal axis and vertical emittance on the vertical axis. The orange plots were measured with nominal settings of the bunch-by-bunch feedback system (BBF) [6], which are set the same way as in normal physics experiments. The setting of the BBF for vertical oscillation during normal physics experiments is intended to dump the frequency of the 0 mode (vertical betatron tune: ν_y). The points circled by the black dashed line in figure 1 indicate that the vertical beam size blow-up occurs when the bunch current is about 0.9 mA or higher. The BBF and vertical beam size blow-up relationship will be explained later.

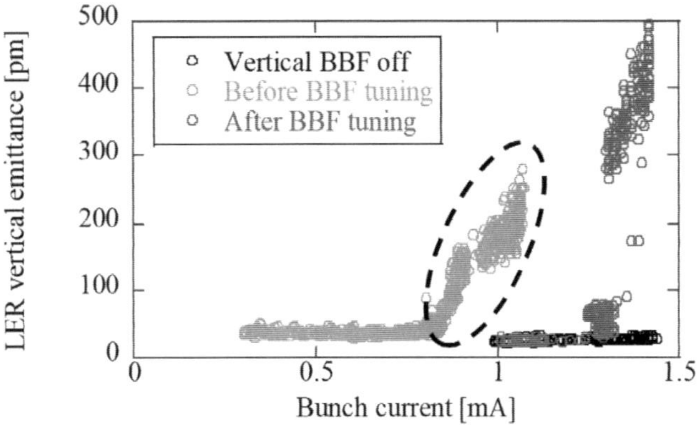

Figure 1. The LER vertical emittance before and after BBF tuning and BBF off versus bunch current.

To measure the tune, a pilot bunch is excited at varying frequencies near the betatron tune. The tune shift was calculated using PyHEADTAIL [7] and the fast Fourier transform (FFT) analysis of the vertical motion in the LER was conducted using the bunch oscillation recorder (BOR) [8], as shown in figure 2. The orange and light blue dashed lines in figure 2 show the frequencies of the 0 mode and -1 mode ($\nu_y - \nu_s$), respectively, with the bunch current on the horizontal axis. ν_y at a bunch current of 0 mA is denoted as ν_{y0}. ν_{y0} is 0.5977 during the measurement in figure 2.

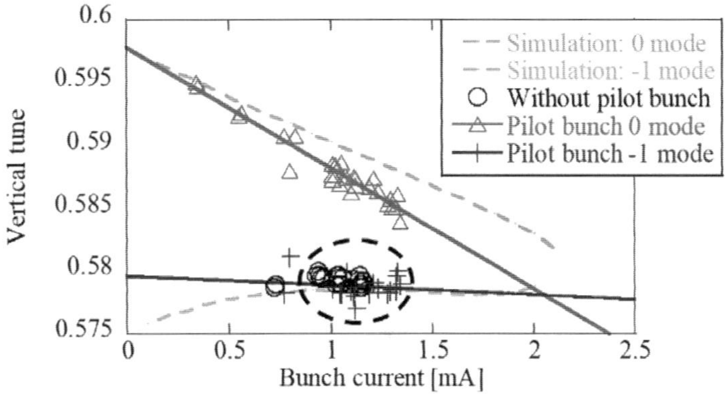

Figure 2. FFT peaks of the LER vertical oscillation versus the bunch current.

The measurement results showed different spectra in the oscillation of the pilot bunch and the non-pilot bunches. The FFT analysis of the pilot bunch revealed two high peaks. These two peaks, corresponding to the 0 mode in red and the -1 mode in blue, are plotted in figure 2. The tune shift of the measured value (red line) differs from the calculated value (orange dashed line) by ap-proximately 20 % in the slope. This difference may be due to an impedance source not yet included in the calculations.

The results of the FFT analysis for the non-pilot bunches showed no peaks at a low bunch current and only one high peak at a high bunch current. The frequency of the single peak point at the high bunch current for the non-pilot bunches is plotted in black in figure 2. This peak was found to have nearly the same frequency as that of the -1 mode for the pilot bunch. At low bunch currents, the peak of the -1 mode for the pilot bunch cannot be measured. Therefore, in the region of low bunch currents, the regression curves of measured values (blue line) do not match simulation results (light blue dashed line).

The measurements circled by the black dashed lines in figure 1 and figure 2 were taken simultaneously. A single peak for the non-pilot bunches in figure 2 was observed around the time when the vertical beam size blow-up started. Interestingly, figure 2 shows that transverse mode coupling instability (TMCI) theory [9] predicts that the vertical beam size blow-up would occur at a bunch current of over 2 mA (the point where the blue and red lines cross), but our experiment showed that it occurred at a bunch current of approximately 0.9 mA.

On reading reference [10], we thought if there was a relationship between the vertical beam size blow-up and BBF. The BBF parameters (number of taps and phases) were changed during our study, and measurements were performed.

Figure 1 also plots the beam size when the BBF is tuned to not enhance the oscillation of the -1 mode as much as possible (hereafter referred to as "BBF tuning") and when the BBF is turned off (BBF off). The bunch current threshold of the vertical beam size blow-up was found to be higher with BBF off (black) or after BBF tuning (red) than before BBF tuning (orange). By the way, the number of bunches cannot be increased due to multi-bunch instability with BBF off.

Next, the growth rate of the dipole oscillation was measured with the BBF off. Figure 3 shows the results of the FFT analysis of the vertical motion in the LER for the pilot bunch (lower plot of figure 3) and the non-pilot bunches (upper plot of figure 3) when the growth rate was measured. We found that even with the BBF off, there was a high peak corresponding to the -1 mode. ν_{y0} is 0.5900 during the measurements in figure 3.

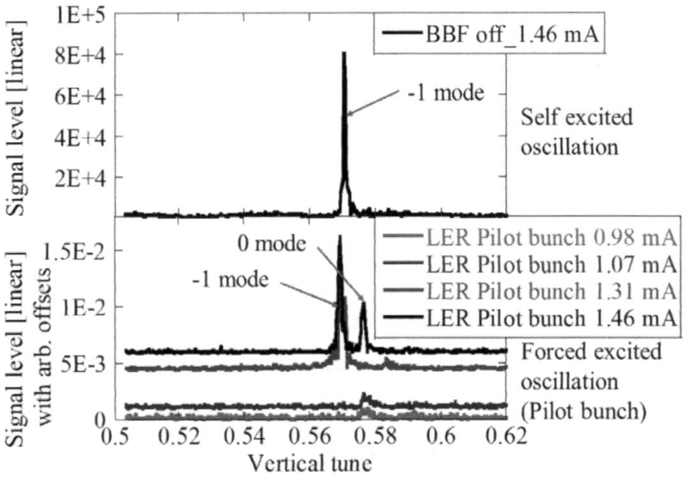

Figure 3. Results of the FFT analysis of the vertical motion in the LER.

Figure 4 shows three bunches each with time on the horizontal axis and vertical amplitudes on the vertical axis. There are bunches with small and large amplitudes. The independent oscillation of other bunches indicates that it is a single-bunch instability. It can also be seen that some bunches grow and then dampen after the oscillations have grown.

During normal physics experiments in the SuperKEKB LER, the accelerator is operated at a chromaticity of about 1.5 units in both planes. Figure 5 shows the growth rate of head-tail instability by chromaticity as a dashed orange line. The growth rate is obtained by calculating the eigenvalues and eigenmodes of a single-bunch oscillation [11]. The black circles in figure 5 represent the measurement points. Figure 5 shows that the measured and simulated results exhibit similar trends.

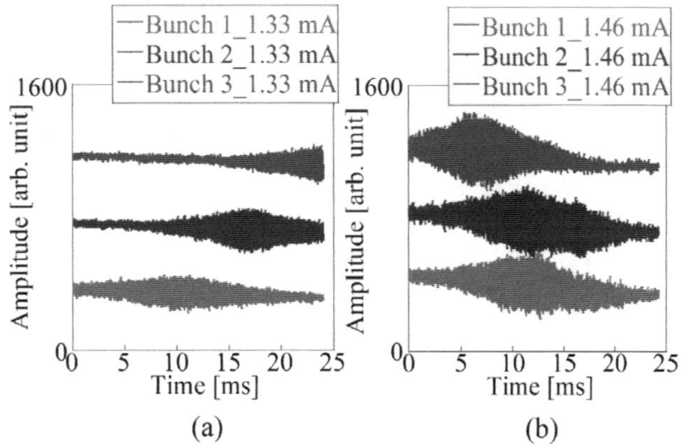

Figure 4. Vertical oscillations in the LER at a bunch current of (a) 1.33 mA and (b) 1.46 mA during growth rate measurement.

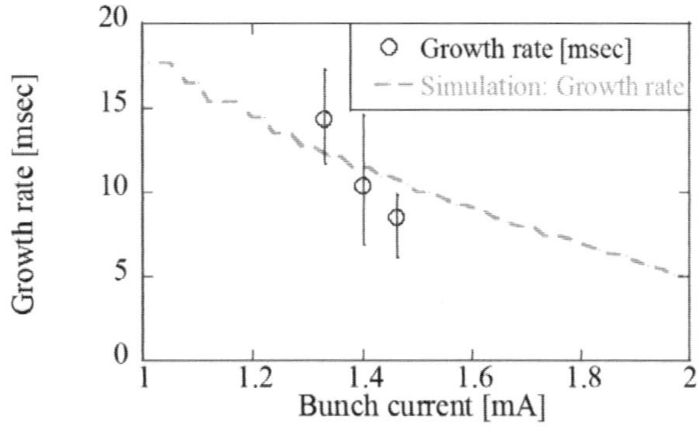

Figure 5. Measured and simulated results of growth rate.

3. Vertical beam size blow-up in the LER during collision condition

We wondered whether the beam-beam effect might mitigate the vertical oscillation in the LER during the collision condition. The following measurements were performed after the BBF tuning.

The lower plot of figure 6(a) and (b) shows the results of the FFT analysis of the vertical motion in the LER for the non-pilot bunches during the collision and single-beam conditions. The maximum bunch current is 1.43 mA in the LER and 0.97 mA in the HER for this study. The results of the FFT analysis of the vertical motion in the HER are also plotted in the upper plot of figure 6(a) under the collision condition. The HER data are included in figure 6(a) because the vertical tune in the HER (dashed black line) affects the results of the FFT analysis of the vertical motion in the LER in the collision condition (the lower plot of figure 6(a)).

In the upper plot of figure 6(b), the results of the FFT analysis of the vertical motion for the pilot bunch in the LER are plotted together for the 0 and -1 modes oscillations and can be distinguished. The frequency of the -1 mode for the non-pilot bunches are the same as the frequency of the -1 mode for the pilot bunches. Comparing the blue circles in the lower plots of figure 6(a) and (b), the signal peaks for the single-beam condition are higher than any other peak in the collision condition. The blue circle is the point where the tune shifted due to high bunch current (about 1.4 mA). From these measurements, it can be inferred that the beam-beam effect mitigates the oscillation in the vertical motion in the LER. ν_{y0} is 0.5890 during the measurement in figure 6.

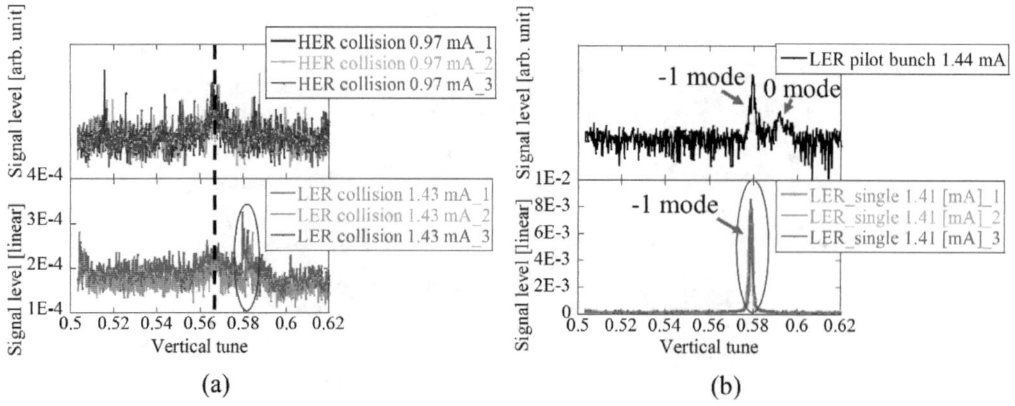

Figure 6. Results of the FFT analysis of the vertical motion during (a) the collision and (b) the single-beam conditions.

Figure 7 shows (a) the specific luminosity and (b) the vertical emittance during the high bunch current collision. Figure 7(b) shows that after BBF tuning, vertical emittance blow-up in the LER is less likely to occur even under high bunch current conditions. For this reason, after BBF tuning, it was possible to perform the collision study at a higher bunch current, which was not possible before BBF tuning, as shown in figure 7(a).

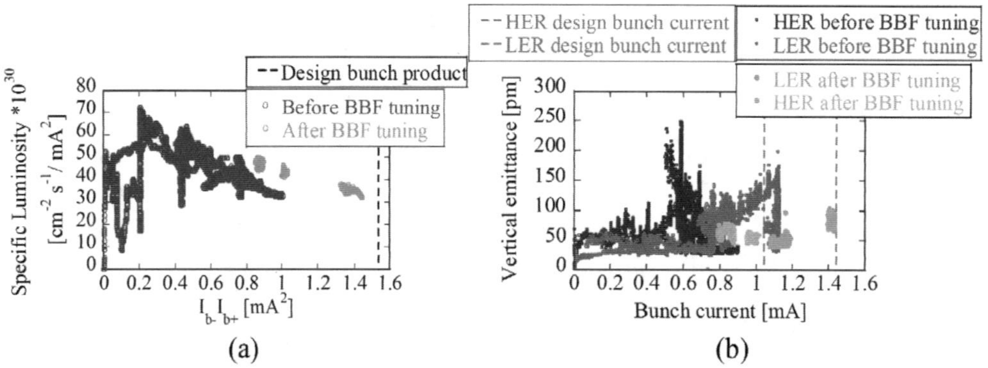

Figure 7. (a) Specific luminosity (b) vertical emittance before and after BBF tuning during high bunch current collision study.

4. Conclusion and future work

The conclusions of the main measurements are as follows:

- The vertical beam size blow-up in the LER during the single-beam condition occurred when the bunch current was higher than approximately 0.9 mA and was attributed to oscillations caused by the -1 mode.
- It was found that the threshold of the beam size blow-up could be increased by BBF tuning such that the oscillation of the -1 mode is not enhanced, as much as possible. After BBF tuning, the collision could be performed at the designed bunch current for the first time.
- From the collision and single-beam condition measurements, it could be inferred that the beam-beam effect mitigates the vertical oscillation in the LER caused by the -1 mode.

Future studies will determine why the oscillation caused by the -1 mode is larger than that caused by the 0 mode when the bunch current increases.

Acknowledgments
We thank all the staff of the SuperKEKB accelerator and Belle II detector and Prof. Mauro Migliorati and Dr. K. Oide for their cooperation.

References

[1] Ohnishi Y *et al.* 2013 *Prog. Theor. Exp. Phys.* **2013** 3
[2] Ohnishi Y *et al.* 2021 *Eur. Phys. J. Plus* **136** 1023
[3] Terui S *et al.* 2021 *Proc. 12th Int. Particle Accelerators Conf. (IPAC'21)* (Campinas, Brazil) pp 3537-40
[4] The GDFIDL electromagnetic field simulator http://www.gdfidl.de
[5] Flanagan J W *et al.* 2008 *Proc. 11th Eur. Particle Accelerators Conf. (EPAC'08)* (Genoa, Italy) pp 1029-31
[6] Tobiyama M 2016 Beam Instrumentation and bunch feedback systems Presented at 21th KEKB Accelerator Review Committee https://www-kekb.kek.jp/MAC/2016/Report/Tobiyama.pdf
[7] PyHEADTAIL code repository https://github.com/ Py-COMPLETE/
[8] Tobiyama M and Flanagan J W 2012 *Proc. 1st Int. Beam Instr. Conf. (IBIC'12)* (Tsukuba, Japan) pp 138-42
[9] Transverse mode coupling instability (TMCI) *Handbook of Accelerator Physics and Engineering* 3rd printing ed A W Chao and M Tigner (Singapore: World Scientific)
[10] Tobiyama M and Ohmi K 2011 *Proc. 10th Eur. Workshop on Beam Diag. and Instr. for Particle Accelerators Conf. (DIPAC'11)* (Hamburg, Germany) pp 212-14
[11] Kuroo N, Ohmi K, Oide K, Zhou D and Zimmermann F 2018 *Phys. Rev. Accel. Beams* **21** 031002

Lattice design of the UVSOR-IV storage ring

E Salehi[1], Y Taira[1], M Fujimoto[1,2], L Guo[2] and M. Katoh[1,3]

[1] Institute for Molecular Science, Okazaki, Japan
[2] NUSR, Nagoya University, Nagoya, Japan
[3] Hiroshima University, Higashi-Hiroshima, Japan

Email: elham@ims.ac.jp

Abstract. We are designing a storage ring lattice for the future plan of UVSOR. As a candidate, we have designed a storage ring of 1 GeV electron energy, which is higher than the present value, 750 MeV. The magnetic lattice is based on a compact double bend achromat cell, which consists of two bending magnets and four focusing magnets, all of which are of combined function. The circumference is 82.5 m. The emittance is 4 nm in the achromatic condition, which becomes lower in the non-achromatic condition. The lattice has moderately large dynamic aperture with four sextupole families. The lattice of 6-fold symmetry has six straight sections of 4 m long and six of 1.5 m long. Undulators can radiate nearly diffraction-limited light in VUV. If we install high field multipole wigglers at the short straight sections, they can provide high flux tender X- rays. We are expecting usage of a laser-based accelerator as the injector, which might be developed in the next decade. As an alternative plan, we have designed a traditional injector, which consists of a linear accelerator and a booster synchrotron that can be constructed inside of the storage ring.

1. Introduction

UVSOR is a low energy synchrotron light source, which had been operated since 1983. After two major upgrades [1-4], now it is called UVSOR-III. The circumference of the storage ring is 53 m and the electron beam energy 750 MeV. It has 8 straight sections and six of them are occupied with undulators of various kinds. One straight section is used for beam injection and another for RF acceleration. It has a moderately small emittance of about 17 nm and provides vacuum ultraviolet light of high brightness.

Nowadays, to meet the demand of diffraction limited light beam in the vacuum ultraviolet and x-ray range from scientific community, several synchrotron light sources, which have exceedingly small emittance less than 1nm, are under consideration, construction, or in operation [5-7]. In such a situation, we have started considering a future plan for UVSOR with an emittance smaller than at least a few nm to provide diffraction-limited light in the vacuum ultraviolet range. As the first step of the investigation, we have analyzed the present magnetic lattice of UVSOR based on tie diagram to explore the possibility to get a lower emittance with some minor changes in the configuration of magnets. We have found a few optics which has significantly (but not drastically) smaller emittance around 10 nm than the present value, 17 nm which may be useful for some special experiments [8]. To reach the low emittance around a few nm, we have started designing a totally new storage ring. The storage ring has a higher electron energy, 1 GeV and a larger circumference, 82.5 m. The lattice consists of twelve double bend cells. Among twelve straight sections, two sections will be used for the injection and RF cavity, and ten sections will be used for insertion devices.

In this paper, we will describe the design of the new lattice for UVSOR-IV and the related beam dynamics studies.

2. Lattice design

The lattice has been designed based on a compact double bend achromat cell (DBA), which consists of two bending magnets and four focusing magnets, all of which are combined function magnets. Two sextupole families are located in between two combined dipoles for the chromaticity correction and two harmonic sextupole families are also employed to correct the high order geometric aberrations. This lattice has twelve DBA cells with six long straight sections about 4 m and six short straight sections around 1.5 m long. These lengths are same as those of UVSOR-III. This may enable us to use the undulators at UVSOR-III in the new ring.

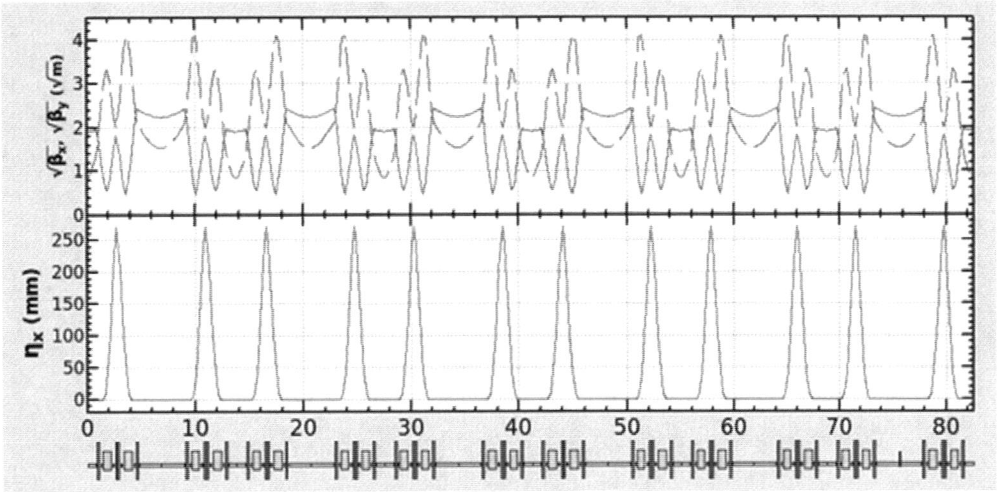

Figure 1. Lattice function of 1 GeV storage ring for UVSOR-IV. The blue and red lines show the horizontal and vertical betatron tune, respectively.

The lengths of the straight sections can be longer if the larger circumference is allowed. It is noted that this lattice has a flexibility on the dispersion function at the straight sections, which enables to realize various operation modes such as of achromatic, lower emittance, or isochronous.

A tune survey was performed to find the linear optics with a low emittance and appropriate optical functions. ELEGANT [9] was used for the calculations. The emittance is around 4 nm in the achromatic condition, which becomes lower in the non-achromatic condition. Figure 1 shows the lattice functions in the achromatic condition. The major parameters are listed in table 1 and are compared with those of UVSOR-III.

Table 1. Lattice parameters.

	UVSOR-III	**UVSOR-IV**
Electron energy	750 MeV	1 GeV
Stored current	300 mA	300 mA
Circumference	53.2 m	82.5 m
RF frequency	90.1 MHz	89.4 MHz
Harmonic num.	16	26
RF voltage	120 kV	100 kV
Coupling ratio	0.01	0.01
Energy spread	0.000526	0.000576
Emittance	16.9 nm	4.3 nm
Betatron tunes	(3.75, 3.20)	(8.63, 3.19)

It is noticeable that if we occupy three of short straight sections with 2 T multipole wigglers in symmetry, the energy loss due to the radiation in the wigglers enhances the damping effect and the emittance reduces to 3.7 nm. These wigglers can provide tender X-ray, which are hardly accessible at the present UVSOR-III.

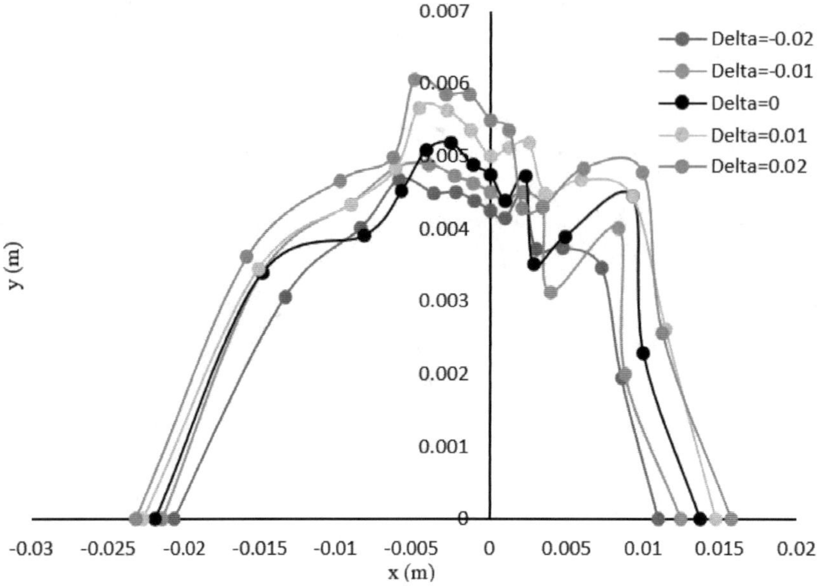

Figure 2. Dynamic aperture of optic for different energy deviation. The particle tracking has been run for 1024 turns passing through the storage ring.

2.1. Dynamic aperture
In a low emittance lattice, strong quadrupoles are generally employed which result into a large negative chromaticity. For chromaticity correction, strong sextupoles are needed. Due to their nonlinear effects, the dynamic aperture, which corresponds to the maximum betatron amplitude where electrons remain bounded, decreases. It should be, anyway, large enough for the beam injection and storage. We use ELEGANT [9] to optimize the dynamic aperture without considering multipole errors and misalignment errors. The strengths of the harmonic sextupoles are selected for maximizing the dynamic aperture. Figure 2 shows the dynamic aperture for on- and off energy (+/- 1,2%) particles at the straight section, which is calculated by tracking 1024 turns. In this optics, the horizontal aperture for the on-momentum electron is sufficiently large, about -23 to 15 mm and vertical aperture is about 6 mm. Further study is ongoing to realize a larger dynamic aperture for off- momentum particles and also with various machine errors.

3. Touschek lifetime and intra-beam scattering effects
The electron beam lifetime had been one of the most important parameters for a synchrotron light source. In recent years, the so called top-up injection is widely employed and the beam lifetime is less important than be-fore. However, to reduce the beam intensity fluctuation and to keep the radiation level sufficiently low for the safety regulation, the lifetime is still an important parameter. Touschek scattering, in which the electrons in a same bunch scatter each other, get large longitudinal momentum, come out of the momentum aperture and get lost, dominates the lifetime in low emittance storage rings, particularly of low energy. Figure 3 shows Touschek life-time vs. RF voltage. The calculation of Touschek lifetime is based on the momentum acceptance (MA). The relevant MA is the minimum of the RF MA (RF bucket height) and the lattice MA. The RF MA is given by the RF voltage and it is independent of the location in the lattice. The local lattice MA which depends on where the scattering

event occurred and varies along the lattice, is obtained with tracking by ELEGANT. It can be seen from figure 3 that the lifetime increases with the RF voltage and then saturates as electrons are lost due to exceeding the lattice MA during their post-scattering trajectories.

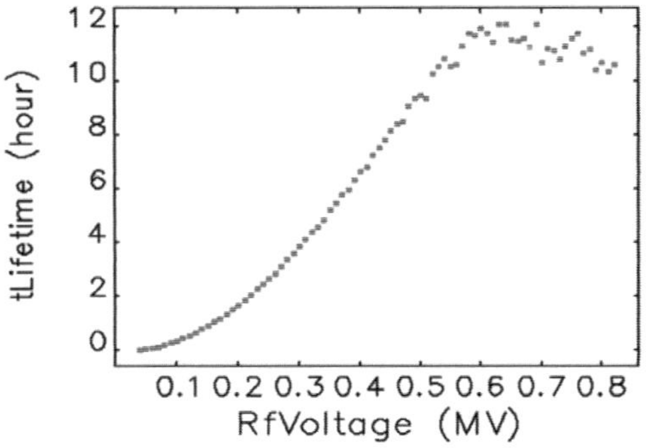

Figure 3. Touschek lifetime vs RF voltage.

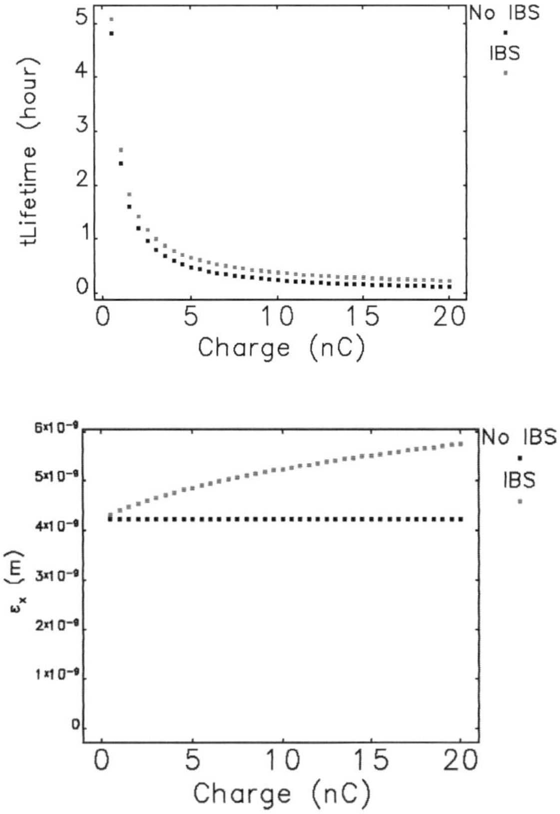

Figure 4. IBS effect on Touschek lifetime (top) and longitudinal emittance (bottom) versus bunch charge.

Intra-beam scattering (IBS), which is the result of the multiple small-angle Coulomb scattering of charged particles in the beam, can increase beam emittance. Figure 4 shows the Touschek lifetime (top) and the emittance (bottom) versus bunch charge with IBS effect (red) and without IBS effect (black). From figure 4 (top), we see that IBS effect increases the Touschek lifetime, while the emittance is increased as shown in figure 4 (bottom). In single-bunch mode, the new lattice design would have a bunch charge of about 11.6 nC, and the Touschek lifetime would be of about one hour. The increase of the emittance due to IBS effect is around 26 % that is far from the target emittance. Fortunately, there are some techniques that can reduce its impact, such as increasing the transverse coupling using some designated skew quadrupoles. Another technique is the addition of a harmonic cavity to produce a bunch lengthening, which is routinely used at UVSOR-III [10]. To check the possibility of using harmonic cavity, we run the simulation code with different values of bunch length. Figure 5 shows the result of IBS effect on the emittance versus bunch length for bunch charge corresponding to 11.6 nC. The plot shows that the lengthening of the bunch due to harmonic cavity up to 80 mm can decrease the effect of IBS on the emittance to an acceptable level with to $\varepsilon_0 = 4.3$ nm.

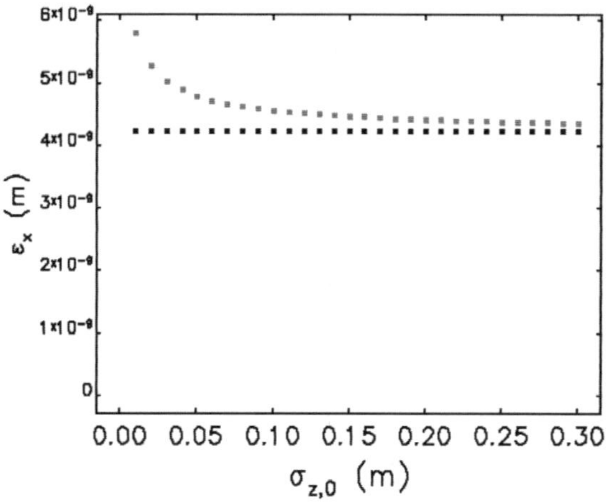

Figure 5. Longitudinal emittance vs bunch length with IBS (red) without IBS (black) at the bunch charge of 10.1nC.

4. Conclusion

A new 1 GeV storage ring lattice with a circumference of 82.5 m has been proposed to upgrade the UVSOR synchrotron facility. The lattice has been designed based on DBA cell to have a low emittance. Optimising the magnet arrangement and the working point, a small natural emittance 4.2 nm is obtained in the achromatic condition. The lattice has twelve straight sections, ten of them are available for insertion devices. Occupying 3 short straight sections with 2 T multipole wigglers can reduce the natural emittance to 3.6 nm-rad. Moreover, these wigglers provide intense tender X-rays.

Dynamic aperture was studied for on- and off-momentum particles without machine errors. Our simulations show that IBS effect on the beam emittance is significant. One possible technique to decrease the IBS effect is lengthening the bunches by using harmonic cavity. Our simulations shows that the increasing of the bunch length by 80 mm can reduce IBS effect on the emittance to an acceptable level.

Acknowledgements

The authors thank Prof. S. Kera, the facility director, for his support and other facility staffs for their cooperation.

References

[1] Katoh M, Hayashi K, Honda T, Hori Y, Hosaka M, Ki-noshita T, Kouda S, Takashima Y and Yamazaki J 2001 *Nucl. Instrum. Methods Phys. Res., Sect.* A **467** 68
[2] Katoh *et al.* 2004 *AIP Conf. Proc.* **705** 49
[3] Adachi M, Zen H, Konomi T, Yamazaki J, Hayashi K and Katoh M 2013 *J. Phys. Conf. Ser.* **425** 042013
[4] Katoh M *et al.* 2010 *AIP Conf. Proc.* **1234** 531
[5] Eriksson M *et al.* 2016 *Proc. 7th Int. Particle Accelerator Conf. (IPAC'16)* (Busan, Korea) pp 11-15
[6] Wanzenberg R *et al.* 2019 *AIP Conf. Proc.* **2054** 030002
[7] Hara T *et al.* 2021 *Phys. Rev. Accel. Beams* **24** 110702
[8] Salehi E and Katoh M 2021 *Proc. 12th Int. Particle Accelerator Conf. (IPAC'21)* (Campinas, Brazil) pp 3970–72
[9] Borland M 2000 elegant: A Flexible SDDS-Compliant Code for Accelerator Simulation Tech. rep. LS-287 Advanced Photon Source
[10] Mochihashi A *et al.* 2018 *Proc. 9th Int. Particle Accelerator Conf. (IPAC'18)* (Vancouver, Canada) pp 3280–82

A method for obtaining 3D charge density distribution of a self-modulated proton bunch

T Nechaeva[1], L Verra[1,2,3], G Zevi Della Porta[2] and P Muggli[1]

[1] Max-Planck-Institute for Physics, Munich, Germany
[2] CERN, Geneva, Switzerland
[3] Technical University Munich, Munich, Germany

E-mail: tatiana.nechaeva@cern.ch

Abstract. The Advanced Wakefield Experiment (AWAKE) at CERN is the first plasma wakefield accelerator experiment to use a proton bunch as driver. The long bunch undergoes seeded self-modulation (SSM) in a 10 m-long plasma. SSM transforms the bunch into a train of short micro-bunches that resonantly drive high-amplitude wakefields. We use optical transition radiation (OTR) and a streak camera to obtain time-resolved images of the bunch transverse charge density distribution in a given plane. In this paper we present a method to obtain 3D images of the bunch by scanning the OTR across the entrance slit of the streak camera. Reconstruction of the 3D distribution is possible because with seeding self-modulation is reproducible. The 3D images allow for checking the axi-symmetry of SSM and for detecting the possible presence of the non-axi-symmetric hosing instability (HI).

1. Introduction

Plasma-based acceleration is a promising alternative to conventional one, as plasma can sustain accelerating fields several orders of magnitude higher than the RF cavities. Acceleration of electron bunches up to 8 GeV [1] and 42 GeV [2] has already been demonstrated in laser wakefield and beam-driven wakefield experiments, respectively.

The Advanced Wakefield Experiment (AWAKE) uses a long relativistic proton bunch as driver. This long bunch, propagating in plasma, undergoes self-modulation instability (SMI). SMI transforms the bunch into a train of short micro-bunches that drive high-amplitude wakefields. In order to control SMI we use a relativistic ionization front (RIF) or a short electron bunch preceeding the proton bunch that creates initial seed wakefields. When seeded, self-modulation (or SSM) yields reproducible outcome [3, 4].

Seeding of the self-modulation that is an axi-symmetric process with the electron bunch relies on the alignment of the electron and proton bunches with respect to each other. When misalignment is present, SSM is accompanied by the non-axi-symmetric hosing instability (HI). HI occurs in the plane of misalignment and, in case this plane is different from the main plane of observation, additional diagnostic is required in order to detect this instability.

In this paper we present a method that allows simultaneous observation of the proton bunch evolution in two perpendicular planes and, by extension, in any plane. We use a streak camera that yields time-resolved images of slices of the proton bunch charge density distribution [5, 6]. From the slices we obtain a 3D charge density distribution, which gives the possibility to detect whether HI is present in the plane perpendicular to the main plane of observation.

2. Optical setup

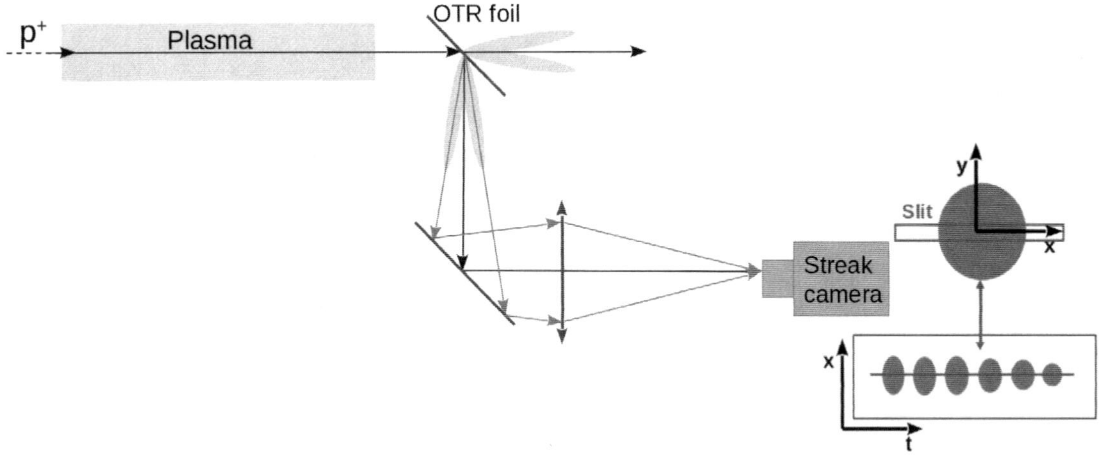

Figure 1. Simplified schematic of OTR transport line.

After exiting the Rubidium (Rb) vapor source, where plasma is created, the proton bunch propagates through a 280 μm-thick Silicon waver coated with 1 μm mirror-finished Aluminium foil, see figure 1 [6]. The emitted backwards optical transition radiation (OTR) contains information of the spatio-temporal proton bunch charge distribution. It is collected and guided in free-space using optical relay imaging to the streak camera room, where it is imaged onto the slit of the streak camera. The field of view of the imaging system is $\sim \pm$ 4 mm [7]. The streak camera yields light intensity distribution as a function of time and position.

The streak camera can operate with various slit widths and in different time windows. The choice of these parameters affects the temporal resolution of the resulting images. We measure the time resolution in two time windows, 73 ps and 210 ps respectively, and in the range of slit widths from 10 μm to 200 μm, using a 120 fs-short laser pulse. We define time resolution as the standard deviation in time of the intensity of the measured much shorter signal. The optimal resolution value is obtained with the slit width of 20 μm with minimum loss of signal, therefore this width is used in all the measurements.

In the measurement presented below we use 1 ns time window in order to capture the full proton bunch charge density distribution. Corresponding time resolution value is $\sim$ 4.79 ps.

The spatial resolution of the imaging system is measured using a 1951 USAF resolution target illuminated with the uniform blue light. Assuming a Gaussian response of a point light source, we do a convolution of the response with the square signal reproducing the selected target element. We find modulation transfer function (MTF) as a function of a ratio between the standard deviation of the Gaussian response and the line width of the selected element. We define spatial resolution as the standard deviation of the response function corresponding to the MTF of the signal. We obtain a resolution value of $\sim$ 0.18 mm.

3. Method and results

The streak camera produces time-resolved images of the transverse proton bunch charge density distribution. Under normal operation conditions the proton bunch OTR is centered on the slit of the camera to obtain the slice of the distribution around the bunch propagation axis. In order to detect potential transverse instabilities it is possible to offset the OTR with respect to the slit, thus to produce images of the proton bunch charge density at different transverse positions across its distribution. In the experiment we change the OTR position on the slit with the last reflecting mirror of the transport line in front of the streak camera (shown in figure 1). The distance between the mirror and the slit is $\sim 0.93\,\mathrm{m}$, while the maximum OTR offset introduced is $\sim 1.1\,\mathrm{mm}$, thus the effect of change in angle (maximum angular offset is $\sim 1.18\,\mathrm{mrad}$) is negligible. We can consider the variation of the OTR position with respect to the slit as equivalent to the change of the slit position across the transverse proton bunch distribution. A graphic overview of the method summarizing the discussion above is given on figure 2a).

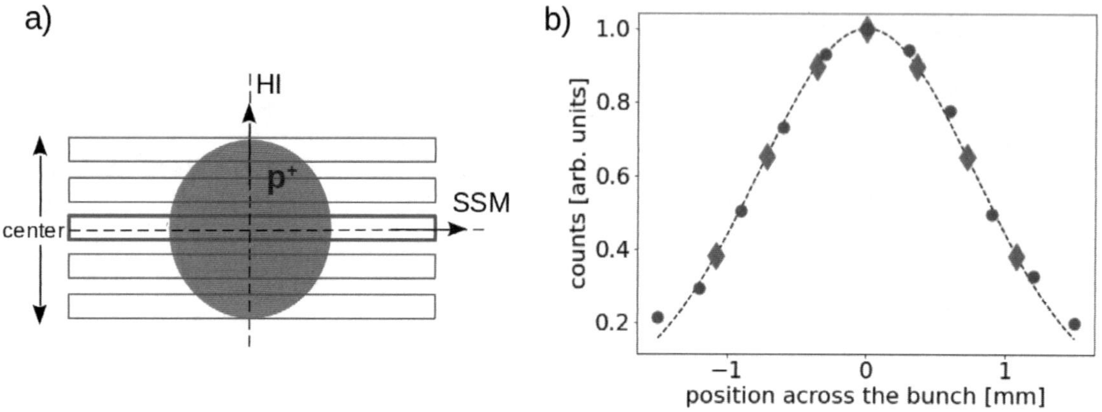

Figure 2. a) Schematic overview of the scanning procedure. In order to detect HI that occurs in the plane perpendicular to the slit it is possible to vary the OTR position with respect to the slit. It is equivalent to change of the slit position across the transverse proton bunch distribution. Blue rectangles indicate various slit positions. b) Average sum of counts of the images of the time-integrated proton bunch charge density distribution as a function of the position across the bunch. Red points – data, black dashed line – gaussian fit, blue diamonds – positions across the bunch where the slices are obtained in the experiment.

We determine the central position of the proton bunch OTR on the slit and locate the edges of the distribution by varying the mirror angle and obtaining corresponding images of the time-integrated charge density distribution that is produced without streaking. We plot the average sum of counts of the images (red circles on figure 2b) as a function of the position across the bunch. The transverse proton bunch distribution is Gaussian, therefore we perform a Gaussian fit. We use the mirror angular offset corresponding to the maximum of the Gaussian as the initial position. The rms size of the proton bunch at the OTR screen is $\sigma_\mathrm{p} = 0.73\,\mathrm{mm}$. We vary the mirror angle so that we record sets of bunch slices every $0.36\,\mathrm{mm}$ or $\sim 0.5\sigma_\mathrm{p}$ (blue diamonds on figure 2b). This step size is also larger than the spatial resolution of the system.

In the experiment we start the measurement from the central position of the proton bunch OTR on the slit and we vary the mirror angle in a way described above to obtain 3 slices above and 3 slices below the central one (see figure 2b). We record 5 events per slice and obtain the average time-resolved image of the proton bunch charge density distribution for each of them. We then produce a 2D distribution containing the average sum of all the slices and a 3D distribution consisting of the slices placed in order in a linear rectangular grid. We test the method by reconstructing the charge density distribution of the un-modulated (incoming) proton bunch propagating in the Rb vapor, that is as if in vacuum. Results are shown on figures 3 and 4.

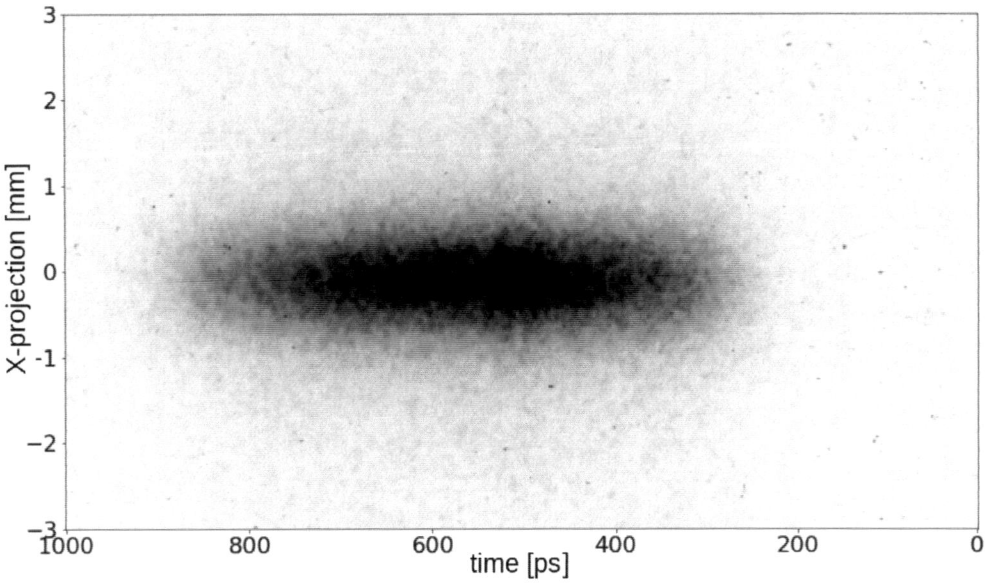

Figure 3. Proton bunch charge density distribution containing an averaged sum of all the slices as a function of time and position in the slit plane. The bunch propagates from left to right.

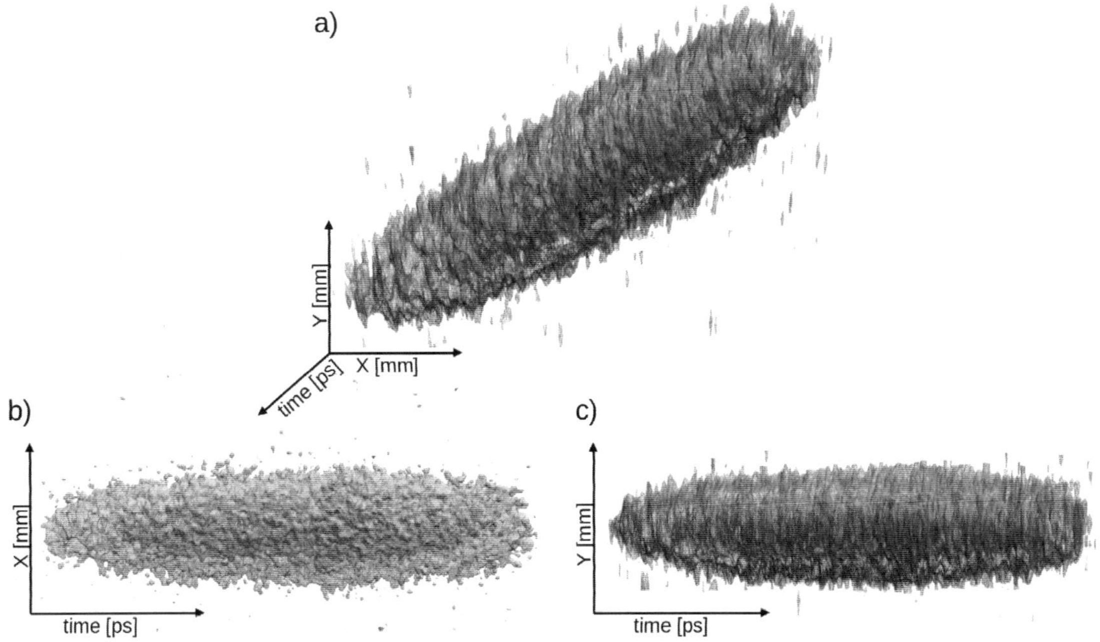

Figure 4. 3D reconstruction of the proton bunch charge density distribution: a) isometric view, b) projection of the slit plane, c) projection of the plane perpendicular to the slit plane. Slices separated by $0.5\sigma_\mathrm{p}$ are placed in a linear rectangular grid. On b) and c) the bunch propagates from left to right.

Conclusion

In the AWAKE experiment a long proton bunch propagating in plasma undergoes SMI. This instability can be controlled by placing a short electron bunch that creates initial wakefields ahead of the proton bunch and on its axis. Misalignment between the two bunches causes HI to occur in addition to SSM. HI develops in the plane of misalignment and SSM occurs in the perpendicular plane. It is important to be able to detect HI (if it occurs) when the system is supposed to be well aligned and therefore to develop a tool that allows simultaneous observation of two planes. We will purposely misalign the two bunches to induce HI and to study its characteristics.

We introduced a method that allows to reconstruct time-resolved 3D charge density distribution of the proton bunch using a streak camera. We tested the procedure by obtaining the distribution of the incoming proton bunch propagating in the Rb vapor. We will use the method in the upcoming measurements of the proton bunch evolution in plasma.

References

[1] Gonsalves A J *et al.* 2019 *Phys. Rev. Lett.*, **122** 084801
[2] Blumenfeld I *et al.* 2007 *Nature* **445** 741–4
[3] Batsch F *et al.* (AWAKE Collaboration) 2021*Phys. Rev. Lett.* **126** 164802
[4] Verra L *et al.* (AWAKE Collaboration) 2022 *Phys. Rev. Lett.* **129** 024802
[5] Rieger K, Caldwell A, Reimann O, Tarkeshian R and Muggli P 2017 *Rev. Sci. Instrum.* **88** 025110
[6] Adli E *et al.* (AWAKE Collaboration) 2019 *Phys. Rev. Lett.* **122** 054802
[7] Bachmann A-M and Muggli P 2020 *J. Phys.: Conf. Series* **1596** 012005

Concrete shielding activation for proton therapy systems using BDSIM and FISPACT-II

E Ramoisiaux[1], C Hernalsteens[1,2], R Tesse[1], E Gnacadja[1], N Pauly[1], M Vanwelde[1] and F Stichelbaut[3]

[1] Service de Métrologie Nucléaire, Université libre de Bruxelles, Brussels, Belgium
[2] CERN, 1211 Geneva 23, Switzerland
[3] Ion Beam Applications (IBA), Louvain-la-Neuve, Belgium

E-mail: eliott.ramoisiaux@ulb.be

Abstract. Proton therapy systems are used worldwide for patient treatment and fundamental research. The generation of secondary particles when the beam interacts with the beamline elements is a well known issue. In particular, the energy degrader is the dominant source of secondary radiation. This poses new challenges for the concrete shielding of compact systems and beamline elements activation computation. We use a novel methodology to seamlessly simulate all the processes relevant to the activation evaluation. A realistic model of the system is developed using Beam Delivery Simulation (BDSIM), a Geant4-based particle tracking code that allows a single model to simulate primary and secondary particle tracking and all particle-matter interactions. The secondary particle fluxes extracted from the simulations are provided as input to FISPACT-II to compute the activation by solving the rate equations. This approach is applied to the Ion Beam Applications (IBA) Proteus®ONE (P1) system and the shielding of the proton therapy research centre of Charleroi, Belgium. Proton loss distributions are used to model the production of secondary neutrals inside the accelerator structure. Two models for the distribution of proton losses are compared for the computation of the clearance index at specific locations of the design. Results show that the variation in the accelerator loss models can be characterised as a systematic error.

1. Introduction

Numerous proton therapy centres have been built for patient treatment and fundamental research over the past two decades [1]. It is well known that proton therapy machines generate a large number of secondary particles, mainly neutrons, when the proton beam interacts with the beamline elements [2]. In particular, the energy degrader is the dominant source of secondary radiation. Those neutrons interact with the concrete shielding via nuclear reactions, mainly neutron capture and spallation, producing radioactive nuclides. Some are long-lived and are responsible for the long-term activation of the shielding.

When designing a new centre or preparing experimental setups, the complete modelling of proton therapy systems from the primary and secondary beam interactions to the beamline and shielding activation is a complex but necessary task. We establish a method, inspired from the Rigorous Two-Step (R2S) [3], coupling Beam Delivery SIMulation (BDSIM) [4] with the code and library database FISPACT-II [5]. BDSIM provides a full 3D model of the proton therapy system and its shielding that includes the particle-matter interactions of Geant4 and the tracking of all the particles through the beamline magnetic elements, vacuum windows and air gaps. FISPACT-II is subsequently used for the activation computation.

Journal of Physics: Conference Series **2420** (2023) 012064 doi:10.1088/1742-6596/2420/1/012064

This methodology is thoroughly described in [6] and was applied to the shielding design of the future proton therapy centre of Charleroi, Belgium. The IBA Proteus®ONE proton therapy system that will be used in the centre was already modelled in BDSIM and validated against experimental data in [7]. The model was then used for the secondary particle generation required for shielding activation studies with FISPACT-II. The shielding activation results were fully validated against the IBA shielding design that was obtained using MCNPX [8].

During the elaboration of the BDSIM model of the future proton therapy centre of Charleroi, it was required to model in detail the proton loss pattern inside the accelerator (S2C2). Indeed, the acceleration of the primary protons is not simulated in BDSIM and therefore, the secondary neutron generation is realised by combining two different simulations. The first simulation considers the transport of the primary particles through the beamline with the secondary neutrons generated by interactions of the beam with the beamline elements. On the other hand, the second simulation simulates the propagation of lost protons inside the S2C2 structure. The S2C2 secondary neutrons generation process is presented in figure 1.

Figure 1. Illustration of the S2C2 secondary neutrons generation process in BDSIM. The protons (shown in blue) and the neutral particles, neutrons and photons (shown in green) are represented. Distribution of protons lost during the acceleration and the extraction processes serve as primary input (left). The resulting secondaries are produced by the interactions of the protons with the S2C2 structure (right).

We propose to use the BDSIM/FISPACT-II methodology to characterise the impact of two S2C2 proton loss distributions on the activation results. Figure 2 represents the BDSIM model of the vault of the proton therapy centre of Charleroi with its concrete shielding design. The concrete shielding was implemented using *Pyg4ometry*, a Python library that enables users to create GDML-based geometry rapidly [9, 10]. The S2C2 and the degrader, the two main elements of the beamline at the origin of most of the secondary particle generation, can be observed. We study the activation inside a cylindrical volume in the North Wall as it is the wall most affected by the S2C2 secondary neutrons. The specific location and geometry of the volume have been chosen to fit with a possible future experimental measurements campaign design that will consist of cylindrical removable concrete cores placed in the most irradiated parts of the shielding vault.

The activity of a compound is determined by its clearance index. The clearance index is defined as the sum A_i/CL_i over all the material radionuclides with A the specific activity and CL the clearance level allowed by the Belgian legislation. If the clearance index exceeds the value of 1, the compound is considered radioactive waste. The main isotopes produced in concrete are listed in table 1 with their corresponding clearance level.

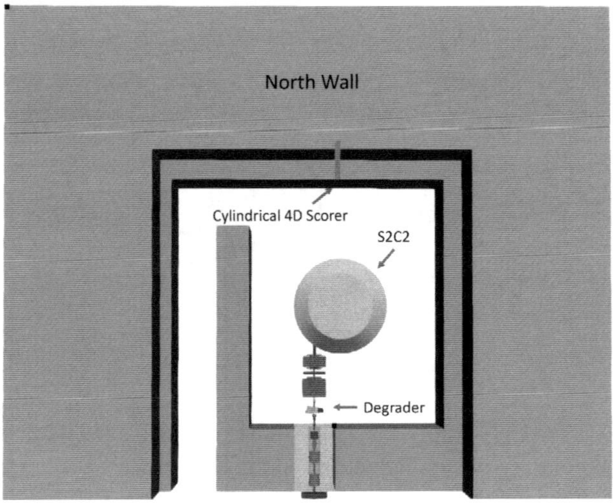

Figure 2. Realistic model of the S2C2 vault shielding. The North Wall is designated as well as the cylindrical mesh position.

Table 1. Clearance levels for the main isotopes produced in concrete [11].

Nuclide	CL (Bq/g)
^{3}H	100
^{7}Be	10
^{22}Na	0.1
^{54}Mn	0.1
^{60}Co	0.1
^{134}Cs	0.1
^{152}Eu	0.1
^{154}Eu	0.1

2. Tools and methods

Simulations with two different proton loss distributions inside the S2C2 have been realised using the BDSIM model. The irradiation condition studied are those used during the centre dimensioning [6]: 300 hours of irradiations per year with an S2C2 current of 150 nA. The degrader has been calibrated for a delivered beam energy of 100 MeV, which is the future centre most-used value.

The two distributions for the S2C2 losses were characterised based on the IBA design [11]. For both loss models, we consider that 25 % of the beam current is lost uniformly along the circumference at top energy (230 MeV) and that 45 % of the beam current is lost at the septum level during the extraction phase, as shown in figure 3.

In the first S2C2 loss model, model A, all the proton losses are simulated in the horizontal mid-plane, with the septum loss emitted in the forward direction. In the second model, model B, a normally distributed contribution of 1 mm (resp. 1 mrad) σ is added to the vertical position (resp. direction) of the circumference lost protons, while the septum lost protons are uniformly distributed in a 30° cone. The two S2C2 loss models are represented in figure 4.

The differential fluence of the secondary neutrons is scored following the predefined energy group structure "CCFE-709" in the cylindrical scorer mesh presented in figure 2 using the 4D-Scoring BDSIM feature [12]. The comparison between the neutron differential fluence in

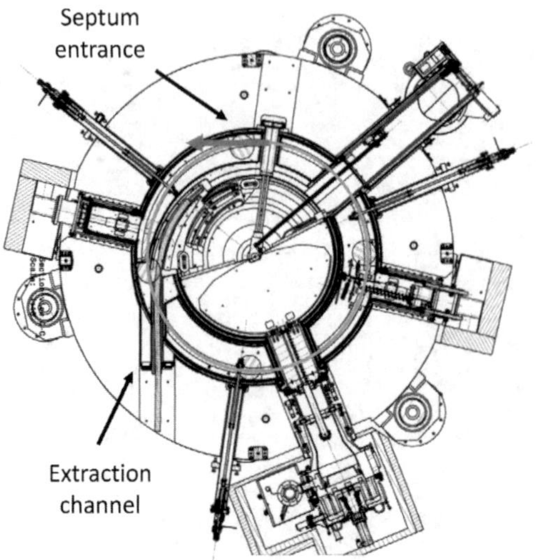

Figure 3. Illustration of the two sources of proton losses in the S2C2 simulated in BDSIM: the uniformly distributed losses along the accelerator circumference (in green) and the losses located at the septum (in red).

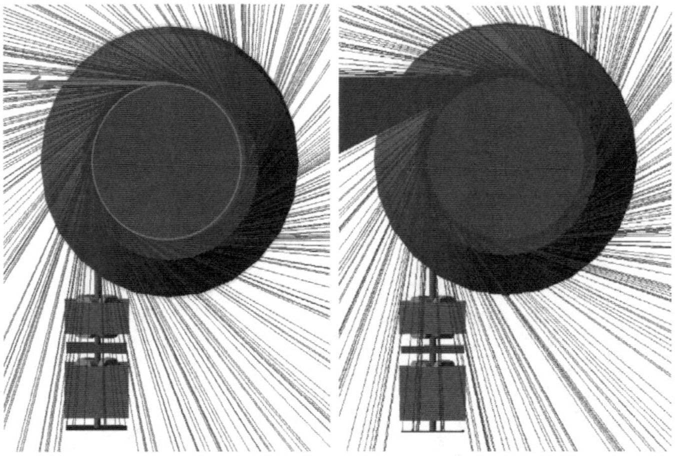

Figure 4. S2C2 loss models. On the left, model A, with the circumference losses simulated in the horizontal mid-plane and the septum losses emitted along with the red arrow. On the right, model B, with a normally distributed contribution of 1 mm (resp. 1 mrad) σ added to the vertical position (resp. direction) of the circumference lost protons and the septum loss uniformly distributed in a 30° cone.

the cylindrical core when using either of the models is presented in figure 5. The differential secondary fluences obtained from the circumference losses are almost identical between the two models. This similarity is expected behaviour as the impact of the random variation in position and angle of the lost protons simulated for the circumference losses is entirely hidden by the multiple scattering and interactions that the protons undergo inside the S2C2 structure. On the other hand, the differential fluence obtained from the septum losses is higher when using model A. This difference is easily explained by the variation of the septum losses distributions in model B, which spreads the secondary neutrons away from the North wall.

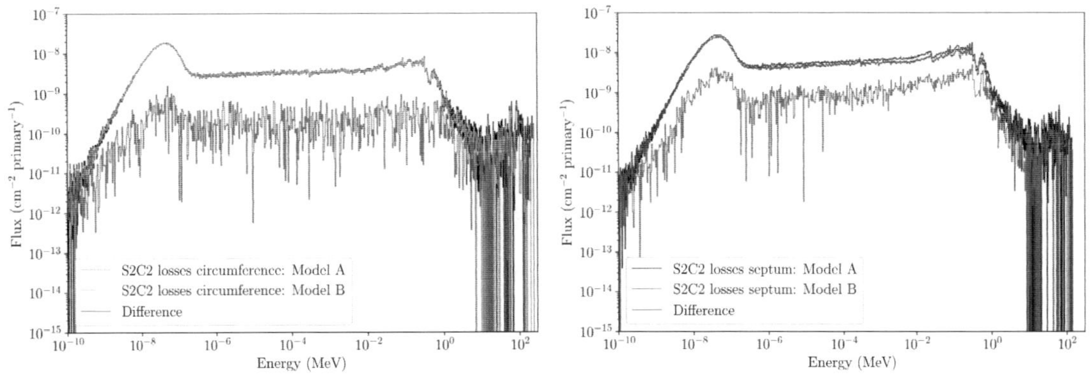

Figure 5. Comparison of the differential secondary neutron fluence extracted from the first 10 cm of the scored volume using the BDSIM 4D Scoring feature between the two S2C2 models. The differential secondary neutron fluence generated from the S2C2 circumference losses on the left and from the septum on the right.

The differential fluences of the secondary neutrons are then provided to FISPACT-II to compute the activation over an irradiation and a cooling period of 20 years. The clearance index is calculated along the cylindrical scored volume for both S2C2 loss models and the beamline losses. In figure 6, the activation associated with model A is observed to be larger than for model B, as expected by the comparison of the related differential fluences distributions. The most substantial difference between the models is 7 %. These results are not critical as the activated thickness - the depth at which the clearance index drops below 1 - is not impacted.

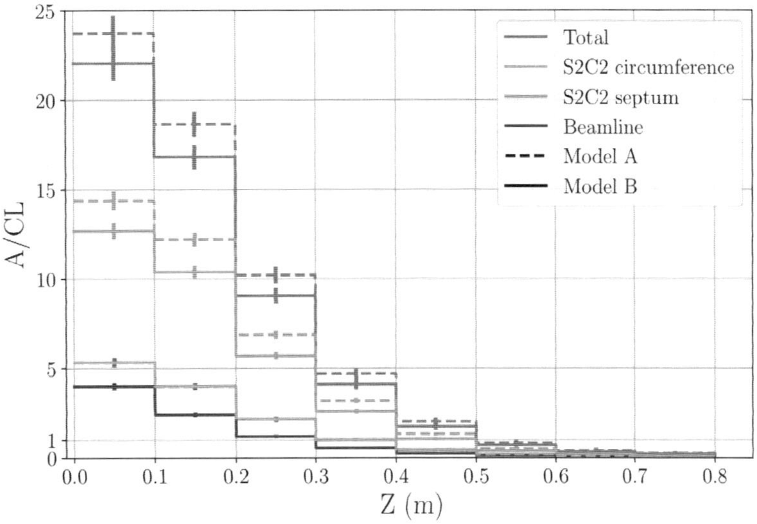

Figure 6. Evolution of the clearance index along with the thickness of the scored volume. The clearance indexes related to the circumference and the septum S2C2 losses are represented in green and orange, to the beamline losses in blue, and the total is represented in red.

FISPACT-II computation results also provide information about the impact of either model on the evolution of the radioactive nuclides concentration in the scored volume. The evolution of the clearance index of the most radioactive part of the core following the main radioactive nuclides is presented in figure 7. The nuclide-related activation varies only slightly between models A and B. Such variation will not significantly impact experimental measurements of the activation. The variation in the S2C2 loss models can be characterised as a systematic error.

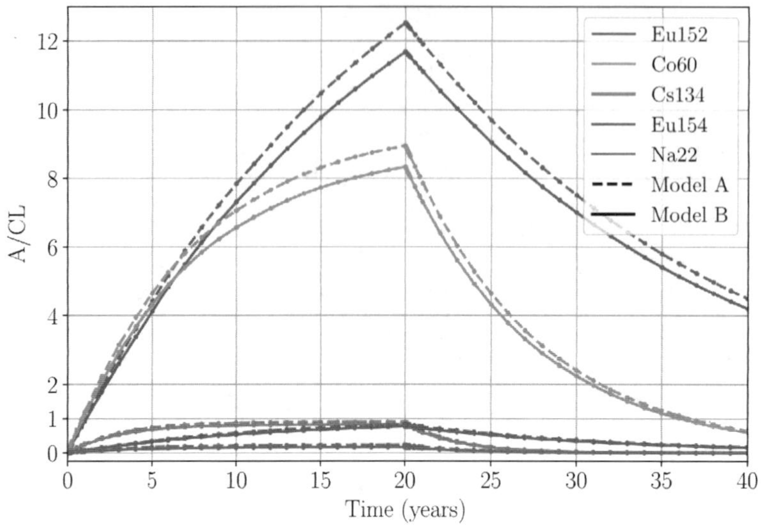

Figure 7. Evolution of the clearance index of the most radioactive part of the core following the main radioactive nuclides.

3. Conclusion and outlooks

The BDSIM/FISPACT-II methodology has been applied to the specific case of determining if the choice of the S2C2 loss patterns impacts the activation results. The activation along a cylindrical scorer in the North Wall of the vault was studied. The results compared the two models and showed a minimal variation for the activation after 20 years of operation and its evolution per nuclides. This variation between models can therefore be characterised as a systematic error. Such models of the S2C2 loss patterns will be used for further studies without impacting their main results.

Acknowledgments

This work has received funding from the Walloon Region (SPW-EER) Win^{2}Wal program under grant agreement No. 1810110 and PIT Prother-wal program under grant agreement No. 7289. The authors thank IBA for the support provided during this work.

References

[1] PTCOG. (2021) Particle therapy co-operative group - particle therapy centers `https://www.ptcog.ch/`
[2] Paganetti H 2011 *Proton Therapy Physics* (Boston, USA: CRC Press)
[3] Chen Y and Fischer U 2002 *Fusion Eng. Des.* **63-64** 107–14
[4] Nevay L J *et al.* 2020 *Comput. Phys. Commun* **252** 107200
[5] Sublet J -C *et al.* 2017 *Nucl. Data Sheets,* **139** 77–137
[6] Ramoisiaux E *et al.* 2022 *Eur. Phys. J. Plus* **137** 889
[7] Hernalsteens C*et al.* 2021 *EPL* **132** 50004
[8] Pelowitz D 2008 *MCNPX 2.6 Manual* (USA: Los Alamos National Laboratory)
[9] Boogert S T *et al.* 2019 *Proc. 10th Int. Particle Accelerator Conf. (IPAC'19)* (Melbourne, Australia) pp 3244–7
[10] Walker S *et al.* 2022 *Comput. Phys. Commun* **272** 108228
[11] Stichelbaut F 2019 Tech. rep. Ion Beam Applications
[12] Ramoisiaux E *et al.* 2021 *Proc. 12th Int. Particle Accelerator Conf. (IPAC'21)* (Campinas, Brazil) pp 1252–5

Characterization of the electron beam visualization stations of the ThomX accelerator

Alexandre Moutardier[1], Christelle Bruni[1], Jean-Noël Cayla[1], Iryna Chaikovska[1], Sophie Chance[1], Nicolas Delerue[1], Hayg Guler[1], Hugues Monard[1], Maher Omeich[1] and Scott David Williams[1,2]

[1] Université Paris-Saclay, CNRS/IN2P3, IJCLab, 91405, Orsay, France
[2] now at the University of Melbourne, Melbourne, Australia

E-mail: `moutardier@ijclab.in2p3.fr`

Abstract. We present an overview of the diagnostics screens stations - named SSTs - of the ThomX compact Compton source. ThomX is a compact light source based on Compton backscattering. It features a linac and a storage ring in which the electrons have an energy of 50 MeV. Each SST is composed of three screens, a YAG:Ce screen and an Optical Transition Radiation (OTR) screen for transverse measurements and a calibration target for magnification and resolution characterisation. The optical system is based on commercial lenses that have been reverse-engineered. An Arduino is used to control both the aperture and the focus remotely, while the magnification must be modified using an external motor. We report on the overall performance of the station as measured during the first steps of beam commissioning and on the optical system remote operations.

1. Overview of the optical diagnostics at ThomX

The ThomX Compact Compton source is a novel accelerator which is being commissioned in Orsay [1]. We report here on the optical diagnostics used in this accelerator.

The overall diagnostics layout is shown in figure 1. There are five screens stations. They are located near the RF Gun, at the end of the linac, in the transfer line (twice) and in the extraction line. The current operation permit from the Nuclear Safety Authority only allows commissioning of the straight part of the accelerator after the photo-cathode (upper part of the drawing), hence this paper will focus on the two first screens stations, named LI/DG/SST.01 and TL/DG/SST.01.

The green dashes show the location of the screens stations.

Each screens station is made of a vacuum vessel with a UHV-compatible actuator to move the screens in vacuum. The visualisation is made using a Basler scA640-70gm CCD camera equipped with a commercial lens. A drawing of these diagnostic stations can be seen in figure 2. There are three types of screen for each station: a YAG:Ce screen, an Optical Transition Radiation (OTR) screen and a calibration target. Those screens are at a 45° from both the beam trajectory and the camera axis. Each screen is separately described in the following sections.

The TL/DG/SST.01 has also a sapphire screen used to generate Cerenkov light and allow longitudinal length measurement. This feature is not yet commissioned and will not be mentioned further in this paper.

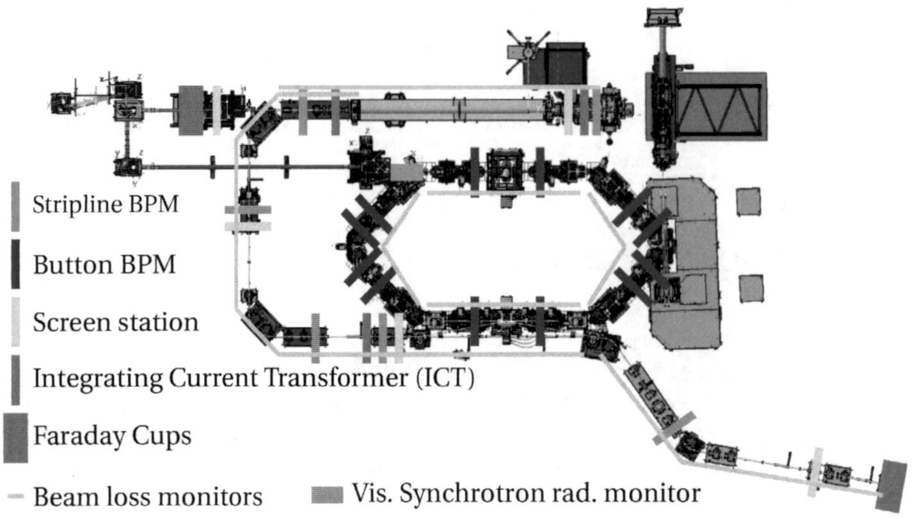

Figure 1. Location of the optical diagnostics on the ThomX accelerator. The beam is produced at the top of the drawing and propagates counter-clockwise. The green dashes show the location of the screens stations.

2. Commercial lens reverse-engineering

2.1. Lens characteristics

The lens used in both SSTs is a Tamron : 18-400mm F/3.5-6.3 Di II VC HLD. Those lens have a focal length of 18 mm to 400 mm and an apperture going from f/3.5 (f/6.3) to f/22 (f/40) at a focal length of 18 mm (400 mm).

2.2. Reverse-engineering

This kind of lens allows auto-focus and aperture control by photographer's cameras, but the ThomX cameras do not have the required outputs. To remotely control both those parameters, an Arduino Uno circuit controlled by a web interface has been developed [2].

A 3D printed structure is under test to remotely control the magnification by adding an external motor.

3. Calibration target

Before using a scintillation screen to evaluate beam size, some calibrations must be done. For that purpose, a specific calibration target, named USAF1951 calibration chart, is used (see figure 3). Nine identical calibrated targets allow the measurement of the magnification and resolution at each target position. Two numbers - group and element - are use to characterise each triplet of vertical and horizontal lines [3]. The width of a line - or a fifth of its length - may be computed using equation (1).

$$d[\text{mm}] = 2^{-1\times(1+group+(element-1)/6)} \tag{1}$$

From the size of the largest element, one may compute the pixel/millimetre ratio needed to compute the beam size - hence the magnification - while the resolution comes out from the smallest element's triplet of lines that one can discriminate. Table 1 show the resolution and magnification computed for both images of figure 3. The magnification computation give a 2x2 matrix used to compute, from horizontal (h) and vertical (v) pixel length on the image, the horizontal (x) and vertical (y) physical length of the beam (in mm) as show in equation (2).

Figure 2. Left: CAD 3D view of one of the diagnostic stations. The beam travels from the front left to the back right on the image. The screen is inserted by the motorized UHV-actuator on the right. The observation camera is located below the station. On some stations such as the one depicted here, a vacuum pump can be inserted on top of the station.
Right: Photo (top) and drawing (bottom) of the screens mechanical arm. Three to four screens can be inserted via these mechanical arms. In this example, the 4th screen is the Cerenkov screen and it has a slightly different angle.

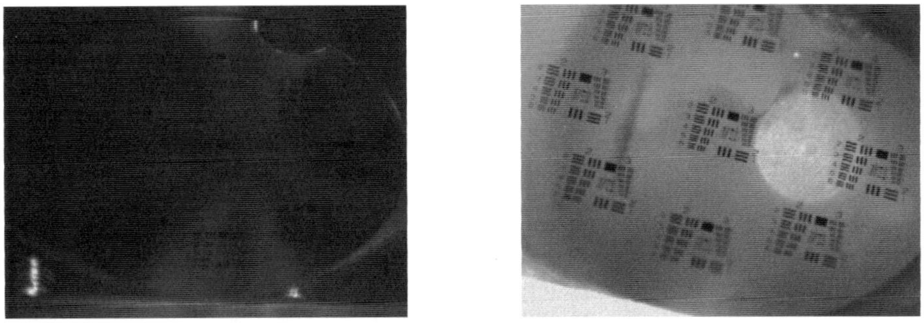

Figure 3. USAF1951 test chart on Li/DG/SST.01 (left) and on TL/DG/SST.01 (right). Nine calibrated targets allow the measurement of magnification and resolution. The bottom and top part of the screen are blurry because of the 45° rotation of the screens and the short depth of field. The illumination and exposure is adapted to allow automatic computation.

$$\begin{pmatrix} x[\mu\text{m}] \\ y[\mu\text{m}] \end{pmatrix} = \begin{pmatrix} HX & VX \\ HY & VY \end{pmatrix} \times \begin{pmatrix} h[pixel] \\ v[pixel] \end{pmatrix} \tag{2}$$

One may notice reflections on figure 3. It comes from the LEDs used to light the screen. To avoid too much reflection preventing from reading the target, several LEDs have been tested.

Table 1. Calibration parameters computed for images of figure 3. Resolution is given for the upper target while over parameters are the mean for all target detected by the analysis's code.

	Resolution [µm]		**Magnification** [µm/px]			
	Horz	**Vert**	**HX**	**VX**	**HY**	**VY**
Li/DG/SST.01	125	99	36	0	1	−36
TL/DG/SST.01	79	79	28	10	−5	27

The less directional the LED light, the better the image. The actual configuration is not perfect but allows nevertheless the use of a python script to automatically compute the calibration of the screens. A dedicated paper about this computation has been presented at IBIC2020 [2].

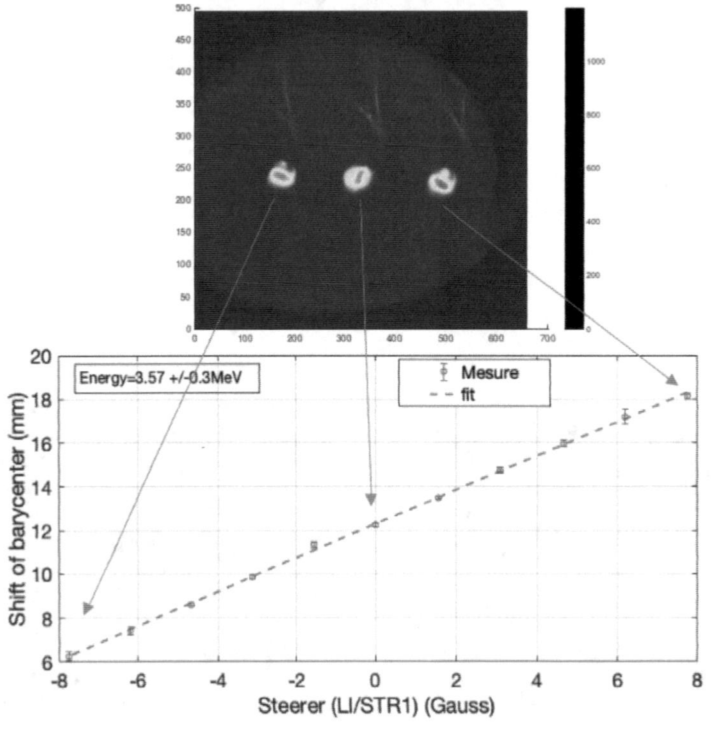

Figure 4. Energy measurement on the YAG:Ce screen of LI/DG/SST.01 using a steerer to deflect the beam. Each spot correspond to the beam position at three different values of the steerer's strength : ≈ -8 G, 0 G, ≈ 8 G. The measured energy was 3.6 ± 0.3 MeV at the exit of the photo-gun.

4. YAG:Ce screen

A YAG:Ce screen is used on ThomX to visualise the transverse projection of the beam. This kind of screen is particularly efficient during commissioning as the yield of photo generated is

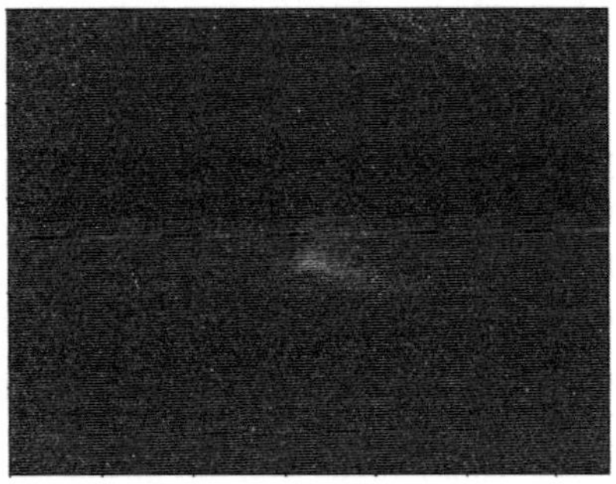

Figure 5. ThomX beam at 30 pC, 50 MeV on TL/DG/SST.01. The left image is one beam taken with a 1 ms exposition time on the YAG:Ce screen. The right image is the subtraction of two images of the OTR screen with a 10 s exposition time at 10 Hz, one with and one without the photo-emitted beam.

considerable, which allows visualisation of a small charge beam (around tenth of pC) or dark current. Those screens are also used to measure the beam energy with a steerer magnet. While the operation permit does not allow the use of the bending dipoles, this is the main method for energy measurement. Figure 4 shows one of those energy measurements on LI/DG/SST.01.

5. OTR screen

The optical transition radiation (OTR) is another way to measure transverse beam characteristics. Beam size measurements are more precise than with YAG:Ce screens as no blooming occurs, but the yield of photon generated is much smaller.

Figure 5 shows this yield difference as one 30 pC, 50 MeV beam alone can be clearly identified on the YAG:Ce screen's image (top) whereas to see the OTR light, an image of one hundred beams and a background noise subtraction is needed. A preliminary computation of OTR light collection by the camera give a better visualisation than those measured. Hence some measurement of the transmittance of the optical system is under process as well as a verification of potential misalignment of the camera to try to increase the OTR light collection.

6. Conclusion

The ThomX accelerator is under commissioning since June 2021. Screens stations are one of the first electron diagnostics used, as even during the RF commissioning YAG:Ce screen allows to visualise the dark current. Those screens are used every day to visualise the beam and to measure its size and energy thanks to calibration targets. Optical radiation transition has been visualised but the very low yield has yet to be completely understood. The reverse engineering on commercial lens provides a better adjustment of the optics and remote control.

Acknowledgments

Work done on behalf of the ThomX collaboration. List of members available on :

- https://thomx.ijclab.in2p3.fr/collaboration-thomx

Work supported by the French "Agence Nationale de la Recherche"as part of the program "investing in the future"under reference ANR-10-EQPX-51 and by grants from Région Ile-de-France.

Journal of Physics: Conference Series **2420** (2023) 012065 doi:10.1088/1742-6596/2420/1/012065

References

[1] Variola A *et al.* 2014 *ThomX Technical Design Report* Tech. rep. Laboratoire de l'Accélérateur Linéaire (in2p3-00971281)
[2] Williams S D *et al.* 2020 *Proc. 9th Int. Beam Instr. Conf. (IBIC'20)* (Santos, Brazil) pp 281–4
[3] Wikipedia contributors 2021 *USAF resolution test chart*

Journal of Physics: Conference Series **2420** (2023) 012066 doi:10.1088/1742-6596/2420/1/012066

Surrogate-based Bayesian inference of transverse beam distribution for non-stationary accelerator systems

H Fujii[1], N Fukunishi[1] and M Yamakita[2]
[1]Nishina Center, RIKEN, 2-1 Hirosawa, Wako, Saitama, 351-0198, Japan
[2]Department of Systems and Control Engineering, Tokyo Institute of Technology, 2-12-1 Ookayama, Meguro-ku, Tokyo, 152-8550, Japan
E-mail: hiroki.fujii@riken.jp

Abstract. Constraints on the beam diagnostics available in real-time and time-varying beam source conditions make it difficult to provide users with high-quality beams for long periods without interrupting experiments. Although surrogate model-based inference is useful for inferring the unmeasurable, the system states can be incorrectly inferred due to manufacturing errors and neglected higher-order effects when creating the surrogate model. In this paper, we propose to adaptively assimilate the surrogate model for reconstructing the transverse beam distribution with uncertainty and underspecification using a sequential Monte Carlo from the measurements of quadrant beam loss monitors. The proposed method enables sample-efficient and training-free inference and control of the time-varying transverse beam distribution.

1. Mathematical background

1.1. Accelerator systems with unknown drift and parameter uncertainties

The dynamical system of the accelerator with measurement error can be written as

$$\dot{\boldsymbol{x}} = f(\boldsymbol{x}, \boldsymbol{u}) \tag{1}$$

$$\boldsymbol{y} = h(\boldsymbol{x}, \boldsymbol{u}) + \boldsymbol{\epsilon}, \tag{2}$$

where $\boldsymbol{x} \equiv \boldsymbol{x}(t) \in \mathbb{R}^n$ is the vector of the time-varying hidden states, $\boldsymbol{u} \in \mathbb{R}^m$ is the control input state, t is the time. $\boldsymbol{y} \in \mathbb{R}^d$ is the measurement, and $\boldsymbol{\epsilon} = [\epsilon_1, ..., \epsilon_d]^T$ is the independent measurement noise.

To produce a quality beam of interest by inferring hidden states $\boldsymbol{x}$, we introduce a surrogate model as

$$\dot{\hat{\boldsymbol{x}}} = \hat{f}(\hat{\boldsymbol{x}}, \boldsymbol{u}) \tag{3}$$

$$\hat{\boldsymbol{y}} = \hat{h}(\hat{\boldsymbol{x}}, \boldsymbol{u}), \tag{4}$$

where $\hat{\boldsymbol{x}} \equiv \hat{\boldsymbol{x}}(t) \in \mathbb{R}^n$, and $\hat{\boldsymbol{y}} \in \mathbb{R}^d$ are the counterpart of the real system equations (1) and (2) built based on the parameterized system model. One may infer the hidden state by simply minimizing the loss function

$$C(\boldsymbol{y}) = \|\boldsymbol{y} - \hat{\boldsymbol{y}}\|_2^2. \tag{5}$$

However, the accurate state $\boldsymbol{x}$ cannot be necessarily obtained due to the model uncertainties and possible multiple hidden states when the measurement is sparse. The goal of the following Approximate Bayesian Computation (ABC) framework [1] is to sequentially filter and track the time-varying hidden state estimates $\hat{\boldsymbol{x}}$ by minimizing the expectation of the cost function in equation (5) throughout the operation of the accelerator complex.

1.2. Markov-Chain Monte-Carlo (MCMC) for the initial posterior inference

Given an observation $\boldsymbol{y}$, the initial posterior distribution of interest for a given fixed control input $\boldsymbol{u}$ can be given by

$$p(\hat{\boldsymbol{x}}|\hat{\boldsymbol{y}}) = \frac{p(\hat{\boldsymbol{y}}|\hat{\boldsymbol{x}})p(\hat{\boldsymbol{x}})}{p(\hat{\boldsymbol{y}})}, \tag{6}$$

here subscripts for denoting the time steps are omitted in this section, e.g., $\hat{\boldsymbol{x}} = \hat{\boldsymbol{x}}_{k=0}$. Since $p(\hat{\boldsymbol{x}}|\hat{\boldsymbol{y}})$ is a probability distribution, we can rewrite this equation as

$$p(\hat{\boldsymbol{y}}) = \int p(\hat{\boldsymbol{y}}|\hat{\boldsymbol{x}})p(\hat{\boldsymbol{x}})d\hat{\boldsymbol{x}}. \tag{7}$$

The probability density function $p(\hat{\boldsymbol{y}})$ given a measurement $\boldsymbol{y}$ with normally distributed sensor noise with standard deviation $\boldsymbol{\sigma}_n = \{\sigma_{n1}, ..., \sigma_{nd}\}$ is expressed as

$$p(\hat{\boldsymbol{y}}) = \prod_j^d \frac{1}{\sqrt{2\pi}\sigma_{nj}} \exp\left(-\frac{|\hat{y}_j - y_j|^2}{2\sigma_{nj}^2}\right) \tag{8}$$

and $p(\hat{\boldsymbol{y}}|\hat{\boldsymbol{x}})$ drawn by surrogate model equation (4). Then, the proposal $p(\hat{\boldsymbol{x}})$ of a general nonlinear system can be sampled using MCMC [2]. In this work, the parallel tempering MCMC (PT-MCMC) algorithm is adopted to efficiently find the optimal and promote mixing across the state spaces. Details on PT-MCMC can be found in [3].

1.3. Particle filtering for tracking hidden states

The sequential importance resampling is a class of particle filter algorithm. The sequence is initialized with a set of N particles representing $p(\hat{\boldsymbol{x}})$, and assign the normalized weights $W_{k=1}^s = 1/N$, for all $s = 1, ..., N$. Then particles at a discrete time step $k-1$ are sequentially propagated for a new time step k to update prior $p(\hat{\boldsymbol{x}}_k^s|\hat{\boldsymbol{x}}_{k-1}^s, \boldsymbol{u}_{k-1})$. The weight update of s-th particle is performed based on the measurement and cosine similarity

$$\begin{aligned} &p(\hat{\boldsymbol{y}}_k^s|\hat{\boldsymbol{x}}_k^s, \boldsymbol{u}_k, \boldsymbol{u}_{k-1}) \\ &\propto \mathrm{sim}_k(\Delta_k\hat{\boldsymbol{y}}^s, \Delta_k\boldsymbol{y}) \prod_j^d \exp\left(-\frac{|\hat{y}_j^s - y_j|^2}{2\sigma_{nj}^2}\right). \end{aligned} \tag{9}$$

Here, $\mathrm{sim}_k(\Delta_k^s\hat{\boldsymbol{y}}, \Delta_k\boldsymbol{y})$ is a similarity measure between the responses of the actual system $\Delta_k\boldsymbol{y} \equiv \boldsymbol{y}_k - \boldsymbol{y}_{k-1}$ and the surrogate model $\Delta_k^s\hat{\boldsymbol{y}} \equiv \hat{\boldsymbol{y}}_k^s - \hat{\boldsymbol{y}}_{k-1}^s$ expressed by

$$\mathrm{sim}_k(\Delta_k^s\hat{\boldsymbol{y}}, \Delta_k\boldsymbol{y}) = \begin{cases} \frac{1+\tau(\Delta_k^s\hat{\boldsymbol{y}}, \Delta_k\boldsymbol{y})}{2} & (\|\Delta_k^s\hat{\boldsymbol{y}}\| > 2\sigma_n) \\ 1 & (\|\Delta_k^s\hat{\boldsymbol{y}}\| \leq 2\sigma_n) \end{cases} \tag{10}$$

and $\tau(\boldsymbol{a}, \boldsymbol{b}) \equiv \frac{\boldsymbol{a}\cdot\boldsymbol{b}}{\|\boldsymbol{a}\|\|\boldsymbol{b}\|}$ is the cosine similarity. The inequality conditions in equation (10) are introduced such that the noise level does not adversely affect the weight updates. Then the sample particles are reweighed according to equation (6) as

$$W_k^s = \frac{p(\hat{\boldsymbol{y}}_k^s|\hat{\boldsymbol{x}}_k^s, \boldsymbol{u}_k, \boldsymbol{u}_{k-1})W_{k-1}^s}{\sum_s p(\hat{\boldsymbol{y}}_k^s|\hat{\boldsymbol{x}}_k^s, \boldsymbol{u}_k, \boldsymbol{u}_{k-1})W_{k-1}^s}. \tag{11}$$

After each update of states and weights, the particles are resampled by systematic resampling to obtain new equally-weighted particles to prevent degeneracy when the effective sample size $ESS = \left(\sum_{s=1}^{N}(W_k^s)^2\right)^{-1}$ is below the user-defined threshold [4].

2. Numerical experiments

2.1. CNN modeling of the transverse beam dynamics with parameterized beam source condition

In the numerical experiment, the system equation (1) is defined to mimic the beam transport line after the AVF cyclotron as in figure 1(a) of [5]. Control inputs $\boldsymbol{u} \in \mathbb{R}^7$ are assigned to applied excitation current for a quadrupole triplet (u_1, u_2, u_3), horizontal and vertical steerers (u_4, u_5), and a subsequent quadrupole doublet (u_6, u_7).

This study's incident particle distribution containing $N = 40,000$ particles is characterized by skewed bi-Gaussian transverse beam distribution with skewness parameters $\alpha_{\mathrm{x,y}}$ [6, 7], while fixing the beam width. Here, the skewness in x-axis α_{x} modifies the standard normal distribution $\phi(\mathrm{x}) = \frac{1}{\sqrt{2\pi}}\exp\left(-\mathrm{x}^2/2\right)$ to be $\phi_\alpha(\mathrm{x}) = 2\phi(\mathrm{x})\Phi(\alpha\mathrm{x})$, where $\Phi(\mathrm{x})$ is the cumulative distribution function of the normal distribution $\Phi(\mathrm{x}) = \int_{-\infty}^{\mathrm{x}}\phi(t)dt = \frac{1}{2}\left[1 + \mathrm{erf}\left(\frac{\mathrm{x}}{\sqrt{2}}\right)\right]$. The same applies to the y-axis. Note that the skewness is a practical problem for the beam extracted by the AVF Cyclotron of Nishina Center as in figure 5 of [5].

The observation function $h(\boldsymbol{x}, \boldsymbol{u}) \in \mathbb{R}^4$ is defined to sum up the transverse two-dimensional histogram $I(n_{\mathrm{x}}, n_{\mathrm{y}}) \in \mathbb{R}^{52\times52}$ as in figure 1(c) overlapped by the beam loss monitors on each side (R: Right, L: Left, U: Up, D: Down) as in figure 1(d), where $n_{\mathrm{x}} \in \mathbb{N}$ and $n_{\mathrm{y}} \in \mathbb{N}$ are indexes of pixels which evenly discretize the field of view at the beam loss monitor location.

As summarized in figure 1, we assume the skewness parameters of the incident beam before the transport slowly vary like a sigmoid function over 60 discrete time steps as in figure 1(b). The transverse beam distributions at $k = 0$ (figure 1(c-i)) and at $k = 60$ (figure 1(c-ii)) differ as shown in figure 1(c-iii) as a result of these time-varying skewness parameter. The subset

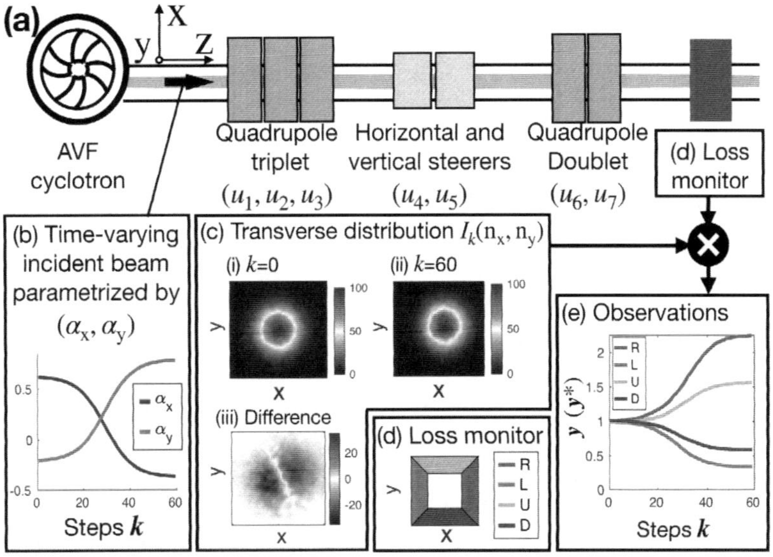

Figure 1. (a) Beam transport system's overview. (b) Time variation of hidden states $\alpha_{\mathrm{x,y}}$ over discrete time steps k. (c) Transverse distribution $I(n_{\mathrm{x}}, n_{\mathrm{y}})$ at the position of the beam loss monitors (d). (e) The number of incoming particles for each loss monitor relative to the initial observation $\boldsymbol{y}^*$.

figure 1(e) describes the time evolution of the observations by loss monitor figure 1(d) at each side relative to the optimized value y_j^* at $k = 0$.

Following the similar procedure in [8, 9], the corresponding simulation-based surrogate model is trained using Convolutional Neural Network to output transverse distribution estimate $\hat{I}(n_\mathrm{x}, n_\mathrm{y}) \in \mathbb{R}^{52\times52}$ as a function of the excitation currents and hidden parameters of the incident beam distribution $\alpha_{\mathrm{x,y}} \in [-3, 3]$. The average reconstruction error of the predicted transverse distribution $\hat{I}(n_\mathrm{x}, n_\mathrm{y})$ defined by the equation below resulted in around 7 % for the test data.

$$Error = \sum_{n_\mathrm{x},n_\mathrm{y}} |\hat{I}(n_\mathrm{x}, n_\mathrm{y}) - I(n_\mathrm{x}, n_\mathrm{y})| / \sum_{n_\mathrm{x},n_\mathrm{y}} |I(n_\mathrm{x}, n_\mathrm{y})|$$

We take into account the input bias in surrogate modeling, e.g., i-th component of the input is assumed to be

$$\hat{u}_i = (1 + \delta_i)u_i, \tag{12}$$

where δ_i is the hidden state representing the scaling error.

In the later sections, the nine-dimensional vector $\hat{\boldsymbol{x}} = [\delta_1,\ldots,\delta_m,\ \alpha_\mathrm{x},\ \alpha_\mathrm{y}]^T$ is considered as the hidden states to be updated by sequential Monte-Carlo (SMC). By computing the likely hidden states $\hat{\boldsymbol{x}}$ from measurements $\boldsymbol{y}$ with the ABC framework, surrogate-based reconstruction of the transverse beam distribution can be made even if invasive diagnostic devices such as wire scanners or scintillators are not insertable remotely due to insufficient room for installation or because of their damage threshold.

2.2. *Process variance for hidden states tracking*

Changes in the hidden parameters are updated by conditioning the probability distribution of states, which is spread by artificial process noise, with the observation at each step k. For simplicity of calculation, a Gaussian uncertainty with a standard deviation σ_{vi} up to 5 % of the maximum i-th training excitation current $\boldsymbol{U}$ is considered for δ_i in equation (12); $\sigma_{vi} = 0.05 \max U_i$. In the sequential filtering process, $\Delta\hat{\boldsymbol{u}} \sim \mathcal{N}(\Delta\boldsymbol{u}, \gamma\boldsymbol{\Sigma}_v)$ is added to the surrogate model input for each move $\Delta\boldsymbol{u}$ on the actual system. Here, γ is the squared average of the control input norm for each move relative to $\max U_i$ expressed as $\gamma \equiv \frac{1}{m}\sum_i^m \frac{\Delta u_i^2}{\max U_i^2}$, and $\boldsymbol{\Sigma}_v = \operatorname{diag} \boldsymbol{\sigma}_v^2$ where $\boldsymbol{\sigma}_v^2 = [\sigma_{v1}^2, \ldots, \sigma_{vm}^2]$.

On the other hand, the process variances of the skewness parameters $\alpha_{\mathrm{x,y}}$ are determined from the variances of the resultant MCMC particle distribution. The MCMC is initialized first by finding $\hat{x}$ which minimizes the cost equation (5) while the initial guesses $\alpha_{\mathrm{x,y}} = 0$ are used for the skewness parameters, and then subsequent proposal moves are accepted for the criterion conditioned by equation (8).

2.3. *Measurement based control while tracking the hidden state space for the underspecified model*

In this section, we aim at matching the observation $\boldsymbol{y}$ to the target observation $\boldsymbol{y}^* \in \mathbb{R}^d$ while tracking the hidden states with SMC. For the weighted sum of the observations on the surrogate model $\hat{\boldsymbol{y}}_k = \sum_s W_k^s \hat{\boldsymbol{y}}_k^s$, the optimal input at time $k+1$ can be determined by the following two steps

$$\boldsymbol{u}^* = \arg\min_u \|(\boldsymbol{y}_k - \boldsymbol{y}^*) - (\hat{\boldsymbol{y}}_k - \hat{\boldsymbol{y}}_{k+1})\|_2^2 \tag{13}$$

$$\boldsymbol{u}_{k+1} = \boldsymbol{u}_k + \lambda(\boldsymbol{u}^* - \boldsymbol{u}_k). \tag{14}$$

Please note that if $\hat{\boldsymbol{y}}_k$ is equal to $\boldsymbol{y}_k$, $\hat{\boldsymbol{y}}_{k+1}$ is made close to $\boldsymbol{y}^*$. Here, the exponential convergence factor λ $(0 < \lambda \leq 1)$ is introduced to regulate the input move within the proximity of the input

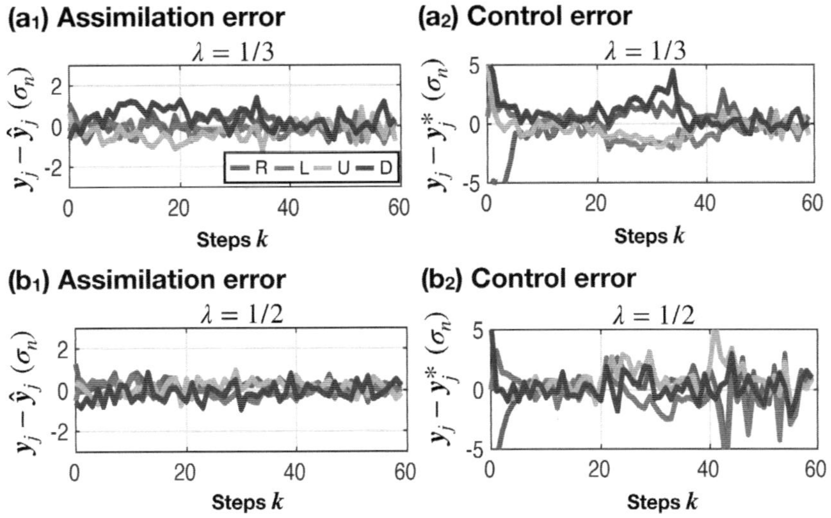

Figure 2. Results of control attempts toward y_j^* ($j \in R, L, U, D$) by equation (13) for (a) $\lambda = 1/3$ and (b) $\lambda = 1/2$.

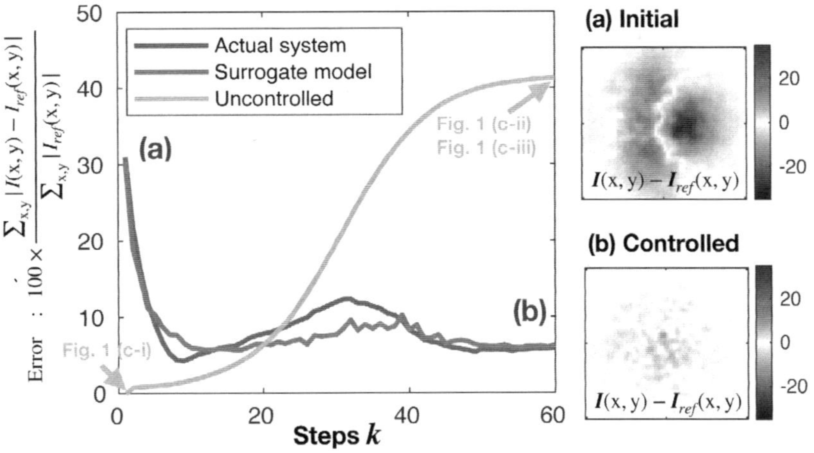

Figure 3. Left: The blue line shows the relative error of $I(n_\mathrm{x}, n_\mathrm{y})$ when the model's state is controlled based on the distribution $\hat{I}(n_\mathrm{x}, n_\mathrm{y})$ represented by the red line toward $I_{ref}(n_\mathrm{x}, n_\mathrm{y})$. The yellow line is the uncontrolled reference when the state evolves as in figure 1. Right: Error in the transverse distribution $I(n_\mathrm{x}, n_\mathrm{y})$ at (a) $k = 0$ and (b) $k = 60$.

state $\boldsymbol{u}_k$. In this study, the optimization problem equation (13) is solved using the simplex method [10] for the measurement estimates $\hat{\boldsymbol{y}}$ inferred by the surrogate model with tracked hidden states $\hat{\boldsymbol{x}}_k = \sum_s W_k^s \hat{\boldsymbol{x}}_k^s$ with $s = 5{,}000$. Here, the standard deviation of the Gaussian noise ϵ_j is set to 5 % relative to y_j^*, while the measurement uncertainty σ_j is designed to be 10 % of the desired values for each output y_j^*. To demonstrate the robust behavior of the proposed scheme to underspecified model, the beam current is gradually enhanced by 10 % with a factor which varies like a sigmoid function similar to $\alpha_{\mathrm{x,y}}$. In figure 2, the control strategy equation (14) for a randomly picked input state $\boldsymbol{u}_0$ is demonstrated for (a) $\lambda = 1/3$ and (b) $\lambda = 1/2$. Figure 2(a) demonstrates the successful stabilization of the time-varying system only with a few samples around the optimal $\boldsymbol{y}^*$ for the noisy sensor measurement, while a slightly delayed response is observed for the higher rate of change in hidden states $|\alpha_{k+1}-\alpha_k|$. Figure 2(b) suggests that large λ can lead to unstable response because too large input move toward optimal $\boldsymbol{u}^*$ can promote quick degeneracy of sample particles.

2.4. Inference based control strategy for the transverse beam distribution

Instead of the control strategy based on observable $\boldsymbol{y}$, we design the control law as to minimize $\sum_{n_x,n_y} |\hat{I}_{k+1}(n_x, n_y) - I_{ref}(n_x, n_y)|^2$ by replacing the cost function in equation (13), where the reference distribution $I_{ref}(n_x, n_y)$ is defined to be the one shown in figure 1(c-i). The left window in figure 3 illustrates the responses of the actual system and surrogate model to this control strategy beginning at a randomly picked control state $\boldsymbol{u}_0$ under the time-varying hidden state as in figure 1(b). In this scenario, the error of the actual system (blue) and surrogate model (red) are close throughout the state space tracking by the particle filter, and the distribution error is minimized as shown in the right panes of figure 3. As a result, the error at the final state (at $k = 60$) is much less than the uncontrolled case (yellow) whose distribution began by $I_{ref}(n_x, n_y)$.

3. Conclusion

We checked the efficacy of the similarity-based hidden state tracking and control using SMC on a simulation-based surrogate model. The method has proven its sample-efficient and robust behavior for uncertainties and under-specifications. The inference-based optimization is demonstrated for the beam distribution estimate of the assimilated surrogate model. Future directions would be the high-dimensional extension of the SMC algorithms potentially with Ref. [11] and [12], error compensation of the surrogate model itself, and global tracking of multiple plausible clusters of possible states. The theoretical bound on the exponential convergence factor λ should be also investigated.

Acknowledgments

The authors gratefully acknowledge valuable discussions with Hiroyasu Tsukamoto.

References

[1] Beaumont M A 2019 *Annu. Rev. Stat. Appl.* **6** 379–403
[2] Marjoram P, Molitor J, Plagnol V and Tavaré S *Proc. National Academy of Sciences* 26 pp 15324–8
[3] Baragatti M, Grimaud A and Pommeret D 2013 *Statistics and Computing* **23** 535–49
[4] Doucet A and Johansen A M 2009 *Handbook of Nonlinear Filtering* 4–6
[5] Fujii H, Scheinker A, Uchiyama A, Gessner S J, Kamigaito O and Fukunishi N *Proc. 18th Ann. Meeting of Part. Acc. Soc. of Japan* pp 146–50
[6] O'Hagan A and Leonard T 1976 *Biometrika* **63** 201–3
[7] Mudholkar G S and Hutson A D 2000 *J. Stat. Plann. Inference* **83** 291–309
[8] Scheinker A, Cropp F, Paiagua S and Filippetto D 2021 (arXiv:2105.03584)
[9] Scheinker A, Cropp F, Paiagua S and Filippetto D 2021 *Sci. Rep.* **11** 19187
[10] Nelder J A and Mead R 1965 *Comput. J.* **7** 308–13
[11] Poterjoy J 2016 *Mon. Weather Rev.* **144** 59–76
[12] Septier F and Peters G W 2016 *IEEE J. Sel. Top. Signal Process.* **10** 312–27

A beam position monitor for electron bunch detection in the presence of a more intense proton bunch for the AWAKE Experiment

Collette Pakuza[1], Philip Burrows[1], Roberto Corsini[2], Wilfrid Farabolini[2], Pierre Korysko[2], Michal Krupa[2], Thibaut Lefevre[2], Stefano Mazzoni[2], Eugenio Senes[2] and Manfred Wendt[2]

[1] John Adams Institute of Accelerator Science, University of Oxford, Oxford, United Kingdom
[2] CERN, Geneva, Switzerland

E-mail: collette.pakuza@physics.ox.ac.uk

Abstract. The Advanced Proton Driven Plasma Wakefield Experiment (AWAKE) at CERN uses 6 cm long proton bunches extracted from the Super Proton Synchrotron (SPS) at 400 GeV beam energy to drive high gradient plasma wakefields for the acceleration of electron bunches to 2 GeV within a 10 m length. Knowledge and control of the position of both copropagating beams is crucial for the operation of the experiment. Whilst the current electron beam position monitoring system at AWAKE can be used in the absence of the proton beam, the proton bunch signal dominates when both particle bunches are present simultaneously. A new technique based on the generation of Cherenkov diffraction radiation (ChDR) in a dielectric material placed in close proximity to the particle beam has been designed to exploit the large bunch length difference of the particle beams at AWAKE, 200 ps for protons versus a few ps for electrons, such that the electron signal dominates. Hence, this technique would allow for the position measurement of a short electron bunch in the presence of a more intense but longer proton bunch. The design considerations, numerical analysis and plans for tests at the CERN Linear Electron Accelerator for Research (CLEAR) facility are presented.

1. Introduction
The AWAKE experiment uses the wakefields generated by a long proton bunch with length of the order of a few hundred ps to accelerate short electron bunches [1]. The set-up is shown in figure 1. The proton bunches arrive every 15-30 s from the SPS and their typical parameter ranges are given in table 1. They then propagate colinearly with a 120 fs long, 780 nm central wavelength laser pulse inside a 10 m long rubidium (Rb) vapour source. The laser is used to singly ionise the Rb vapour to a plasma with the same density as the vapour. This density can be chosen in the range $1\text{-}10\times10^{14}\ \text{cm}^{-3}$ required for the generation of wakefields of the order of 1 GV/m [2]. To effectively drive large amplitude wakefields, the drive bunch should have transverse and longitudinal sizes of the order of the plasma wavelength which for the given plasma density range is $\sim$1 mm. Since the proton bunch is several cm long, the generation of large amplitude wakefields relies on a process called seeded self-modulation (SSM). Here, the proton bunch is divided into a train of micro-bunches with longitudinal size less than and period equal to the plasma wavelength [3]. The relativistic ionisation front of the laser pulse seeds the self-modulation

process creating a reference phase for the correct injection of electrons in order for them to be focussed and accelerated. During Run 1 (2016-2018), the self-modulation of an SPS proton bunch into a train of over 20 micro-bunches and the acceleration of electrons from 19 MeV to 2 GeV in a 10 m plasma cell was successfully demonstrated [1]. The goals for Run 2 (2021-2024) include SSM via electron bunch seeding and the addition of a density step in the vapour source for maintaining wakefield amplitudes at maximum level over longer distances [4].

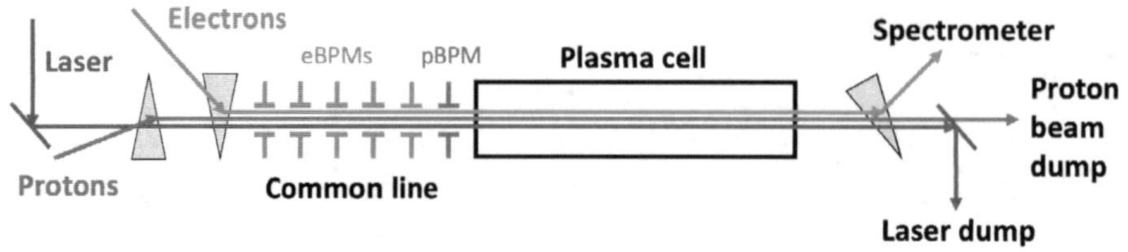

Figure 1. Schematic of the AWAKE experiment.

Amongst the diagnostics for measuring the occurrence of SSM in the proton bunch and the electron beam parameters after acceleration, the measurement of the beam position of the electron and proton bunches before the plasma cell for alignment purposes is a crucial aspect of the experiment. The beam position monitoring system for the protons is composed of 21 dual plane button-style beam position monitors (BPMs) between the extraction point from the SPS to downstream of the plasma cell [5]. For the electrons, there are 5 shorted stripline BPMs in the common beam line as shown in figure 1. The electron BPMs operate at 404 MHz with position resolution of 10 μm in both planes [6].

Table 1. Proton and electron parameters at AWAKE [7].

Parameter	Protons	Electrons
Energy/GeV	400	0.01-0.02
Bunch length/ps	200-400	0.3-10
Bunch charge/nC	48	0.1-1

2. Motivation for a high frequency electron beam position monitor

In the common beam line at AWAKE, both proton and electron bunches are present. The electron BPM system operating at 404 MHz are able to detect electrons when the protons are not present. If both beams are present, the proton signal dominates and prevents the position measurement of the electron bunch. This is shown via the spectra of the bunches in figure 2 assuming Gaussian bunches. Currently, the beam position of the electron bunches is determined by utilizing the different repetition rates of both beams. The electron bunches are produced at a rate of 10 Hz while the protons arrive every 15-30 s. The position of the electron bunch when the protons are present is extrapolated from shots before and after the proton bunch. In order to make a more meaningful measurement, a BPM to measure the electron bunch in the presence of a proton bunch is required. This is done by exploiting the different bunch lengths of the particle beams and operating at a high enough frequency where the electron signal dominates.

3. A BPM based on Cherenkov diffraction radiation

3.1. Theory of Cherenkov diffraction radiation

The theory of Cherenkov radiation developed by Frank and Tamm in 1937 [8] describes the generation of radiation as a charged particle traverses a medium at a velocity greater than the phase velocity of light in that medium. A coherent wavefront is formed at the Cherenkov angle. Cherenkov diffraction radiation

(ChDR) refers to a charged particle travelling in close proximity to, but not inside, a dielectric target and polarises the atoms on its surface [9]. The radiation emitted from this interaction, known as ChDR, propagates through the medium at the Cherenkov angle. Due to its non-invasive nature and well-defined angle of emission, ChDR could provide a useful tool for particle beam diagnostics [10,11]. In the context of beam position monitoring, dielectric buttons, also referred to as radiators, can be used to generate ChDR to couple to the beam field for detection.

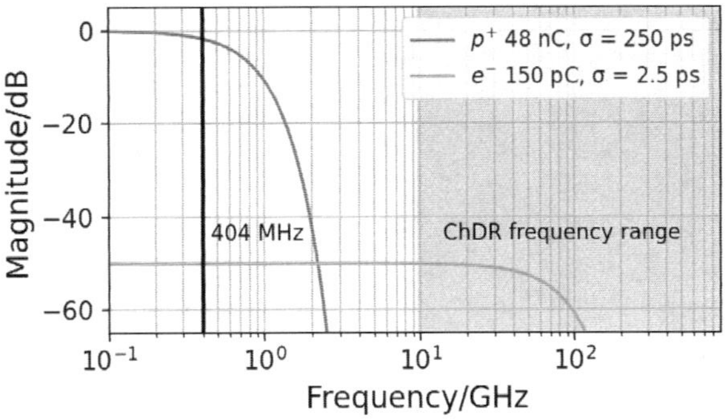

Figure 2. Spectra of the proton and electron bunches at AWAKE assuming Gaussian longitudinal distributions.

3.2. Design considerations

To utilise the already limited space available in the common line at AWAKE, the ChDR radiators were designed to fit into the button housing that are compatible with the existing proton BPM (pBPM) body. This was in view of converting the pBPM before the plasma cell into a ChDR BPM and installing an additional ChDR BPM in the furthest drift section upstream of the plasma cell in the common line. Due to the geometry of the pBPM buttons and the radiation tolerance, alumina with a relative dielectric permittivity of 9.4 and Cherenkov angle of 71° was chosen as the radiator material. The design is shown in figure 3.

From previous streak camera measurements of the proton bunch profile at AWAKE, it was found that the spectral content of the proton bunches extends to frequencies higher than expected for an ideal Gaussian bunch [12]. For typical electron parameters, the frequency range in which the electrons dominate is between 20-40 GHz. If the filtering of the electron signal is to be done at the radiator, then this would require radiator diameters less than a few mm. This poses limitations in the power output and increases the fragility of the radiator. As a result, a diameter of 6 mm corresponding to a cut-off frequency of 9.6 GHz was chosen to partially filter the signal whilst the remainder of the high-pass filtering is realised by commercially available WR28 rectangular waveguides operating in the Ka-band (26.5-40 GHz) with a cut-off of 21.1 GHz. Due to the availability of in-house components, an operating frequency of 30 GHz was chosen.

Since the signal needs to be transmitted efficiently from the circular alumina waveguide to the WR28 waveguide at the chosen operating frequency, a transition was designed based on a quarter-wave impedance transformer for maximum power transmission between the two dominant modes of the waveguides. The design involves a ∅6 mm, 9.9 mm long fused quartz piece inserted between the alumina and WR28 waveguide. This corresponded to 99% power transmission at 30 GHz. The transition can be attached to the extended end of the alumina shown in figure 3.

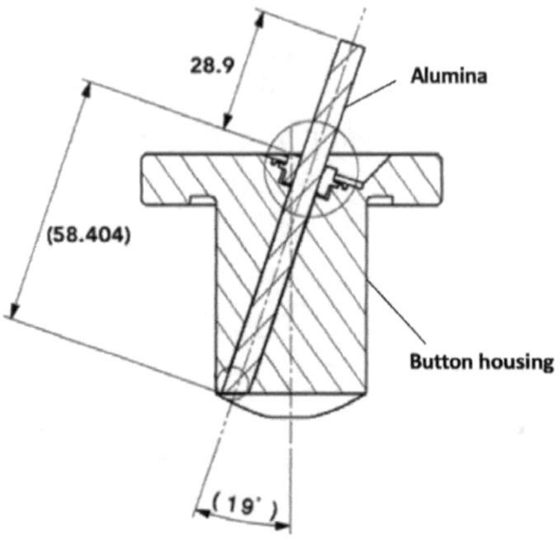

Figure 3. Schematic of ChDR BPM button with radiator angled at 71°.

3.3. Numerical simulations of the ChDR BPM

To better understand the behaviour of the propagation of ChDR inside the radiator, one arm of the pick-up (PU) was modelled and simulated in CST Studio Suite Wakefield solver [13]. The model is shown in figure 4. It includes the ∅60 mm AWAKE beampipe, one ∅6 mm alumina angled at 71°, and the electron beam shown by the blue and orange arrows. The electron bunch parameters are similar to those at AWAKE. The background material is set as a perfect electrical conductor. From the time evolution of the field shown in figure 4, it can be seen that the ChDR is generated as the particle beam passes the surface of the radiator and propagates at the Cherenkov angle inside the radiator. In addition, radiation generated from the finite geometry of the PU is also coupled out of the PU and can be seen by the succeeding wavefronts that are not propagating at the Cherenkov angle. As these fronts travel along the alumina, they undergo multiple reflections.

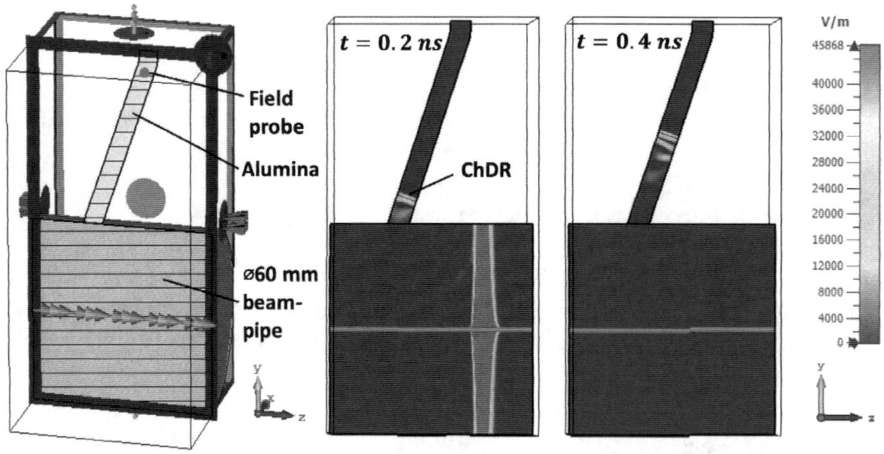

Figure 4. 3D CST model of one arm of the monitor (left) and time evolution of the radiation generated as the beam passes the radiator presented as a contour plot of the absolute E-field (right).

A field probe was placed at a location 86 mm inside the alumina to measure the E-field as a function of time. This was recorded for varying vertical beam offsets. The peak field was plotted as a function of the beam offset and is shown in figure 5. Assuming the system is symmetric, this behaviour was mirrored for the opposite arm of the PU. By taking the difference over sum of the measured peak fields as a function of the beam offset, a position sensitivity of 5.8 %/mm was obtained from the linear region as shown in figure 6.

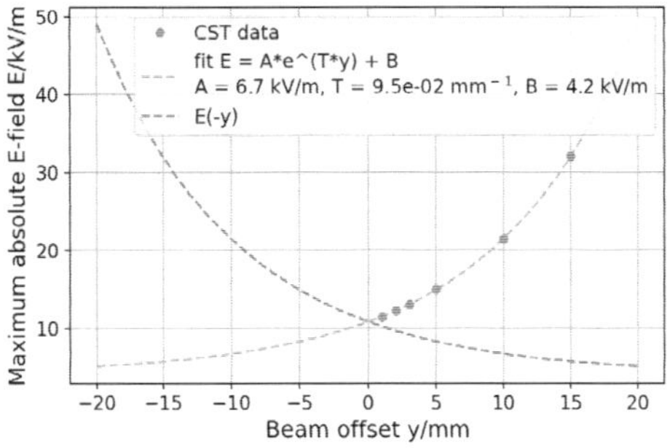

Figure 5. Peak absolute E-field from CST for different vertical beam offsets. The dotted lines are the behaviour of both arms of the plane assuming a symmetric system.

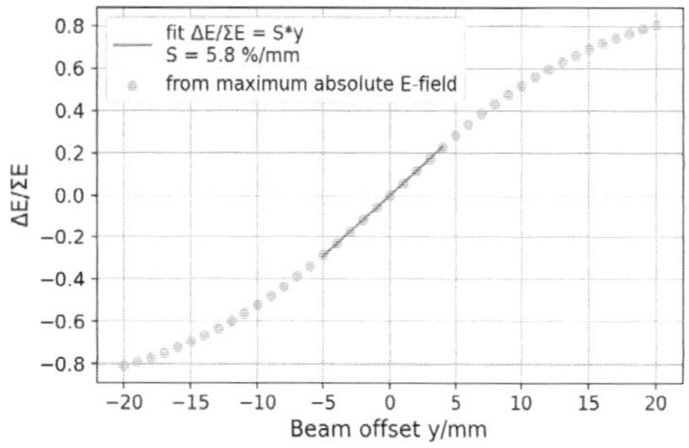

Figure 6. Difference over sum of the E-field measured from the two arms in the vertical plane of the monitor as a function of vertical beam position. The position sensitivity is 5.8 %/mm in the linear region.

3.4. A ChDR monitor for tests at CLEAR

Beside the first successful beam tests at AWAKE following the recent installation of one ChDR BPM [14], a BPM with two ChDR buttons in the horizontal plane is foreseen for dedicated beam studies at CLEAR this year [15]. The monitor will be installed in vacuum approximately 7 m upstream of the beam dump. It will be mounted on a translation stage for position scan measurements. The detection system for each arm will comprise a 30 GHz, 300 MHz bandwidth (BW) band-pass filter (BPF), a variable attenuator and a Ka-band zero-bias Schottky diode detector. The signals will then be measured via a scope. The goal is to scan the parameter space of the electron bunch, measure the signals, and compare them to what is expected from analytical and numerical models. This would provide important information for the continued commissioning of the two ChDR BPMs at AWAKE.

4. Conclusion

The chosen design of the ChDR BPM was modelled and simulated in a numerical electromagnetic solver. The results showed the generation of ChDR at the Cherenkov angle inside the dielectric medium as expected from theory. It also provided the position sensitivity of an ideal ChDR BPM system with ∅6 mm alumina buttons angled at the Cherenkov angle which is 5.8 %/mm. To benchmark the simulation results, in-vacuum tests will be performed at CLEAR this year.

References

[1] Aldi E *et al.* 2018 *Nature* **561** 363
[2] Muggli P *et al.* 2016 *Proc. North American Particle Accelerator Conf. (NAPAC2016)* (Chicago, IL, USA) p 687-9
[3] Muggli P *et al.* 2018 *Plasma Phys. Control. Fusion* **60** 014046
[4] Muggli P and for the AWAKE Collaboration 2020 *J. Phys.: Conf. Ser.* **1596** 012008
[5] Mazzoni S *et al.* 2017 *Proc. 8th Int. Particle Accelerator Conf. (IPAC'17)* (Copenhagen, Denmark) pp 404-7
[6] Gorgisyan I *et al.* 2018 *J. Phys.: Conf. Ser* **1067** 072015
[7] Pepitone K *et al.* 2018 *Nucl. Instrum. Methods Phys. Res., Sect.* A **909** 102
[8] Frank I M and Tamm I E 1937 *Dokl. Akad Nauk SSSR* **14** 107

[9] Shevelev M V and Konkov A S 2014 *J. Exp. Theor. Phys.* **118** 501
[10] Alves D *et al.* 2019 *Proc. 8th Int. Beam Instrum. Conf. (IBIC2019)* (Malmö, Sweden) p 660-4
[11] Curcio A *et al.* 2020 *Phys. Rev. Acc. Beams* **23** 022802
[12] Senes E 2020 PhD Thesis University of Oxford
[13] CST Studio Suite `https://www.cst.com/`
[14] Senes E *et al.* 2022 *Proc. 13th Int. Particle Accelerator Conf. (IPAC'22)* (Bangkok, Thailand) pp 343-6
[15] Gamba D *et al.* 2018 *Nucl. Instrum. Methods Phys. Res., Sect.* A **909** 480

6D Phase space diagnostics based on adaptively tuned physics-informed generative convolutional neural networks

Alexander Scheinker[1], Daniele Filippetto[2] and Frederick Cropp[3]

[1] Los Alamos National Laboratory, 87544, Los Alamos, USA
[2] Lawrence Berkeley National Laboratory, 94720, Berkeley, USA
[3] University of California Los Angeles, 90095, Los Angeles, USA

E-mail: `ascheink@lanl.gov`

Abstract. A physics-informed generative convolutional neural network (CNN)-based 6D phase space diagnostic is presented which generates all 15 unique 2D projections (x, y), (x, y'),...,(z, E) of a charged particle beam's 6D phase space (x, y, z, x', y', E). The CNN is trained by supervised learning over a wide range of input beam distributions, accelerator parameters, and the associated 6D beam phase spaces at multiple accelerator locations. The CNN is applied in an un-supervised adaptive manner without knowledge of the input beam distribution or accelerator parameters and is robust to their unknown time variation. Adaptive feedback automatically tunes the low-dimensional latent space of the encoder-decoder CNN to predict the 6D phase space based only on 2D (z, E) longitudinal phase space measurements from a device such as a transverse deflecting RF cavity (TCAV). This method has the potential to provide diagnostics beyond the existing state of the art at many accelerator facilities. Studies are presented for two very different accelerators: the 5-meter-long ultra-fast electron diffraction (UED) HiRES compact accelerator at LBNL and the kilometer long plasma wakefield accelerator FACET-II at SLAC.

1. Introduction

Particle accelerators are large complex systems with many coupled components. Accelerator beams are complex objects with dynamics governed by nonlinear collective effects such as space charge and coherent synchrotron radiation. Because of their complexity, particle accelerator controls and diagnostics can greatly benefit from advanced machine learning (ML) [1], and control theory techniques.

Supervised learning techniques are being applied at CERN for the reconstruction of magnet errors in the incredibly large (thousands of magnets) LHC lattice [2]. Bayesian methods have been developed for online accelerator tuning of the LCLS [3], Bayesian methods with safety constraints are being developed at the SwissFEL and the High-Intensity Proton Accelerator at PSI [4], and at SLAC Bayesian methods are being developed for the challenging problem of hysteresis [5] and surrogate models are being developed for the beam at the injector [6]. Convolutional neural networks (CNN) have been used to generate incredibly high resolution virtual diagnostics of the longitudinal phase space (LPS) of the electron beam in the EuXFEL [7]. A laser plasma wakefield accelerator has also been optimized by utilizing Gaussian processes at the Central Laser Facility [8].

A limitation of standard ML methods is the requirement of re-training whenever a system changes. Because accelerators are changing continuously and detailed beam measurements usually interrupt operations repetitive re-training is not a feasible solution. Recently, powerful model-independent feedback control methods, known as extremum seeking (ES), have been developed which can handle unknown and quickly time-varying nonlinear systems in which the direction of the controller's input is unknown and quickly time-varying [9, 10]. For example, it is possible to use ES for RF cavity resonance control based only on ambiguous reflected power measurements [11]. A limitation of model-independent feedback is the possibility of getting stuck in a local minimum.

Due to the complimentary strengths and weaknesses of ML and model-independent feedback, efforts are being made to combine the two fields via adaptive ML (AML) which provides the best of both worlds: an ability to learn directly from large complex data, while maintaining robustness to time variation and distribution shift. The first demonstration of the AML approach was the use of neural networks together with ES for automatic control of the time-varying longitudinal phase space distribution of the LCLS beam [12]. AML methods have also combined CNNs and ES to track time-varying input beam distributions at the HiRES UED [13], and preliminary results have shown an ability to adaptively tune the low-dimensional latent space of encoder-decoder CNNs to track all 15 unique 2D projections of beam's 6D phase space despite unknown and time-varying input beam distributions and accelerator and beam parametes [14].

2. AML for 6D diagnostics

In this work we present simulation-based AML studies at the HiRES UED [15], for predicting all 15 unique 2D projections of a charged particle beam with unknown and time-varying input beam conditions at the photocathode, unknown beam charge and injector solenoid magnet strength, and demonstrate that this method has the capability to accurately predict beyond the span of the training set data.

Time-varying systems, or systems with distribution shift, are an open problem and an active area of research in the ML community [16–20]. In this work we tackle the problem of distribution shift by incorporating model-independent adaptive feedback directly within the architecture of an encoder-decoder CNN which takes beam distributions and parameters (charge and solenoid current) as inputs and generates 15 256×256 pixel 2D projections ($\sim 10^6$ dimensions) of the beam's 6D phase space downstream from the injector, as shown in figure 1, where the CNN's predictions are compared to the ground truth for a beam with unknown input distribution and unknown charge passing through an accelerator with an unknown solenoid strength. Our CNN is applied in an unsupervised adaptive way by squeezing down to a 2D latent space between the encoder and decoder sections which is adaptively tuned using ES with time-varying cost

$$C(t) = \int\int |\rho_{z,E}(t) - \hat{\rho}_{z,E}(t)|\, dEdz, \tag{1}$$

a comparison between the CNN's longitudinal phase space (LPS) prediction $\hat{\rho}_{z,E}$ and the measurement of the LPS as provided by a TCAV $\rho_{z,E}$. By forcing the CNN to simultaneously generate all 15 projections of the 6D phase space we introduced observational biases directly through data embodying the underlying physics, allowing the CNN to learn functions that reflect the physical structure of the data [21].

3. Tracking phase space

Figure 2 shows the results of changing the input beam distribution, beam charge, and solenoid current far beyond the span of the training data. The black lines show the change relative to the initial starting condition. The green lines show the CNN's errors if assuming known beam distribution, charge, and solenoid strength, with catastrophic failure beyond the span of the

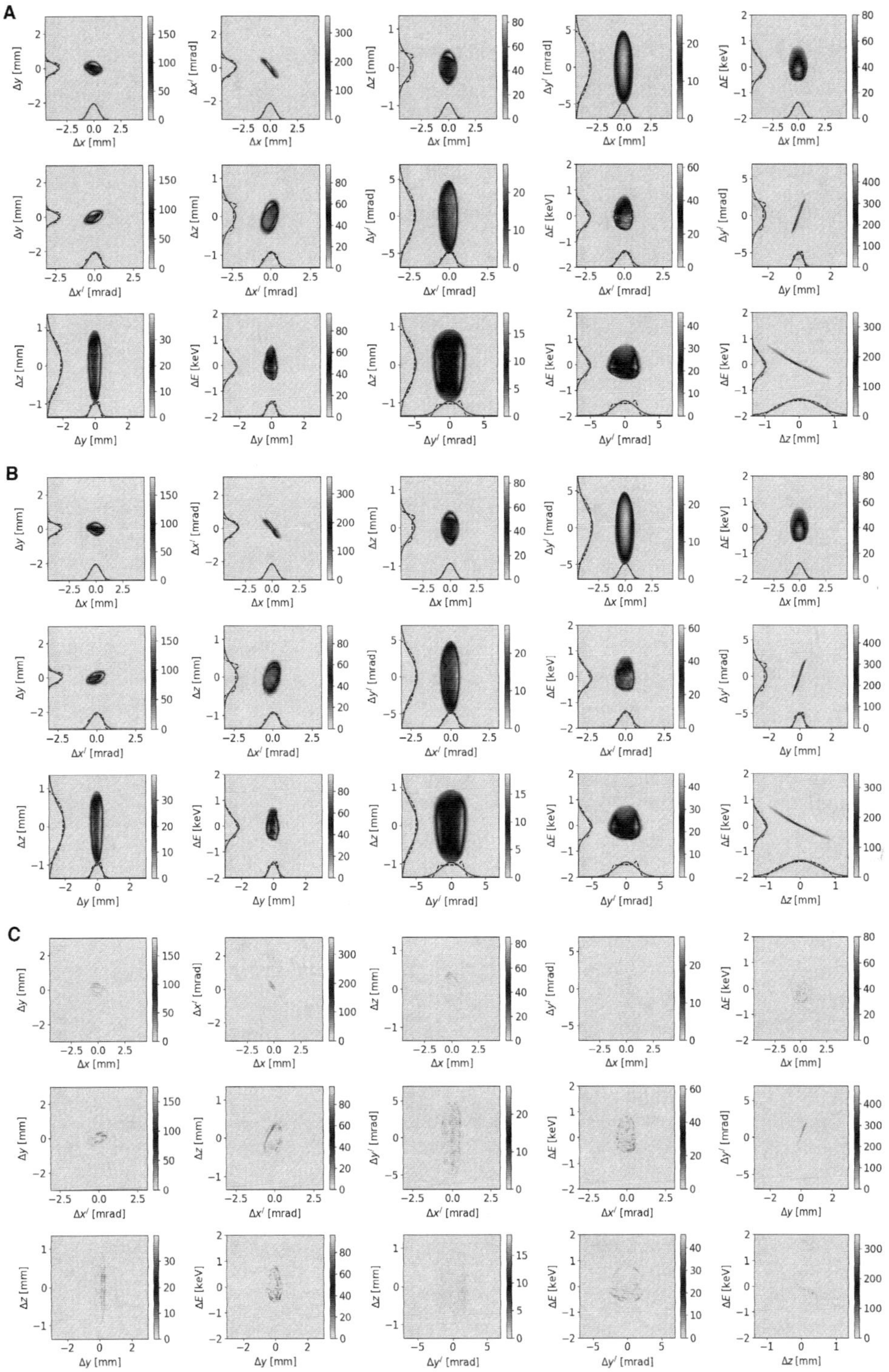

Figure 1. **A:** 15 2D projections of a beam's 6D phase space, ρ_{ij}. **B:** The same 15 2D projections generated by the CNN, $\hat{\rho}_{ij}$. **C:** Absolute difference $|\hat{\rho}_{ij} - \rho_{ij}|$. Color scale of each projection set by maximum value of actual projection: max $\{\rho_{ij}\}$.

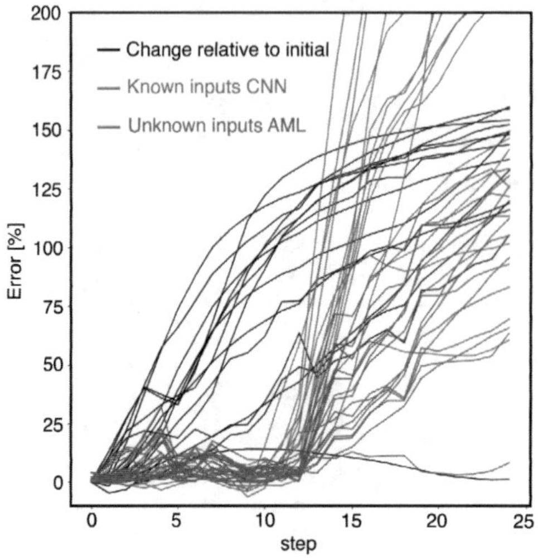

Figure 2. Error is shown in percent for each of the 15 projections of the beam's 6D phase space as the input beam distribution, charge, and solenoid strength are moved beyond the span of the training set.

training set where the CNN's predictions are far worse than simply doing nothing. Finally, the red lines show the error when we do not have access to the unknown beam distribution, charge, and solenoid strength, but with the use of adaptive feedback which has access to the (z, E) projection to be used as feedback within the latent space by continuously minimizing the cost function , equation (1).

4. Conclusions

We have demonstrated preliminary studies of a physics-informed AML method for tracking all 15 projections of a charged particle beam with unknown and time-varying initial distribution and charge at the photocathode and unknwon and time-varying solenoid strength at the injector based only on TCAV measurements of the (z, E) LPS.

Acknowledgments

Work funded by US Department of Energy, DOE Office of Science Graduate Student Research (SCGSR) contract numbers 89233218CNA000001 and DE-AC02-05CH11231 and by the NSF under Grant No. PHY-1549132.

References

[1] Gentile U and Serio L 2017 *Proc. 2017 Eur. Conf. on Electrical Engineering and Computer Science (EECS)* (Bern, Switzerland) pp 90–95
[2] Fol E, Tomás R and Franchetti G 2021 *Eur. Phys. J. Plus* **136** 365
[3] Duris J, Kennedy D and Ratner D 2018 *Proc. 61st ICFA Adv. Beam Dyn. Workshop High-Intensity High-Brightness Hadron Beams (HB2018)* (Daejeon, Korea)
[4] Kirschner J, Mutný M, Krause A, de Portugal J C, Hiller N and Snuverink J 2022 *Preprint* ArXiv:2203.13968
[5] Roussel R J and Hanuka A *Proc. 12th Int. Particle Accelerator Conf. (IPAC'21)* (Campinas, Brazil) pp 2159–62
[6] Gupta L, Edelen A, Neveu N, Mishra A, Mayes C and Kim Y K 2021 *Mach. Learn.: Sci. Technol.* **2** 045025
[7] Zhu J, Chen Y, Brinker F, Decking W, Tomin S and Schlarb H 2021 *Phys. Rev. Appl.* **16** 024005
[8] Shalloo R *et al.* 2020 *Nat. Commun.* **11** 1–8
[9] Scheinker A and Krstić M 2012 *IEEE Trans. Autom. Control* **58** 1107–22

[10] Scheinker A and Scheinker D 2016 *Automatica* **69** 250–57
[11] Scheinker A 2016 *IEEE Trans. Control Syst. Technol.* **25** 1521–28
[12] Scheinker A, Edelen A, Bohler D, Emma C and Lutman A 2018 *Phys. Rev. Lett.* **121** 044801
[13] Scheinker A, Cropp F, Paiagua S and Filippetto D 2021 *Sci. Rep.* **11** 1–11
[14] Scheinker A 2021 *J. Instrum.* **16** P10008
[15] Filippetto D and Qian H 2016 *J. Phys.* B*: At. Mol. Opt. Phys.* **49** 104003
[16] Quiñonero-Candela J, Sugiyama M, Schwaighofer A and Lawrence N D 2008 *Dataset shift in machine learning* (Massachusetts: MIT Press)
[17] Moreno-Torres J G, Raeder T, Alaiz-Rodríguez R, Chawla N V and Herrera F 2012 *Pattern Recognit.* **45** 521–30
[18] Sugiyama M and Kawanabe M 2012 *Machine learning in non-stationary environments: Introduction to covariate shift adaptation* (Massachusetts: MIT Press)
[19] Ovadia Y, Fertig E, Ren J, Nado Z, Sculley D, Nowozin S, Dillon J, Lakshminarayanan B and Snoek J 2019 *Advances in neural information processing systems* **32**
[20] Subbaswamy A and Saria S 2020 *Biostatistics* **21** 345–52
[21] Karniadakis G E, Kevrekidis I G, Lu L, Perdikaris P, Wang S and Yang L 2021 *Nat. Rev. Phys.* **3** 422–40

A machine learning approach to electron orbit control at the 1.5 GeV synchrotron light source DELTA

D Schirmer

Center for Synchrotron Radiation (DELTA), TU Dortmund University, Germany

E-mail: detlev.schirmer@tu-dortmund.de

Abstract. Machine learning (ML) methods have found their application in a wide range of particle accelerator control tasks. Among other possible use cases, neural networks (NNs) can also be utilized for automated beam position control (orbit correction). ML studies on this topic, which were initially based on simulations, were successfully transferred to real accelerator operation at the 1.5–GeV electron storage ring of the DELTA accelerator facility. For this purpose, classical fully connected multi-layer feed-forward NNs were trained by supervised learning on measured orbit data to apply local and global beam position corrections. The supervised NN training was carried out with various conjugate gradient backpropagation learning algorithms. Afterwards, the ML-based orbit correction performance was compared with a conventional, numerical-based computing method. Here, the ML-based approach showed a competitive orbit correction quality in a fewer number of correction steps.

1. Introduction

Stable electron orbit control is an important task especially for modern synchrotron light sources. For this purpose, singular value decomposition (SVD) of the orbit response matrix is a standard numerical tool at storage rings worldwide. An alternative concept applies machine learning techniques as an heuristic method, inspired by the pioneering work done at NSLS/BNL [1]. Since 2018, machine learning (ML) based orbit correction (OC) methods have extensively been studied and applied at DELTA [2–5], a 1.5–GeV electron storage ring operated as a synchrotron light source [6] and a new facility for ultrashort pulses in the VUV and THz regime [7, 8].

2. Machine operation

The OC hardware setup of the DELTA storage ring consists of 54 beam position monitors (BPMs) which determine the position in both orbit planes simultaneously as well as 30 horizontal and 26 vertical corrector magnets (steerers) [9]. The ML methods, which were successfully tested by means of simulations on a DELTA storage ring OC model, were applied to the real storage ring.

2.1. ML-Data acquisition

Compared to the uncoupled case in previous real machine studies [3], now a dedicated data acquisition script randomly varies all steerer strengths in both planes at once within intervals from typically ±200 mA up to ±500 mA. The interval limits are a compromise between risk of beam losses and minimizing relative measurement errors due to the limited steerer strength resolution of 2.4 mA. After each perturbation, the steerer strength changes and the emerging closed orbit differences in both planes are measured, error-cleaned (e.g., deleting of hardware and software related readout errors) and recorded. The data pool noise level was estimated to approximately 2% caused by the combined error of BPM and steerer strength readback accuracies, mainly dominated by the limited steerer strength granularity.

2.2. Weighted beam position monitors

To increase the impact of orbit deviations at more important storage ring positions (e.g., synchrotron radiation source points or the injection region) each BPM can be assigned with an individual weight factor. With $\tilde{w}^{\mathrm{BPM}}_{x,z}$ as a diagonal matrix of BPM weight factors, the weighted closed-orbit error $\chi^{w}_{x,z}$ can be evaluated as a scalar quantity for both planes (x, z) by

$$\chi^{w}_{x,z} = \left\| \tilde{w}^{\mathrm{BPM}}_{x,z} \cdot (\overrightarrow{\Delta d}_{x,z} + \tilde{R}_{x,z} \cdot \overrightarrow{\Delta I}_{x,z}) \right\|_2 . \tag{1}$$

The goal for an orbit correction algorithm is to minimize the residual closed-orbit error $\chi^{w}_{x,z}$ for arbitrary orbit deviations $\overrightarrow{\Delta d}_{x,z}$ with respect to any desired reference orbit. The product of the response matrix $\tilde{R}_{x,z}$ and the steerer strength changes $\overrightarrow{\Delta I}_{x,z}$ in equation (1) can be determined by means of a reverse NN (see figure 1). It can also be trained with the experimental data patterns, but now each squared network error $e^2_{pj} = (o^{\mathrm{BPM}}_{pj} - t^{\mathrm{BPM}}_{pj})^2$ must be weighted by an individual BPM weight factor w^{BPM}_{pj} as follows:

$$E^{R}_{\mathrm{mse}} = \frac{1}{P}\sum_{p=1}^{P}\frac{1}{N}\sum_{j=1}^{N} w^{\mathrm{BPM}}_{pj}(o^{\mathrm{BPM}}_{pj} - t^{\mathrm{BPM}}_{pj})^2 . \tag{2}$$

The reverse mean squared NN error E^{R}_{mse} sums up the squared differences between all numbers of neurons N at the NN target t and output o for a specific quantity of data patterns P. Thus, the reverse trained NN, as a representation of the orbit response matrix $\tilde{R}_{x,z}$, is able to determine orbit deviations $\overrightarrow{\Delta d}_{x,z}$ at all BPMs for given steerer strength changes $\overrightarrow{\Delta I}_{x,z}$. Afterwards, the weighted orbit error $\chi^{w}_{x,z}$ in equation (1) can be minimized using a numerical optimizer, e.g., the BFGS Quasi-Newton method [10]. In addition, the optimizer itself has also been replaced by a pre-trained NN. For this purpose, the optimizer has to pre-calculate the optimum $\chi^{w}_{x,z}$-values for all measured orbit deviations $\overrightarrow{\Delta d}_{x,z}$. These data pairs again serve as labeled input/target data to train a dedicated NN as an optimizer substitute.

2.3. Neural network training

Since each corrector strength variation normally affects the beam displacement at all BPM positions in the storage ring, a fully connected feed-forward neural network (FFNN) was specified as the NN connection architecture. Thus, NNs to be trained by the experimental data are composed of three layers with a total of 272 neurons (108/108/56) and approximately 17700 weights and biases (see figure 1). The supervised network training was tested with various conjugate gradient backpropagation learning algorithms. Most effective learning was achieved with scaled conjugate gradient (scg) algorithms [11]. Typical forward and reverse training curves for pure training data (blue) and the related validation data sets (green) are shown in figure 2. In both cases, the network output error is reduced continuously, without over- or under-fitting issues, mainly limited by the data noise level. The best forward validation performance with an E_{mse}-value of $7.3 \cdot 10^{-4}\,\mathrm{A}^2$ was reached with no significant improvement beyond approx. 1600

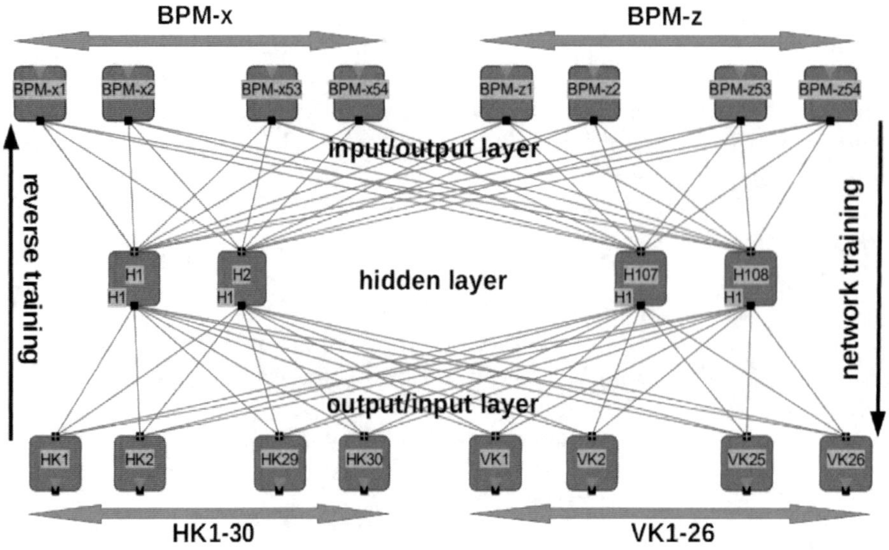

Figure 1. Schematic layout of the neural network topology for a full x,z-coupled orbit correction at the storage ring DELTA. The input layer represents 54 BPMs for each plane, fed by measured orbit deviations. It is connected via a 'hidden' layer with the output values for 56 corrector magnets (HK1-30, VK1-26). The correction considers both transverse coordinates (x,z) as well as their coupling. In total, the network consists of approx. 17700 connections (red lines, only partially shown). For reverse training, the input and output layers are swapped.

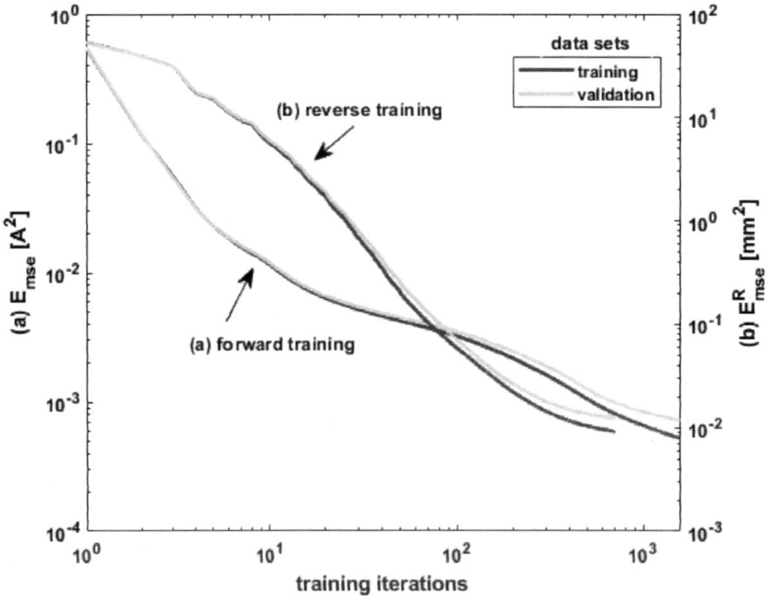

Figure 2. Supervised NN training, based on a scaled conjugate gradient (scg) backpropagation algorithm, with experimental data (blue) and verification of the NN performance with an additional 'unseen' validation test data set (green).

scg-training iterations applying a full batch training. The reverse training reduced the network output error E^R_{mse} to $1.2 \cdot 10^{-2}\,\mathrm{mm}^2$ after about 700 iterations. In both cases, the network's fitness gains approximately three orders of magnitude compared to the starting values, which indicates sufficient fitness performance. Hence, conventional NNs are able to learn and generalize the correlation between orbit deviations and the related steerer strength variations.

3. Benchmark results

The performance of the ML-based OC program was benchmarked against a recently implemented numerical approach [12,13] at different comparison terms. The tests considered actual reference settings, which include the currently valid reference orbit and the corresponding weight factors (wf) for all K BPMs in operation. Various arbitrary steerer-induced orbit errors have been enforced (a–d) in both orbit planes, respectively (hk/vk). The benchmark results are compared in figure 3a (ML-based) and figure 3b (numerical) separately for each orbit plane.

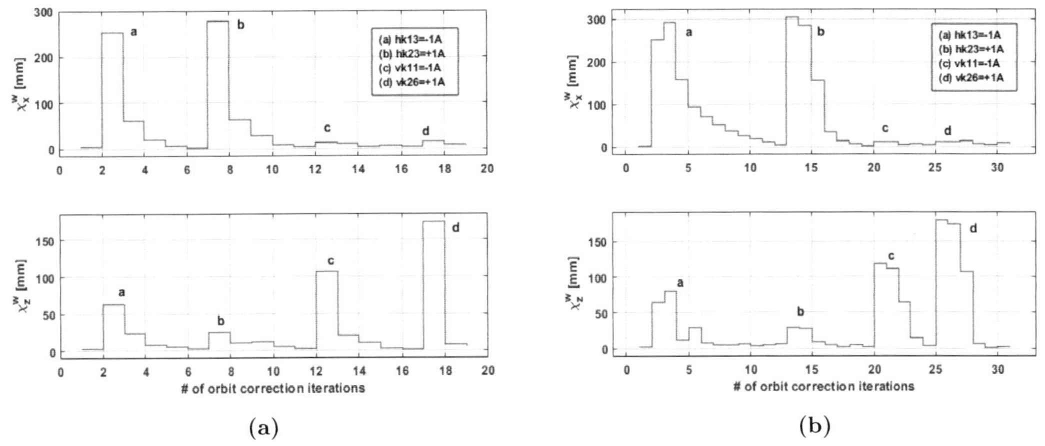

Figure 3. (a): Weighted rms orbit error $\chi^w_{x,z}$ for different steerer-induced (hk/vk) orbit disturbances (a-d) compensated iteratively by the ML-based OC program. Due to x,z-coupling, the provoked deviations appear in both orbit planes. A total of 18 steps were required to compensate for all disturbances (a-d). In comparison to figure 3b, similar final residual orbit errors are achieved in significantly fewer iteration steps. (b): Weighted rms orbit error $\chi^w_{x,z}$ to benchmark the orbit correction quality and convergence for a conventional numerical method [12,13]. The correction was performed iteratively for the same steerer-based (hk/vk) orbit distortions (a-d) as presented in figure 3a. In total, 30 steps were required to compensate for all disturbances (a-d).

In addition, typical but not pre-trained sources for orbit disturbances have been provoked. This includes unmatched closed orbit bumps (e.g., injection dc-bump), strong sextupole strength changes (SF/SD) or ramping of insertion devices (e.g., the undulator U250). The benchmark results are shown in figure 4a (ML-based) and figure 4b (numerical) for each plane separately. In all plots, the residual orbit quality $\chi^w_{x,z} := \left(\frac{1}{K}\sum_{i=1}^{K}\left(w^{\mathrm{BPM}}_{x,z}\cdot\Delta_{x,z}\right)^2_i\right)^{1/2}$ is scored by the weighted root-mean-squared value (wrms) over all K BPMs in both orbit planes (x, z).

As can be seen from the graphs, both OC programs worked similarly stable and they were able to compensate all enforced orbit disturbances without any beam losses. The ML-based OC needs approximately 2 to 4 orbit correction iterations to equalize the individually provoked orbit deviations. Even distortions (e–j), which have not trained during supervised learning were compensated with analogous quality. After each provoked error, the residual weighted orbit error converged to less than 3 mm, which corresponds to several ten μm in real beam offset.

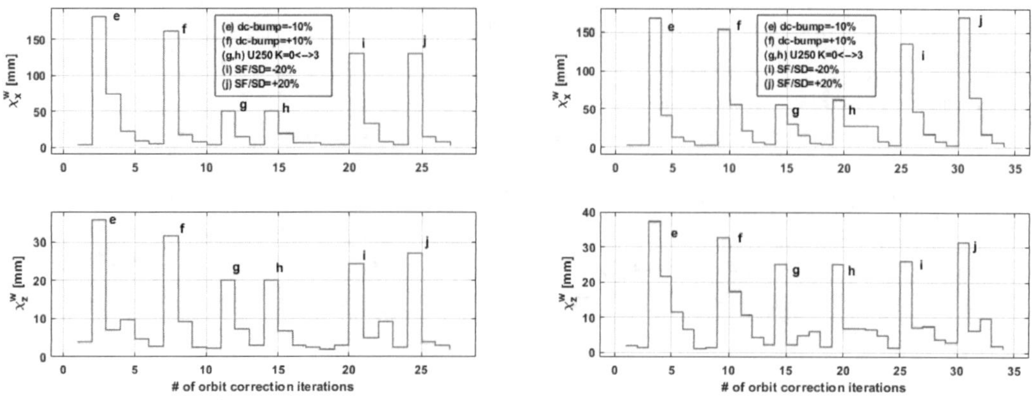

(a) (b)

Figure 4. (a): Individual ML-based correction steps to compensate orbit deviations caused by different error sources (e-j) which were not included in the training data sets. Although these types of error sources were not considered during NN training, all disturbances (e-j) could be compensated in 26 steps. (b): Correction for different scenarios of enforced orbit deviations (e-j) performed with a conventional numerical orbit correction program [12, 13]. In this case, in total 33 steps were needed to compensate the same perturbations (e-j) as depicted in figure 4a.

In direct comparison, the standard numerical OC method requires more orbit correction steps to compensate the same provoked orbit disturbances. The residual weighted orbit error converged also to less than 3 mm, but overall, in these benchmark examples, the standard OC version needs in total 63 steps (see figure 3b and figure 4b) compared to 44 steps for the ML-based implementation (see figure 3a and figure 4a).

4. Summary

A machine learning-driven orbit correction algorithm has successfully been implemented at the 1.5-GeV electron storage ring of the DELTA accelerator facility. The achieved remaining orbit error is similar to the results of standard SVD-like correction approaches. In general, ML-based methods require fewer correction steps which leads to faster orbit correction convergence. Because ML techniques based on training with real machine data, this method automatically incorporates storage ring imperfections (e.g., alignment errors) and non-linearities (e.g., magnetic fringe fields). Furthermore, it has been demonstrated that NNs were also able to cope with beam disturbances that had not been trained before. In addition, even changes of BPM offsets, e.g. due to realignments of individual lattice magnets, can be taken into account, simply by editing the related orbit reference data file. Therefore, retraining of the entire neural network is not required. In general, once successfully trained, the NN-based application showed high reliability, numerical stability and robustness.

References

[1] Bozoki E and Friedman A 1994 *Proc. 4th Eur. Particle Accelerator Conf. (EPAC'94)* (London,UK) pp 1589–91
[2] Schirmer D 2018 *Proc. 9th Int. Particle Accelerator Conf. (IPAC'18)* (Vancouver, Canada) pp 4855–58
[3] Schirmer D 2019 *Proc. 17th Int. Conf. on Accelerator and Large Experimental Physics Control Systems (ICALEPCS'19)* (New York, USA) pp 1426–30

[4] Schirmer D, Althaus A, Hüser S, Khan S, Schüngel T 2021 *17th DELTA Annual Report 2021* (Dortmund, Germany) pp 3–4

[5] Schirmer D, Althaus A, Hüser S, Khan S, Schüngel T 2021 *Proc. 18th Int. Conf. on Accelerator and Large Experimental Physics Control Systems (ICALEPCS'21)* (Shanghai, China) pp 631–35

[6] Tolan T, Weis T, Westphal C and Wille K 2003 *Synchrotron Radiat. News* **16(2)** pp 9–11

[7] Khan S *et al.* 2011 *Synchrotron Radiat. News* **24** pp 18–23

[8] Khan S *et al.* 2013 *Synchrotron Radiat. News* **26** pp 25–29

[9] Grewe M 2005 SVD-basierte Orbitkorrektur am Speicherring Delta Ph.D. thesis Dortmund University Germany

[10] Broyden-Fletcher-Goldfarb-Shanno (BFGS) algorithm `https://en.wikipedia.org/wiki/Broyden-Fletcher-Goldfarb-Shanno\_algorithm`

[11] Møller M F 1993 *Neural Networks* **6** pp 525–533

[12] Kötter S, Riemann B and Weis T 2017 *Proc. 8th Int. Particle Accelerator Conf. (IPAC'17)* (Copenhagen, Denmark) pp 673–75

[13] Kötter S, Glassl A, Isbarn B D, Rohde D, Sommer M and Weis T 2018 *Proc. 9th Int. Particle Accelerator Conf. (IPAC'18)* (Vancouver, Canada) pp 3507–10

A data-driven anomaly detection on SRF cavities at the European XFEL

A Sulc, A Eichler and T Wilsken

Deutsches Elektronen-Synchrotron DESY, Notkestr. 85, 22607 Hamburg, Germany

E-mail: sulcan@desy.de

Abstract. The European XFEL is currently operating with hundreds of superconducting radio frequency cavities. To be able to minimize the downtimes, prevention of failures on the SRF cavities is crucial. In this paper, we propose an anomaly detection approach based on a neural network model to predict occurrences of breakdowns on the SRF cavities based on a model trained on historical data. We used our existing anomaly detection infrastructure to get a subset of the stored data labeled as faulty. We experimented with different training losses to maximally profit from the available data and trained a recurrent neural network that can predict a failure from a series of pulses. The proposed model is using a tailored architecture with recurrent neural units and takes into account the sequential nature of the problem which can generalize and predict a variety of failures that we have been experiencing in operation.

1. Introduction

The superconducting radio-frequency (SRF) cavities are responsible for accelerating beams which are used in the European X-ray Free Electron Laser (EuXFEL) to obtain extremely brilliant X-ray photon light.

Particle accelerators use the cavity resonators operating in radio-frequency spectra to accelerate particles by synchronization with their frequency. The cavities accelerate and energize particles by the induced alternating electric field.

For superconductivity, it is necessary to maintain the cavities cooled to very low temperatures, usually near absolute zero, with a cryogenic system. The cryogenic system maintains the temperature to preserve the superconductivity. The superconductivity minimizes the losses through the wall to a minimum and thus almost all RF power can be transmitted to the passing beam.

EuXFEL is currently operating 784 SRF cavities and it is necessary to use automated algorithms to prevent failures. One kind of failure we are particularly interested in are quenches. A quench is when cavity walls lose their superconductivity due to temperature breakdown. It leads to a loss of superconductivity and energy is dissipated through the cavity walls (the surrounding helium bath is heated up) thus the quality factor decreases, i.e. efficiency decreases. Although the quench limits are experimentally tested and set in the firmware to hard limits, there are numerous situations where the cavity can quench due to e.g. degradation which lowers quench limits.

EuXFEL SRF cavities are operating in pulsed mode, therefore we have available a sequence of waveforms with a fixed length. The current quench detection system at EuXFEL is observing

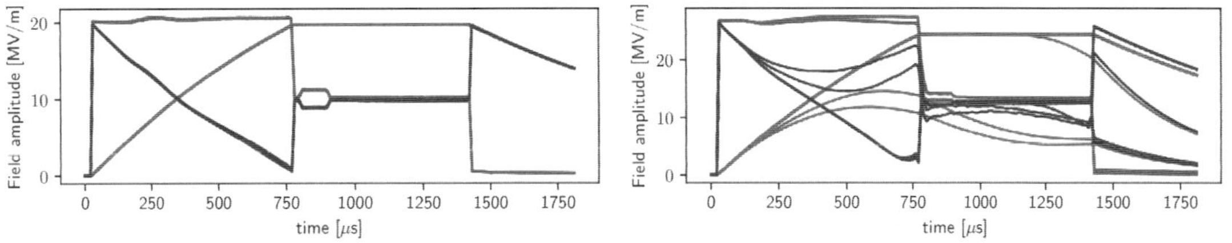

Figure 1. Two examples of amplitudes of healthly (left) and quenching amplitudes (right). Probe $\mathbf{p}$, forward $\mathbf{f}$ and $\mathbf{r}$ signals are depicted red, green, and blue respectively.

the quality factor [1]. It is one of the classic methods for the detection and prevention of such quenches. In [2] an online approach for quench detection based on the calculation of detuning and bandwidth on superconducting cavities is presented which is specially tailored to continuous wave operation. A model-based approach for anomaly detection on cavities is shown in [3]. In [4, 5] the parity space method is used to detect anomalies on SRF cavities. Recently data-driven machine learning approaches [6, 7] are used for cavity breakdown prediction on cavities.

Our currently deployed quench detection server [1] provides a daily overview of probable quenches. Recently, EuXFEL created a dataset of events that are probable faults. The availability of such a labeled dataset allowed us to experiment with data-driven machine learning models. This paper presents a study of data-driven anomaly detection to detect faults on RF cavities tailored to the case of EuXFEL. We demonstrate that vanilla data-driven machine learning methods can be trained to predict potential failures with very limited access to training data labeled as faulty.

The structure of this paper is the following: First, we describe the procedure of preprocessing data. Then, we present details of the proposed architecture used for the prediction of faults. Lastly, we show the results of our approach on a test set using different data-driven approaches.

2. Method

2.1. Notation

At a time moment t we observe a pulse which consists of three types of complex-valued waveforms: probe $\mathbf{p}$, forward $\mathbf{f}$, reflected $\mathbf{r}$, see figure 1. Each event consists of a series of pulses, where all waveforms are 1.82 ms long and sampled at 1 MHz. This yields 1820 values per waveform.

Processing the entire 1820 values of each waveform is unnecessarily redundant and computationally and data-intensive, therefore we further subsampled each waveform to 182 values. Furthermore, for case of data handling, we further transformed each type of waveform from amplitude and phase to the IQ coordinates. In summary, we stack a series of transformed IQ waveforms, i.e. 1092 values per pulse $\mathbf{x}$. We further perform normalization of each waveform independently to the $(0, 1)$ range.

2.2. The Model

We experiment with two models. One is based on semi-supervised anomaly loss (SAL) the other with binary cross-entropy loss (BCE). For both models, we use identical architecture with recurrent units, but the final layers are slightly different. The proposed model consists of stacked Long Short-Term Memory (LSTM) units with a linear unit, see figure 2. The first LSTM layer encodes the input $\mathbf{x}$ into 256-dimensional vector $\mathbf{h}_1$, which is further passed another LSTM layer $\mathbf{h}_2$ with identical dimensionality as $\mathbf{h}_1$. The choice of architecture is not arbitrary.

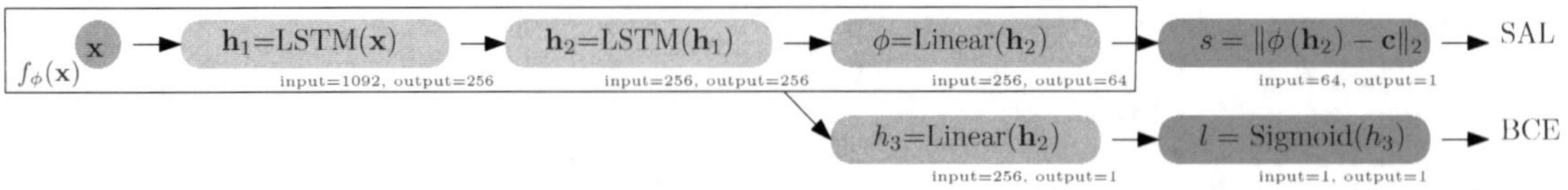

Figure 2. The architecture consists of two LSTM layers. The final linear (green) layer has either 64-dimensional output for a model trained with SAL or 1-dimensional for a model trained with BCE loss. As an input, the network receives a series of signals where a pulse is a 1092 dimensional vector $\mathbf{x}$. For the SAL, the anomaly score s is obtained by calculation of the L_2 distance of the output of the feature layer $f_\theta(\mathbf{x})$ from $\mathbf{c}$. For the model trained with BCE, the likelihood l of a healthy signal is obtained after the application of the Sigmoid function.

Most of the available training data is recorded with less than 250 pulses. Therefore to cover the entire time range of the labeled faults, we designed a two-layer network where input and hidden recurrent neural units have 256 hidden units. This should provide sufficient freedom to train the temporal relations that are contained in individual training events.

2.2.1. Semi-supervised anomaly detection. Values from $\mathbf{h}_2$ are presented into the final linear layer that produces a 64-dimensional vector ϕ. We refer to transformed inputs into the final linear layer as features $f_\theta(\mathbf{x})$. Calculation of anomaly score s is performed by measuring the L_2 distance of the input's features $f_\theta(\mathbf{x})$ from a common centre $\mathbf{c}$.

2.2.2. Classifier. Unlike the model trained with SAL, the classifier trained with BCE has a single binary value that signifies if the output is faulty or not, therefore the last vector $\mathbf{h}_2$ is fed into a linear unit with just one output h_3. A sigmoid unit can be used to obtain a likelihood of a healthy signal l.

2.3. Optimization

The critical component of our model is a proper loss function. Since we have only a few labeled training data as a fault, the BCE may suffer from biases toward healthy data. For this purpose, we adopted SAL [8] defined over N training samples as

$$L(\theta) = \frac{1}{N}\sum_{i}^{N} \|f_\theta(\mathbf{x}_i) - \mathbf{c}\|_2 + \eta\|f_\theta(\mathbf{x}_i) - \mathbf{c}\|_2^{y_i}. \tag{1}$$

The first term is a one-class loss [9] that fits the input data regardless of label. The second term is the SAL [8] which takes into account the category of the input data $\mathbf{x}$ by exponentiating by y. Variable y signifies if $\mathbf{x}$ is an anomaly or not. Intuitively, if y is not an anomaly, then the value is set to 1 and the optimization trains the network and $\mathbf{c}$ to move as close as possible. Contrarily, if y is a (known) anomaly, the value is set to -1 and it intuitively moves the network $f_\theta(\mathbf{x})$ and $\mathbf{c}$ away from each other thus increases the anomaly score for such samples. Since we have a small set of partially labeled data that contains various types of faults, the trained model mostly adapts to standard operations in healthy data with a noticeable emphasis on unhealthy data which are moved away from each other. The benefit of SAL is that features $f_\theta(\mathbf{x})$ reveal information about various faults in the feature space since it optimizes distances in it. It potentially allows further analysis with very little supervision because the network itself plays a role of a bottleneck. It is important to note that unlike [8], we update $\mathbf{c}$ during optimization.

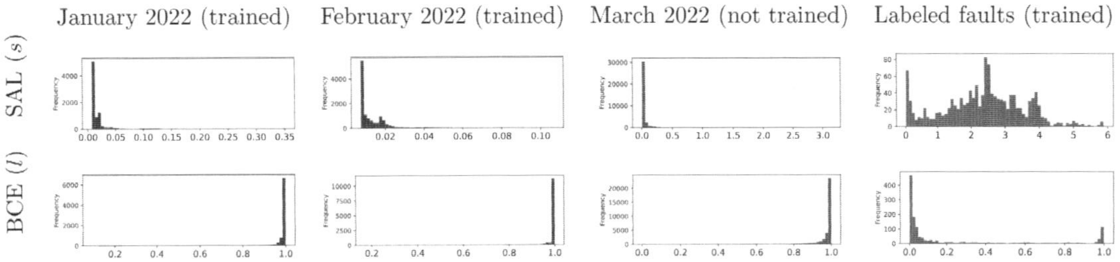

Figure 3. Histograms of anomaly scores s trained with SAL (Top) and likelihoods l trained with BCE (Bottom) for different datasets. (Top) The events sampled from trained months (Jan, Feb) have very low anomaly score s and the highest scores are always below 0.35. Events sampled in March have an increase in anomaly scores since there are seasonality effects that might be labeled as an anomaly. (Bottom) Likelihood of a healthy last pulse l of a model trained with BCE loss.

We also evaluated the LSTM model with BCE loss. It requires minor changes in architecture and the output of $\mathbf{h}_2$ is replaced with a single-valued binary output h_3. After applying sigmoid, we have a likelihood l of a healthy pulse. This architecture has an important merit because evaluation provides a likelihood of whether the input is an anomaly or not and avoids the need to specify a threshold.

Each event consists of a sequence K of pulses $(\mathbf{x}_1, \ldots \mathbf{x}_K)$. For healthy events, the loss is calculated for all pulses in the event, in the case of faulty events only the last event is trained.

3. Experiments

Faulty data was detected by an available quench detection server [10]. Faulty events consist of 250 pulses or less with usually 200 pulses before the event and 50 afterward. The healthy data were equally sampled and downloaded from our DAQ system [12]. Healthy events usually have 250 or 500 pulses. The training data are sparsely sampled from all available stations over a period of five months.

We have 81922 healthy events and 1331 labeled faulty events available. The test set is randomly sampled with 7803 healthy and 102 faulty events. Additionally, we created one test set with only healthy events over a period of March 2022 to test how models react to novelty.

Table 1. Evaluation of model trained models on two test sets. The test set consists of samples randomly sampled from available data. March 2022 is sampled over the entire month to test how the model responds to long-term untrained events.

	Test set				March 2022	
Method	**TP**	**TN**	**FP**	**FN**	**TN**	**FP**
SAL	**103**	**7691**	**0**	**9**	**35034**	**952**
BCE	96	7685	6	16	34869	1117

3.1. Evaluation

We evaluated both approaches after 50 epochs trained with ADAM. Table 1 and figure 3 show their comparisons.

The model trained with SAL performs better on the test set, where none of the healthy events was wrongly identified (FP) and only 9 faulty events were not identified (FN), see table 1. The threshold for labeling an event as an anomaly was all scores exceeding one times the standard

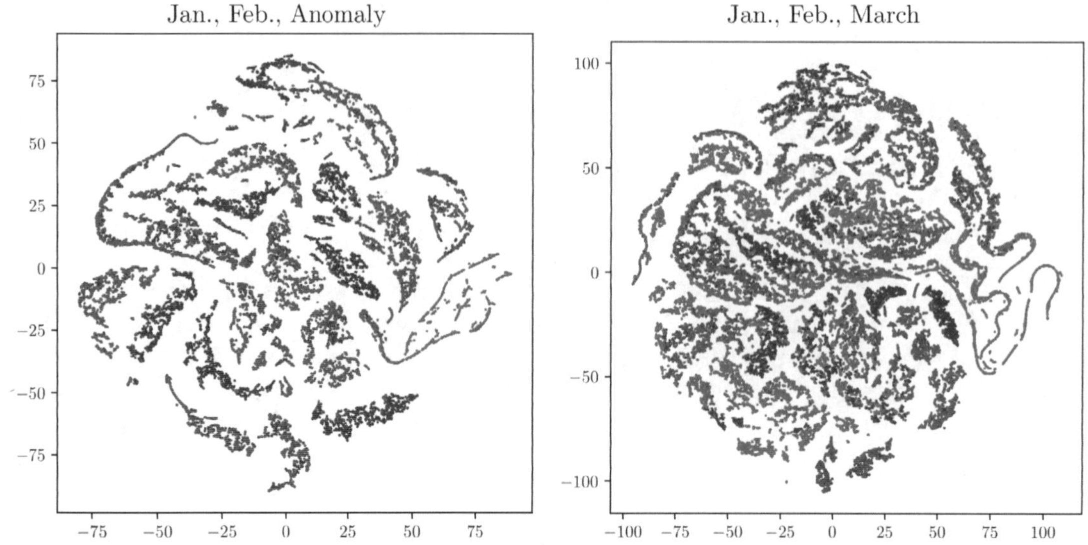

Figure 4. T-SNE Embedding [11] of feature outputs $f_\theta(\mathbf{x})$ of SAL on the last pulse in the event. (Left) Training and testing events from Jan 2022 (green), Feb 2022 (blue) and anomaly (red). (Right) Training and testing events from Jan. 2022 (green), Feb 2022 (blue) and March 2022 (red).

deviation of all scores in the test set. Slightly better performance than BCE is noticeable on the healthy test set sampled in March 2022, where 952 events were identified as a fault. This can further be identified in the right image in figure 4, where a part of the healthy events of March 2022 noticeably deviates from the trained datasets for January and February 2022.

The model trained with BCE wrongly identified 6 healthy events (FP) as faulty and 16 faulty events were not identified (FN). The model also performs slightly worse on the March 2022 test set by identifying 1117 healthy events as faulty.

4. Conclusion and Future Work

In this paper, a data-driven and model-free approach to detecting cavity anomalies is shown. We show a training model which can bypass the disproportionally many healthy training samples by using a SAL [8] and compare it with BCE. The SAL allowed us to train the proposed model with an abundance of healthy data. As a byproduct, the model is trained to project inputs to a feature space that reveals the potential for further classification of different types of faults.

Experiments show that our method can identify a large part of faults in our test set. One of the major limitations is that waveforms may vary over longer periods. This causes a noticeable increase in false positives for events from different time periods.

In the future, since the lower-dimensional features of the SAL model still carry the information about a fault, we would like to experiment with different models to achieve better interpretability of the features and distinguish between different types of faults. Since the dimensionality of features is much smaller, this should also require smaller training datasets. Additional insights can also be revealed by using generative models for anomaly detection [13] instead of discriminative ones.

Acknowledgments
We thank Julien Branlard for his helpful suggestions and corrections on the paper. This work was supported by HamburgX grant LFF-HHX-03 to the Center for Data and Computing in Natural Sciences (CDCS) from the Hamburg Ministry of Science, Research, Equalities and Districts.

References
[1] Branlard J, Ayvazyan V, Hensler O, Schlarb H, Schmidt C, and Cichalewski W 2013 *Proc. 14th Int. Conf. on Accelerator and Large Experimental Physics Control Systems (ICALEPCS'13)* (San Francisco, CA, USA) pp 1239–41
[2] Bellandi A, Butkowski Ł, Dursun B, Eichler A, Gümüş Ç, Kuntzsch M, Nawaz A, Pfeiffer S, Schlarb H, Schmidt C, Zenker K and Branlard J 2021 *IEEE Trans. Nucl. Sci.* **68** 385–93
[3] Nawaz A S, Pfeiffer S, Lichtenberg G and Rostalski P 2018 *Proc. 9th Int. Particle Accelerator Conf. (IPAC'18)* (Vancouver, Canada) pp 2502–04
[4] Nawaz A S, Pfeiffer S, Lichtenberg G and Rostalski P 2018 *IFAC-PapersOnLine* **51** 1379–86
[5] Eichler A, Branlard J and Timm J H K 2022 Anomaly detection at the european xfel using a parity space based method *Preprint* arXiv:2202.02051
[6] Tennant C, Carpenter A, Powers T, Shabalina A S, Vidyaratne L and Iftekharuddin K 2020 *Phys. Rev. Accel. Beams* **23** 114601
[7] Obermair C, Apollonio A, Cartier-Michaud A, Catalán Lasheras N, Felsberger L, Millar W L, Pernkopf F and Wuensch W 2021 *Proc. 12th Int. Particle Accelerator Conf. (IPAC'21)* (Campinas, Brazil) pp 1068–71
[8] Ruff L, Vandermeulen R, Görnitz N, Binder A, Müller E, Müller K and Kloft M 2019 Deep semi-supervised anomaly detection *Preprint* arXiv:1906.02694
[9] Ruff L, Vandermeulen R, Goernitz N, Deecke L, Siddiqui S A, Binder A, Müller E and Kloft M 2018 *Proc. 35th Int. Conf. on Machine Learning (ICML2018)* (Stockholm Sweden) *PMLR* **80** pp 4393–4402
[10] Timm J H K, Branlard J, Eichler A and Schlarb H 2021 *Proc. 12th Int. Particle Accelerator Conf. (IPAC'21)* (Campinas, Brazil) pp 3344–46
[11] Hinton G and Roweis S 2002 *Advances in Neural Information Processing Systems* vol 15 ed S Becker, S Thrun and K Obermayer (Massachusetts: MIT Press)
[12] Hinsch K *et al.* 2017 *Proc. 16th Int. Conf. on Accelerator and Large Experimental Physics Control Systems (ICALEPCS'17)* (Barcelona, Spain) pp 1–5
[13] Schlegl T, Seeböck P, Waldstein S M, Schmidt-Erfurth U and Langs G 2017 *International conference on information processing in medical imaging* ed M Niethammer *et al.* (Springer)

Journal of Physics: Conference Series **2420** (2023) 012071 doi:10.1088/1742-6596/2420/1/012071

Online optimization of NSLS-II dynamic aperture and injection transient

X Yang, G Wang, V Smaluk, L H Yu, T Shaftan, Y Li, F Plassard, S Buda, Y Hidaka, D Durfee, Y Tian, Y Hu, K Ha, A Derbenev, B Bacha, C Danneil and D Padrazo

National Synchrotron Light Source II, Brookhaven National Laboratory, Upton, NY 11973, USA

Email: xiyang@bnl.gov

Abstract. The goal of the NSLS-II online optimization project is to improve the beam quality for the user experiments. To increase the beam lifetime and injection efficiency, we have developed a model-independent online optimization of nonlinear beam dynamics using advanced algorithms, such as Robust Conjugate-Gradient Algorithm (RCDS). The optimization objective is the injection efficiency and optimization variables are the sextupole magnet strengths. Using the online optimization technique, we increased the NSLS-II dynamic aperture and reduced the amplitude-dependent tune shift. Recently, the sextupole optimization was successfully applied to double the injection efficiency up to above 90% for the high-chromaticity lattice being developed to improve the beam stability and to increase the single-bunch beam intensity. Minimizing the beam perturbation during injection is the second objective in this project, realized by online optimization of the injection kickers. To optimize the full set of kicker parameters, including the trigger timing, amplitude, and pulse width, we upgraded all kicker power supplies with the capability of tunable waveform width. As a result, we have reduced the injection transient by a factor of 29, down to the limit of 60 μm.

1. Introduction

The Facility Improvement Project "Methods of online optimization of NSLS-II storage ring concurrent with user operations" was dedicated to developing a set of software tools for online optimization of beam dynamics and minimization of the injection transients. The online optimization approach is based on the use of the measured beam and machine parameters to evaluate performance functions, which are optimized using advanced algorithms designed to work reliably in noisy environments.

In the frameworks of this project, we planned to explore and improve the performance of the optimization algorithms. We use beam-based model-independent online methods to improve the injection efficiency and beam lifetime by direct optimization of sextupoles. For the minimization of the injection transients [1-4], the specific plan was to optimize the matching of Storage Ring (SR) injection kickers [5-7].

We expect the NSLS-II user community to benefit from the reduced top-off injection frequency, improved beam stability, and transparent injection. Since nonlinear beam dynamics is a subject of high scientific interest, we believe the results of this project will be beneficial not only for NSLS-II but also for other present and future light sources.

2. Study results

2.1. Optimization of sextupoles to increase dynamic aperture

To test the RCDS optimization algorithm [8], we loaded the high chromaticity 7 lattice. The injection efficiency was decreased to around 20%. The RCDS online optimization of sextupoles recovered the injection efficiency up to 90% at the second iteration. In this experiment, we found two to three iterations were adequate to achieve the optimal solution. Extra iterations were not so effective, the injection efficiency went from mild improvement to no improvement at all. The evolution of the power supply currents of the NSLS-II sextupole families is shown in the right plot of figure 1.

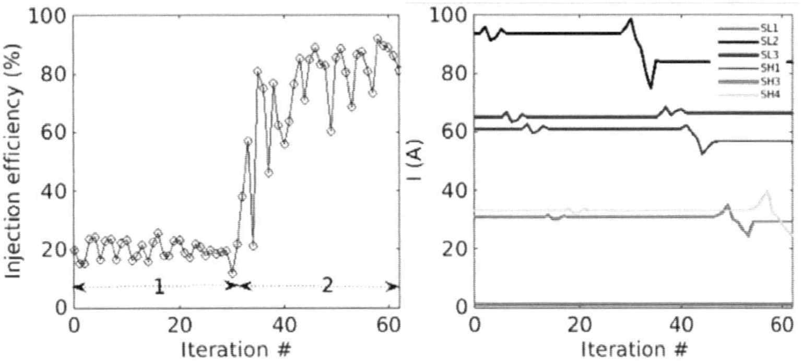

Figure 1. Two iterations of the RCDS online optimization of sextupoles for the high chromaticity 7 lattice: injection efficiency (left) and the sextupole settings (right).

We characterized the dynamic aperture (DA) for the NSLS-II operational lattice. We injected a beam of 2 mA current in 50 bunches, which were well aligned on the flat top of the pinger pulse. Then, we gradually increased the pinger voltage and recorded the beam current measured by the DCCT. Figure 2 shows the horizontal (left) and vertical (right) DA before (red) and after (blue) RCDS online optimization. We have achieved a more than 20% increase in the horizontal DA with no change in the vertical DA. We also analysed the measured betatron tune shift with amplitude and observed a reduction of the horizontal tune shift with amplitude by a factor of two.

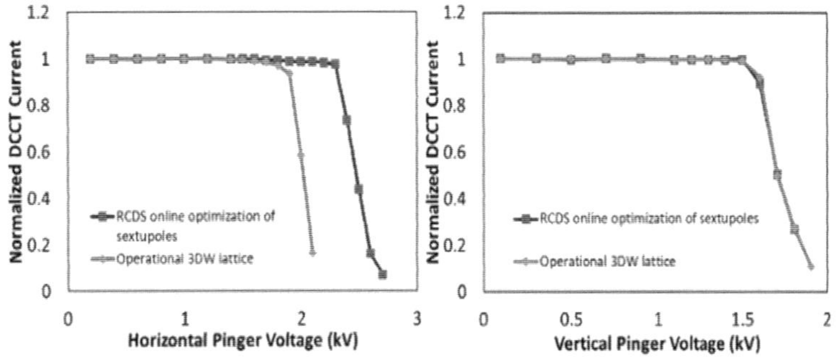

Figure 2. Horizontal (left) and vertical (right) DA before (red) and after (blue) sextupole optimization in the 3DW lattice.

We successfully applied the sextupole optimization to the lattice with high chromaticity, +7, in both horizontal and vertical directions. High chromaticity provides fast damping of betatron oscillations allowing us to reduce the bunch-by-bunch feedback gain. This helps to improve the beam size stability in the operations with small vertical emittance. High chromaticity is also useful to increase the single-bunch beam intensity, so the other application of this lattice is the new operation mode with high-intensity bunches for time-resolved experiments.

2.2. Minimization of the injection transient

For the injection, four pulse kickers create a horizontal orbit bump in the injection straight section [6,7]. The orbit outside the bump is unperturbed if the kickers are perfectly matched. Any mismatch of the kicker pulses results in the residual beam oscillation around the ring perturbing the user beam. The online optimization was applied to provide the top-off injection with minimized perturbations of the beamline user operations. Optimizing the matching of four injection kickers helps to reduce the perturbation of the stored beam orbit and achieve a more transparent injection.

Initially, all the kickers were set to the operational values with the same kicker amplitude. We did the optimization in three steps with different combinations of the knobs [9]. First, we varied amplitudes of 3 kickers and monitored the residual betatron oscillation during the injection in one fixed RF bucket. This step showed very fast convergence (within one hour) and minimized the injection transient from 1.2 mm down to 0.15-0.2 mm peak-to-peak, which is dominated by the kicker shot-to-shot variation. Then, with the optimized kicker amplitudes, we varied the timing delays of 3 kickers and repeated the same optimization process. As predicted, these two steps worked very well to minimize the injection transient of a certain injected bunch, but other bunches in the train still have a large transient. In the last step, we varied both amplitude and timing with excitation of all the bunches from 0 to 1000 with a step of 100, in the same way as the top-off injection. The objective function is the average RMS value of each bunch. Although the knobs were set close to the optimal values after the first two steps, the optimization took a much longer time, about 4 hours, to reach the minimum. We observed a dependence of the residual oscillation on the number of the injection bucket, illustrated by figure 3.

The RMS values of the measured oscillations are shown before the optimization (black), after the optimization with the first 20 buckets filled with 1.6 mA (red), and after the global optimization of 1000 buckets filled with 22 mA (magenta). The global optimization of the pulse amplitude and timing of the injection kickers reduces the injection transient by a factor of six from 1.2 mm down to 0.2 mm.

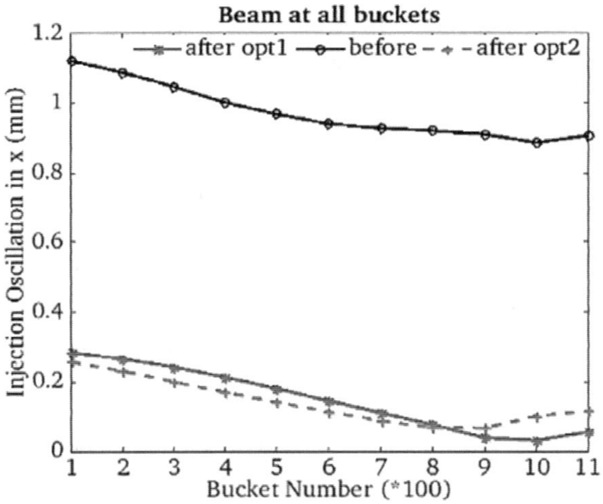

Figure 3. RMS value of BPM turn-by-turn data vs the injection bucket: before optimization (black), after optimization of the first 20 buckets (red), after global optimization of 1000 buckets (magenta).

2.3. Reduction of timing jitter

The timing jitter caused by the timing system only should be smaller than 15 ps. Using the event generator transistor-to-transistor logic output channels providing the triggers for the injection kickers, we confirmed that the jitter is mainly caused by the kicker system itself.

The time jitters of all 4 injection kickers were similar, about 2 ns in RMS and 4-5 ns in FWHM depending on where the measurement was performed. A smaller jitter of about 4 ns was measured right at the pulser, and jitter of 5 ns was measured after the delay line. Each kicker has 10 IGBTs connected in series and triggered by fiber optical receivers. Based on the diagnostic measurements, we concluded the time jitter comes from the fiber optics receiver. A new trigger board was designed, manufactured, and tested. It provides a significant reduction of the timing jitter by a factor of 10, from 2 ns down to 200 ps. For each of the 4 SR injection kickers, 11 boards (1 master fanout and 10 stage trigger boards) are needed. All 44 trigger boards were replaced sequentially during the maintenance days in May-June 2019. The performance was evaluated before and after the installation of new trigger boards by analysis of 180 kicker waveforms to extract the timing and amplitude jitters:

- A reference waveform is obtained by averaging 180 injections.
- Each waveform is subtracted from the reference waveform.

After the installation of new trigger boards for injection kicker 1, 180 waveforms were taken for all 4 kickers. The processed data show the cosine-like patterns for kickers 2, 3, and 4, which indicates the timing jitter of these kickers, except kicker 1. To extract the timing jitters, the cosine-like patterns were normalized by (amplitude ω) [10]. As shown in figure 4, the timing jitter of kicker 1 is on the noise level; the timing jitters of the other three kickers are about 2 ns peak-to-peak. After the upgrade of new trigger boards for all 4 kickers, their timing jitters have gone down to the noise level, as shown in figure 5.

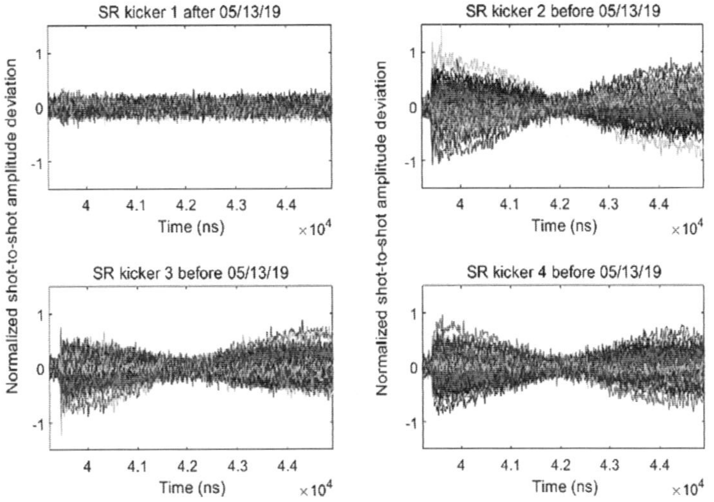

Figure 4. Data were taken at 05/13/19 with upgraded k1.

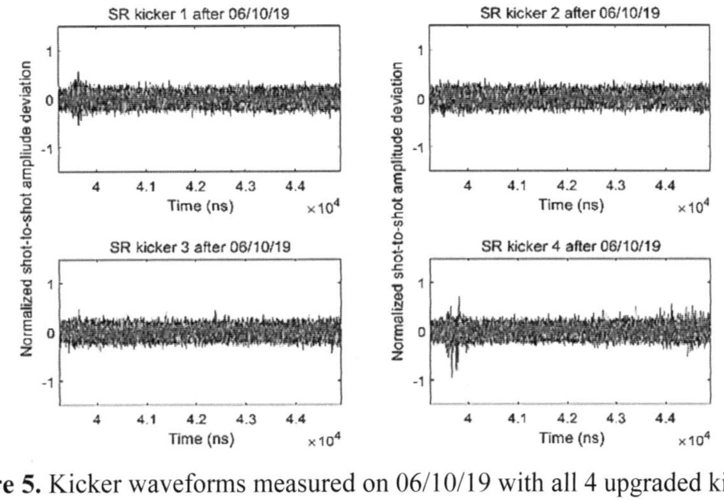

Figure 5. Kicker waveforms measured on 06/10/19 with all 4 upgraded kickers.

Journal of Physics: Conference Series **2420** (2023) 012071 doi:10.1088/1742-6596/2420/1/012071

2.4. Implementation of tunable kicker waveform

Mode analysis of the SR injection kickers provides the information of the amplitude deviation, timing offset, and waveform width deviation [10]. Four pulsers have been designed, manufactured, tested, and installed. The tuning slot is activated by a stepper motor to vary the inductance and, therefore, the width of the kicker pulse. The motion of the stepper motor is precisely controlled by the integrated motion control functions of a SIEMENS 1215C DC/DC/DC PLC. The achieved tuning range of 70 ns meets the specification with nanosecond precision, as shown in figure 6. Then, the associated control program was developed, tested, and installed.

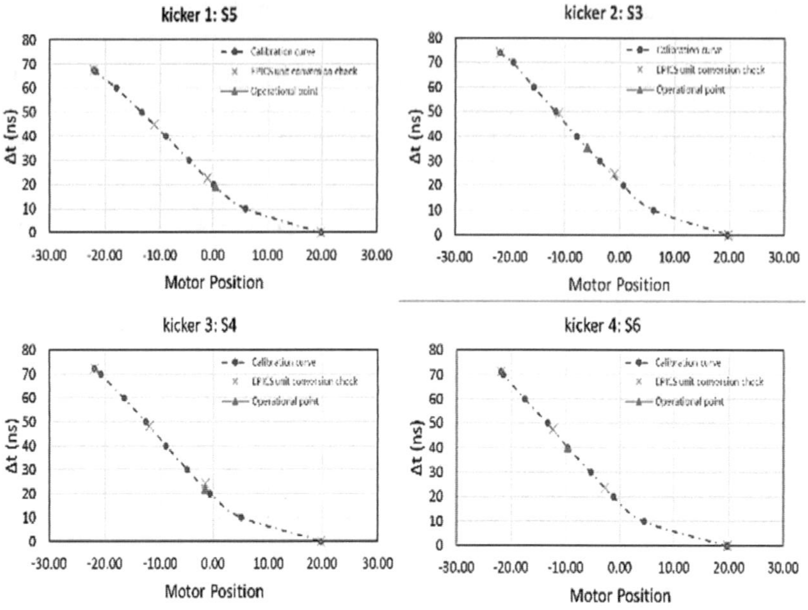

Figure 6. Unit conversions from the motor position to the variation of the kicker pulse width.

2.5. Optimization with full set of kicker parameters

The kicker hardware upgrade reduces the trigger timing jitter down to 1 ns in RMS and provides a tunable pulse width with a 1 ns precision in a range of 70 ns. The upgrade allows us to optimize the injection transient with a full set of parameters, including the amplitude, trigger timing, and pulse width of all 4 injection kickers.

However, there were some issues related to the BPM turn-by-turn readback, which distorted the measurement of injection transients. We found that the BPM turn-by-turn data had a shot-to-shot variation of the injection transient larger than usual for the RF buckets in the range of 0-300 and 800-1200. The misalignment of the BPM timing in the turn-by-turn mode was identified at the top of the list of leading causes. Then, all BPMs were realigned with respect to timing event 47 for all RF buckets from 0 to 1200. As a result, we have an improved BPM turn-by-turn readback with a timing error significantly smaller than the duration of a bucket (2 ns). Now, the BPMs can be used as a diagnostic tool for the online optimization of the injection transient.

Figure 7 presents the results of the injection transient optimization, the RMS residual oscillations are shown as a function of the injection bucket number before (black) and after (blue) the first RCDS online optimization, after fixing the trigger timing jitter (magenta), after implementing the adjustable width of the kicker waveform (red), and after fixing the BPM alignment issue (green and cyan). With the reliable BPM turn-by-turn readback for the buckets from 1 to 1000, we have achieved the targeted injection transient of 70 μm, which is close to the ultimate limit of 60 μm determined by the design of the off-axis injection scheme [5-7].

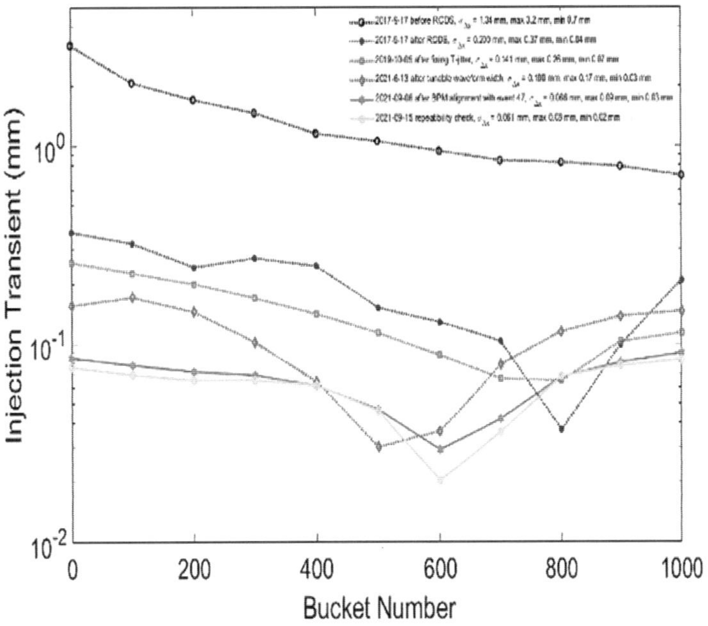

Figure 7. The measured injection transient as a function of bucket number. The injection transient is reduced to 60 μm, with the total improvement by a factor of 20.

3. Conclusion

By applying RCDS online optimization, we increased the horizontal dynamic aperture by 20% for the operational lattice. Using the same approach, we improved the injection efficiency from 40% to 90% for the newly developed lattice with high chromaticity.

Ultimately, we have minimized the injection transient by matching the injection orbit bump via online optimization of the Storage Ring injection kickers including the timing delay, pulse amplitude, and pulse width. First, we reduced the timing jitter of the injection kickers from 4-5 ns down to 1 ns by the upgrade of the fiber-optical fanout trigger boards with the new design. Then, simulation studies indicated that for the minimum injection transient of 60 μm, the pulse widths of all kickers need to be adjusted with nanosecond precision. We have completed the hardware modifications of the injection kicker power supplies and achieved the desired range of 70 ns to adjust kicker pulse width, the nanosecond tuning precision has been demonstrated too. The goal was to reduce the injection transient down to a level that would be nearly transparent to most of the beamline users, about 60 μm. We have minimized the RMS injection transient down to 70 μm, which is close to the target value. We plan a few experiments with NSLS-II beamlines to check the residual injection transient effects on the X-ray beam quality. After these final tests, the optimization results are ready to be implemented in regular user operations.

Acknowledgements

This manuscript has been authored by Brookhaven Science Associates, LLC under Contract No. DE-SC0012704 with the U.S. DOE.

References

[1] Keil J, Kube G, Obier F, Sahoo G K and Wanzenberg R 2018 *Proc. 9th Int. Particle Accelerator Conf. (IPAC'18)* (Vancouver, Canada) pp 1467–70
[2] Mitsuhashi T and Ueda A 2005 *Proc. Particle Accelerator Conf. (PAC'05)* (Knoxville, TN USA) pp 2717–19
[3] Liuzzo S M *et al.* 2016 *Proc. 7th Int. Particle Accelerator Conf. (IPAC'16)* (Busan, Korea) pp 3420–23

[4] Leng Y B, Yan Y B, Yang Y, Yuan R X and Zhang N 2013 *Proc. 2nd Int. Beam Instrumentation Conf. (IBIC'13)* (Oxford, UK) pp 746-48
[5] Dierker S *et al*. 2007 National Synchrotron Light Source II Preliminary Design Report Tech. rep. BNL-94744-2007 Brookhaven National Laboratory
[6] Shaftan T V *et al*. 2015 *Proc. Particle Accelerator Conf. (PAC'05)* (Knoxville, TN USA) pp 3408–10
[7] Shaftan T V *et al*. 2010 *Proc. 1st Int. Particle Accelerator Conf. (IPAC'10)* (Kyoto, Japan) pp 2672–74
[8] Huang X and Safranek J 2015 *Phys. Rev. Accel. Beams* **18** 084001
[9] Wang G M, Cheng W X, Choi J, Shaftan T V and Yang X 2017 *Proc. 8th Int. Particle Accelerator Conf. (IPAC'17)* (Copenhagen, Denmark) pp 4683–85
[10] Yang X *et al*. 2021 Online Optimization of NSLS-II Dynamic Aperture and Injection Transient (Facility Improvement Project) Tech. rep. NSLSII-ASD-TN-372; BNL-222523-2021-TECH Brookhaven National Laboratory

Dispersion-free steering beam based alignment at SwissFEL

Eugenio Ferrari[1], Marco Calvi[1], Romain Ganter[1], Christoph Kittel[1,2], Eduard Prat[1], Sven Reiche[1] and Thomas Schietinger[1]

[1]Paul Scherrer Institut, CH-5232 Villigen PSI, Switzerland.
[2]Faculty of ICT, University of Malta, Msida MSD 2080, Malta.

E-mail: eugenio.ferrari@psi.ch

Abstract. Micron-level alignment of the undulator line is required for successful operation of linear accelerator based high gain free electron lasers to produce powerful radiation at X-rays' wavelengths. Such precision in the straightness of the trajectory allows for an optimal transverse superposition between the electrons and the photon beam. This is extremely challenging and can only be achieved via beam-based techniques. In this paper we will report on the dispersion-free steering approach implemented at SwissFEL, that helped achieving improved performance for both the hard and soft X-ray beamlines.

1. Introduction

SwissFEL [1] is a free-electron laser (FEL) user facility, based on the SASE principle, delivering ultrashort photon pulses in the X-ray regime. For a schematic representation of the facility, see figure 1. The electrons needed for FEL production are generated in a normal conducting photocathode gun and accelerated through a normal conducting linac. The gun and booster sections, before the first bunch compressor, employ S-band technology. All the other accelerating cavities, divided for convenience into 4 linacs, are at C-band frequency. Two magnetic chicanes are used to compress the pulse duration. An X-band harmonic cavity is also installed and used to linearise the phase space before compression.

Two different photoinjector lasers are used to generate two electron bunches with temporal separation of 28 ns. This allows parallel operation of two distinct FEL beamlines. The electrons of bunch 1 are accelerated to the required energy and injected in the Aramis undulator, for hard-X-ray generation in the 2–12 keV photon energy range. The typical electron beam energies are 3–6 GeV, depending on the requirement of the specific user experiment. Bunch 2 electrons are extracted after Linac 2, using a combination of resonant kicker and septum, and they are injected at 3 GeV in the Athos branch for soft-X-ray production. The photon energy range of the second beamline is 0.25–1.9 keV. Both beamlines can operate in parallel at a full repetition rate of 100 Hz.

The Aramis undulator [1] consists of 13 in-vacuum variable-gap modules, each 4 m-long, interleaved with 0.75 m break sections containing a sub-μm resolution BPM, a quadrupole magnet and a phase shifter. Athos [2] on the other hand consists of 16 out-of-vacuum Apple-X modules of 2 m length with break lengths of 0.8 m, housing a CHIC-chicane instead of the phase

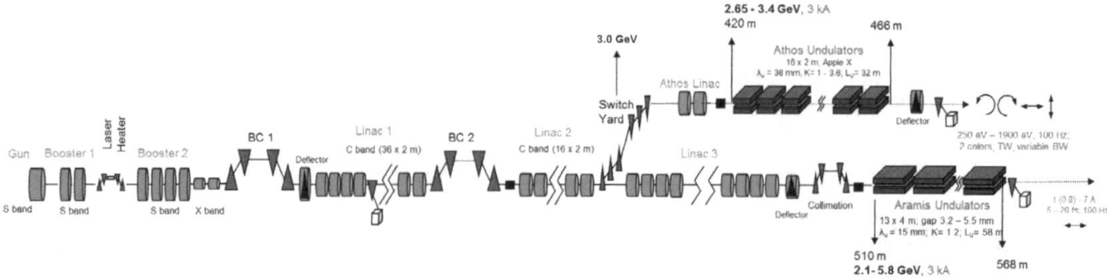

Figure 1. Schematic overview of the SwissFEL accelerator and FEL lines. Design parameters are also shown, as well as the typical operation energies for the different sections. Two bunches, separated by 28 ns, are generated at the gun using two different photoinjector lasers. The first bunch is sent to the Aramis beamline (hard X-rays), while the second bunch is extracted after Linac 2 using a resonant kicker and septum and, through a switchyard, injected into the Athos beamline (soft X-rays). Image adapted from [3].

shifter [4]. Each quadrupole/BPM pair is mounted on a remotely controlled plate with travel range of 1 mm in the horizontal and vertical planes for alignment purposes in both the Aramis and Athos undulator beamlines. The BPM offsets relative to the centre of the quadrupoles can be corrected via software. The undulator modules can also be remotely aligned, in the horizontal and vertical plane, both in offset and angle.

To achieve best lasing performance in a reproducible way, the alignment of the quadrupole magnets is critical. The electron and photon beams must be transversally superimposed inside the undulator for efficient FEL amplification. The maximum allowed transverse misalignments should be better than 10 μm. Such extreme tolerances are not achievable using traditional alignment techniques (tunnel survey), hence it is necessary to implement beam-based alignment (BBA) techniques.

2. Procedure

Originally developed for the LCLS [5] and successfully demonstrated at PAL [6] and EuXFEL [7], the dispersion-free steering BBA is a procedure that allows finding a straight electron beam trajectory along the undulator line. The straightness of the trajectory inside each of the undulator module relies instead on the correct compensation of first and second field integrals, corrected using dedicated coils set according to feed-forward tables. The procedure allows the reconstruction of the position of the quadrupoles with respect to a straight line as well as the offsets of the BPMs by measuring and minimising the dispersion along the beamline. It is based on measuring the electron beam trajectories for widely different energies in order to distinguish between magnetic kicks, caused by quadrupole offsets or other spurious kicks, which scale linearly with energy, and BPM offsets, which are energy independent [5, 6]. Critical for the success of the procedure is that the transport optics till the entrance of the beamline stays constant when changing energy, while nothing must be changed in the undulator beamline.

In the case of SwissFEL, for such wide energy changes we follow different strategies depending on the beamline. In the case of Aramis, we put off timing the RF of successive accelerating cavities in Linac 3, see figure 1, such that the beam energy is varied from 5.9 GeV to 3 GeV. We usually consider four energy steps, equally spaced in inverse electron beam energy, such that the strength of the quadrupole magnets varies by the same amount (while staying at the same current). In the case of Athos, we change the acceleration settings of Linac 2, see figure 1, while compensating the energy change using Linac 3 to maintain lasing in Aramis. This allows for the beam based feedbacks, mostly acting on bunch 1, to keep stable the transverse and longitudinal beam properties for both bunches in the common part of the machine.

In the case of Athos we vary the beam energy between 3 GeV and 2 GeV. In the future, it will be possible to use the dedicated C-band station at the beginning of the Athos beamline also for the BBA, although with reduced energy variation (±250 MeV). For both beamlines we go from higher to lower energy settings. At each energy, we save the machine settings to have a faster set-up in successive iterations of the BBA procedure. Together with appropriate cycling of the magnets, such a strategy allows for better reproducibility of the measurement conditions, which in turn help the convergence of the procedure. The BBA is now routinely performed with the gaps of the undulators closed. It was only done with open undulators at the beginning, to establish reasonable starting conditions and minimise the deposited radiation dose on the devices.

For a successful BBA, injection offset and angular errors, coming from the upstream Linac, must be evaluated and corrected. Our approach is to use two pairs of steering magnets, one for each plane, installed on two quadrupoles preceding the part of the undulator beamline for which the BBA is applied, to correct the trajectory along the full undulator line. This is performed either temporarily modifying the settings of the trajectory feedback or using dedicated trajectory correction tools. We ignore the readings of the two BPMs attached to these two quadrupole magnets in this phase.

At the end of the BBA, the offsets of the launch BPMs are adapted so that they read zero and the corresponding quadrupoles are moved to minimise the corrector current, thus achieving an appropriate injection into the FEL beamline. The trajectory feedback configuration is finally restored: the correction loop runs along the full undulator line to compensate for spurious kicks from the undulators or the chicanes, when their parameters are modified, like changing the undulator gap to tune the photon energy. The full procedure requires a few hours for each of the beamlines.

3. Results

The typical convergence of the procedure is achieved in 3 or 4 loops, with the initial one giving the largest corrections. We usually repeat the procedure until the correction amplitude is below 5 μm in the case of Aramis and 10 μm in the case of Athos.

In figure 2, a comparison of the results of two different BBAs, performed one year apart on the Aramis beamline, are reported. One can clearly see that the BPM offsets are very similar in the two cases (orange and red curves), once the launch region is excluded. This indicates that the relative offset between quadrupoles and BPMs is quite stable. The quadrupole positions are instead different in the two cases. This is probably due to the different quadrupole strengths in the two cases, as well as different undulator parameters. In the case of Athos, the vertical plane shows a smooth behaviour of the positions, whereas in the horizontal plane we typically observe an alternating up-down behaviour. This could be explained by residual kicks present between two quadrupoles, either due to a sign error in the correction tables for the first- and second-order integral of the undulators or the CHIC chicanes. We plan to further investigate this effect by performing dedicated BBAs with different correction signs, as well as for different polarisation settings of the undulator modules.

In figure 3, we report the average trajectory for the four BBA energies (a, b) and root-mean-square (rms) variation (c, d) for the Athos beamline. Panels (a, c) refer to the case before the BBA, (b, d) to the case after the BBA. The locations of the 16 undulator modules are shown as red boxes. The reduction of the trajectory variation for the different energies is clear, in particular comparing the rms before and after the BBA. The average of the rms trajectory variation is decreased from 290 μm to 4 μm for the horizontal plane, and from 93 μm to 2 μm in the vertical plane. The BPM with average significantly different from zero and rms larger than 20 μm corresponds to the location between the two kicker magnets used to correct the injection

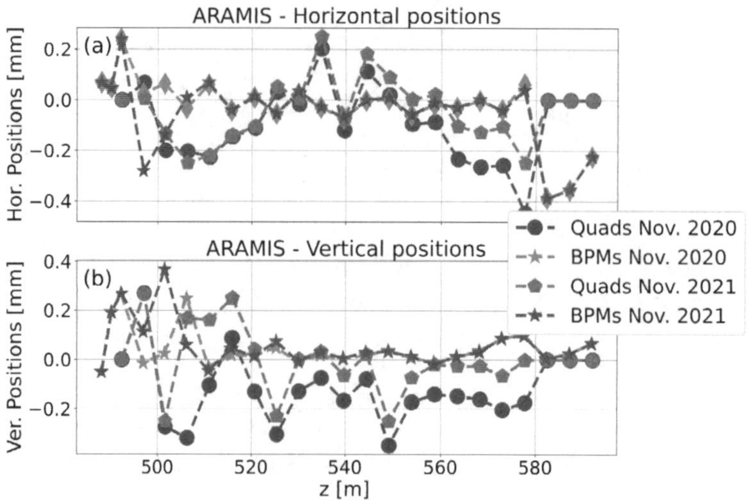

Figure 2. Horizontal (a) and vertical (b) positions of the elements along the Aramis beamline at the end of two BBA procedures, performed one year apart. One can clearly see that the BPM offsets are very similar in the two cases (orange and red curves), if the launch region is excluded. This indicates that the relative offset between quadrupoles and BPMs is quite stable. The quadrupole positions, however, are different in the two cases.

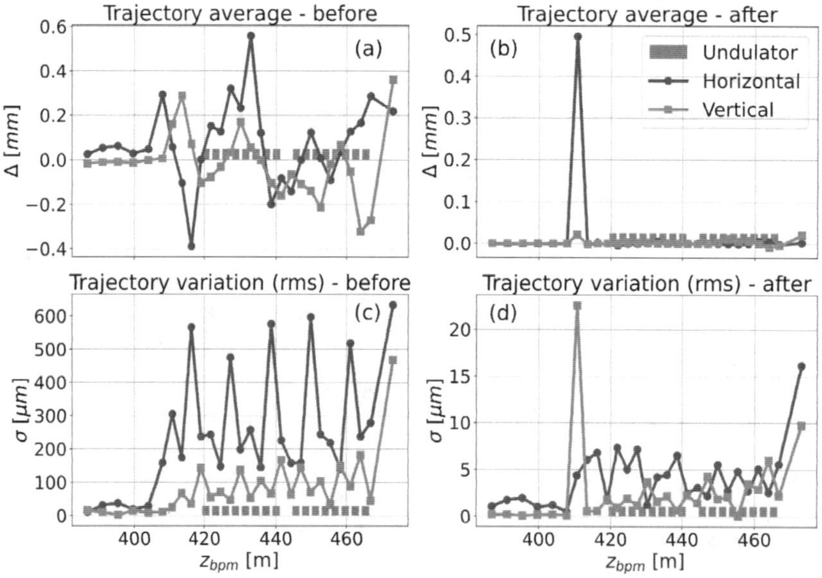

Figure 3. Average trajectory for the four BBA energies (a, b) and root-mean-square (rms) variation (c, d) for the Athos beamline. (a, c) refer to the case before the BBA, (b, d) to the case after the BBA. The locations of the 16 undulator modules are shown as red boxes. One can clearly see the reduction in the trajectory change for large energy changes, indicating a significant reduction in the dispersion, and hence in the straightness, of the beamline.

error in the beamline, which can be different for each energy. Such results clearly demonstrate a significant reduction of the dispersion, and hence to the improved straightness of the beamline.

After proper alignment of the beamline components, a pointing correction towards the user stations can be required. It is usually performed by physically shifting all the components as a line, relying on new targets for the trajectory feedbacks for small final adjustments.

4. Maintaining relative alignment of quadrupoles and undulators

As reported in [6], the BBA procedure could find a bow-like solution in case of large misalignment of the undulator modules when constraints on the offsets are imposed (bounded solution). The unconstrained solution is instead immune to such bow-like solutions [6], although being characterised by larger uncertainties in the reconstruction. The behaviour is due to the quadratic dependence on energy (not linear as for quadrupolar kicks) when passing off-axis in an undulator and can lead to sub-optimal performance.

To minimise such an effect we implemented an EPICS based server, monitoring the positions of the quadrupoles (undulator follower), that adapts the position of the undulator girders, in order to maintain their relative alignment established either during the tunnel survey phase ($\sim$100 μm), or improved by photon based techniques [8].

During the December 2021 Athos BBA, we actually observed the bow formation when we tried to disable the automatic undulator follower. By comparing the bounded and unbounded solutions for the BBA inversion problem, using the same input trajectories, we noticed a significant bow-like trend for the quadrupole offsets, in particular for the horizontal plane. This was also confirmed looking at the pointing of groups of undulators on a photon imager, which confirmed the presence of the bow. This confirmed the necessity of following the movement of the quadrupoles induced by the BBA and eventual pointing corrections with the undulators, e.g., keeping the undulator follower enabled. Such an approach maintains the relative alignment between undulators and quadrupole magnets. Using the unbounded approach in January 2022, we were able to recover a bow-free trajectory, with improved transverse profile and overall better FEL performance. When comparing with the results reported in [6], the usage of short 2 m modules for Athos has the advantage of relaxing the required alignment precision, allowing us to utilise the unconstrained solution despite its larger uncertainty when compared to the constrained one.

5. Conclusion

We presented the BBA procedure and results for SwissFEL, for both the Aramis (hard X-rays) and the Athos (soft X-rays) beamline. The BBA helped establishing robust and reliable operations for both FEL beamlines. Both Aramis and Athos deliver >1 mJ photon pulse energy in their respective energy ranges. The procedure works also in the presence of spurious trajectory coupling, as observed between the horizontal and vertical plane in Athos, even though it is not explicitly included in the model. The BBA helped in obtaining good performance, as well as in improving the transverse photon beam shape. It has been fundamental in reducing or eliminating beam clipping due to the small aperture of the Athos vacuum chamber.

References

[1] Prat E *et al.* 2020 *Nat. Photonics* **14** 748
[2] Abela R *et al.* 2019 *J. Synchrotron Rad.* **26** 1073
[3] Ganter R (Ed.) 2017 Athos Conceptual Design Report Tech. rep. PSI Rep. 17-02 Paul Scherrer Institut
[4] Prat E *et al.* 2016 *J. Synchrotron Rad.* **23** 861
[5] Emma P, Carr R and Nuhn H-D 1999 *Nucl. Instrum. Methods Phys. Res., Sect.* A **429** 407
[6] Kang H-S and Loos H 2019 *Phys. Rev. Accel. Beams* **22** 060703
[7] Scholz M, Decking W and Li Y 2019 *Proc. 39th Int. Free-Electron Laser Conf. (FEL'19)* (Hamburg, Germany) pp 592–95
[8] Tanaka T *et al.* 2012 *Phys. Rev. Spec. Top. Accel Beams* **15** 110701

Journal of Physics: Conference Series **2420** (2023) 012073 doi:10.1088/1742-6596/2420/1/012073

New generation of very low noise beam position measurement system for the LHC transverse feedback

Daniel Valuch[1] and Viera Stopjakova[2]

[1]Accelerator Systems department, CERN, Geneva, Switzerland
[2]UEF FEI, Slovak University of Technology, Bratislava, Slovakia

E-mail: daniel.valuch@cern.ch

Abstract. Recent studies showed that the transverse feedback system noise floor in the Large Hadron Collider (LHC) must be reduced by at least factor of two in order to operate the machine with large beam-beam tune shift as foreseen in the High Luminosity (HL) LHC. Also, the future feedback system foreseen to suppress the LHC Crab Cavity noise relies on improved noise performance of the beam position measurement system. An upgrade program was launched to lower the LHC transverse feedback system noise floor during the LHC Long Shutdown II. A new generation, very low noise beam position measurement module was developed and tested with beam. Innovative methods in the RF receiver, digital signal processing, thorough optimization of every element in the signal chain from pickup to the kickers allowed to achieve a significant reduction of the system noise floor. This unprecedented noise performance opens also new possibilities for auxiliary instruments, using the position data from the transverse feedback. The paper presents the new system, notable implementation details and measured performance.

1. Transverse feedback in LHC

The Large Hadron Collider (LHC) relies on a transverse feedback (TFB) to suppress coupled bunch instabilities and to damp injection oscillations conserving the injected beam emittance. Apart of these primary functions, the TFB in the LHC provides a large number of auxiliary functions, for example all kinds of beam excitation for abort and injection gap cleaning [1, 2], controlled transverse emittance blow-up for aperture and collimation measurements [3], or controlled transverse losses for special measurements [4].

During LHC Run I, the LHC TFB system started to provide very valuable bunch-by-bunch, turn-by-turn beam position data for diagnostics purposes. The data quickly became the second most important "commodity" delivered by the transverse feedback. A high performance computing system, called ADTObsBox was introduced [5] to collect and analyze these data in real time, allowing for example a real time detection of transverse instability onset, assisted tune measurements, collimator impedance measurements and many more. The beam position measurement quality is equally important also for this secondary application.

The beam position is sensed by four stripline type pickups in the arc around LHC point 4. Symmetrically at both sides of the former interaction point. The pickups are referenced as Q9 and Q7 (by the quadrupole magnets they belong to), Q8 and Q10 respectively. Signals of both pickup electrodes (referenced as electrode A and B) are transported to the surface. A and B signals are combined by a 180-degree, hybrid coupler to produce the analogue Σ and the delta Δ signals. These raw, impulse like signals are then feeding the Beam Position Measurement module (BeamPos), which downconverts and digitizes them. A normalized bunch position is calculated digitally by a field programmable gate array (FPGA). One data point per bunch per turn is sent over an optical link to Digital Signal Processing Unit where all feedback related calculations are performed. An analogue correction/excitation signal is sent back to the tunnel, amplified by tetrode amplifiers to a peak amplitude of up to 10 kV and fed to electric field kickers.

2. Performance of the old system

Though the TFB performance during the LHC Runs I (2009-2013) and II (2013-2018) was greatly sufficient for the LHC operation, recent studies showed that the TFB noise floor in the LHC must be reduced by at least factor of two in order to operate the machine with large beam-beam tune shift as foreseen in the High Luminosity (HL) LHC [6] and suppress the risk of loss of Landau damping by noise [7]. Improvement by factor of 4 is required to recover an emittance growth rate in the order of 2% per hour as in the present LHC. Also, the future feedback system foreseen to suppress the LHC Crab Cavity noise in HL-LHC [8] relies on improved noise performance of the upgraded TFB.

With the majority of the signal processing chain of the LHC TFB being digital, the beam position measurement subsystem is considered to define the noise performance of the whole LHC TFB system, therefore the effort to lower it focuses on improving the beam position measurement subsystem. The noise floor of the TFB system can be determined only indirectly: either beam based by measurement of transverse emittance blow-up [6], or by analysis of the beam position measurement hardware [9].

For latter, a complete numeric model of the receiver and the signal processing chain was constructed. The model was fed by a noiseless input signal corresponding to a full scale bunch position. Then a real BeamPos module in LHC was set to nominal operating conditions, but without beam and noise from the analogue to digital converters was measured. The signal was superimposed to the ideal position data and processed by the model. Distribution of the measured position was recorded.

Noise floor of the LHC TFB beam position measurement system used in the Run I/II was found to be 1.03-1.40 μm_{RMS}, depending on pickup, settings and operating conditions.

3. New beam position module

The Beam Position Measurement module used by LHC TFB for Runs I and II [10] was developed within the first large scale digital low-level RF project at CERN, circa 2006. Very little was known about how could it be used beyond the basic feedback loop functionality, or what should be the optimal electrical parameters. No special optimization was done for noise performance, or operational flexibility at the time. For development of the new Beam Position Module, the noise performance and the operational flexibility were the primary design parameters. Thermal stability, or accuracy of absolute position measurements are not of a large concern. A multi-domain simulation model of the complete system was created starting with the beam-pickup interaction (Particle studio), transmission line system and signal processing in the RF domain (Microwave office), downconverters and intermediate frequency amplifiers (SPICE), digitization, noise and digital signal processing (Python).

Four different position measurement methods, compatible with the existing system were evaluated:

(1) Direct sampling of the A and B signals by high speed digitizers. The pickup electrode signals are typically a 1 ns long pulses with not more than a 2 dB difference in amplitude

for a full scale bunch position movement. Effective number of bits (ENOB) of currently available multi-Gsps digitizers is too low for our application.

(2) Direct sampling of a band-pass filtered A and B signals by high resolution, medium speed digitizers. At the time of development, state of the art analogue to digital converters were delivering 13-14 bits ENOB, at a sampling frequency of low hundreds Msps. Noise performance of this method was found to be about the same as the old system.

(3) Direct sampling of a band-pass filtered Σ and Δ signals by high resolution, medium speed digitizers. The bunch repetition frequency in LHC is 40 MHz. Convenient frequencies to measure would therefore be integer multiples of the bunch repetition frequency, e.g. 200, or 400 MHz. The LHC TFB uses a bunch synchronous sampling and data processing. Convenient data rates would be 40, or 120 MHz. Theoretical analysis of this option showed, that if a 200 MHz wavelet is generated by the analogue FIR filter, and it is sampled at 240 Msps, an improvement of noise floor of factor of 2 to 3 might be in reach. This option was implemented and successfully tested in LHC towards end of Run II [11]. Though the method was successfully tested, the aim of the upgrade project was to obtain the best possible noise performance with the technology available to date.

(4) Continue using the super-heterodyne principle and high resolution, medium speed digitizers similar to the old beam position measurement system, but "super-optimize everything possible". Every component of the signal chain was modelled, or measured, thoroughly analyzed and optimized for the ultimate noise performance. It was possible to connect up to six, 16-bit, 125 Msps dual analogue to digital converters (ADCs [12]) to the Artix7 FPGA used by the LHC TFB modules, setting the baseline for this option.

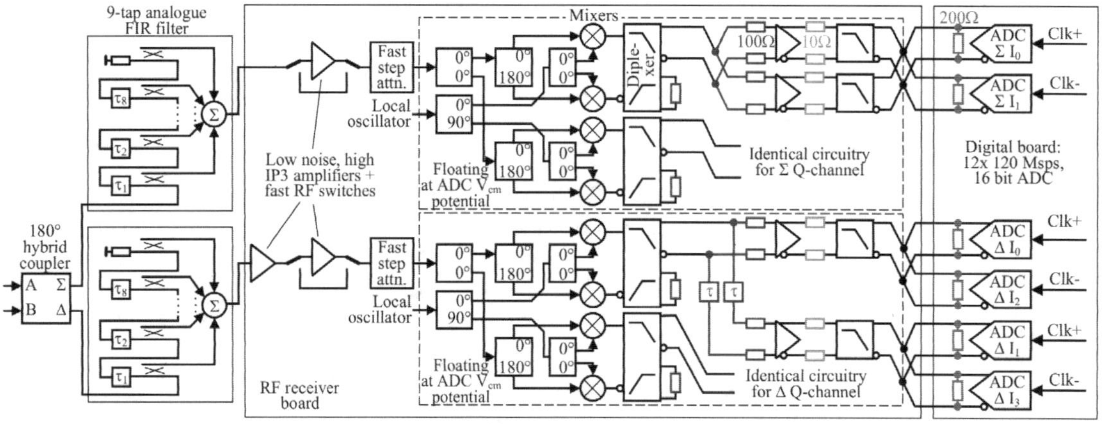

Figure 1. Diagram showing the new receiver topology, with important sub-circuits which influence the overall noise performance. Black lines are 50 Ω characteristic impedance, blue elements and lines are 100 Ω.

3.1. New concept of a fully differential RF receiver

The new generation beam position measurement module [13, 14] profits from a high signal power available from the pickup. Ideally the whole signal chain from the pickup to the ADC could be completely passive. This required development of a fully differential, quadrature RF receiver. The result is shown in figure 1. The wavelet forming analogue FIR filter was designed as a fully fused, 4-layer printed board made on microwave substrate. A very tight electrical parameter control was possible, all 40 produced filters were identical without any need of tuning. Broad side coupled directional couplers with progressively increasing coupling extract most of the energy at the frequency of interest (400 MHz) from the through line, forming a 9-RF period wavelet with very clean and flat impulse response. ADCs typically use a differential input, we have designed the whole RF receiver as fully differential to use all the available signal power.

The input RF signal is split to 0°/180° (differential) components and downconverted by two Level 13 mixers using a local oscillator signal split to 0°/0° (common mode). High LO-IF isolation of the mixer, together with the common mode rejection of the IF amplifier and the ADC suppresses the LO leakage below the ADC resolution. The main noise contribution comes from the IF amplifier. Thanks to available high IF output voltage from the mixers, its gain could be kept at the very minimum (3-6 dB), contributing not more than 0.5 LSB (least significant bit) to the ADC noise floor.

Majority of the RF receiver is floating at a common mode potential of ADC (0.9 V) to prevent back-biasing of the mixers through the differential IF amplifiers. Many "analogue" tricks are implemented in the new receiver allowing to form e.g. delay lines, pulse responses, or IF amplifier paralleling with the aim not to waste any signal power in termination resistors.

3.2. Interleaved sampling and ADC noise floor

Each of the IF signals is sampled by multiple ADCs. The Δ signal is unique for each bunch and each passage, so we have to treat it as a single shot phenomena. Four interleaved ADCs are used to sample it at a total rate of 4 x 120 = 480 Msps. Two of the ADCs are drive by non-inverted and the other two by inverted clock. A further interleaving is obtained by time-delaying two of four input analogue signals by quarter of a sampling period. The Σ signal is not expected to change much withing few turns, so only two ADCs are used providing a sampling frequency of (240 Msps). More Σ samples can be acquired from successive turns and averaged. A total of 12 ADCs is used. The datasheet ADC transition noise is specified at 3.4 LSB_{RMS}. IF amplifiers were designed to contribute not more than 0.5 LSB_{RMS}. As the real ADC performance was found a bit better, the achieved total ADC noise floor (including the drivers) is about 3.6 LSB_{RMS}.

4. Final noise performance

Using the same method as for the old module (numerical model plus superimposed real noise data), the noise floor of the new generation beam position module was determined to be 0.22 μm_{RMS}, with Gaussian distribution as shown in figure 2. An improvement by more than factor 6 with respect to the first generation. From Run III (2022-), the LHC TFB will also use four pickups instead of two, further lowering the overall TFB system noise floor. The obtained noise performance is close to the limit of available technology. There is still a bit of potential to further lower the noise floor by digital signal processing of the Σ signals.

The new module supports bunch by bunch gain control which is transparent for operation. For ultimate performance, the receiver gain can closely follow the machine cycle, e.g. large acceptance for injected bunch train, increase after the injection transient is damped, and compensate for lost bunch intensity during collisions. The new system also allows to run at optimal gain for circulating bunches with very different intensities, e.g. for machine development sessions.

To validate the technology, one prototype of the new generation beam position module per beam per plane was installed to LHC for the beam test in October 2021. LHC was operated with upgraded TFB during the whole two week period. Figure 3 shows an injection transient captured by the old and new beam position measurement modules.

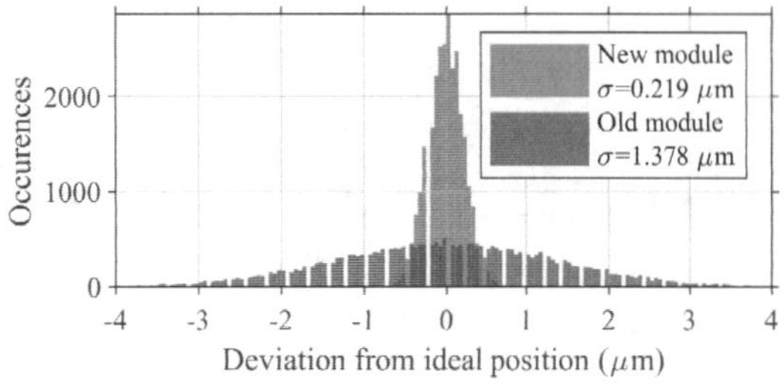

Figure 2. Comparison of position measurement quality for the old and new BeamPos modules (27000 samples). Same operating conditions, such the performance can be directly compared.

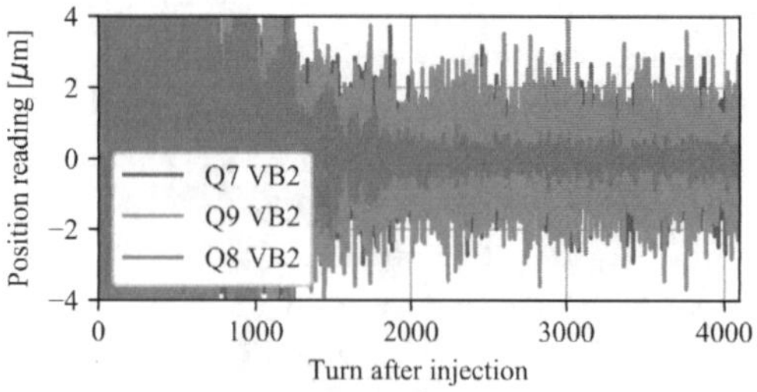

Figure 3. Injection into LHC captured by the old (Q7,Q9) and new (Q8) beam pos. meas. module. Data are notch filtered and β-normalized, so they can be directly compared.

5. Conclusion

For Run III, the LHC TFB was upgraded with a new generation, very low noise beam position measurement modules. The measurement subsystem noise floor was lowered by more than a factor 6 with respect to the old modules. Together with the doubled number of pickups and more advanced operation possibilities it is expected the overall TFB noise floor will improve even further.

Acknowledgments

Authors would like to thank X. Buffat and S. Furuseth for their extensive theoretical work on noise related phenomena in LHC, which triggered and stimulated many new ideas for the LHC TFB. Their studies allowed us to set concrete engineering constraints and requirements on the TFB hardware developments. We would also like to thank M. Söderén for all software and high performance computing work for LHC TFB. And finally, G. Kotzian, for his long-standing inspiration and support of the LHC TFB.

References

[1] Uythoven J A *et al.* 2014 *Proc. 5th Int. Particle Accelerator Conf. (IPAC'14)* (Dresden, Germany) pp 138–40
[2] Gianfelice-Wendt E *et al.* 2012 *Proc. 3rd Int. Particle Accelerator Conf. (IPAC'12)* (New Orleans, LA USA) pp 496–98
[3] Höfle W *et al.* 2012 *Proc. 3rd Int. Particle Accelerator Conf. (IPAC'12)* (New Orleans, LA USA) pp 4059–61
[4] Sapinski M *et al.* 2013 *Proc. 4th Int. Particle Accelerator Conf. (IPAC'13)* (Shanghai, China) pp 3025–27
[5] Soderen M and Valuch D 2019 *EPJ Web Conf.* **245** 01036
[6] Buffat X, Herr W, Pieloni T and Valuch D 2020 *Phys. Rev. Accel. Beams* **23** 021002
[7] Furuseth S V and Buffat X 2020 *Phys. Rev. Accel. Beams* **23** 114401
[8] Baudrenghien P and Mastoridis T 2015 *Phys. Rev. Spec. Top. Accel Beams* **18** 101001
[9] Valuch D 2018 On improving the ADT pick up resolution Presented at HL-LHC WP2 meeting 10.4.2018
[10] Valuch D and Baudrenghien P 2007 Beam phase measurement and transverse position measurement module for the LHC Poster presented at LLRF07 workshop (Knoxville, USA)
[11] Buffat X *et al.* 2019 Noise studies with new ADT pickups (MD4143) Tech. rep. CERN-ACC-NOTE-2019-0026
[12] Analog Devices, LTC2185 datasheet `https://www.analog.com/media/en/technical-documentation/data-sheets/218543f.pdf`
[13] LHC ADT beam position module, digital board. EDA-04230 in CERN design archive `https://edms.cern.ch/item/EDA-04230-V3-1/0`
[14] LHC ADT beam position module, RF receiver board. EDA-04231 in CERN design archive `https://edms.cern.ch/item/EDA-04231-V3-0/0`

Online correction of laser focal position using FPGA-based ML models

J Einstein-Curtis[1], S J Coleman[1], N M Cook[1], J P Edelen[1], S Barber[2], C Berger[2] and J van Tilborg[2]

[1] RadiaSoft LLC, Boulder, CO, USA

[2] Lawerence Berkeley National Lab, Berkeley, CA, USA

E-mail: joshec@radiasoft.net

Abstract.
Ultrafast lasers play an increasingly critical role in the generation, manipulation, and acceleration of electron beams for High Energy Physics applications. Laser plasma accelerators enable order of magnitude improvements in accelerating gradient and promise compact tunable GeV electron beam sources, while novel photocathode systems permit fundamental advances in electron beam manipulation for accelerator and radiation applications Advances in fast feedback systems are required to stabilize laser performance at kHz repetition rate operation against environmental fluctuations. A field programmable gate array (FPGA) based digital control system, coupled with responsive optics, can provide rapid and precise stabilization of ultrafast lasers. A collaboration between RadiaSoft and the Lawrence Berkeley National Laboratory BELLA Center to develop, test, and deploy these systems across a range of beamlines operating at >1 Hz repetition rate, including 1 kHz systems, was created.

1. Introduction

Laser plasma accelerators (LPAs) rely upon accurate control of ultrafast lasers, typically Ti:Sapph and Nd:Yag amplifier systems [1, 2]. The BELLA Center at Lawerence Berkeley National Laboratory (LBNL) features several ultra-short pulse, high-energy beamlines to develop LPAs. These accelerators require highly repeatable, stable interaction points to generate high-quality electron beams, which necessitates a collection of active and passive controls to mitigate environmental, mechanical, and component variations.

Recent work has primarily focused on enhancing transverse beam stability [3–5]. This paper describes a a strategy to address focal position stability, leveraging a machine learning (ML) enhanced wavefront diagnostic in tandem with a Field Programmable Gate Array (FPGA) controller to correct focal position at a kHz-scale rate. By building a model of wavefront at the interaction point, it is possible to use a non-perturbative measurement to calculate the focal position [6].

2. Facility and equipment

The initial model was created for the BELLA HTU laser system, shown in figure 1. This beamline operates with 1 kHz seed pulses and a 1 Hz full-power pulse. A HASO FIRST Shack-Hartmann wavefront sensor was used as the ground-truth imaging device of the interaction and post-interaction region, with the pre-interaction region sensor a Thorlabs WFS20-7AR. A Xilinx Zynq ZCU104 FPGA evaluation kit was used for testing to provide flexibility during the

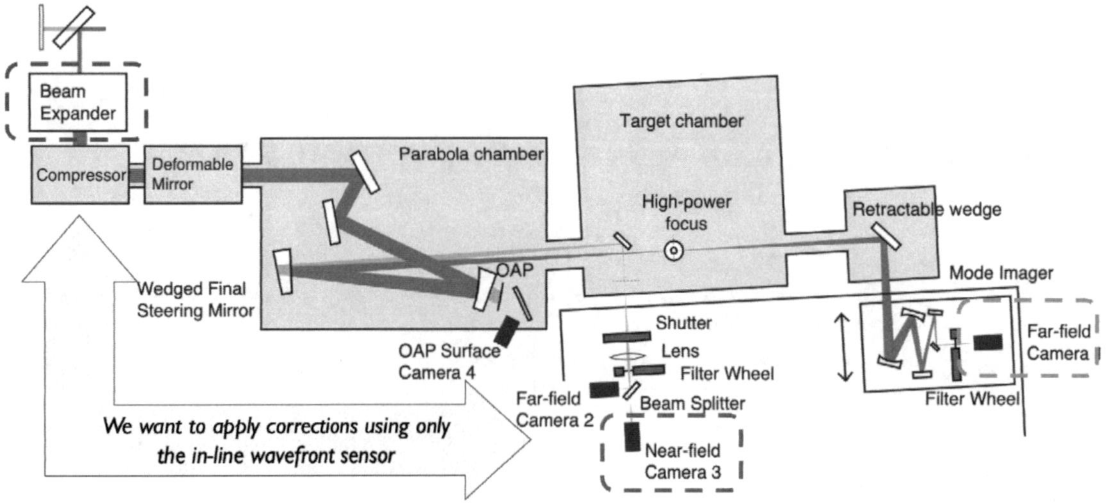

Figure 1. Diagram of HTU laser system at LBNL, highlighting the proposed correction scheme. Machine learning techniques are used to correlate a fast, non-perturbative sensor (2) with a high-quality, but perturbative wavefront sensor (1) which cannot be used for online correction. The resulting online diagnostic is used to deduce variations away from the desired focal position, which is then corrected for prior to the next shot by changes made to a transmissive lens beam expander (3).

Table 1. Optimal lens movement vs focal shift and beam size change. Focus shift is per mm lens translation. Beam size change is change per mm lens translation.

Element	Shift	Size Change
Transmissive Amp3-in	2mm	x1.348
Transmissive Amp4-in	2mm	x1.046
Reflective Amp4-out	1mm	x1.002

prototype phase, including a variety of customizable I/O, well-supported manufacturer-provided software, and a variety of processing options in support of ML operations.

3. Focal position investigations

To determine the optimal lenses to move for a focal shift, we looked at the magnitude of the shift at final focus and the (unwanted) increase in beam size throughout the optical chain. Table 1 summarizes these parameters for three different lenses in the telescope.

From these simulations we determined that the reflective Amp4-out is not ideal as a motorized correction optic for focal location because it is more weakly responsive, shifting the focus by only 1 mm per mm translation. Moreover, the off-axis reflective geometry introduces beam centroid kicks, even in response to relatively mild beam size variations. Ultimately, we determined the Amp4-in telescope is the best choice.

To verify our model, we measured the focal location vs lens separation at high power. Our measurement used a comparable method of capturing leakage from the final steering mirror thus measuring raw focal location without the need for further calibration or renormalization. The inset of figure 2 provides details of the measured focal position and radius of curvature taken from the wavefront sensor.

When comparing measurements to the simulation, we note that the focus shift per mm stage

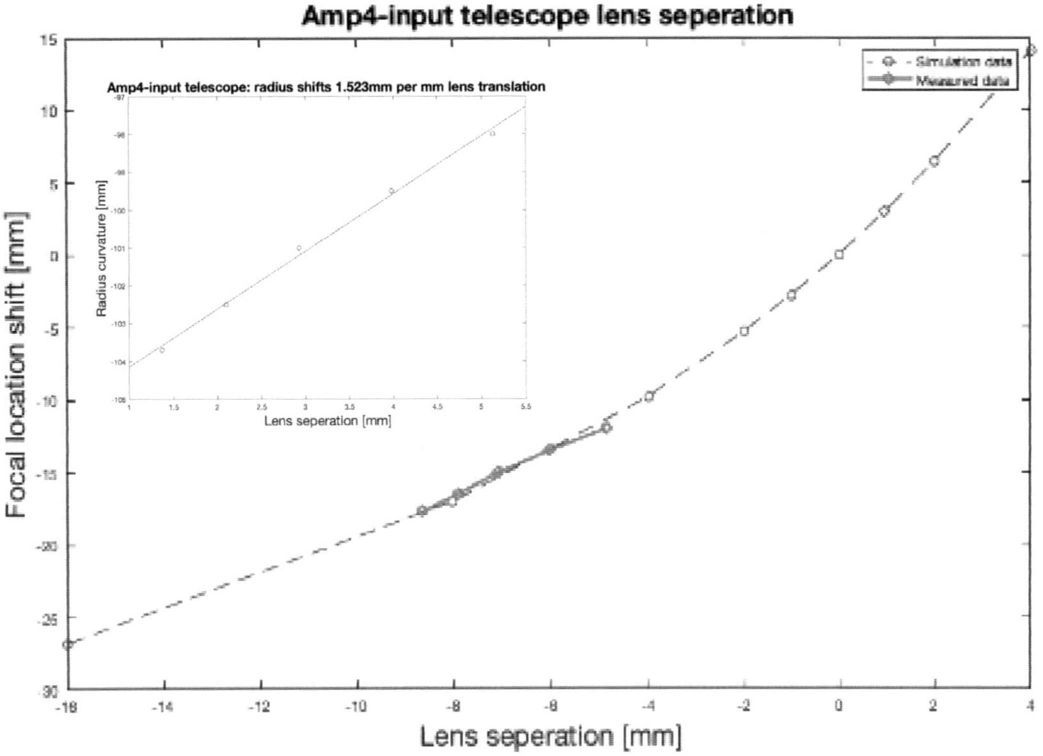

Figure 2. Focal location vs lens separation.

motion depends on the nominal Amp4-in lens separation. For a perfectly collimated beam entering the $-f_1/+f_2$ telescope, and for a perfectly collimated beam leaving the telescope (lens separation is $f_2 - f_1$), the slope change is 2mm focus shift per 1mm change in lens separation.

However, for the situation where the lens separation is NOT equal to $f_2 - f_1$, for example because the input beam has a divergence or the output beam is not perfectly collimated, this slope will have a different value.

By overlapping the experimental data (red circles) with the simulation (blue circles), we find a good agreement for one very specific initial lens separation offset (circa -6 mm). The slope at this separation is 1.52mm focal position shift for every 1mm lens motion. Figure 2 confirms this result.

This validates the use of a telescoping optic configuration for making controlled adjustments to the laser focal position. This design was validated through simulation and experimental measurement.

4. Model development

Several datasets were collected to examine changes in focal position on a shot-by-shot basis. The intra- and inter-shot variation over time, as shown in figure 3, show millimeters of variation in the calculated radius of curvature, highlighting the need for correction schemes.

Examining the extrapolated focal positions from each dataset reveals significant discrepancies between the two sensor measurements. Figure 4 shows a correlation plot between the two sensors, for which the raw correlation, as measured by the Pearson's coefficient between computed radius

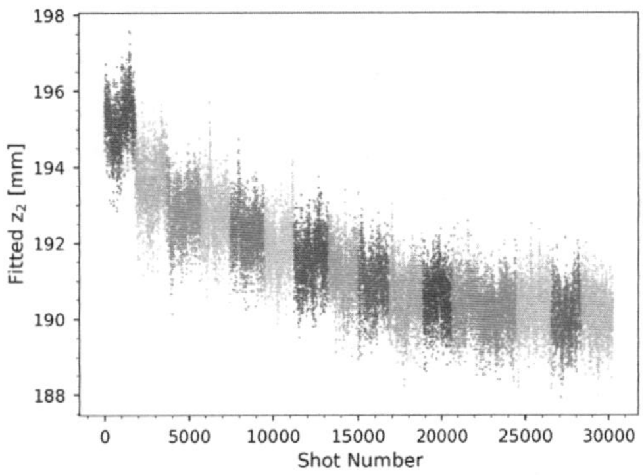

Figure 3. Representative dataset variation in fitted Z_2 calculation.

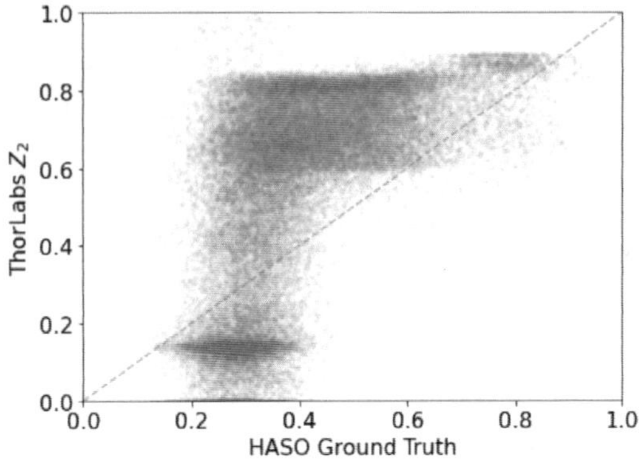

Figure 4. Raw HASO/ThorLabs correlation.

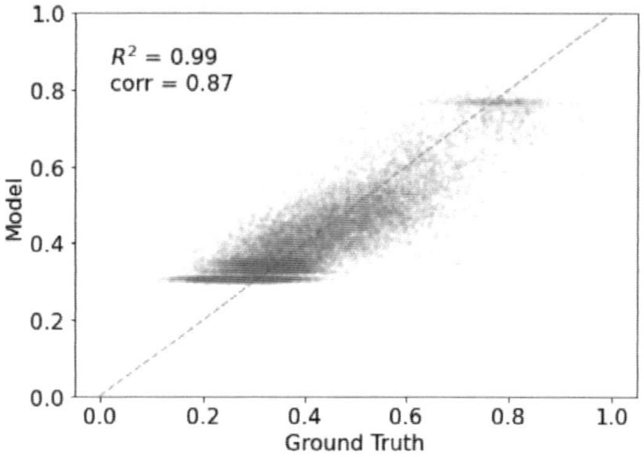

Figure 5. Feed-forward network accuracy.

of curvature, is only 0.45.

Due to the lack of correlation between the two sensors using raw pixel data, it became necessary to develop a pre-processing flow and system model to accurately capture systematic differences in the two measurements. We thus developed and trained a set of neural network models with the aim of improving the correlation between the two devices. The output for the trained network was a prediction of the radius of curvature, to be compared against the HASO WFS measurement.

Our initial efforts considered two different types of neural networks – convolutional neural nets (CNN) and more general feedforward neural nets (FFNN). Each network was trained using PyTorch, an open source library for developing machine learning models.

CNNs are designed to operate on images as inputs and are useful for computer vision applications. Our CNNs were trained using 12×12 pixel-by-pixel image data of the wavefront from the Thorlabs WFS. We found that the CNN produced only modest improvements in the correlation, to upwards of 0.63 from the initial value of 0.45. We thus transitioned to exploring more general FFNNs.

Our FFNN architecture featured 2-4 fully connected hidden layers, ReLU activation functions, and implemented a robust scaler on inputs and outputs. These features were chosen to be fast-executing and compatible with our FPGA deployment strategy. We concluded that the FFNN consistently produced better results than the CNN, but did exhibit tradeoffs between input space size and network complexity, with implications for performance at high repetition rate.

The best correlation was found by augmenting the pixel data with additional Zernike polynomial fitting terms; we explored several different strategies for generating the fit. Using

Table 2. Correlation values for different data.

Dataset	Pearson Correlation
Zernike fits only	0.45
CNN - pixel data only	0.63
FFNN - pixel data only	0.82
FFNN - pixel data & Zernike fits	0.87

the Thorlabs toolkit to produce a 5th order fit provided an additional 16 terms to include in the input space of the network, and improved dataset correlation to as high as 0.87, using only two hidden layers, as shown in figure 5.

Using external fitting libraries, such as the Mahotas library [7], permitted higher order fits, such as a 28-value, 6th-order fit. However, increasing the fit complexity showed diminishing returns, as correlations did not improve significantly, while speed of execution declined. Using a 6th-order fit does enable a network to be trained using only fitting data (28 inputs), and can result in comparable performance to that of the full set of pixel values. Table 2 summarizes the correlation performance for each of our approaches.

5. Implementation

A full correction implementation was prototyped using the FPGA system and was tested on the bench to meet-or-exceed the operational requirements of the HTU beamline. This implementation utilized the Xilinx Vitis AI toolkit in conjunction with the Xilinx Deep Learning Processor (DPU) to minimize the use of custom FPGA designs and software.

Due to driver limitations of the Thorlabs WFS20 sensor, in particular being limited to a Windows-based platform, the sensor was unable to be directly connected to the processing platform. This necessitated the use of an alternative data communication channel. This channel was created in Python using ZeroMQ, and tested to transfer wavefront data to the processing platform at the limit of the sensor capture rate (about 0.9 kHz). Validation data from model development was used to test the model processing performance, and achieved a better than 5 kHz throughput, with well-understood bottlenecks and limitations. Due to the minimal number of outputs, additional output data processing should not over-burden the system, enabling performance that meets the 1 kHz seed pulse rate on the HTU beamline.

6. Conclusion

We have demonstrated a model of the BELLA Center HTU beamline interaction region and developed a correction method for the focal position. This model, in conjunction with slow controllers, corrects for measured system variations in simulation. This method has been demonstrated in prototype hardware using simulated data and meets-or-exceeds the necessary performance requirements with room for expansion and increased model complexity as needed.

Limitations exist in that variations between seed and full-power pulses might require multiple models for proper correction. Plans exist to continue this work on additional beamlines to develop a flexible, plug-and-play framework for additional LPAs.

Acknowledgments

This material is based upon work supported by the U.S. Department of Energy, Office of Science, Office of High Energy Physics, under Award Number DE-SC0021680.

Journal of Physics: Conference Series **2420** (2023) 012074 doi:10.1088/1742-6596/2420/1/012074

References

[1] Esarey E, Schroeder C B Leemans W P 2009 *Rev. Mod. Phys.* **81** 1229
[2] Dann S J D *et al.* 2019 *Phys. Rev. Accel. Beams* **22** 041303
[3] Isono F, van Tilborg J, Barber S K, Geddes C G R, Tsai H, Schroeder C B, Leemans W P 2019 *Update on BELLA Center's Free-Electron Laser Driven by a Laser-Plasma Accelerator* (CLEO: Science and Innovations SF31)
[4] Barry M 2015 Evaluation of Laser Stabilization and Imaging Systems for LCLS-II Tech. rep. SLAC-TN-15-012 SLAC National Accelerator Laboratory
[5] Nguyen L *et al.* 2019 *Proc. North American Particle Accelerator Conf. (NAPAC'19)* (Lansing, MI USA) pp 938–41
[6] Shalloo R J *et al.* 2020 *Nat. Commun.* **11** 6355
[7] Coelho L P 2013 *J. Open Res. Software* **1** e3

Journal of Physics: Conference Series **2420** (2023) 012075 doi:10.1088/1742-6596/2420/1/012075

Commissioning and first results of an x-band LLRF system for TEX test facility at LNF-INFN

L Piersanti, D Alesini, M Bellaveglia, S Bini, B Buonomo, F Cardelli, C Di Giulio, E Di Pasquale, M Diomede, L Faillace, A Falone, G Franzini, A Gallo, G Giannetti, A Liedl, D Moriggi, S Pioli, S Quaglia, L Sabbatini, M Scampati, G Scarselletta, A Stella, S Tocci and L Zelinotti

INFN - Laboratori Nazionali di Frascati, Via Enrico Fermi 54, 00044, Frascati, Italy

E-mail: luca.piersanti@lnf.infn.it

Abstract. In the framework of LATINO project (Laboratory in Advanced Technologies for INnOvation) funded by Lazio regional government, the commissioning of the TEst stand for X-band (TEX) facility has started in 2021 at Frascati National Laboratories of INFN. Born as a collaboration with CERN to test high gradient accelerating structures, during 2022 TEX aims at feeding the first EuPRAXIA@SPARC_LAB X-band structure prototype. During 2021 the commissioning has been successfully carried out up to 48 MW. The power unit is driven by an X-band low level RF system, that employs a commercial S-band (2.856 GHz) Libera digital LLRF (manufactured by Instrumentation Technologies), with an up/down conversion stage and a reference generation and distribution system able to produce coherent frequencies for the American S-band and European X-band (11.994 GHz), both designed and realized at LNF. The performance of the system, with a particular focus on amplitude and phase resolution, together with klystron and driver amplifier jitter measurements, will be reviewed in this paper. Moreover, considerations on its suitability and main limitations in view of EuPRAXIA@SPARC_LAB project will be discussed.

1. Introduction

The TEX facility [1] is one of the pillars of the LATINO project [2], funded by Lazio regional government, that aims to provide companies and the scientific community with the advanced technologies and skills developed in the field of particle accelerators for research, medical and industrial applications. Its commissioning has started in November 2021 at INFN National Laboratories of Frascati with the Site Acceptance Test (SAT in the following) of the RF power unit [3]. During these preliminary tests also the LLRF system has been commissioned, showing some promising results, but also highlighting some critical aspects concerning, for instance, sample rate, front-end and back-end bandwidth, which are particularly important for a future LLRF system development for an X-band driven linac.

2. TEX test facility

The TEX facility main goal is to test at high power X-band RF components at LNF. Waveguide devices and accelerating structures could be conditioned at this frequency in the facility, which is supplied by a VKX8311A 50 MW, 1.5 μs klystron from CPI LLC (USA), and a K400 450 kV solid state modulator from Scandinova (Sweden). The repetition rate will be at least 50 Hz. A new control room has been realized and equipped, and a concrete shielded bunker has been built to host the structures under test. A Memorandum of Understanding has been also signed with CERN, to profit from the well established experience on X-band technology acquired with the operation of the XBOX test stands. In fact, the test-facility will be also used to test and condition CLIC structures. TEX will be also an R&D and test facility for X-band RF components, LLRF systems, beam diagnostics, vacuum technologies and control system in view of INFN future accelerator EuPRAXIA@SPARC_LAB [4, 5] that will be built at LNF. This project foresees in fact the construction of a X-band linac booster up to 1 GeV, working with a target accelerating gradient of 60 MV/m.

A sketch of the TEX area is reported in figure 1, where the modulator cage, the bunker and the waveguide network CAD design is reported.

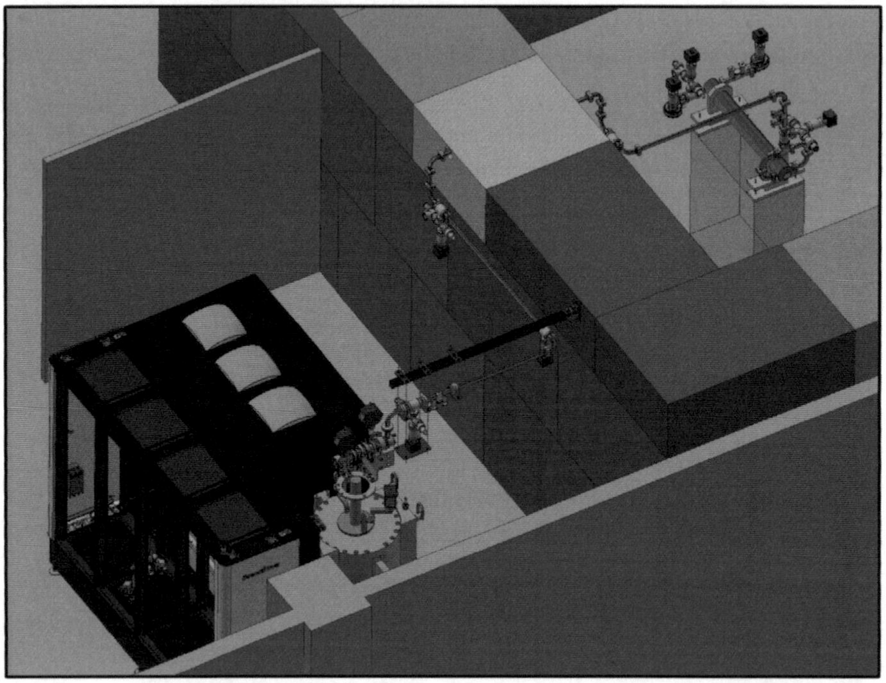

Figure 1. 3D CAD drawing of the TEX facility experimental area.

The SAT of the RF power unit took place in November 2021, once the building air conditioning and civil engineering works have been completed. The entire waveguide network is now in place up to the final RF splitter right before the accelerating structure input coupler (where two 25 MW RF loads have been connected in parallel for the high power tests instead). Using this RF layout, it was possible to reach 48 MW, with a pulse length of 150 ns at 50 Hz repetition rate during the SAT. In figure 2 is shown the waveguide network outside (left) and inside (right) the bunker.

To ensure a safe and repeatable accelerating structure test, aiming to reach the nominal accelerating gradient minimizing the operation time, a complex automatic conditioning/interlock algorithm has been foreseen, implemented and preliminarily tested during the modulator SAT [6]. In its final version, it will include also an online measurement of the produced dark current and of the breakdown rate, to safely guide the conditioning process. An EPICS based Telegram integration has been also deployed for the alarms handling of the facility [7].

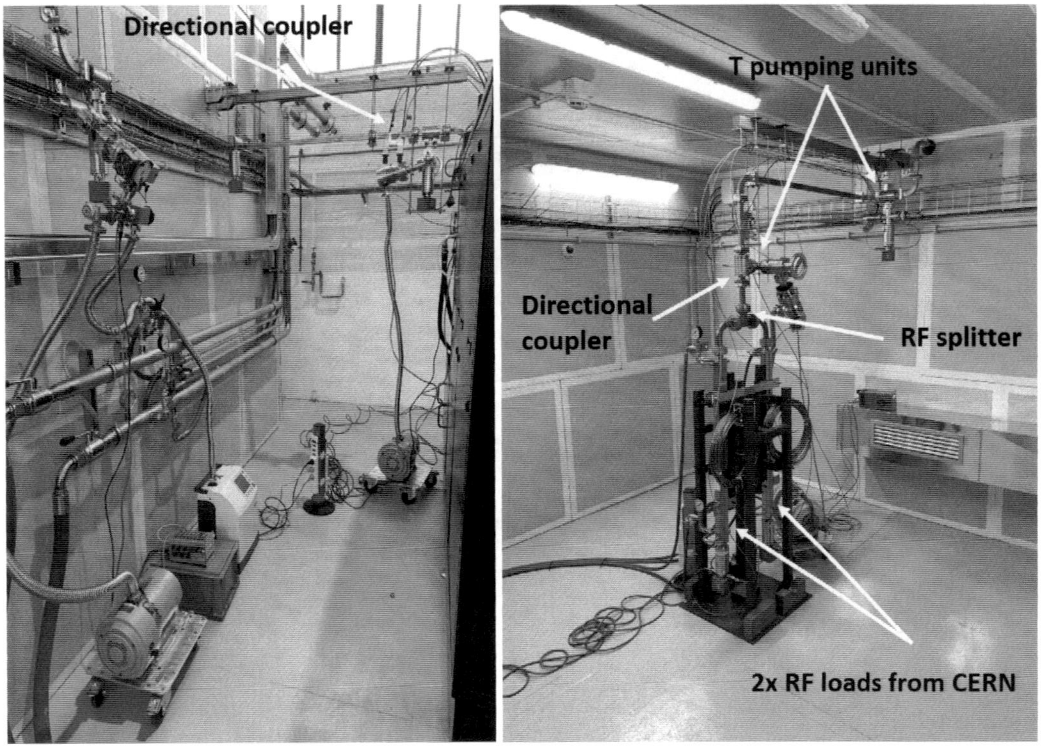

Figure 2. Pictures of the waveguide network in the klystron area (left) and inside the bunker (right).

3. LLRF system commissioning

A block diagram of the whole RF layout including the BOC pulse compressor (that will allow the simultaneous conditioning of two structures in parallel, and will be installed in a second phase) is shown in figure 3 with a particular emphasis on the LLRF system and all the RF signals picked-up for diagnostics and feedback.

A detailed description of the development of the TEX X-band LLRF system is reported in [8]. The system RF source is an ultra low phase noise Rohde&Schwarz synthesizer (SMA100B). The core of the system is based on a commercial S-band (2.856 GHz) Libera LLRF (purchased from Instrumentation Technologies, and whose performance has been already reviewed in [9, 10] for a similar architecture). In addition, a custom up/down converter and reference generation stages, that translate the carrier frequency to the European X-band (11.994 GHz) and viceversa, have been both developed at LNF. It has been installed and commissioned in Fall 2021 prior to the modulator SAT. The LLRF system is hosted in the rack room adjacent the control room. A picture of the Low-Level RF rack is shown in figure 4

As visible from the picture, in addition to the up/down converter and the reference generation some ancillary systems have been also developed, in particular: (i) a splitter stage for klystron forward and reverse power signals, which are needed by the RF mask digitizer (12-bit, 3.2 GS/s Wavecatcher [11]) employed for pulse to pulse breakdown detection [12, 13]; (ii) a trigger and interlock distribution, to safely inhibit the RF pulses generation whenever the machine protection system detects a possible threat.

The RF signals picked-up by the directional couplers are routed to the LLRF system front-end by means of Andrew FSJ1-50 1/4 inch phase stable cables. Even though the phase stability

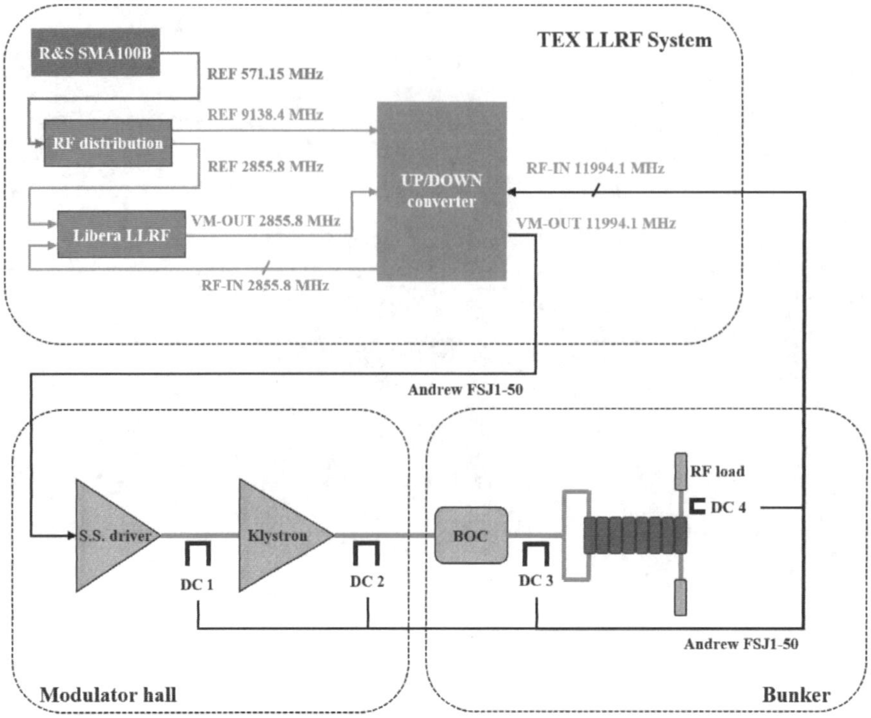

Figure 3. TEX RF system block diagram. It includes also the BOC pulse compressor that will be added in a second phase.

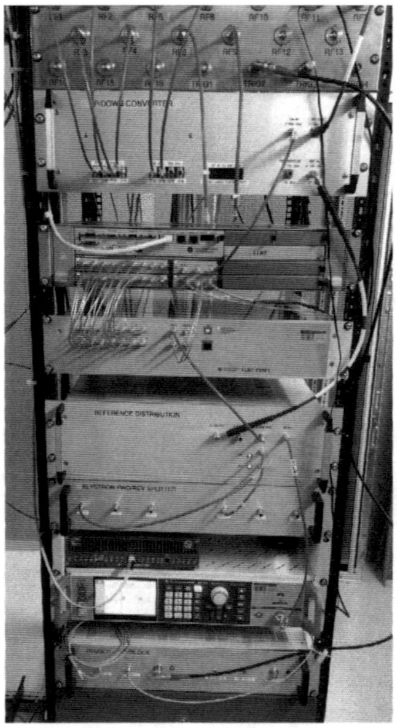

Figure 4. TEX LLRF rack.

Table 1. RF jitters preliminary measurement.

	Amp. jitter	Phase jitter	Time jitter
KLY FWD	0.04 %	0.0894 deg	20.7 fs

requirement is not strictly necessary in a test facility, i.e. where no beam is going to be accelerated, we have chosen such cables to test their performance (e.g. in terms of long term drifts) also in view of the EuPRAXIASPARC_LAB TDR.

During the SAT, signals from the DC1 and DC2 have been acquired and all the LLRF system functionalities have been successfully tested. Some preliminary measurements have been carried out aiming at characterizing the amplitude and phase jitter on the klystron output power, which is one of the most important parameters to qualify the stability of the power source. In figure 5 the outcome for 20 MW, 300 ns pulse amplitude (left) and phase (right) stability measurements are shown, while table 1 summarizes the results. Very promising jitter values have been obtained: 0.04 % and 20.7 fs rms for amplitude and phase respectively. These performance are compatible with state of the art solid state power sources, and represent a strong groundwork for the future X-band linac facility EuPRAXIASPARC_LAB.

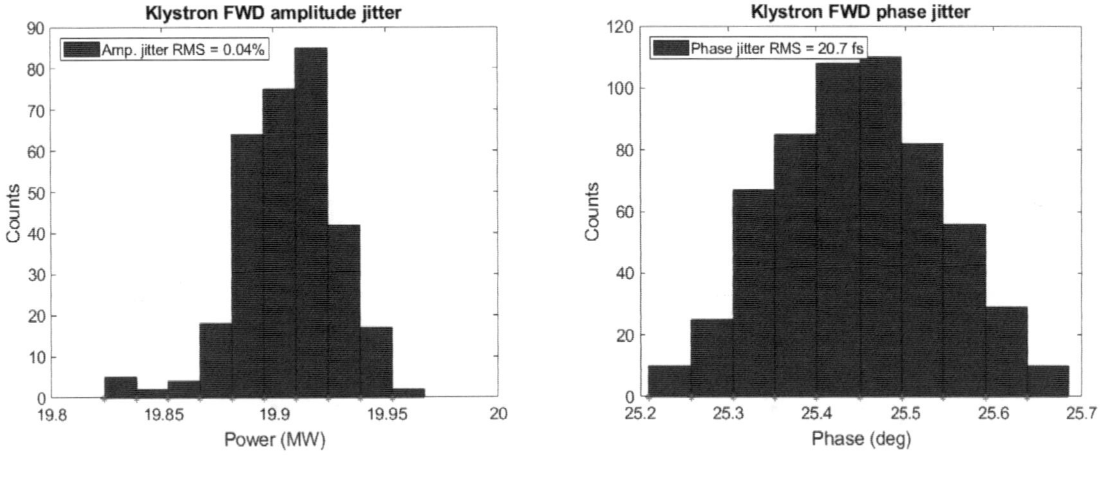

Figure 5. Measured amplitude and phase jitter of klystron output forward power. The measurements refer to 20 MW, 300 ns at 50 Hz repetition rate RF pulses.

As soon as the facility will start its experimental activity, an extensive measurement campaign will be carried out to confirm and optimize these results at full power (maximizing the signal dynamics on the front-end), to identify all the phase jitter contributions (LLRF, driver, klystron) and to evaluate the Libera LLRF noise floor.

The LLRF system reviewed in this paper, certainly suitable for a test stand, it does not seem the optimal choice for an X-band based linac for various reasons. First of all, the high conversion loss ($\approx$ 14 dB) of the down-converter stage limits the dynamic range of the front-end. In fact, even if high LO power RF mixers have been employed, the maximum RF level at the LLRF front-end is 6 dBm, while the Libera LLRF could accept signals up to 10 dBm. Then, considering that the filling time of X-band structures usually ranges between 150 ns and 200 ns, the available measurement window of 8.6 μs results too wide and the sampling frequency too low (one ADC sample every 8.4 ns). This, together with a back-end and front-end bandwidth respectively of 16 MHz and 5 MHz, represents a major limitation with such short pulses, especially if fast pulse modulations are required.

4. Conclusions

In Fall 2021 the LLRF system of the TEX facility has been successfully commissioned and the first RF pulses have been delivered to the RF loads to condition the waveguide network and to test the RF power unit during the modulator SAT. A final power of 48 MW, 150 ns at 50 Hz repetition rate has been reached before the end of the SAT. Some preliminary performance of the X-band LLRF system have been evaluated. Promising results have been obtained, with a measured amplitude and phase jitter of the klystron forward output power of 0.04 % and 20.7 fs respectively. As soon as the facility will start its experimental activity, more detailed tests on the LLRF system will be carried out.

Acknowledgments

LATINO is a project co-funded by Regione Lazio within POR-FESR 2014-2020 program.

References

[1] Pioli S *et al.* 2021 *Proc. 12th Int. Particle Accelerator Conf. (IPAC'21)* (Campinas, Brazil) pp 3406–09
[2] Sabbatini L *et al.* 2019 *Proc. 10th Int. Particle Accelerator Conf. (IPAC'19)* (Melbourne, Australia) pp 2277–79
[3] Cardelli F *et al.* 2022 *Proc. 13th Int. Particle Accelerator Conf. (IPAC'22)* (Bangkok, Thailand) pp 1148–51
[4] Alesini D *et al.* 2018 EuPRAXIA@SPARC_LAB Conceptual Design Report Tech. rep. INFN-18-03/LNF
[5] Ferrario M *et al.* 2018 *Nucl. Instrum. Methods Phys. Res., Sect.* A **909** 134
[6] Pioli S *et al.* 2022 *Proc. 13th Int. Particle Accelerator Conf. (IPAC'22)* (Bangkok, Thailand) pp 751–53
[7] Pioli S *et al.* *Proc. 13th Int. Particle Accelerator Conf. (IPAC'22)* (Bangkok, Thailand) pp 1145–47
[8] Piersanti L *et al.* 2021 *Proc. 12th Int. Particle Accelerator Conf. (IPAC'21)* (Campinas, Brazil) pp 3371–74
[9] Piersanti L *et al.* 2016 *Proc. 7th Int. Particle Accelerator Conf. (IPAC'16)* (Busan, Korea) pp 407–10
[10] Piersanti L *et al.* 2018 *Proc. 9th Int. Particle Accelerator Conf. (IPAC'18)* (Vancouver, Canada) pp 2162–65
[11] Breton D *et al.* 2014 *Proc. 19th IEEE-NPSS Real Time Conf. (RT2014)* (Nara, Japan) pp 1–8
[12] Alesini D *et al.* 2017 *Proc. 8th Int. Particle Accelerator Conf. (IPAC'17)* (Copenhagen, Denmark) pp 3662–65
[13] Pioli S *et al.* 2017 *Proc. 8th Int. Particle Accelerator Conf. (IPAC'17)* (Copenhagen, Denmark) pp 1822–23

A Python Framework for High-level Applications in Accelerator Operations

J Chrin, V Erçağlar and T Schietinger

Paul Scherrer Institut, 5232 Villigen PSI, Switzerland

E-mail: jan.chrin@psi.ch

Abstract. A Python graphical framework providing reusable components to facilitate the development of accelerator applications, that meet the basic requirements of experts and operators alike, is presented. Such a collective approach serves to bridge the gap between the expert developer and the operational team, resulting in applications that are inherently cohesive, durable and easily navigable. The operational advantages and underlying principles are exemplified in a reference application that provides executable examples of customary practices, and further highlights a number of composite and control system-enabled widgets.

1. Perspective and motivation

The development of high-level applications crosses the domain of several groups, each possessing a distinctive skill set and ambition. Typically the 'expert' application is developed by the scientist, engineer, whose primary interest is in providing an interface that details the hardware capabilities of the system. The 'operator' application on the other hand may require only a condensed view and one that offers automated procedures geared towards the everyday operation of the accelerator. Other, beam dynamics, applications will undertake specific measurements crucial to the optimization of the accelerator, and may further require interaction with accelerator models, message reporting capability, and the means to store and retrieve data resulting from their analyses. Many such measurements and procedures are first deployed by the expert in the commissioning phase of the accelerator before ultimately being delegated to machine operators once routine operation is established. The diversity of developers, however, inevitably leads to a variety of frameworks and appearances, with some duplication in that similar functionalities are instigated in different manners and with a non-uniform behaviour. These subtle details of cross-domain application development are acknowledged to add to the tasks and challenges faced by the operator [1, 2].

This inadvertent disparity between applications, however, can be alleviated by supplying a controlled, coordinated, and configurable graphical interface wherein common functionality is provided through predetermined inputs with well defined responses. Applications then become inherently homogeneous, and developers are further relieved of implementing peripheral tasks, releasing time to prioritize their particular area of interest.

In the following, a Python graphical framework is described that provides a base class that can be inherited by the application developer, and adapted to specific needs. The focus is on the components and methodology that constitute the framework, rather than any specific application that is built upon the infrastructure, as applied within the context of SwissFEL and the Swiss Light Source (SLS).

2. A python graphical framework

The Python programming language is presently enjoying a high profile within the accelerator community being the preferred language for application development at facilities of various size [3, 4]. While scripting is regarded as gratifyingly intuitive and powerful, our applications are inherently graphical and this in itself adds a new level of complexity. To this end, PyQt, a Python graphical user interface (GUI) module that connects to the Qt C++ framework [5,6], is the prevailing choice. Significantly, in this work, Qt modules are imported through the QtPy abstraction layer [7] allowing our framework to be used effortlessly across Qt versions.

The Qt library provides much functionality, with numerous classes. Its specific application domain of windows, widgets, layouts, colours, shapes, and more, however, presents the developer with expansive possibilities to interrogate and discern before converging on a finely-tuned appearance. Furthermore, and rather critically, any complex physics analysis, or any other resource intensive operation having a long running time, needs to be delegated to a separate thread. Here, Qt's dedicated thread support is a fitting option given that our applications interact with other components of the Qt library. In this way, the main thread remains responsive at all times, widgets with read backs continue to be updated, and user interaction, where permitted, is not interrupted. Such considerations are a prerequisite in gaining a satisfactory user experience. Once these challenges are overcome, the programmed solutions may be presented to the developer in a reusable form.

2.1. Apps4Ops: A characteristic style

The GUI implementation class follows Qt's customary 'main-window-style' approach that offers predefined options for user input in the form of a menu bar, toolbar, status bar, a central widget, and dock windows. The adoption of such a ready-made, integrated approach, at the outset, is an important factor in achieving a polished, intelligible, and navigable user interface. The static visualization of data is accomplished through the comprehensive Matplotlib library [8, 9]. For the display of continuous, real-time data, however, the PyQtGraph [10] library, based on Qt's GraphicsView framework, is preferred for its speedier response time. The principle components that constitute the framework, and other practicalities that play a part in achieving the desired homogeneity among applications, coupled with an optimized user experience, are elaborated.

2.1.1. Control system-enabled widgets. The low-level control system is built on the EPICS (Experimental Physics and Industrial Control System) framework [11], which comprises an extensive set of software tools tailored to the needs of particle accelerators and large-scale experiments. Interaction with the EPICS-based control system is established through a proven C++ abstraction layer [12–14], for which high-performance bindings to Python have been provided using the Cython technology [15]. Several control system-enabled Qt widgets, and composites, have been made available. These connect to a gateway module that handles their connectivity management, propagates updates, and ensures their display conforms to the recognized style.

2.1.2. Configuration. The provision of an application framework that spans across different accelerators requires a multi-faceted configuration mechanism that allows the generic, the accelerator-specific, and the application's user behaviour to be configured via a hierarchy of configuration files. Their tasks may be numerous, hence only certain representative usage is outlined here.

At the core level, a Qt style sheet file defines the appearance of GUI elements, including fonts, sizes, and colour schemes, in accordance with the stipulated style guide. In addition, a configuration file storing simple data structures in JSON (JavaScript Object Notation) format [16], provides information such as default menu options, and the destination of various output files. The second level configuration file provide handles to several accelerator-specific parameters, e.g., exposing the EPICS channels that constitute the accelerator-specific header widget. The application's user configuration file may be as extensive as the user requires. It can also be used to enact a number of predefined widgets and associated tasks, e.g., checkboxes that control simulated runs, debug options, etc., and also to formulate further GUI components that provide input parameters that dynamically feed into the application and its analysis tasks. A destination subpanel for analysis results, typically in graphical form, can also be entered.

The configuration files, and other small resources files such as icons and help pages, are registered and safe guarded in an accelerator-dependent resources file from which a Python module is generated. The resource module is then imported by the application, from where the files can be accessed directly. No assumptions need then be made about the location of the files, or the application's working directory; with help files also incorporated into the resources module, no external web help pages need be summoned. The default generic and accelerator level configurations can also be replaced by user-supplied files through command line interface flags.

2.1.3. Initialization. Applications are expected to initialize as speedily as is feasible. Faster times can be achieved by assigning any time-intensive 'loading' method to a single-shot timer with zero-timeout; the loading method is then executed when the event queue next allows, i.e., once the main window initialization is complete. Another helpful approach is to inform the application at startup of its associated channels. The underlying EPICS client library allows connections to be established in unison with a single call. Having connections established in advance of widget creation allows pertinent information concerning the channels to be readily available to any associated widget and allow its configuration to be determined dynamically and spontaneously. These data are further stored locally and can be recalled in the event of a channel being unavailable at start-up time. The widget is automatically reconfigured once a connection is established. Likewise, other application's settings, such as the window dimensions and positions, will have been previously optimized, remembered locally, and recalled on startup, providing consistency between sessions. Nevertheless, despite these optimizations, where complex applications may still take a few seconds to startup, which is invariably the case, a so-called 'splash screen' is shown ahead of the forthcoming main window display to reassure the user that the application is indeed loading. A progress bar on the splash screen further informs the user of the time remaining for the application to initialize.

2.1.4. Finalization. Best practice for a tidy application exit dictates that any pending actions are first completed or at least interrupted gracefully. This may entail checking if any analysis threads are still running and that control system parameters are restored to their pre-measurement values. Explicitly releasing application resources, such as those managing EPICS connectivity, reassures the developer of sound memory management in the underlying libraries.

2.1.5. Menu bar and toolbar. Menus and toolbars furnish a GUI with a convenient graphical presentation of user commands that perform specific tasks. Placements in the main window's menu bar are grouped into submenus consisting of a list of logically related pull-down items. A dynamically created menu bar is provided for operations on files, as necessitated for the display of recently used files, while other static menu bars are provided for performing peripheral tasks such as:

- initiating output actions, e.g., saving data to HDF5 files [17], reporting to the electronic logbook [18],
- capturing images and screenshots,
- connectivity to printers,
- interrogating the standard output file, clearing entries in the message log window,
- displaying procedural help pages and information that discloses 'behind-the-scenes' details, such as software versions and the application responsible,
- enabling a tidy exit.

Actions to application-specific items can be customized by the user by resetting the base class implementations. The application menu items are selected by the application's user configuration file.

The toolbar, located immediately below the horizontal menu bar, is populated with icons that provide quick access to the most commonly used commands.

2.1.6. Central widget. An accelerator-specific header widget takes ownership of the top area of the central widget. It displays the most pertinent information related to the current status of the machine. The remainder of the central widget is partitioned into dedicated subpanels that serve as destination locations for operator/expert input parameters to measurement procedures, the visualization of results, and the display of log messages. The partition of the central widget can be configured to the application's needs through the application-specific configuration file.

2.1.7. Message reporting. The use of a carefully considered message reporting interface ensures a consistent set of data is delivered to the log window. Mandatory information includes the origin and timestamp of the message, coupled with a severity level. The data are supplemented by a number of optional fields that are filled at the discretion of the user. In the event of a fatal condition that prevents the application from continuing, the provision of the developer to supply a meaningful and helpful message, that may also propose a solution, can hasten the return to normal operation.

2.1.8. Analysis module. An example analysis module which inherits Qt's thread class demonstrates the proposed analysis procedures. Input data from the operator or expert panels are automatically propagated to the analysis class. These, together with the analyzed data, are encapsulated into dictionaries for subsequent display and storage. Stored data can be recalled into the application and reanalysed, thereby demonstrating the seamless integration of logged and real-time accelerator data into the same framework. Interthread communication, as required for progress updates and message reporting, is accomplished through Qt's distinctive signal and slots mechanism.

2.2. Deployment and experience

A generic, skeleton application exemplifying features of the graphical framework acts as a reference for developers. An application that makes typical use of the infrastructure is shown in figure 1. The emphasis has been on creating fastidious modules to help ensure excellent operational ability. Even the simplest of push buttons must execute its task impeccably and be equipped to act with integrity in the event of an anomaly. The adoption of the framework ensures that applications have recognizable buttons and icons that lead to predictable actions, and accelerates the creation of user interfaces by relieving developers of implementing peripheral tasks, thereby releasing time to concentrate on their particular area of interest. The involvement of both graduate and undergraduate students in research and creative activities is also an integral

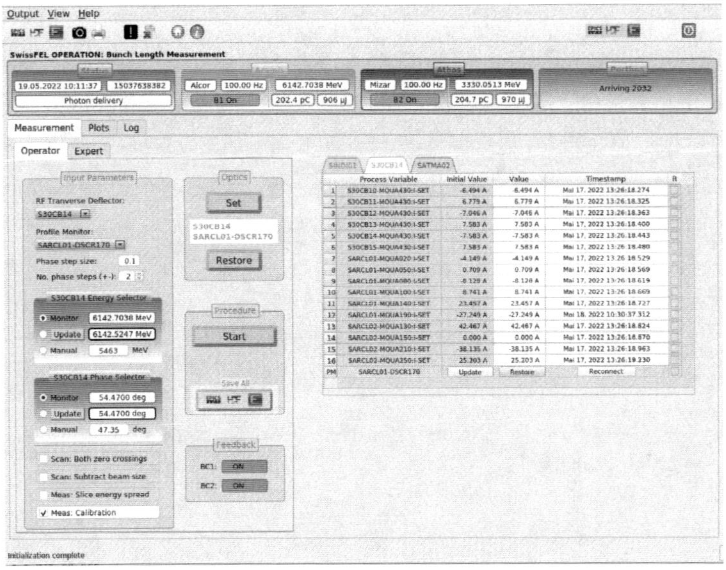

Figure 1. An example of an application using the 'main-window-style' framework. It comprises a menu bar, a toolbar, an accelerator-specific header widget, customized epics-enabled widgets, a status bar, and partitions for operator and expert inputs and procedures, the visualization of data, and the display of log messages.

part of our organization; their use of the framework increases productivity and readies their work for operational use. It further promotes sustainability in that only a single framework need be maintained and extended for basic application needs.

The deployment of the framework has proven to be an iterative process as essential feedback is returned from a focus group of users. The sharing of their experience, the problems encountered, and subsequent proposals, feature in a collective decision on the evolution of an application. The collaborative approach brings flexibility into our section, promotes inter-department activities, and equally broadens our own horizon.

3. Concluding remarks

Several common, peripheral requirements for high-level applications have been identified and incorporated into a dedicated graphical framework that serves as a foundation for the development of applications that are collectively cohesive, functional, durable, and navigable. The *Apps4Ops* software architecture has encompassed several renewed efforts as our understanding of operator requirements, and the capabilities of the underlying software, has advanced with experience. It has now been applied to a number of beam dynamics applications developed by accelerator physicists achieving a level of homogeneity and consistency that is shaping the evolution of high-level applications for operations at the existing SwissFEL and the upcoming SLS 2.0 [19, 20].

Acknowledgments

We are grateful to our many colleagues in the Accelerator Operation and Development Dept. of the Large Research Facilities (GFA) Division, for their valuable feedback, especially concerning enhancements to the user experience. The Python environment within GFA is provided by the Electronics and Control Systems Dept. One of us (JC) thanks Ivan Sinkarenko, CERN, for fruitful Pythonic discussions.

References

[1] Jacquet D 2016 *Proc. 7th Evian Workshop on LHC Beam Operation (Evian Les Bains)* pp 157–60
[2] Deghaye S 2016 *Proc. 7th Evian Workshop on LHC Beam Operation (Evian Les Bains)* pp 161–66
[3] Zhang T, Chen J H, Liu B and Wang D 2015 *Proc. 12th Int. Computational Accelerator Physics Conf. (ICAP2015)* (Shanghai, China) pp 23–25
[4] Elson P, Baldi C and Sinkarenko I 2021 *Proc. 18th Int. Conf. on Accelerator and Large Experimental Physics Control Systems (ICALEPCS 2021)* (Shanghai, China) pp 236–41
[5] Qt `https://www.qt.io`
[6] PyQt `https://riverbankcomputing.com/software/pyqt`
[7] QtPy `https://pypi.org/project/QtPy`
[8] Hunter J D 2007 *Computing in Science and Engineering* **9** 90
[9] Matplotlib `https://matplotlib.org`
[10] PyQtGraph `https://www.pyqtgraph.org`
[11] EPICS `https://epics-controls.org`
[12] CAFE `http://cafe.psi.ch`
[13] Chrin J 2015 *Proc. 15th Int. Conf. on Accelerator and Large Experimental Physics Control Systems (ICALEPCS 2015)* (Melbourne, Australia) pp 1013–16
[14] Chrin J, Aiba M and Snuverink J 2019 *J. Phys.: Conf. Ser.* **1350** 012155
[15] Chrin J 2016 *Proc. 11th Int. Workshop on Personal Computers and Particle Accelerator Controls (PCaPAC 2016)* (Campinas, Brazil) pp 21–24
[16] JSON `https://www.json.org`
[17] HDF® `https://www.hdfgroup.org`
[18] ELOG `https://elog.psi.ch/elog`
[19] Streun A, Garvey T, Rivkin L, Schlott V, Schmidt T, Willmott P and Wrulich A 2018 *J. Synchrotron Radiat.* **25** 631
[20] Streun A 2022 *Proc. 13th Int. Particle Accelerator Conf. (IPAC'22)* (Bangkok, Thailand) pp 925–28

Journal of Physics: Conference Series **2420** (2023) 012077 doi:10.1088/1742-6596/2420/1/012077

Reconstruction of transverse phase space from transverse feedback data for real time extraction of vital LHC machine parameters

G Kotzian[1], M Söderén[1], P S Solvang[1], D Valuch[1] and V Stopjakova[2]

[1] Accelerator Systems department, CERN, Geneva, CH

[2] UEF FEI, Slovak University of Technology, Bratislava, SK

E-mail: gerd.kotzian@cern.ch

Abstract. The LHC transverse feedback system (ADT) provides bunch by bunch, turn by turn, normalized and digitized beam position signals from four pick-ups per plane and for each beam. Together with already existing powerful computer-based observation systems, this data can be used to reconstruct in real-time the transverse phase space coordinates of the centre-of-charges, for each individual bunch. Such information is extremely valuable for machine operation, or transverse instability diagnostics.

This paper aims on discussing and evaluating methods of combining four position signals for such analysis in the presence of noise and with active transverse feedback. Comparisons are made based on the extraction of vital parameters like the fractional tune or transverse activity. Analytical and numerical results are further benchmarked against real beam data.

1. Transverse feedback and ATDObsBox

During the LHC Long Shutdown II the LHC transverse feedback system (ADT) [1] was subject to an upgrade of its Beam Position Monitor (BPM) hardware, aiming for an improvement of the system's noise floor [2]. New low-noise beam position hardware is now available for LHC Run III, providing independent processing of 16 dedicated pick-ups. This allows for a total of four beam position streams per plane and per beam, representing the transverse centre-of-charges of each individual bunch.

This data is available in real-time to the LHC Transverse Feedback system for damping and stabilizing the beam, as well as to the computer-based observation system ADTObsBox [3]. The ADTObsBox is capable of recording and processing all digital data streams from the available pick-ups at full data rate (bunch-by-bunch or 40 Msps, with 16 bit resolution), for instance, to combine the data in real-time and extracting valuable metrics of the ADT performance (e.g. transverse activity) or the beam itself (bunch-by-bunch fractional tune).

One considerably useful representation of transverse motion is the use of phase space coordinates, readily described in Ref. [4] as an analytic signal $x[n]$, expressed as,

$$x[n] = y[n] + jy'[n]. \tag{1}$$

Here, the transverse normalized phase space coordinates $y[n]$ and $y'[n]$ represent normalized position data respectively the corresponding normalized slope values at turn index n.

From the notation in equation (1) we can immediately derive a measure of transverse beam activity, denoted by $A[n]$, as the modulus – or amplitude – of the complex vector,

$$A[n] = \text{abs}\left\{x[n]\right\}. \tag{2}$$

If $A[n]$ decreases over time then the transverse activity is considered to be damped, whereas growing values provide an indication of transverse instability.

Equivalently, the fractional tune, $Q[n]$, follows from phase space data by observing the phase component $\varphi[n] = \arg\left\{x[n]\right\}$ of equation (1), specifically, the relative phase advance between consecutive turns, described as,

$$Q[n] = \frac{1}{2\pi}\arg\left\{\frac{x[n-1]}{x[n]}\right\} = \frac{1}{2\pi}\left(\varphi[n-1] - \varphi[n]\right). \tag{3}$$

For the analytic evaluation, we implicitly assume that the beam is centred in the pick-ups and we observe betatron oscillations, i.e. $(y)^2 + (y')^2 = \text{const}$. Furthermore, for our assessment, we shall use a damped, complex-valued harmonic oscillator as beam model, which provides for the kth beam position monitor at the nth turn the phase space coordinates as follows,

$$x_k[n] = A_0 e^{-j\phi_k}\left(\alpha \cdot e^{-j\omega_0}\right)^n. \tag{4}$$

Here, A_0 and ϕ_k are initial conditions, α accounts for an amplitude decay and $\omega_0 = 2\pi Q_f$ represents the angular frequency at the fractional betatron tune Q_f.

2. Phase space reconstruction

In order to satisfy equation (1), and by acknowledging that the ADT Beam Position Monitors readily provide normalized readings, we are looking for indirect methods to obtain slope samples from beam position measurements.

In the following, we evaluate two methods for combining beam position data of four LHC pick-ups, identified as a *spatial* and a *temporal phase shift* in Ref. [5]. The first combines the information of several beam position monitors based on their longitudinal distribution in the accelerator, the later relies on processing the history of recorded beam position data using filter kernels.

2.1. Spatial method

This method relates the data measured by two or more independent beam position monitors at different longitudinal azimuths. Figure 1 outlines the case for $N = 2$ signal sources. For this method, the beam position monitors should have a betatron phase advance ideally between $60^o < (\phi_2 - \phi_1) < 120^o$.

For the case of the LHC ADT, the data provided by individual beam position monitors $y_k[n]$ are meticulously time-aligned during setting-up, with index n corresponding to the same bunch data at the same turn. Therefore, using the "Pick-up Vector Sum" algorithm from Ref. [5] it can be shown that, choosing appropriate mixing coefficients a_{11} and a_{12}, the position signal $y_I[n]$ in figure 1 of a virtual beam position monitor can be constructed, such that

$$y_I[n] = a_{11}y_1[n] + a_{12}y_2[n]. \tag{5}$$

Equivalently, applying the same approach of pick-up signal mixing, and by taking into account an additional 90° phase advance compared to the virtual beam position used for equation (5), we obtain the representation of the slope, $y_Q[n]$, described as

$$y_Q[n] = a_{21}y_1[n] + a_{22}y_2[n]. \tag{6}$$

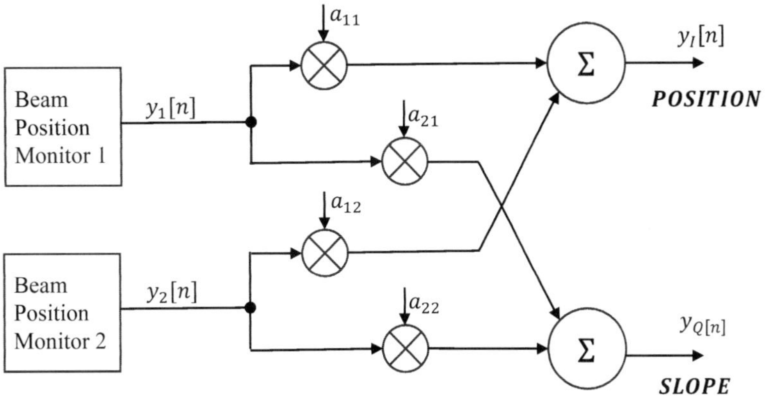

Figure 1. Spatial phase space reconstruction: in-phase term $y_I[n]$ and quadrature component $y_Q[n]$ calculated from vectorial rotation and combination of two beam position sequences $x_1[n]$ and $x_2[n]$.

The pair of equations (5) and (6) represent Cartesian coordinates, describing the reconstructed phase space for an arbitrary longitudinal position. It is a convenient practice to chose the ADT kicker position as reference for the phase space reconstruction. By using the same longitudinal reference this technique is further expandable for using multiple beam position monitors – four in the case of the LHC ADT – as shown in figure 2.

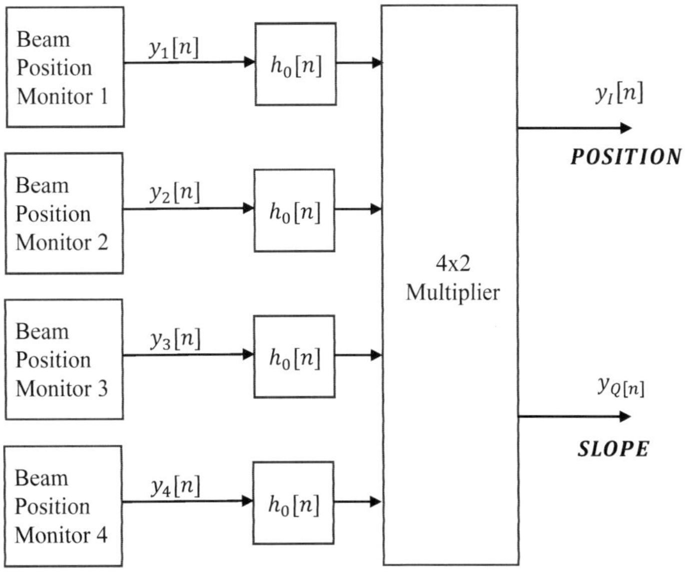

Figure 2. Spatial combination of four beam position streams.

We therefore obtain an analytic signal $x_S[n] = y_I[n] + jy_Q[n]$ as the reconstructed normalized transverse phase space based on a scalar combination of real-valued position sequences, $y_k[n]$, using

$$x_S[n] = h_0[n] * \sum_k (a_{1k} + ja_{2k}) \cdot y_k[n]. \tag{7}$$

The asterisk operator ($*$) represents the discrete-time convolution of the weighted BPM signals with an additional finite impulse response (FIR) filter of impulse response $h_O[n]$, allowing shaping of the input noise. When no filter is used, this method has zero group delay and therefore it is suitable for applications requiring low latency.

2.2. Temporal method

Referring to [4], this reconstruction method obtains the slope signal as the 90° phase rotated version of the position data utilizing digital filters.

As outlined in figure 3, bunch-by-bunch data $y_k[n]$ provided from individual Beam Position Monitors passes through a pair of matched filter kernels, denoted as $h_I[n]$ as $h_Q[n]$. The filters' even and odd symmetric impulse responses generate in-phase (I) and quadrature output signals (Q), which can be combined afterwards thanks to an identical group delay.

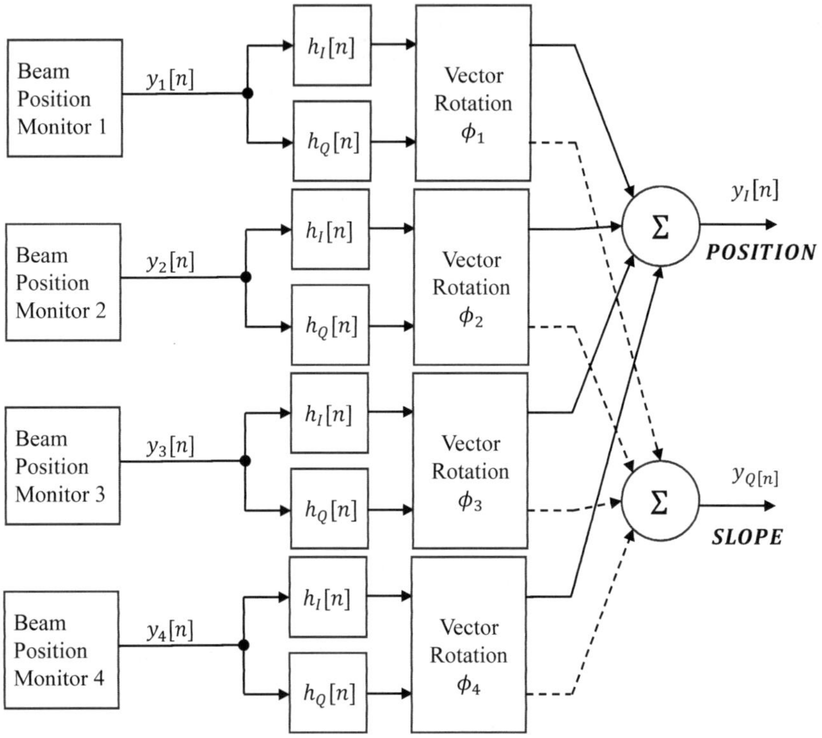

Figure 3. Temporal phase space reconstruction using matched digital filters $h_I[n]$ and $h_Q[n]$ on the input data stream $y_k[n]$ of four Beam Position Monitors.

It is worth noting that the calculated FIR filter output pairs readily represent phase space coordinates at the longitudinal position of the corresponding monitor. Therefore, to aggregate four pick-ups to an arbitrary longitudinal reference location, the individual output vectors need to be aligned before summing their contributions. This is done by a vector rotation, ϕ_k, towards a common longitudinal position (e.g. to the location of the ADT kicker).

The analytic signal $x_T[n] = y_I[n] + jy_Q[n]$ obtained from the temporal method can therefore be described as,

$$x_T[n] = \sum_k \left(y_k[n] * (h_I[n] + jh_Q[n])\, e^{j\phi_k} \right). \tag{8}$$

With a short 5 taps kernel length, these filters are tuned to exhibit nominal transmission at the fractional tune for the corresponding plane, rendering them applicable for a tune range exceeding ± 0.02 around the target tune [4].

3. Results

To evaluate the expected performance we model the inherent noise from each Beam Position Monitor as an independent additive white Gaussian noise process, $e[n]$ ($\overline{e[n]} = 0$, noise power $\overline{e^2[n]} = \sigma_e^2$), which is added to the desired noise-free position signal, $y_p[n]$,

$$y[n] = y_p[n] + e[n]. \tag{9}$$

The convolution in equation (7) and equation (8) with selective FIR filters lets signals close to the target tune pass unaltered in amplitude, but shapes the noise power for out-of-band signals. For the example of the in-phase filter (equivalent for the quadrature component), the noise power follows from,

$$\sigma_I^2 = \sigma_e^2 \sum_m |h_I[m]|^2 . \tag{10}$$

The overall noise contribution is further reduced by the superposition of $N = 4$ Beam Position Monitors, leading to a final RMS noise amplitude,

$$\sigma_A = \sigma_e \sqrt{\frac{\sigma_I \sigma_Q}{N}}. \tag{11}$$

It can be shown that the activity defined in equation (2)with noise present according to equation (9) follows the well known Rice probability distribution. The expected RMS amplitude noise follows the RMS input noise, reduced by filtering and by the combination of 4 BPMs. For example, in the case the filter kernels listed in [4] are used for the temporal method, then the expected total process gain is 9.2 dB.

The measurement noise also affects the observable fractional tune (equation (3)) from the reconstructed phase space data $x_T[n]$, with the RMS phase noise given as,

$$\varphi_{\mathrm{rms}} = \mathrm{atan}\frac{\sigma_A}{A_0} \tag{12}$$

At this point it is worth noting that, for large amplitudes $\overline{A[n]} >> 10\sigma_A$, the reconstructed values for amplitude and phase exhibit measurement noise following the assumed normal distribution (standard deviations σ_A respectively φ_{rms}), however, due to the filtering their spectral density is no longer white.

Furthermore, for $\overline{A[n]} << 10\sigma_A$ the observed phase vector resembles a uniform distribution, with an arbitrary phase value between $[-\pi, \pi]$, providing no useful information on the fractional tune. Numerical simulations using equation (4) and a BPM noise floor of $\sigma_e = 0.219\,\mu$m (taken from [2]) indicate that a minimum signal to noise ratio (SNR) of $20\log_{10}(A[n]/\sigma_e) \geq 30$ dB is required to obtain an RMS tune variation of σ_Q better than 10^{-3}.

This value is confirmed by measurements of LHC injection transients, shown in figure 4 as an example, where the turn-by-turn fractional tune of one individual bunch is reconstructed from 4 LHC BPMs using the method described by equation (8). The red trace indicates the 16-turns moving average of the instantaneous fractional tune $Q_T[n]$ (in gray), and with the RMS tune error depicted in black. Between 2000 and 2500 turns after injection the oscillation amplitude has decayed to an SNR of approx. 30 dB, and the measured RMS tune variation $\sigma_Q = 1.1 \cdot 10^{-3}$, which is in excellent agreement with the numerical prediction.

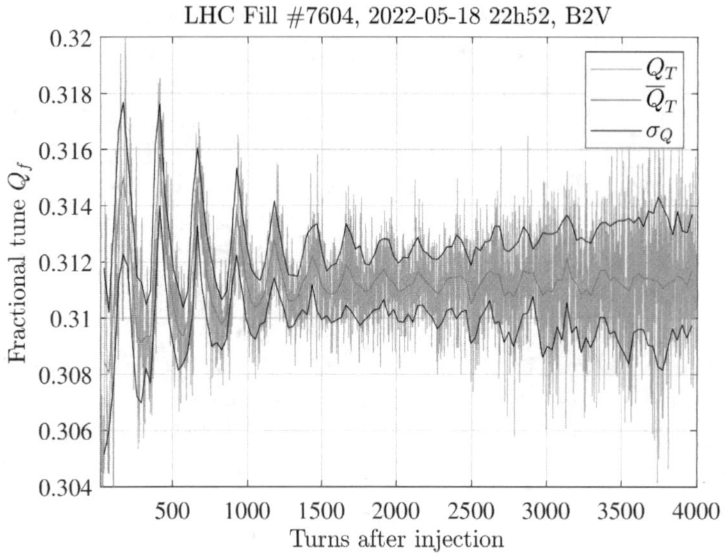

Figure 4. Fractional tune from reconstructed transverse phase space using 4 LHC BPMs using the temporal method.

4. Conclusions

The suggested methods are both valid candidates for reconstructing the transverse phase space in real-time. With the spatial method being attractive for the analysis of fast beam transients, for example during injection transients and with 5 turns damping time, both methods profit from the suppression of out-of-band noise using filter kernels which improves the SNR.

Potentially, more elaborate filter response functions can be designed to lower the SNR even further, for instance by processing the beam position data of 100 to 1000 turns.

References

[1] Söderén M *et al.* 2019 *Proc. 9th LHC Operations Evian Workshop* (Evian Les Bains, France) pp 165–71
[2] Valuch D and Stopjakova V 2022 *Proc. 13th Int. Particle Accelerator Conf. (IPAC'22)* (Bangkok, Thailand) pp 849–52
[3] Söderén S and Valuch D 2019 *EPJ Web Conf.* **245** 01036
[4] Kotzian G 2017 *Proc. 8th Int. Particle Accelerator Conf. (IPAC'17)* (Copenhagen, Denmark) pp 1920–23
[5] Kotzian G 2017 *Proc. 8th Int. Particle Accelerator Conf. (IPAC'17)* (Copenhagen, Denmark) pp 1924–27

Journal of Physics: Conference Series **2420** (2023) 012078 doi:10.1088/1742-6596/2420/1/012078

The new eddy current type septum magnet for upgrading of fast extraction in main ring of J-PARC

T Shibata[1], K Ishii[1], S Iwata[1], T Sugimoto[1], H Matsumoto[1] and K Fan[2]

[1] J-PARC Center, Accelerator Laboratory, KEK, High Energy Accelerator Research Organization, 1-1 Oho Tsukuba Ibaraki 305-0801 Japan

[2] Huazhong University of Science and Technology, 1037 Luoyu Road Wuhan Hubei 430074 China

E-mail: tatsunobu.shibata@j-parc.jp

Abstract. As part of the goal of increasing the beam power of the Main Ring for Fast eXtraction (FX) in J-PARC to 750 kW, we have been evaluating new FX low-field septa of induced eddy current type (Eddy-septum) since 2014. To reduce the leakage field in the circulating line of the eddy-septa, new pure iron duct-type magnetic shields were produced in July 2021, and in March 2022 we verified that the leakage field was greatly reduced . The construction of the two eddy-septa and their power supplies was completed in May 2022. Beam operation with a new 1.3-s cycle will be started in June 2022.

1. Introduction

The J-PARC proton accelerator experiment laboratory consists of a 400-MeV LINAC, a 3-GeV Rapid Cycling Synchrotron (RCS) and a 50-GeV Main Ring (MR). The MR is a proton synchrotron that receives the 3-GeV proton beam from the RCS, accelerates it to an energy of 30 GeV, and extracts it to the neutrino facility (NU) or the hadron facility. The present operation cycle is 2.48 s and the maximum beam power for fast extraction mode (FX) is 515 kW, which was achieved in March 2021 [1]. To realize the original design value of 750 kW, we are working towards operating the MR with a cycle of 1.3 s in June 2022 [2], referred to as "1 Hz operation". The cycle will also be reduced to 1.16 s to increase the beam power to 1.3 MW by 2028 (1.3 MW beam) [2]. The magnets for FX (FX magnets) in the MR, which are used for switching the direction of the proton beam to the NU or the abort dump (ABT), have been in the process of being upgraded for 1 Hz operation and 1.3 MW beam for a decade, and this will be completed by summer 2023. The FX magnets consist of five kickers, two low-field septa (LF-septa) and four high-field septa [3]. The upgrade includes an improvement of the power supply for the kickers and replacing the septa with new ones. During a long-term shutdown of the MR in 2021–2022, the two LF-septa was removed from the beam line and the new septa were installed. The construction of two sets of power supplies for the new septa was completed in May 2022. This article focuses on the upgrade of the LF-septa system.

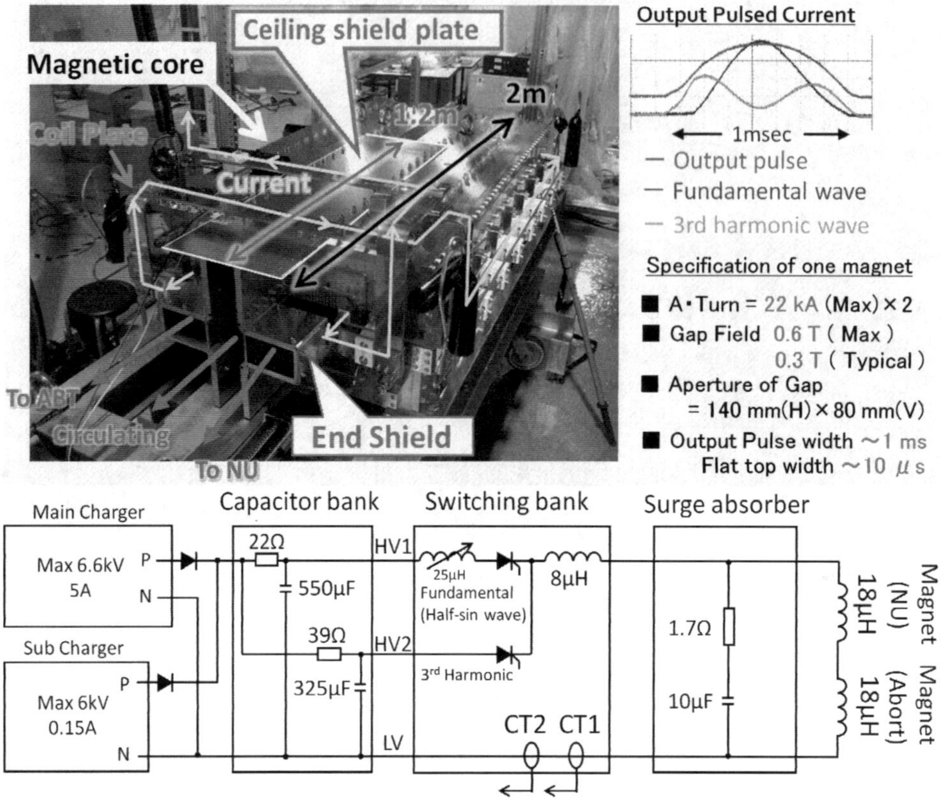

Figure 1. Photograph of eddy septum magnet (upper left), waveform of output pulsed current and specifications of eddy septum system (upper right), and circuit diagram of pulse power supply (lower).

2. FX low-field septum magnet

The previous LF-septa, removed in 2021, had a conventional configuration with a thin septum coil. The typical gap field for the previous LF-septa was 0.3 T, and the applied current waveform had a time width of ~1.5 s, the same as that for the high-field septa. Each previous LF-septum consisted of two septa, located symmetrically with respect to the beam axis, one used for NU extraction and the other for ABT extraction, with a circulating beam passing through the space between the two septa. These septa were mounted in a vacuum chamber, because the extraction lines were not sufficiently separate from the circulating line, and there was no space to mount a beam duct.

The new LF-septa are induced eddy current type (eddy-septa) (figure 1) [4, 5]. We will use two eddy-septa, mounted in two vacuum chambers, as in the case of the previous LF-septa. The eddy-septa have developed in some electron and proton accelerators in the world except J-PARC: e.g., the injection/extraction of synchrotron SOLEIL [6, 7], the injection of SSRF storage ring [8, 9], the extraction of SPS in CERN [10]. The upstream and downstream eddy-septa are called Eddy-1 and Eddy-2, respectively. The most significant feature of an eddy-septum is that it can greatly reduce the leakage field compared to the previous LF-septa. Since the configuration of an eddy-septum is based on a C-type bending magnet, it does not have a septum coil but has a thin copper plate on the opening side of the magnetic core (septum plate). The applied current waveform for an eddy-septum is a short-pulse half-sinusoidal wave. When the short-pulse current is applied to the eddy-septum, the pulsed leakage field penetrates the septum plate, inducing an eddy current in the septum plate that cancels the leakage field. Other advantages of eddy-septa for high-power beam operation include: (1) a zero-Gauss leakage field during beam injection and acceleration because they are excited only during beam extraction, (2) a wider horizontal aperture of the magnetic gap due to the lack of a septum coil, which constrains the aperture, (3) lower possibility of vibration damage to the insulator on the septum coil and (4) lower joule heating in the magnetic coil and core due to the short-pulse current. However, the power supply (PS) must be replaced because that for the previous LF-septa cannot supply the short-pulse current. The PS has been under development since 2014 in the electric building along with the other power supplies for the FX magnets [3, 11, 12, 13, 14]. One septum PS consists of two main chargers, a sub-charger, a capacitor bank, a switching bank and a surge absorber (figure 1). The main chargers charge the capacitor bank with a maximum charging voltage of 6 kV and the switching bank then discharges the capacitor bank with a maximum output power of 6 kV×22 kA. The output short-pulse current has a full and flat top width of ~800 µs and ~10 µs , respectively. It is a composite of a fundamental and a 3rd harmonic sinusoidal wave, to optimize the flatness of the flat top region of the waveform by adjusting the difference between each discharge time.

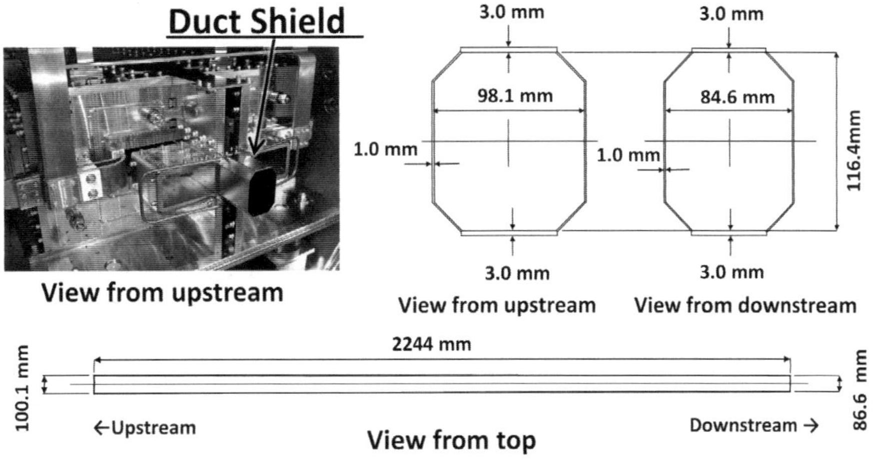

Figure 2. Location and illustration of duct shield for Eddy-1.

3. History of development of eddy-septa

The most significant technical goals regarding the magnets are achieving the expected gap field and an extremely low leakage field in the circulating line. We measured the field integral (BL) of the gap field with a charging voltage of 3 kV, which is half the maximum value, and obtained 0.449 T·m. Since the BL for one of the previous LF-septa for 30-GeV beam extraction was 0.44 T·m, we determined that the optimal charging voltage for 30-GeV beam extraction is ≈3 kV. In addition, we confirmed good linearity between the BL and the charging voltage in the range from 2 to 4 kV without any saturation. When the bending angle is fixed, the extraction energy is proportional to the BL, that is, the charging voltage. Therefore, we can estimate the optimal charging voltage for any extraction energy. In 2019, we measured the uniformity of the BL along the horizontal direction in the gap space, and obtained a gradient of 0.15%/47 mm, which was smaller than the value of 0.8%/44 mm for the previous LF-septa [14], showing that the uniformity was improved. We started taking measures against field leakage in the circulating line in 2014. First, a copper ceiling plate with a thickness of 5 mm was mounted on top of the magnetic core to shield against the induced field generated in the bus-bar coils over the magnetic cores [3]. We measured the longitudinal distribution of the leakage field in the circulating line and found that a leakage field of 3–4 Gauss and 10 Gauss still existed in the magnetic core region and the end-fringe of the magnetic core (end-fringe field), respectively [14], indicating that we need to further reduce the leakage field by using an additional shield.

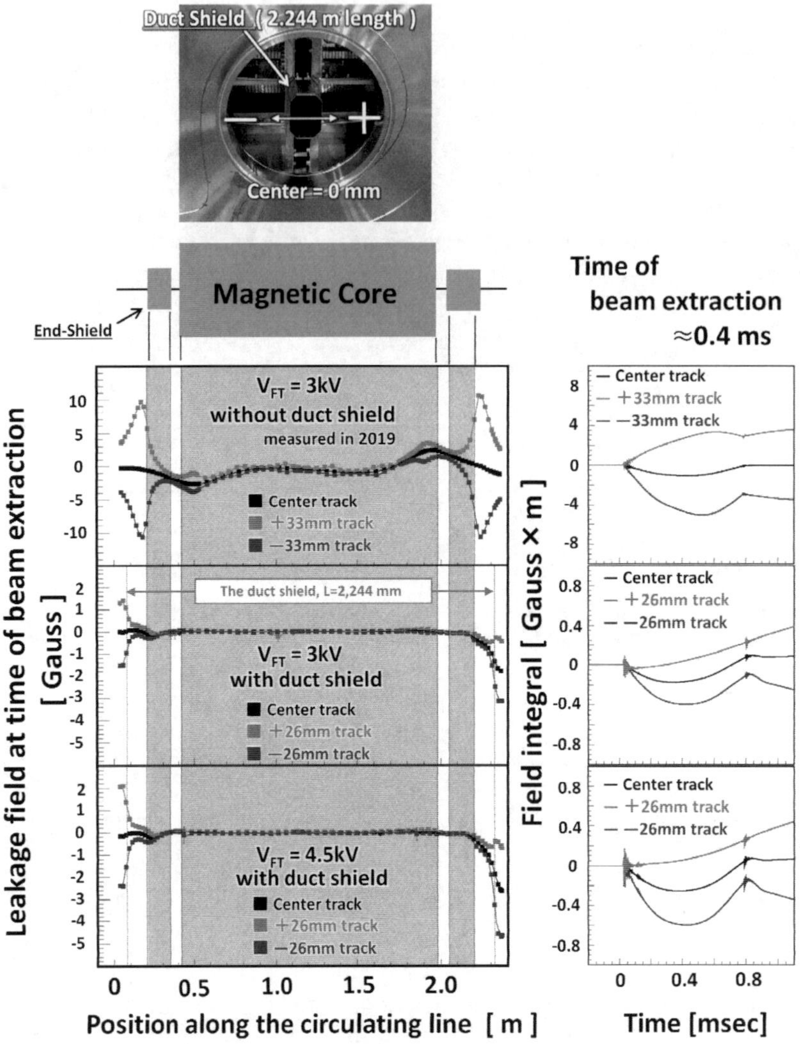

Figure 3. Leakage field and BL without any duct shields measured in 2019 (top) and with duct shield measured in 2022 (middle and bottom).

4. Reduction of leakage field in circulating beam line

First, to reduce the end-fringe field, in 2019 we produced a rectangular trial duct made of pure iron with a length of 0.3 m. The duct was mounted on the end-fringe of the magnetic core, which allowed the end-fringe field to be reduced to below 4 Gauss. Then, in 2020 we produced a pure iron rectangular duct with a length of 2.3 m to reduce the leakage field for the entire circulating line. To prevent beam loss, we optimized the thickness of the side to 1 mm and that of the top and bottom to 3 mm. The leakage field at the center of the circulating line with the duct shield was reduced to ≈0 Gauss and the end-fringe field to 2 Gauss. The time-dependent BL is ≈0 Gauss·m at beam extraction. Based on these results, we decided to use a pure iron duct shield for beam operation. The actual duct shields for Eddy-1,2 were designed in 2021. A length of 2.244 m was used, where the length of the vacuum chamber is 2.254 m, The thickness is the same as that for the trial version. The cross-sectional shape was chosen to be an octagon to fit the shape of the circulating line. The actual duct shields are tapered along the longitudinal direction because the two eddy-septum magnetic cores have horizontal apertures that become narrower from the entrance to the exit. We optimized the inner dimensions of the duct shield using the SAD [15] beam simulation program. The inner width at the entrance and exit for Eddy-1 were set to 98.1 mm and 84.6 mm, and those for Eddy-2 were set to 83.4 mm and 74.4 mm, respectively. The inner height of the duct shield was set to 116.4 mm, larger than the size of the circulating beam. Figure 2 shows the duct shield for Eddy-1.

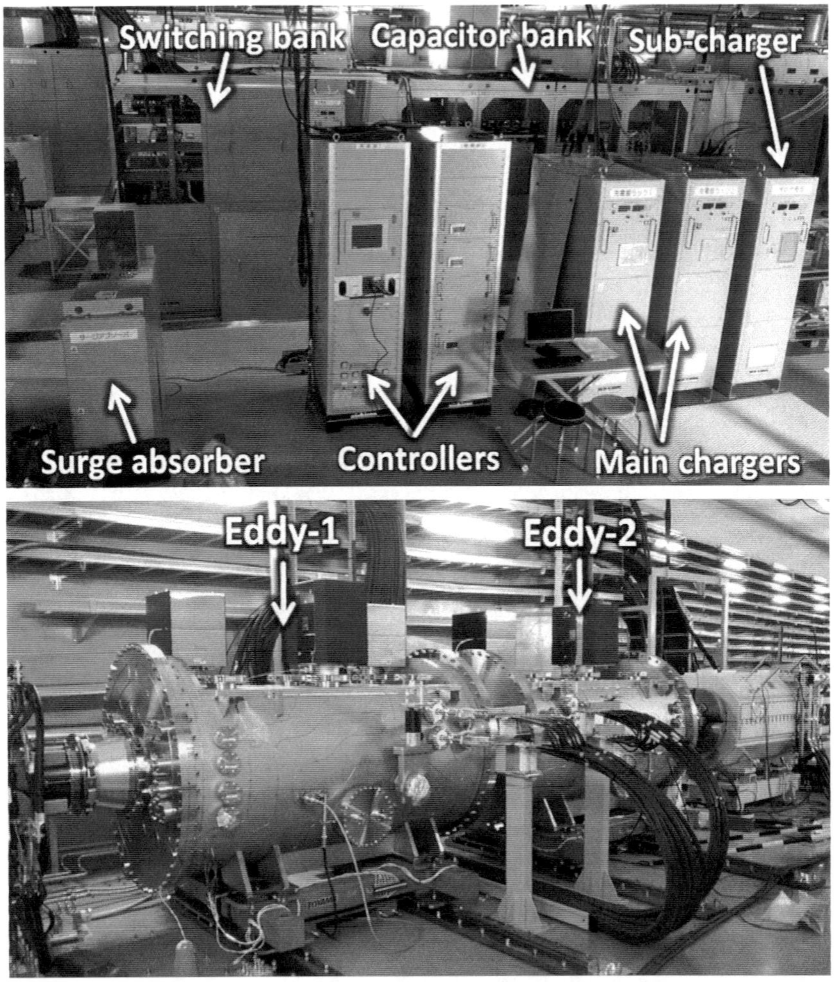

Figure 4. Photograph of PS for Eddy-1 (upper) and Eddy-1,2 installed in the MR (lower).

The production of the duct shields was competed in July 2021 and the shield for Eddy-2 was mounted in the circulating line in January 2022. In March 2022, the longitudinal distribution of the leakage field in the duct shield with V_{FT} of 3 kV and 4.5 kV was measured along three tracks: in the center of the duct shield (center track), parallel to the beam direction, and at ±26 mm from the center track in the horizontal direction (±26-mm tracks). The measurement results together with the leakage field without the duct shield measured in 2019 are shown in figure 3. The field strength in the region of the magnetic cores on all tracks were reduced to below 0.1 Gauss. In addition, the end-fringe field of ≈10 Gauss was also reduced to a few Gauss, although we found that the end-fringe field had an asymmetric distribution with respect to 0 Gauss due to the magnetic field generated by the current passing through the bus bars over the magnetic cores. The time-dependent BL on the center track without any duct shield reached 1 Gauss×m, and was reduced to ≈0.2 Gauss×m using the duct shield. The maximum BL values with V_{FT} of 3 kV and 4.5 kV on the −26-mm track were 0.4 and 0.6 Gauss×m, respectively. Thus, we verified that the duct shield strongly reduced the leakage field. The duct shield for Eddy-1 was mounted in April 2022.

5. Installation of eddy-septa system in MR

Both eddy-septa were mounted in the vacuum chambers and installed in the MR, and construction of the two PS sets for the eddy-septa was also completed in the electric building by the end of April 2022. We conducted the first operational tests of the Eddy-2 system, followed by the Eddy-2 system, with a maximum V_{FT} of 4.5 kV. Upgrading of the LF-septa system, including the construction of all of the eddy-septa systems, was completed in May 2022. Figure 4 shows photographs of the PS and the vacuum chambers that the septa are mounted in. Beam operation with a cycle of 1.3 s will be started in June 2022, and a high-power beam will be provided to the NU from November 2022.

6. Summary

As part of the upgrade of the FX septa to increase the beam power of the MR in J-PARC, the low-field septa are being replaced with eddy-septa. New additional duct shields were produced and mounted in the circulating line in 2022, and we verified that the leakage field was greatly reduced. The installation of the eddy-septa and the power supplies was completed at the end of May 2022. Beam operation with the new 1.3-s cycle will be started in June 2022.

References

[1] J-PARC Group 2021 Rep. KEK report 2021-2
[2] Igarashi S *et al.* 2021 *Prog. Theor. Exp. Phys.* **2022** 033G01
[3] Shibata T *et al.* 2016 *Proc. 7th Int. Particle Accelerator Conf. (IPAC'16)* (Busan, Korea) pp 1340-2
[4] Kumagai K *et al.* 1991 *Proc. 8th Symp. on Accelerator Science and Technology* (Saitama, Japan) pp 220-2
[5] Barnes M J *et al.* 2010 Rep. CERN-2010-004 pp 167-184
[6] SOLEIL `https://www.synchrotron-soleil.fr/en`
[7] Lebasque P *et al.* 2006 *Proc. 10th European Particle Accelerator Conf. (EPAC'06)* (Edinburgh, UK) pp 3511-13
[8] SSRF `https://lssf.cas.cn/en/index.jsp`
[9] Kang W *et al.* 2001 *Proc. 2nd Asian Particle Accelerator Conf. (APAC'01)* (Beijing, China) pp 663-5
[10] Schröder G H *et al.* 2000 *Proc. 7th European Particle Accelerator Conf. (EPAC 2000)* (Vienna, Austria) pp 2370-2
[11] Shibata T *et al.* 2017 *Proc. 8th Int. Particle Accelerator Conf. (IPAC'17)* (Copenhagen, Denmark) pp 573-5
[12] Shibata T *et al.* 2018 *J. Phys.: Conf. Ser.* **1067** 052020
[13] Shibata T *et al.* 2019 *J. Phys.: Conf. Ser* **1350** 052020
[14] Shibata T *et al.* 2021 *JPS Conf. Proc.* **33** 011033
[15] SAD `https://acc-physics.kek.jp/SAD/`

Spallation target optimization for ADS by Monte Carlo codes

M Mumyapan, M Ghergherehchi, D H Ha, H Namgoong and J S Chai

Department of Electrical and Computer Engineering, Sungkyunkwan University, Suwon, Gyeonggi-do, Korea

E-mail: jschai@skku.edu

Abstract. Accelerator Driven System are advanced system for the use of Thorium as fuel, aiming to reduce nuclear waste through transmutation. The spallation target, which is responsible for producing neutrons, is one of the main parts of the ADS system. In this research, neutronic parameters of spallation targets consisting of several materials LBE, Mercury, and Lead, on the cylindrical and conic shapes using Monte Carlo codes (FLUKA, PHITS, MCNPX) were investigated. Energy deposition and spallation neutron yield of spallation target with different shapes and dimensions have been calculated to optimization of the target. According to the results, the neutron yield values from MCNPX and PHITS are similar and it's close to the experimental result. On the other hand, the error rate of the values in FLUKA is higher.

1. Introduction

Accelerator Driven Systems (ADS) are up-and-coming tools that provide reliable energy and transmute long-lived radioactive waste. The simultaneous operation of ADS's passively safe subcritical core and accelerator distinguishes it from conventional fission reactors. So, the unique feature of this system is that the reaction stops shortly after the proton beam from the accelerator is turned off. In this way, it provides safety that will significantly reduce the risk of nuclear accidents. In addition, nuclear waste is a big problem for which no solution can be found [1]. ADS aims to transmute long-lived radioactive waste like Pu, and Np isotopes [2]. Thus, it can figure out this problem and contribute to the sustainability of nuclear energy [3].

Powerful particle accelerators with high proton beam energy, spallation target, and sub-critical system are the main components of ADS [4]. The particle accelerator continuously delivers a dense beam of accelerated particles toward the target. As a result, these neutrons can be multiplied in the subcritical nuclei enclosing the target [5].

The spallation target is the part that connects the accelerator and the subcritical reactor. Therefore, the spallation target design has a key point for ADS design. There are some difficulties for design spallation target; selecting the most suitable material, heat removal, radiation damage, and distribution of power density. To handle these issues, subjects such as neutronic analysis, and energy deposition are required for spallation target design [6].

The neutron yield per incident proton is the principal data that has been used in the ADS study [7]. Neutron yield calculation with MCNPX has been investigated by several researchers [8-10]. However, a comparison of neutron yield with different Monte Carlo codes such as FLUKA and PHITS was not well reported.

Three well-proved Monte Carlo codes (FLUKA, PHITS, and MCNPX) were used to calculate and compare the neutron yield to determine the most suitable material and select the optimum size and shape for the spallation target.

2. Materials and methods

In this paper, by using FLUKA, PHITS, and MCPNX, the neutron yield of the cylindrical target was calculated according to different proton beam energies and this process was repeated for 4 different target materials (LBE, Mercury, Lead, and Tungsten). The fixed cylinder target with a radius of 20 cm and height of 80 cm was chosen based on the literature [10]. To compare the different shapes, the dimensions of the Conical target were selected the same (figure 1). Proton beam (2.35 mm spatial FWHM) with energies ranging from 600 to 1500 MeV, in the z-direction (0, 0, -2) injected into the target.

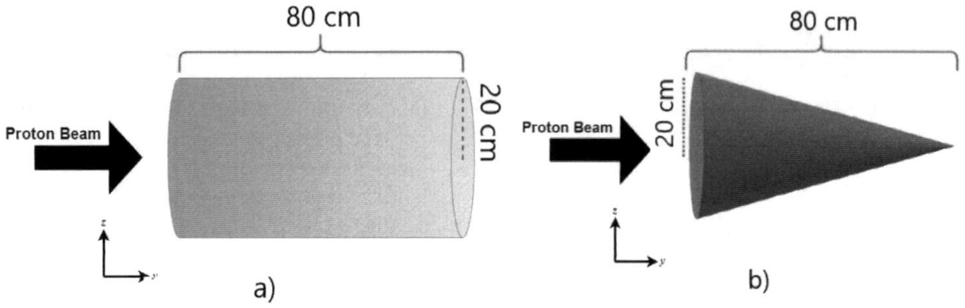

Figure 1. Fixed cylindrical and conical spallation target.

To determine the optimum target shape and dimensions, the neutron yield of the cylindrical target as well as the conical target were calculated and repeated for different materials (Tungsten, LBE, Mercury, and Lead) and dimensions. The length of the cylindrical target has been changed to 20, 40, 60, 80 and 100 cm, and then the radius of the target has been changed to 5, 10, 15, 20, 25 and 30 cm. Conical target length has been varied from 40, 60, 80, 100 cm and radius 10, 20, 30, 40 and 50 cm. Equation (1) is used to calculate the neutron yield per incident proton [11]. Proton beam energy, target material, shape, and dimensions of the target are the factors affecting the neutron yield. Sp is the number of primary protons and Sn is the number of primary neutrons.

$$Spallation\ neutron\ yield\ per\ proton\ \ Yn/p = \frac{Sn}{Sp} \tag{1}$$

3. Result and discussion

Neutron yields at different proton beam energies using four different target materials for a fixed radius and length cylinder target were simulated. Energy from 600 to 1500 MeV with a step of 200 MeV was selected. Figure 2 shows that by increasing proton energy increases, the neutron yield increases in direct proportion, and the highest value throughout the proton energy belongs to the Tungsten target, 43.6803 n/p at 1500 MeV. Next, the highest value was followed by Mercury 43.33 n/p. On the other hand, the lowest yield was seen in Boron, Nickel and Cobalt, which values under 20 n/p at 1500 MeV.

Neutron yield values for LBE and Lead targets are very close to each other and are approximately 41 n/p at 1500 MeV. Nickel, Cobalt and Boron target materials have not been used in the next stages because their yield values are quite low.

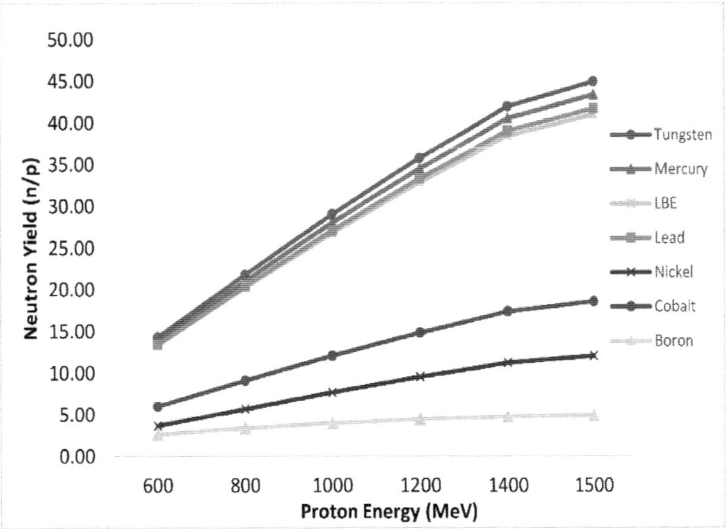

Figure 2. Neutron yield versus proton energy for different target materials by MCNPX.

Figure 3 shows the neutron yield with a different target radius for a cylindrical target at fixed energy, 1 GeV proton beam. Regardless of the target material, it was observed that neutron increases with the increase of the target radius. While there is a rapid increase up to 20 cm, after that there is a gradual increase. It means after 20 cm, the rate of increase is insignificant and almost saturated. Therefore, a 20 cm radius were selected as an optimum target radius and the neutron yield were calculated for the target length varying from 20 cm to 100 cm (figure 4). Calculation was repeated for four different materials. According to result, neutron yield increased for the Tungsten target up to 40 cm, no change was observed after 60 cm for target length. For Mercury Target, this optimum value appears to be between 40 cm and 60 cm target length, while for LBE and Lead target the optimum length was calculated as 60 cm. So, after this optimum value, there is no advantage in terms of neutron yield.

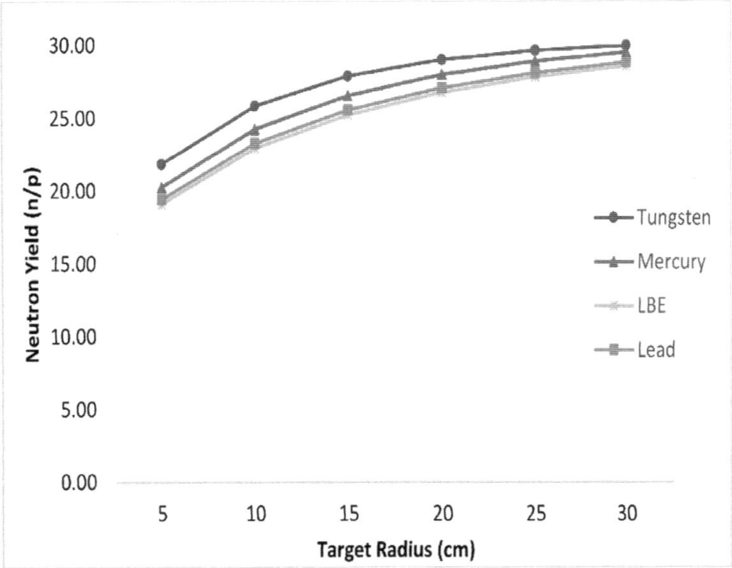

Figure 3. Neutron yield versus target radius at 1 GeV by MCNPX.

Table 1. Comparison of neutron yield for conical and cylinder target.

Target Materials	Conical Target (R:20cm L:80cm)			Cylindrical Target (R:20 cm L:80 cm)			Average Relative Differences (%)
	MCNPX (n/p)	PHITS (n/p)	FLUKA (n/p)	MCNPX (n/p)	PHITS (n/p)	FLUKA (n/p)	
LBE	24.26	25.48	27.52	26.75	28.62	30.95	11.03
Mercury	26.07	27.86	29.72	27.99	28.45	32.47	8.29

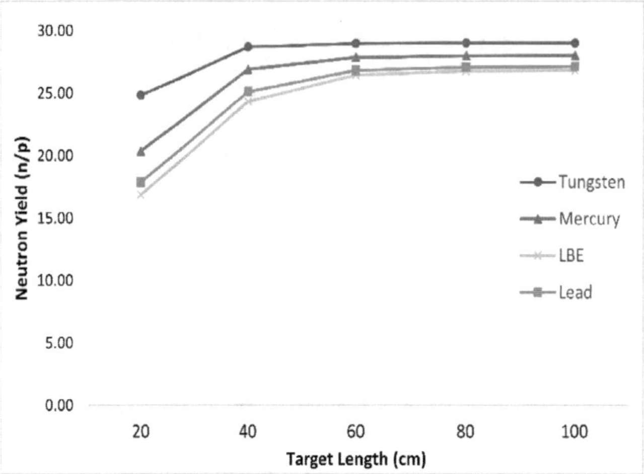

Figure 4. Neutron yield versus target length at 1 GeV by MCNPX.

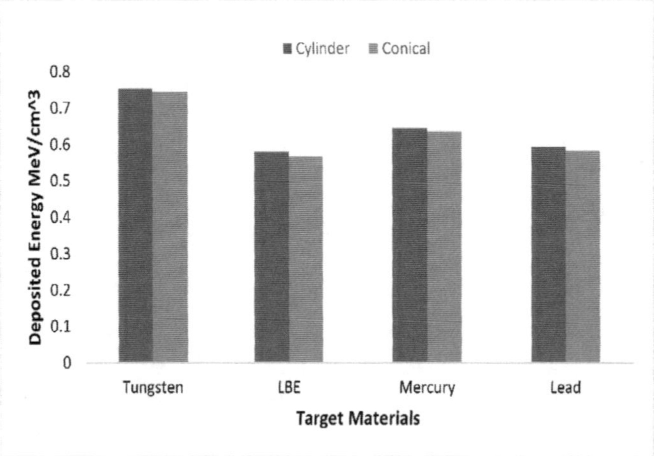

Figure 5. Deposited energy of 4 different materials for cylinder and conical targets.

To see the effect of different target shapes on neutron yield, the yield of Cylinder target and Conical target (table 1) were selected. For the same radius, it was observed that neutron yield for cylinder target is higher than conical target. FLUKA has the highest neutron yield for various target shapes and materials, while MCNPX has the lowest.

Table 2. Comparison of our work and experimental data for lead target (10.2 cm x 61 cm) (diameter x length) and tungsten target (10.2 cm x 40 cm)

Material	Energy (MeV)	Experimental result (n/p) [12]	MCNPX (n/p)	PHITS (n/p)	FLUKA (n/p)	Relative Differences (%)
Lead	1000	17.5	19.50	19.58	21.12	10.25 / 10.62 / 17.14
	1400	26.3	27.60	26.86	29.55	4.71 / 2.08 / 10.99
Tungsten	1000	20.5	21.95	22.62	24.08	6.60 / 9.37 / 14.86
	1400	28.5	30.90	31.99	34.48	5.28 / 10.9 / 17.34

In addition, conical target (34.64 x 80 cm) and cylinder target (20 x 80 cm) neutron yields with the same volume for four different materials were simulated by FLUKA. There is a more or less ±1 percent difference between the conical target and cylinder target with the same volume. As a result, it was observed that geometry has an effect on neutron yield.

The deposited energy for Tungsten, LBE, Mercury and Lead target materials because of the collision of 1 GeV protons for cone and cylinder was calculated with FLUKA (figure 5). As can be seen, the amount of deposited energy varies depending on the target material, but is nearly identical for the same target shape. So, deposited energy does not depend on geometry. In addition, Tungsten target has the highest energy deposition (0.752 MeV/cm^3), whereas LBE target has the lowest (0.586 MeV/cm^3).

To validate the simulation results, our results calculated by Monte Carlo codes were compared with the experimental data (table 2). In the experiment conducted by BNL, the neutron yield of two different target materials (Tungsten and Lead) data was used for different proton beam energies (1000 and 1400 MeV) [8, 12]. In this study, the obtained data calculated by Monte Carlo Codes has an average uncertainty of less than 0.8 %. The average relative differences between MCNPX, PHITS, FLUKA and experimental result for neutron yield were 6.71 %, 8.24 %, and 15.08 %, respectively.

4. Conclusion

In this study, we investigated different spallation target materials with different target shapes and dimension by using 3 Monte Carlo Codes. We confirmed that proton beam energy, target materials, target geometry and dimension have an effect on neutron yield. The conical target and cylindrical target were compared in terms of neutron yield and deposited energy. Although LBE neutron yield is relatively low compared to other target materials, it generates less heat. Obtained results by Monte Carlo codes were confirmed with experimental data. PHITS and MCNPX results are very close to each other, with an average relative difference of 2.61 %.

The next step is to design reflector and shielding for ADS system by using Monte Carlo codes. Later we will focus on LBE target cooling to remove heat in the spallation target. To overcome this problem, it is planned to use the Computational Fluid Dynamics (CFD) based approach.

Acknowledgements
This work was supported by the Korea Medical Device Development Fund grant funded by the Korea government, the Ministry of Science and ICT, Project Number: 1711135001, KMDF_PR_20200901_0042

References

[1] IAEA 2018 *IAEA Nuclear Energy Series* NW-T-1.14
[2] Zhivkov P, Stoyanov C and Furman W 2018 *EPJ Web Conf.* **194** 08002
[3] Lattefer A P 2008 *Nuclear Waste Research* (Nova Publishers) p 44
[4] Zhiguang W 2019 *Strategic Study of CAE* **21** pp 1-13

[5] Ghorbanpour E and Ghasemizad A 2015 *Indian Journal of Science and Technology* **8** pp 1-6
[6] Bauer G S 2010 *J. Nucl. Mater.* **398** pp 19-27
[7] Minh T T 2021 *Atoms* **9** 95
[8] Amirkhani M A , Hassanzadeh M and Safari S A 2019 *Radiation Physics and Engineering* **3** pp 29-36
[9] Frolova T A 2017 *Nucl. Energy Technol.* **3** pp 60-64
[10] Didi A *et al.* 2018 *Moscow University Physics Bulletin* **73** pp 612-17
[11] Kadi Y and Revol J P 2003 Design of an accelerator-driven system for the destruction of nuclear waste Lect. note at Workshop on Hybrid Nuclear Systems for Energy Production
[12] Kumar V *et al.* 2003 *Pramana* **60** pp 469-81

Status of the engineering design of the IFMIF-DONES high energy beam transport line and beam dump system

D Sánchez-Herranz[1,2], O Nomen[2], M Sanmartí[2], B K Singh[2], F Arranz[3], I Podadera[3], C Oliver[3], P Cara[4], V Hauer[5] and F Ogando[6]

[1] University of Granada, C/ Gran Vía de Colón, 48, 18010, Granada, Spain
[2] Catalonia Institute for Energy Research, Jardins de les Dones de Negre 1, 2ª pl., 08930, Sant Adrià del Besòs, Barcelona, Spain
[3] Centre for Energy, Environmental and Technological Research, Av. Complutense, 40, 28040, Madrid, Spain
[4] Fusion for Energy BA/IFMIF, Boltzmannstraße 2, 85748, Garching bei München, Germany
[5] Karlsruhe Institute of Technology, Hermann-von-Helmholtz-Platz 1, 76344, Eggenstein-Leopoldshafen, Germany
[6] National Distance Education University, Bravo Murillo, 38, 28015, Madrid, Spain

Email: dsherranz@ugr.es

Abstract. IFMIF-DONES plant [1] (International Fusion Materials Irradiation Facility – DEMO Oriented Neutron Source) is currently being developed under the framework of a work package of the EUROfusion Consortium. It will be a facility located at the south of Spain in Granada. Its objective and main goal is the testing and qualification of fusion materials by the generation of a neutron flux with a broad energy distribution covering the typical neutron spectrum of a (D-T) fusion reactor. This is achieved by the Li (d,xn) nuclear reactions occurring in a liquid lithium target where a 40 MeV at 125 mA deuteron beam with a variable beam footprint between 200 mm x 50 mm and 100 mm x 50mm collides. The Accelerator Systems is in charge of providing such high energy deuterons to produce the required neutron flux. The High Energy Beam Transport line (HEBT) is the last subsystem of the IFMIF-DONES accelerator, and its main functions are to guide the deuteron beam towards the liquid lithium target and to shape it with the required rectangular reference beam footprints. The HEBT system includes a Beam Dump devoted to stop the beam during commissioning and start-up phases. The present work details the present status of the HEBT engineering design, including beam dynamics, vacuum configuration, radioprotection, beam diagnostics devices and remote handling analyses performed detailing the layout and integration of required components through-out the beamline.

1. Introduction

The linear accelerator for the IFMIF-DONES facility [2] will produce a continuous wave deuteron beam of 125 mA, accelerated up to 40 MeV that will impinge on a liquid lithium target producing a fusion-like neutron flux for the assessment of materials damage in future fusion reactors. IFMIF-DONES linear accelerator is divided in seven subsystems: Injector, Radio Frequency Quadrupole, Medium Energy Beam Transport line, Superconducting RF LINAC, High Energy Beam Transport line, Radio Frequency Power and Accelerator Systems Ancillaries.

IFMIF-DONES accelerator is based on the design of LIPAC [3,4], currently undergoing its commissioning phase. The IFMIF-DONES High Energy Beam Transport Line and Beam Dump System (HEBT & BD) design has been developed considering as main requirements the ones shown in table 1.

Table 1. Requirements HEBT&BD system.

Requirement	Value
Particle Type	D+
Beam energy	40 MeV
Nom. beam peak current	125 mA
Nom. duty cycle	100%
Nom. transmission ratio	>99.8%
Beam dynamics length	48.8 m
Achromatic bending	9º
Beam footprint at target	20x5cm^2[10x5cm^2]
Beam Dump peak power	5 MW
Beam Dump avg. power	50 kW
RMS beam size BD cone	≈ 40 mm
Divergence at BD cone	14-16 mrad
Off axis	<3 mm
Pressure at SRF Linac	$<5\cdot10^{-6}$ Pa
Pressure at Li target	$>10^{-2}$ Pa

A preliminary design of the HEBT system was initiated following the requirements specified [5], establishing a first layout. It shall provide a continuous wave (CW) deuteron beam with adjustable rectangular footprint. Due to the development of the facility and its systems associated, multiple aspects have been either modified or developed for the HEBT main or beam dump transport lines. Next sections are devoted to give an overview of the current design and configuration updates. Main points of the HEBT engineering design which have been detailed from the previous preliminary design correspond to: Detailed design of magnets, recalculation of vacuum subsystem and adaptation to aluminum of beam facing components along the transport line, considering limiting factors for beam pipes the maximum outer diameter given by fixed aperture of magnets and minimum one due to beam envelope obtained in simulations. An important consequence of this material change to aluminum is the necessity to implement new type of flange connections, which is currently under-study. Figure 1 on the following page depicts configuration of all integrated components along the HEBT system.

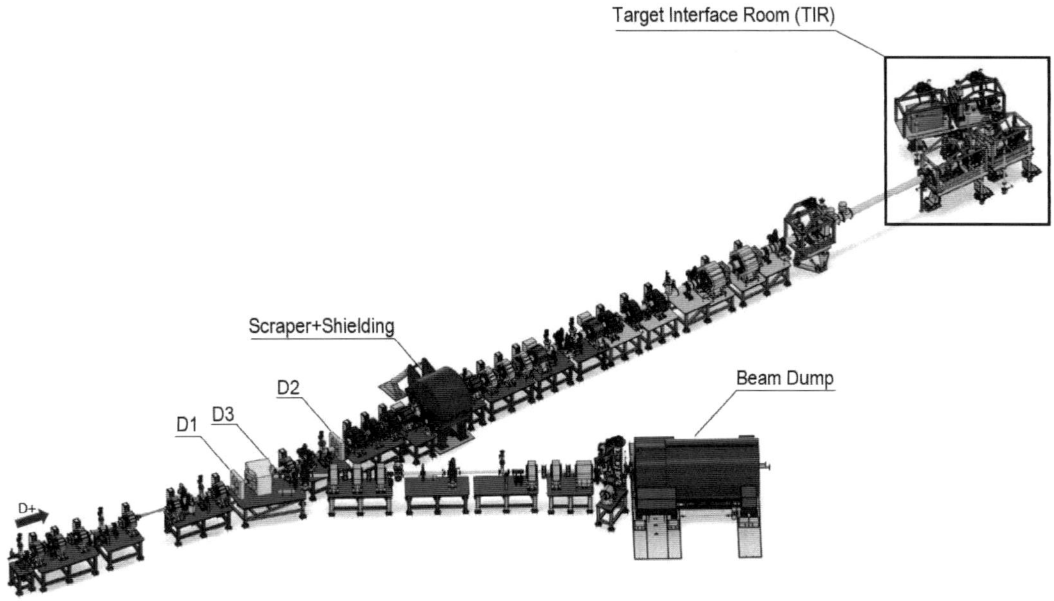

Figure 1. General distribution of HEBT mainline and beam dump transport line.

2. HEBT layout and design

2.1. Beam dynamics characteristics

HEBT section has been designed to handle a 40 MeV beam at 125 mA coming out from the SRF Linac, trans-porting it to the lithium target securely and with the beam size and characteristics required by the project to optimize the irradiation field [6]. Beam dynamics optimization has been performed modifying its original layout with the aim to guarantee enough space for components integration, adapt the main line and beam dump transport line to building modifications and increase space around individual steerers to minimize the magnetic field aberrations. Beam sizes present vary from apertures of 20 mm at the HEBT entrance up to 200 mm where beam enters the target area.

2.2. Magnets and steerers

Transportation and shaping of the beam throughout HEBT line is done by means of multiple magnets. Twenty-four quadrupoles grouped in four types (QD-A, QD-B, QD-C and QD-D) have been identified to keep the beam pipe aperture as small as possible. Two types of dipoles are also needed for the beam bending. The first type, namely DP-A type, covering the D1 and D2 dipoles in the main HEBT line, will have a small bending angle of 4.5^{o}. The DP-B type is used for D3 in the BDTL line having a larger bending angle of 30°.

Non-linear magnets have been designed according to the nominal tuning requirements. A couple of octupoles OP-A and OP-B will be implemented, each of them with different characteristics, whereas one type DDP is expected for the two dodecapoles needed.

Horizontal and vertical dipolar correctors are needed to steer beam trajectory deviations appearing due to up-stream elements or magnets misalignments and field errors. Twenty pairs of correctors with an integrated field of 40 G.m are recommended, all grouped in three types (ST-A, ST-B and ST-C) depending on aperture of section where they are located.

A margin of 5 mm between the beam-stay-clear radius that must be exclusively reserved for the beam to travel avoiding any losses and magnets' mechanical aperture has been considered to properly accommodate vacuum pipe thicknesses and alignment tolerances. Table 2 presents type of magnets and their main characteristics.

Table 2. Magnets characteristics.

Magnet type	Aperture [mm]	Pole field [T]	Units
QD-A	55	0,550	9
QD-B	85	0,595	10
QD-C	130	0,390	2
QD-D	70	1,050	3
DP-A	65	0,410	2
DP-B	55	0,850	1
DDP	65	0,360	2
OP-A	65	0,400	1
OP-B	85	0,360	1

2.3. Diagnostics devices

Given the very high current and power to be transported by IFMIF-DONES high energy beam transport line, beam diagnostics devices are of paramount importance to properly tune, characterize and monitor the accelerated beam. Multiple types of diagnostics have been integrated all along the HEBT, each of them with different functionality and characteristics [7].

Twenty-two Beam Position Monitors (BPMs) are necessary to detect possible beam deviation from the central axis and correct the dipolar errors created by misalignments of the different accelerator elements, mainly non-linear magnets.

To measure beam current in both CW and pulsed modes, sixteen current transformers are placed at multiple locations. Three types of current monitors will be in-stalled: ACCT, used for pulsed mode, DCCT for CW and CWCT capable of measure in both operation modes.

Beam Profile Monitors are required for both magnets tuning and beam characterization. Due to the limited space, a minimum set of fourteen have been considered. A combination of non-interceptive and interceptive instruments has been implemented by means of Beam Profile Chambers (BPCs), compromising a vacuum chamber where a pair of Fluorescence Profile Monitors (FPMs) plus a single Secondary Emission Monitor (SEM Grid) are installed. These BPCs shall be as close as possible to the non-linear magnets for an optimum octupole and dodecapole tuning to sustain proper beam shaping at downstream sections where no magnetics elements are present.

Last type of diagnostic devices implemented are six-teen Beam Loss Monitors (BLoMs), which are primary diagnostics for beam tuning and equipment protection via measurement of the amount and location of beam losses along the beam transport line. LIPAc Ionization chambers type is the present choice for those monitors.

2.4. Vacuum system modelling

Vacuum system of HEBT has been recalculated, leading to modification on pump types integrated and pumping speeds. These changes are required due to, first, beam pipe material replacing stainless steel grade 316L for aluminum 6061 which implies changes on thermal out-gassing rates, secondly and Argon injection inside Target Vacuum Chamber (TVC) required to prevent boiling or significant vaporization from the lithium surface ensuring a total pressure in TVC above 10^{-2} Pa. Two cases for argon injection during nominal beam operation were simulated with slightly different pump speeds and active pumping points. In both cases, argon flow rate was kept constant at a value of $5 \cdot 10^{-4}$ Pa. m^3/s. Figure 2 shows HEBT and BD pressure profiles for argon injection in both alternatives.

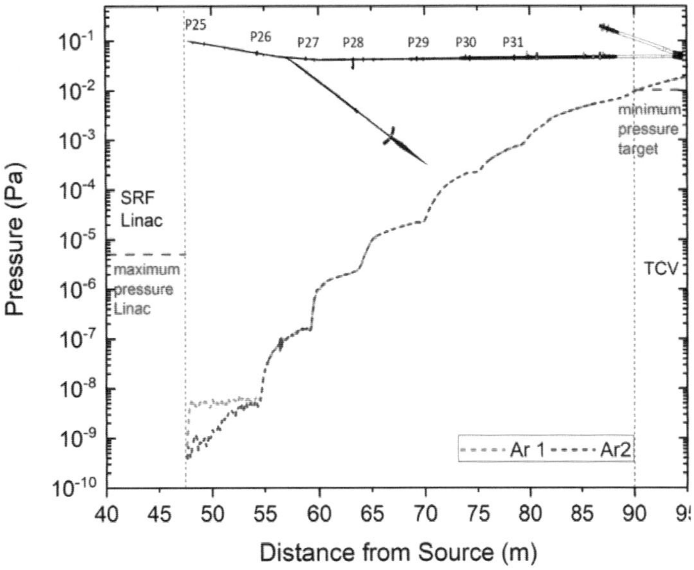

Figure 2. Pressure profile along HEBT (Case Ar1 + 2).

As it can be seen, in both cases the pressure falls far be-low the maximum permitted pressure of $5 \cdot 10^{-6}$ Pa at the SRF Linac side, it also stays higher than 10^{-2} Pa on TVC side. Thus, either of the two pump configurations results obtained will be suitable. New pumping configuration listing pump types and speeds is listed in table 3.

Table 3. HEBT and BD pumping configuration.

Point	Type	Pump speed [Ar1, l/s]	Pump speed [Ar2, l/s]
P25	NEG/SIP	0	240
P26	CP	1000	1100
P27	TMP	1100	1100
P28	TMP	1100	1100
P29	TMP	1100	1100
P30	TMP	1100	1100
P31	TMP	1100	1100

2.5. RH and radioprotection

HEBT spans along two different maintenance areas, established based on radiation dose rates calculated to workers. Most of it lays inside the first area where limited access is allowed for hands-on operations. The second area is named Target Interface Room (TIR), where maintenance is required to be carried out by remote handling [8]. To meet manual servicing inside the first area, it has been identified that the use of stainless steel as beam facing material leads to production of Co-56 with $T_{1/2}$=77 days, ruling out any possibility of hands-on maintenance during beam stop period, hence material for beam facing components was changed to aluminum. Activated devices, scraper and beam dump cartridge must be taken out from their radiation protection shielding by remote handling means [9]. Once removed, vacuum pumps, diagnostics devices and other components can be handled manually.

3. Conclusion

Engineering design of HEBT is achieving a mature stage and adapted to requirements and changes that arise as IFMIF-DONES facility overall design evolves.

Current configuration explained, portrays progress made from the previous preliminary stage towards a more detailed system in which mechanical and functional aspects of the components have been further specified. Magnets' characteristics are now defined, and their mechanical implementation performed, vacuum system has been successfully adapted to the required argon injection modification together with subsequent integration of newly required vacuum pumps, and finally adaptation of the beam line components material to aluminum has been identified and is presently being applied, paying attention to constraints such as, beam pipe thickness increment and flange type.

Future steps and required updates of the HEBT will be intrinsically and greatly linked to LIPAc results currently being obtained.

Acknowledgments

This work has been carried out within the framework of the EUROfusion Consortium, funded by the European Union via the Euratom Research and Training Programme (Grant Agreement No 101052200 — EUROfusion). Views and opinions expressed are however those of the author(s) only and do not necessarily reflect those of the European Union or the European Commission. Neither the European Union nor the European Commission can be held responsible for them..

References

[1] Królas W *et al.* 2021 *Nucl. Fus.* **61** 125002
[2] Podadera I *et al.* 2021 *Proc. 12th Int. Particle Accelerator Conf. (IPAC'21)* (Campinas, Brazil) pp 1910–13
[3] Dzitko H *et al.* 2021 *Fusion Eng.Des.* **168** 112621
[4] Cara P *et al.* 2021 *Proc 28th IAEA Fusion Energy Conf. (FEC'20) TECH/4* (Virtual) p 650
[5] Nomen O *et al.* 2020 *Fusion Eng. Des.* **153** 111515
[6] Oliver C *et al.* 2021 *Proc. 12th Int. Particle Accelerator Conf. (IPAC'21)* (Campinas, Brazil) pp 3251–53
[7] Podadera I *et al.* 2019 *Proc. 8th Int. Beam Instrumentation Conf. (IBIC'19)* (Malmö, Sweden) pp 200-04
[8] Sánchez-Herranz D *et al.* 2021 *Fusion Eng.Des.* **168** 112636
[9] Coloma S *et al.* 2020 *Fusion Eng. Des.* **165** 112216

13th International Particle Accelerator Conference (IPAC'22) IOP Publishing
Journal of Physics: Conference Series **2420** (2023) 012081 doi:10.1088/1742-6596/2420/1/012081

The reduction of the leakage field of the injection septum magnet in main ring of J-PARC

T Shibata, N Matsumoto, K Ishii, T Sugimoto and H Matsumoto

J-PARC Center, Accelerator Laboratory, KEK, High Energy Accelerator Research Organization, 1-1 Oho Tsukuba Ibaraki 305-0801 Japan

E-mail: tatsunobu.shibata@j-parc.jp

Abstract. A new injection septum magnet 1 (Inj-SM1) was installed in the Main Ring (MR) of J-PARC in 2016 as an upgrade. We confirmed that the leakage field upstream from the circulating duct of Inj-SM1 was sufficiently smaller than that for the previous Inj-SM1. Nevertheless, we attempted to further reduce the leakage field by using a new magnetic shield to achieve a higher proton beam intensity. In 2017, the first version of the new magnetic shield was produced and installed. The strength of the leakage field with the shield was ≈30% of that without the shield; however, the magnetic field of a quadrupole magnet (QM) upstream of Inj-SM1 was reduced to below the permissible limit, ≈0.1% of the total field integral of the QM, from the point of view of beam optics. Consequently, the first version failed. A second shield, for which the defects of the first version were fixed, was produced in 2018. The strength of the leakage field of Inj-SM1 with the second shield was reduced to ≈30% of that without the shield, while that of the QM was maintained. Thus, we decided to use the second shield for beam operation starting in November 2019. We confirmed that the impact of the leakage field on the circulating beam was 1/20 of that of the previous Inj-SM1.

1. Introduction

The J-PARC Main Ring (MR) provides a high-intensity proton beam with an energy of 30 GeV to the neutrino facility [1] or the hadron facility [2]. The present operation cycle is 2.48 s and the maximum beam power to the neutrino facility is 515 kW, which was achieved in March 2021 [3]. To realize the original design value of 750 kW, we will start the operation of the MR with a cycle of 1.3 s in June 2022; this is referred to as 1-Hz operation. The cycle will be further reduced to 1.16 s by 2028 to increase the beam power to 1.3 MW [4]. Over the last decade, the injection magnets in the MR have also been upgraded for 1-Hz operation and a 1.3-MW beam. The high-field injection septum magnet 1 (Inj-SM1), which has a typical field strength of 1.4 T, is used for injecting the proton beam from the rapid cycling synchrotron into the circulating orbit in the MR. The first Inj-SM1 started operation in 2008 (start of the J-PARC MR) and was replaced with the new septum magnet for 1-Hz operation in 2016 [5,6]. The replacement of the magnet was conducted because it was extremely difficult to further reduce the leakage field of ≈20 gauss around the entrance and exit of the circulating beam duct beside the septum plate. The reduction of the leakage field is an important target for the new septum magnet.

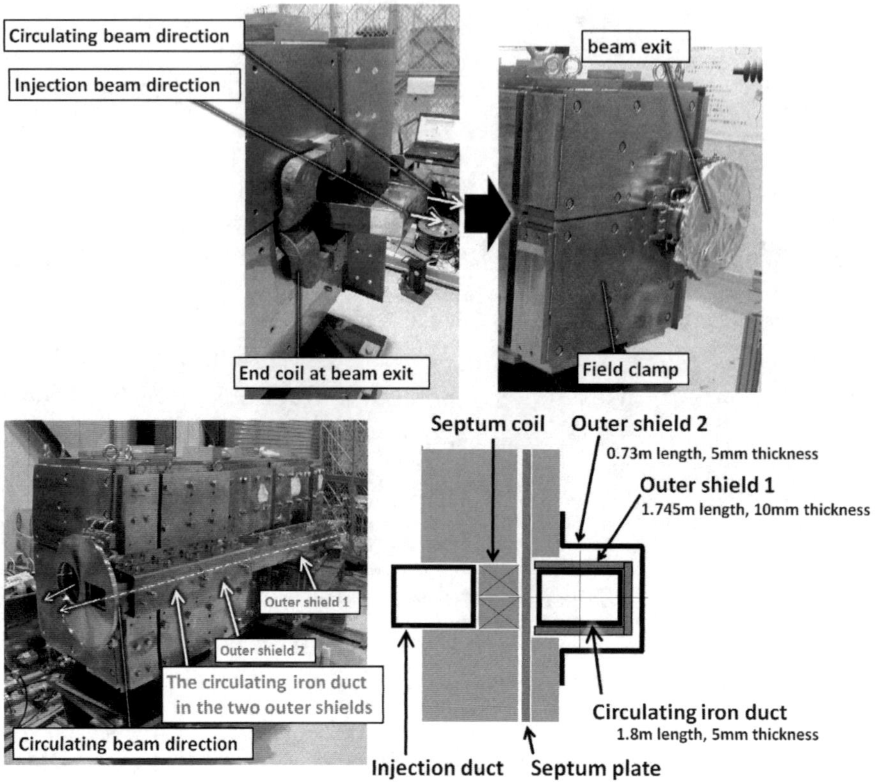

Figure 1. End coil and field clamp at beam exit (top) and several magnetic shields of Inj-SM1.

2. Measures against leakage field in 2015-2016

To minimize the leakage field, magnetic shields, vacuum chambers made of magnetic stainless steel [7], or an eddy current can be used to cancel the leakage field [8–10]. To reduce the leakage field of Inj-SM1, we applied several magnetic shields made of pure iron. The most significant contribution to the leakage field is from an end coil at the beam exit because it is located very close by the circulating beam line. To reduce the leakage field, we covered the entire end coil by a magnetic shield made of silicon-steel (field clamp) (figure 1 (top)). This reduced the leakage field to a few gauss [5, 6]. The second most significant contribution is the leakage field that passes through the septum plate. An ideal septum magnet has zero leakage on the outside of the septum plate; however, an actual septum magnet has a small leakage with an order of magnitude that is 0.1% of the gap field. For example, for a gap field of 1.4 T, the strength of the leakage field is ≈14 gauss. Therefore, we covered the circulating duct beside the septum plate by two magnetic shields made of pure iron (outer shields 1 and 2) (figure 1 (bottom)). Furthermore, we fabricated a 5-mm-thick circulating duct made of pure iron, which had a dc permeability of ≈8000 emu, in March 2016 [6] (figure 1 (bottom)), We measured the time-dependent field integral (BL) of the leakage field along the beam axis in the circulating duct with an applied current of 2,700 A, which is close to the optimal current for beam operation (2,590 A). The maximum value of BL during 0.2 s from the start of beam acceleration was 2 gauss·m. The bend angle of the circulating beam caused by the leakage field was then measured with the 3-GeV circulating beam in December 2016. Based on the measured angle, we obtained a BL value of ≈2.5 gauss·m, which is consistent with the field measurement [6]. The BL of the leakage field of the first Inj-SM1 was ≈30 gauss·m [11], which is 10 times greater than that of the second

Inj-SM1. Because our ultimate target for the leakage field is zero gauss·m, we attempted to further reduce the leakage field.

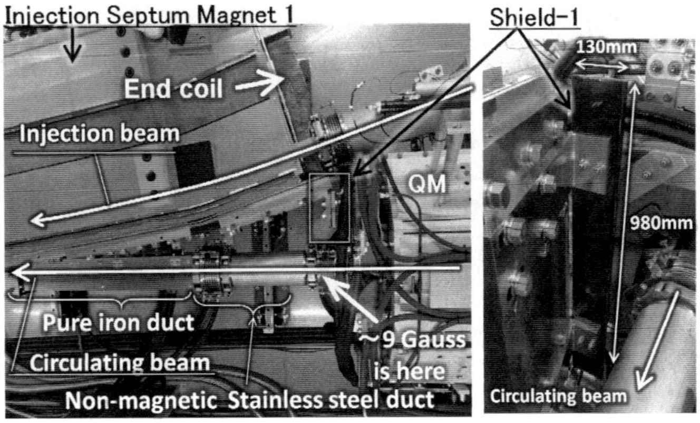

Figure 2. Shield-1 mounted on Inj-SM1 in 2017.

3. Reduction of leakage field in 2017

There was a leakage field of ≈9 gauss upstream from the pure iron circulating duct, where no shield was mounted and the circulating ducts were made of non-magnetic stainless steel [6] (figure 2). The origin of the field was the end coil at the injection beam entrance of Inj-SM1. Unfortunately, it was impossible to cover the end coil by a field clamp because of a lack of space. In addition, we could not cover the stainless steel circulating ducts by a magnetic shield because the magnetic field in the circulating ducts includes not only the field of Inj-SM1 but also that of a quadrupole magnet (QM) located upstream of Inj-SM1. The large loss of the leakage field of the QM affects beam optics, which must be corrected using a trim coil mounted in the QM [12]. This correction was not ideal. We thus mounted an additional magnetic shield in the space between the end coil and the circulating ducts. The detailed structure of the additional magnetic shield (Shield-1) was designed using Opera-3D-Tosca in 2017 [13]. The dimensions of Shield-1 were set to 130 mm (width) × 980 mm (height) × 13 mm (thickness) (figure 2). We measured the leakage field of each of the magnets with and without Shield-1 using a Hall probe (F.W. Bell, model 8030 [14]). First, we measured the leakage field of Inj-SM1 at three positions from the stainless steel circulating ducts to the outside of the ring. The distances from the center of the beam axis were all ≈160 mm, and those from the surface of the end fringe of the QM, z, were 18, 38, and 82 cm. We had to remount the probe whenever we changed the measurement position because we had only one probe. The reproducibility of the measurement position added some systematic uncertainty (a few percent) to the measured values. The applied current was the optimal pattern for beam injection. The rise, flat top, and fall time were all 0.2 s and the flat-top current was 2,598 A. The measured waveforms of the leakage field are shown in figure 3 (top). They indicate that the leakage field at $z = 18$ and 38 cm was reduced from ≈9 to ≈3 gauss and from ≈3 to ≈1 gauss, respectively. No reduction was found at $z = 82$ cm because the origin of the leakage found at this position was the induced eddy current field generated on the septum plate, and Shield-1 was not located between this position and the plate. Next, we measured the leakage field of the QM. The results are shown in figure 3 (bottom). The leakage field at $z = 18$ cm on the top of the duct with Shield-1 was 6∼7% lower than without the shield. We found no difference at the three positions on the inside because they were far from Shield-1. The leakage field at $z = 25$ and 39 cm on the outside (close to Shield-1) was reduced to 33% and 41%, respectively. From these results, the total loss of the BL of the QM caused by Shield-1 was estimated to be 0.3%, which is larger than the permissible limit of ≈0.1% [15]. Thus, we concluded that Shield-1 cannot be used for beam operation.

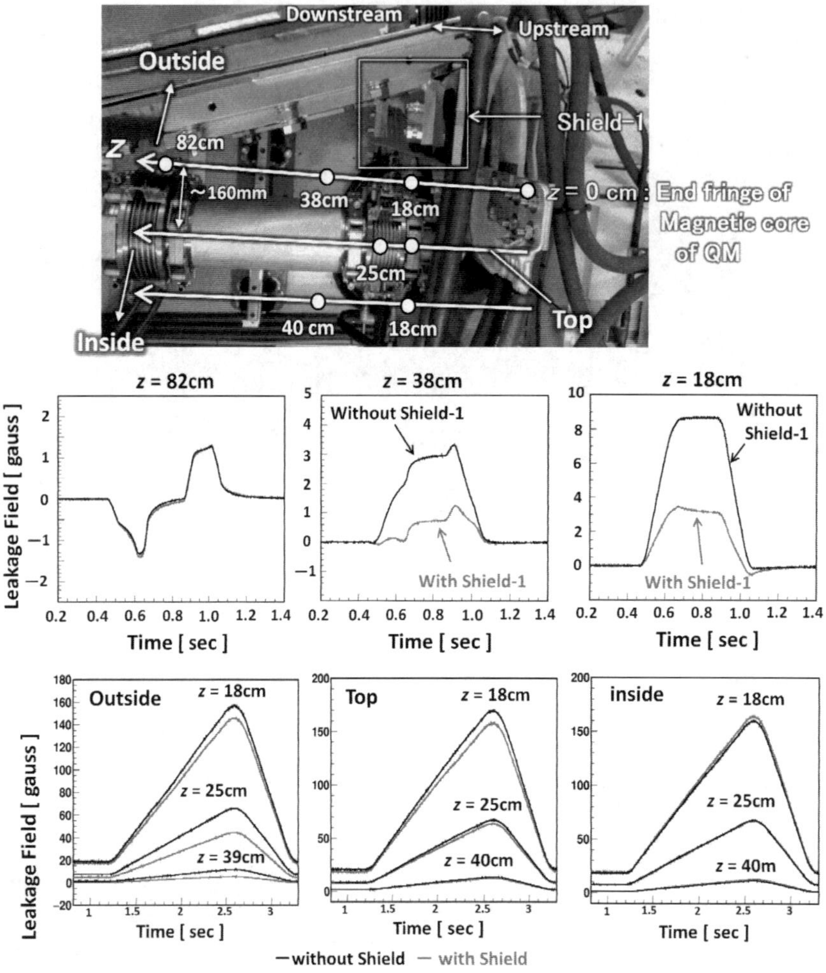

Figure 3. Waveforms of leakage field of Inj-SM1 (top) and quadrupole magnet (bottom) with and without Shield-1 measured in 2017.

4. Reduction of leakage field in 2019

We considered that the cause of the large loss of the leakage field of the QM was Shield-1 being located in front of the magnetic coil of the QM, which obstructed the field of the QM, as shown in figure 2. Therefore, we fabricated a new pure iron shield that did not interfere with the QM (Shield-2) in 2018. The dimensions of Shield-2 are 110 mm (width) × 600 mm (height) × 20 mm (thickness) (figure 4). Shield-1 was mounted perpendicular to the septum plate, whereas Shield-2 was mounted parallel to the septum plate so as not to reduce the leakage field of the QM (figure 4). In 2019, we measured the leakage field of Inj-SM1 and the same QM as that measured in 2017. To eliminate the systematic uncertainty caused by the reproducibility of the measurement position, we used eight Hall probes (F.W. Bell, model BH-200), one for each measurement position. The measured waveforms of the leakage field of Inj-SM1 are shown in figure 5 (top). The measurement positions on the outside of the circulating ducts were $z = 20.5$,

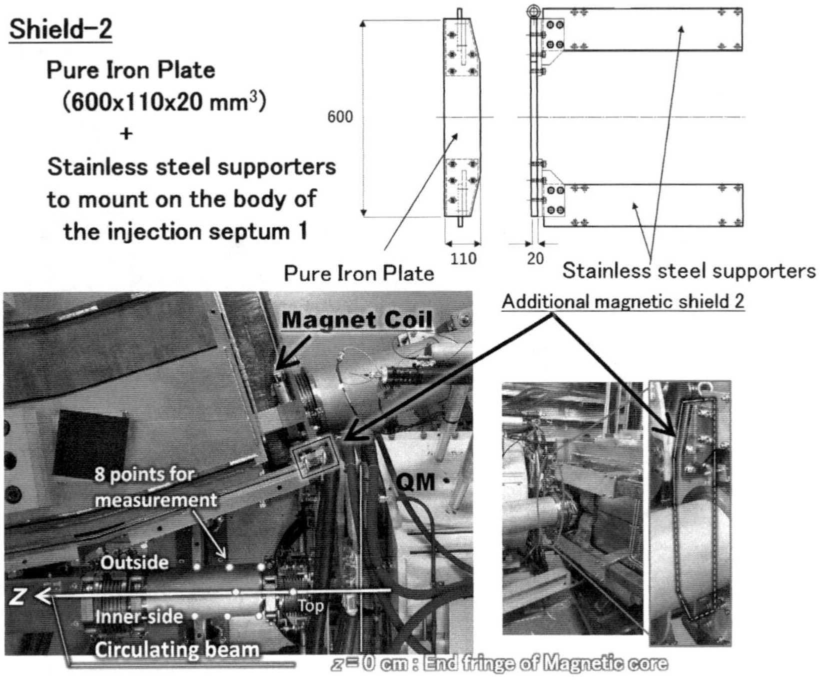

Figure 4. Shield-2 mounted on Inj-SM1 in 2019.

33, and 52.5 cm. We verified that the leakage field at $z = 33$ and 20.5 cm was reduced to ≈30% of that without Shield-2. Almost no reduction was found at $z = 52.5$ cm for the same reason as that given above. Next, we measured the leakage filed of the QM at the eight positions on the top, outside, and inside of the circulating ducts (figure 5 (bottom)). There was no difference between the measurements made with and without Shield-2 at all positions except $z = 20.5$ cm on the inside, where the probe was removed once by accident. From the above results, we found that the leakage field of only Inj-SM1 was reduced by Shield-2. Thus, we concluded that Shield-2 can be used for beam operation. We started to use Shield-2 in November 2019.

5. Beam study with new shield

Finally, the impact of the leakage field of Inj-SM1 on the 3-GeV circulating beam was measured in November 2019. The time-dependent shift of the beam position with respect to the horizontal direction while Inj-SM1 is excited is shown in figure 6. The shift was ≈0.25 mm, which corresponds to a BL of 1.6 gauss·m caused by the leakage field. Thus, we confirmed that the impact on the circulating beam was further reduced by ≈1/3 by Shield-2. The shift of the beam position by the leakage field of the previous Inj-SM1 was ≈5 mm [11], which is ≈20 times higher.

6. Summary

The new Inj-SM1 installed in 2016 reduced the leakage field to 1/10 of that of the previous one with the use of several shields. In 2019, the leakage field was further reduced by ≈1/3 by an additional magnetic shield that does not reduce the leakage field of the QM upstream. The impact of the leakage field on the 3-GeV circulating beam was measured. We confirmed that it was ≈1/20 of that of the previous Inj-SM1.

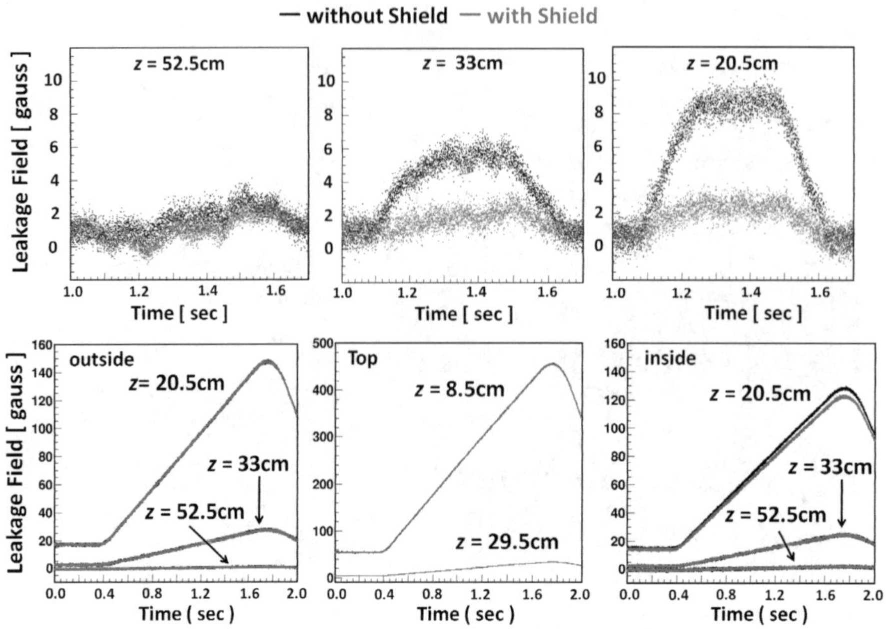

Figure 5. Waveforms of leakage field of Inj-SM1 (top) and QM (bottom) with and without Shield-2 in 2019.

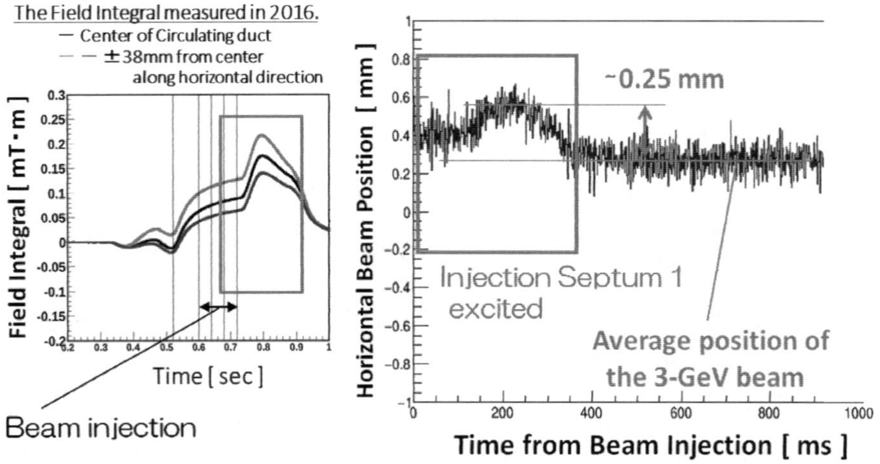

Figure 6. Shift of 3-GeV beam position with respect to horizontal direction caused by leakage field of Inj-SM1 measured in 2019.

References

[1] https://j-parc.jp/Neutrino/en/index.html
[2] https://j-parc.jp/Hadron/en/index.html
[3] J-PARC Group 2021 KEK report 2021-2
[4] Igarashi S *et al.* 2021 *Prog. Theor. Exp. Phys.* **2017** 033G01
[5] Shibata T *et al.* 2016 *Proc. of 7th Int. Particle Accelerator Conf. (IPAC'16)* (Busan, Korea) pp 1337–39
[6] Shibata T. *et al.* 2017 *Proc. of 8th Int. Particle Accelerator Conf. (IPAC'17)* (Copenhagen, Denmark) pp 576–78
[7] Yoshimoto M *et al.* 2008 *IEEE Trans. Appl. Supercond.* **18** pp 297–300

[8] Shuman D *et al.* 2005 *Proc. of Particle Accelerator Conf. (PAC'05)* (Knoxville, Tennessee USA) pp 3718–20
[9] Lian-Hua O *et al.* 2010 *Chinese Physc. Conf.* **34** pp 389–93
[10] Shibata T *et al.* 2016 *Proc. of 7th Int. Particle Accelerator Conf. (IPAC'16)* (Busan, Korea) pp 1340–42
[11] Fan K *et al.* 2013 *Proc. 10th Annual Meeting of Particle Accelerator Society of Japan (PASJ2013)* (Nagoya , Japan) pp 54–57
[12] Sato K C *et al.* 2015 *Proc. of 15th Int. Conf. on Accelerator and Large Experimental Physics Control Systems (ICALEPCS2015)* (Melbourne, Australia) pp 353–55
[13] `https://www.3ds.com/ja/products-services/simulia/products/opera/`
[14] `https://fwbell.com/`
[15] Igarashi S 2017 private comm.

Journal of Physics: Conference Series **2420** (2023) 012082 doi:10.1088/1742-6596/2420/1/012082

The new high field septum magnet for upgrading of fast extraction in main ring of J-PARC

T Shibata[1], K Ishii[1], S Iwata[1], T Sugimoto[1], N Matsumoto[1], H Matsumoto[1] and K Fan[2]

[1] J-PARC Center, Accelerator Laboratory, KEK, High Energy Accelerator Research Organization, 1-1 Oho Tsukuba Ibaraki 305-0801 Japan

[2] Huazhong University of Science and Technology, 1037 Luoyu Road Wuhan Hubei 430074 China

E-mail: tatsunobu.shibata@j-parc.jp

Abstract. The upgrade of the beam power of J-PARC Main Ring to 750 kW by reducing the cycle from 2.48 s to 1.3 s is underway, and the upgrade of the four high-field (HF) septa will be completed by summer of 2023. The operation test of one HF septum, SM31, was conducted in 2020. First trial was 1 Hz operation test, and we confirmed the joule heating at the magnetic coil was lower than limit. Second was magnetic field measurement. We obtained a good linearity between the applied current and the gap field without saturation. The field integral of the gap field was measured to estimate the optimal current for beam extraction, and we found it is 3,400 A. We compared the gap field between in the neutrino line and that in the beam abort line, then we found the longitudinal distributions have a small discrepancy but it was acceptable. The large leakage field was observed around the end-fringes, therefore, in order to reduce the leakage field, we produced an additional duct shield for mounting in the circulating duct in 2022. We verified that the additional duct shield reduced the leakage field to $\approx$2% of unshielded leakage field. The new SM31 and other two HF septa have been installed in MR by end of May 2022. The beam operation with the new HF septa will be started in June 2022.

1. Introduction

The J-PARC Main Ring (MR) provides the high-intensity proton beam with an energy of 30 GeV to the neutrino facility (NU) or the hadron facility. The present operation cycle is 2.48 s and the maximum beam power for Fast eXtraction mode (FX) is 515 kW which was achieved in March 2021 [1]. To realize original design value of 750 kW, we are going to start operation of MR with the cycle of 1.3 s, it is referred to as "1 Hz operation"; moreover, the cycle will be shorten to 1.16 s for increasing the beam power to 1.3 MW by 2028 (1.3MW beam) [2]. The magnets for FX in MR, which are used for switching the proton beam direction to NU or abort dump (ABT) [3], also have been upgrading for 1 Hz operation and 1.3 MW beam, and will be completed by summer of 2023. The four high-field septum magnets (SM) with magnetic field of above 1 T are used, and these magnets are called SM30, SM31, SM32 and SM33 from upstream. To upgrade, the replacement of these septa with new ones is underway. We produced the new septa in 2015, and conducted the operation test of the new SM30 and measured its magnetic field at first in 2018-2019 [3, 4]. Next, we conducted the test operation of the new SM31 in autumn 2020. The aim of the operation test was 1 Hz operation and measurement of the gap field and leakage field. After operation test, two large-sized vacuum flanges made of pure Titanium were weld to the vacuum ducts of the new SM31 in November 2020. In 2021, we produced two additional duct shields which can be mounted in the circulating duct of the new SM30 and SM31 in order to reduce the leakage field further. Regarding the new SM32, unfortunately, the installation in MR was postponed because of a fatal defect in the magnetic coil found in 2021. The new SM33 have been reconstructed with the new beam ducts and removed magnetic cores of SM32 and SM33. The installation of the new SM30, SM31 and SM33 in MR have been completed in May 2022 (figure 1). The beam operation with new cycle is going to be started in June 2022. This article focuses on the new SM31. The new SM31 is a bipolar septum magnet as same as the new SM30 [4]. Figure 2 shows the photograph, the specification and applied pattern current of the new SM31. The detail of the operation test in 2020 and the reduction of the leakage field by using the additional duct shield in 2022 will be described.

Figure 1. The high-field septa installed in MR in May 2022.

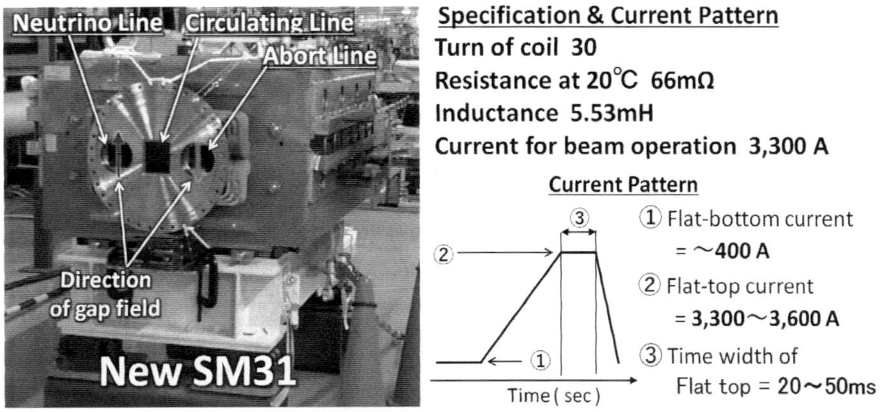

Figure 2. The photograph, the specification and the applied pattern current of the new SM31.

2. 1 Hz operation test

In the 1 Hz operation test, the operation cycles which we tested were 1.32 s, 1.28 s and 1.16 s, and the flat-top current was 3,600 A. We measured the temperature of cooling water flowed in the magnetic coil by using a radiation thermometer during the operation. The temperature increased approximately 13°C in all cycles which were corresponded with our expected values and it was lower than limit value of 20°C. Consequently, it was verified that the 1 Hz operation has no problem.

3. Measurement of gap field

The correlation between the applied current and gap field in the NU line which was measured by a Hall probe are shown in figure 3. We found no saturation at high-field region, thus we confirmed it has good linearity, however the linear-fit value of the slope was 3.472±0.004(stat) which was 0.8 % lower than design value of 3.501. The possible cause of the lower slope is small leakage of magnetic flux from the magnetic core.

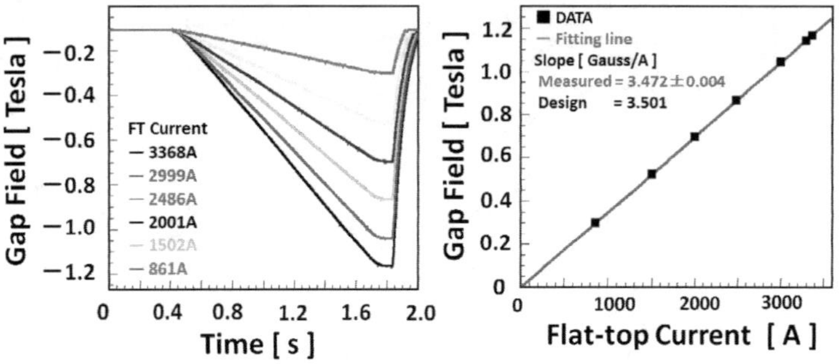

Figure 3. The waveforms of gap field in the NU line (left), and the correlation between the flat-top of the applied pattern current and the gap field (right).

To decide the optimal current for 30-GeV beam extraction, we measured the field integral (BL) along the center of the NU line with the flat-top current of 3,300 A, and the BL was 1.745 T·m. Since the supposed BL for the beam extraction is 1.808 T·m, we estimated the optimal current for beam operation at 3,400 A. Next, we measured the transverse distribution of the BL along the horizontal direction, and the result is shown in figure 4. It indicates the BL has an asymmetric sextupole field component, and the difference of the value of the BL between maximum and minimum in the region of ±20 mm which is size of the extraction beam [5] was ≈0.17%. It corresponds to the bending angle of 0.05 mrad for the 30-GeV proton beam. However, since the bending angle is almost same scale of fine-tuning of position and direction of the extraction beam by changing the field strength of the HF-septa, the difference is not serious problem.

Finally, we compared the longitudinal distribution of the gap field between in the NU and in the ABT line with the flat-top current of 3,300 A and 3,400 A (figure 5), and found an obvious discrepancy. We presumed the cause of the discrepancy is slight difference of the magnetic cores such as volume, shape and processing. The large asymmetric structure should make the leakage field larger, but that is enough small and the discrepancy is acceptable.

4. Measurement of leakage field

The longitudinal distribution of the leakage field in the circulating duct was measured along the several straight tracks: on the center of the beam-line (center track) which is parallel to the beam direction, on ±10 mm from center track to horizontal direction (10-mm tracks), on ±25 mm from

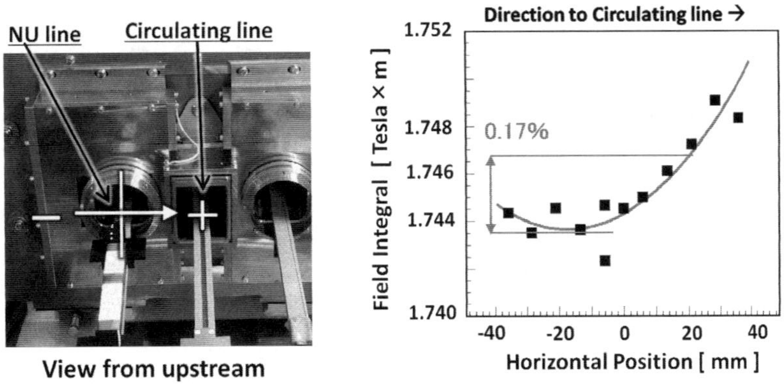

Figure 4. The view of the new SM31 from upstream (left), and the transverse distribution of the BL in the NU line, and the red curve is quadratic fit (right).

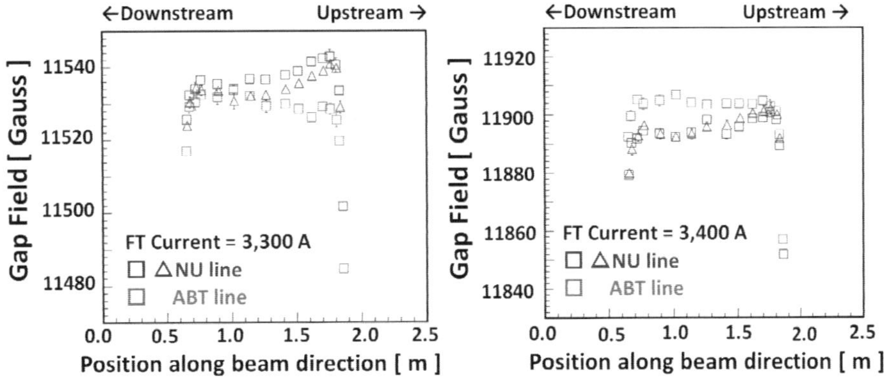

Figure 5. Comparison of longitudinal distribution of the gap field between NU and ABT line with 3,300 A (left) and 3,400 A (right).

center track (25-mm tracks), and on 18 mm from the inside wall of the duct (18-mm tracks) (figure 6). The 18-mm tracks are not parallel to the beam direction because the circulating duct is tapered along the longitudinal direction of which the horizontal size of entrance and exit are respectively 88 mm and 218 mm. The flat-bottom and flat-top of the applied pattern current were 408 A and 3,300 A, respectively. The results shown in figure 6 indicate the large leakage field still exists around the end-fringe of the magnetic cores (end-fringe field), especially, extremely large end-fringe field of which the maximum value was 170 Gauss were observed on the 18-mm tracks. We investigated the end-fringe field in detail, and found the cause was magnetic saturation of the septum plates made of iron. Next, the horizontal distribution of the BL which used flat-bottom current (fb-BL) and flat-top current (ft-BL) were also measured along the several tracks (figure 7). We found a dipole and quadrupole field component in the both of results, especially, the dipole field component of 1.2 Gauss·m in fb-BL and 1.6 Gauss·m in ft-BL were large and they affect the circulating beam. To reduce the end-fringe field and the dipole field component, the best measures is installation of an additional magnetic shields in the circulating duct.

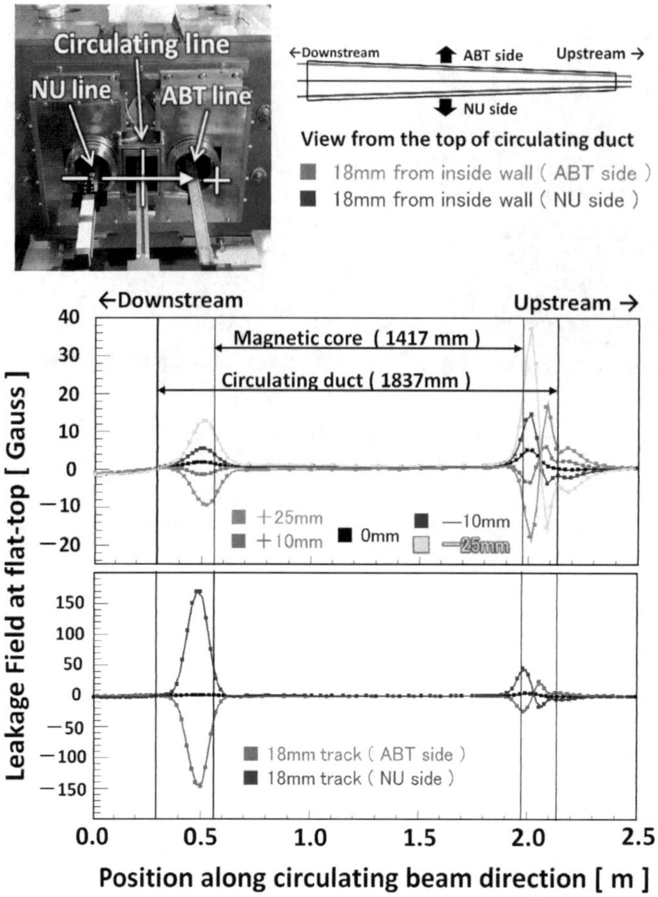

Figure 6. The longitudinal distribution of the leakage field in the circulating duct along the center track, 10-mm tracks, 25-mm tracks, and 18-mm tracks.

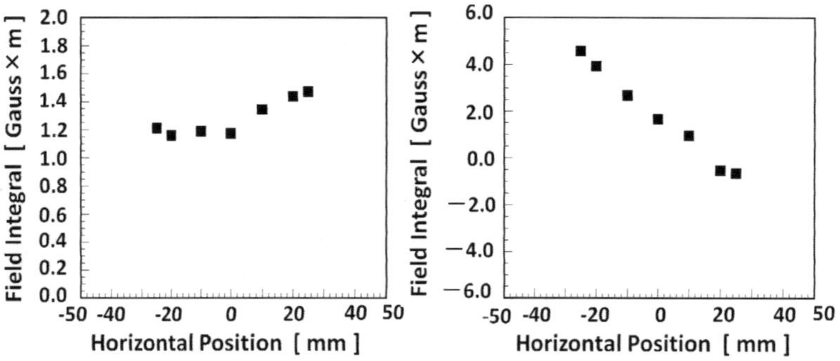

Figure 7. The horizontal position dependence of the BL: fb-BL (left) and ft-BL (right).

5. An additional inner shield

In order to reduce the leakage field in entire area of the circulating duct of 1.837 m length, we decided to produce a duct shield of 1.935 m length and 3 mm thickness made of pure iron (inner shield). The inner shield is also tapered as same as the circulating duct of which the cross section of entrance and exit are respectively 82.2 mm×99.4 mm and 219.6 mm×99.4 mm. The production of the inner shield was completed in March 2022, and we mounted in the circulating duct. After installation of the new SM31 in MR, we have measured the end-fringe field with the inner shield along a 18-mm track of ABT side in May 2022. Figure 8 shows a photograph of the inner shield mounted in the new SM31, and the longitudinal distribution with the flat-top current of 3,960 A. We found that the large end-fringe field was reduced to ≈2% of unshielded end-fringe field, thus we concluded the reduction of the end-fringe field by using the inner shield is a most effective measures.

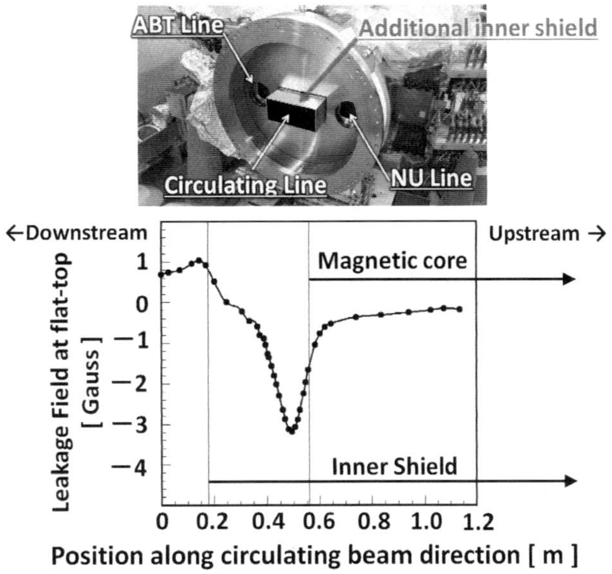

Figure 8. The additional inner shield mounted in the new SM31, viewing from downstream (upper). The longitudinal distribution of the end-fringe field along the 18-mm track with the inner shield (lower).

6. Summary

We are working on upgrade of the FX high-field septa for increasing the beam power of MR in J-PARC. In 2020, we conducted operation test of the new SM31. The high repetition operation had no problem, and we obtained a good linearity between the applied current and gap field. The longitudinal distributions of the gap field in the neutrino line and abort dump line have small discrepancy, but it was acceptable. On the other hand, we found large leakage field around the end-fringe, and we produced an additional duct shield which is mounted in the circulating duct. The inner shield was installed in March 2022, and we verified the leakage field was reduced to ≈2%. The beam operation with the new high-field septa will be started in June 2022.

References

[1] J-PARC Group 2021 KEK report 2021-2
[2] Igarashi S *et al.* 2021 *Prog. Theor. Exp. Phys.* **2017** 033G01
[3] Shibata T *et al.* 2019 *Proc. of 10th Int. Particle Accelerator Conf. (IPAC'19)* (Melbourne, Australia) pp 2001–04
[4] Shibata T *et al.* 2021 *J. Phys. Conf. Ser.* **33** 011034
[5] Iwata S 2021 Private comm.

Role of surface chemistry in conditioning of materials in particle accelerators

G Sattonnay, S Bilgen, B Mercier, D Longuevergne, S Della-Negra and I Ribaud

Université Paris-Saclay, CNRS/IN2P3, IJCLab, Orsay France

Email: gael.sattonnay@ijclab.in2p3.fr

Abstract. For the vacuum scientists and the accelerator community, finding solutions to mitigate pressure rises induced by electron, photon and ion desorption, and also beam instabilities induced by ion and electron clouds is a major issue. Along the time, changes in the surface chemistry of vacuum chambers are observed during beam operations, leading to modifications of: outgassing rates, stimulated desorption processes and secondary emission yields (SEY). To understand the role of the surface chemistry of air exposed materials in the electron conditioning process, typical air exposed materials used in particle accelerators : thin film coatings (NEG, i.e a Ti-Zr-V alloy, and carbon), copper (and its oxides Cu_2O and CuO) and niobium were conditioned by low energy electron irradiation for a better understanding of electron-cloud effect. First, SEY was measured to understand the changes of surface conditioning upon particle irradiation; then, surface chemistry evolution after electron irradiation was investigated by both XPS and TOF-SIMS analyses using the ANDROMEDE facility at IJCLab. Finally, the relationship between the surface chemistry and the conditioning phenomenon will be discussed.

1. Introduction

Electron emission is a major phenomenon involved in the formation of the electron-cloud (EC) in the LHC. EC effects have been identified among the major performance limitations for the Large Hadron Collider (LHC). The EC is induced by the electron multipacting process related to the secondary electron yield (SEY) of the surfaces exposed to electron irradiation. The SEY depends strongly on the nature of materials and on the surface chemistry, and consequently the EC formation is determined by the surface properties. Nevertheless, electron bombardment of the inner surface of the beam pipes is also responsible for a beneficial effect called "surface conditioning" (or the "scrubbing effect") leading to a reduction of SEY and therefore inducing a mitigation of the multipacting process and electron cloud build up).

The aim of this study was first to validate the set-up developed in our laboratory and devoted to SEY measurements, by reproducing the results that exist elsewhere for Cu and Nb for instance. Then, we proposed an alternative method for the analysis of the surface chemistry to the X-ray Photoelectron Spectroscopy which is traditionally used: Time-of-Flight Secondary Ion Mass Spectrometry with high energy gold nanoparticles to probe the surface. We investigated the conditioning of copper beam screen induced by electron bombardment and identified the role played by the modifications of the surface chemistry in this phenomenon.

The experimental set up for SEY measurements is an in-house build set-up developed at IJCLab. It consists of a single UHV chamber (base pressure: 7×10^{-10} mbar) equipped with an electron gun providing a pulsed electron beam (with a pulse length of 30 ms) in the energy range 10 to 4000 eV, with an intensity from few nanoamperes to 50 μA. The sample is carried by a single manipulator allowing for a precise positioning of the sample in the chamber.

The SEY was measured by the sample bias method that is a two-step SEY measurement. First, the primary current I_p is acquired for each primary electron energy E by applying a positive sample bias (V=+50 V). In this case, the current measured on the sample is $I_M(+50V) = I_P$. The emitted secondary electrons (SE) are trapped and recaptured by the sample. The sample polarity is then switched to a negative value (V=-20 V) and I_M is acquired while shooting with the electron gun on the sample with the same energy settings as during I_p acquisition. The SE current (I_{SE}) is given by:

$$I_{SE} = I_p - I_M(-20V) \tag{1}$$

The SEY δ is then obtained by:

$$\delta = \frac{I_{SE}}{I_p} = \frac{I_p - I_M(-20V)}{I_p} = 1 - \frac{I_M(-20V)}{I_M(+50V)} \tag{2}$$

In the present work, we used the ANDROMEDE facility [1] to analyse samples of copper beam screen. 12-MeV Au_{400}^{4+} ion beams are accelerated by a NEC Pelletron® 4MV electrostatic accelerator to bombard samples. The emitted secondary ions (both positive and negative ions) are analysed with the ToF spectrometer EVE In this set-up, it is possible to record a mass spectrum with only one single ion impact. Consequently, the secondary ion emission yield is strongly enhanced and the sensitivity of chemical surface analysis is improved. A high detection efficiency for molecules deposited in very small quantities on metal surfaces (lower than the monolayer) is reached. Moreover, the limited number of primary ions (about 107 nanoparticles/cm2) required to analysis the samples prevents excessive damage of the surface. Only the extreme surface (i.e. the molecules deposited on the surface and the materials surface corresponding to a depth of ≈10 nm) is analysed.

The experimental system used for XPS analysis in the present work is the K-ALPHA system from Thermo-Fisher, equipped with a monochromatic Al kα X-ray source (hν=1486.6 eV) and an Ar^+ ion gun for XPS sputter depth profiling. For each analysis, a full energy spectrum is acquired for identification of the constituent elements and additional sweeps are then performed in narrower energy regions to obtain high energy-resolution spectra. Spectra displayed in the following in normalized intensity were obtained after normalization of the signal to its maximum intensity.

The following samples were analysed : an OFE-Cu, a Cu_2O sample provided by the NEYCO company, CuO produces via a chemical route on Cu [2], a Ti-V-Zr coating deposited on Cu (non-activated NEG) and a Nb sample (provided by the company Tokyo Denkai).

2. Secondary electron yield evolution

To illustrate the SEY evolution during electron-conditioning, results obtained for an as-received copper beam screen sample during its irradiation at room temperature by 500 eV electrons is shown in figure 1. SEY is observed to decrease over the full considered energy range when the irradiation dose increases. The maximum SEY, initially equal to 2.00, decreases down to 1.10 after a dose of 6 10^{-2} C/mm^2. For larger irradiation doses, no further evolution of SEY is observed. These results are in good agreement with those previously published in the literature: the initial value of the SEY for an as-received copper is reported to be between 2.5 and 2.0 and for a conditioned sample the SEY decreases down to a value between 1.2 and 1.1 [3-5].

Typical conditioning curves, i.e. evolution of the maximum SEY with respect to electron irradiation dose, of air exposed samples (namely Cu, CuO, Cu_2O, Nb, NEG coating and Carbon-coating) are given in figure 2. It is seen that the maximum SEY of metallic sample (Cu and NEG) decreases from an initial value around 2.1 down to an ultimate SEY (maximum SEY at the saturation of the decrease) of about 1.1. The saturation of the SEY decrease is obtained for doses higher than 10^{-3} - 10^{-2} C/mm^2. As shown in other studies, the SEY of the conditioned carbon deposit is well below one. This is why it can be used as an antimultipacting layer in accelerator beamlines. A surprising result was obtained for CuO, since its SEY in a fully conditioned state is also less than one: it reaches a SEY value close to 0.8 after electron-conditioning. This result is in agreement with a study that shows a SEY for a CuO nanolayer lower than one [6]. However, it was recently mentioned that CuO should not condition easily [5]. This discrepancy could come from the influence of the thickness of the CuO layer, or from the effect of a possible difference in the stoichiometry of the CuO samples used in the different studies.The SEYs for the other materials are consistent with the literature.

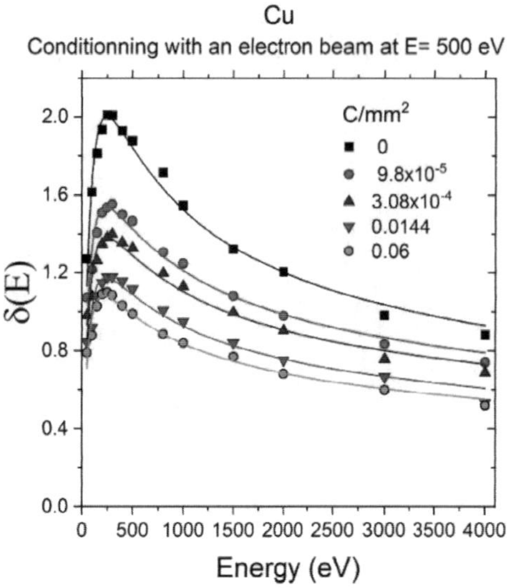

Figure 1. Secondary electron yield curves as a function of primary electron energy, for an as-received copper beam screen conditioned with a 500 eV electron beam at different irradiation doses.

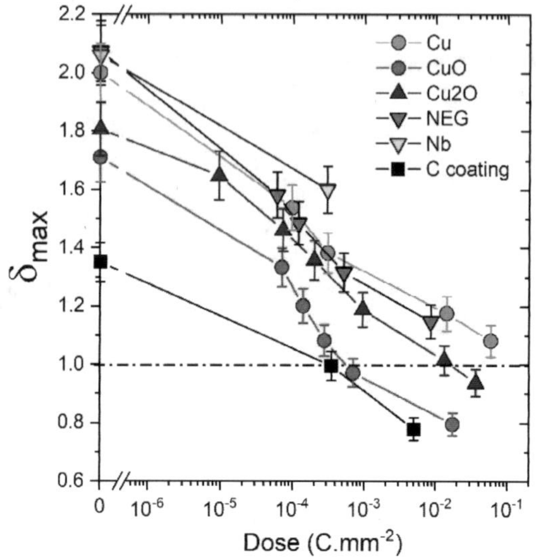

Figure 2. δ_{max} as a function of the dose for different impinging electron energies on several materials conditioned with a 500 eV electron beam: Cu, CuO, Cu_2O, Nb, NEG coating and Carbon-coating.

3. Surface chemistry evolution

Evolution of the Cu-surface after electron-conditioning was more particularly investigated. In the as-received state, the C 1s line shows a main peak at 284.8 eV ascribed to C-C bonds (sp3 hybridization). A second peak at 286 eV corresponding to C-O bonds, and a third at 288.2 eV related to O-C=O groups are visible (figure 3). These latter are due to an adventitious carbon contamination and comes from the different compounds forming the adsorbed layer. In the fully conditioning state (figure 3b), the XPS analysis indicates that the decrease of the secondary electron yield corresponds to significant modifications of the surface chemistry. The disappearing of the peak at 288.2 eV on the C 1s line implies the removal of O-C=O groups. The main C 1s peak is shifted towards 284.8 eV, indicating the transformation of the adventitious carbon layer (sp3) into a more graphitic form (sp2), as already observed in previous studies, for instance in [4]. This phenomenon is called sometimes "graphitization". Figure 3c shows the XPS analysis of a carbon coating layer provided by the CERN: we found the main C 1s peak related to the sp2 hybridization.

From the analysis of the negative secondary ion spectra recorded with ANDROMEDE, a significant increase of carbon (C_n^-) and carbon cluster (C_nH^-) emission for the conditioned Cu sample is observed (figure 4). Comparison of the conditioned sample spectrum with a graphene spectrum obtained under the same conditions suggests that a film of hydrogenated graphene covers the irradiated surface as the carbon cluster distribution is similar. Thus, in agreement with the XPS results, the ToF-SIMS analysis shows that a carbon layer whose bonds are that of graphene (sp2 hybridization) is formed on copper surface under electron bombardment. It is worth noting that this carbon layer is not composed only of carbon, but it contains also high amounts of hydrogen and oxygen.

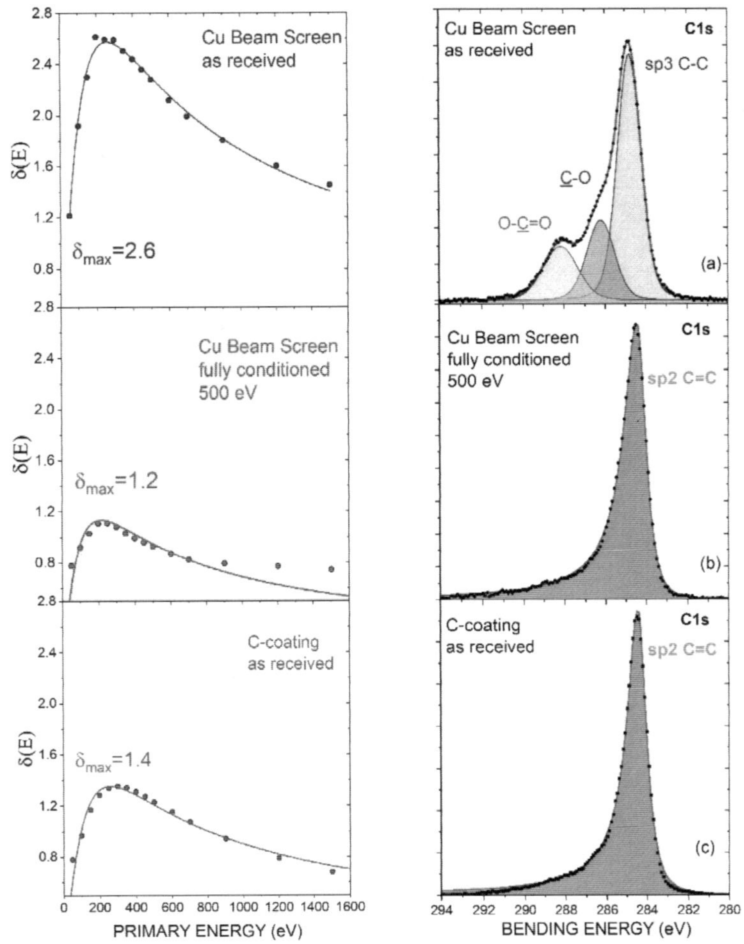

Figure 3. SEY curves and C1s XPS spectra measured on a LHC Cu beam screen sample: (a) "as received"; (b) fully conditioned; (c) on an as received amorphous carbon coating.

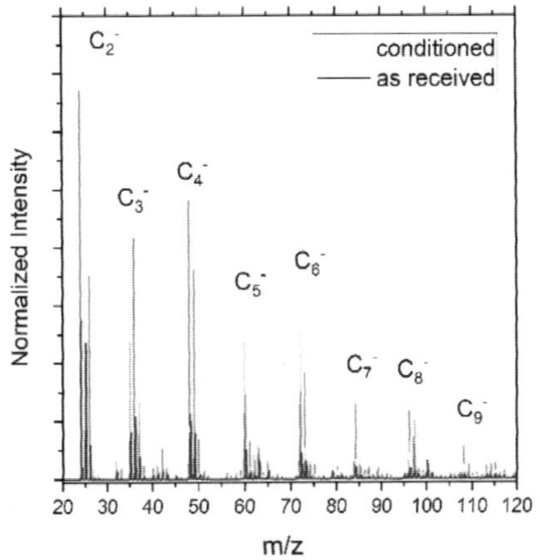

Figure 4. Negative secondary ion mass spectrum for a copper beam screen after electron conditioning process displaying intense carbon cluster peaks.

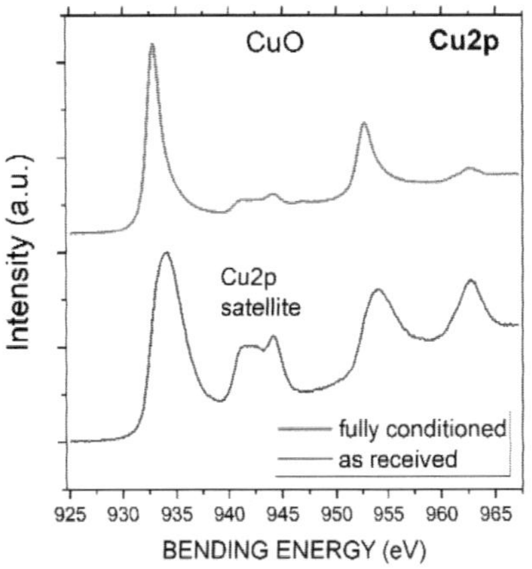

Figure 5. Cu2p XPS spectra measured on a CuO layer deposited on Cu.

Evolution of the CuO surface chemistry during electron-conditioning was also investigated by XPS. Figure 5 shows the XPS Cu 2p lines for an as-received and a fully conditioned CuO sample. The Cu 2p satellite shape on the as received sample spectrum is the signature of a CuO sample. After electron conditioning, the intensity of the Cu 2p satellite strongly decreases, indicating that the conditioned area was partially transformed into Cu_2O. Thus, the electron irradiation produced a partial reduction of CuO into Cu_2O.

4. Conclusion

Conditioning effect was investigated for several materials by measuring the variation of their SEY with the electron dose. Study of the copper beam screen conditioning under irradiation with 500 eV electrons evidenced two processes: (i) a surface cleaning by electron stimulated desorption; (ii) the formation of a graphene-like layer. Both processes lead to the decrease of the copper SEY down to value close to unity. Carbon transformation upon electron irradiation should be also an effective phenomena to explain the SEY decrease with the electron dose for the other metallic materials, for which the SEY of a fully conditioned state is close to one. A SEY value below 1 after conditioning for CuO represents also an interesting result since it is close to the SEY of carbon-coating, this latter being known for its antimultipacting properties. Further investigations are needed to explain the conditioning of CuO.

References

[1] Lai A T L *et al.* 2020 *J. Vac. Sci. Technol.* B **38** 044008
[2] Welegergs G G *et al.* 2021 *Mater. Today:. Proc.* **36** 509
[3] Cimino R *et al.* 2012 *Phys. Rev. Lett.* **109** 064801
[4] Larciprete G R *et al.* 2013 *Appl. Surf. Sci.* **328** 356
[5] Petit V *et al.* 2019 *Phys. Rev. Accel. Beams* **22** 083101
[6] Aguilera L *et al.* 2013 *J. Phys. D: Appl. Phys.* **46** 165104

Few-nanosecond stripline kickers for top-up injection into PETRA IV

G Loisch, V Belokurov and F Obier

Deutsches Elektronen-Synchrotron DESY, Germany

E-mail: gregor.loisch@desy.de

Abstract. PETRA IV is the planned ultralow-emittance upgrade of the PETRA III synchrotron light source at DESY, Hamburg. The current baseline injection scheme is an off-axis, top-up injection with few-nanosecond stripline kickers, which would allow for accumulation and least disturbance of experiments during injection. Besides the requirements on kick-strength, field quality, pulse rise-rate, and heat management, two mechanical designs with different apertures are necessary, as the devices will be used for injection and the transverse multi-bunch feedback system. In this contribution we will present the current status of 3D finite element simulations of electromagnetic fields as well as the mechanical design and first pulse electronics tests.

1. Introduction

Several major lightsources have recently been upgraded or are currently undergoing an upgrade to so-called 4th generation light sources, which approach the diffraction limit of the produced synchrotron light with ultra-low emittances in the range of a few 10 pm to 150 pm. One difficult aspect of the beam optics of such low-emittance storage rings is a very limited dynamic aperture. This hinders top-up injection of bunches into the machine and has led to the choice of a full-charge swap-out injection scheme in most designs.

The PETRA IV project aims at upgrading the PETRA III 6 GeV synchrotron radiation facility into a 4th generation light source [1]. Simulations indicate that in PETRA IV top-up injection is nevertheless possible and it is currently pursued as the baseline injection scheme. To limit the impact of an injection event on the stored beam, the kickers in PETRA IV are designed to have a kick duration equal to the target bunch separation of 2 ns. This is only achievable with short stripline kickers, which are being developed for this purpose at DESY. First simulations of the electromagnetic design of these devices have been presented before [2]. In this contribution we will give an update on the kicker magnet requirements for single-bunch top-up injection, show the updated electromagnetic design, and go into some details of the mechanical design and pulse electronics development.

2. Injection kicker requirements

The stripline design had to be adjusted to the new baseline top-up injection scheme from the original swap-out injection of 80 ns bunch trains. To not constrain the dynamic aperture of the storage ring, the free aperture of the striplines has to be at least 16 mm in diameter. Stripline length and high voltage (HV) pulse duration define the kick strength on neighbouring bunches, which are not being injected into. This residual kick shall be minimised for least disturbance

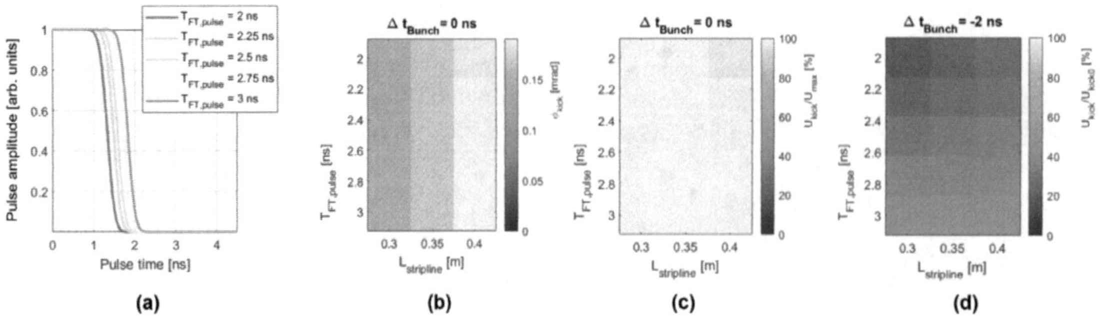

Figure 1. Analytical calculation of stripline length and pulse shape constraints. Assumed pulse shapes are shown in (a), the absolute kick strength at 12 kV/6 GeV is shown in (b). The kick strength relative to the maximum kick strength possible for the given length (peak amplitude during full passage duration of the kicked bunch) is shown in (c) and (d) shows the kick strength onto a bunch with 2 ns delay relative to the nominal strength given in (b).

of user experiments during injection. On the other hand, shortening the striplines also reduces the absolute kick strength of one kicker module and ultimately results in a very large number of modules, making injection increasingly complicated. Additionally, the length of every kicker module should be used efficiently. Figure 1 shows analytical calculations for all these criteria. In figure 1 (a) half of the assumed, time-symmetric input HV pulses is depicted for several assumed pulse lengths. The absolute kick angle for different lengths is shown in figure 1 (b). Efficiency of length usage is judged upon by the relation between actual kick angle and the maximum possible kick angle at a certain voltage and stripline length, which is shown in figure 1 (c). Finally, the kick strength on a neighbouring bunch at 2 ns separation relative to the nominal kick strength is shown in figure 1 (d).

Based on these calculations, a stripline length of 0.35 m with a flat top pulse length of 2 ns was chosen. The assumed rise and fall times are 0.5 ns. With these theoretical parameters a total of 16 stripline kicker modules (plus 4 spare modules) would be required for a 4-kicker bumped orbit at the Lambertson-type injection septum in the current lattice design. Pulse voltage has to be adjustable within a range of (8-14) kV in order to provide the required kick strength at the four different positions as well as a certain adjustment range and redundancy.

3. Electromagnetic design

Simulations of the electromagnetic behaviour of the striplines are performed using CST Microwave Studio. The simulations have been integrated into a Bayesian optimisation routine to allow complex geometry optimisation [2]. The cross-section geometry is shown in the lower right of figure 2. A symmetric aperture of 16 mm in diameter is kept free between the stripline blades and the side fenders, which are introduced to reduce capacitive coupling between the blades. The same cross-section geometry is also planned to be used for the striplines of the transverse multi-bunch feedback system of PETRA IV.

In the end sections of the stripline blades, not only the optimal electromagnetic design has to be taken into account, but also constraints as e.g. possibilities for mechanical adjustment of the stripline positions have to be considered. A first optimisation of this has been performed for a prototype of the PETRA IV injection kickers. Simulated reflected pulses at the rising and falling edge of the input current of this prototype exhibit a maximum of 4 % of the incoming amplitude. Further optimisation of the setup is still ongoing, especially with respect to the longitudinal and transverse beam impedance of the kickers as a single module and in an arrangement of several modules in a row. One possible approach that is considered here, which has been presented for feedback striplines, but not for injection kicker setups so far, is the introduction of additional capacitance at the stripline blade ends, to suppress the excitation of higher order modes [3].

4. Mechanical design

The mechanical design of the first prototype of the PETRA IV injection kickers is shown in figure 2. Due to several changes of the mechanical requirements in the last months, this prototype still exhibits an aperture of 14 mm in diameter and as this setup is envisaged to be used for measurements as an injection kicker and as a feedback kicker, the length between the two contacts of the electrodes, i.e., the distance between the central pins of the vacuum feedthroughs of one stripline blade is 0.3 m. The basic approach for the design was to use the

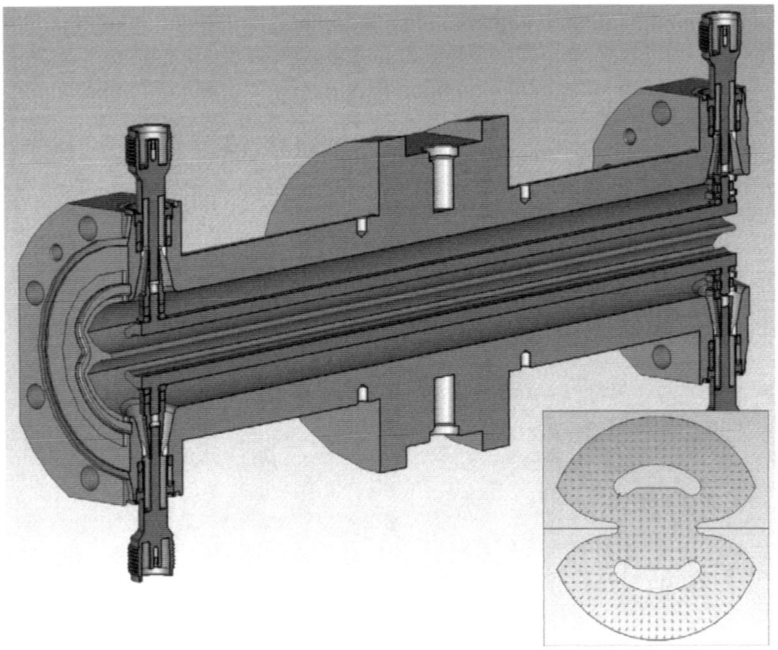

Figure 2. Mechanical design of the prototype PETRA IV stripline kicker. Inset in the lower right shows the cross section geometry.

outer conductor of the stripline as a vacuum vessel, unlike other recent designs [4–6]. This less complex mechanical design was chosen to enable a simple combination of different numbers of stripline kicker modules at the various kicker positions. Vacuum levels are planned to be ensured by non-evaporable getter (NEG) coating of the inner surfaces of the kicker. The outer tank has a DN63 CF flange connector on either end, which is machined from bulk steel material and vacuum feedthroughs are integrated into it. This reduces the gap between two sets of electrodes to ¡ 14 mm, depending on the excess length of the electrodes extending outwards from the feedthrough connectors.

Weldable feedthroughs from CeramTec have been acquired and their DC breakdown voltage has been measured. A maximum hold-off voltage of 15 kV was found in these tests at a pressure of ca. 10^{-7} mbar. Operation voltage was thus constrained to 14 kV, with an extra safety margin due to the short duration of the HV pulses used during operation. Dielectric losses were measured at 1 GHz, 300 W and only a minor temperature increase of ca. 20 °C was found. Time dependent reflectometry (TDR) measurements were performed and showed tolerable levels of reflection.

A sliding contact has been introduced at one end of each electrode, to prevent excessive mechanical stress on the vacuum feedthroughs when the electrodes are heating up during vacuum baking or due to induced currents from the high average current, stored beams. Electrodes will be made of tungsten to minimise thermal expansion while maintaining good electrical and thermal conductivity. By using a two-threaded nut and an RF spring, the stripline blades can be adjusted in height . No additional adjustment in transverse direction is foreseen at the moment other than the exact placement of the vacuum feedthroughs.

5. Pulse electronics

The changed requirements on the PETRA IV injection kicker from bunch train to single-bunch top-up injection has also changed HV pulse requirements. Ground-symmetric HV pulses have to be supplied to the stripline electrodes, each of which are matched to an electrical impedance of 50 Ω in the odd mode, i.e., opposite polarity of the two electrodes. A development project has been initiated with the high power electronic systems group at the ETH Zurich, to explore the pulse rise and fall time limits of available semiconductor HV switch components and different circuit topologies. Preliminary results indicate that Marx generator and linear transformer driver (LTD, also called inductive adder) topologies enable similar switching times in the range of ca. 3 ns to 12 ns with some slight disadvantages in the rise times and advantages in the fall times for the LTD due to additional parasitic inductances in the switching circuit [7,8]. Further studies include the application of advanced gate circuitry to these switches to probably increase switching speed slightly [9, 10], whereas the required switching times of ca. 0.5 ns seem to be inaccessible. Therefore, a test pulse system from FID GmbH based on drift step recovery diodes

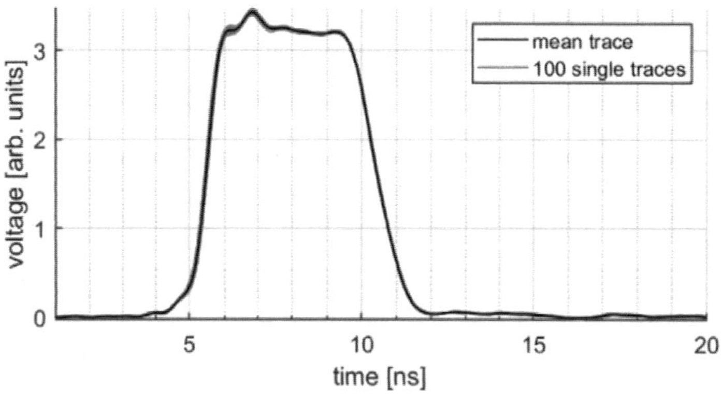

Figure 3. Single and averaged current traces measured with a 4 ns test pulse from FID GmbH.

(DSRDs) [11] has been received and tested for pulse stability, rise and fall times, and also for kick tests with an existing, 2 m long stripline kicker at the European XFEL. This pulser was designed for a pulse length of 4 ns and exhibited rise and fall times of ca. 1 ns and 2 ns, respectively. Figure 3 shows 100 current traces measured with a 5 GHz bandwith, 10 GS/s Tektronix oscilloscope using an FID 40 dB attenuator and an API/Weinschel 53-40-33 40 dB attenuator at the upstream end of a European XFEL-type, 2 m long stripline kicker.

The measured timing jitters were around (30–40) ps with no significant dependence on temperature or operation time. It should be noted that the measured timing jitter is of the same order of magnitude as the measurement resolution.

Beam positions of bunches deflected using this combination of an existing kicker and the FID test pulser are shown in figure 4. In the recorded positions at 3 different downstream beam

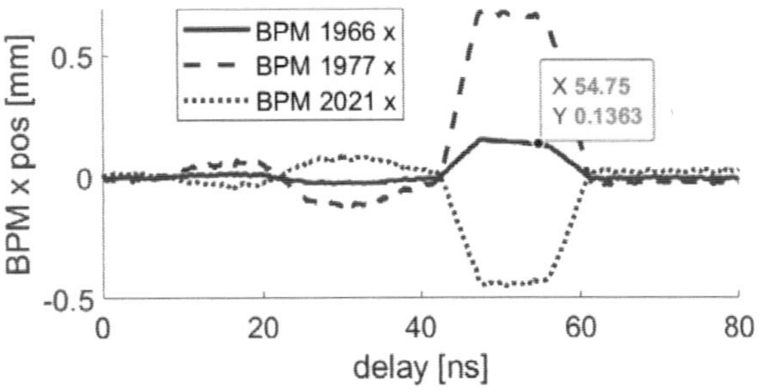

Figure 4. Measured transverse bunch positions for bunches deflected using a 2 m long stripline kicker at the European XFEL and the HV pulses shown in figure 3. The positions are measured at three different positions downstream the kicker position at 1960 m in the XFEL beamline, with the exact measurement positions given in the legend. For the two positions further downstream, several quadrupole magnets were used in between the kicker and the measurement position.

position monitors (BPMs) the same behaviour is seen: the deflection rises at a timing delay between ca. (60–40) ns and levels at a peak deflection with only minor dependence on the delay. This minor dependence is due to the bunches being deflected at a position more upstream in the stripline and therefore a larger drift distance to the BPM. This slope therefore also becomes less pronounced when measuring at BPMs further downstream.

At later other delays (¡ 40 ns) two more deflection peaks were measured. The exact cause of these peaks is yet to be understood, but probably internal reflections due to insufficient electrical matching are responsible for this behaviour. Such reflections have been observed in TDR measurements and are likely caused by several ceramic support stems and vacuum ports in the stripline setup.

6. Summary and outlook

Stripline kickers are being developed to enable single bunch top-up injection in PETRA IV. The simulated residual kick strengths on bunches upstream and downstream of the kicked bunch are below 25 % of the nominal kick strength. A total of 3 bunches out of ¿3000 stored bunches are expected to be disturbed. It is also planned to use a similar kicker design for the transverse multi-bunch feedback system of PETRA IV. Currently, a first prototype is being built and measurements are planned for the early second half of 2022. Based on the experiences with this first setup, dedicated prototypes of the final injection and feedback kicker will be built and shall be tested under operational conditions in an existing storage ring before production of the required ca. 20 injection kicker modules and up to 16 feedback kicker modules.

Acknowledgments

The authors would like to thank M. Jebramcik, J. Kahl, I. Agapov from DESY and M. Sack from KIT for useful discussions. The provision of a high voltage test pulser by FID GmbH is also highly appreciated.

References

[1] Schroer C *et al.* 2018 *J. Synchrotron Rad.* **25** pp 1277–90
[2] Loisch G *et al.* 2021 *Proc. 12th Int. Particle Accelerator Conf. (IPAC'21)* (Campinas, Brazil) pp 2863–65
[3] Olsson D *et al.* 2017 *Proc. 8th Int. Particle Accelerator Conf. (IPAC'17)* (Copenhagen, Denmark) pp 4285–88
[4] Wang L *et al.* 2021 *Nucl. Instrum. Methods Phys. Res., Sect.* A **992** 165040
[5] Waldron W *et al.* 2021 *Proc. 12th Int. Particle Accelerator Conf. (IPAC'21)* (Campinas, Brazil) pp 3487–89

[6] Paraliev M *et al.* 2021 *Proc. 12th Int. Particle Accelerator Conf. (IPAC'21)* (Campinas, Brazil) pp 2889–92
[7] Risch R and Biela J 2021 *Proc. IEEE Pulsed Power Conf. (PPC2021)* (San Antonio, Texas, USA) pp 1–12
[8] Risch R, Hu A and Biela J 2020 *Proc. 11th Int. Conf. Integrated Power Electr. Sys. (CIPS2020)* (Berlin, Germany) pp 1–6
[9] Hochberg M *et al.* 2016 *Proc. PCIM Europe 2016* (Nuremberg, Germany) pp 1578–83
[10] Sack M *et al.* 2021 *Proc. PCIM Europe digital days* (Nuremberg, Germany) pp 40–46
[11] Grekhov I V *et al.* 1985 *Solid State Electr.* **28** pp 597–99

Journal of Physics: Conference Series **2420** (2023) 012085 doi:10.1088/1742-6596/2420/1/012085

Ultra-fast generator for impact ionization triggering

A A del Barrio Montañés [1,2], V Senaj[1], T Kramer[1], Y Dutheil[1] and M Sack[2]

[1] CERN, Geneva, Switzerland
[2] KIT, Karlsruhe, Germany

E-mail: alicia.ana.del.barrio.montanes@cern.ch

Abstract. Impact ionization triggering can be successfully applied to standard thyristors, thus boosting their dI/dt capability by up to 1000x. This ground-breaking triggering requires applying significant overvoltage on the anode-cathode of a thyristor with a slew rate $> 1\,\mathrm{kV/ns}$. Compact pulse generators based on commercial off-the-shelf (COTS) components would allow the spread of this technology into numerous applications, including fast kicker generators for particle accelerators. In our approach, the beginning of the triggering chain is an HV SiC MOS with an ultra-fast super-boosting gate driver. The super boosting of a 1.7 kV rated SiC MOS allows to reduce the MOS rise time by a factor of > 26 (datasheet tr $= 20\,\mathrm{ns}$ vs. measured tr $<$ 800 ps), resulting in an output voltage slew rate $> 1\,\mathrm{kV/ns}$ and an amplitude $> 1\,\mathrm{kV}$. Additional boosting is obtained by a Marx generator with GaAs diodes, reaching an output voltage slew rate $> 11\,\mathrm{kV/ns}$. The final stage will be a Marx generator with medium size thyristors triggered in impact ionization mode with sufficient voltage and current rating necessary for the triggering of a high power thyristor. This paper presents the impact ionization triggering of a small size thyristor.

1. Introduction

In beam transfer facilities for particle accelerators, magnets generally require multi kV/kA pulse generators. Presently this is achieved using thyratrons. However, these devices present several disadvantages such as erratic firing, their need of complex triggering and biasing electronics, their rarity, approaching obsolescence and high cost. These factors are driving their replacement by semiconductor switches.

Recent investigations have shown that switching semiconductors using impact ionization mode is a viable option to replace thyratrons. Impact ionization is achieved when a high dU/dt creates an ionization wavefront in the semiconductor structure. This ultra fast switching technique is even feasible in cheap commercially available thyristors [1]. Thyristors are cheaper and faster than thyratrons (when triggered in impact ionization mode), and offer a higher current density compared to thyratrons.

The following conditions are necessary to achieve impact ionisation on a thyristor:

- A voltage applied by the triggering circuit on the anode-cathode of the thyristor more than double the static breakdown voltage
- Said triggering voltage requires a slew rate of $>1\,\mathrm{kV/ns}$ in order to create the right conditions in the semiconductor structure

- All this while providing enough current to charge the parasitic capacitance of the thyristor with the required dU/dt.

In this paper, we present a Marx generator based topology that achieves impact ionization triggering in a small size (D2PAK) thyristor.

2. Methodology

Traditionally, as seen in [2], the triggering generators for impact ionization used drift step recovery diodes (DSRD) and semiconductor opening switch (SOS) diodes. Generators based on both of these components are bulky, require relatively long pre-charging and more importantly, DSRD and SOS themselves are not commercially available. This is why we propose an alternative approach, based on three main stages.

The first stage has been described in [3] and [4]: ultra-fast MOSFET triggering based on the upgraded gate boosting technique outlined by M. Hochberg *et al.* [5]. M. Azizi *et al* presented in [6] a gate-boosting driving method optimized for a SiC MOSFET to reduce its turn-on time. Our SiC MOSFETs were more aggressively gate-boosted, resulting in an acceleration of their rise time by a factor of >26, depending on the device.

In order to more aggressively boost the MOS device, as per the depicted method in [4], we developed a driver that can deliver 430 V/ns onto a 50 Ω load, with a maximum output voltage >300 V. The resulting output voltage is presented in figure 1, indicated by the blue lines.

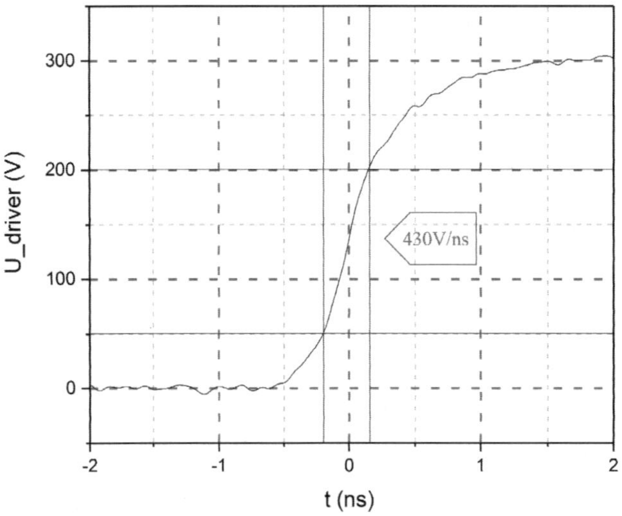

Figure 1. Output voltage of our MOS driver.

This SiC MOSFET, the second stage of the series, was tested onto a 50 Ω load (by mistake up to 2.1 kV, and survived) at the nominal voltage of 1.3 kV and it produced a 1.2 kV/ns slew rate, as seen in figure 2. Once these parameters are achieved, we can use this pulse generator for impact ionization triggering in the third stage.

The third stage is an ultra-fast Marx generator with GaAs diodes switched in impact ionization mode acting as switches. As seen in figure 3, this generator is made up of multiple stages of 900 V rated GaAs diodes [7]; their number depends on the required output voltage amplitude. The triggering of this Marx generator is conducted by a super-boosted SiC MOSFET

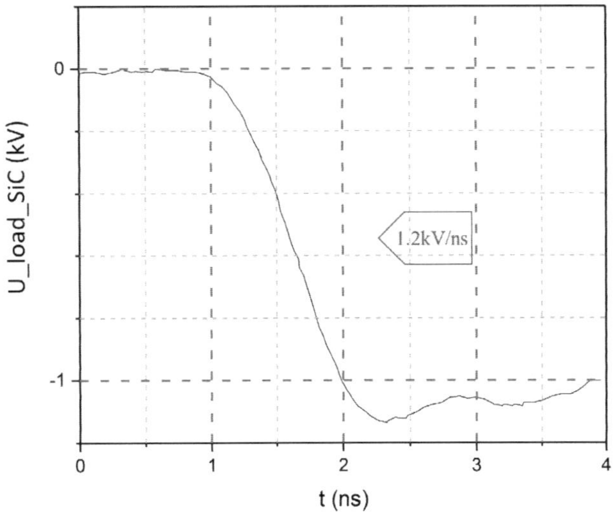

Figure 2. SiC MOSFET voltage output on 50 Ω load.

(Infineon IMBF170R450M1), thanks to the method that has been described above. The negative voltage generated by the MOSFET is applied on the anode of the GaAs diodes.

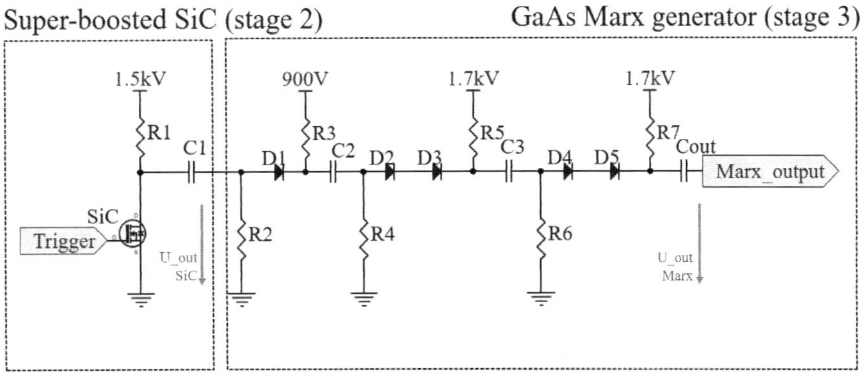

Figure 3. Schematic of stages 2 and 3.

The best results were obtained using GaAs diodes switched in impact ionization mode from the wavefront initiated by the SiC MOSFET (figure 2). However, the same principle can be applied to small thyristors.

With the pulse generated at the output of the three series stages of the Marx generator, a negatively biased thyristor is triggered. The chosen component is VS-16TTS12S-M3 from Vishay, a 1.2 kV rated device. The configuration is shown on figure 4.

3. Results

The different tests have been done using a 6 GHz oscilloscope with a series of 10-18 GHz, 50 Ω attenuators used as output loads.

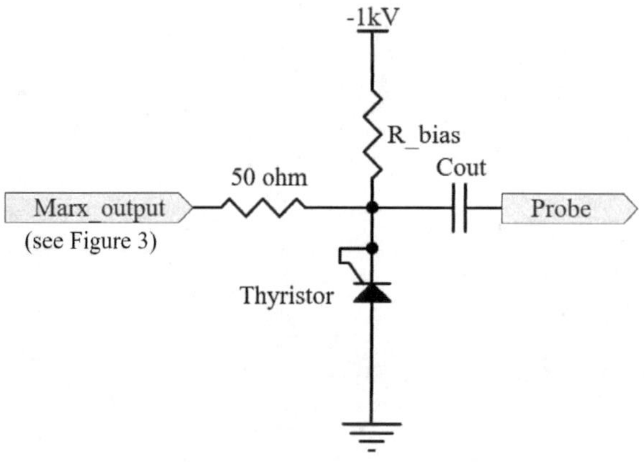

Figure 4. Schematic of Marx generator triggering a thyristor.

Figure 5 shows the output voltage of the Marx generator. The maximum voltage reached is 4.1 kV (negative) and the slew rate, more than 11 kV/ns on a 50 Ω resistive load.

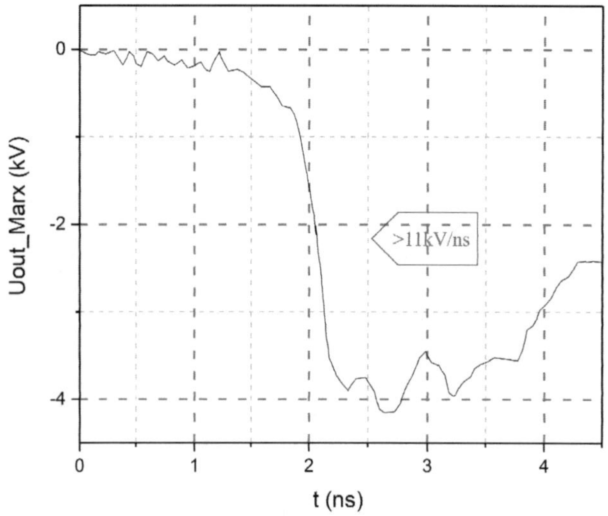

Figure 5. Output waveform of the three stage Marx generator on 50 Ω load.

It can be expected that the magnitudes and output pulse will diminish when triggering a capacitive load, which is what a thyristor represents to the generator. However, it is considered that for a 1.2 kV rated thyristor, the generator output is high enough to ensure that impact ionization will also occur on this thyristor.

As represented on figure 6, this assumption is correct. The phenomenon of impact ionization happening on the thyristor is clearly seen, in blue in figure 6, with an ultra-fast thyristor commutation of ≈ 200 ps, this being on the limit of resolution of our instrumentation.

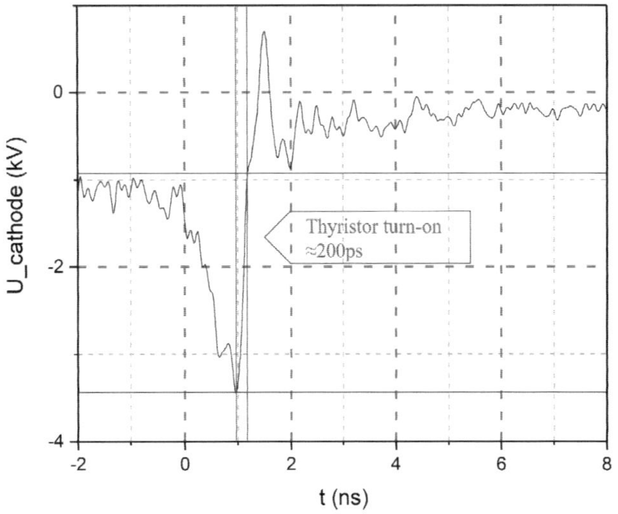

Figure 6. Impact ionization triggering of a thyristor in our laboratory.

To summarise, an innovative Marx generator is presented in this publication. With three stages of GaAs diodes, a slew rate of 11 kV/ns, and a peak output voltage on a resistive load of 4.1 kV are reached thanks to the applied ultra-fast impact ionization process.

Therefore, this Marx generator is considered to be sufficiently fast to also trigger a small-sized thyristor in impact ionization mode.

Consequently, a 1.2 kV rated, D2PAK thyristor was connected at the output of the Marx generator; with a biasing of 1 kV. The output voltage reaches an ultra-fast commutation slew rate higher than 17 kV/ns and a commutation rise time < 200 ps.

4. Conclusion

Impact ionization was achieved, proving the feasibility of this triggering method for thyristors. Furthermore, this triggering technology enables the use of conventional thyristor switches for the construction of a fast HV pulse generator – again in a Marx architecture.

With this research, we lay the foundations for a future fast multi-stage thyristor Marx generator needed to achieve the high voltage and current necessary to trigger larger thyristor stacks in impact ionisation mode. Finally, this could replace thyratrons in beam transfer facilities for particle accelerators.

Acknowledgments

This work has been sponsored by the Wolfgang Gentner Programme of the German Federal Ministry of Education and Research

References

[1] Gusev A *et al.* 2019 *Proc. 2019 IEEE Pulsed Power and Plasma Science (PPPS)* (Orlando, FL USA) pp 1–4

[2] Grekhov I V and Mesyats G A 1999 Physical basis for high power semiconductor nanosecond opening switches Tech. rep. Digest of Technical Papers 12th IEEE International Pulsed Power Conference (Cat. No.99CH36358) pp 1158–61 vol.2

[3] Senaj V, Kramer T, M. Sack M and del Barrio Montañés A A 2021 *Proc. 12th Int. Particle Accelerator Conf. (IPAC'21)* (Campinas, Brazil) pp 4454–56

[4] del Barrio Montañés A A, Senaj V, Kramer T and Sack M 2021 *Euro-Asian Pulsed Power Conference (EAPPC-BEAMS 2020)* (Biarritz, France)

[5] Hochberg M, Sack M and Mueller G 2016 *Proc. 2016 IEEE Int. Power Modulator and High Voltage Conf. (IPMHVC)* pp 171–75

[6] Azizi M, van Oorschot J J and Huiskamp T 2020 *IEEE Trans. Plasma Sci.* **48** 4262

[7] 3-5 Power Electronics Diskrete Bauelemente `http://3-5pe.com/de/rodukte-services/diskrete-bauelemente/`

Journal of Physics: Conference Series **2420** (2023) 012086 doi:10.1088/1742-6596/2420/1/012086

Branch module for an inductive voltage adder for driving kicker magnets with a short circuit termination

J Ruf[1,2], M J Barnes[1], Y Dutheil[1], T Kramer[1] and M Sack[2]
[1]CERN, 1211 Geneva, Switzerland
[2]KIT, 76131 Karlsruhe, Germany

E-mail: johannes.ruf@cern.ch

Abstract. For driving kicker magnets terminated in a short circuit, a branch module for an inductive voltage adder has been designed and assembled. The module has been designed for a maximum charging voltage of 1.2 kV and an output current of 200 A considering the current doubling due to the short circuit termination. It features three consecutive modes of operation: energy injection, freewheeling, and energy extraction. Therefore, the topology of the branch module consists of two independently controlled SiC MOSFET switches and one diode switch. In order not to extend the field rise time of the kicker magnet significantly beyond the magnet fill time, the pulse must have a fast rise time. Hence, the switch for energy injection is driven by a gate boosting driver featuring a half bridge of GaN HEMTs and a driving voltage of 80 V. Measurements of the drain source voltage of this switch showed a fall time of 2.7 ns at a voltage of 600 V resulting in a voltage rise time of 5.4 ns at the output terminated with a resistive load. To meet both the rise time and current requirements, a parallel configuration of four SiC MOSFETs was implemented.

1. Introduction

At present, installations at CERN, for driving kicker magnets equipped with a short-circuit termination, comprise pulse generators based on transmission lines [1]. For some of these transmission-line based generators, a future replacement by an inductive voltage adder is currently under investigation. The inductive voltage adder comprises branch modules which are paralleled in layers for sharing the load current and stacked for adding up the voltage. The inductive voltage adder usually feeds into a matched load. Thereby, the match between the power to be fed into the load impedance and the power delivered by one module is made by an appropriate selection of the number of modules per layer and the number of stacked layers. When driving a transmission line type kicker magnet terminated in a short circuit, with a low impedance source like an inductive adder, the following mode of operation can be used to deal with the reflections from the termination [2]. Initially, the generator outputs a voltage that travels as a wave along the connecting cable and the transmission line type kicker magnet. The wave is reflected at the short circuit termination and travels back to the pulse generator, causing a doubling of the current. Thereby, energy is fed into the system comprising the connecting cable and the kicker magnet and is stored as magnetic energy. Once the reflected wave reaches the pulse generator, the generator stops feeding energy into the system. Hence, the voltage over the output of the pulse generator settles to zero, disregarding losses. Finally, at the end of the pulse, the energy is extracted from the load.

2. Circuit topology of the branch module

In order to implement the functionality described above, the branch modules of the inductive voltage adder need to be equipped with an appropriate circuitry. Figures 1 and 2 show a simplified schematic with an appropriate arrangement of MOSFET switches, and a capacitor C_{pulse} to store initial energy. The circuit elements in figures 1 and 2 can each represent multiple paralleled components.

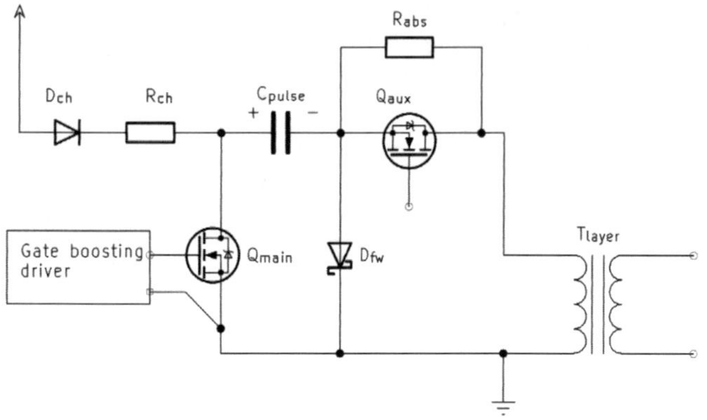

Figure 1. Simplified circuit diagram of one branch module with a resistor R_{abs} for dissipating the energy. ($R_{abs} = 10\,\Omega$; $Q_{main} = 25\,\mu F$; Q_{main}: 4 × G3R160MT17J; Q_{aux}: 2 × C2M0045170P; D_{fw}: 2 × GB25MPS17-247).

Initially, the capacitor is charged via the circuit elements D_{ch}, R_{ch}, and D_{fw}. Thereby, D_{ch} prevents undesired discharging. To transfer energy from the capacitor to the load, consisting of the cable and the kicker magnet, both MOSFET switches Q_{main} and Q_{aux} are closed. Thereby, the transformer T_{layer} is part of the inductive adder and serves for combining the current originating from the branch modules of one layer. The output wave from T_{layer} travels through the kicker magnet to the short circuit. When the wave reflected from the short circuit reaches the module, switch Q_{main} opens, and the current commutates into the freewheeling diode D_{fw}. For extracting the energy at the end of the pulse switch Q_{aux} opens, and the energy stored in the load dissipates in the resistor R_{abs}.

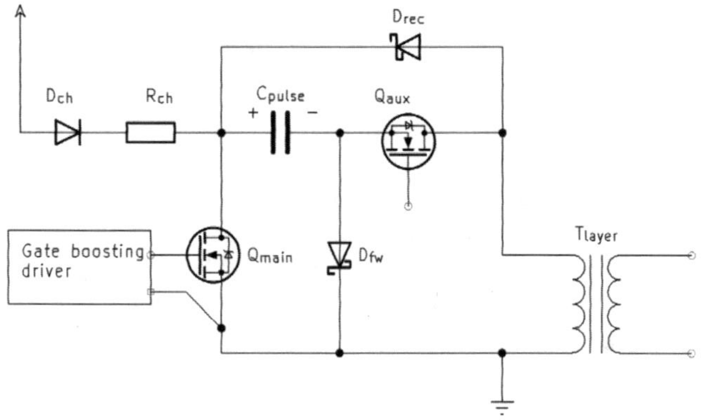

Figure 2. Simplified circuit diagram of one branch module with a diode D_{rec} for recuperating the energy. (D_{rec}: 2 × GB25MPS17–247. For other main circuit parameters see figure 1).

The magnet fill time describes the time the travelling wave needs to propagate twice, i.e., forward and backward, through the short-circuit terminated kicker magnet. The field rise time comprises the magnet fill time and the rise time of the pulse. As a short field rise time is desirable, a fast rise of the pulse is crucial. Therefore, the MOSFET Q_{main} is driven by a gate-boosting circuit. It closes slightly after the switch Q_{aux} has closed and, therefore, defines the rise time of the initial pulse. In addition, its turn off defines the time for the current to commutate from Q_{main} to D_{fw}. MOSFET Q_{aux} is driven by a commercial gate driver and, hence, has a slower switching speed. Moreover, the gate driver for Q_{aux} features an insulation versus the ground reference for Q_{main}. The branch module has been designed for a voltage of 1.2 kV and a peak current of 200 A with a pulse repetition rate of 10 Hz. Switch Q_{aux} and the diode D_{fw} need to be capable of carrying this current. Hence, Q_{aux} has been implemented as two SiC MOSFETs of the type C2M0045170P [3] in parallel configuration. Similarly for the diode D_{fw} two paralleled diodes of the type GB25MPS17-247 [4] have been selected. The switch Q_{main} carries only the initial current before current doubling due to reflection of the wave front at the short circuit, i. e. 100 A. For this switch Q_{main} four paralleled MOSFETs of type G3R160MT17J [5] have been selected (see figure 3).

3. Gate-boosting driver

Figure 3 shows a simplified schematic of the implemented gate-boosting driver circuit. It features an additional capacitance (C_{gb}) in series to the gate, compensating the parasitic inductance of the gate circuit, and forming a capacitive divider with the gate capacitance [6]. A GaN-HEMT half bridge (Q_{gb1}, Q_{gb2}) is fed by a voltage of 80 V to compensate for the effect of the capacitive voltage divider. The resistors $R_{d,b}$ ensures equal current distribution among the four paralleled MOSFETs of Q_{main}. The resistors $R_{d,a}$ and $R_{d,b}$ also provide sufficient damping for operation without an overshoot of the gate voltage.

$$C_{gb} = \frac{C_g}{\frac{V_{gd}}{V_{g,nom}} - 1} \tag{1}$$

The capacitance C_{gb} can be calculated according to equation (1) based on effective gate capacitance C_g, the desired gate voltage $V_{g,nom}$, and the available gate driving voltage V_{gd}.

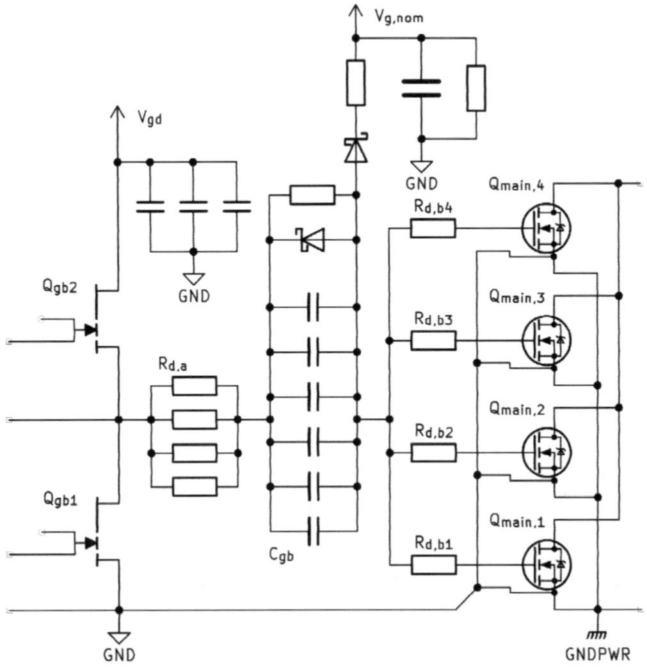

Figure 3. Simplified circuit diagram of the gate-boosting driver for Q_{main}. ($V_{gd} = 80$ V; $V_{g,nom} = 15$ V; $Q_{gb1,2}$: GS66506T; $R_{d,a} = 0.75\,\Omega$; $C_{gb} = 2.52$ nF; $R_{d,b1..4} = 3\,\Omega$; $Q_{main,1..4}$: G3R160MT17J).

4. Measurement results

For initial tests, one module has been directly connected to a resistive load without a transformer. Figure 4 shows a photo of the test setup. The value of the resistive load has been selected to be 10.8 Ω; this results in a current corresponding to the designed operating point of the module in an inductive voltage adder before the arrival of the reflected wave front. The current through the load has been measured by means of a shunt resistor with a bandwidth of 1.2 GHz. For voltage measurements voltage probes of type 10076C [7] and tpp0500 [8] have been employed. The measurements were performed with an oscilloscope with a bandwidth of 500 MHz. This results in a rise time of the voltage measurement setup of approximately 1 ns [9] which is considered to be sufficiently fast for the measurements.

Figure 4. Photo of the test setup.

The tests have been performed at a voltage of 600 V as well as at the rated voltage of 1.2 kV. Figure 5 shows the voltage at the gate of one of the MOSFETS comprising Q_{main}, measured where the gate and kelvin-source leads enter the housing: it rises within 2.8 ns to a voltage of 15 V. The voltage between the drain and source of one of the two center MOSFETs exhibits a fall time between 90 % and 10 % of 2.7 ns for a charging voltage of 600 V and 3.4 ns for a charging voltage of 1.2 kV as shown in figure 6. Figures 7 and 8 show the voltage and current measurements at the output of the module. The rise time (10 % to 90 %) for both voltage and current has been measured to be 5.4 ns for a charging voltage of 600 V and 5.0 ns for a charging voltage of 1.2 kV as shown in figure. 7. For the tests the total pulse length has been adjusted to 400 ns. The pulse to pulse amplitude stability with this initial test setup has been measured to 461 ppm over 1032 pulses at 1.2 kV.

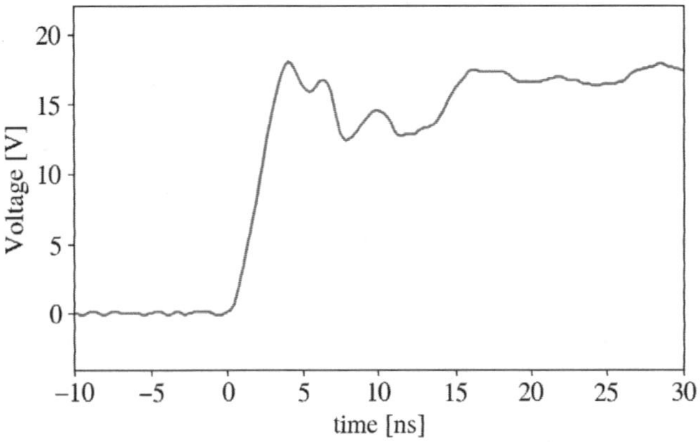

Figure 5. V_{gs} of MOSFET Q_{main}.

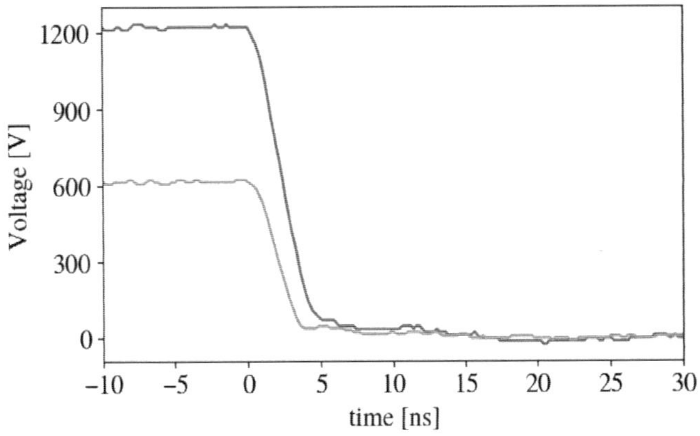

Figure 6. V_{ds} of MOSFET Q_{main} for a charging voltage of 600 V (orange) and 1.2 kV (blue) both with a resistive load of 10.8 Ω.

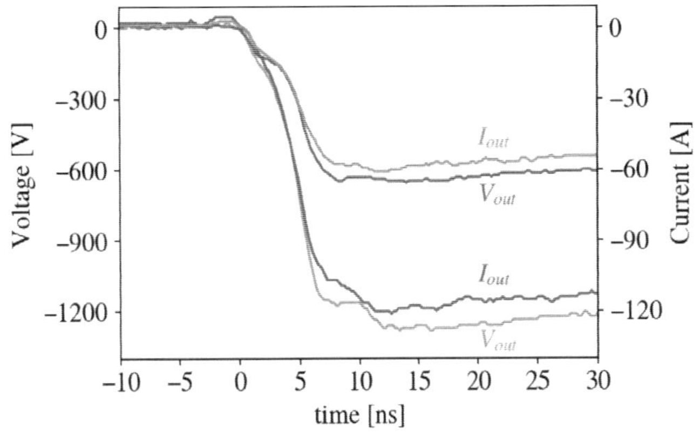

Figure 7. Output voltage and load current of the branch module for a charging voltage of 600 V (blue and orange) and 1.2 kV (cyan and red), both with a resistive load of 10.8 Ω.

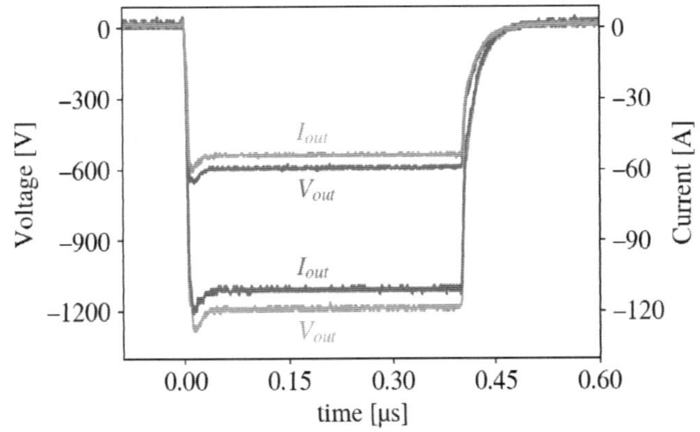

Figure 8. Output voltage and load current of the branch module for a charging voltage of 600 V (blue and orange) and 1.2 kV (cyan and red), both with a resistive load of 10.8 Ω.

5. Conclusion

A branch module for an inductive voltage adder for driving a kicker magnet terminated in a short circuit has been designed and built. It features two active SiC MOSFET switches and a diode switch. A gate driver for MOSFET Q_{main} featuring gate-boosting allows for operation with a fast rise time. The module has been tested connected to a resistive load. At a voltage of 1.2 kV and a current of 110 A a rise time of 5.0 ns has been measured. As the next step the module will be tested with a replacement load emulating a kicker magnet terminated in a short circuit.

Acknowledgments
This work has been sponsored by the Wolfgang Gentner Programme of the German Federal Ministry of Education and Research.

References
[1] Barnes M J 2018 *Proc. CAS–CERN Accelerator School* vol. 5/2018 CERN-2018-008-SP
[2] J. Ruf, M.J. Barnes, T. Kramer, and M. Sack 2021 *Proc. 12th Int. Particle Accelerator Conf. (IPAC'21)* (Campinas, Brazil) pp 3521–24
[3] CREE 2021 Datasheet C2M0045170P Rev. -, 04-2018
[4] GeneSiC 2020 Datasheet GB25MPS17-247 Apr. 20 Rev 1.4
[5] GeneSiC 2021 Datasheet G3R160MT17J Rev 21/Feb
[6] Sack M, Hochberg M, Herzog D, and Mueller G 2021 *Proc. PCIM Europe digital days 2021; International Exhibition and Conference for Power Electronics, Intelligent Motion, Renewable Energy and Energy Management* p 1–7
[7] Keysight 2021 InfiniiVision Oscilloscope Probes and Accessories Rev May 18, 2021
[8] Tektronix, TPP0500 and TPP1000 500 MHz and 1 GHz 10X Passive Probes Instructions Revision A
[9] A. Schwab 1981 *Hochspannungsmesstechnik* zweite Auflage (New York:Springer)

Proposal of a VHEE Linac for FLASH radiotherapy

L Giuliano[1], F Bosco[1], M Carillo[1], D De Arcangelis[1], L Ficcadenti[1], D Francescone[1], A De Gregorio[1], G Franciosini[1], M Migliorati[1], A Mostacci[1], L Palumbo[1], V Patera[1], A Sarti[1], D Alesini[2], M Behtouei[2], L Faillace[2], A Gallo[2], B Spataro[2], A Vannozzi[2], P Cirrone[3], G Cuttone[3], G Torrisi[3], M G Bisogni[4], J H Pensavalle[4], F Di Martino[5], V Favaudon[6], S Heinrich[6] and A Patriarca[6]

[1]Sapienza University, Rome, Italy, also INFN-Roma, Rome, Italy
[2]INFN-LNF, Frascati, Italy
[3]INFN-LNS, Catania,Italy
[4]INFN-Pisa, Pisa, Italy
[5]Azienda Ospedaliera Pisana, Pisa , Italy
[6]Curie Institute, Orsay, France

E-mail: lucia.giuliano@uniroma1.it

Abstract. Translation of electron FLASH radiotherapy in clinical practice requires the use of high energy accelerators to treat deep tumours and Very High Electron Energy (VHEE) could represent a valid technique to achieve this goal. In this scenario, a VHEE FLASH linac is under study at the University La Sapienza of Rome (Italy) in collaboration with the Italian Institute for Nuclear Research (INFN) and the Institut Curie (France). Here we present the preliminary results of a compact C-band system aiming to reach an high accelerating gradient and an high pulse current necessary to deliver high dose per pulse and ultra-high dose rate required for FLASH effect. We propose a system composed of a low energy high current injector linac followed by a modular section of high accelerating gradient structures. CST code is used to define the required LINAC's RF parameters and beam dynamics simulations are performed using TSTEP and ASTRA.

1. Introduction

In 2014 Favaudon et al. [1] discovered a new effect, called FLASH, that can change the scenario of the radiotherapy. Several pre-clinical studies demonstrated that a high dose per pulse (1-10 Gy) with ultra high dose-rate ($> 10^6$ Gy/s) and a total irradiation time < 100 ms, decreases dramatically the toxicity in the healthy tissue while keeping the same efficacy in cancer cure [2]. The technology of the accelerators actually used in the radiotherapy does not allow to explore this novel approaches in terms of dose and dose-rate, therefore a new low energy LINAC (7 MeV) [3,4] has been developed by our team in collaboration with S.I.T. - Sordina IORT Technologies S.p.A. [5] and has been installed at Institut Curie (Orsay-France) in August 2020, in order to investigate the FLASH effect.

A strong interest has also grown on high energy machines (60-150 MeV) able to treat deep tumours. Recent studies [6–8] have investigated the potential of the VHEE for clinical applications, showing that 60–150 MeV beams could become an alternative modality in radiotherapy.

The proposed VHEE source is based on a C-band LINAC, working at the frequency of 5.712 GHz, which allows more compactness compared to the traditional S-band cavities. The irises of the cavities have been designed with a radius such to get a high accelerating gradient and at the same time ensure a good transmission efficiency of electrons, necessary to transport the high peak current required for the high dose and dose rate in the FLASH irradiation.

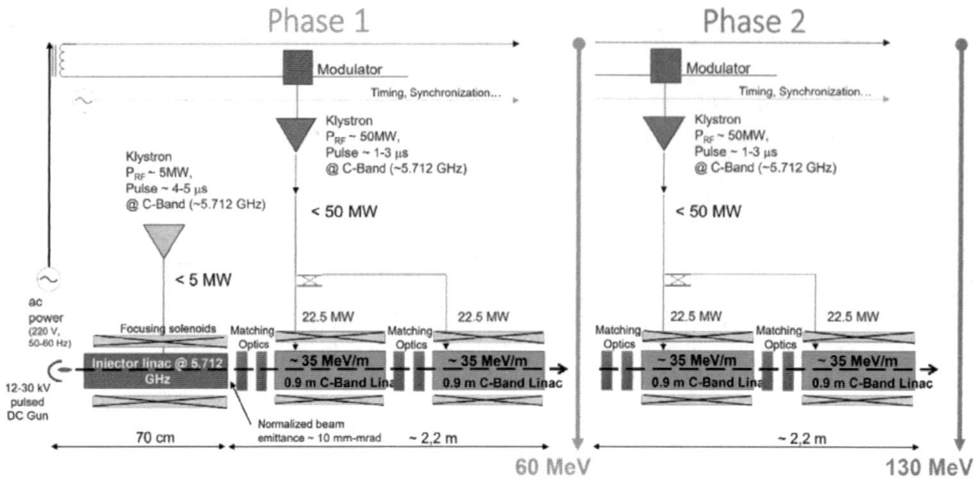

Figure 1. Layout of the VHEE Linear Accelerator for VHEE FLASH radiotherapy with the injector and four TW high-gradient accelerating structures.The maximum expected beam energy is about 130 MeV.

The whole system is thought to be realize in two successive phases (phase 1 and phase 2) as show in figure 1. In phase 1, we use an injector (standing wave LINAC) and two traveling wave accelerating structures 90 cm long. The injector accelerates, at an energy of 10 MeV, a current of 200 mA generated by a pulsed DC gun. The RF design of the C-band injector is essentially based on the experience gained with above mentioned S band low energy LINAC, scaling properly the whole geometry. The electron beam is injected into a compact traveling wave (TW) C-band structure characterized by a high accelerating gradient ($\sim$ 40 MeV/m without the beam) able to bring the energy of the electron beam up to about 60 MeV. In phase 2, with two addition accelerating structures the system can reach an energy up to 130 MeV. The injector is powered by a C-band klystron of 5 MW, upstream of a circulator necessary to prevent reflected waves from the accelerating structure that can damage the klystron itself. A 50 MW klystron, foreseen for phase 1, feeds the first two accelerating structures. The power is split such to feed two LINACs by means of a waveguide network. A second 50 MW C-band klystron provides power to the other two LINACs for the phase 2.

2. RF design optimization

The RF accelerator design is performed by using CST 3D code [9] with eigen and frequency domain solver. The electrons are emitted from a triode electron gun with spherical emission area (see figure 2) with a diameter of about 6 mm. The nominal operation energy is around 12 keV adjustable up to 30 keV. The first cavity of the injector is composed of a half cell with an input plate required to adapt the electric field from the gun to the linac (figure 3). The first linac is a Standing Wave (SW) bi-periodic structure, working in $\pi/2$mode with a frequency of operation of 5.712 GHz. The first part of the linac is composed of three SW cavities called

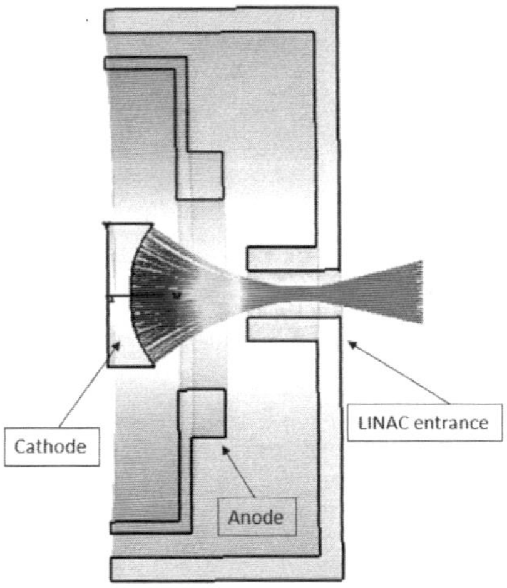

Figure 2. Transverse view of the DC gun. The electrons are emitted from the cathode, pre-accelerated in the gun at energies of about 15 keV and injected into the linac.

bunching section, aimed to generate the bunches and maximize the beam charge capture which results to be around 45%, corresponding to 225 mA.

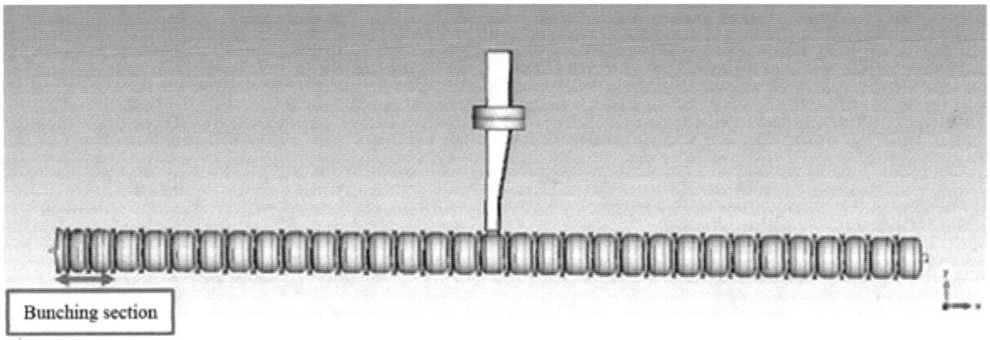

Figure 3. 12 MeV injector linac in C-band with bunching section. The output pulsed beam current is 200 mA.

After the bunching section, the accelerating cells are designed for $\beta \simeq 1$ and the particles reach the required energy after 22 cells.

The high gradient section is a travelling wave (TW) structure operating at 5.712 GHz with a phase advance per cell of $2/3\ \pi$ that guarantees the best efficiency for this type of accelerating structure. A previous experience with the ELI accelerating TW structures was useful to guide in the design of these structures [10–12]. Following preliminary results of beam dynamics, the single cell has been simulated with values of the iris radius "a" of 5 mm and a thickness "t" of 3 mm. Eventually, we performed an optimization of the elliptical shape of the irises, obtaining the best value at $r_1/r_2 = 1.55$. The cell and its geometric parameters are shown in figure 4.

After the single cell's optimization, the final TW structure with same cells was designed integrating the couplers (figure 5). The design procedure is based on an analysis of the phase of the reflection coefficient S11 (figure 6) for different lengths of the short-circuited structure [13].

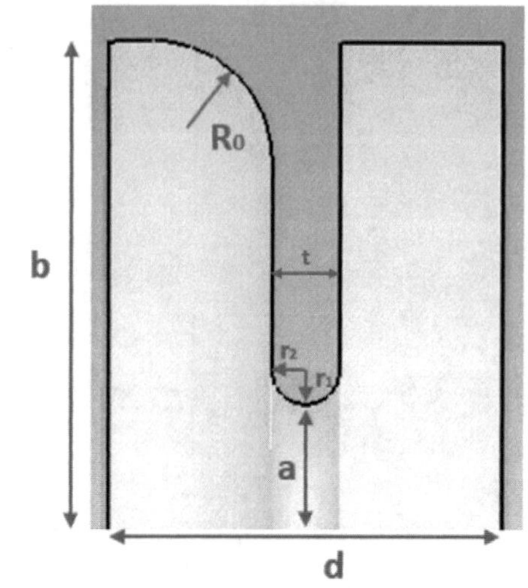

Figure 4. Cell and its geometric parameters: a = 5 mm; b = 20.908 mm; t = 3 mm; d = 17.45 mm; R0 = 5.25 mm; $r_1/r_2 = 1.55$

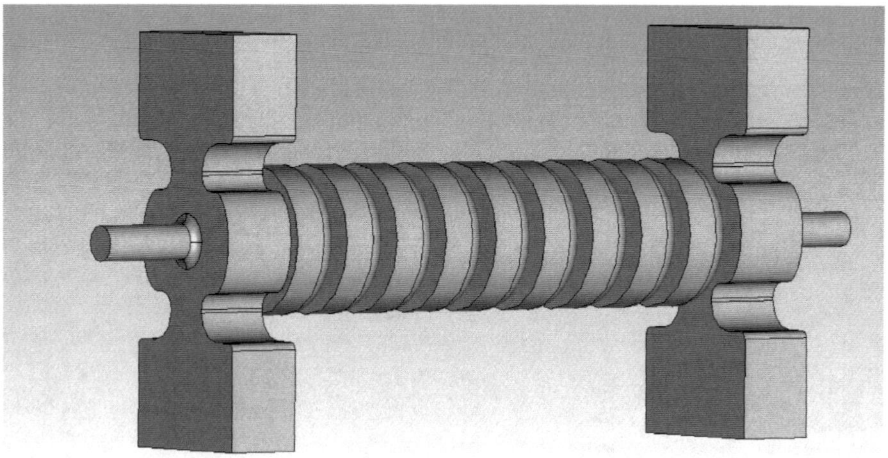

Figure 5. VHEE high gradient Linear Accelerator prototype for VHEE FLASH radiotherapy.

3. Beam dynamics studies

After finishing the layout system, we performed the beam dynamics simulations with the TSTEP [14] and ASTRA [15] codes. In figure 7, it is shown the beam energy gain at full RF power and on-crest operation for each linac. In this case, the maximum achievable energy is about 80 MeV in Phase 1 and 150 MeV in Phase 2, which are well above the minimum required energies for the two phases which are 60 MeV and 130 MeV, respectively. As a result, our proposed system has a very large range of flexibility in the output energy, by easily varying the delivered RF power and phase to each accelerating section. The plot starts from the cathode, located at z = 0 m up to the TW linacs operating at 35 MV/m loaded gradient with 200 mA beam current. In addition, beam dynamics simulations show that no focusing solenoids are required around the accelerating structures. Nevertheless, in our proposed layout, we introduce possible solenoids to distribute along the beamline if necessary, especially from a radiosafety

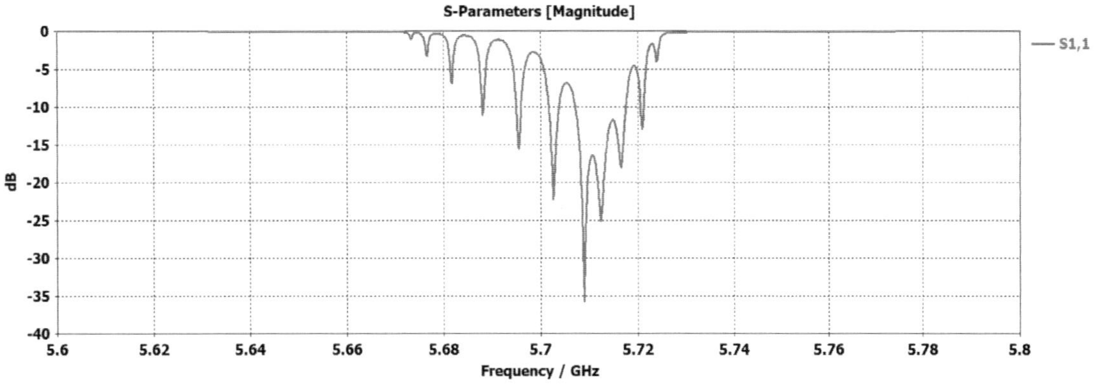

Figure 6. S11 of the TW high gradient structure calculated using CST Studio SUITE code.

point of view.

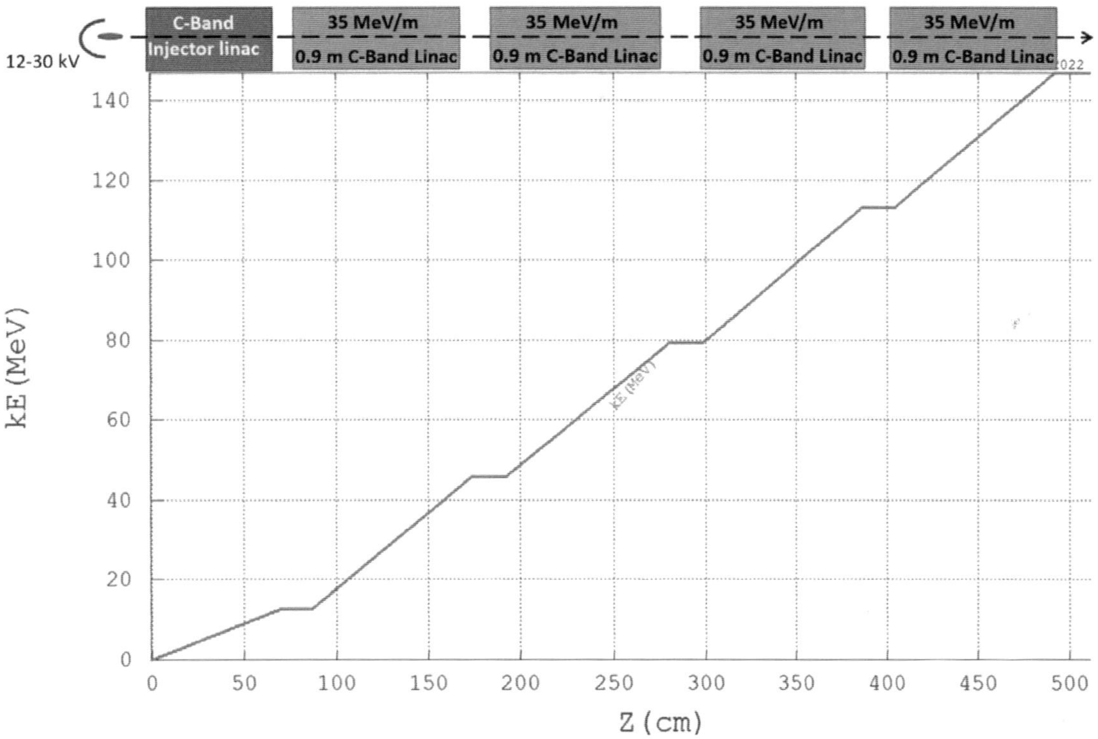

Figure 7. Beam energy gain. The 10 MeV electron beam, from the injector to the last 90 cm long TW linacs with an accelerating loaded gradient of 35 MV/m. The beam current at the exit is 200 mA.

The beam phase-space plots at the linac exit are given in figure 8. The bunch length is around 20 ps FWHM. The energy spread FWHM value is <0.2%. The beam spot shows a FWHM value of 1.8 mm with the total transverse distribution concentrated in about a 4 mm diameter. The normalized transverse beam emittance is in the order of 10 mm-mrad.

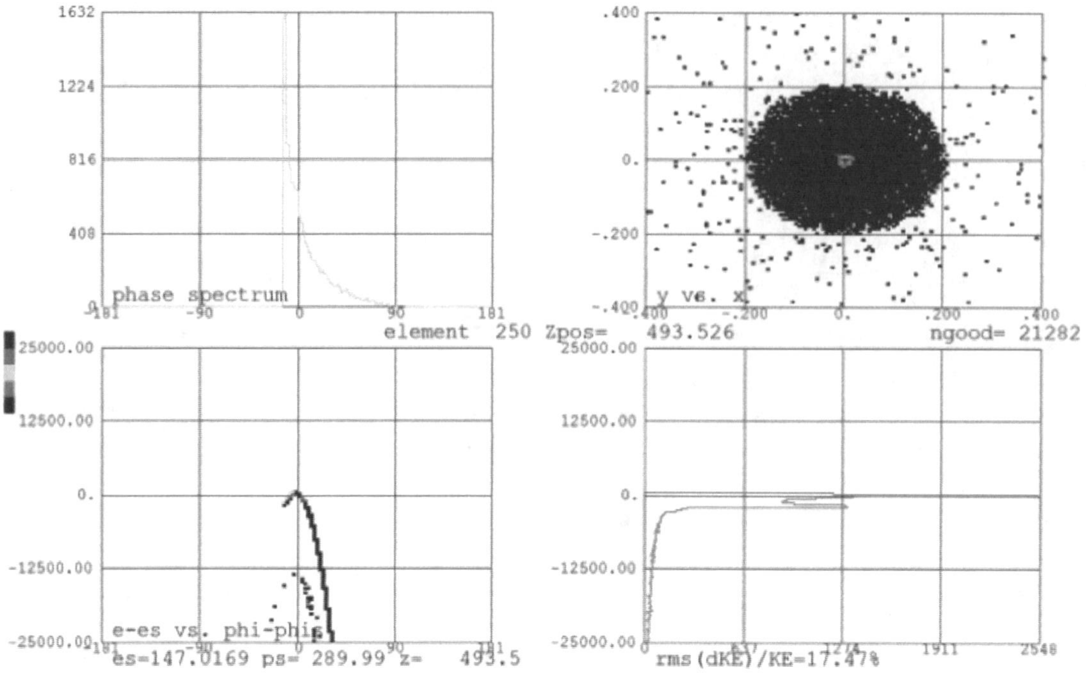

Figure 8. TSTEP output electron beam parameters. In clockwise order: longitudinal phase distribution in degrees; transverse beam spot size in cm; beam energy spectrum in MeV; beam longitudinal phase-space (energy vs. RF phase)

4. Conclusion

In this paper, the design of the VHEE FLASH linac-based machine for a VHEE FLASH-RT Research Laboratory proposed at Sapienza University has been reported. The accelerator is able to deliver high-dose in single pulse thanks to the high current per pulse (> 200 mA), suitable to investigate the FLASH fundamental mechanisms in pre-clinical and radio-biological experiments opening the way to the clinical transfer of the technique. Different strategies for the transverse beam spot size magnification are under study, in order to irradiate large surface as required in the clinical practise. We have developed the RF design and the beam dynamics characterization of the VHEE system optimized to operate with two nominal output electron beam energies: 60 and 130 MeV. Studies are on going to complete the characterization of the machine and to manufacture and test the RF prototypes.

Acknowledgments

The research work has been partially carried out in the framework of the FRIDA project funded by the INFN CSN5.

References

[1] Favaudon V *et al.* 2014 *Sci. Transl. Med.* **6** 245ra293
[2] Esplen N *et al.* 2020 *Phys. Med. Biol.* **65** 23TR03
[3] Martino F Di *et al.* 2021 *Front. Phys.* **8** 630534
[4] Faillace L *et al.* 2021 *Phys. Rev. Accel. Beams* **24** 050102
[5] SIT `https://www.soiort.com`
[6] Schuler E *et al.* 2017 *Med. Phys.* **44** 2544
[7] Subiel A *et al.* 2014 *Phys. Med. Biol.* **59** 5811

[8] Sarti A *et al.* 2021 *Front. Phys.* **11** 777852
[9] CST Studio Suite `www.cst.com`
[10] Alesini D *et al.* 2013 *J. Instrum.* **8** P05004
[11] Alesini D *et al.* 2013 *J. Instrum.* **8** P10010
[12] Alesini D *et al.* 2017 *Phys. Rev. Accel. Beams* **20** 032004
[13] Alesini D, Gallo A, Spataro B, Marinelli A and Palumbo L 2007 *Nucl. Instrum. Methods Phys. Res., Sect. A* **580** 1176
[14] TSTEP `https://tstep.lmytechnology.com/TSTEP.html`
[15] ASTRA `https://www.desy.de/~mpyflo/`

Origin and mitigation of the beam-induced surface modifications of the LHC beam screens

V Petit, P Chiggiato, M Himmerlich, S Marinoni, H Neupert, M Taborelli and L Tavian

European Organization for Nuclear Research, Geneva, Switzerland

E-mail: `valentine.petit@cern.ch`

Abstract. All over Run 2, the LHC beam-induced heat load on the cryogenic system exhibited a wide scattering along the ring. Studies ascribed the heat source to electron cloud build-up, indicating an unexpected high Secondary Electron Yield (SEY) of the beam screen surface in some LHC regions. The inner copper surface of high and low heat load beam screens, extracted during the Long Shutdown 2, was analysed. On the low heat load ones, the surface was covered with the native Cu_2O oxide, while on the high heat load ones CuO dominated at surface, and it exhibited a very low carbon coverage. Such chemical modifications increase the SEY and inhibit a proper conditioning of the affected surfaces. Following this characterisation, the mechanisms for CuO build-up in the LHC beam pipe were investigated on a newly commissioned cryogenic system allowing electron irradiation, surface chemical characterisation by X-ray Photoelectron Spectroscopy and SEY measurements on samples held below 15 K. In parallel, curative solutions against the presence of CuO in the LHC beam screens were explored, which could be implemented in-situ to recover a proper conditioning and lower the beam-induced heat load.

1. Introduction

An unexpected high beam-induced heat load was measured in the beam screen cooling circuit of half of the eight LHC arcs all over Run 2 (2015-2018). This puzzling data was detected first after the Long Shutdown 1 (LS1), a period of maintenance and upgrade including the venting of the cryogenic beam vacuum system to air [1]. Surface analyses were performed on the copper layer of beam screens extracted from the accelerator during the LS2 (2019-2021). The measurements evidenced the presence of native Cu_2O oxide on the surface of the low heat load beam screens, while the high heat load ones were oxidized to non-native CuO oxide and showed an exceptionally low carbon coverage [2]. In addition, the Secondary Electron Yield (SEY) of the beam screens after extraction was found higher when CuO was present. Finally, the room temperature conditioning by electron irradiation of CuO-rich surfaces was observed to be moderate compared to the one of Cu_2O, but not completely inhibited, and CuO was observed to be partially reduced to Cu_2O [2]. This reduction could be hindered or slowed down in the operating conditions of the beam screens, namely between 5 and 20 K, leading to the survival of a high-SEY oxide in some LHC beam screens, to an intense electron cloud activity and to the related high beam-induced heat load. Here, the behaviour of CuO under electron irradiation in cryogenic conditions and the origin of its build-up in some LHC beam screens are assessed. In addition, possible techniques for in-situ CuO reduction are reviewed.

2. Consequence and origin of CuO build-up in the LHC

The cryo-conditioning tests performed in this work were carried out on a new system enabling X-ray photoelectron spectroscopy (XPS), electron irradiation and SEY measurement (part still under commissioning) of samples below 15 K. Irradiation was performed by 250 eV electrons.

2.1. Cryo-conditioning of CuO

The fraction of CuO at the surface of beam screen samples was deduced from XPS measurements of the Cu $2p_{3/2}$ state [3]. CuO fraction evolution during electron irradiation at 250 eV, both at room temperature and 15 K, is shown in figure 1.

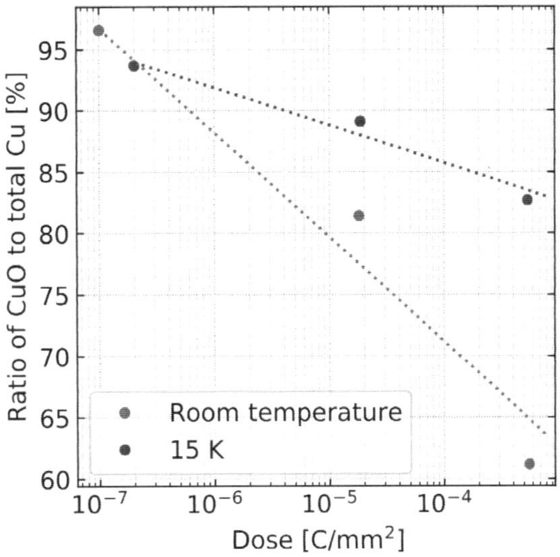

Figure 1. Effect of temperature on CuO reduction under electron irradiation at 250 eV (lines are guide to the eye).

CuO reduction is significantly hindered at 15 K, possibly because of the reduced diffusivity of Cu and O ions in the oxide layer at cryogenic temperature. Therefore, the build-up and the stability of this high-SEY oxide at cryogenic temperature together with the low content of carbon, which limits the graphitic coverage essential for SEY decrease [4, 5], could be responsible for the high beam-induced heat load occurring in some regions of the LHC arcs.

2.2. Origin of CuO build-up

Copper hydroxide $Cu(OH)_2$, which spontaneously grows on air-exposed copper surfaces [6], has been proposed as a precursor for CuO build-up in the LHC [2]. However, room temperature conditioning of airborne $Cu(OH)_2$ layers led to their conversion to Cu_2O, i.e. a reduction to Cu^+, which has a sufficiently low SEY [5]. Nevertheless, also in this case, temperature is expected to play a major role in electron irradiation-induced reactions by affecting diffusion kinetics. Another factor which influences the species formed or decomposed by electron irradiation is the presence of carbon on oxidized Cu surfaces. The absolute carbon coverage is known to influence the chemical transformation during electron bombardment [7] and could thus be decisive to determine the build-up or not of CuO. The observed modifications of air-exposed copper samples during conditioning at room and cryogenic temperatures are reported in figure 2, and the ones for conditioning at 15 K of samples with different carbon coverages are shown in figure 3. In these figures, the evolution of the chemical state of copper during conditioning, as monitored by XPS, is followed for different doses through a Wagner-like plot [2, 3], where the usual Cu $2p_{3/2}$ binding energy axis is replaced by the centre of gravity of this line, to account for the presence of multiple compounds [2]. The position of reference CuO, $Cu(OH)_2$, Cu_2O and Cu(0) samples is shown for comparison.

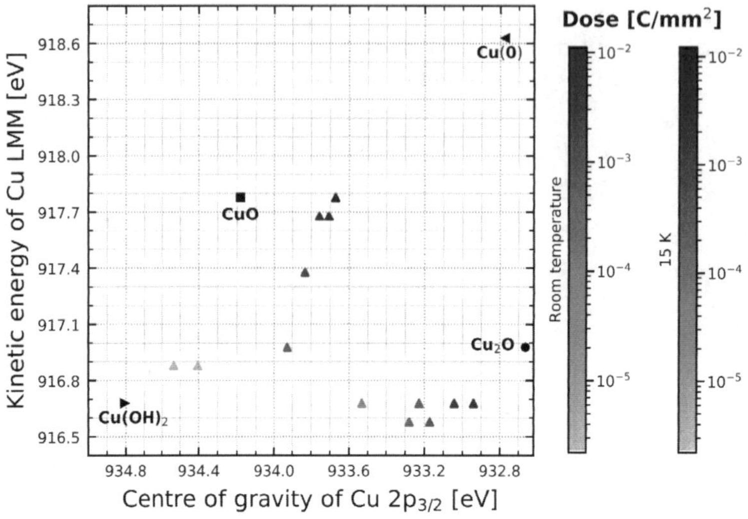

Figure 2. Conditioning of twin air-exposed copper samples at room temperature and 15 K.

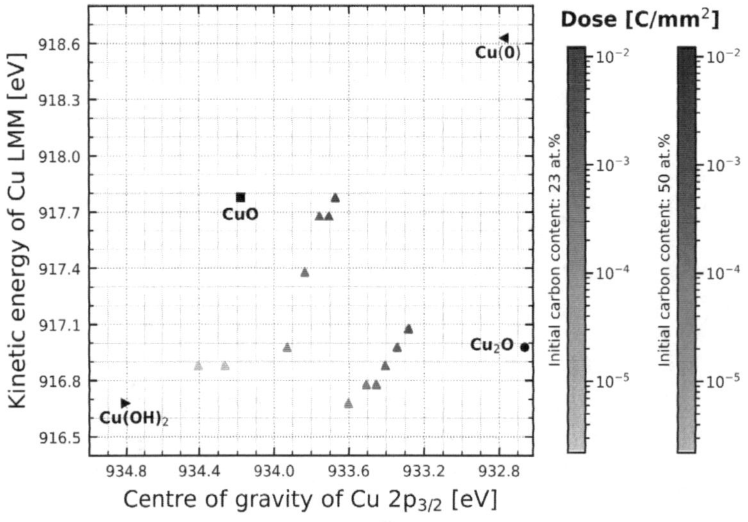

Figure 3. Conditioning at 15 K of air-exposed copper samples with different initial carbon coverages.

Figure 2 clearly shows the influence of temperature on the type of oxide dominating the surface after conditioning. While the conditioning at room temperature provokes the transformation of $Cu(OH)_2$ into Cu_2O, the conditioning at 15 K of a twin sample deviates towards CuO. The formed CuO-like product is similar to the one observed in the LHC-extracted beam screens, where the oxide observed after extraction was even closer to reference CuO in extreme cases [2].

Figure 3 shows the conditioning behaviour at 15 K of two copper samples with initially identical amount of copper hydroxide, but different carbon coverages (23 and 50 at.%). In the presence of high carbon coverage, conditioning leads to significant, but still incomplete, reduction of $Cu(OH)_2$ to Cu_2O, i.e. the dehydration of $Cu(OH)_2$ to CuO is partially avoided. In addition, the carbon content drops to 35 at.%, as expected for a consumption of carbon species during the oxide reduction [7]. At low carbon coverage, $Cu(OH)_2$ is converted to CuO and the amount of carbon drops to 7.5 at.%, which is extremely low and is compatible with the amount of carbon found on the CuO-rich LHC-extracted beam screens [2]. To reproduce the Run 1 – LS1 – Run 2 LHC scheme and propose a scenario leading to CuO build-up only after LS1 in the accelerator, a copper sample exposed to air for more than a year and showing 40 at.% of carbon has been first conditioned at 15 K and then reconditioned at 15 K after 2.5 months of humid air exposure. The corresponding Wagner-like plot is given in figure 4.

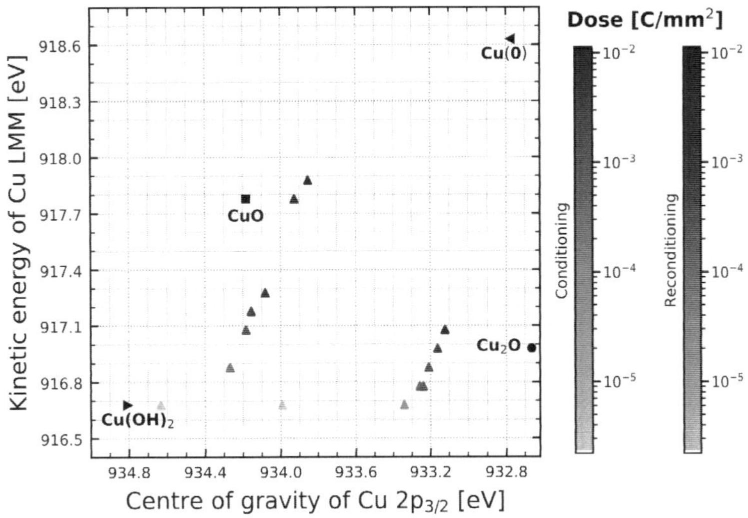

Figure 4. Conditioning and reconditioning at 15 K of an air-exposed copper sample.

Thanks to the initial carbon coverage, the transformation of $Cu(OH)_2$ into CuO is partially inhibited during the first conditioning. In parallel, the carbon content decreases and stabilizes to 31 at.% at the end of this conditioning. After 2.5 months of humid air exposure, $Cu(OH)_2$ coverage is significantly higher than for a non-conditioned sample stored in the same conditions. Indeed, in the current case, the formed layer is even similar to the one of a reference $Cu(OH)_2$ sample, pointing towards an increased surface reactivity after conditioning due to electron irradiation. Subsequently, the second conditioning leads to the transformation of $Cu(OH)_2$ to CuO. This difference of oxidation product formed under electron irradiation could be due to the higher amount of $Cu(OH)_2$ before the second irradiation, but also, as discussed before, to the lower amount of carbon, already partially depleted during the first conditioning. The carbon decrease had stabilized at the end of the first conditioning, but continued to decrease down to 13 at.% during the second irradiation, again compatible with carbon consumption by reaction with oxygen from the oxide or hydroxide [7]. In summary, this conditioning – air exposure –

Journal of Physics: Conference Series **2420** (2023) 012088 doi:10.1088/1742-6596/2420/1/012088

reconditioning sequence and the resulting surface modifications are compatible with the observed heat load increase in some arcs comparing LHC Runs 1 and 2.

3. Mitigation techniques

In parallel to the investigations on the origin of CuO build-up in the LHC, in-situ curing methods have been explored. All of them are based on the removal of CuO from the surface by a reaction with a reducing gas promoted by various sources of energy. For this purpose, samples have been exposed to UV light and to atmospheric pressure plasma, both in 1 bar of N_2-H_2 (95%-5%) mixture. Other samples were exposed to atomic hydrogen generated on a hot tungsten filament (H_2 pressure: 10^{-2} mbar). The Cu $2p_{3/2}$ spectra of a CuO reference sample before treatment, as well as the ones of three samples treated by the different techniques are shown in figure 5. In all cases, a complete CuO reduction is obtained, as witnessed by the complete disappearance of the CuO satellite between 939 eV and 945 eV. These techniques seem therefore promising for an in-situ deoxidation of the LHC beam screens.

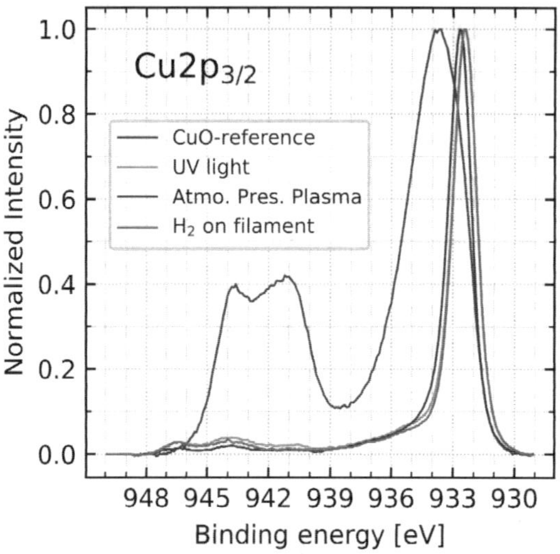

Figure 5. Cu $2p_{3/2}$ spectra of CuO sample before and after various treatments.

4. Conclusion

Thanks to a newly commissioned cryogenic system equipped for XPS analysis and electron irradiation, it is demonstrated that CuO is not reduced during conditioning by electron irradiation at 15 K. The build-up and the stability of this high-SEY oxide together with a low carbon coverage in some part of the LHC arcs can thus be responsible for high beam-induced heat load via intense electron cloud activity. In parallel, conditioning tests at 15 K of $Cu(OH)_2$-covered copper surfaces with various amounts of carbon allowed to propose a scenario explaining the observed oxidation state and the associated high heat load of some LHC beam screens. The following sequence is proposed: at the start of LHC Run 1, the beam screens exhibited either a low $Cu(OH)_2$ coverage and/or a large carbon content, which led to a normal (CuO-free) conditioning during Run 1, with possible start of carbon depletion → the high surface reactivity after electron irradiation during Run 1 enhanced hydroxide formation due to air-exposure during LS1, possibly influenced by the local ambient conditions → a local high $Cu(OH)_2$ and low carbon coverage then led to CuO build-up during Run 2. CuO-free beam screens could have resulted from conditioning of surfaces with low $Cu(OH)_2$ coverage or high carbon content at the end of LS1.

Techniques for in-situ deoxidation of the beam screens have been explored on a laboratory test bench and reduction of CuO is possible by exposure to reducing gas activated either by UV light or in an atmospheric pressure plasma, as well as by exposure to atomic hydrogen formed on a hot tungsten filament. The compatibility of these techniques with an in-situ implementation is under investigation and a further step is required to assess and avoid possible recurrence of CuO on the treated surfaces.

Acknowledgments
The authors would like to thank the CERN Beam-Induced Heat Loads Task Force for the support along this study.

References
[1] Iadarola G, Bradu B, Mether L, Paraschou K, Petit V, Rumolo G, Sabato L, Skripka G, Taborelli M and Tavian L 2021 *Proc. 12th Int. Particle Accelerator Conf. (IPAC'21)* (Campinas, Brazil) pp 1273–78
[2] Petit V, Taborelli M, Zanin D A, Himmerlich M, Neupert H, Chiggiato P and Iadarola G 2021 *Commun. Phys.* **4** 1–10
[3] Biesinger M C 2017 *Surf. Interface Anal.* **49** 1325–34
[4] Cimino R, Commisso M, Grosso D R, Demma T, Baglin V, Flammini R and Larciprete R 2012 *Phys. Rev. Lett.* **109** 064801
[5] Petit V, Taborelli M, Neupert H, Chiggiato P and Belhaj M 2019 *Phys. Rev. Accel. Beams* **22** 083101
[6] Barr T L 1978 *J. Phys. Chem.* **82** 1801–10
[7] Losev A, Kostov K and Tyuliev G 1989 *Surf. Sci.* **213** 564–79

Evaluation of geometrical precision and surface roughness quality for the additively manufactured radio frequency quadrupole prototype

T Torims[1], A Cherif[2], N Delerue[6], M Foppa Pedretti[3], D Krogere[1], T Otto[7], G Pikurs[1], M Pozzi[3], A Ratkus[1], M Thielmann[4], M Vedani[5], M Vretenar[2] and P Wagenblast[4]

[1] Riga Technical University, Riga, Latvia
[2] CERN, Geneva, Switzerland
[3] Rösler Italian S.R.L, Concorezzo, Italy
[4] TRUMPF Laser und Systemtechnik GmbH
[5] Politecnico di Milano, Italy
[6] IJCLab, Orsay, France
[7] TalTech, Tallinn, Estonia

E-mail: toms.torims@rtu.lv

Abstract. A multidisciplinary collaboration within the I.FAST project teamed-up to develop additive manufacturing (AM) technology solutions for accelerators. The first prototype of an AM pure-copper Radio Frequency Quadrupole (RFQ) has been produced, corresponding to ¼ of a 4-vane RFQ. It was optimised for production with state-of-the-art laser powder bed fusion technology. Geometrical precision and roughness of the critical surfaces were measured. Although the obtained values were beyond standard RFQ specifications, these first results are promising and confirmed the feasibility of AM manufactured complex copper accelerator cavities. Therefore, further post-processing trials have been conducted with the sample RFQ to improve surface roughness. Algorithms for the AM technological processes have also been improved, allowing for higher geometrical precision. This resulted in the design of a full 4-vane RFQ prototype. At the time of the paper submission the full-size RFQ is being manufactured and will undergo through the stringent surface quality measurements. This paper is discussing novel technological developments, is providing an evaluation of the obtained surface roughness and geometrical precision as well as outlining the potential post-processing scenarios along with future tests plans.

1. RFQ by additive manufacturing

The RFQ is a crucial accelerator component down-stream the particle source, providing simultaneous and efficient focusing, bunching and acceleration of a particle beam before injection into the linear accelerator [1]. Conventionally it is manufactured from highly conductive materials and consists of four complex shape segments with modulated vanes. To ensure optimum beam optics and radio-frequency properties, all internal surfaces of the RFQ are designed with high-precision tolerances and fine surface roughness parameters. Consequently, the standard RFQ manufacturing routine is complex; each sequence is costly, time consuming and requires expensive technological operations. Among other things, this involves series of multiple high-precision machining operations followed by specialized brazing and material stress-release treatment [2, 3]. Thus, there is an opportunity to re-evaluate established manufacturing approaches and to consider other emerging technologies.

Additive Manufacturing (AM) is a natural candidate, as one of the enabling technologies that has reached the required maturity level to be applied within the accelerator community. The comprehensive study carried out by the I.FAST project confirms the opportunities provided by AM technology [4, 5]. Initial results of the survey have revealed that metal AM activities within the community started already in 2007 and since, the number of applications is growing exponentially. This is explained by the virtues of AM technology: it is accessible, has wide applications within aerospace, medical and automotive industries and is changing the technological paradigm and way of thinking within the engineering wing of the accelerator community. Indeed, AM is offering valuable opportunities, such as the capability to work with a wide range of materials and to provide new design options [6]. Proliferation of the AM has led to the standardisation of this technology. For instance, today the ISO 52900:2021 standard [7] provides for general terms and common language for the growing AM community. Subsequently, series of the standards are being compiled within the Guide to AM Standards [8].

Furthermore, it has been demonstrated [9] that AM can contribute to the manufacturing routines of the RFQ's, e.g., is enabling for design optimisation and improved general properties. The biggest advantage is that AM allows for the RFQ to be built in a single piece, thereby avoiding the complex technological operations, and permitting the design of more efficient internal cooling channels. Recent attempt to produce with AM the pure copper drift tube linac components, is showing that cavity parameters (e.g. quality factor Q and shunt impedance per unit length R_L) are of comparable values to the conventionally machined elements [10].

At the same time, it is important to note that the pure copper, applied for RFQs, is a challenging material for laser-based AM processes itself. It requires state-of-the-art equipment and expertise which is not yet self-evident within the community. For instance, the high thermal conductivity of copper also results in rapid heat diffusion away from the melt area, generating local thermal gradients, hence the risk of the significant residual stresses and part distortion [9].

The limited energy absorption provided by infrared laser radiation on copper is another issue to consider [11]. In the case of laser powder bed fusion (L-PBF) technology, due to the high reflectivity of copper, most of the energy provided by the laser during manufacturing is reflected instead of being absorbed by the powder. This may lead to incomplete fusion of the powder when employing conventional infrared fibre laser sources [12 – 14]. It has been recognized [15 – 17] that increasing the laser power above 500 W can lead to increased part densities. However, high-power lasers can also generate unstable melt tracks due to the different absorptivity of the powder and the molten material. In addition, the large amount of reflected energy that irradiates the optical mirror of the machine during the printing process may damage it [18].

There have been three promising strategies to solve energy absorption issue: increase laser power up-to 1 kW; increase the absorption of the particles through coating; or switch to a green laser source where absorption is almost three times higher. Recent research results [11, 19] are showing that the use of green laser sources are effectively enabling higher absorbance for copper. In case of L-PBF, this leads to relative material densities exceeding 99.8% and which almost eradicates the indicated drawbacks.

2. The additive manufacturing process and equipment

The above-mentioned approach has led to the development of the TruPrint1000 Green Edition by TRUMPF - this machine is based on a modified L-PBF system with an integrated TruDisk1020 1kW frequency doubled 515nm thin disk green laser as energy source for the copper powder bed fusion of 10-45μm particle size (m4p PureCu). At 1 μm wavelength the absorptivity of copper powder is below 10%, whereas at around 500μm it rises 6x to ~60% [20].

With short wavelength, the fusion process becomes more stable and productive in the resulting quality of the parts, compared to a 1 μm laser process, yielding density of > 99,5% and electrical conductivity of >100% International Annealed Copper Standard - IACS (corresponding to >58 MS/m). It is important to note that this matches the properties of conventionally manufactured annealed electrolytic-tough pitch (ETP) copper [21] and falls well within the requirements set for the RFQ, i.e. >90% IACS. The very first demonstration of an AM pure-copper RFQ, corresponding to ¼ of a 4-vane, has been performed on TruPrint1000 Green Edition at IWS Fraunhofer [9].

2.1. Design and manufacturing of the full-size RFQ
Subsequently, collaboration within the I.FAST project continued with the design of full size 4-vane RFQ. This demonstrator has been manufactured by TRUMPF on the extended AM system prototype like the one described in [22], yet with a much larger build volume of 300 mm in diameter and 400mm in height. This meant a full RFQ segment could be built, with the cross-section of Ø148 mm and 248 mm in length (see figure 8).

The system is based on the general design of a TruPrint 5000 and is similarly equipped with a TruDisk 1020. Process chamber inertisation down to 100ppm of oxygen and an efficient gas flux system for the removal of processing by-products are key enablers to achieving the high material quality in a highly productive process with melting rates of typically 16-21 cm^3/h.

Thanks to the availability of the state-of-art AM technological equipment and experience acquired with ¼ sample RFQ sections design, production and measurement results the following major technological improvements were done (see figures 1 and 2):

- High resolution modelling and increased mesh-size;
- Flanges for the vacuum tests;
- Orifices for the RF tests;
- Enhanced internal lattice structures;
- Design optimisation of the cooling channels.

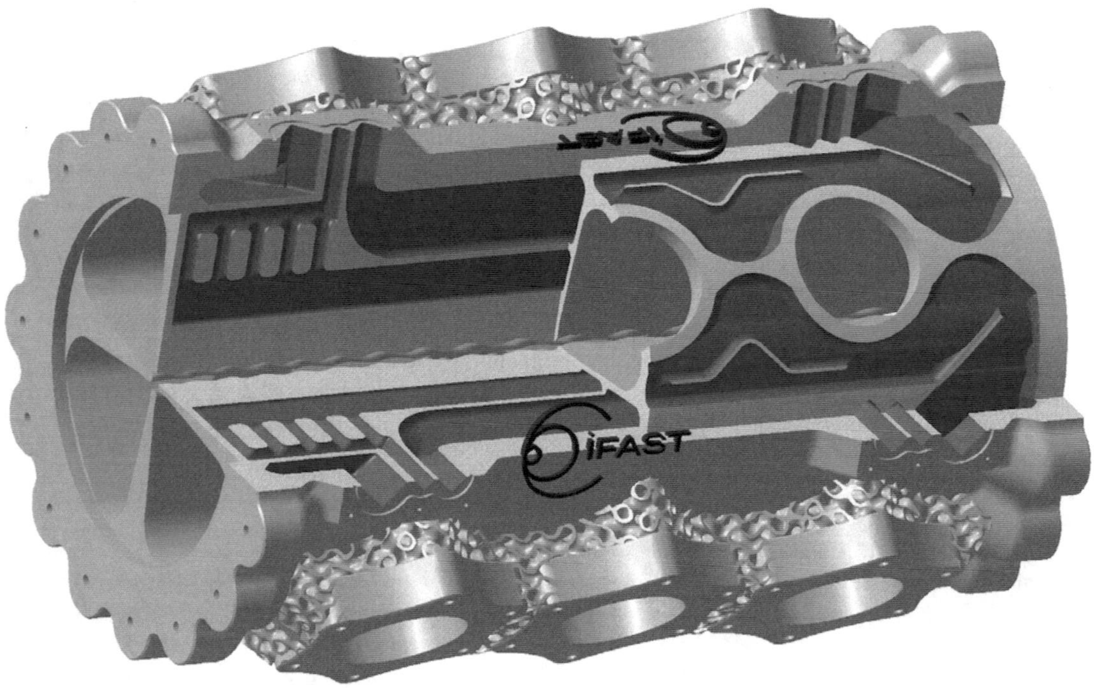

Figure 1. Full-size optimised RFQ.

The overall build duration for the full RFQ demonstrator at TRUMPF amounted to 110 hours, with 6400 layers of 40μm layer thickness and the laser power of 650W with scan velocity at 800 mm/s.

3. Post-processing of RFQ

The initial results of [9] revealed that high-precision accelerator components require post-processing technological operation to ensure the stringent surface roughness requirements of R_a=0.4 μm.

Two sets of the ¼ of a 4-vane were manufactured, measured and then one of them was cut into the smaller segments to allow for the post processing with the three different technologies.

3.1. Conventional surface mass finishing

This process was performed by Rösler in vibrating and rotating machines, where samples have been inserted together with abrasive media, water, and compounds. The surface finishing has been obtained mainly by the mechanical abrasion of the media on the component, due to the relative speed among them. For the finishing of ¼ RFQ a three-step process was designed:

1. Grinding: 12h with Rösler media RXX 07/14 ZS and compound ZF 113 at 1% vol.
2. Smoothing: 6h with Rösler media RKH/4 10/20 DK and compound ZF 113 at 1% vol.
3. Polishing: 1h with Rösler media RP 3/5 ZS and compound ZF 113 at 1% vol.

3.2. Chemically assisted surface finishing

Differently from conventional processes, the surface machining was obtained through the combination of the mechanical abrasion on the surface and the chemical reactions mediated by the compound. Likewise, a three step process was applied - grinding: 1h with Rösler media RXX 07/14 ZS and compound CMP 03/21 L + steps 2 and 3 - idem to the above - see the figure 2.

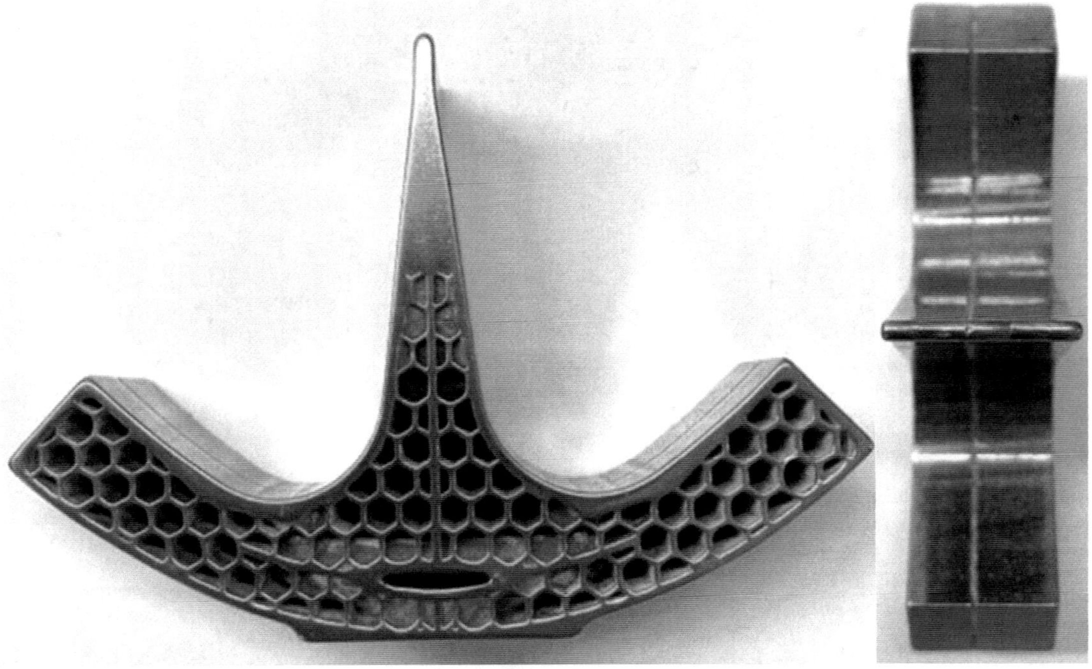

Figure 2. ¼ RFQ sample - chemically assisted finishing.

3.3. High precision surface finishing with MMP TECHNOLOGY®

This technology is based on a mechanical-physical-catalyst surface treatment and was applied to ¼ RFQ which was placed inside a treatment tank [23].

4. Evaluation of geometrical precision and surface roughness

4.1. Geometrical accuracy

The metrology measurements were performed at Fraunhofer IWS with the 3D scanner ATOS GOM Core 135 (see figure 3) and subsequently at CERN by the EN-MME-MM group using Creaform METRASCAN Elite Black and ZEISS Prismo Ultra 12-18-10 (see figure 4). Two sets of measurements were taken: before and after of the post-processing steps.

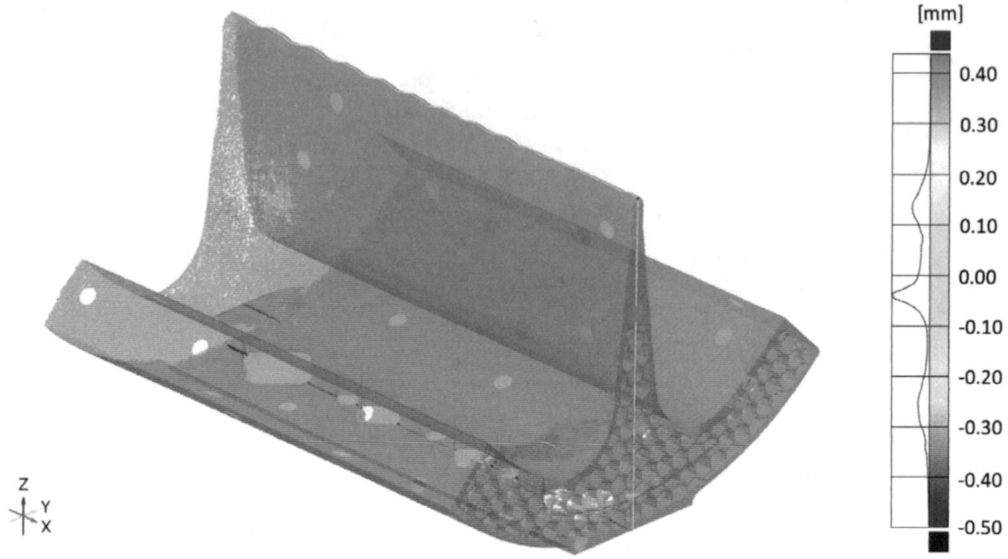

Figure 3. Deviation measurements of ¼ RFQ sample No.1 – prior post-processing.

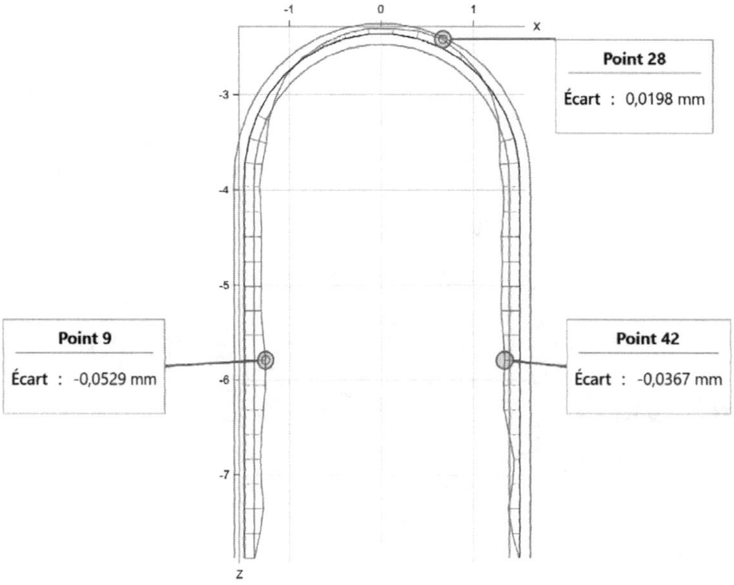

Figure 4. Vane-tip and side measurements of ¼ RFQ sample No.1 – prior post-processing.

Post-processed ¼ RFQ samples were consequently measured in the Rösler and CERN metrology labs with GOM ATOS compact scan and Creaform METRASCAN respectively (see figure 5).

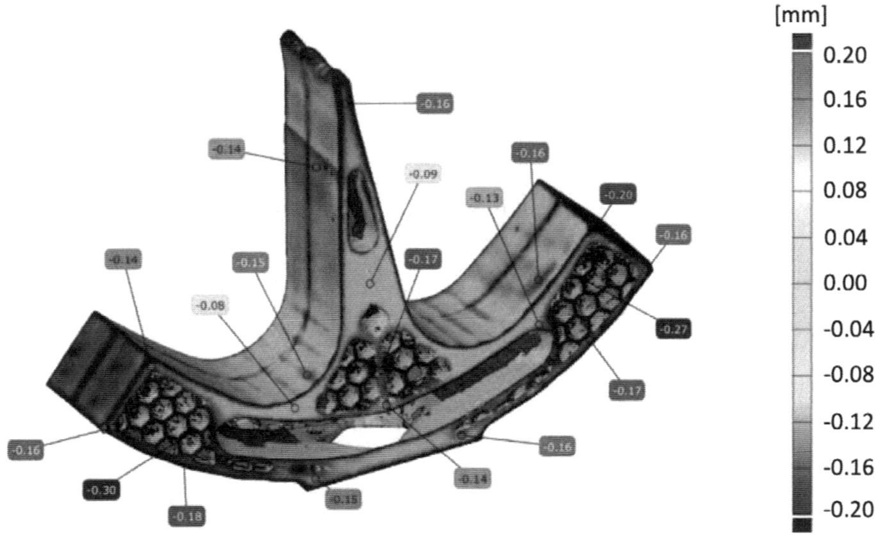

Figure 5. Measurements of ¼ RFQ segment sample No.1 - post-processed.

4.2. Surface roughness

Correspondingly, the surface roughness measurements were carefully planned and were performed in the critical parts of the RFQ. These measurements were done prior and after post-processing operations at CERN and at Rösler by the Mitutoyo 3000 stylus surface roughness measuring equipment, ensuing the comparable protocols (see figure 6).

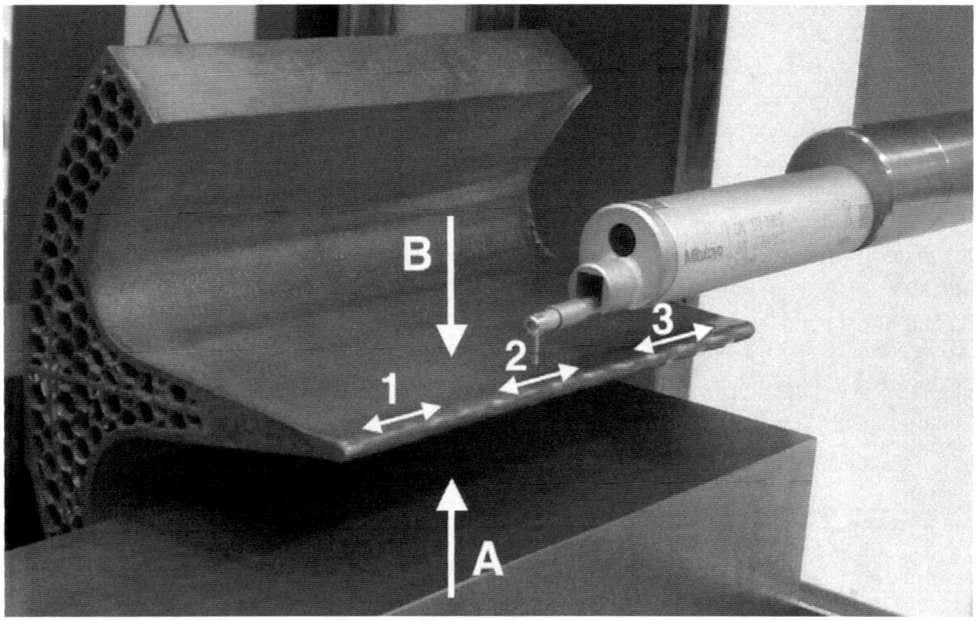

Figure 6. Measurements of ¼ RFQ segment sample No.2 - post-processed with MMP TECHNOLOGY®

To assess the overall results, the obtained surface roughness amplitude values were comprehensively evaluated in an aggregated way: R_a - average, of profile height deviations from the mean line and R_z - maximum peak to valley height of the profile, within a single sampling length.

Table 1. Surface roughness of ¼ RFQ before and after post-processing.

Post processing method	Side	R_a, µm	R_z, µm
Before post-processing		13.82	48.86
Trad. mass finishing	A	0.09	0.83
	B	0.07	0.58
Chemically assisted	A	0.07	0.67
	B	0.12	0.97
MMP TECHNOLOGY®	A	0.30	3.24
	B	0.11	1.03
Target roughness		0.4	not set

5. Conclusions and way forward

5.1. Attained geometrical accuracy

The I.FAST AM collaboration has previously con-firmed [8] that even without surface finishing operations, the surface precision of the AM manufactured RFQ's is approaching the required precision of 20 µm on the vane-tip and fully reaching 100 µm on other surfaces. Certain accuracy deviations exceeding 20 µm, on the vane-tip, are attributed to the features of the standard AM techno-logical process and could be effectively eradiated by enhancing (increasing source file mesh-size) of the RFQ vain-tip design and optimising the AM technological settings (process parameters).

Still, even without the RFQ design and AM process improvements, impact of the applied post-processing operations has proved to be beneficial for attainment of the required geometrical accuracy. Measurements of the geometrical precision revealed encouraging results – post-processing has improved surface accuracy by eliminating excess peaks and is opening valuable surface engineering features. At the same time, it must be noted that - post processing has removed extra surface layer below pre-set values. Chemically assisted finishing has removed surplus of 115-135 µm and finishing with MMP TECHNOLOGY® has reduced surface volume by 40-70 µm. This must be considered and appropriately compensated within the design and technological process for AM manufactured and post-processed parts.

5.2. Post-processing and achieved surface roughness

Required R_a values < 0.4µm were successfully accomplished by the three different post-processing methods. For the conventional surface mass finishing, processing time was longer, however, material removal was smaller, in comparison with chemically assisted process, where process time was smaller and material removal was higher finishing (see figure 7).

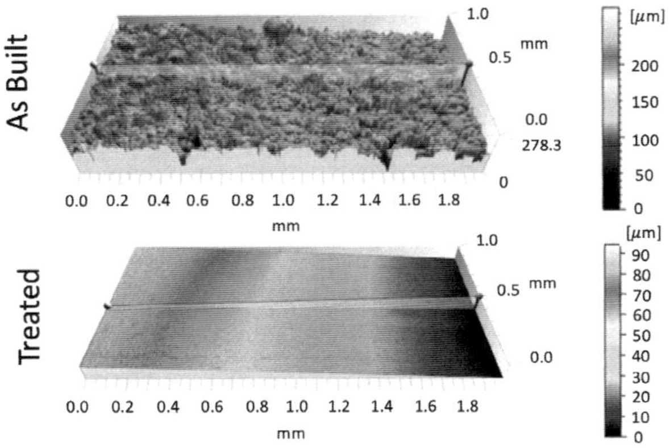

Figure 7. 3D surface roughness of the post-processed ¼ RFQ sample.

The limiting factor for mass finishing processes is the presence of deep valleys on the surface. It is because the mechanical abrasion provided by the media occurs preferentially on peaks and they remain on the surface also at the end of the process. Instead, chemically assisted surface and presumably also MMP TECHNOLOGY®, is enabling much higher material removal. In this case, the external surface layers, where defects are most prevailing, could be completely removed and thus issues related to the presence of valleys shall be mitigated.

Although feasibility of the pure-copper AM production and its surface treatment approach has been fully con-firmed, other methods of the surface finishing, such as laser smoothening will be considered in the future re-search. This will be studied in the context of the 3D sur-face roughness parameters and surface texture requirements.

5.3. AM produced full-size RFQ and the next steps

I.FAST AM collaboration has developed a full-size RFQ design which was optimised for AM and improved features. The RFQ demonstrator has been AM printed on TruPrint 5000 machine by TRUMPF (see figure 8). This segment has reached dimensions of 250 mm in length and cross-section of Ø148 mm. This falls within the real size limits of the compact linear accelerator RFQ's [24]. This is confirming the feasibility of large-size pure-copper AM printing capabilities for the accelerator community at large.

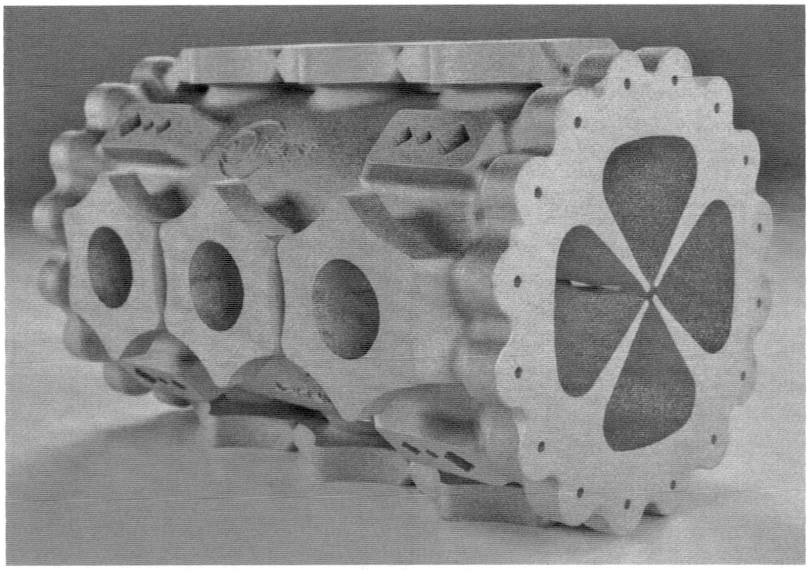

Figure 8. Full-size AM produced RFQ of pure copper.

This demonstrator will undergo the same geometrical accuracy and surface roughness measurements as ¼ of a 4-vane RFQ. This will be complemented with vacuum, water-tightness, and RF tests which are planned at IJCLab, and are integral part of the I.FAST project AM package. An AM produced RFQ demonstrator has been designed and equipped with the flanges and orifices enabling these tests. Separately, pure copper AM sample High Voltage Holding tests are planned at CERN.

Even though required geometrical accuracy and surface roughness of ¼ RFQ have been achieved, on the inner and outer surfaces, attention will remain on the vane-tip as a crucial and most challenging element of the RFQ. Post-processing will be applied to the full-size RFQ demonstrator with the subsequent analysis routines described in this paper.

Acknowledgments
This project has received funding from the European Union's Horizon 2020 Research and Innovation programme under grant agreement No 101004730 and is supported by the Latvian Council of Science under grant agreement VPP-IZM-CERN-2020/1-0002.

References

[1] Lombardi A 2005 The Radio Frequency Quadrupole (RFQ) Tech. rep. CAS, CERN, Geneva, Switzerland
[2] Mathot S 2008 *Proc. 11th Eur. Particle Conf. (EPAC'08)* (Genoa, Italy) pp 1494-96
[3] Scibor K *et al.* 2020 *Proc. euspen's 20th Int. Conf. and Exhibition* (Geneva, Switzerland)
[4] Pikurs G and Delerue N 2022 *presentation at the I.FAST Workshop on AM applications* (Geneva, Switzerland)
[5] I.FAST project https://ifast-project.eu
[6] Barlier C and Bernard A 2015 *Collection Technique & ingénierie. Mécanique et matériaux* p 401
[7] ISO/ASTM 52900:2021 Additive manufacturing - General principles - Fundamentals and vocabulary
[8] Guide to Additive Manufacturing Standards https://www.amable.eu/fileadmin/amable/documents/publications/AMable_Standardisation-in-AM_Status2022_v0100.pdf
[9] Torims T, Pikurs G, Gruber S, Vretenar M, Ratkus A, Vedani M, López E and Brückner F 2021 *Instruments* **5** 35
[10] Mayerhofer M *et al.* 2022 *Rev. Sci. Instrum.* **93** 023304
[11] El-Wardany T I *et al.* 2018 *J. Electron. Packag.* **140** 20907
[12] Gruber S *et al.* 2021 *Materials* **14** 3642
[13] Lykov P A *et al.* 2016 *Mater. Sci. Forum* **843** pp 284-88
[14] Silbernagel *et al.* 2019 *Addit. Manuf.* **29** 100831
[15] Trevisan F *et al.* 2017 Selective laser melting of chemical pure copper powders *Proc. Int. Powder Metallurgy Congress and Exhibition (Euro PM 2017)* (Milan, Italy)
[16] Colopi M *et al.* 2018 *Procedia CIRP* **74** pp 59-63
[17] Ikeshoji T T *et al.* 2018 *JOM* **70** pp 396-400
[18] Jadhav S D *et al.* 2021 *Addit. Manuf.* **42** 101990
[19] Jadhav S D *et al.* 2019 *J. Mater. Process. Technol.* **270** pp 47-58
[20] Becker D 2014 Selektives Laserschmelzen von Kupfer und Kupferlegierungen *Apprimus Wissenschaftsverlag* Aachen http://publications.rwth-aachen.de/record/444890
[21] Jiang Q *et al.* 2021 *Coatings* **11** pp 1-30
[22] Wagenblast P *et al. Proc. SPIE 11271, Laser 3D Manufacturing VII* (San Francisco, CA USA) p 112710J
[23] BinC Industries http://mmptechnology.com/site/mmp-technology/
[24] Vretenar M *et al.* 2014 *Proc. 27th Linear Accelerator Conference (LINAC2014)* (Geneva, Switzerland) pp 935-38

Tuning of superconducting cavities using the FFT of transmitted power

E Laface, C G Maiano, P Pierini and M Wang
European Spallation Source ERIC, Lund, Sweden
E-mail: emanuele.laface@ess.eu

Abstract. We implemented a method to tune the ESS superconducting cavities based on the spectral analysis of the high resolution data available from the Low Level RF system (LLRF) for the transmitted power, without the need of connecting a network analyzer or any other dedicated instrumentation along the RF chain. A frequency peak up to 4 MHz off from the resonating frequency can be detected and used to control the stepper motor of the tuner until the cavity is stretched to the proper length to reach the correct operation frequency. Experience of its use at the ESS Test Stand 2 (TS2) facility at Lund during cryomodule acceptance testing is presented.

1. Introduction

The European Spallation Source superconducting linac is composed by 36 elliptical cavities with geometric β of 0.67 and 84 elliptical cavities with geometric β of 0.86. The cavities are assembled in groups of 4 per each cryogenic module as described in [1, 2] and operate at the frequency of 704.42 MHz, accelerating the proton beam from 216 MeV up to the 2 GeV.

Superconducting cavities are fabricated with a lower frequency than the operational one and, once cooled down in a bath of helium at 2.0 K, the thermal contraction brings them close to the goal frequency, but not with the needed accuracy imposed by the small cavity bandwidth. The final adjustment needs to be done with a tuner that stretches the cavity through the action of a step motor. In order to follow the tuning process, one of the standard techniques consists in monitoring the S21 cavity transmission with a Vector Network Analyzer (VNA). The use of VNA S21 requires either to modify the high power RF wave-guide distribution system (RFDS) inserting matched transitions and disconnecting the pickup cables, or to inject the signals from the strongly attenuated ports in the directional couplers of the RFDS, with additional amplifier stages. Both methods introduce risks in the linac installation (RFDS leaks, missing cable re-connections) and is time consuming.

In all modern LLRF systems the cavity pick up (PU) for the transmitted signal is available with a high sampling rate (~10 mega samples per second) along the pulse. Its analysis in the spectral domain provides sufficient accuracy to identify the narrow cavity resonance in the noisy pattern of the highly detuned (~200 band widths) system. The pulsed klystron power, which has a carrier frequency of 704.42 MHz, has Fourier components that fall into the cavity resonance, and these appear with a clear signature peak in the spectral analysis of the transmitted signal. A similar approach was used to tune the 1.3 GHz SRF cavities at the European XFEL, DASY, Hambourg, Germany [3].

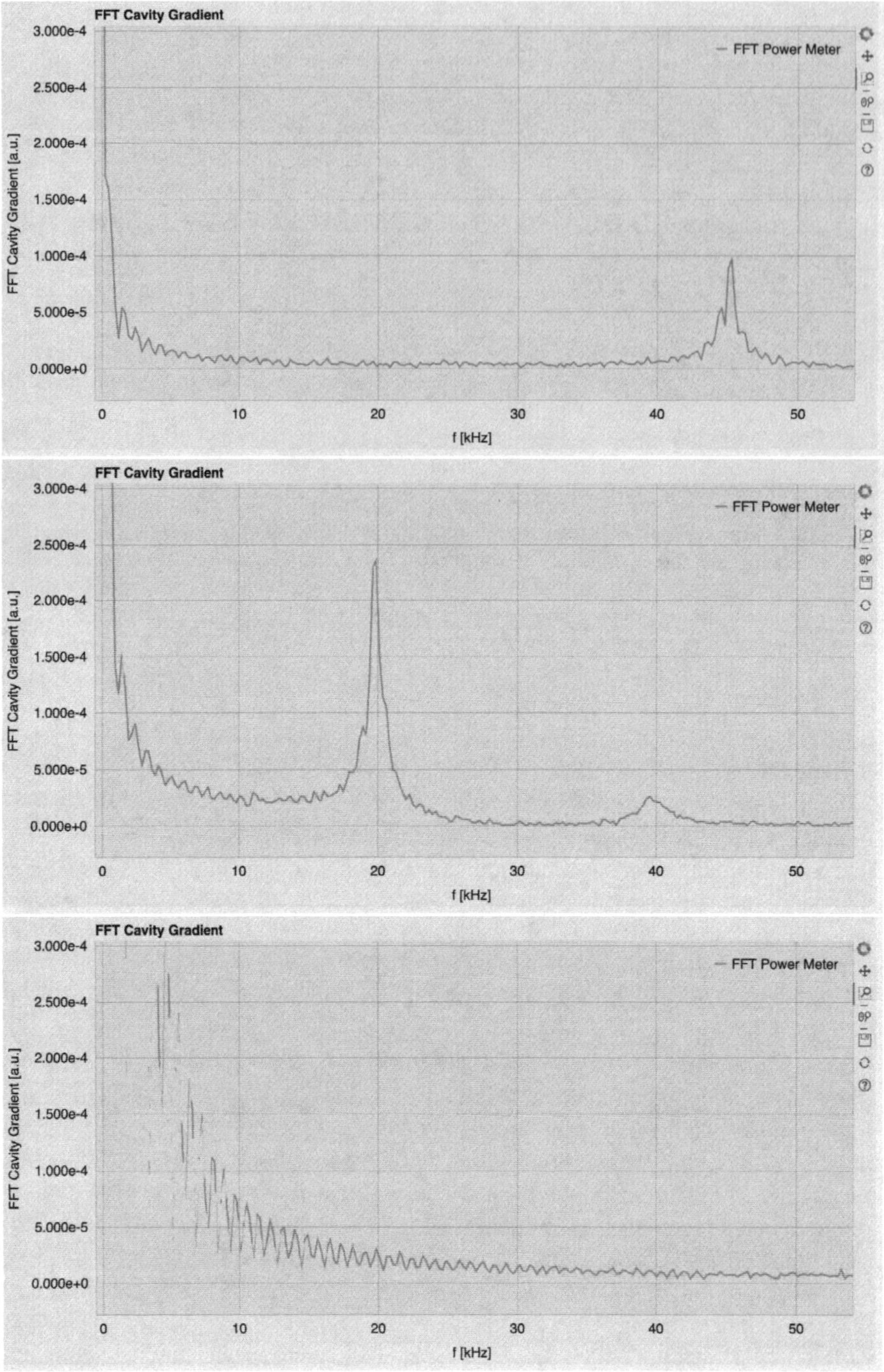

Figure 1. Fast Fourier Transform of the transmitted power in three moments of the tuning process for the 3rd cavity in the cryomodule CM01. The peak moves toward the zero (carrier RF frequency) when the cavity is approaching the driver frequency of the klystron, and in the last step it is no longer visible.

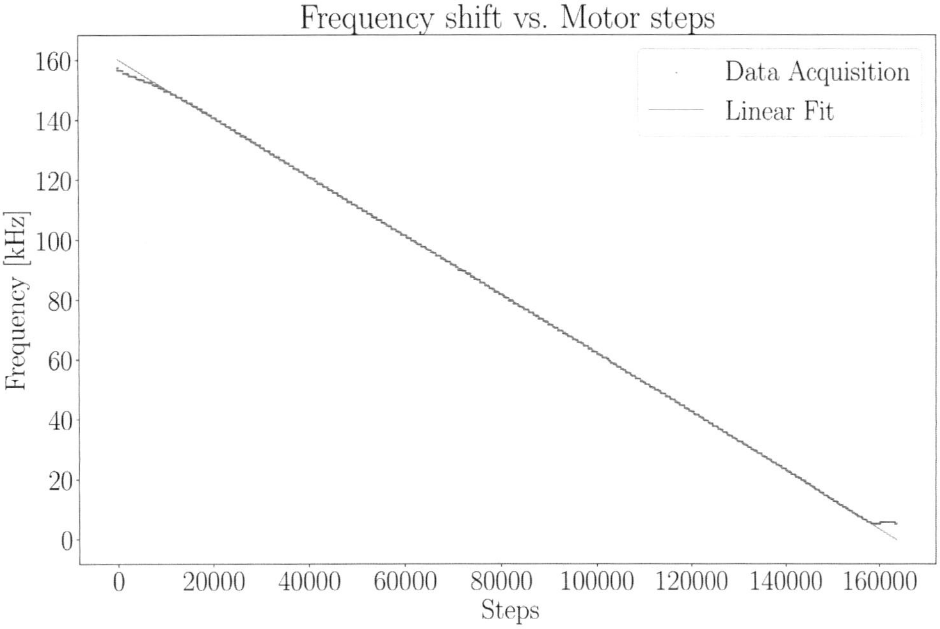

Figure 2. Frequency versus motor steps.

2. The FFT technique for tuning

The technique is illustrated in figure 1, where the FFT of the transmitted power signal is shown at three different positions during the motion of the tuner. The signature peak of the detuned frequency in the upper plot is shifted of 45 kHz with respect to the 704.42 MHz design frequency, and in the lower plots it moves towards the klystron frequency due to the tuner motion.

The method has been initially implemented at the ESS TS2 as a high level application that acquires (here at 1 Hz) the down-sampled transmitted power amplitude from the LLRF system, digitized in an array of 35000 samples with a sampling rate of 8.386 MHz in a 4.174 ms time window. The data is then subjected to an FFT transformation that has an un-aliased reach of approximately 4 MHz from the carrier frequency. DC component minimization is performed by subtraction of the mean value of the signal.

The analysis of the FFT data can isolate the cavity frequency signature that can be followed during the process and correlated to the motor position information to derive and verify the design frequency tuning coefficient.

2.1. Far tuning

The cavity frequency depends on the tuner motor position, which is available in the control system as an EPICS Process Variable (PV). We are then able to evaluate the correlation of the frequency versus the number of steps and verify the linearity of the system over almost the full range [4], as shown in figure 2.

The position of the stepper motor is indicated by the number of full steps performed by the motor controller. A full turn of the motor axis corresponds to 200 full steps. The motor is equipped with a 1:100 gearbox to rotate the main tuner screw with a 1.5 mm pitch that actuates a double leverage system with ratio 1:17 which produces the cavity elongation. The combined effect of the cavity frequency sensitivity (217 kHz/mm) and the mechanical reduction results in

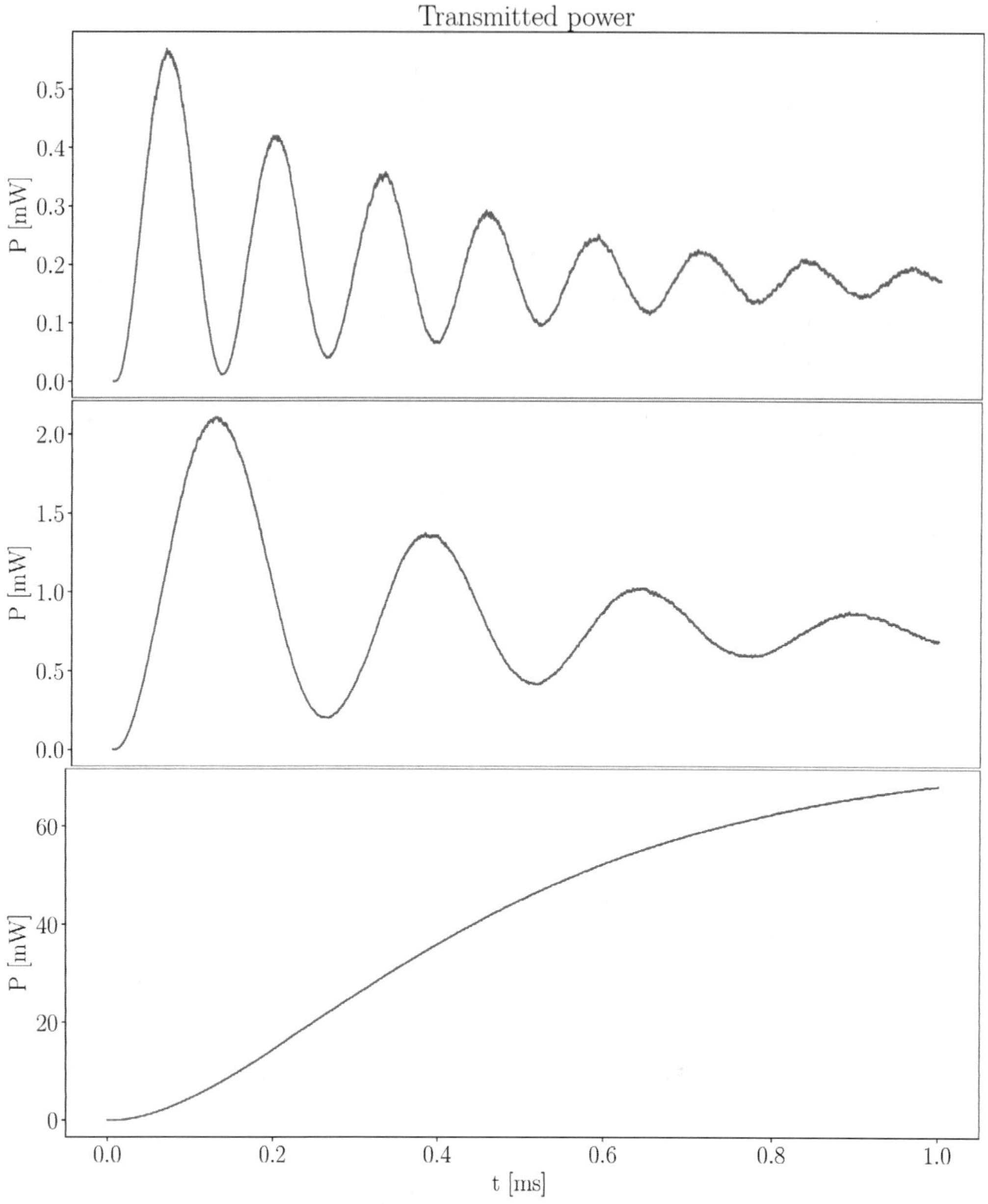

Figure 3. Amplitude of the transmitted power close to the resonance, in the first ms of the RF pulse. Top: the 8 peaks indicate a 8 kHz detuning. Middle: After correction of approximately 4000 steps the cavity has a remaining detuning of 4 kHz. Bottom: After a further tuning of 4000 steps the cavity is on resonance and the signal does not show oscillation, but the usual filling characteristic of an SRF cavity, with a long fill time.

a nominal design sensitivity of 190 Hz/motor turn (200 full steps), that is 0.95 Hz/full step.

The function that relates the frequency to the steps fits very well with an RMS error on the frequency of ± 10 Hz except for the two regions at the opposite extremes. At the beginning, when the tuning range is between 0 and 10000, the tuner did not establish a full mechanical coupling between the cavity and the tuner. At the ending of the range during the final tuning

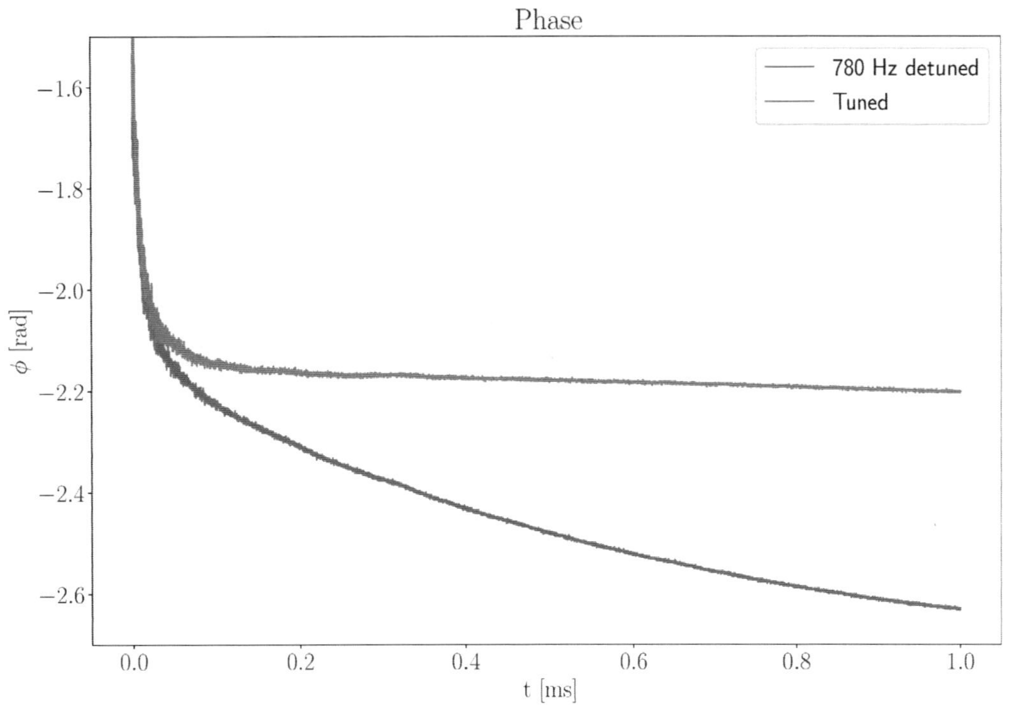

Figure 4. Fine tuning of the cavity in the last bandwidth fraction from the resonance. In blue the phase is not flat, it means that the input and the transmitted frequencies are not the same and the motor can be adjusted until the phase is flat as in the red plot. The detuning from the blue and the red plots was of 780 Hz.

process between 158000 and 164000, the data can deviate from the linear behaviour due to the small mechanical hysteresis shown by the tuning leverage upon inversion of motion and due to the loss of accuracy in the peak identification when it reaches the carrier DC component.

The slope of the linear interpolation gives −0.98 Hz per full motor step that is comparable with the expected design value for the tuner of 0.95 Hz per full motor step [4].

2.2. Approach to resonance

Once the detuning frequency is below 10 kHz the FFT analysis becomes unreliable because the detuned peak is too close to the DC component and the mechanical position may be affected by a small hysteresis if inversion occurs in the last steps of the tuning process.

When the cavity frequency is close to the drive klystron, the transmitted power starts to build up in the cavity and the small frequency difference between the cavity frequency and the driver frequency appears as a "beating" modulation over the transmitted power.

The number of beating peaks present in the transmitted power in 1 ms gives a visual indication of the approximate detuning frequency and allows a further adjustment with the sensitivity of 1 Hz/step. An example of three moments of this tuning process is represented in figure 3.

2.3. Fine tuning

When the cavity frequency approaches the klystron frequency within a fraction of its bandwidth (∼1 kHz), the last tuning action can be performed by flattening the transmitted power phase. When the input and the transmitted power are at the same frequency the phase is flat, as it can be seen in figure 4.

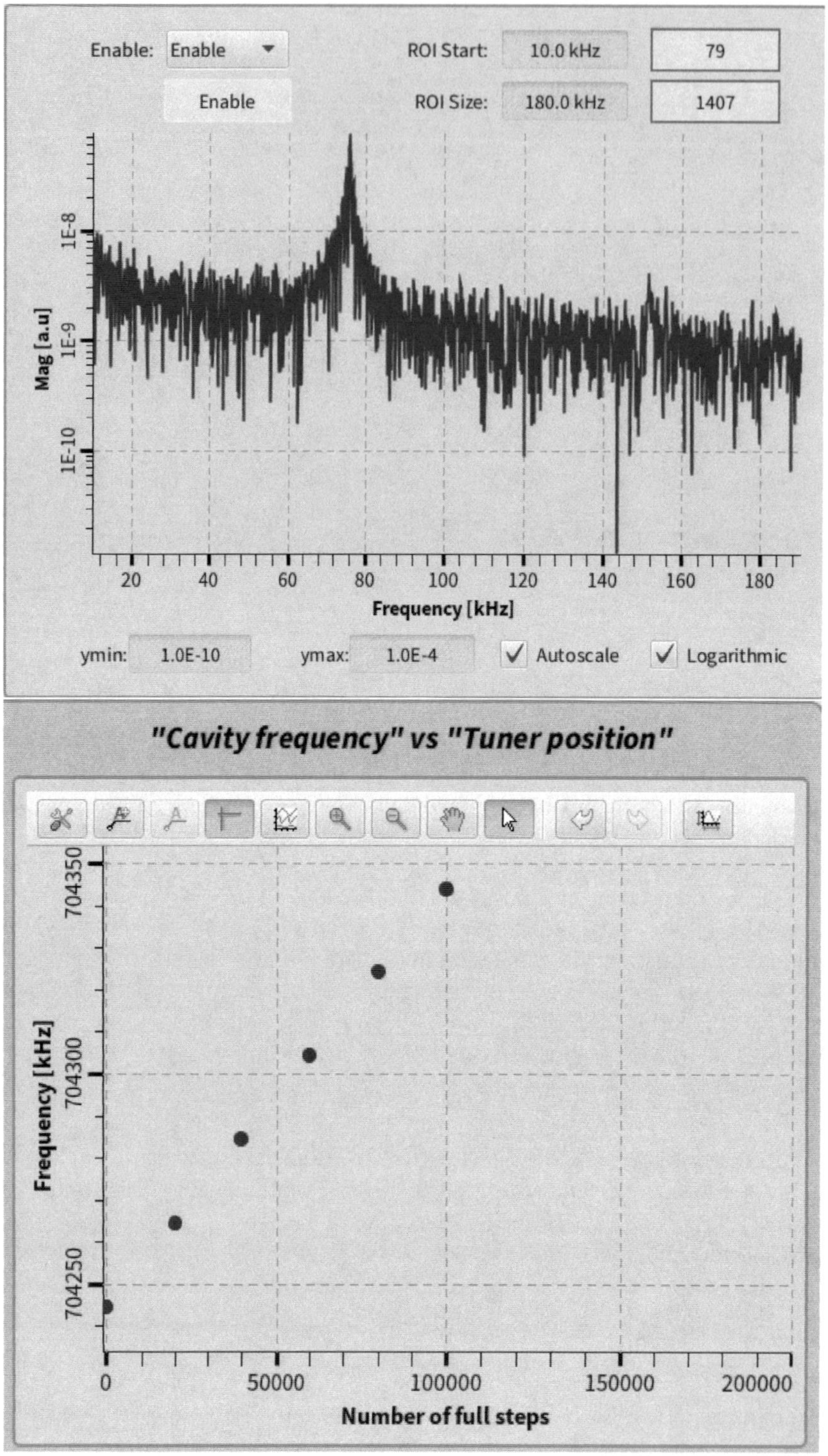

Figure 5. The FFT method implemented in the OPI for normal operations. Top: the frequency spectrum. Bottom: Frequency versus motor steps.

3. EPICS implementation

The method described above was developed during the TS2 testing of the first cryomodule (CM01) of the ESS series, after a direct frequency check with a VNA. The procedure was firstly implemented as a high level scripting application in Python.

After this successful test the method was ported to a full EPICS IOC module (based on the AreaDetector generic module) and a full EPICS graphic interface developed for the operations.

This version is now routinely used in the test facility to tune the 36 medium β and 84 high β cavities with the same conceptual procedure described in this paper, integrated in the operational screens as shown in figure 5. A further version is being developed for the regular commissioning of the superconducting linac.

4. Conclusions

We presented a procedure to tune the superconducting cavities of the ESS linac based on the FFT analysis of the transmitted power. The FFT is used during the far tuning to bring the system within reach of the klystron frequency from the parking position starting the cavity excitation. The frequency beating between the two systems is then used to approach the cavity bandwidth to the klystron frequency. Finally, the phase of the cavity is finely flattened in order to have the cavity to resonate exactly at the klystron frequency.

References

[1] Sertore D, Bertucci M, Bonezzi M, Bosotti A, D'Ambros A, Grimaldi A, Michelato P, Monaco L, Pagani C and Paparella R 2021 *Proc. 12th Int. Particle Accelerator Conf. (IPAC'21)* (Campinas, Brazil) pp 1420–23

[2] Cenni E, Baudrier M, Carbonnier P, Devanz G, Hanus X, Maurice L, Plouin J, Roudier D and Sahuquet P 2019 *Proc. 19th Int. Conf. on RF Superconductivity (SRF'19)* (Dresden, Germany) pp 1143–46

[3] Branlard J *et al.* 2017 *Proc. 8th Int. Particle Accelerator Conf. (IPAC'17)* (Copenhagen, Denmark) pp 3638–41

[4] Devanz G 2021 CEA in-kind parthner Private Comm.

X-Rays energy measurements during the RFQ conditioning at the European Spallation Source

E Laface[1], C G Maiano[1], R Zeng[1] and O Piquet[2]
[1]European Spallation Source ERIC, Lund, Sweden
[2]CEA Paris-Saclay - DRF/Irfu/DACM, France

E-mail: emanuele.laface@ess.eu

Abstract. The Radio Frequency Quadrupole (RFQ) was conditioned at the European Spallation Source during spring 2021. We used part of the conditioning time to estimate the accelerating potential within the RFQ analyzing the x-rays bremsstrahlung radiation emitted by the electrons released and accelerated in the RFQ. The results of these measurements are in good agreement with the theoretical prediction.

1. Introduction

The Radio Frequency Quadrupole, [1] section 3.2 and [2], is an accelerating, focusing and bunching structure located after the Ion Source and Low Energy Beam Transport at the European Spallation Source in Lund, Sweden. This structure has the goal of accelerating the 62.5 mA of proton beam from 75 keV to 3.6 MeV at a frequency of 352.21 MHz with a repetition rate of 14 Hz for a pulse of 2.86 ms. In order to achieve this goal, the RFQ needs several weeks (in our case eight) of conditioning, where the power, the repetition rate and the length of the pulse are gradually increased up to the nominal level. The conditioning phase removes the residual contaminants deposited on the surface, as well as surface imperfections, making the device stable in terms of operations.

A side effect of the RFQ powering is that some electrons are generated and accelerated within the RFQ itself and, when those electrons hit the surface they release the energy as x-rays that can be measured outside the RFQ structure. The distribution of energies among the electrons follows the spectrum of bremsstrahlung, and in particular the maximum reachable energy (the asymptotic behaviour) is expected to be the potential produced withing the RFQ. This gives a method to measure the potential produced in the RFQ for each applied power [3–5].

2. The X-rays detector

The x-rays measurements were performed using CdTe detector from Amptek: Model XR-100-CdTe is a high performance x-rays detector, pre-amplifier, and cooler system using a $5 \times 5 \times 1$ mm^3 Cadmium Telluride (CdTe) diode detector mounted on a two-stage thermoelectric cooler. Cadmium Telluride (CdTe) crystal is the sensor material for direct conversion of the x-rays radiation [6]. This high-Z semiconductor material (Cd with Z=48, Te with Z=52) provides excellent stopping power and thus high efficiency in the desired energy range (see figure 1).
The detector and Digital Pulse processor, (figure 2) were installed on a flange outside the fourth slot of the fifth RFQ module at a distance of 3.03 m from the beginning of the RFQ (figure 3).

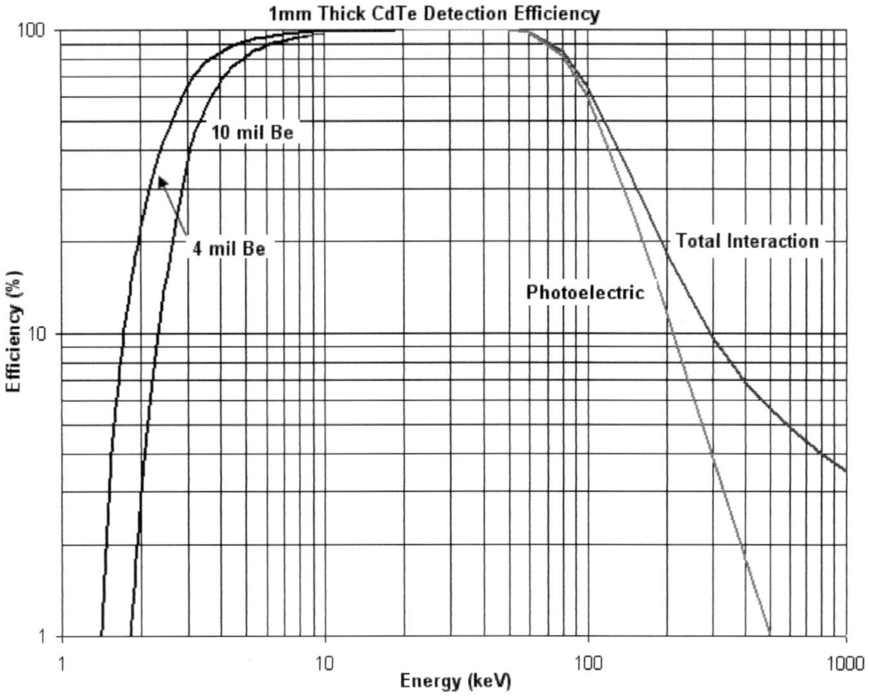

Figure 1. Detector efficiency (1 mm thick crystal).

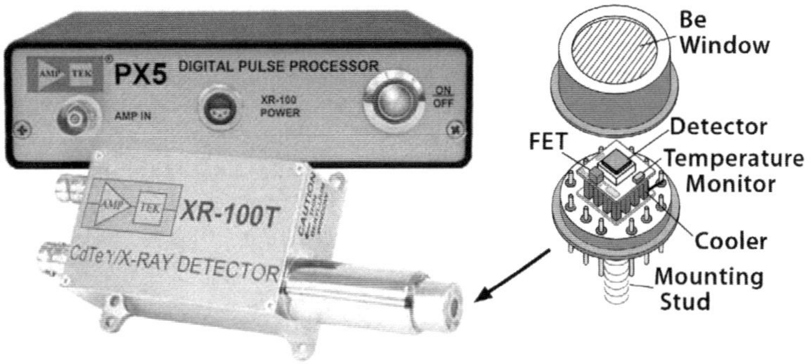

Figure 2. Detector.

After some tests of setup it was decided to avoid the use of a shielding liner for the installation of the x-ray detector because the copper of the RFQ and the stainless steel of the flange was sufficient to shield the low energy x-rays avoiding the saturation of the detector.

2.1. Calibration

Before starting to acquire the measurements we did a calibration of the detector, using isotopes with lines in the energy region of interest, which is from 0 to 120 keV. The selected sources for the calibration were:

- ^{241}Am with lines at 13.81, 17.70, 20.70, 26.34 and 59.54 keV as in figure 4
- ^{152}Eu with lines at 39.52, 45.41 and 121.78 keV as in figure 5

Figure 3. Detector installed on the RFQ flange.

The linear function between channels and energy is obtained with a best fit between the calibrated points (figure 6). The residual error of the fit, that is of 0.3 keV, is used as error in the energy measurement of the instrument.

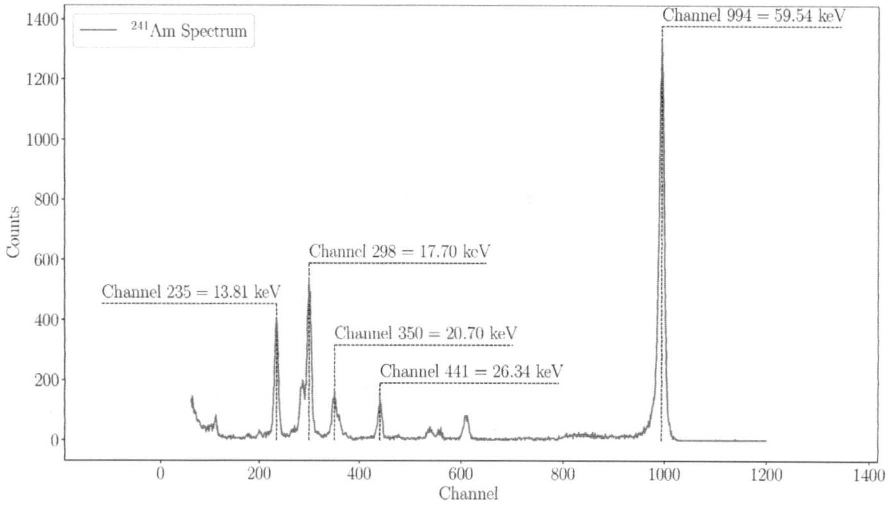

Figure 4. Calibration with ^{241}Am source.

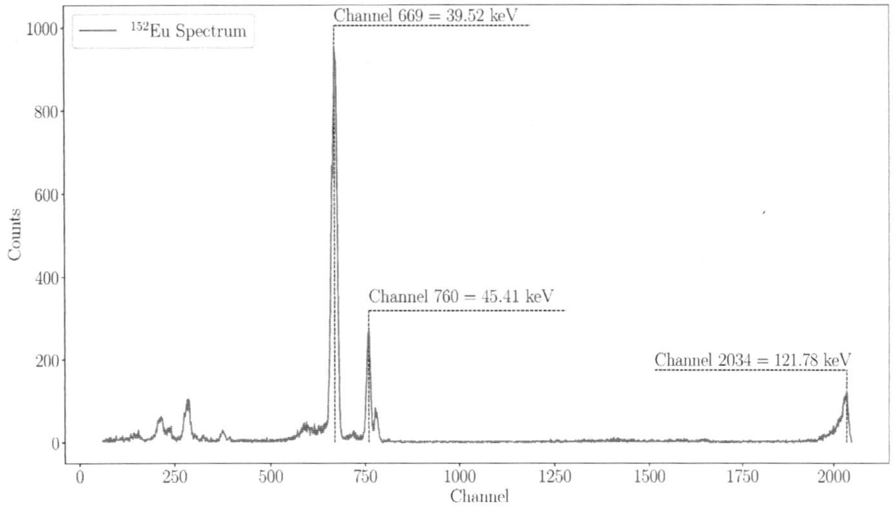

Figure 5. Calibration with ^{152}Eu source.

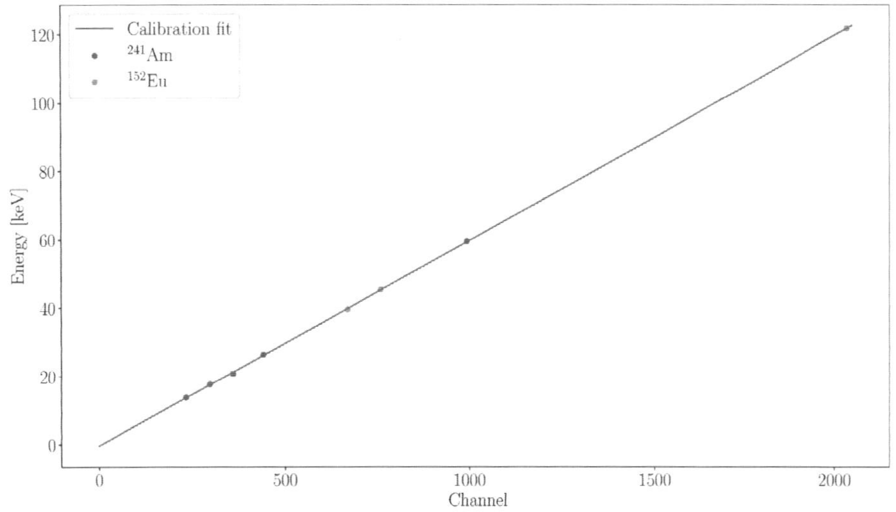

Figure 6. Calibration fit.

2.2. Configuration
The detector configuration was selected starting from Amptek general guideline, fine tuned for our specific case. Parameters like detector voltage and and peak time were unchanged, since the energy resolution had been already optimized at the vendor: nominal: 25 mm^2 crystal shows $<$ 1.5keV FWHM at ^{57}Co 122 keV line. The total number of channels of 2048 and the gain of 12 were chosen to focus on the energy detection window from 0 to slightly above 120 keV (the maximum energy peak visible in the spectrum was ^{152}Eu line at 121.78 keV). No pile-up rejection mode (PUR) option was activated since the rate of events was sufficiently low: no dead time was measured at any power level.

3. Measurements and analysis

The x-rays spectrum was acquired for each value of the forward power from 450 to 800 kW with steps of 50 kW. The pulse that was sent to the RFQ was, for all the acquisitions, the nominal one: 2.86 ms of flat top pulse with a repetition rate of 14 Hz. For each set-point the data were acquired for a time long enough to see a clear signal over the noise background. This means that the time of acquisition at low power was much longer than the time of acquisition at high power because the rate of detected x-rays was lower at lower power. For this reason, the counts per channel are normalized with time in the data analysis, such that the comparison is homogeneous. A background measurement, that is at zero power in the RFQ, was acquired during a shutdown, normalized to the acquisition time and subtracted from the power measurements. The measured background is due to natural radioactivity of the materials surrounding the detector. Isotopes like ^{232}Th and ^{40}K are , with high probability, contained in the concrete walls and Compton continuum extends to low energy region where our ROI sits.

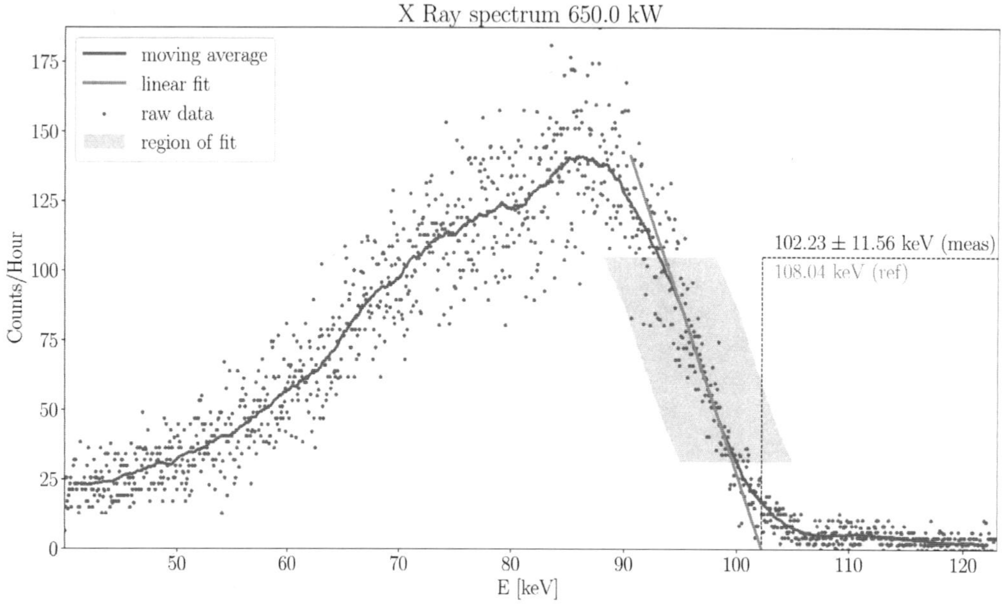

Figure 7. Example of a spectrum: 650 kW of inward power at 14 Hz, 2.86 ms. The error reported here in the measurement is only the residual of the linear fit, the full error must account also the instrument error from the calibration procedure as previously described. The background was subtracted from a background measurement done when the power in the RFQ was zero. The residual background from natural activity and cosmic rays is visible as the counts after the bremsstrahlung region.

The acquisitions showed a clear bremsstrahlung pattern in the performed measurement, as an example we show the acquisition spectrum at 650 kW in figure 7.

The spectrum was analyzed to identify the limit in energy of the electrons. This limit is, in first approximation, when the linear component of the spectrum intercept the zero of the counts. The first step is to identify the linear region applying a movable average with a windows of 20 channels. The size of this window was chosen after several test with different sizes in order to reduce the noise maintaining, at the same time, the underlying physics. The linear fit is then performed on the most linear region (the largest identified that minimize the quadratic error between the linear fit and the data). Finally the value of the zero intersection is evaluated as the maximum energy of the bremsstrahlung electrons, with the error estimated as the square root of the residuals.

The measured value was compared with the theoretical estimation based on the model of the RFQ. The values are reported in table 1. The point at 500 kW was rejected because we discovered during the analysis that the power in the RFQ was not kept constant during the measurement, generating a spectrum that was not possible to use for the linear extrapolation.

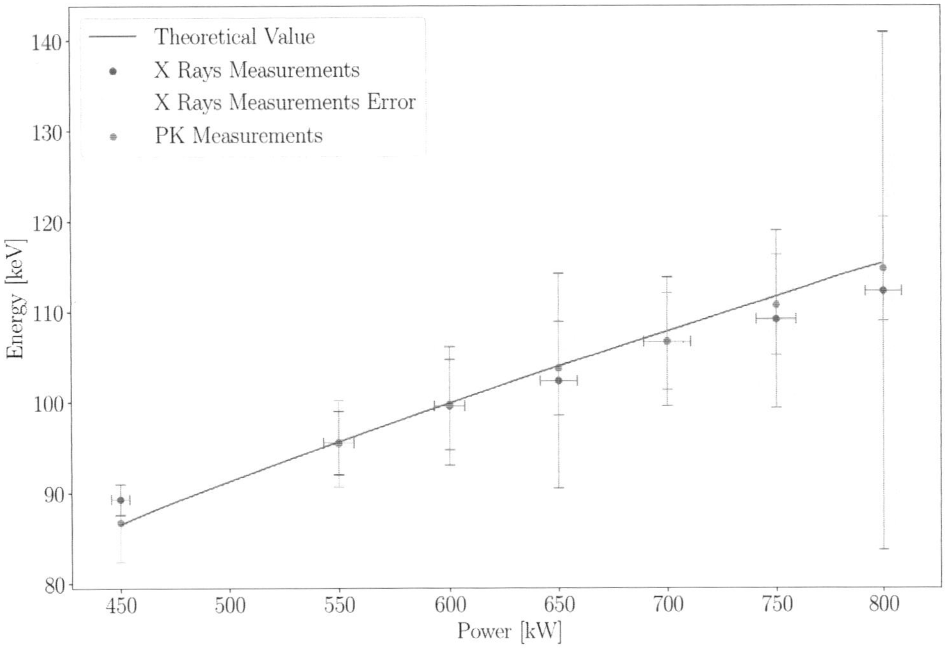

Figure 8. Energy vs. power plot. The solid line is the theoretical value, the blue dots are the x-rays measurements with the error bars in green; the orange dots are the measured voltage in the RFQ from the PKs, reported with their errors for comparison.

Table 1. Energy vs. Power in the RFQ. Theoretical estimation and values measured from the bremsstrahlung electrons. The point at 500 kW was rejected because the power was changed during the measurement and the final value cannot be estimated.

Power [kW]	Theoretical [keV]	Measured [keV]
450 ± 4.18	86.5	89.27 ± 1.74
550 ± 6.87	95.63	95.51 ± 3.51
600 ± 6.90	99.89	99.57 ± 6.56
650 ± 8.45	103.96	102.33 ± 11.86
700 ± 10.71	107.81	106.67 ± 7.12
750 ± 9.00	111.67	109.14 ± 9.81
800 ± 8.40	115.33	112.26 ± 28.62

The errors reported in table 1 for the measured values are the square root of the residuals of the linear fit combined with the error of the detector. The errors on the power readings are the errors between the set-point and the real value of power applied to the RFQ. The overview of the measurements with errors and the theoretical value is in figure 8 where, for comparison, was it added also the value of the voltage measured in the power meters (PK) installed in the RFQ scaled at the position of the x-ray detector. The error on the PK signal is the nominal power meter calibration error , around 0.2 dB, that corresponds to 5%.

4. Conclusions

The analysis of the bremsstrahlung x-rays produced by the electrons in a RF device such as the RFQ can be exploited to estimate the maximum accelerating field generated within the device. We proved that the obtained values from the measurements at different powers, from 450 to 800 kW in the RFQ, are in good agreement with the theoretical estimation from the model and from the readings obtained by the power meter PKs. The main source of errors in the analysis is the noise in the fit that can be reduced with long acquisition time.

We also proved that the ESS RFQ is producing an accelerating field according to the design and it is ready for the next phase of operations that is the beam commissioning. A similar technique of x-ray analysis was applied to the superconducting cavities at the ESS, using different kind of detectors, and the results are available in [7].

References

[1] Garoby R *et al.* 2017 *Phys. Scr.* **93** 014001

[2] Piquet O *et al.* 2019 *Proc. 10th Int. Particle Accelerator Conf. (IPAC'19)* (Melbourne, Australia) pp 855–57

[3] Letchford A, Bailey C, Church R, Duke J, Findlay D, Murdoch G, Stevens A, Schempp A and Vormann H 2000 *7th Eur Particle Accelerator Conf. (EPAC'00)* (Vienna, Austria) pp 833–35

[4] Ferdinand R, Beunard R, Di Giacomo M, Desmezieres V, Robillard P, Cazaux S, Desmons M, France A, Leboeuf D, Piquet O *et al.* 2006 *10th Eur Particle Accelerator Conf. (EPAC'06)* (Edinburgh, UK) pp 282–84

[5] Schneider P, Born D, Droba M, Lorey C, Meusel O, Noll D, Podlech H, Schempp A, Thomas B and Wagner C 2016 *28th Linear Accelerator Conf.(LINAC'16)* (East Lansing, MI, USA) pp 394–97

[6] Redus R 2021 Efficiency of XR-100T-CdTe Detectors Application Note (AN-CdTe-001 Rev. 1) `https://www.amptek.com/internal-products/xr-100t-cdte-cadmium-telluride-detector-efficiency-application-note`

[7] Maiano C G, Elias N, Laface E, Pierini P, Sagliano L, Wang M and Cenni E 2022 *Proc. 13th Int. Particle Accelerator Conf. (IPAC'22)* (Bangkok, Thailand) pp 1265–67

Journal of Physics: Conference Series **2420** (2023) 012092 doi:10.1088/1742-6596/2420/1/012092

Design studies on a high-power wide-band RF combiner for consolidation of the driver amplifier of the J-PARC RCS

H Okita[1], F Tamura[1], M Yamamoto[1], M Nomura[1], T Shimada[1], M Yoshii[1], C Ohmori[1], Y Sugiyama[1], K Hasegawa[1], K Hara[1] and M M Paoluzzi[2]

[1] J-PARC Center, JAEA & KEK, Tokai-mura, Japan
[2] CERN, Geneve, Switzerland

E-mail: hidefumi.okita@j-parc.jp

Abstract. A power upgrade of the existing 8-kW solid-state driver amplifier is required for the acceleration of high intensity proton beams on the J-PARC 3-GeV Rapid Cycling Synchrotron. The development of a 25-kW amplifier with gallium nitride (GaN) HEMTs, based on 6.4-kW modules is ongoing. The combiner is a key component to achieve such a high output power over the wide bandwidth required for multi-harmonic RF operation. This paper presents the preliminary design of the combiner. The circuit simulation setup and results, including the realistic magnetic core characteristics and frequency response of the cable, are reported.

1. Introduction

In the J-PARC 3-GeV Rapid Cycling Synchrotron (RCS), the consolidation of the amplifier chain of the RF system is desired [1]. The solid state amplifier is one of the important components of the amplifier chain. The output power and bandwidth of the existing solid state amplifier are 8 kW and 450 kHz–5.1 MHz, respectively. For the high power beam acceleration, improvements in both the output power and bandwidth are desired.

The development of the 25-kW amplifier with gallium nitride (GaN) HEMTs is ongoing. The target bandwidth is 100 kHz–10 MHz. The 25-kW amplifier consists of four 6.4-kW amplifier modules and one 25-kW combiner as shown in figure 1. In the current design, the bandwidth of the 6.4-kW amplifier module sufficiently covers our requirements. Hence, the realization of the high power combiner with the target bandwidth is the key of the development of the amplifier.

Transmission line transformers (TLTs) are the primary component of the 25-kW combiner. TLTs have wide bandwidths and high transmission efficiencies. A typical TLT consists of magnetic cores and transmission lines. To realize a wide bandwidth, it is necessary to design TLTs with a short cable length and a large number of winding turns.

Beyond 1 kW, the TLT design is challenging. Since high power coaxial cables are generally thick and stiff, the cable length and winding turn of the coil are inevitably longer and fewer. Furthermore, the power loss in the magnetic core should also be taken into account. We develop the TLT model using the electric circuit simulator, which can evaluate the bandwidth and power loss in the core considering the actual cable characteristics and complex permeability. A preliminary design of the 25-kW combiner is presented.

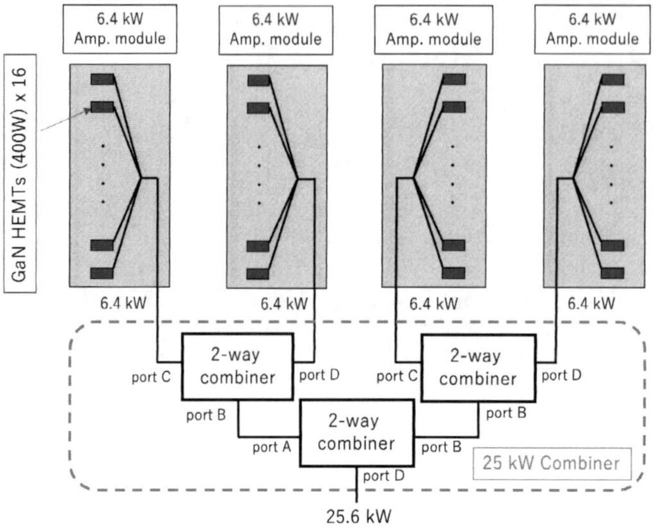

Figure 1. The configuration of 25-kW GaN amplifier.

2. Overview of the 25 kW combiner

The 25-kW combiner consists of three 2-way combiners as shown in figure 1. The 2-way combiner has four ports and consists of six TLTs as shown in figure 2. When P_A and P_B are input to ports A and B, ports C and D output $1/2(\sqrt{P_A} - \sqrt{P_B})^2$ and $1/2(\sqrt{P_A} + \sqrt{P_B})^2$, respectively. When P_C and P_D are input to ports C and D, ports A and B output $1/2(\sqrt{P_C} - \sqrt{P_D})^2$ and $1/2(\sqrt{P_C} + \sqrt{P_D})^2$, respectively.

The bandwidth of the 25-kW combiner depends on the TLT design. The cable attenuation due to the skin effect degrades the high frequency response. Cables with a low attenuation and short length should be employed. The isolation of the TLTs is associated with the low frequency response. If the isolation is insufficient, common mode signals are excited in the coaxial cables which causes the degradation of the low frequency response. Thus, a large number of turns and low loss magnetic core should be preferred to realize the required low frequency response.

3. Circuit simulation model of TLT

A circuit simulation model of TLTs is developed with LTspice [3]. Our model contains the skin effect, common mode propagation, and complex permeability, which are not trivial in LTspice simulations, to evaluate the bandwidth and core loss. The details of the TLT model are described below.

The developed TLT model consists of the coaxial cable model and core model. Figure 3 depicts the coaxial cable model. To simulate the skin effect, the parallel circuit consists of multiple resistors ($R1$–$R6$) and inductance ($L2$–$L6$) is introduced [4]. This parallel circuit represents a conductor line divided into multiple concentric circles. The number of stages in the parallel circuit varies depending on the accuracy and the upper frequency limit. In this case, the number of the stages is six. The values of each resistance and inductance are determined by the

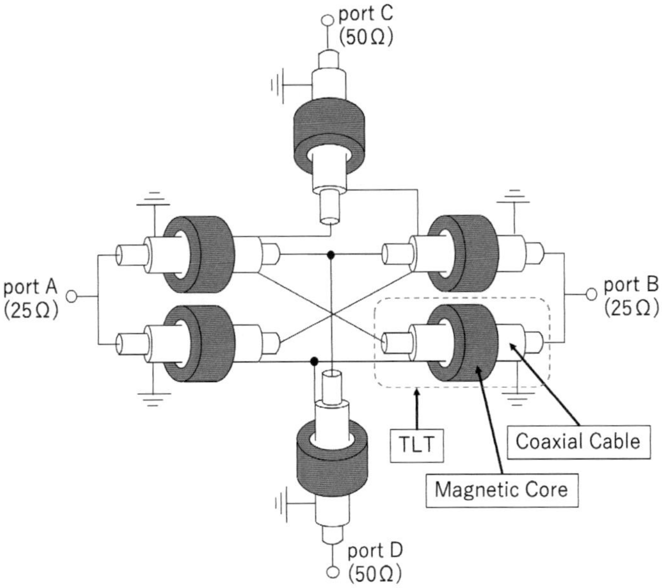

Figure 2. The configuration of the 2-way combiner [2].

property of actual cables. Details in derivation of each resistance and inductance are described in Ref. [4], and Ref. [5] provides the computational programs.

The coaxial cable model also includes the technique to simulate the common mode propagation. LTspice provides "TLINE" for simulating the signal propagation in a transmission line. However, "TLINE" only supports a single transmission line mode. To calculate the common mode signal propagation, we use two "TLINE" as shown in figure 3, which is recommended in the example simulation code provided by LTspice.

Magnetic cores with the complex permeability ($\mu = \mu' - j\mu''$) can be represented by an equivalent circuit, which consists of resistance and inductance varying with frequencies. In LTspice, "arbitrary behavioral voltage source" is used to simulate the frequency-dependent resistance and inductance. Figure 4 shows the toroidal coil model considering the complex permeability. Toroidal cores are represented by the "arbitrary behavioral voltage sources" [6]. The behavior of the voltage source $V(s)$ can be controlled using the Laplace variable $s = j\omega$. $V_L(s)$ and $V_R(s)$, corresponding to the inductance and resistance of toroidal cores, are

$$V_L(s) = s\mu' \frac{\mu_0 t N^2}{2\pi} \left(\ln \frac{b}{a}\right) (I_1(s) + I_2(s)) \quad \text{and} \tag{1}$$

$$V_R(s) = s\frac{-s}{|s|}\mu'' \frac{\mu_0 t N^2}{2\pi} \left(\ln \frac{b}{a}\right) (I_1(s) + I_2(s)), \tag{2}$$

where μ_0 is the magnetic permeability in vacuum, a, b and t are the inner and outer radius and thickness of toroidal cores, respectively, $I_1(s)$ and $I_2(s)$ are the current inner and outer conductors of coaxial cables, and N is the number of turns.

The parameters μ' and μ'' are introduced to the simulation model using the fitting function $g(x)$, which is given in Ref. [6] as

$$g(x) = \frac{10^{k_2 \log_{10} x + l_2}}{1 + 10^{(k_2 - k_1)\log_{10} x + (l_2 - l_1)}}, \tag{3}$$

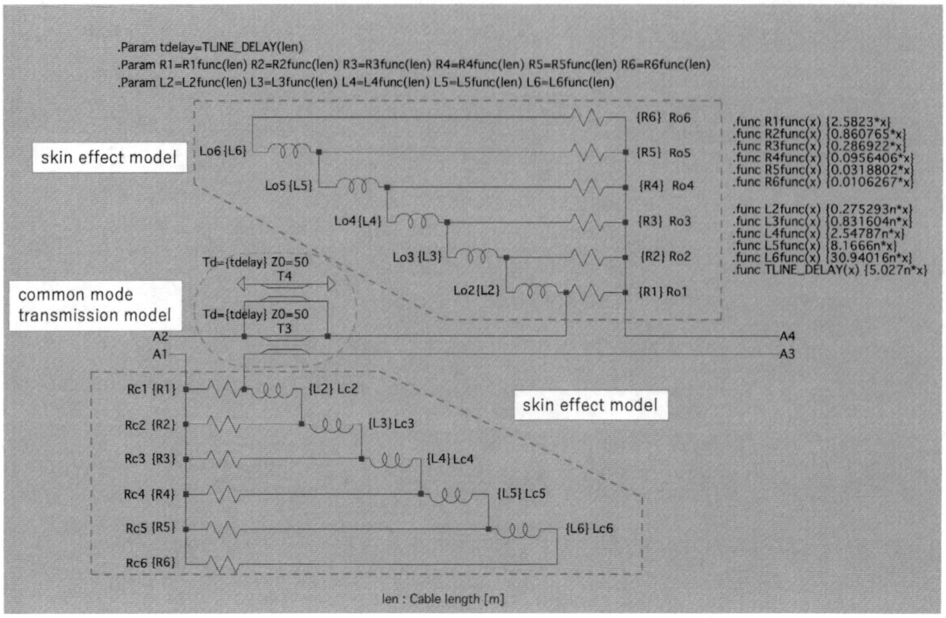

Figure 3. The coaxial cable model considering the skin effect and common mode signal propagation with LTspice. Parameters shown in this figure is for LMR-400-FR cable.

where k_1, k_2, l_1, l_2 are the parameters derived from the curve fitting of complex permeability data. Figure 4 shows the magnetic core model for FINEMET (FT-3L) core [7]. The curve fitting is performed on the complex permeability data of FT-3L in the frequency range of 10 kHz–10 MHz.

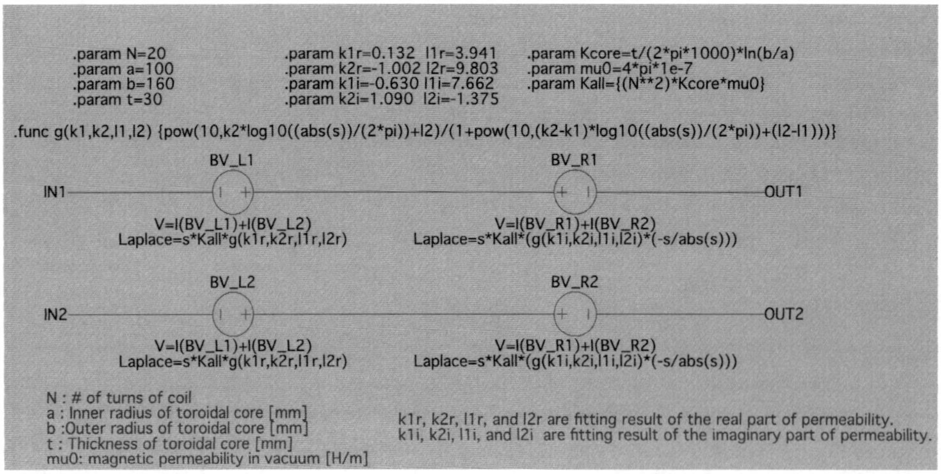

Figure 4. The toroidal core model considering the complex permeability with LTspice. Parameters shown in this figure are for FINEMET (FT-3L).

4. Design of 25-kW combiner

The overview of the 25-kW combiner is shown in figure 5. Two types of TLTs are designed, corresponding to each transmission power. For the TLT at the output port (TLT-A), the transmission power is 25.6 kW. For the other TLTs (TLT-B), the transmission power is less than 6.4 kW.

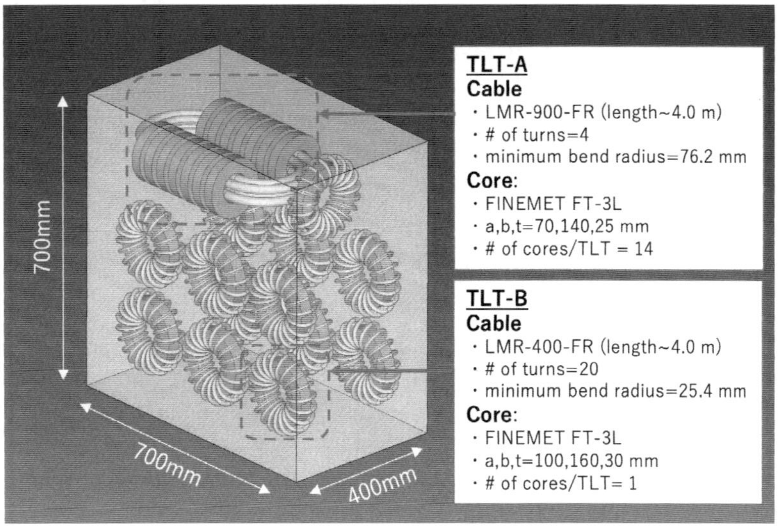

Figure 5. Overview of the 25 kW combiner.

For coaxial cables, LMR series from Amphenol [8], which have a high power handling and flexibility, are employed. LMR-900-FR and LMR-400-FR are employed for the TLT-A and TLT-B, respectively. For the magnetic core, FT-3L, which has a high relative permeability with a low core loss, is employed.

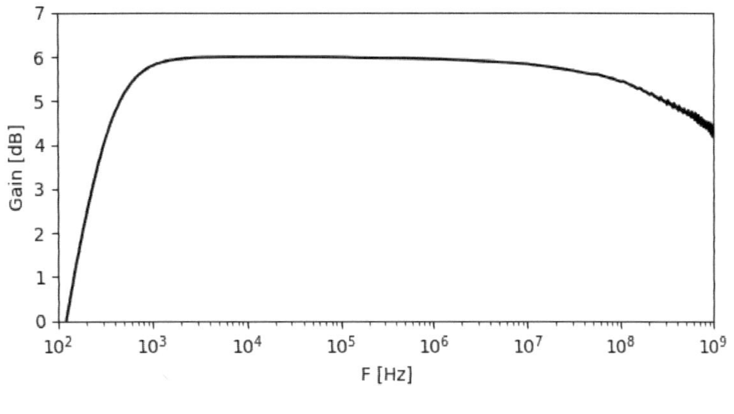

Figure 6. Bandwidth of the designed 25-kW combiner calculated using LTspice.

For the TLT-B, a toroidal coil with a single core is designed thanks to the small minimum bend radius (25.4 mm) of LMR-400-FR. For the TLT-A, the minimum bend radius of LMR-900-FR is not sufficiently small (76.2mm) to implement a single core configuration. Therefore, a coil with multiple magnetic cores is designed. The cable length is set to 4 m. The number of turns for each TLT is 4 and 20 for the TLT-A and TLT-B, respectively.

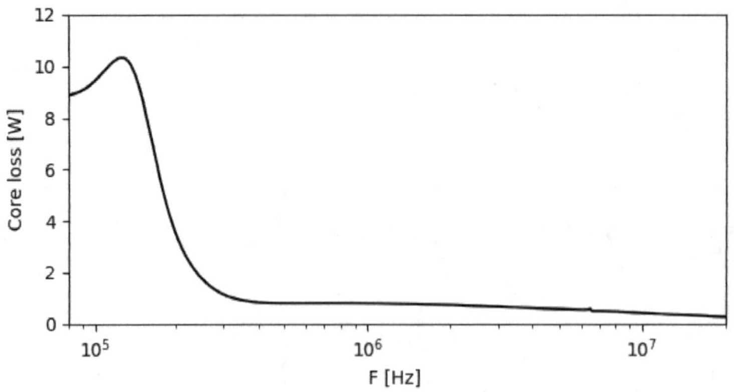

Figure 7. Core loss of the designed 25-kW combiner calculated using LTspice.

The LTspice model of the 25-kW combiner is developed using the TLT model described in the previous section. The simulated bandwidth of the combiner is shown in figure 6. To observe the bandwidth clearly, the frequency range is set to 100 Hz – 1 GHz. In the frequency range out of 10 kHz – 10 MHz, for the complex permeability, the extrapolation of the fitting results is used. The drop in the gain at the higher frequency side is due to the skin effect, while the low frequency side is due to the common mode propagation. The bandwidth of the designed 25-kW combiner sufficiently covers our requirement, from 100 kHz to 10 MHz. Figure 7 shows the core loss. The core loss is expected to be very low, approximately 10 W maximum.

5. Summary

The development of the 25 kW amplifier with GaN HEMTs is ongoing for the consolidation of the RF system in the RCS. Design studies for the 25-kW combiner, which is a key component of the 25-kW amplifier, are conducted. Circuit simulator models that include the skin effects, common mode propagation, and complex permeability, which are essential for evaluating bandwidth and core loss of the combiner, have been developed. The preliminary design of the 25 kW combiner is established. The circuit simulation result shows that the designed 25-kW combiner has a wide bandwidth and a low core loss, which covers our requirements. A prototype of the 25 kW combiner is to be constructed.

References

[1] Tamura F *et al.* 2011 *Nucl. Instrum. Methods Phys. Res., Sect.* A **999** 165211
[2] Barabas U 1979 *IEEE Trans. Microwave Theory Tech.* **MTT-27**
[3] Analog Devices, LTspice version 17.0.21.0 `https://www.analog.com/en/design-center/design-tools-and-calculators/ltspice-simulator.html`
[4] Yen C-S, Fazarinc Z and Wheeler R K 1982 *Proc. of IEEE* **70** 750
[5] Transmission Line (Cable) SPICE Model `http://www.mogami.com/e/cad/spice-tline.html`
[6] Nomura K *et al.* 2016 *Proc. IEEE Applied Power Electronics Conf. and Exposition (APEC)* (Long Beach, California USA) pp 3146–52
[7] Hitachi Material Ltd `https://www.hitachi-metals.co.jp/e/products/elec/tel/p02_21.html`
[8] LMR cable `https://www.timesmicrowave.com/DataSheets/Categories/high-power-10-17.pdf`

CERN Linac4 Chopper Dump: operational experience and future upgrades

C J Sharp, R Franqueira Ximenes, M Calviani, D Grenier, C Y Mucher, E Grenier-Boley, C Torregrosa, J R Hunt, L S Esposito, G Costa, P Andreu Muñoz and A M Krainer

European Organization for Nuclear Research (CERN), CH-1211 Geneva, Switzerland

E-mail: calum.james.sharp@cern.ch

Abstract. The Chopper Dump in the Linac4 accelerator at CERN is a beam-intercepting device responsible for the absorption of the 3 MeV H^- ion beam produced by the Linac4 source and deflected upstream by an electromagnetic chopper. It allows a portion of the beam, which would otherwise fall into the unstable region of the radiofrequency buckets in the Proton Synchrotron Booster, to be dumped at low energy with minimal induced radiation. It may also be used to absorb the entire beam. With peak currents of 25 to 45 mA and shallow penetration, this results in large deposited energy densities, thermal gradients and mechanical stresses. Additional constraints arise from geometric integration, vacuum and radiation protection requirements. Material selection, beam-matter interaction studies and thermo-structural analyses are important aspects of the design process. The Chopper Dump underwent modification in 2019 following observed material degradation in the original version of the device. The experience gained, modifications made and observations noted since then are detailed herein. Against this background, the design and analysis of an upgraded device, intended to cope with future operational conditions, is outlined and discussed.

1. Introduction

Chopping of the 3 MeV H^- beam in CERN's Linac4 accelerator allows bunches that would fall into the unstable region of the radiofrequency buckets of the PS booster to be dumped at low energy [1]. A chopping system consisting of electrostatic plates and quadrupole magnets is used to deflect particles vertically onto a conical dump in the medium energy beam transport (MEBT) line. Chopping at low energy minimises the required chopping element strength and induced radiation, but results in more demanding low-penetration conditions for the absorbing surface. The dump must absorb the entire beam when required and dissipate the resulting heat so as not to exceed material limits on temperature or stress. Additional challenges in the design arise from tight geometric constraints, vacuum and radiation protection.

Material degradation in a previous version of the dump prompted installation of a modified version during 2019 [2]. This has operated successfully at 25 mA peak dumped current but is predicted not to withstand future conditions up to 45 mA. Activation due to ^{65}Zn production also restricts long term use. A project to consolidate the dump is ongoing.

2. Operational experience

2.1. Nickel-coated device

The original dump, first operated in 2013, consisted of a GlidCop® AL-60 core (figure 1a) – a dispersion-strengthened copper-based metal-matrix composite desired for its high thermal conductivity and good mechanical properties – with the internal surface coated in nickel to reduce activation [3].

2.2. Observed degradation

In 2019, material degradation in the form of delamination and local melting of the nickel was observed (figure 1b), a concern for device integrity and potential introduction of debris in the beamline. Subsequent analysis concluded that the degradation was a result of a loss of local heat conduction due to subsurface cracks or voids in the coating, potentially caused by plastic deformation and low-cycle fatigue during the pulse-cooldown cycle [2]. Hydrogen implantation and blistering was identified as another possible cause.

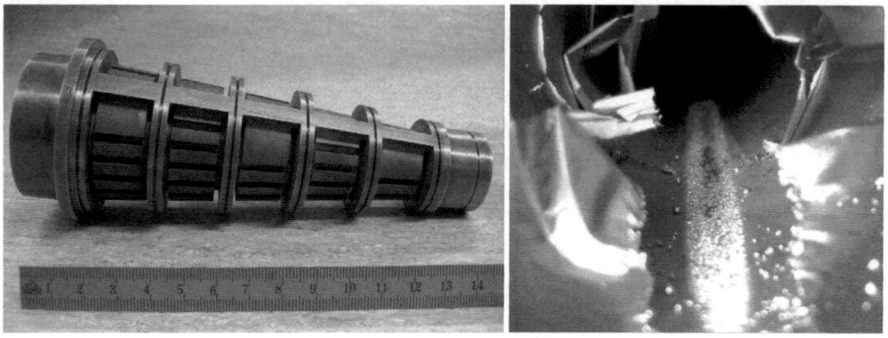

Figure 1. (a) GlidCop® core, (b) damaged Ni coating [2].

2.3. Modifications

A modified version of the dump with the nickel coating removed (except for a small region at the aperture) was installed in August of that year. With the beam impinging directly on the GlidCop® substrate, this provides better heat conduction and increased resilience owing to the material's superior thermo-mechanical properties. Device activation is increased but acceptable in the short term.

2.4. Operation and subsequent inspection

The modified device has operated during Linac4 commissioning and the start of Run 3 during 2019-2022. An inspection in the 2021 year-end technical stop observed no sign of melting or delamination problems as experienced in the previous device. Some minor changes were noted, however are not thought to have a significant impact on continued operation in 2022. These included discolouration (darkening) approximately in the region of beam impact (figure 2) thought to be caused by migration of intrinsic carbon to the surface of the copper. This phenomenon, particularly under annealing temperatures and in the presence of hydrogen (of which the beam is a source), is detailed by studies on the chemical vapour deposition of graphene [4, 5].

Reflective surface marks could indicate some minor deterioration of ridges left behind after machining. However, no worsening of damage to the remaining nickel at the aperture of the dump was observed. Finally, it was noted that the beam impact position appears further downstream

Journal of Physics: Conference Series **2420** (2023) 012093 doi:10.1088/1742-6596/2420/1/012093

than the nominal case, potentially putting at risk downstream equipment and supporting the inclusion of a more comprehensive alignment system in future.

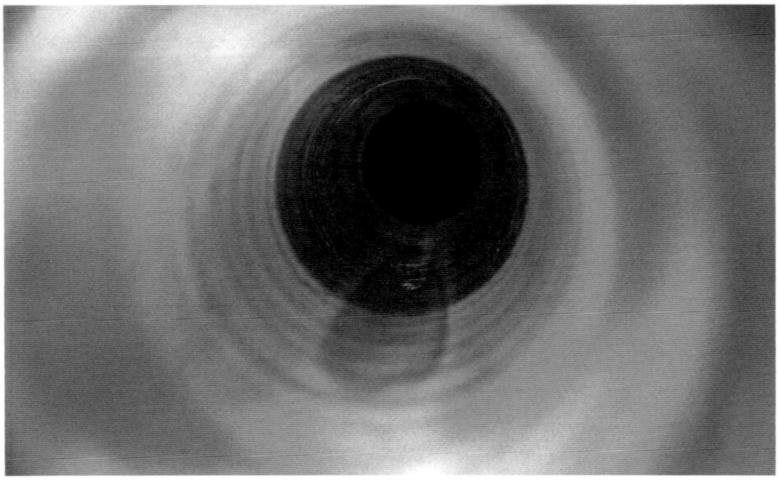

Figure 2. 2021 inspection - GlidCop® surface.

3. Upgrade

Whilst operating successfully at 25 mA dumped current, finite element studies indicate the dump cannot withstand higher intensities planned for the future [6]. Radiological constraints on production of ^{65}Zn from direct exposure of copper to the beam also prohibit its use in the long term.

The upgrade currently underway is specified to withstand peak dumped currents of 27 mA (Run 3) and 45 mA (Run 4) with dumping of the entire 800 µs pulse at 0.83 Hz. This results, respectively, in 81 kW and 135 kW pulse power, and 54 W and 90 W average power (an 80% increase on the existing 25 mA specification which yields 75 kW pulse power and 50 W average power). The exclusion of highly conductive copper as a material option makes managing the concentrated energy deposition an even greater challenge.

3.1. Conceptual design

The initial idea was to create a device similar to the existing one, but using Inconel® 625 (a nickel superalloy) as the core material. However, whilst this offered advantageous properties, mechanically and radiologically (as isotopes produced in nickel are short-lived) [7], its low thermal conductivity resulted in unacceptably high temperatures.

This prompted a more comprehensive redesign (figure 3) with R4550® isostatically-formed graphite [8], which is used extensively in beam-intercepting devices [9, 10]. However, its porosity means it cannot serve as the vacuum chamber nor host water-cooling channels. Hence, a shrink-fitted stainless steel jacket with a helical cooling channel was introduced with a precise interference to ensure good thermal contact.

A 3-point adjustable support platform and fiducial markers were included to facilitate full control of the alignment.

3.2. Beam-matter interaction studies

The Fluka Monte-Carlo simulation code [11, 12] was used to simulate the beam-material interaction. This provided a distribution of deposited energy density throughout the material,

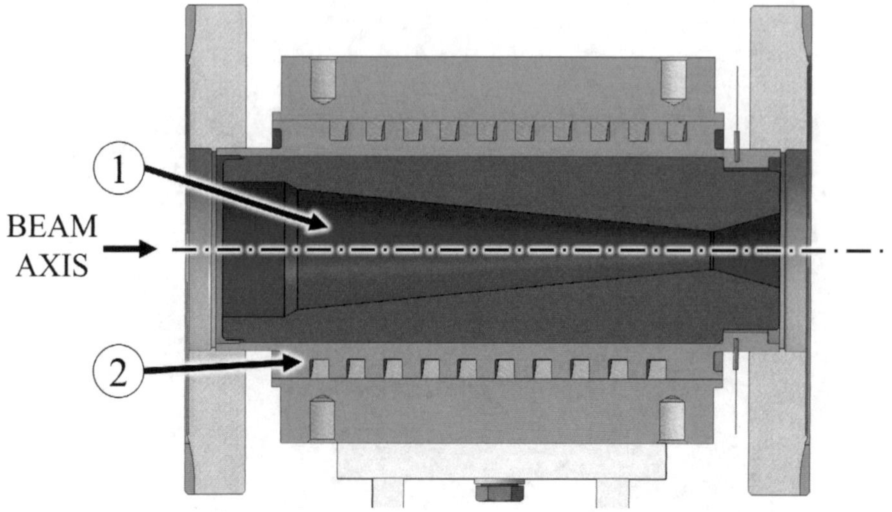

Figure 3. CAD model vertical cross section - graphite core (1) with cooling jacket (2).

highlighting the shallow penetration of around 20 µm into the conical surface and high deposited energy densities up to 34.2 kJ/cm^3 per pulse in the 45 mA case.

3.3. Finite element studies

Thermal (steady-state and transient) and structural (steady-state and quasi-static) finite element analysis (FEA) was completed in ANSYS® Workbench 2020 R2 [13].

A two-scale modelling approach consisted of a coarse global model (figure 4) of the core and cooling jacket plus a finer sub-model of the beam-intercepting surface.

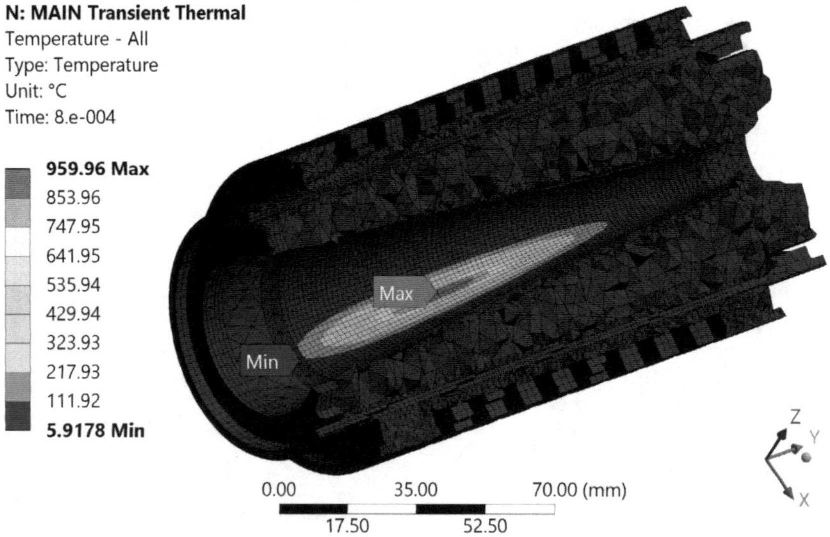

Figure 4. Global model - temperature plot (end of pulse).

Meshing created a global model with 432 000 elements and a sub-model with 78 000 elements with sizing controls used to match the array of energy deposition data. Element aspect ratio was unavoidably high; however, with small deformations and well-formed elements, this was deemed acceptable and confirmed by basic mesh convergence studies.

Thermal boundary conditions in the global model consisted of a convection condition applied on the water-cooling channel with 25 °C fluid temperature and a heat transfer coefficient of $7000\,\mathrm{W/m^2K}$, plus the beam loading applied as a 2D surface heat flux (flattened from the 3D energy deposition array). Then, in the sub-model, temperatures from the global model were transferred to the cut boundary surfaces, and the beam loading was applied with a higher-fidelity three-dimensional distribution of heat generation.

Structurally, the shrink-fitting pre-stress was accounted for by including the real interference in the global model geometry and resolving the contact to a specified penetration tolerance at the beginning of the simulation. A fixed support boundary condition was applied on the downstream jacket edge which will be welded to the downstream flange. In the sub-model, cut-boundary displacements were transferred from the global model. In both cases, the temperature results from thermal simulations were applied to simulate the thermal stress.

The sub-model results (table 1) indicated, in the 45 mA case, a peak temperature of 1054 °C. Although high, this remains below the limits imposed by the onset of sublimation.

Table 1. Selected thermo-structural FEA results - full dumping of 800 µs pulse.

Result		27 mA	45 mA
Steady state max. temperature	[°C]	35.2	42.2
Transient peak temperature	[°C]	629.8	1054.2
Minimum principal stress	[MPa]	(-)46.9	(-)75.7

Stresses in the core are generally compressive owing to the shrink-fitting stress. In the region of maximum stress, bi-axial compression dominates. The failure criterion is therefore considered simply as when the minimum principal stress exceeds the compressive strength. For bi-axial compression, this is consistent with the Mohr-Coulomb criterion implemented in ANSYS®. Comparing the worst case minimum principal stress with an ultimate compressive strength of 130 MPa yields an acceptable factor of safety of 1.7.

The calculation to specify the shrink-fitting interference between the core and cooling jacket was carried out analytically in the first case, using the Lamé equations for thick cylinders, and later with finite element simulations. Manufacturing limitations required that calculations be made at the limits of the tolerances so as to ensure sufficient interface pressure and acceptable stresses in all cases. The diametric interference specified was 30-70 µm.

3.4. Fatigue

A phenomenological S-n curve assessment of fatigue due to the 0.83 Hz pulse-cooldown cycle highlighted slim safety margins, although uncertainties were large owing to a lack of material data. A second calculation considered that the cyclic stresses are extremely superficial, and that even if a crack were to exist on the inner core surface penetrating to a depth of 0.3 mm (where the stress is effectively non-cyclic), the crack tip stress intensity factor would likely not exceed the fracture toughness and the crack would not propagate.

3.5. Cooling system performance

Cooling system calculations were performed with analytical equations using the empirical Dittus-Boelter correlation for calculation of the Nusselt number and thus the heat transfer coefficient.

With the 6 mm square channel specified, this yielded an estimated heat transfer coefficient of 7000 W/m^{2}K which was then applied in the FEA studies.

Preliminary CFD analysis broadly confirmed the analytical findings, although velocities near the inner wall may be reduced by centrifugal effects. However, the system is not considered particularly sensitive to this as critical temperatures are dominated by local conduction on a short timescale.

3.6. Microstructural effects
Whilst temperatures in the homogeneous FEA were well below sublimation point, there is some concern that isolated grains in the microstructure of graphite with reduced capacity for conduction could reach higher temperatures.

An analytical micro-model of an isolated 10 µm grain of graphite with the peak deposited energy density applied and radiation to the surrounding bulk material modelled by the Stefan-Boltzmann law indicated temperatures may indeed reach sublimation point.

In a parallel approach, integrating the specific heat capacity over temperature up to 3800 K yields an energy requirement of 7050 kJ/kg. Filtering the energy deposition map by this threshold gives a region (accounting for up to 45% or 48 J of energy deposition) in which isolated grains may adiabatically reach this temperature. If a sublimation energy requirement of 41 900 kJ/kg is now considered (adding latent heat to the previous figure), and assuming that 1% of the material consists of isolated grains, then the sublimation rate is estimated as 1.14×10^{-8} kg per pulse.

If sustained, this equates to 0.034 g per hour, which would be unacceptable. However, as isolated grains sublimate, the vulnerable portion of material is expected to deplete, meaning sublimation may cease after an initial "bake-out" period. Re-condensation and deposition may also occur.

Experimental tests in a 45 keV test stand are planned to assess this potential for surface degradation under low energy beam conditions.

3.7. Device production
Trials have been carried out to verify the assembly process and associated precision manufacturing requirements. Shrink-fitting requires that the core and jacket be heated to 250 °C such that different degrees of thermal expansion generate a clearance to allow insertion of the core. Metrology analysis of the interface surfaces will ensure compliance with tolerances and reduce uncertainty over the interface pressure, thermal contact and stresses in the core.

The graphite will also undergo vacuum firing treatment to reduce outgassing at high operational temperatures, followed by residual gas analysis to assess vacuum compliance.

4. Conclusions
Building on past operational experience, a new design for CERN's Linac4 Chopper Dump has been proposed and analysed. Whilst further studies on the integrity of graphite under shallow beam penetration are ongoing, the device is foreseen to be installed in 2022/23.

References
[1] Lombardi A 2014 Commissioning of the Low-Energy Part of Linac4 Tech. Rep. CERN-ACC-2014-394 CERN, Geneva, Switzerland
[2] Torregrosa Martin C, Ciccotelli A and Esposito L S 2021 Thermo-mechanical analysis and reconditioning of the Linac4 chopper internal dump Tech. Rep. EDMS:2384553 CERN, Geneva, Switzerland
[3] Arnaudon L *et al.* 2006 Linac4 Tech. Rep. CERN-AB-2006-084, CARE-Note-2006-022-HIPPI CERN Geneva, Switzerland
[4] Khaksaran M H and Kaya I I 2019 *ACS Omega* **4** 9629
[5] Çelik Y, Escoffier W, Ming Y, Flahaut E and Suvacı E 2016 *Carbon* **109** 529

[6] Widorski M, Torregrosa Martin C, Calviani M and Lallement J 2021 Linac4 medium energy beam transport line dump Tech. Rep. EDMS:2435672 CERN, Geneva, Switzerland
[7] Bruno L, Magistris M and Silari M 2004 Conceptual design and radiological issues of a dump for the 3 MeV test facility Tech. Rep. CERN-SC-2004-008-RP-TN CERN Geneva, Switzerland
[8] SGL Carbon 2019 Sigrafine® R4550 Data Sheet TDS R4550.00
[9] Leitão I and Maglioni C 2014 Design specifications for the Linac4 main and LBE dump cores Tech. Rep. EDMS:1256894 CERN, Geneva, Switzerland
[10] Maglioni C 2013 The Linac4 commissioning dump Tech. Rep. EDMS:1182648 CERN, Geneva, Switzerland
[11] Ahdida C *et al.* 2022 *Front. Phys.* **9** 788253
[12] Battistoni G *et al.* 2015 *Annals of Nuclear Energy* **82** 10
[13] ANSYS® Ansys Mechanical Enterprise 2020 R2

TURBO: A novel beam delivery system enabling rapid depth scanning for charged particle therapy

J SL Yap[1], S L Sheehy[1,2], A F Steinberg[1,3,4], H XQ Norman[1,3,4] and R B Appleby[3,4]

[1] School of Physics, University of Melbourne, Melbourne, VIC 3010, Australia
[2] Australian Nuclear Science and Technology Organisation (ANSTO), Lucas Heights, NSW 2234, Australia
[3] Cockcroft Institute, Daresbury, Warrington WA4 4AD, United Kingdom
[4] Department of Physics and Astronomy, University of Manchester, Manchester M13 9PL, United Kingdom

E-mail: jacinta.yap@unimelb.edu.au

Abstract. Charged particle therapy (CPT) is a well-established modality of cancer treatment and is increasing in worldwide presence due to improved accelerator technology and modern techniques. The beam delivery system (BDS) determines the overall timing and beam shaping capabilities, but is restricted by the energy variation speed: energy layer switching time (ELST). Existing treatment beamlines have a $\pm 1\%$ momentum acceptance range, needing time to change the magnetic fields as the beam is delivered in layers at various depths across the tumour volume. Minimising the ELST can enable the delivery of faster, more effective and advanced treatments but requires an improved BDS. A possibility for this could be achieved with a design using Fixed Field Alternating Gradient (FFA) optics, enabling a large energy acceptance to rapidly transport beams of varying energies. A scaled-down, novel system – Technology for Ultra Rapid Beam Operation (TURBO) – is being developed at the University of Melbourne, to explore the potential of rapid depth scanning. Initial simulation studies, beam and field measurements, project plans and clinical considerations are discussed.

1. Introduction

The use of charged particle beams as a therapeutic modality has several recognised advantages over conventional treatments as the characteristic 'Bragg Peak' enables a precise amount of radiation to be delivered with a greater radiobiological effect. High facility capital costs are prohibitive and new developments in delivery techniques, accelerators and related technologies, along with improved clinical and biological outcomes can better exploit CPT benefits and treatment accessibility. The emergence of methodologies such as ultra-high dose rate 'FLASH' [1], Arc [2] and multi-ion [3] therapies also place greater demands on the capabilities of the BDS: significant improvements are needed to overcome existing technical limitations [4].

Although there are many factors which contribute to the delivery process and treatment time, the ELST is a bottleneck constraint and this deadtime whilst waiting for magnetic field adjustments to transport different beam energies has several implications, impacting treatment efficiency and efficacy [5]. As most facilities offer state-of-the-art treatments with active pencil beam scanning, the ELSTs accumulate and longer beam delivery times translate to higher costs, lower patient throughput and can decrease treatment quality: physical uncertainties and sensitivities with motion cause inferior dosage i.e. 'interplay effects' [6]. Numerous mitigation approaches [7] currently used in the clinic to manage uncertainties can be complex to apply and further extend treatment times. Shorter ELSTs, thus faster delivery with the possibility of improved dosimetric quality and robust plans are also particularly beneficial for treating mobile sites [8].

The need for better immobilisation has led to upright treatments being revisited [9] and could offer practical and clinical benefits, notably enabling fixed beamline: gantry-less treatments [10]. This presents the opportunity of not only added cost reduction [11], but in combination with a large energy acceptance BDS with fast ELST, can further improve the effectiveness of standard pencil beam scanning and accommodate new delivery techniques. The applicability for heavier ions could be even greater than for protons given the impracticality of a gantry due to the complexities and infrastructure cost; the increased radiobiological factor may also favour fewer fields and fractions [12]. The clinical advantages and requirements will direct the development of TURBO as a scaled-down, proof-of-concept demonstrator to de-risk these concepts and explore the feasibility of a large energy acceptance BDS for CPT.

2. TURBO

TURBO comprises several interchangeable modules (A–F as shown in figure 1) adapted for the UniMelb low energy ion '*Pelletron*' accelerator [13] (0.5–3.5 MeV protons & He at 100's nA to several µA). The FFA section will be curved to accommodate the system requirements for a clinical setting, followed by a fast scanning system downstream to replicate active pencil beam scanning. A self-developed control system will manage all component operation, where precision and synchronicity is required to enable rapid beam delivery.

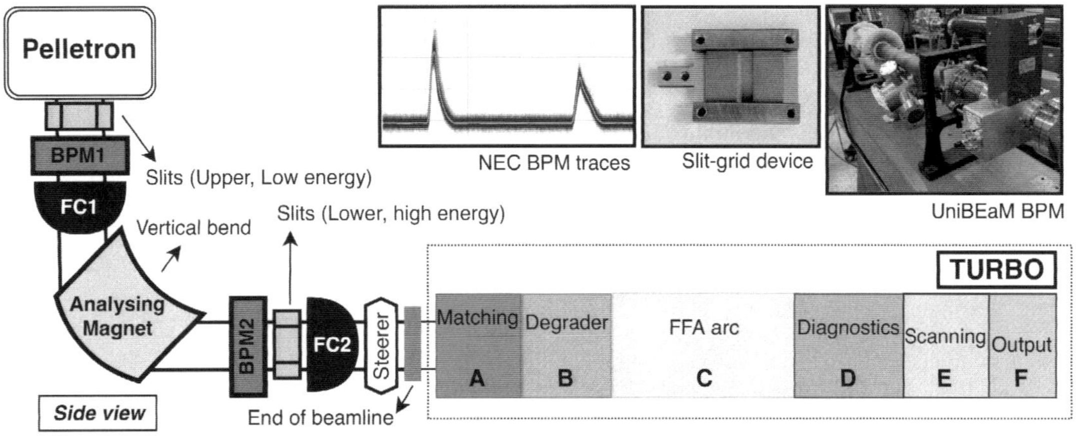

Figure 1. Schematic of the Pelletron and TURBO with key components labelled. Two pairs of vertical and horizontal slits (typically 2.54 mm apart) collimate the beam tails and the vertical bending magnet ensures the desired beam energy. NEC BPM traces, the slit-grid device (with tantalum sheets) and the UniBEaM detector integrated into the beamline are shown.

Initial measurements and beam characterisation studies have been initiated to establish the suitability of the Pelletron for TURBO. Two NEC helical wire beam profile monitors [14] (BPM1,2) and Faraday cups are situated together with the upper and lower slits. These BPMs measure both the horizontal and vertical beam distributions simultaneously but do not give the beam centre position. To determine the beam profile and position, a UniBEaM fibre scanner system [15, 16] was integrated into the beamline ~2.10 m downstream of the lower slits. A single motorised (200 µm diameter) optical fibre linearly translated across the beam path, provides a signal at each position from detected scintillation light. Three sets of initial measurements were recently obtained with the fibre scanner in the vertical plane (figure 2).

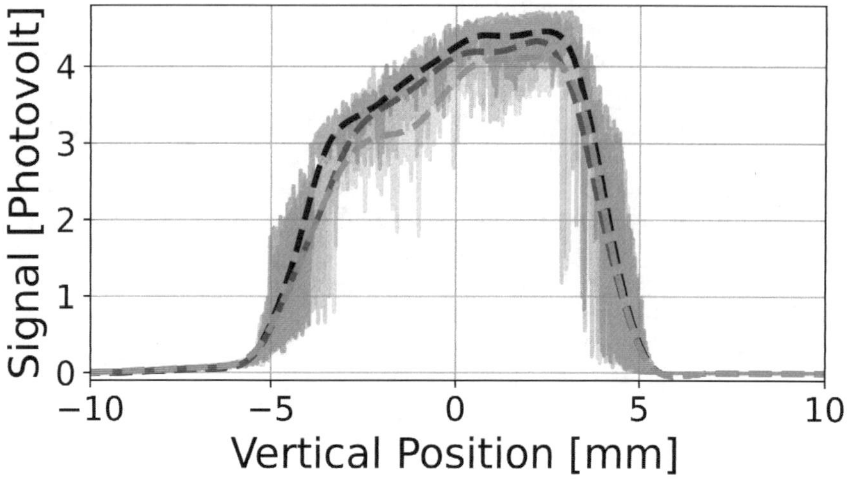

Figure 2. UniBEaM vertical beam profiles. The raw data (shaded), are passed through a low-pass filter to reduce noise.

As full characterisation has not been previously carried out on this machine, slit-grid measurements (figure 1) are planned to resolve information about the Pelletron beam phase space and optical parameters [17], and energy measurements using a dipole as a simple spectrometer. Initial modelling suggests a drift of ~1 m between the dipole and UniBEaM should produce an energy resolution <10% at 3.5 MeV. Supplementary imaging diagnostics are also anticipated using (30 & 50 µm thick) YAG screens.

A degrader is necessary to rapidly change the beam energy, requiring a design specific for the low energy range and beam conditions. As protons are easily stopped, varying layers of different thickness materials were considered; effects of Kapton® film were investigated by initial simulations using the Monte Carlo code TOPAS (v3.7.0) [18]. An ideal beam with no angular spread but a uniform distribution over the aperture of the lower slits was generated. A beam of 3.5 MeV (mean) protons can be completely degraded to 0.1 MeV with 150 µm of Kapton and various energy steps can be attained by combining various standard thickness films: 31 µm generates 3.14 MeV protons; 50 µm, 2.74 MeV; 75 µm, 2.30 MeV; 100 µm, 1.77 MeV; 133 µm, 1.45 MeV and 131 µm, 0.90 MeV. The energy spread grows linearly with increasing Kapton thickness due to multiple coulomb scattering: the beam distribution becomes more Gaussian but the transmission decreases. Subsequent simulations with realistic beam parameters will indicate the need for collimators or additional mitigation to control the spread.

3. Optics

Although 'FFA optics' have previously been demonstrated [19], they have never been integrated into an existing accelerator. For TURBO, the beamline is constrained by the beam parameters at either end of the FFA arc: the orbit offsets are not a function of energy at either end, so the optics must constitute a closed-dispersion arc. Also, having zero dispersion at the output makes quality assurance and beam scanning easier. For the input from the Pelletron, we are currently investigating whether a variable matching section is required to ensure that the beam parameters are energy-independent.

Three methods of creating a large energy acceptance arc with zero dispersion at the start and end are discussed in table 1. One is to use an 'adiabatic transition', where the strength of the bending field is smoothly increased over many cells, reaching a maximum at the midpoint of the arc before decreasing symmetrically. Another would be to construct specialised matching cells, to map the input beam optics and offsets to a linear arc by controlling beam trajectories with nonlinear fields and shaped magnet edge angles. The final method is to use nonlinear magnets throughout the arc, setting multipole strengths as are required for stability. Based on the considerations in table 1, a closed dispersion arc utilising nonlinear optics throughout should provide the most control over the beam dynamics for the full range of energies, while keeping the arc short enough to be viable.

Table 1. Comparison of methods to produce a closed dispersion FFA arc.

Method	Advantages	Disadvantages
Adiabatic Transition	Uses only linear magnets	Smooth transition requires a long beamline
	Simple to design and construct	Matching is always approximate
Matching Cells	Allows for periodic linear optics in arc	Errors in matching magnets amplified in arc
	Nonlinearity limited to a short region	Dynamics in arc hard to control for all energies
Full Nonlinearity	Fully control dynamics for all energies	Correcting magnet errors over full arc is difficult
	Tuneable dispersion throughout the arc	Aperiodicity requires unique magnets at each point

Fixed field accelerators have been operated using both electromagnets and permanent magnets [19, 20]. For the FFA arc, commercially available permanent magnet blocks are arranged in a plastic 3D printed holder, producing a Halbach array [21] with the desired fields; this is a simplified version of the magnets used for CBETA [22]. An advantage of this method is that is it straightforward to design and construct each unique magnet in the arc rapidly, ensuring that they have the right multipole fields. If an array is no longer required, it is easy to remove and repurpose the magnetic blocks. To verify our plans, a prototype dipole has been modelled, constructed, and measured. Using a maximum of 44 magnetic blocks ($10\times5\times25$ mm^3), the field was compared to modelling in Magpylib [23]. For this prototype (figure 3), the internal field was maximised by using 14 sets of 3 magnet blocks in a Halbach cylinder. Although the field quality is poor ($\pm1.65\%$ in a good field region of ±15 mm along the midplane), it will be improved with an increased quantity of magnetic blocks in future.

4. TURBO concept for heavy ions

Alongside development of the scaled-down TURBO beamline, work is underway towards a full clinical energy design for transportation of heavier ions. To realise the design, superconducting magnets are required for increased magnetic fields; Canted Cosine Theta (CCT) is a potential arrangement. The coil geometry produces pure dipole fields whilst cancelling unwanted solenoidal fields [24]; higher order multipoles can be adjusted by modifying the coil windings. Other single-pass optics studies [25, 26] have presented curved, alternating gradient (AG)-CCT configurations, consisting of focusing-defocusing quadrupole layers inside CCT dipoles; [25] reports a momentum acceptance of $\pm25\%$. Similarly, an AG-CCT arrangement could achieve strong focusing in TURBO to transport a wide range of rigidities without magnet ramping during treatment. Additional clinical specifications being considered for TURBO are described in [4], with magnetic requirements listed in table 2. The magnetic field ranges are calculated firstly for a bending radius of 2 m. To transport the particles to full treatment energies, three modes are considered. To maintain a similar momentum acceptance from accelerating p^+ to $^{16}O^{8+}$ requires: B_{dip} = (0.62–1.23) T for $E_{p^+,\text{min}}$ to $E_{\text{He}^{2+},\text{min}}$ ($\frac{\delta P}{P} = \pm33\%$); B_{dip} = (1.25–2.20) T for $E_{\text{He}^{2+},\text{min}}$ to $E_{\text{C}^{6+},\text{mid}}$ ($\frac{\delta P}{P} = \pm33\%$); B_{dip} = (2.20–3.65) T for $E_{\text{C}^{6+},\text{mid}}$ to $E_{\text{O}^{8+},\text{max}}$

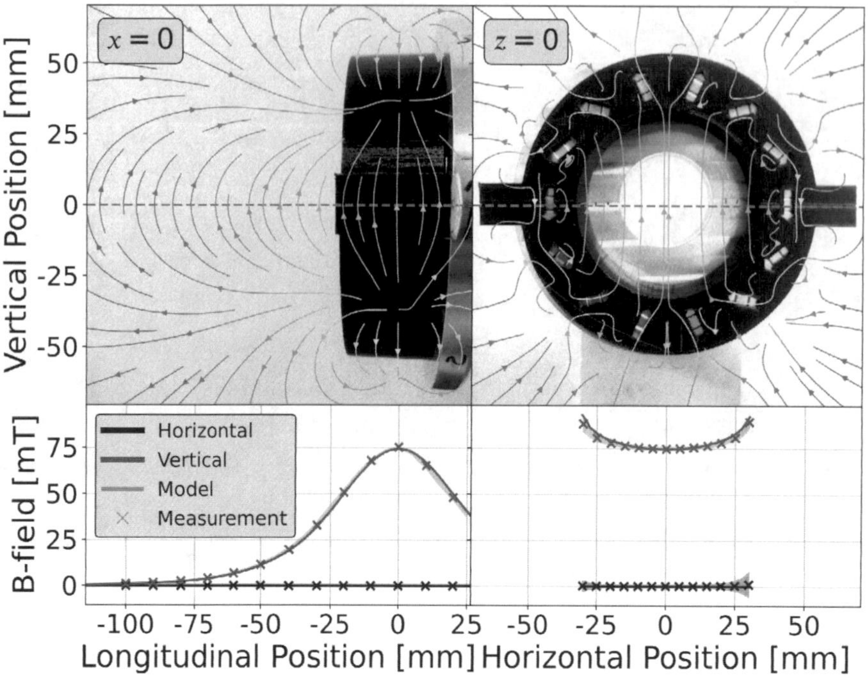

Figure 3. Prototype Halbach dipole with simulated fieldmap. Measurements along the midplane are compared to the predicted field in the lower part of the figure.

($\frac{\delta P}{P} = \pm 28\%$). The bore radius of the magnet must also accommodate a suitable treatment field size at patient isocentre (minimum 10×10 cm^2 [27]). The exact design parameters will depend on the optics to determine what higher order multipoles are tolerable. Future modelling of a CCT magnet for TURBO will consider unwanted longitudinal field components in beam dynamics studies [28] and detailed particle tracking through the full 3D field map.

Table 2. Key clinical parameters for p^+ to $^{16}O^{8+}$.

Species	E [MeV/u]	$B\rho$ [Tm]	$\pm\frac{\delta P}{P}$	B_{dip} [T]
p^+	70–250	1.23–2.43	33%	0.6–1.22
$^4He^{2+}$	70–250	2.43–3.32	33%	1.22–1.66
$^{12}C^{6+}$	70–430	2.50–6.63	45%	1.25–3.32
$^{16}O^{8+}$	70–500	2.50–7.26	49%	1.25–3.63

The modularity of TURBO allows the flexibility to explore and iterate between different concepts: further experimental, simulation and design work will progress the development of a laboratory implemented demonstrator, to improve beam delivery for existing and future clinical CPT facilities.

References

[1] Favaudon V *et al.* 2014 *Sci. Transl. Med.* **6** 245ra93
[2] Ding X, Li X, Zhang J M, Kabolizadeh P, Stevens C and Yan D 2016 *IJROBP* **96** 1107–16

[3] Ebner D K, Frank S J, Inaniwa T, Yamada S and Shirai T 2021 *Front. Oncol.* **11** 1–8
[4] Yap J S L, Higgins E R and Sheehy S L 2021 *Proc. 12th Int. Particle Accelerator Conf. (IPAC'21)* (Campinas, Brazil) pp 1256–59
[5] Yap J, Franco A D and Sheehy S 2021 *Front. Oncol.* **11**
[6] Bert C, Grözinger S O and Rietzel E 2008 *Phys. Med. Biol.* **53** 2253–65
[7] Rietzel E and Bert C 2010 *Med Phys* **37** 449–60
[8] Van De Water S, Kooy H M, Heijmen B J and Hoogeman M S 2015 *Int. J. Radiat. Oncol. Biol. Phys.* **92** 460–68
[9] Rahim S, Korte J, Hardcastle N, Hegarty S, Kron T and Everitt S 2020 *Front. Oncol.* **10** 1–5
[10] Mazal A *et al.* 2021 *Front. Oncol.* **10** 1–14
[11] Bortfeld T R, de Viana M F and Yan S 2020 *Z. Med. Phys.* 6–8
[12] Ebner D K and Kamada T 2016 *Front. Oncol.* **6** 6–11
[13] Colman R and Legge G 1993 *Nucl. Instrum. Methods. Phys. Res. B* **73** 561–69
[14] Hortig G 1964 *Nucl. Instrum. Methods* **30** 355–56
[15] Auger M, Braccini S, Carzaniga T S, Ereditato A, Nesteruk K P and Scampoli P 2016 *J. Instrum.* **11** P03027
[16] Potkins D E, Dehnel M P, Kubley T, Toader O F and Lobanov N R 2017 *Proc. 6th Int. Beam Instrum. Conf. (IBIC'17)* (Grand Rapids, MI, USA) pp 335–37
[17] Forck P 2011 Lecture notes on beam instrumentation and diagnostics: Joint university accelerator school
[18] Perl J, Shin J, Schümann J, Faddegon B and Paganetti H 2012 *Med Phys* **39** 6818–37
[19] Machida S *et al.* 2012 *Nat. Phys.* **8** 243–47
[20] Bartnik A *et al.* 2020 *Phys. Rev. Lett.* **125**(4) 044803
[21] Halbach K 1980 *Nucl. Instrum. Methods* **169** 1–10
[22] Brooks S J, Cintorino J, Jain A K and Mahler G J 2017 *Proc. 8th Int. Particle Accelerator Conf. (IPAC'17)* (Copenhagen, Denmark) pp 4118–20
[23] Ortner M and Coliado Bandeira L G 2020 *SoftwareX*
[24] Meyer D I and Flasck R 1970 *Nucl. Instrum. Methods* **80** 339–41
[25] Brouwer L *et al.* 2020 *Nucl. Instrum. Methods Phys. Res., Sect.* A **957** 163414
[26] Benedetto E, Harbi N A, Brouwer L, Tommasini D, Prestemon S, Riboni P and Amaldi U 2021 A carbon-ion superconducting gantry and a synchrotron based on canted cosine theta magnets *Preprint* ArXiv:2105.04205
[27] Robin D S, Arbelaez D, Caspi S, Sun C, Sessler A, Wan W and Yoon M 2011 *Nucl. Instrum. Methods Phys. Res., Sect.* A **659** 484–93
[28] Veres D, Vaszary T, Benedetto E and Barna D 2022 *IEEE Trans. Appl. Supercond.* **32** 1–14

Upgrade of a proton therapy eye treatment nozzle using a cylindrical beam stopping device for enhanced dose rate performances

E Gnacadja[1], C Hernalsteens[1,2], N Pauly[1], E Ramoisiaux[1] R Tesse[1] and M Vanwelde[1]

[1] Service de Métrologie Nucléaire, Université libre de Bruxelles, Brussels, Belgium
[2] CERN, 1211 Geneva 23, Switzerland

E-mail: eustache.gnacadja@ulb.be

Abstract. Proton therapy is a well established treatment method for ocular cancerous diseases. General-purpose multi-room systems which comprise eye-treatment beamlines must be thoroughly optimized to achieve the performances of fully dedicated systems in terms of depth-dose distal fall-off, lateral penumbra, and dose rate. For eye-treatment beamlines, the dose rate is one of the most critical clinical performances, as it directly defines the delivery time of a given treatment session. This delivery time must be kept as low as possible to reduce uncertainties due to undesired patient movement. We propose an alternative design of the Ion Beam Applications (IBA) Proteus Plus (P+) eye treatment beamline, which combines a beam-stopping device with the already existing scattering features of the beamline. The design is modelled with Beam Delivery SIMulation (BDSIM), a Geant4-based particle tracking and beam-matter interactions Monte-Carlo code, to demonstrate that it increases the maximum achievable dose rate by up to a factor 3 compared to the baseline configuration. An in-depth study of the system is performed and the resulting dosimetric properties are discussed in detail.

1. Introduction

The use of proton beams to treat ocular tumors is well established [1–3]. The optimization of high energy multi-room proton therapy systems to allow ocular tumors treatment poses many challenges compared to dedicated low energy facilities. The key point of the design of such systems is the trade-off that must be found between a small depth-dose distal fall-off[1] (DFO, typically below 2 mm) and a high dose rate (often required to be at least equal to 15 Gy/min) in order to minimize treatment delivery time. Indeed, as eye tumors depths often vary from 5 to 35 mm, the beam nominal energy of high energy systems must be reduced from $\sim$250 MeV to $\sim$70 MeV or less prior to additional range shifting inside of a dedicated eye nozzle. As this energy degradation unevitably leads to a significant energy spread, momentum slits must be positioned at a location of high dispersion to intercept the more energetic particles and reduce the DFO of the dose deposition profiles. Such particles interception directly affects the transmission of the beamline, leading to a huge limitation on the dose rate performances of the system.

[1] The distal fall-off is defined as the difference between the 20 % and 80 % dose points on the distal side of the Bragg peak: DFO = $R_{80} - R_{20}$.

The Ion Beam Applications (IBA) eye treatment beamline, so-called "eyeline", is an example of such a high energy system, in which the beam is produced at a nominal energy of 230 MeV. The IBA eyeline has been previously designed [4] and experimentally optimized with an energy of 105 MeV at the nozzle entrance [5], leading to a maximum dose rate of 30 Gy/min, but with a DFO of only 3.2 mm. More recently, the different steps toward a numerical optimization for a smaller DFO (below 2 mm) has been presented in detail in [6], using Beam Delivery SIMulation (BDSIM) [7]. BDSIM is a geant4 based particle tracking and beam-matter interactions simulation code, that has already proven its ability to model low energy proton therapy systems [8]. This smaller DFO requirement forced the reduction of the nominal energy at the nozzle entrance from 105 to 80 MeV, leading to a smaller maximum achievable dose rate. Table 1 summarizes the clinical performances required during the optimization presented in ref. [6].

Table 1. The clinical requirements (minimum range (R_{min}), maximum range (R_{max}), DFO, Flatness, Lateral Penumbra (LP) and Dose Rate (DR)) imposed on the dosimetric properties of the IBA passive scattering eyeline.

R_{min} (mm)	R_{max} (mm)	DFO (mm)	Flatness (%)	LP (mm)	DR (Gy/min)
5	35	< 2	< 2	< 1.5	> 15

As discussed in [6], two designs of the IBA eyeline were studied. The first design only uses the scattering and range shifting features of the system to optimize the nozzle, with at the end a maximum achievable dose rate of 17 Gy/min. The second, alternative design combines a cylindrical, lead material beam stopper, with the scattering features of the nozzle to achieve a transversally flat dose profile. Figure 1 illustrates the BDSIM model of the nozzle as designed with this cylindrical beam stopping device.

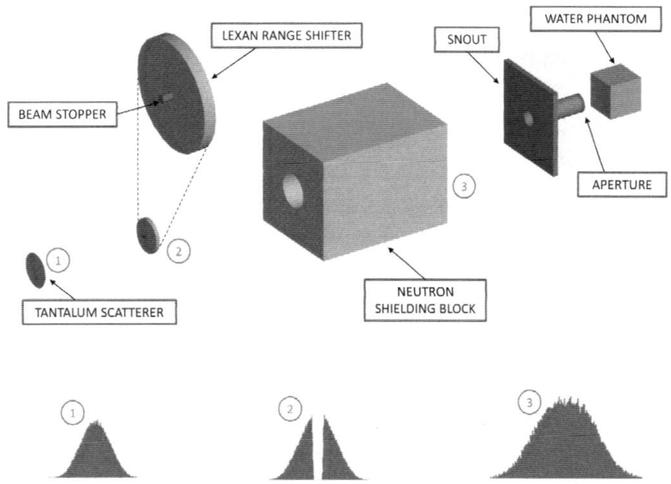

Figure 1. BDSIM model of the IBA eye treatment nozzle as designed with a cylindrical beam stopping device.

The beam first interacts with a thin 150 µm-thick Tantalum foil. Thanks to the high Z and high density of this Tantalum foil, a significant, randomly distributed transverse angle is given

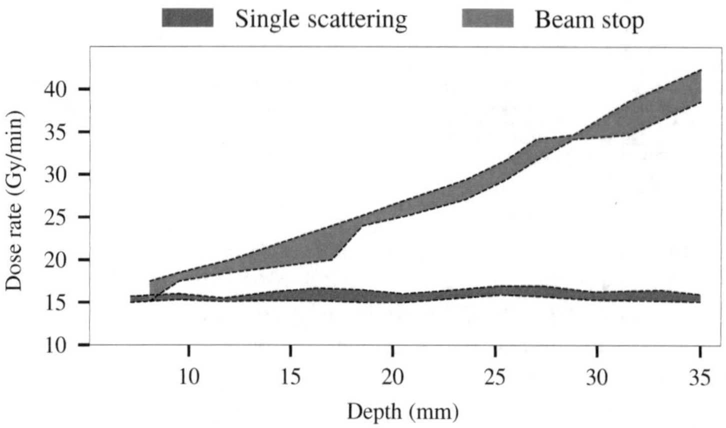

Figure 2. Evolution of the dose rate in function of clinical range, as calculated for the single scattering (blue) and the beam stop (red) designs in [6]. The maximum achievable dose rate with the beam stopper design is 42.4 Gy/min, while the single scattering maximum value is around 17 Gy/min.

to the particles. As the beam propagates along the nozzle after this interaction, the transverse angles are converted into position and the beam size increases. After a distance of 35.5 cm, the beam crosses the lead beam stopper, which has a thickness of 10 mm. Such thickness is enough to stop all particles that cross the beam stopper, meaning that a hole is created in the middle of the beam distribution just after this interaction. Then the remaining particles continue to propagate on a distance of 165 cm up to isocenter. As the beam propagates, the hole is progressively filled by the particles due to their transverse angle, and a transversally flat profile is obtained at isocenter.

The design parameters obtained in [6] (scattering foil, beam stopper radius and position) when optimizing the nozzle with the beam stopper are summarized in table 2.

Table 2. The design parameters of the nozzle when used with a beam stopping device. A 12 mm thick Lexan range shifter, placed just downstream of the beam stopper, is used to fix the maximum clinical range to 35 mm.

Ta thick. (mm)	Le thick. (mm)	Radius (mm)	Thick. (mm)	D_{FS-BS} (cm)
0.15	12	3.6	10	35.5

Figure 2 shows the results of the dose rate optimization using both designs, as obtained in [6]. We can observe that the beam stopper design allows achieving a maximum dose rate of 42.4 Gy/min, for the maximum clinical range of 35 mm. On the other hand, as the range decreases, the dose rate also decreases due to the fact that range shifting also induces scattering, leading to beam losses along the nozzle.

The aim of this paper is to investigate the possibility of further increasing the dose rate values at shallow depths (below 20 mm) by changing the nominal energy at the nozzle entrance from 80 to 90 MeV in this interval. The goal is to benefit from the difference in beamline transmission between these two energies (1.15 % at 80 MeV, 2.23 % at 90 MeV) to restore dose rate values that are higher than 20 Gy/min across all the treatment range interval. After discussing the detailed numerical model of the nozzle, the design parameters that allow obtaining a transversally flat profile at isocenter are presented. Then the main simulation results (lateral and depth dose profiles, penumbra and dose rate in function of depth) are discussed in detail. Finally, the impact of the energy switching on the DFO is assessed for a full modulated Spread-Out Bragg Peak (SOBP) which has a maximum range of 20 mm.

2. Simulation results

Figure 3 shows a comparison between the single scattering and the beam stopper designs, for the same nominal energy of 80 MeV at the entrance of nozzle. As can be seen, the maximum dose is almost 3 times higher for the beam stopper design compared to the single scattering solution.

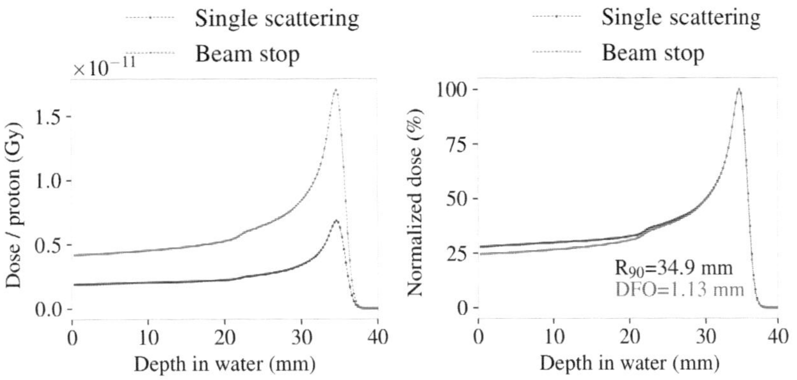

Figure 3. Bragg peaks simulated for a clinical range of 35 mm with the single scattering (blue) and the beam stopper (red) designs. The dose per primary proton (left) illustrates the dose rate improvement by a factor of 3 with the beam stopper design. After normalization (right), the beam stopper design exhibits a dose at skin which is almost 5 % smaller than with the single scattering design.

Figure 4 shows a comparison of the lateral profiles obtained for the single scattering and the beam stopper design.

We observe that the clinical properties are the same for both designs in the irradiation field. At skin, the lateral dose tails (delivered outside of the irradiation field) are more significant for the single scattering design than with the beam stopper, meaning that the latter allows giving a smaller dose to shallow healthy tissues.

Figure 5 compares the evolution of the lateral penumbra with the clinical range for the two designs. Both exhibit similar behaviour, showing an increase of the penumbra with the range, as the protons are scattered in water.

In order to maintain the dose rate higher than 20 Gy/min over all the treatment range interval, a nominal energy of 90 MeV was selected for clinical ranges from 5 to 20 mm. Figure 6 shows a comparison between the estimated dose rate values with only 80 MeV and the results obtained when switching to 90 MeV for ranges smaller than 20 mm. We can clearly observe a significant jump in the dose rate curve at 20 mm for the solution that mixes 80 and 90 MeV. The jump is then followed by a decrease but the absolute minimum of the dose rate remains higher than 20 Gy/min.

On the other hand, the DFO is expected to inevitably increase as we need more range shifting inside of the nozzle to set the Bragg peaks at the same ranges as with a nominal energy of 80 MeV. Such DFO increase can be observed in figure 7, which compares the SOBPs obtained for the two

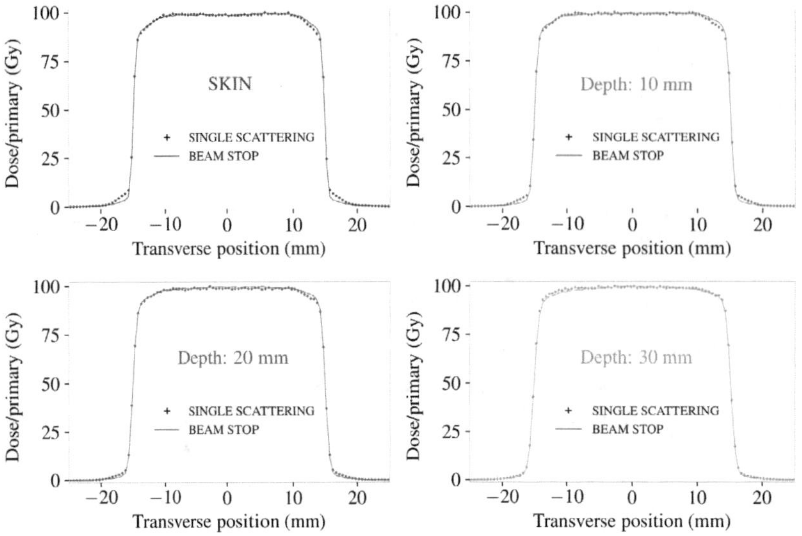

Figure 4. Lateral dose profiles at 4 different depths (skin, 10, 20 and 30 mm) in water. The single scattering and the beam stopper designs have similar clinical properties in the irradiation field. At skin we can observe slightly smaller lateral dose tails for the beam stopper design.

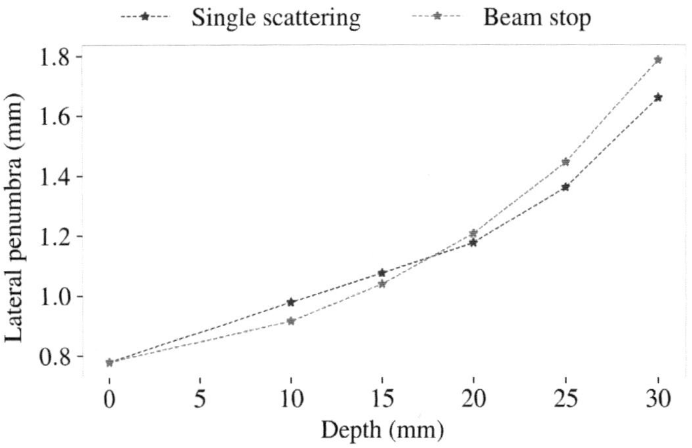

Figure 5. Lateral penumbra in function of the clinical range for the single scattering (blue) and the beam stopper (red) designs. An increase of the penumbra from 0.8 to around 1.8 mm is observed for both designs.

different energies for a maximum clinical range of 20 mm. The DFO increases from 1.13 mm at 80 MeV to 1.5 mm at 90 MeV. However, it remains smaller than the limitation of 2 mm presented in table 1, meaning that the proposed upgrade does not have a clinically significant impact on the DFO of the system.

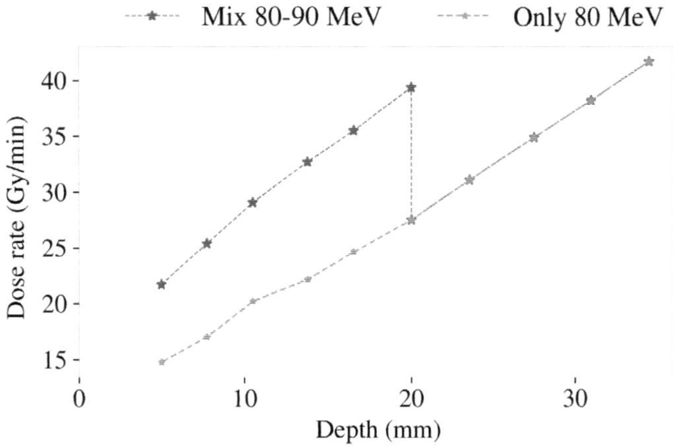

Figure 6. Dose rate values evaluated in function of the clinical range. The results when using a nominal energy of 80 MeV from 5 to 35 mm are shown in orange, while the green curve shows the results obtained when switching to 90 MeV for shallow depths (below 20 mm).

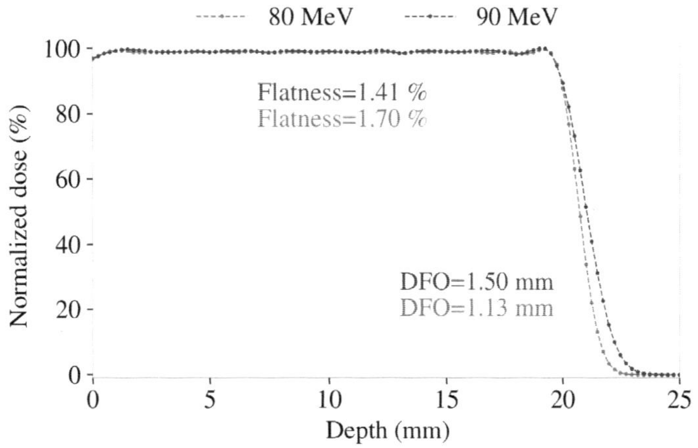

Figure 7. The full-modulation SOBPs for a clinical range of 20 mm, at 80 MeV (red) and at 90 MeV (blue). The flatness of both curves remains below the 2 % limitation. At 90 MeV, a slightly higher DFO is observed due to the energy spread induced by the additional range shifting required inside of the nozzle.

3. Conclusion

We present in this paper an upgrade of the design of the IBA passive scattering ocular tumors treatment beamline using a beam stopping device. This design combines the existing scattering features of the beamline with a cylindrical, 10 mm thick beam stopper to provide a transversally flat dose deposition profile at isocenter. Starting from the previous results presented in [6] with a single nominal energy of 80 MeV at the entrance of the nozzle, we demonstrate that switching from 80 to 90 MeV for shallow depths (<20 mm) allows to further optimize the dose rate and restore values obtained at larger depths (>20 mm). A minimum dose rate of 20 Gy/min is obtained across all the required treatment range interval (from 5 to 35 mm). Moreover, we show that the upgrade only induces an increase of 0.4 mm to the DFO, which reaches a maximum of 1.5 mm at a depth of 20 mm. As this value remains smaller than the limitation of 2 mm, the upgrade still fulfils all the targeted clinical requirements for the system.

Acknowledgments
The authors thank IBA for the support given to this work.

References
[1] Jovanovic P *et al.* 2013 *Int. J. Clin. Exp. Pathol.* **6** 1230–44
[2] Damato B *et al.* 2013 *Saudi J. Ophthalmol.* **27** 151–57
[3] Gragoudas E S *et al.* 1977 *Am. J. Ophthalmol.* **83** 665–73
[4] Prieels D 2006 Calculations and design considerations of the nozzle dedicated to eye irradiation Tech. rep. Ion Beam Applications
[5] Slopsema R *et al.* 2014 *Med. Phys.* **41** 011707
[6] Gnacadja E *et al.* 2022 *Phys. Rev. Research* **4** 013114
[7] Nevay L J *et al.* 2020 *Comput. Phys. Commun.* **252** 107200
[8] Hernalsteens C *et al.* 2020 *EPL* **132** 50004

Journal of Physics: Conference Series **2420** (2023) 012096 doi:10.1088/1742-6596/2420/1/012096

Achromatic gantry design using fixed-field spiral combined-function magnets

R Tesse[1], C Hernalsteens[1,2], E Gnacadja[1], E Ramoisiaux[1] N Pauly[1] and M Vanwelde[1]

[1] Service de Métrologie Nucléaire, Université libre de Bruxelles, Brussels, Belgium
[2] CERN, 1211 Geneva 23, Switzerland

E-mail: robin.tesse@ulb.be

Abstract. Arc-therapy and flash therapy are promising proton therapy treatment modalities as they enable further sparing of the healthy tissues surrounding the tumor site. They impose strong constraints on the beam delivery system and rotating gantry structure, in particular in providing high dose rate and fast energy scanning. Fixed-field achromatic transport lattices potentially satisfy both constraints in allowing instant energy modulation and sufficient transmission efficiency while providing a compact footprint. The presented design study uses fixed-field magnets with spiral edges respecting the FFA scaling law. The cell structure and the layout are studied in simulation and integrated in a compact gantry. Results and further optimizations are discussed.

1. Introduction

The treatment of cancer is a major societal challenge for which numerous treatment techniques are actively developed. Novel proton therapy treatment modalities, such as flash and arc therapy, aim at increasing the clinical efficiency and reducing the treatment time. These techniques pose new challenges for the accelerators and beam delivery systems as they require tighter control of the beam spot sizes at the isocenter, better control of beam losses and higher beam currents [1–3].

Gantry designs providing achromatic transport with fixed-field magnets meet these challenges and allow arbitrarily fast energy scanning. We report on a preliminary design of the CASPRO ("Compact Achromatic System for Proton Therapy") project, for an achromatic proton gantry in the energy range 70-230 MeV using normal conducting magnets. The objective is to retain the reduced footprint of modern variable-field compact gantries, such as the IBA Proteus One [4] or the ProNova SC360 superconducting gantry [5].

The adopted strategy is to divide the system into three distinct blocks. The proposed block diagram, presented in figure 1, is as follows. A proton beam with an energy of 230 MeV enters the first block that consists of a fixed dipole that deflects the beam and is sent on an energy degrader mounted directly on the gantry, to make the system more compact. The beam transport line following the degrader follows a straight scaling fixed-field lattice design. The second block, detailed in this paper, bends the beam towards the isocenter and is at the heart of the design to reduce the footprint by providing a small bending radius and achromatic properties across the entire energy range. The studied concept is based on Fixed-Field Alternating Gradient (FFAGs) cells using normal conducting magnets. The starting point is the spiral FFAG cells of the RACCAM design [6–9]. The third block is composed of the scanning magnets that are positioned downstream of the last cell to limit the opening of the FFA magnets.

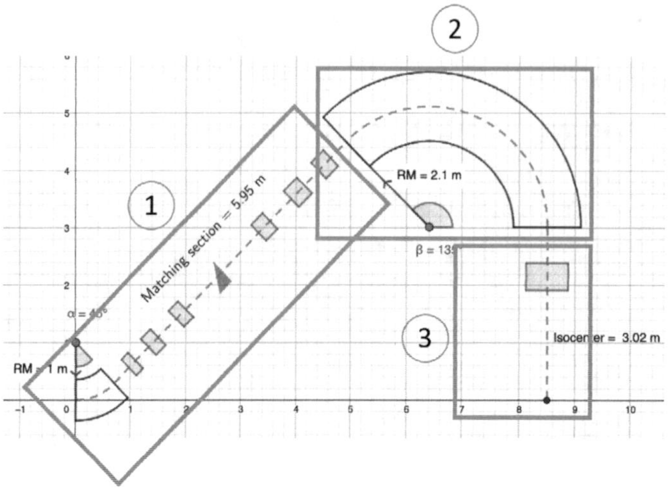

Figure 1. The proposed structure for the CASPRO achromatic gantry. The three main blocks are highlighted in orange.

2. Spiral FFA cell

We adapt the design of the RACCAM original cell aiming to minimize the size of the magnets. The parameters before and after optimisation are presented in table 1. An additional constraint arise from the the fact the gantry design uses downstream scanning magnets to scan the beam at isocenter. To minimize the required in-plane aperture of the scanning magnets, the orbit excursion with energy has to be minized. The aim is to keep the excursion span over the full energy range below 20 cm.

Table 1. Periodic cell parameters of the RACCAM design and the preliminary design of the gantry cell.

Parameters (units)	RACCAM	CASPRO
Mean radius R_0 (m)	3.48	2.1
Angle (degrees)	45	45
Field index	4.415	6.0
Spiral angle (degrees)	50.36	65
Field at R_0 (T)	1.5	1.6
Gap (cm)	3	3

The simulations are performed with the ray-tracing code Zgoubi [10,11] using the Zgoubidoo Python interface [12]. Zgoubidoo provides many additional functionalities, such as the ability of running multiple instances of Zgoubi for parallel tracking, analyzing tracking results, and plotting the beamline geometry in a global coordinates system with particle trajectories. With these different modules, Zgoubidoo is suited to study non-linear accelerators such as FFAs [13]. Figure 2 shows the reduction in footprint after modifications of the original cell.

First, we determine the periodic closed-orbit solutions for the cell in the energy range 70-230 MeV. The results are shown in figures 3 and 4. The periodic trajectory solution, visible in

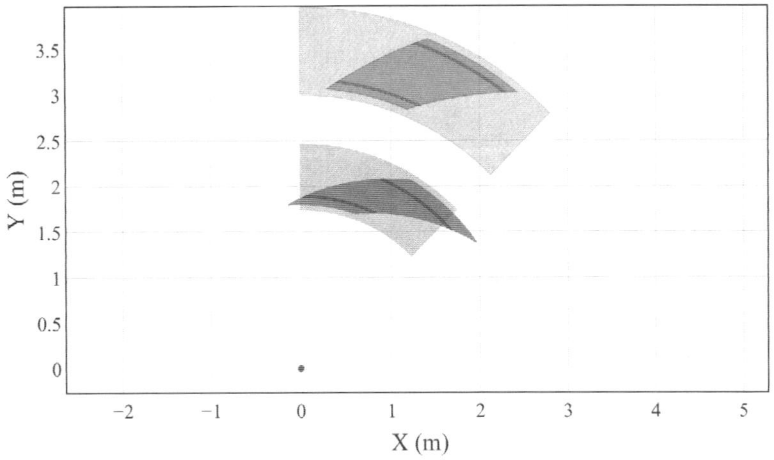

Figure 2. Footprint comparison between the RACCAM cell (in red) and the gantry cell (in blue). We observe the reduction of the footprint with respect to the original design.

figure 3, indicates that a solution is found over the full energy range and that the constraint on excursion is satisfied. The positions and angles as a function of the energy are shown in figure 4.

It is also essential to ensure that the system can remain normal-conducting. For this purpose,

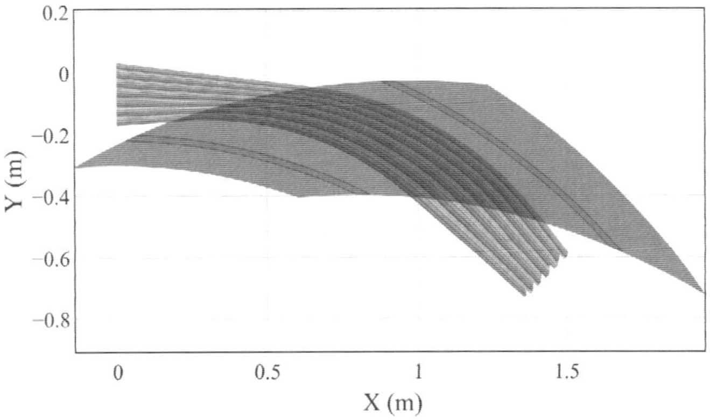

Figure 3. Representation of the cell in a global coordinate system and periodic closed-orbits in the energy range 70-230 MeV. The geometry of the magnet is shown in blue. One clearly see the edge of the magnet which has a spiral curvature and the beam pipe in gray.

we analyze the field along the particle trajectory at different energies in a cell. As we observe in figure 5, the field remains well below 2.3 T. The next study concerns the computation of the Twiss parameters with the beta functions. In figure 6, the evolution of the beta function inside the FFA cell is shown, and we observe the periodicity of these functions. The initial beta values for both planes at the cell's entrance as a function of the momentum deviation are shown in figure 7.

Based on the single cell periodic optics, we study the details of *block 2* (see figure 1). Stable trajectories are found for all energies. Figure 8a shows the results in a global coordinate system. The reduction of the trajectory-energy-excursion towards the isocenter is observed, a desired feature of the design. Figure 8b details the trajectory excursions along the cell for different energies.

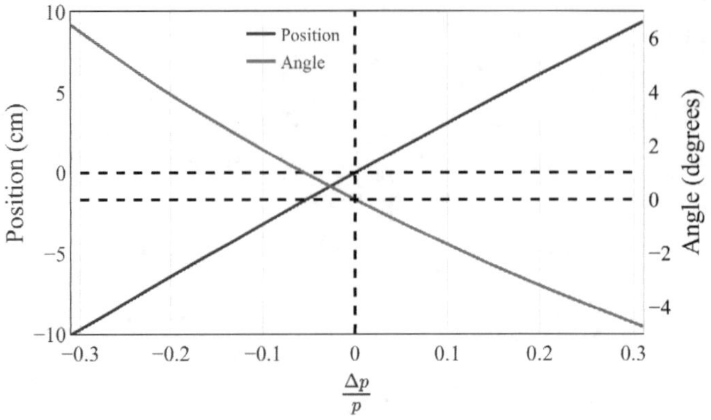

Figure 4. Position (in blue) and angle (in red) of the stable orbit as a function of the momentum deviation at the cell entrance.

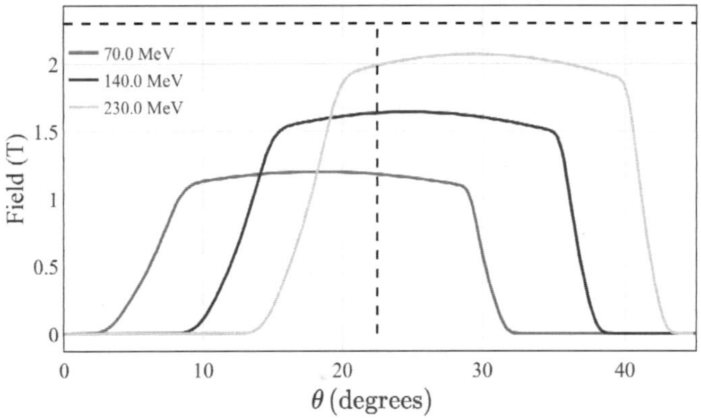

Figure 5. Magnetic field along the particle trajectory at different energies in a cell. For all trajectory, the field is below 2.3 T which ensure normal-conducting magnets.

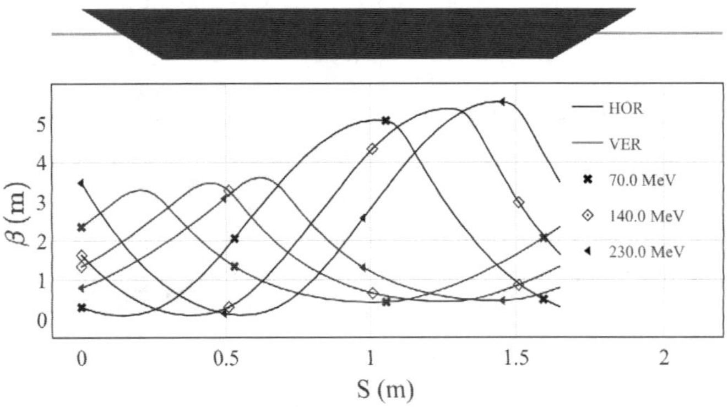

Figure 6. Evolution of the β-functions inside the FFA cell.

Figures 9a and 9b represent the beam size in the horizontal and vertical planes for a beam energy of 140 MeV in the cell, assuming an emittance of 10 mm · mrad. In the first approximation, we compute the relation $\sigma = \sqrt{\beta\epsilon}$. We observe that the beam size in both

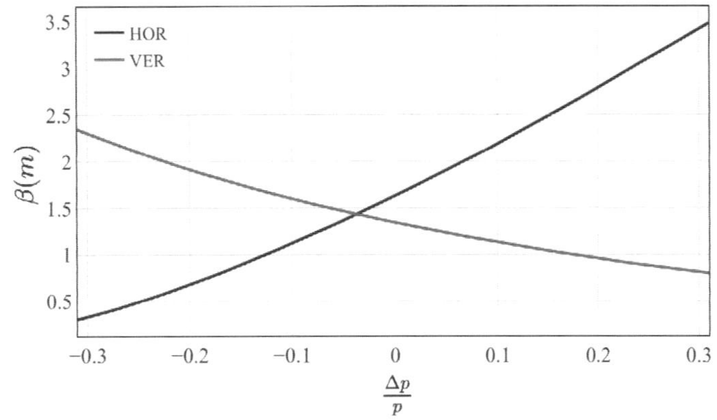

Figure 7. Initial β at the entrance of the cell as a function of the momentum deviation.

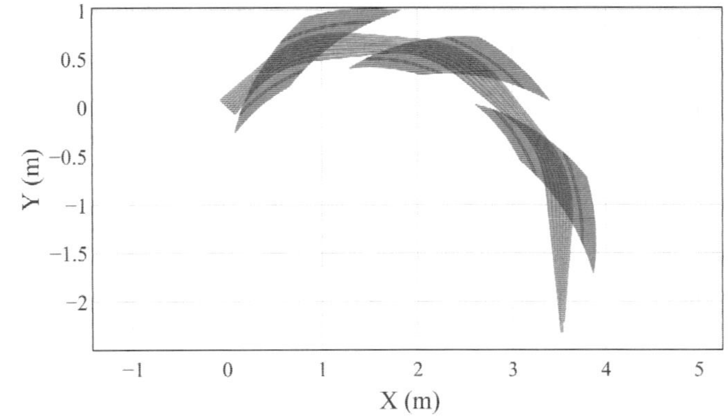

(a) In a global reference frame.

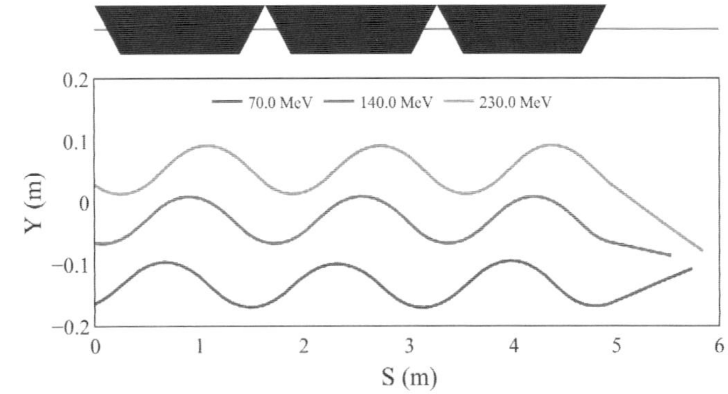

(b) In the Frenet-Serret reference frame.

Figure 8. Tracking of particles along three FFAs cells. We observe a focusing of the different energies at the isocenter. By construction, the orbits are stable in the cells.

planes has the same order of magnitude and does not exceed 8 mm across the three cells. The scanning system will be designed to make the beam size as small as possible at the isocenter.

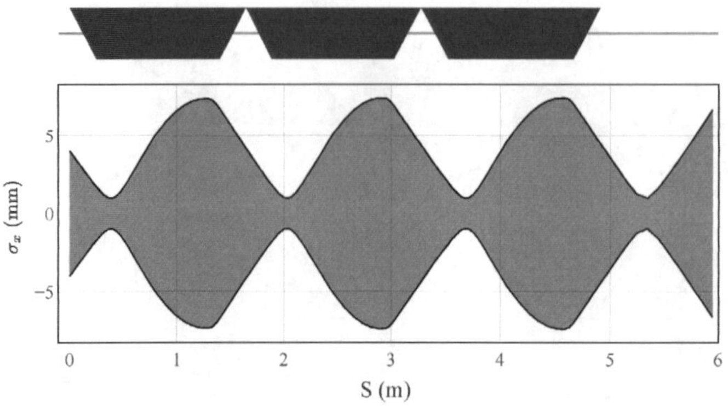

(a) Horizontal plane.

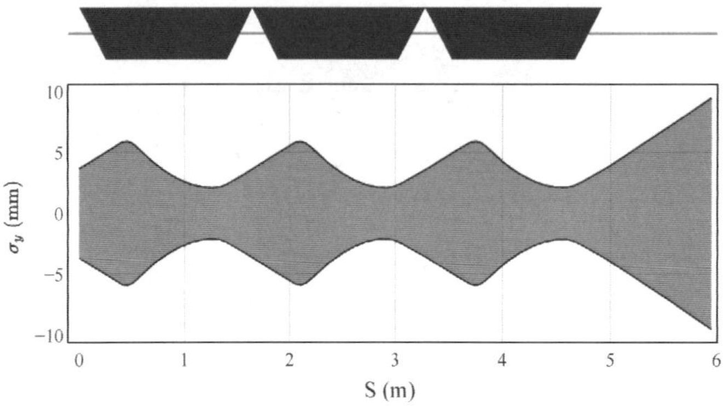

(b) Vertical plane.

Figure 9. Beam size along the line in the horizontal (a) and vertical (b) plane for an beam energy of 140 MeV and assuming an emittance of 10 mm·mrad. The beam size across the three cells doesn't exceed 8 mm in both planes.

3. Conclusion

This paper has illustrated that RACCAM-based cells can be used to design an achromatic gantry for bending the beam towards the isocenter. These cells have the advantage of providing an achromatic beam transport across the entire energy range. By modifying the original cell design, we have reduced the radius of curvature to make the system as compact as possible. The first studies have shown encouraging results, especially concerning the orbit excursion, which has been reduced from 66 to below 20 cm which is suitable for the scanning magnets. Moreover, we verified that the maximum field in the FFAs cells is below 2.3 T which guarantees that all the magnets remain normal-conducting. At the exit of three cells, we have shown that all energies are focused at the isocenter, which simplifies the scanning magnets' design. The next step of this project will focus on a detailed study of the beam size along the line, mainly by changing the emittance with the energy.

Acknowledgments

This work has received funding from the Walloon Region (SPW-EER, Belgium) Win^{2}Wal program under grant agreement No. 1810110. The authors thank IBA for the support given to this work.

References

[1] Toussaint L *et al.* 2019 *Acta Oncol.* **58** pp 1451–56
[2] Li X *et al.* 2019 *Radiother. Oncol.* **137** pp 130–36
[3] Hughes J R and Parsons J L 2020 *Int. J. Mol. Sci.* **21** 6492
[4] Hernalsteens C *et al.* 2020 *Europhys Lett* **132** 50004
[5] Derenchuk V 2014 The ProNova SC360 Gantry, in: Modern Hadron Therapy Gantry Developments Tech. rep. Cockcroft Institute Daresbury UK
[6] Fourrier J *et al.* 2008 *Nucl. Instrum. Methods Phys. Res., Sect.* A **589** pp 133–42
[7] Méot F 2019 RACCAM: An Example of Spiral Sector Scaling FFA Technology Tech. rep. Brookhaven National Laboratory USA BNL–211536-2019-NEWS
[8] Antoine S *et al.* 2009 *Nucl. Instrum. Methods Phys. Res., Sect.* A **602** pp 293-305
[9] Planche T *et al.* 2009 *Nucl. Instrum. Methods Phys. Res., Sect.* A **604** pp 435–42
[10] Méot F 2014 *Nucl. Instrum. Methods Phys. Res., Sect.* A **767** pp 112–25
[11] Méot F 1999 *Nucl. Instrum. Methods Phys. Res., Sect.* A **427** pp 353–56
[12] Hernalsteens C, Tesse R and Vanwelde M 2022 Zgoubidoo `https://ulb-metronu.github.io/zgoubidoo/`
[13] Vanwelde M, Gnacadja E, Pauly N, Ramoisiaux E, Tesse R, and Hernalsteens C 2022 *Proc. 13th Int. Particle Accelerator Conf. (IPAC'22)* (Bangkok, Thailand) pp 118–21

A carbon minibeam irradiation facility concept

M Mayerhofer[1], V Bencini[2], M Sammer[1] and G Dollinger[1]

[1] Universität der Bundeswehr München, 85579 Neubiberg, Germany
[2] University of Oxford, OX1 2JD Oxford, United Kingdom

E-mail: michael.mayerhofer@unibw.de

Abstract. In minibeam therapy, the sparing of deep-seated normal tissue is limited by transverse beam spread caused by small-angle scattering. Contrary to proton minibeams, helium or carbon minibeams experience less deflection, which potentially reduces side effects. To verify this potential, an irradiation facility for preclinical and clinical studies is needed. This manuscript presents a concept for a carbon minibeam irradiation facility based on a LINAC design for conventional carbon therapy. A quadrupole triplet focuses the LINAC beam to submillimeter minibeams. A scanning and a dosimetry unit are provided to move the minibeam over the target and monitor the applied dose. The beamline was optimized by TRAVEL simulations. The interaction between beam and these components and the resulting beam parameters at the focal plane is evaluated by TOPAS simulations. A transverse beamwidth of $< 100\,\mu m$ (sigma) and a peak-to-valley (energy) dose ratio of > 1000 results for carbon energies of 100 MeV/u and 430 MeV/u (~ 3 cm and 30 cm range in water) whereby the average beam current is ~ 30 nA. Therefore, the presented irradiation facility exceeds the requirements for hadron minibeam therapy.

1. Introduction

Compared to proton or heavy ion radiotherapy, minibeam radiation therapy has the potential to further spare normal tissue [1, 2]. The dose required for tumor control is applied through minibeams with an inital transverse width in the submillimetre range. Pencil or planar minibeams are arranged in a grid or an array with a center-to-center distance (ctc) of a few millimeters [3, 4]. The resulting transverse dose distribution with dose minima between minibeam channels spares normal tissue compared to broad beams, reducing side effects. The lateral width of the beams increases with penetration depth due to small-angle scattering. With a suitable ctc, the individual minibeams superimpose in the tumor resulting in a homogeneous dose [5]. The potential of minibeam therapy has been confirmed by preclinical experiments for protons [1, 2, 6–9]. However, it has been shown that the normal tissue sparing decreases with increasing transverse beam width [8]. Therefore, especially the sparing of the deeper normal tissue is limited due to the spreading of the proton beam. In comparison, heavier hadrons such as helium or carbon are less affected by small-angle scattering, which offers the possibility to enhance normal tissue sparing [5, 10, 11]. To evaluate the potential of heavy ions for minibeam therapy in (pre)clinical studies, an irradiation facility needs to be established. Here, we focus on a carbon minibeam facility that could easily be adapted to a helium minibeam facility. Thereby the following requirements should be fulfilled: 1. a carbon ion range in water from 3 cm to 30 cm. 2. a beam current > 1 nA to allow single-session treatments. 3. a transverse beam width of $100\,\mu m$ and a peak-to-valley dose ratio (PVDR) of > 540 which causes no normal tissue reaction for proton minibeams [8] 4. a possibility for beam scanning up to X = Y = ± 15 mm that would be sufficient for preclinical experiments as well as for first clincial trials. This manuscript shows a LINAC-based concept for a carbon (helium ion) minibeam irradiation facility that fulfils these requirements.

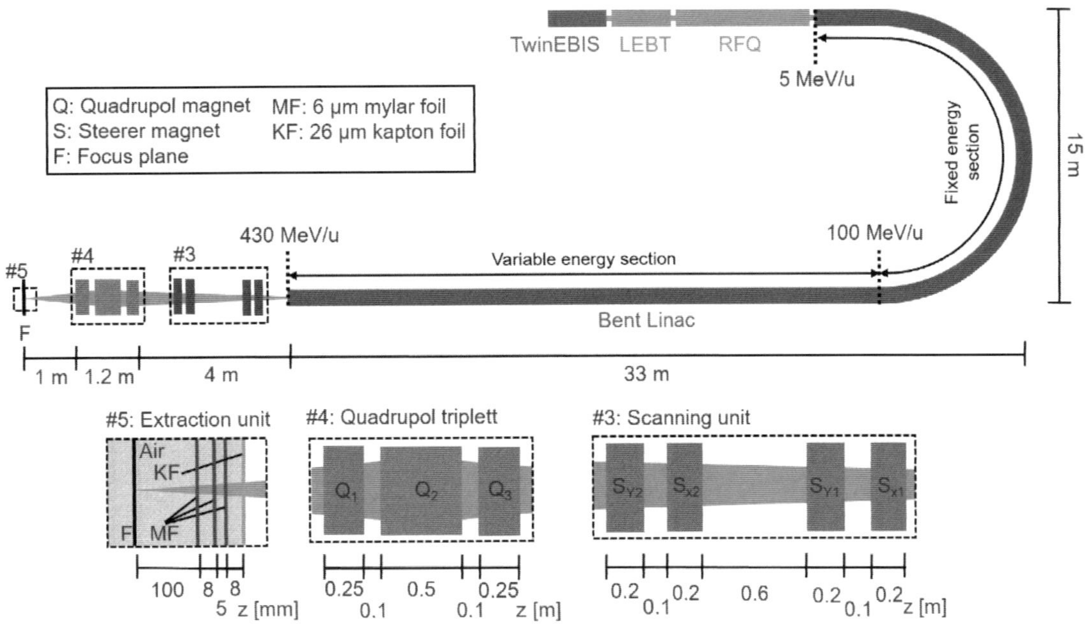

Figure 1. Concept of a carbon minibeam irradiation facility to form submillimeter $^{12}C^{6+}$ -ion beams with energies between 100 MeV/u and 430 MeV/u. Bent-LINAC concept (TwinEBIS, LEBT, RFQ and bent LINAC) is adopted from [13].

2. Hadron minibeam irradiation facility concept

Figure 1 shows the concept we suggest for a hadron minibeam irradiation facility. It is based on a bent carbon LINAC concept currently being developed at CERN and funded by the CERN Knowledge Transfer Office as part of the NIMMS project [12,13]. This LINAC concept consists of 5 substructures. A TwinEBIS source [14,15] delivers $^{12}C^{6+}$ -ions. The ions are initially accelerated and transferred to a subsequent RFQ accelerator by the Low Energy Beam Transport (LEBT). The RFQ accelerates the carbon ions to 5 MeV/u [16]. It operates at a frequency of 750 MHz, allowing a bunch-to-bunch injection into the subsequent 3 GHz bent LINAC structure which can be divided into two parts. The fixed energy section accelerates the ions to 100 MeV/u and does not allow an energy variation due to the dipole magnets in the "arc part". The subsequent variable energy section allows an energy variation of the carbon ions from 100 MeV/u to 430 MeV/u at a mean beam current of 30 nA. The advantage of a bent accelerator over a straight accelerator chain is the reduced space required for future irradiation facilities.

For (pre)clinical minibeam application, several components are added downstream of the bend-Linac which have already been designed similarly for a proton minibeam irradiation facility [17,18]: To focus the beam to a transverse width in the submillimeter range (minibeams) on the target (F in figure 1), a quadrupole triplet is used (see #4, figure 1). To scan the minibeam over the target, a scanning unit (see #3, figure 1) is placed upstream of the quadrupole triplet. It consists of two steerer magnets which first deflect the beam in X- and Y-direction (SX1 and SY1) and two steerer magnets to bend the beam back in the opposite direction (SX2 and SY2). This arrangement allows the beam to pass the quadrupole triplet as close as possible to the optical axis, minimizing abberations. The beam extraction to air is achieved via a 25 µm thick Kapton foil. An air-filled dosimetry chamber (ionization chamber) is provided for dose monitoring. It consists of two 6 µm thick aluminium-coated Mylar foils. The side not shielded by the beam tube is covered by another aluminium-coated Mylar foil.

3. Carbon minibeam simulations

The quadruple triplet (#4 in figure 1) and the scanning unit (#3 in figure 1) are designed using the beam dynamics simulation software Trace 3-D [19] and TRAVEL [20]. Subsequently, the Geant4 toolkit TOPAS [21] was used to verify the beam dynamics and evaluate the interaction of the beam with the extraction window, the dosimetry chamber, and the air gap to the target. The 100 MeV/u and 430 MeV/u $^{12}C^{6+}$ beam of the bent LINAC serves as starting point for all simulations. Figure 2 shows exemplarily the transverse phase space of the 430 MeV/u beam at the end of the LINAC. The energy spread corresponds to 0.1% for both energies.

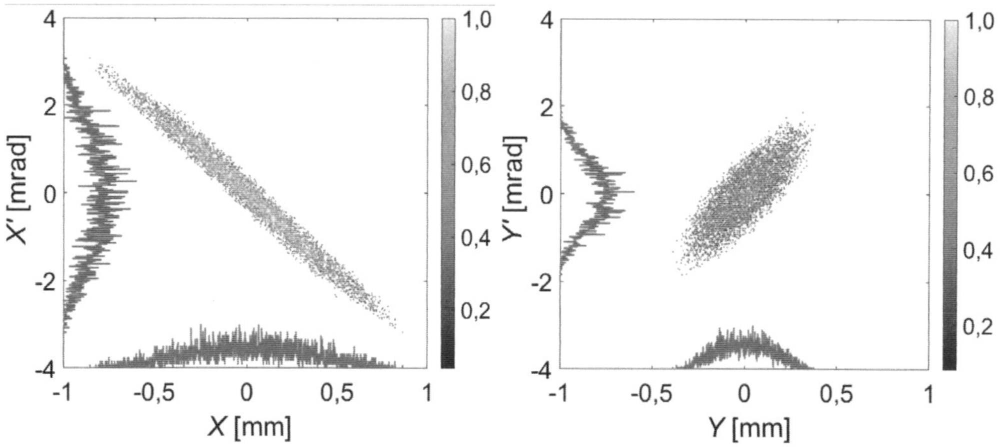

Figure 2. Transverse phase space of the 430 MeV/u bent LINAC $^{12}C^{6+}$ beam.

Figure 3 shows exemplarily the (x)-projection of the $^{12}C^{6+}$-ion distribution at the focal plane (F, figure 1) as simulated by TOPAS. Red represents the 100 MeV/u and blue the 430 MeV/u beams. The associated dashed lines represent the (x)-projection of the maximum deflected 430 MeV/u and 100 MeV/u beams at positions X = Y = +17 mm and X = Y = +38 mm, respectively. The field strengths of the qaudrupole triplet magnets Q1, Q2 and Q3 (see #4 figure 1) required to focus the beams are given in table 1. The magnetic fields of the steerer magnets S_{x1}, S_{y1}, S_{x2} and S_{y2} (see #3 figure 1) which allow the maximum deflection are given in table 2. A transverse beam width of (87 ± 1) µm and (64 ± 1) µm as well as a beam divergence of (2.8 ± 0.1) mrad and (3.1 ± 0.1) mrad results for the 100 MeV/u and the 430 MeV/u beam, respectively. Thereby, the given values represent the standard deviations of the particle distributions (σ). For maximum deflection, the beam spot width increases by less than 10%.

The sparing of normal tissue in minibeam therapy is limited by the dose applied between the individual minibeams and therefore by secondary particles created during the interaction of the beam with the matter. Figure 4 shows the (x)-projection at the target of the 430 MeV/u $^{12}C^{6+}$-ion distribution and additionally that of the secondary electrons and other secondary particles (protons, neutrons, etc.) resulting from fragmentation processes in exit window, dosimetry

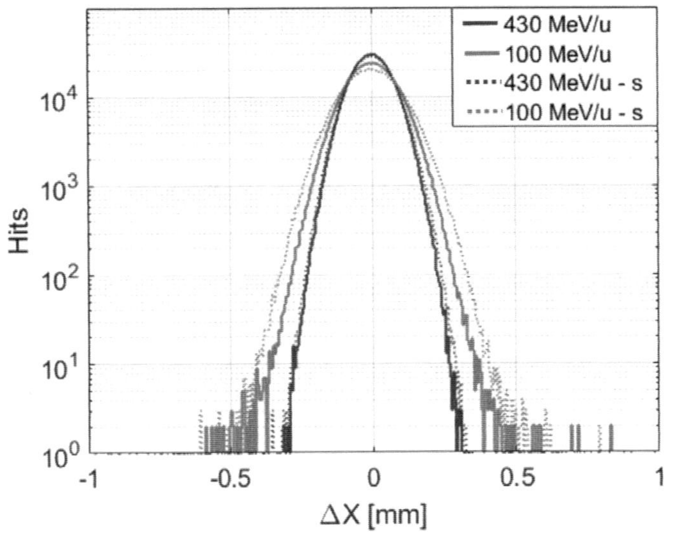

Figure 3. (x)-projection of the 100 MeV/u (red) and 430 MeV/u (blue) $^{12}C^{6+}$-ion distribution at the focal plane. The corresponding dashed lines (s) represent the maximum deflected beams on the position X = Y = +17 mm and +38 mm.

Table 1. Field strengths of the qaudrupole triplet magnets Q1, Q2 and Q3 (see #4, figure 1) for focusing the 100 MeV/u and 400 MeV/u beam.

Energy [MeV/u]	Field strengths [T/m]		
	Q_1	Q_2	Q_3
100	-18.10	15.37	-17.99
430	-40.50	34.29	-39.28

Table 2. Field strengths of the deflecting magnets S_{x1}, S_{y1}, S_{x2} and S_{y2} (see #3 figure 1) for deflecting the 430 MeV/u and 100 MeV/u beam to X = Y = +17 mm and X = Y = +38 mm, respectively.

Steerer magnet	Magnetic fields [mT]
S_{x1}	133
S_{y1}	81
S_{x2}	400
S_{y2}	350

chamber and air gap. The dose between the individual minibeams at the target entry results from the broadly distributed secondary particles of all minibeams in a minibeam array.

For clinically relevant minibeam arrays, the resulting reduction of the Peak to valley (energy) dose ratio (PV(E)DR) is evaluated with TOPAS. Therefore, the interaction of a water phantom positioned downstream to F with an array of 99 x 99 quadratic arrangement minibeams is simulated. Figure 5 shows the dose distribution at phantom entrance for the minibeam unit cell in the centre of the array exemplary for the 430 MeV/u beams. $N_{\mathrm{MB}}X$ and $N_{\mathrm{MB}}Y$ indicate the minibeam number in X and Y direction, respectively. The PV(E)DR is calculated as the ratio of

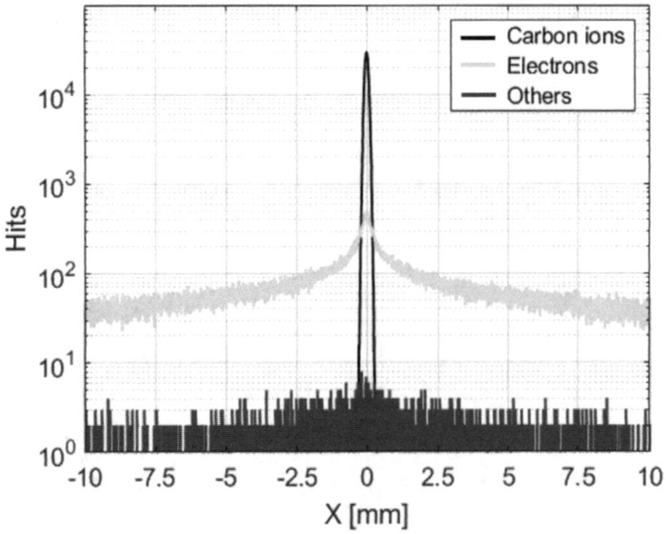

Figure 4. Transverse distribution of 430 MeV/u carbon ions and secondary particels at the focal plane.

maximum dose in the channel (green cross) and averaged dose in an area between the channels (green circle). The ctc was chosen to result in a homogeneous transverse energy dose at the Bragg peak (in water). For beam energies of 100 MeV/u and 430 MeV/u with the associated ctc's of 1.1 mm and 3.4 mm, a PV(E)DR of around 1200 and 2700 results, respectively. The array of 99 minibeams corresponds to a transverse tumor dimension > 100 mm.

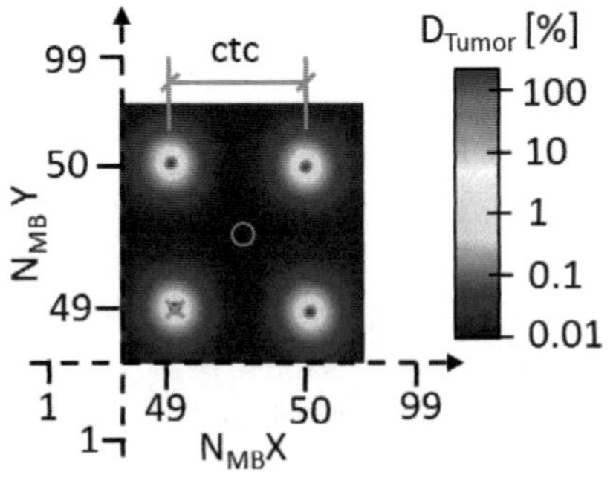

Figure 5. Unit cell of a 99 x 99 minibeam array. The green cross marks the highest and the green circle the lowest dose.

4. Conclusion

The bent carbon LINAC concepts currently under development [13] offer a high brillant beam with a constant average beam current of 30 nA for beam energies between 100 MeV/u and 430 MeV/u (approx. 3 to 30 cm range in water). Using the presented quadrupole triplet, transverse beam sizes below 100 µm (σ) at the patient are achieved although the beam scatters due to the extraction foil, dosimetry unit and air gap. The scanning unit enables a beam deflection of > ± 15 mm. A PV(E)DR of > 1000 is achieved at the patient entrance. However, RBE corrected biological dose distribution in particular in the valley dose regions requires further investigations. The presented concept is a promising (and first) approach for a hadron minibeam irradiation facility. It exceeds all requirements for the planned (pre-)clinical studies defined in the introduction. Besides $^{12}C^{6+}$, theoretically, all ions with the identical mass-to-charge ratio can be accelerated by the bent-Linac. In particular, the presented system also has the potential for $^{4}He^{2+}$ minibeam therapy.

References

[1] Zlobinskaya O *et al.* 2013 *Radiat. Environ. Biophys.* **52** pp 123–33
[2] Prezado Y *et al.* 2013 *Med. Phys.* **40** 031712
[3] Datzmann G *et al.* 2020 *Front. Phys.* **8** 471
[4] Meyer J *et al.* 2018 *BJR* **92** 20180466
[5] Sammer M *et al.* 2021 *Sci. Rep.* **11** pp 1–16
[6] Girst S *et al.* 2016 *Int. J. Radiat. Oncol. Biol. Phys.* **95** pp 234–41
[7] Prezado Y *et al.* 2017 *Sci. Rep.* **7** 14403
[8] Sammer M *et al.* 2019 *PLoS One* **7** e0221454
[9] Sammer M *et al.* 2019 *PLoS One* **14** e0224873
[10] Dilmanian A *et al.* 2012 *Int. J. Radiat. Oncol. Biol. Phys.* **14** e0224873
[11] Martínez-Rovira I *et al.* 2017 *Med. Phys.* **44** pp 4223–29
[12] Vretenar M *et al.* 2021 *Proc. 12th Int. Particle Accelerator Conf. (IPAC'21)* (Campinas, Brazil) pp 1240–43
[13] Bencini V 2020 Design of a novel linear accelerator for carbon ion therapy PhD Thesis Università di Roma
[14] Breitenfeldt M *et al.* 2017 *Nucl. Instrum. Methods Phys. Res., Sect.* A **856** pp 139–46
[15] Pahl H *et al.* 2018 *J. Instrum.* **13** P08012
[16] Bencini V *et al.* 2020 *Phys. Rev. Accel. Beams* **23** 122003
[17] Mayerhofer M *et al.* 2021 *Med. Phys.* **48** pp 2733–49
[18] Mayerhofer M *et al.* 2021 *PLoS One* **16** e0258477
[19] Crandall K *et al.* 1987 Trace 3-D documentation Los Alamos National Laboratory
[20] Perrin A *et al.* 2003 Travel v4.06 user manual CERN internal note
[21] Perl J *et al.* 2012 *Med. Phys.* **39** pp 6818-37

Beam optics studies for a novel gantry for hadron therapy

E Felcini[1], G Frisella[1], A Mereghetti[1], M G Pullia[1], S Savazzi[1], M T F Pivi[2] and E Benedetto[3]

[1]National Center for Oncological Hadrontherapy (CNAO), Pavia, Italy
[2]EBG MedAustron, Wr. Neustadt, Austria
[3]SEEIIST, Geneva, Switzerland

E-mail: enrico.felcini@cnao.it

Abstract. The design of smaller and less costly gantries for carbon ion particle therapy represents a major challenge to the diffusion of this treatment. Here we present the work done on the linear beam optics of possible gantry layouts, differing for geometry, momentum acceptance, and magnet technology, which share the use of combined function superconducting magnets with a bending field of 4 T. We performed parallel– to–point and point–to–point optics matching at different magnification factors to provide two different beam sizes at the isocenter. Moreover, we considered the orbit distortion generated by magnet errors and we introduced beam position monitors and correctors. The study, together with considerations on the criteria for comparison, is the basis for the design of a novel and compact gantry for hadrontherapy.

1. Introduction

In the context of the European project HITRI*plus* [1], the study and design of a novel superconducting gantry has been undertaken. Following up the studies performed by the TERA Foundation on superconducting gantries [2, 3], a wide explorative study on dozens of gantry layouts and optics configurations was performed [4,5]. This manuscript presents the assumptions made and methods used to analyse and compare the different optics solutions. Furthermore, the two most promising layouts are described in more detail.

2. Gantry design choices

Before entering in the details of the optics studies, it is important to highlight the design choices that defined the perimeter of the explored parameters space. The first fundamental choice was to position the scanning magnets after the last bending section. Although this approach poses relevant challenges on the design of the scanning magnets, it allows to operate with standardized dipoles in the whole gantry, avoiding an increase of aperture in last bending section.

The second fundamental choice was the selection of the bending field. While state of the art gantries operate with fields up to 3.5 T [6–8], we decided to push toward 4 T superconducting dipoles [9], aiming for a further reduction of the gantry weight. Depending on the layouts, these dipoles may need a superimposed gradient that can be obtained through an asymmetric assembly of the coils [10]. Even if it would be possible to wind the magnet with separate circuits for dipole and quadrupole (nested magnets), this option was discarded in the context of this study to favor the easiness of construction and operation of the superconducting magnets. To standardize the comparison between the different gantry layouts, two options were considered for the magnet aperture: 70 mm and 90 mm in diameter. Optics layouts that require an aperture larger than 90 mm are considered not suitable for the proposed study. In the same way, two options were considered for the dipoles angle length: 30° and 45°.

3. Gantry beam optics

The main constraints to the beam optics of the gantry, resulting from the continuous interaction with medical doctors and physicist of CNAO (Italian National Center for Oncological Hadron Therapy) and MedAustron (Center for Ion Therapy and Research, Austria), were identified:

- operation with $^{12}C^{6+}$ up to 430 MeV/u kinetic energy, equivalent to 31 cm of range in water;
- beam characteristics at the isocenter independent on the angle of rotation;
- two different beam sizes at the isocenter: 8 mm and 12 mm (FWHM) at the minimum extraction energy;
- global achromaticity to avoid distortion of beam size and position due to the beam momentum spread.

A more thorough and detailed list of medical constraints and requirements is reported in [11]

3.1. Matching procedures at the isocenter

In order to obtain a beam size independent of the gantry angle, two different types of telescopic matching were implemented between the coupling point and the isocenter [12]:

- "point–to–point" matching, where particles with the same initial position in (x, y) end up in the same position independently of the initial divergence (x', y'). In this case, an input round beam in (x, y) maintain its size at the isocenter independently from the gantry rotation angle;
- "parallel–to–point" matching, where particles with the same initial divergence (x', y') end up in the same position independently of the initial position (x, y). In this case, an input round beam in (x', y') maintain its size at the isocenter independently from the gantry rotation angle.

For both matching methods, it is possible to impose a magnification factor (MF), i.e. the factor by which the input beam size is increased to obtain the output beam size at the end of the beam line.

3.2. Beam size regulation at the isocenter

As already specified, for a given beam energy, two spot sizes are required at the isocenter, which can be obtained each with two operating modes:

- varying the MF, keeping the same input beam parameters: the gantry optics is flexible enough to accommodate the change of output beam parameters varying the quadrupole gradients along the line;
- keeping constant the MF, varying the input beam parameters: the gantry gradients are fixed, and all the changes required to attain the output beam parameters are demanded to the beam line upstream of the gantry.

In the first case, the gantry is an active optic element and it is used for beam size control. Although the optics optimization for different MF may be more complex, this results in a simpler overall operation, where the gantry can be operated independently from the upstream line. In the second approach, changing the quadrupoles upstream the gantry to change beam size would requires re-steering the orbit both in the upstream line and in the gantry.

3.3. Achromaticity

The gantry optics must be at least globally achromatic, i.e. for an input beam free of dispersion, the output beam at the end of the beam line is still free of dispersion. Such a requirement is necessary otherwise the characteristics of the delivered beam spot at the isocenter would depend on the momentum distribution of the beam particles and on the gantry angle. The gantry optics can also be made locally achromatic, i.e. the beam will stay free from dispersion in all non–bending sections. This option can be used to reduce the maximum value of dispersion along the line and in general to increase the gantry momentum acceptance. However, given the reduced sizes of the gantries, a locally achromatic optics implies large gradients in the quadrupoles between dipoles, strongly increasing the beam dimension in the non–bending plane and generating large orbit distortions for small alignment errors.

3.4. Momentum acceptance

Momentum acceptance refers to the maximum $\Delta p/p$ that can be transported by the gantry beam line with a specific optics without beam losses. Having a small momentum acceptance implies that magnet currents must be set for each beam energy necessary for the treatment; on the contrary, a large momentum acceptance allows using fewer set points for magnet currents reducing the AC losses on the superconductors. It was estimated that at the minimum energy (120 MeV/u for carbon ions) a momentum acceptance of $\Delta p/p = 1\%$ would allow to effectively irradiate two consecutive tumor slices without changing the magnetic field of the dipoles. The number of slices increases with increasing energies, reaching up to eight slices at the maximum energy. However, a large momentum acceptance comes at the expenses of the magnet aperture. It is worth noting that detailed studies are needed to evaluate the effects of off–momentum operations on the beam size.

3.5. Errors and corrections

The possibility to compensate for alignment and magnetic errors and restore an acceptable orbit and optics was evaluated as well. Alignment errors were distributed according to a Gaussian distribution truncated at $3\,\sigma$ with $\sigma = 0.3$ mm (0.3 mrad) for the position (angle) error. Errors in the magnetic fields were distributed according to a Gaussian distribution truncated at $3\,\sigma$ with $\sigma = 5 \cdot 10^{-5}$ to both dipole and quadrupole excitation errors (ΔB/B and Δg/g), describing errors coming from power supply (and thus common to both dipole and quadrupole components). In addition, a Gaussian distribution truncated at $3\,\sigma$ with $\sigma = 2.5 \cdot 10^{-4}$ was added in order to consider construction errors, e.g. a displacement in the length of the filament, and thus an error in the magnetic length. Using monitors and correctors, one can correct the orbit deviation at the isocenter, but a residual orbit deviation (R_{corr}) is till present along the line (usually in the order of 5 mm–10 mm).The total (radial) space occupied by the beam (R_{tot}) is given by:

$$R_{\text{tot}} = R_{\text{env}} + R_{\text{corr}} \tag{1}$$

$$R_{\text{corr}} = \sqrt{R^2_{\text{corr,x}} + R^2_{\text{corr,y}}} \tag{2}$$

$$R_{\text{env}} = \sqrt{R^2_{\text{env,x}} + R^2_{\text{env,y}}} \tag{3}$$

$$R_{\text{env,x}} = \sqrt{\beta_x \varepsilon_x} + D_x \Delta p/p \tag{4}$$

where the terms of equation (4) are summed linearly since we are interested in the space occupied by the beam during off–momentum operations, typical of the multi–energy extraction [13]. In the case of $\Delta p/p = 0.1\%$, the linear sum takes into account a small variation in momentum during the extraction. We considered R_{corr} as the maximum of the residual orbit deviation on each point along the line. The space occupied by the beam (R_{tot}) defines the required Good Field Region (GFR) for the magnets. As commonly done in designing accelerator magnets, in this study we consider that the GFR is 2/3 of the magnet aperture.

4. Most promising solutions

Taking into account all the previous considerations, two optics layout were selected as the most promising out of more than 30 analysed configurations [4]. The overall parameters of the two layouts are summarized in table 1, while in the following sections they are analysed in more detail.

Table 1. Main Parameters of the 45° and 30° layouts. Both optics are matched as parallel to point.

Parameters	**45° Layout**		**30° Layout**	
Length | Height | Dipole Angle	14.1 m | 5.75 m | 4x45°		12.9 m | 6.4 m | 7x30°	
Dipole Quadrupolar Gradient	0.07 T/m		4.5 | -4.9 | 0.6 T/m	
Dipole Quad. Gradient to change MF	Fixed		±0.3 T/m	
Max R_{tot} along the gantry	$\frac{\Delta p}{p} = 0.1\%$	$\frac{\Delta p}{p} = 1\%$	$\frac{\Delta p}{p} = 0.1\%$	$\frac{\Delta p}{p} = 1\%$
MF=1.5, 2 for $\beta_{x,y} = 5m$	21 mm	54 mm	22 mm	30 mm
$\beta_{x,y} = 2.5, 5$ m for MF=1.5	22 mm	56 mm	24 mm	29 mm
Dipole Aperture (diameter)	70 mm			90 mm

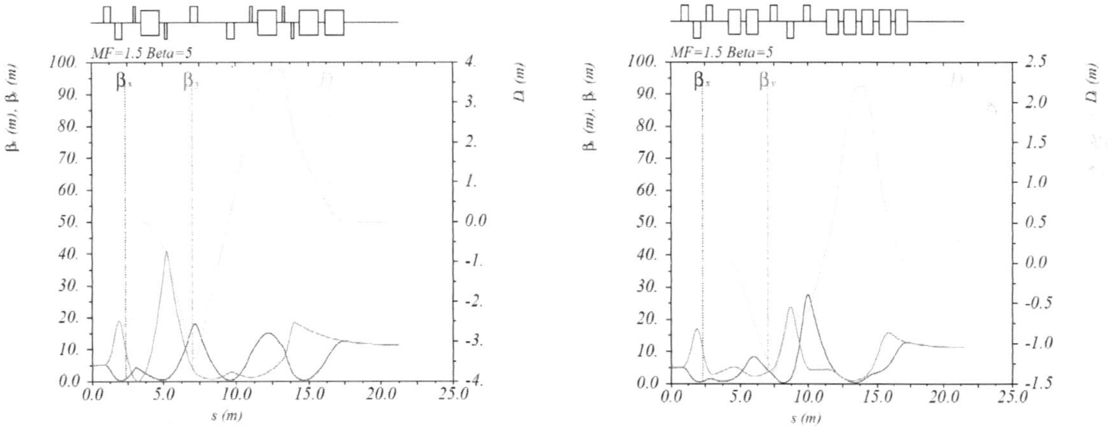

(a) (b)

Figure 1. β (x in black and y in red) and Dispersion (in green) functions for the layout with 45° dipoles (a) and for the layout with 30° dipoles (b).

4.1. Layout with 45° dipoles

The first selected solution is composed of four identical 45° dipoles (i.e. one dipoles' family), with a modest superimposed quadrupolar gradient of 0.07 T/m. This solution, matched as parallel–to–point, allows the change of MF without acting on the gradients of the dipole, but relying on the use of the normal and superconducting quadrupoles. To optimize the gantry dimensions, the superconducting quadrupoles are considered to be directly bolted on the heads of the superconducting bending dipoles (like the spool–piece magnets of the LHC [14]). This layout, operated both changing MF and input β functions, requires a magnet aperture in diameter within 70 mm with a $\Delta p/p = 0.1\%$. Considering a 90 mm aperture, this layout can operate up to $\Delta p/p = 0.4\%$. However, if operated with a $\Delta p/p = 1\%$, the space required by the beam significantly exceeds the limit of 90 mm of aperture; therefore, this layout, cannot operate with a $\Delta p/p = 1\%$. Figure 1(left) shows the optics functions along the gantry, together with the schematics of the line.

4.2. Layout with 30° *dipoles*
The second selected layout operates with seven dipoles subdivided in three families: (i) three dipoles with equal positive quadrupolar gradient (4.5 T/m), (ii) three dipoles with equal negative quadrupolar gradient (−4.9 T/m), (iii) one dipole with a slightly negative quadrupolar gradient (−0.6 T/m). To change magnification factor, a limited variation of the gradients (±0.3 T/m) in the dipoles is required. This would imply the use of nested magnets, that have been excluded from this study. On the other hand, this layout can effectively operate by varying the input betas while maintaining the required aperture within 90 mm for both $\Delta p/p = 0.1\%$ and $\Delta p/p = 1\%$. Figure 1(right) exhibits the optics functions and the schematics of this layout.

5. Conclusion
In this contribution, we reported the main beam optics studies done for the design of a new superconducting gantry for carbon ions. Among dozens of analysed configurations, here we presented the most promising solutions. The first layout is composed of four equal 45° dipoles with an embedded quadrupolar component defined by the coil geometry. This solution can operate both changing the magnification factor or the input beam parameters, but the momentum acceptance is limited to $\Delta p/p = 0.1\%$ for a magnet aperture of 70 mm ($\Delta p/p = 0.4\%$ for 90 mm).

The second layout is based on seven 30° dipoles, subdivided in three families differing by the amount of superimposed gradient. Given the limits we assumed on the magnet technology, this solution can change the beam size at the isocenter only by varying the input beam parameters. Considering a 90 mm magnet aperture, this second layout can operate with a momentum spread up to $\Delta p/p = 1\%$.

The presented work set the basis for future studies on the gantry design, that will have to include a complete and detailed integration between optics, mechanics of the structure, magnet technology, cryogenics and clinical environment.

Acknowledgments
This study was (partially) supported by the European Union Horizon 2020 research and innovation program under grant agreement no. 101008548 (HITRI*plus*).

References
[1] HITRIplus website `https://www.hitriplus.eu/`
[2] Amaldi U *et al.* 2021 A superconducting ion gantry with riboni's unconventional mechanics Tech. rep. CERN, Geneva, Switzerland CERN–ACC–NOTE–2021–0014, NIMMS– Note–002
[3] Benedetto E, Harbi N A, Brouwer L, Tommasini D, Prestemon S, Riboni P and Amaldi U 2021 A carbon-ion superconducting gantry and a synchrotron based on canted cosine theta magnets *Preprint* arXiv:2105.04205
[4] Pullia M G *et al.* 2022 *Proc. 13th Int. Particle Accelerator Conf. (IPAC'22)* 13 (Bangkok, Thailand) pp 2966–69
[5] Pullia M, Felcini E, Savazzi S and Mereghetti A 2022 Explorative beam optic studies for the design of a novel superconducting ion gantry Tech. rep. CERN, Geneva, Switzerland CERN-ACC-NOTE, NIMMS-Note-007

[6] Galonska M, Scheloske S, Brons S, Cee R, Hoppner K, Mosthaf J, Peters A and Haberer T 2013 *Proc. 4th Int. Particle Accelerator Conf. (IPAC'13)* (Shanghai, China) pp 3636–38
[7] Iwata Y *et al.* 2012 *Phys. Rev. Spec. Top. Accel Beams* **15** 044701
[8] Takayama S *et al.* 2022 *IEEE Trans. Appl. Supercond.* **32** 1–4
[9] Rossi L *et al.* 2022 *IEEE Trans. Appl. Supercond.* **32** 1–6
[10] Karppinen M, Ferrentino V, Kokkinos C and Ravaioli E 2022 *IEEE Trans. Appl. Supercond.* **32** 1–5
[11] Pullia M G *et al.* 2022 *Proc. 13th Int. Particle Accelerator Conf. (IPAC'22)* 13 (Bangkok, Thailand) pp 2966–69
[12] Pavlovic M, Griesmayer E and Seemann R 2005 *Nucl. Instrum. Methods Phys. Res., Sect.* A **545** 412–26
[13] Mizushima K, Katagiri K, Iwata Y, Furukawa T, Fujimoto T, Sato S, Hara Y, Shirai T and Noda K 2014 *Nucl. Instrum. Methods Phys. Res., Sect.* B **331** 243–47
[14] Baynham D, Coombs R, Ijspeert A and Perin R 1994 *IEEE Trans. Magn.* **30** 1823–26

Explorative studies of an innovative superconducting gantry

M G Pullia[1], E Benedetto[8], L Dassa[2], E De Matteis[4], M Donetti[1], E Felcini[1], G Frisella[1], M Karppinen[2], C Kurfürst[3], S Mariotto[4,5], A Mereghetti[1], A Mirandola[1], A Pella[1], D Perini[2], L Piacentini[7], M T F Pivi[3], M Prioli[4], A Ratkus[7], L Rossi[4], L Sabbatini[6], S Savazzi[1], M Stock[3], D Tommasini[2], T Torims[7], S Uberti[9], A Vannozzi[6], J Vilcans[7] and M Vretenar[2]

[1]National Center for Oncological Hadrontherapy (CNAO), Pavia, Italy
[2]European Organization for Nuclear Research (CERN), Geneva, Switzerland
[3]EBG MedAustron, Wr. Neustadt, Austria
[4]National Institute for Nuclear Physics (INFN–Milano), LASA Laboratory, Segrate, Italy
[5]University of Milan, Milan, Italy
[6]National Institute for Nuclear Physics (INFN–LNF), Frascati, Italy
[7]Riga Technical University, Riga, Latvia
[8]South East European International Institute for Sustainable Technologies (SEEIIST), Geneva, Switzerland
[9]Università di Brescia, Brescia, Italy

E-mail: marco.pullia@cnao.it

Abstract. The Heavy Ion Therapy Research Integration plus (HITRI*plus*) is an European project that aims to integrate and propel research and technologies related to cancer treatment with heavy ion beams. Among the ambitious goals of the project, a specific work package includes the design of a gantry for carbon ions, based on superconducting magnets. The first milestone to achieve is the choice of the fundamental gantry parameters, namely the beam optics layout, the superconducting magnet technology, and the main user requirements. Starting from a reference 3 T design, the collaboration widely explored dozens of possible gantry configurations at 4 T, aiming to find the best compromise in terms of footprint, capital cost, and required R&D. We present here a summary of these configurations, underlying the initial correlation between the beam optics, the mechanics and the main superconducting dipoles design: the bending field (up to 4 T), combined function features (integrated quadrupoles), magnet aperture (up to 90 mm), and angular length (30° – 45°). The resulting main parameters are then listed, compared, and used to drive the choice of the best gantry layout to be developed in HITRI*plus*.

1. Introduction

In the framework of the Heavy Ion Therapy Research Integration plus (HITRI*plus*) project [1], a new design of a gantry for carbon ions is being developed, based on superconducting (SC) magnets. The new design shall represent the next generation of gantries for hadron therapy, targeting small dimensions, an affordable cost and a credible time scale for construction.

An international panel gathering experts in accelerator design and clinicians at the forefront of research in hadron therapy suggested [2] to start the development from a recent gantry design proposed by a joint TERA–CERN team [3]. The reviewers put particular emphasis on reviewing and confirming the clinical requirements with the clinical staff of the hadron therapy centers participating to the project and on decreasing the overall dimensions and weight of the gantry. For this reason, the panel suggested to push the field on the bending magnets beyond the state of the art of normal [4] and superconducting [5, 6] carbon–ion gantries up to 4 T. Prototyping the magnets was also indicated as a key achievement, in order to prove the feasibility of the design.

Starting from the clinical requirements, this contribution presents the main layout and optics solutions identified during a first exploratory phase of the development; key parameters are compared and the two most promising layouts are chosen for further development. Details on the chosen optics can be found at [7].

2. Clinical requirements

The clinical requirements were extensively discussed with medical physicists and doctors at CNAO [8] and MedAustron [9]. The most relevant ones are:

- the beam with the largest magnetic rigidity is $^{12}C^{6+}$ at 430 MeV per nucleon kinetic energy, corresponding to 6.62 Tm and 31 cm of range in water;
- the scanned area shall be as large as possible, indicatively 350 mm × 350 mm, at least 200 mm × 300 mm; the larger dimension is parallel to the gantry rotation axis. The minimum scanning speed shall be 20 m/s;
- the minimum set of beam sizes at the isocenter shall be 8 mm and 12 mm (FWHM) at the minimum extraction energy; at larger energies, the beam size will reduce following the adiabatic damping of the beam emittance;
- a source–to–axis distance (SAD) of at least 2 m – 2.5 m;
- possibly 360° rotation, minimum 220°;
- volumetric imaging at the isocenter.

3. Layout exploration

All the layouts are based on the following assumptions:

- scanning magnets located downstream of the last bending section, to relax constraints on the aperture of the SC magnets;
- 4 T SC dipoles, corresponding to a bending radius of 1.65 m for $^{12}C^{6+}$ at 430 MeV per nucleon. Depending on the optics, a quadrupole field may be necessary: this would not be obtained via dedicated trimmable coils, but by an asymmetric arrangement of the coils (left–right taken on the magnet cross section). In addition, the magnet aperture should be 70 mm – 90 mm. The maximum ramp rate $\dot{B}$ should be 0.4 T/s;
- achromatic optics solutions at the isocentre, implying that the spot size and position at the isocentre is independent of the beam momentum distribution;
- beam dimensions at the isocentre independent of the gantry angle of rotation. This implies that either a "rotator" [10, 11, 12, 13] is installed upstream of the gantry or that the incoming beam is round, i. e. has the same dimension or divergence in the two transverse planes, depending on the matching method (see later).

Such parameters pose important challenges to the design of the SC dipoles, especially concerning handling the high magnetic energy stored by the dipole for a very small curvature radius and the $\dot{B}$. Both in the framework of HITRI*plus* and other European R&D collaborations, several magnet options have been investigated, both in terms of conductors (i. e. Nb–Ti, Nb_3Sn, REBCO,Bi–2212, MgB_2), and in terms of coil layouts (i. e. CosineTheta, Canted Cosine Theta

Table 1. Key gantry parameters among different layouts (numbered from 1 to 10) and optics: gantry length, radius, distance between end of last dipole and isocentre, number of main SC dipoles and bending angle, number of SC dipole families, and number of NC and SC quadrupoles. The required magnet aperture when matching with varying MF for a given value of input β function or when varying the input β function for a given value of MF are indicated; two values of beam momentum spread ($\Delta p/p$) are considered. The results in the table consider a parallel–to–point matching method but for layouts 3, 4, 7 and 10, for which point–to–point matching is performed.

Geometry parameters

Layout	Length [m]	Radius [m]	After last bending [m]	SC dipoles N × angle	SC dipole families	Quads. NC, SC
1	12.9	6.4	4	7 × 30°	3	0, 6
2	14.05	5.75	3.5	4 × 45°	1	5, 4
3	11.7	5.75	3.5	7 × 30°	3	0, 6
4	13.2	7.3	4	9 × 30°	3	6, 2
5	10.4	6	3.6	9 × 30°	9	0, 2
6	14.2	5.55	3	4 × 45°	1	6, 4
7	14.2	5.55	3	4 × 45°	1	6, 4
8	13.5	5.25	3	4 × 45°	1	5, 4
9	14.05	5.75	3.5	4 × 45°	2	5, 4
10	14.2	5.1	3	4 × 45°	2	0, 8

Aperture [mm]

Layout	$\Delta p/p = 0.1\%$ var. MF	var. β	$\Delta p/p = 1\%$ var. MF	var. β
1	90	90	90	90
2	70	70	163	168
3	90	93	105	111
4	90	90	174	129
5	70	70	90	93
6	70	70	156	168
7	90	90	111	123
8	70	70	162	165
9	93	96	120	135
10	90	90	165	165

(CCT) or race–track [14, 15]). For design and construction of a prototype the HITRI*plus* collaboration has chosen as baseline Nb–Ti wound as CCT. The use of CCT magnets have been already proposed for high momentum acceptance proton gantries [18] operating in steady–state; however, in the analysed HITRI*plus* gantry we are investigating low loss former and coil technology to optimise the performances for a ramped magnet. In addition, the higher magnetic field, different aperture and curvature radius require a complete re–design of the CCT.

The choice of locating the scanning magnets downstream of the last bending section is another source of important challenges. In fact, on one hand, the scanning magnets must be normal conducting (NC) to guarantee the necessary scanning speed; on the other hand, they must provide quite substantial magnetic fields in the gap in order to bend beams of the maximum magnetic rigidity by some tens of mrad, in order to scan an area of some hundreds of mm about 2.5 m downstream. The specific challenges pertaining the scanning system are not addressed in this contribution.

As done in the optics design of other gantries [10, 11, 12, 13, 16], the gantry is considered as a 1:MF telescope (MF stands for "Magnification Factor"). The two beam sizes as per the clinical requirements can be obtained either by varying the MF for a given set of optics functions in input to the gantry ("var. MF", see table 1), or by varying the input optics functions for a specific MF ("var. β", see table 1). Similarly, the optics can be matched "point–to–point", by which the beam dimension in input to the gantry is directly mapped onto the beam dimension at the isocentre, or "parallel–to–point", by which the beam divergence in input to the gantry is mapped onto the beam dimension at the isocentre.

For every explored layout, monitors and correctors were inserted in the layout, to estimate the accuracy in correcting the beam orbit in presence of magnet misalignment and magnetic field errors. For each optics, the residual orbit was calculated and taken into account together with the *spread* and *offset* in beam momentum to evaluate the beam envelope (including the betatron part), which identifies the so–called "good field region" (GFR). As commonly done in magnet design, the GFR sets the magnet aperture, obtained as 3/2 of the GFR as a first approximation. Two values of momentum spread have been considered for each layout and optics, i. e. $\Delta p/p = 0.1\%$ and $\Delta p/p = 1\%$; the former value is close to what operationally attained with CNAO beams, whereas the latter considers the possibility of operating with off–momentum beams with a fixed current in the superconducting magnets for a few tumor "slices".

A detailed study on the mechanics rotating the gantry is being carried out. First results [17] favour statically–balanced solutions to ease operation and safety requirements, versus the solution initially proposed by TERA [3, 16] with no counterweight and lower mass. Moreover, a wide range of solutions has been considered; optimisation is on–going focussing on a few figures of merit like acceleration and breaking, safety and space required.

4. Layout comparison

Table 1 summarises the key parameters of the most relevant layouts and optics solutions. The number of bending magnets is indicated, along with the bending angle and the number of dipole families; the number of quadrupoles, both NC and SC, is reported as well. The required aperture for different matching configurations is indicated for two values of beam momentum spread.

Layouts with a distance between the last bending section and the isocentre smaller than 3 m are discarded, because there would be not enough space for the scanning magnets and the desired SAD. Moreover, solutions requiring a magnet aperture beyond the allowed range (i. e. 70 mm – 90 mm) are discarded as well. Few solutions require separate circuits for dipoles and quadrupoles in the superconducting magnets (nested magnets), which have been discarded too.

As shown, only "Layout 1" allows transporting beams with a 1% beam momentum spread. This layout operates with combined–function dipoles, that would require varying the quadruple gradient independently from the dipole field, i. e. a nested magnet, for changing magnification factor. Therefore, "Layout 1" can be operated only in the scenario for which the input β functions are varied. As complementary layout, i. e. with SC combined–function dipoles of 45°, working with a beam momentum spread of 0.1%. "Layout 2" is optically matched by varying the MF without the need of changing the quadrupole gradient in the dipole. This layout is the only one with enough room downstream of the last bending section and requiring an aperture of 70 mm. Both layouts are optically matched parallel–to–point; their schematics are shown in figure 1.

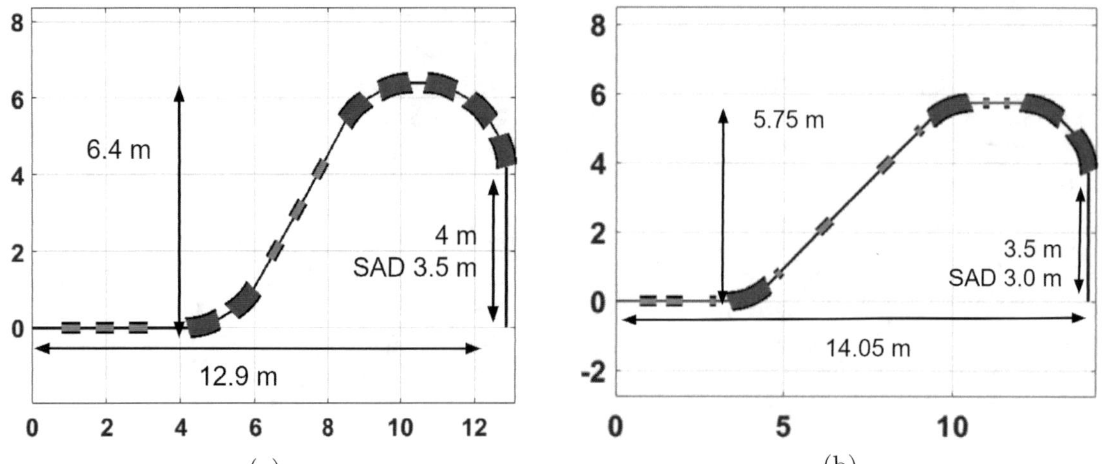

Figure 1. Schematics of the chosen layouts: layout 1 (a) and layout 2 (b). Arrows indicate essential dimensions; the radius is the distance between the gantry rotation axis and the beam height.

The total cost of the gantry has been estimated based on the experience of the teams responsible for the development of the different aspects of the gantry design. The total cost estimation considers superconducting magnets (R&D and construction), NC magnets, power supplies, ancillary systems, rotation mechanics and building (only gantry room). There are small variations among the cited items for all the four configurations, apart from procuring the hardware of the dipoles, more relevant for Layout 1 than for Layout 2, as expected. In general, Layout 1 results to be more expensive than Layout 2, with an estimated cost difference in the order of 10%. Moreover, the difference in total cost between the two layouts in the 220°and 360°configurations is about 15%.

5. Conclusion

In the context of the HITRI*plus* project, a new design of a SC ion gantry is under development. The new gantry shall transport ion species of present and future interest for hadrontherapy, with the maximum beam rigidity set by fully stripped carbon ions at 430 MeV per nucleon. The design is based on SC bending dipoles with the main field at 4 T, corresponding to a radius of 1.65 m. The aperture should be in the range 70 mm – 90 mm, marking a world record in terms of stored magnetic energy for such a small curvature radius, though expected to be at reach; the fast ramping rate of $\dot{B}$ =0.4 T/s is another parameter requiring particular attention in the magnet design. Among dozens of possible layouts and optics solutions, the two most promising ones have been discussed, highlighting differences and challenges of the two designs. An overall cost estimation was made for these two layouts, as well as for the complete and partial rotation options. The layout that will be selected by the HITRI*plus* collaboration will be further developed and optimised.

Acknowledgments

This study was (partially) supported by the European Union Horizon 2020 research and innovation program under grant agreement no. 101008548 (HITRI*plus*).

Journal of Physics: Conference Series **2420** (2023) 012099 doi:10.1088/1742-6596/2420/1/012099

References

[1] The HITRI*plus* project `https://www.hitriplus.eu/`
[2] *The GaToroid and SIGRUM ion therapy gantries: report of the review committee*, private communication
[3] Amaldi U *et al.* 2021 SIGRUM – A superconducting Ion Gantry with Riboni's Unconventional Mechanics Rep. CERN
[4] Galonska M, Scheloske S, Brons S, Cee R, Hoppner K, Mosthaf J, Peters A and Haberer T 2013 *Proc. 4th Int. Particle Accelerator Conf. (IPAC'13)* (Shanghai, China) p 3636
[5] Iwata Y *et al.* 2012 *Phys. Rev. Spec. Top. Accel Beams* **15** 044701
[6] Takayama S *et al.* 2022 *IEEE Trans. Appl. Supercond.* **32**
[7] Felcini E, Frisella G, Mereghetti A, Pullia M G and Savazzi S 2022 *Proc. 13th Int. Particle Accelerator Conf. (IPAC'22)* (Bangkok, Thailand) pp 2962–5
[8] CNAO website `https://fondazionecnao.it/`
[9] MedAustron website `https://www.medaustron.at/en/`
[10] Pavlovic M 2021 *Transport of Ion–Therapy Beams in Rotating Gantries* (Hauppauge, New York: Nova Science) p 68
[11] Badano L, Benedikt M, Bryant P J, Crescenti M, Holy P, Maier A T, Pullia M, Rossi S and Knaus P 2000 Proton–Ion Medical Machine Study (PIMMS) Rep. CERN vol 1
[12] Bryant P J *et al.* 2000 Proton–Ion Medical Machine Study (PIMMS) Rep. CERN vol 2
[13] Lante V *et al.* 2012 Conceptual design for a carbon ion gantry – ULICE Rep. D. JRA6.3
[14] Rossi L *et al.* 2022 *IEEE Trans. Appl. Supercond.* **32** n 1 p 1
[15] Rossi L *et al.* 2022 *IEEE Trans. Appl. Supercond.* **32** n 6 p 1
[16] Benedetto E, Al Harbi N, Brouwer L, Tommasini D, Prestemon S, Riboni P and Amaldi U 2021 *Nucl. Instrum. Methods Phys. Res., Sect.* A submitted
[17] Piacentini L, Dassa L, Perini D, Ratkus A, Torims T, Vilcans J, Vretenar M 2002 Comparative Study on Scenarios for Rotating Gantry Mechanical Structures Rep. CERN
[18] Brouwe L *et al.* 2020 *Nucl. Instrum. Methods Phys. Res., Sect.* A **957** 163414

Journal of Physics: Conference Series **2420** (2023) 012100 doi:10.1088/1742-6596/2420/1/012100

Study of coil configuration and local optics effects for the GaToroid ion gantry design

E Oponowicz[1], L Bottura[1], Y Dutheil[1], A Haziot[1] and A Gerbershagen[1,2]

[1]CERN, Geneva, Switzerland

[2]Particle Therapy Research Center (PARTREC), Department of Radiation Oncology, University Medical Center Groningen, University of Groningen, The Netherlands

E-mail: ewa.oponowicz@cern.ch

Abstract. GaToroid, a novel gantry configuration for hadron therapy, is based on superconducting coils that generate a toroidal magnetic field to deliver the beam to the tumour location. Designing the complex GaToroid coils requires careful consideration of the local beam optical effects. We present a Python-based tool for charged particle transport in complex electromagnetic fields. The code implements fast tracking in arbitrary three-dimensional field maps, and it does not require a definition of specific or regular reference trajectories, as is generally the case for accelerator physics codes. The tool was used to characterise the beam behaviour inside the GaToroid system: to automatically determine the reference trajectories in the symmetry plane and analyse three-dimensional beam dynamics around these trajectories. Beam optical parameters in the field region were compared for various magnetic configurations of GaToroid. This paper introduces the new tracker and shows the benchmarking results. Furthermore, first-order beam optics studies for different arrangements demonstrate the main code features and serve for the design optimisation.

1. Introduction

A number of well-established codes for beam tracking studies are widely available [1, 2]. However, most of them are typically designed for long magnets of small aperture, i.e., magnetic configurations generating conventional fields or superposition of such, which can be easily represented with multipole expansion theory. Beam optics studies in complex magnetic fields are either not implemented, or require complex post-processing [3].

The development of a fast and compact code for particle tracking in arbitrary magnetic (or electric) field was motivated by the beam optics studies of GaToroid, a novel gantry for hadron therapy [4, 5]. The GaToroid concept is based on an axis-symmetric toroidal fixed-field magnet which, in combination with an upstream vector magnet, is capable of delivering the beam onto the patient. Characterisation of the local beam dynamics properties of this system is required to optimise its design. This work demonstrates the transport of carbon ions (C^{6+} of kinetic energy in the range 120 – 430 MeV/u) through several examples of GaToroid configurations.

2. Method

2.1. Coil geometry

The principle of GaToroid is based on two parts: the main toroidal magnet to bend the beam onto the patient (figure 1) and an upstream *vector magnet* (VM, modelled here as single kick) for beam steering. The angular kicks imposed at the VM are polar, depending on the beam kinetic energy, and/or azimuthal, determined by the desired irradiation direction. The azimuthal variations are discrete and limited to the number of coil pairs constituting the toroid. For a given irradiation angle, the beam passes in between a set of two coils tilted symmetrically around the toroid axis (figure 1a). Due to the periodic axis symmetry, beam dynamics studies concentrate on the region bounded by two coils forming the "beam channel".

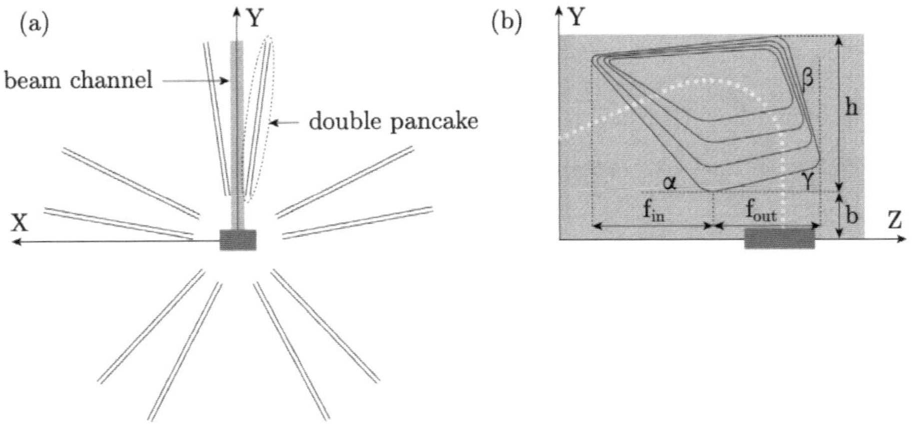

Figure 1. Schematic drawings of GaToroid with patient (green rectangle) in the axis of the toroid (treatment position), and the conductor windings (purple curves): (a) projection of a full gantry consisting of 5 beam channels; each channel is the space between two adjacent coils, (b) projection of the single coil geometry onto the bending plane together with and example beam trajectory (yellow dashed curve). f_{in} - entrance leg length, f_{out} - exit leg length, , h - maximum coil height, b - bore radius, α - entrance face angle, β - back leg angle, γ - exit face angle. The configuration consists of 4 coil grades.

The geometry of the coil (figure 1b) substantially impacts beam dynamics. In a typical toroidal configuration, the magnetic field decreases with the radius of the coil; higher energy particles require a larger kick at VM and hence experience a lower magnetic field and their bending is weaker. For treatment purposes this defocusing effect must be compensated to ensure it is an achromatic system. The field profile is modified by so called grading, i.e., introducing spacing between the coil winding in the coil for an appropriate distribution of the current density. For a detailed explanation the reader is referred to [5, 4]. The main parameters defining the geometry of the coil, are: number of grades, their relative positions and reduction factors; angles (α and γ) and lengths (f_{in} and f_{out}) of the entrance (vector magnet side) and exit (isocentre side) faces; maximum height, and bore size. Additionally, the angle of the back leg β can be optimised to reduce the amount of conductor used, in particular in the region not passed through by the beam.

2.2. Tracking algorithm

A tracking tool was developed in Python based on the *Boris* algorithm [6]. It is validated against a well-established Fortran ray tracing code `Zgoubi` [7]. Here a single charged particle travels through a complex static magnetic field, and the trajectories – read out from the two tracking codes – are compared. The distance between the two positions in a 3D coordinate system increases in the regions of higher magnetic field (stronger bending), but remains in a sub-millimeter range. The difference is assumed to result from the numerical approximation, since it is smaller for a finer mesh of the field map, e.g., max. 0.04 mm for a 25 mm mesh size, and max. 0.01 mm for 10 mm.

2.3. First order transport matrix
Transport of a particle from point A to point B can be described as

$$\begin{pmatrix} x_B \\ x'_B \\ y_B \\ y'_B \\ s_B \\ \frac{\delta p}{p}\Big|_B \end{pmatrix} = M^{AB} \begin{pmatrix} x_A \\ x'_A \\ y_A \\ y'_A \\ s_A \\ \frac{\delta p}{p}\Big|_A \end{pmatrix}, \quad (1)$$

where x_i - horizontal position, x'_i - horizontal angle, y_i - vertical position, y'_i - vertical angle, s_i - longitudinal position, $\frac{\delta p}{p}\Big|_i$ - momentum spread, $i = A, B$, and M^{AB} is the 6x6 linear transport matrix. The matrix can be hence determined by tracking a set of finite (and low) number of particles. In addition to the reference particle of input parameters $(x_0, x'_0, y_0, y'_0, t_0, \frac{\delta p}{p}\Big|_0)$, at least 12 other particles must be tracked: each with only one coordinate different from the respective coordinate of the reference particle. The matrix coefficients are then calculated using the method described in [8]. This can be performed at any step of the reference trajectory, provided all the beam parameters are represented in the local reference system of the pilot particle, i.e. in the reference plane. If trajectories of all other particles include steps before and after the reference plane, and within the tracking region (here, magnetic field region), their positions and momenta are linearly interpolated to the reference plane. Coordinates of a particle which does not reach the reference plane are extrapolated under the assumption of zero field outside of the tracking region; the particle momentum remains unchanged, and the extrapolated position is a translation of its last known position, based on the momentum vector, onto the reference plane.

3. Results

3.1. Reference trajectories
To analyse beam dynamics in the toroidal gantry, the initial step is to ensure that the beam is delivered to the patient location, i.e., isocentre. Ideally, to provide with a full energy acceptance, beams of any kinetic energy within the treatment spectrum should be deflected at the VM such that it reaches the isocentre with a sub-millimeter precision [9] and normal to the irradiation plane (except for specific cases, e.g., in the vicinity of organs at risks, dealt with by the treatment planning systems). However, the isocentre location relative to the coil geometry is arbitrary, and so is the position of the vector magnet (VM). The trajectories of reference particles of various energies, backward-tracked from an example isocentre vertically up, end up in various locations on the isocentre plane; this end position can move by as much as ± 1 m for an average distance between the isocentre and VM of around 11 m. This behaviour depends on the selected isocentre location.

To determine (for a given coil configuration) the positions of the VM and of the isocentre, together with the azimuthal kick angles imparted at VM, an iterative process, restricted to a two-dimensional symmetry plane between two coils, of following steps is performed:

(i) *Find VM position:* The pilot particles of various energies are tracked backward and vertically up from a given isocentre to the the isocentre plane. As their final positions vary as a function of kinetic energy, the VM position is chosen as the average value of the positions of different-energy particles.

(ii) *Optimise azimuthal angles:* The particles are forward-tracked from this VM location and receive azimuthal kicks. For each energy, the kick angle is optimised such that the particle arrives as close as possible to the previously fixed isocentre, and as vertically as possible.

(iii) *Calculate particle angle at isocentre:* Based on the optimisation results, the particles are forward-tracked from the VM to calculate the angle of the particle at the isocentre.

The whole procedure is repeated for multiple isocentre positions and the final choice corresponds to the case of minimum spread of the incoming beam angles for the whole range of the treatment energies.

3.2. Focusing terms

Once central trajectories are known, particle tracking in a 3D field map is performed. Linear transport matrix can be extracted from the tracking data for a segment between any two tracking points of the reference trajectory. Subsequently, this segment can be modelled with a thin quadrupole approximation. The focusing term kl derived form the matrix, further normalised by the beam rigidity $B\rho$ and the segment length L, can be expressed as quadrupole gradient:

$$g\ (\mathrm{T/m}) = \frac{\mathrm{kl} \cdot \mathrm{B}\rho}{\mathrm{L}} \left(\frac{\mathrm{m}^{-1} \cdot \mathrm{Tm}}{\mathrm{m}} \right). \tag{2}$$

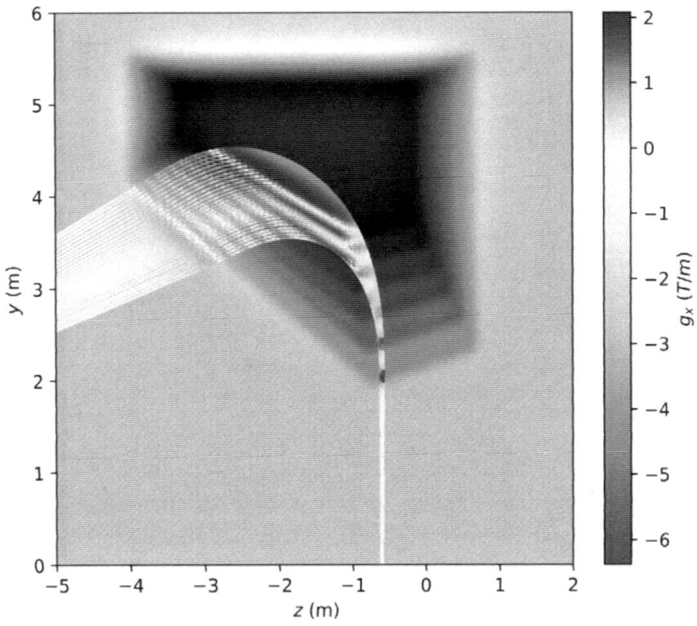

Figure 2. Color map representing the quadrupole gradient (T/m) at different steps of reference trajectories for the entire range of treatment energies, plotted on top of the main magnetic field component of a GaToroid configuration (black and white).

The gradients experienced by beams of various energies (figure 2) give a more detailed, quantitative understanding of the beam dynamics: a strong defocusing effect in the horizontal plane over the full length of propagation. The location of strong quadrupolar behaviour follows the position of the coil grades.

3.3. Coil exit angle

the beam dynamics in the GaToroid system is affected by the set of geometry parameters defining the coil. To demonstrate it, we introduce modification of the exit angle γ (isocentre side), which resembles the edge angle in a conventional bending magnet. Varying the γ angle alone, while keeping the bore size fixed, leads to magnetic configurations for which the beam travelling from the same point (isocentre) experiences different bending strengths, i.e., in the hard edge approximation the magnetic field starts further from the starting location for higher tilt angles γ. Hence, for each magnetic configuration (tilt angle), the entire coil is shifted such that the beam starts being bent in the same location, to decouple the effects of focusing and bending. In each case, the input positions and angles of 300 MeV/u carbon ion beams are optimised such that their trajectories are comparable (figure 3). The focusing coefficients for the entire system (from VM to isocentre) for each configuration are presented in figure 4. They change slightly in the vertical plane and more substantially in the horizontal plane, decreasing with higher angles and changing into negative values (focusing) for 20°.

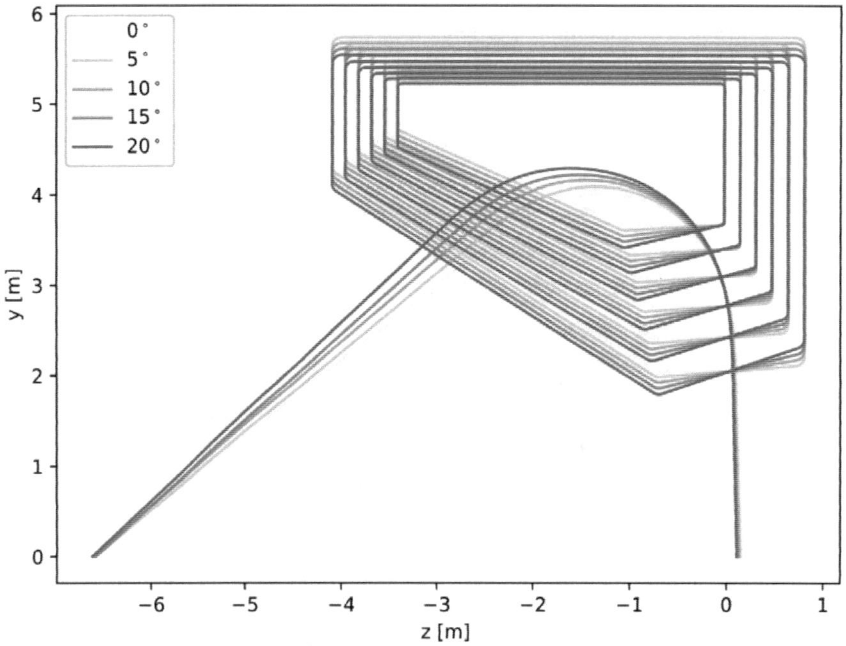

Figure 3. Configurations of the GaToroid coil for several isocentre-side angles γ $(0-20°)$. Different configurations are shifted such that the beam arriving from the isocentre start experiencing the magnetic field at the same location. For each configuration a carbon ion beam of the same energy (300 MeV/u) is tracked such that it arrives with minimum angle to the same isocentre.

4. Conclusion

A new tool for tracking particles, based on Boris algorithm, in any arbitrary three-dimensional magnetic field is presented in this work. Its implementation was motivated by the complexity

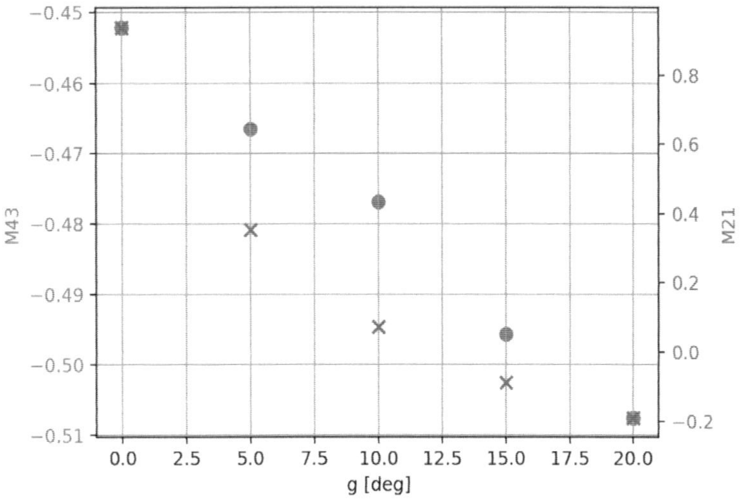

Figure 4. Focusing coefficients in the vertical plane M_{43} (green) and in the horizontal plane M_{21} (red) as a function of the tilt angle γ of the GaToroid isocentre side of the coil.

of the magnetic field generated by GaToroid, a novel toroidal configuration proposed for beam delivery in hadron therapy. The features of the new tracker analyse the beam dynamics in the gantry in order to couple it to the coil design. Several examples, including tracking in two- and three-dimensional magnetic field maps of GaToroid, demonstrate the steps to be undertaken towards the ultimate goal, i.e., optimisation of the GaToroid design. Whilst this work focused on linear beam dynamics, higher orders are to be determined in the next steps. In addition, for further optimisation, GaToroid should be treated as a complete system from the accelerator extraction point to the isocentre.

Acknowledgments
Project co-supported by the CERN Budget for Knowledge Transfer to Medical Applications.

References
[1] Sagan D 2006 *Nucl. Instrum. Methods Phys. Res., Sect.* A **558** pp 356–59
[2] Deniau L, Grote H, Roy G and Schmidt F 2017 The MAD-X program. User's Reference Manual
[3] Gerbershagen A, Meer D, Schippers J M and Seidel M 2016 *Z. fur Med. Phys.* **26** pp 224—37
[4] Bottura L, Felcini E, De Rijk G and Dutoit, B 2020 *Nucl. Instrum. Methods Phys. Res., Sect.* A **983** 164588
[5] Felcini E, Bottura L, Van Nugteren J, De Rijk G, Kirby G and Dutoit B 2020 *IEEE Trans. Appl. Supercond.* **30** pp 1–5
[6] Qin H, Zhang S, Xiao J, Liu J, Sun Y and Tang W M 2013 *Phys. Plasmas* **20** 084503
[7] Méot F 2012 Zgoubi user's guide No. BNL-98726-2012-IR. Brookhaven National Lab., Upton, NY USA
[8] Yoon M and Robin D S 2013 *IEEE Trans. Nucl. Sci.* **60** pp 3837–42
[9] Reimoser S A and Pavlovic M 2001 *Nucl. Instrum. Methods Phys. Res., Sect.* A **456** pp 390–410

Slow extraction modelling for NIMMS hadron therapy synchrotrons

R Taylor[1,2], E Benedetto[2,3], M Sapinski[2,3] and J Pasternak[1,4]

[1]Imperial College London, UK
[2]CERN, Geneva, Switzerland
[3]SEEIIST, Geneva, Switzerland
[4]STFC Rutherford Appleton Laboratory, Didcot, UK

E-mail: rebecca.taylor@cern.ch

Abstract. The Next Ion Medical Machine Study (NIMMS) is an umbrella R&D programme for CERN accelerator technologies targeting advanced accelerator options for proton and light ion therapy. In collaboration with the European programme HITRIplus, one area of study is slow extraction which is required to deliver a uniform beam spill for radiotherapy treatment. Several techniques use the third-order resonance to extract hadrons; these include betatron core driven extraction and radiofrequency knock-out. Flexible simulation tools using these techniques were prepared and initially benchmarked with results from the literature that used the Proton-Ion Medical Machine Study (PIMMS) design. The limits of the current PIMMS design were then pushed to evaluate its compatibility to deliver 10x higher intensity ion beams, and using increased extraction rates.

1. Motivations

Flexible modelling techniques have been developed to facilitate the design of NIMMS [1]. The NIMMS study has a variety of designs available including normal and superconducting options [2]; all aiming to provide a variety of light ion beams for clinical treatment and radiobiological research.

Resonant slow extraction is used to ensure a continuous and stable dose delivery during treatment. It is performed by setting the horizontal tune of the machine close to the third-order resonance (Q_x =1.666) which is excited with strong resonant sextupole magnets. The particles are then driven to the resonance via excitation methods, leave the stable triangle phase-space and are extracted at the electrostatic septum (ES).

Simulations of slow extraction were used to assess if the PIMMS [3, 4] design, as utilized by CNAO and MedAustron therapy facilities, can be adapted to suit advanced accelerator options. The feasibility study was performed, first by modelling the PIMMS lattice, then changing excitation methods to ensure it can be upgraded to meet two requirements: adapting to extract higher emittance beams, and providing the option of increased extraction rates.

1.1. Higher emittance

Multi-Energy Extraction (MEE) uses multiple flat-tops for each required treatment energy, rather than having one energy per magnet duty cycle. This technique decreases wait-time during delivery of different energies, and was proposed and implemented at the NIRS HIMAC center [5].

To deliver a dose of 2 Gy to a 1-liter tumour within one cycle of MEE, the synchrotron must store 1 to 2×10^{10} carbon ions [2], which is 20× higher than what current European hadron therapy synchrotrons can deliver. With an ion source of 600 µA, a multi-turn injection method of 30 turns needs to be performed to achieve this higher intensity, assuming 50% injection efficiency. This method would consequently generate a horizontal, normalised rms emittance increase of $\varepsilon_{x_{\mathrm{rms}}} < 6\pi$ mm mrad. Slow-extraction studies have been performed to assess the fesibility of extracting a beam with these larger emittances.

1.2. Increased extraction rates
Treating with dose delivery rates of 50-100 Gy s^{-1} reduces toxicities to healthy tissue compared to conventional radiotherapy [6]. This technique is known as FLASH therapy. With a suggested threshold dose of 10 Gy, the beam should be provided within 100 ms [7]. The NIMMS synchrotron should provide a hadron beam which can meet the recommended dose within this time frame. The PIMMS revolution frequency for high-energy carbon is 2.8 MHz [4] so 100 ms requires 280,000 turns. This is a challenge compared to nominal slow extraction which delivers over millions of turns. This study will observe whether a high quality beam spill can be provided within this increased extraction rate.

2. Method
Simulations were performed to reproduce slow extraction results from the existing PIMMS synchrotrons, then these were adapted to meet the two advanced options. The lattice was implemented into MAD-X [8] and matched for the extraction conditions for two excitation methods: betatron core and radiofrequency knock-out (RKFO).

The main lattice parameters relevant to slow extraction are horizontal and vertical tune Q_x, Q_y, chromaticity Q'_x, virtual sextupole strength S_{virt} and beam momentum spread $\frac{\Delta p}{p}$, shown in table 1 for both methods.

Table 1. Lattice extraction parameters.

	Q_x	Q'_x	Q_y	S_{virt}	$\Delta p/p$
Betatron	1.666	-4.041	1.720	28.4	0.4%
RFKO $1\varepsilon_x$	1.675	-0.004	1.695	32.1	0.1%
RFKO $6\varepsilon_x$	1.680	-0.007	1.695	32.1	0.1%

The resulting lattice is converted into a high order Polymorphic Tracking Code (PTC) map [9] using the Maptrack module [10, 11], written in python. Maptrack's ability to split a 1-turn map into segments was employed, and custom elements were written in-between map segments, creating dynamic simulations of betatron core and RFKO excitation. The tracked beam was a 400 MeV/u energy carbon beam with a gaussian beam profile corresponding to normalised rms emittance of $\varepsilon_{x_{rms}} = 1\pi$ mm mrad. Each turn, the beam was observed at the position of the electrostatic septum.

Visualisation tools were developed to observe the phase-space evolution and tune change of the beam throughout slow extraction. An interactive demonstration of these tools is in development [12]. The tune is calculated by measuring the fundamental frequency of the particle's position in the x or y plane, over a window of 256 turns, using the Python NAFF module [13].

2.1. Betatron core

PIMMS synchrotrons use betatron core excitation, which decreases the momentum of the particles each turn to push their tune towards the resonance [3]. The dynamics was modelled by changing the particle momentum by a step corresponding to the tune distance to resonance, divided by chromaticity and total number of turns: $(Q_{x\text{res}} - Q_{x\text{min}})/(Q'_x \cdot N_T)$.

In operation, this method requires a debunched beam, therefore is difficult to combine with MEE, and so the excitation method of the NIMMS synchrotron must be adapted accordingly.

2.2. RFKO

RFKO excitation consists of two electrostatic plates which provide a transverse kick at frequencies close to the betatron tune, to increase the amplitude of the particles. It is the method in use at HIMAC [14] and HIT [15] and is currently being implemented at CNAO [16] and considered for MedAustron [17]. The method allows for extraction to start and stop rapidly, so it is ideal for MEE. Equation (1) [18] shows how the voltage V_p relates to the angular kick to a beam of charge-mass ratio Z/A and relativistic particle velocity $\beta = v/c$. The kick strength depends on the properties of the exciter: length L and plate distance a.

$$\theta_k = \frac{Z}{A}\frac{e}{m_p}\frac{V_p}{a}\frac{L}{\beta^2 c^2} \tag{1}$$

The RFKO exciter installed at CNAO is located in-between the injection septum and the first chromatic sextupole. It has a maximum voltage of 367 V, corresponding to a kick of 2.8 µrad for 250 MeV protons and 1.0 µrad for 400 MeV/u carbon ions [16].

Equation (2) is used to incorporate this effect into simulations, with the kick in particle angle, $\delta x'$ relating to amplitude θ_k, fractional tune of the beam q_x and turn number T.

$$\delta x' = \theta_k \cos(2\pi q_x T) \tag{2}$$

A gaussian range of frequencies around q_x is considered, to excite the tune spread of particles. An amplitude curve is added throughout extraction which exponentially increases with time to overcome the increased density of the particles at the core, the function of which is in use in operation at HIT [18].

2.3. High emittance simulations

Conventional betatron core can accommodate larger amplitudes, however RFKO is amplitude-driven, therefore an increase in amplitude requires a shift in slow extraction settings to avoid the initial beam being affected by the extraction region.

Two emittance increases were considered, $\varepsilon_x = 3$ and 6π mm mrad. Extracting a 3π mm mrad beam is compatible with nominal slow extraction settings, however a 6π mm mrad beam is too large for the stable phase-space region, and therefore the distance to the tune had to be increased to accommodate for this, see table 1. Alternatively, it is possible to keep the Q_x constant and decrease S_{virt}, however this would affect the spiral step - defined as final step of the particles before extraction.

3. Simulations

Using nominal betatron core extraction with a momentum shift of -3.2×10^{-9} per turn, the particles are considered extracted when they exceed the electrostatic septa aperture limit of $x_{\text{ES}} = 35$ mm. The intensity drop of the beam and the spill showing number of particles extracted as a function of number of turns is represented in figure 1, which shows 5,000 particles extracted throughout 2,800,000 turns - equivalent to 1 s of spill.

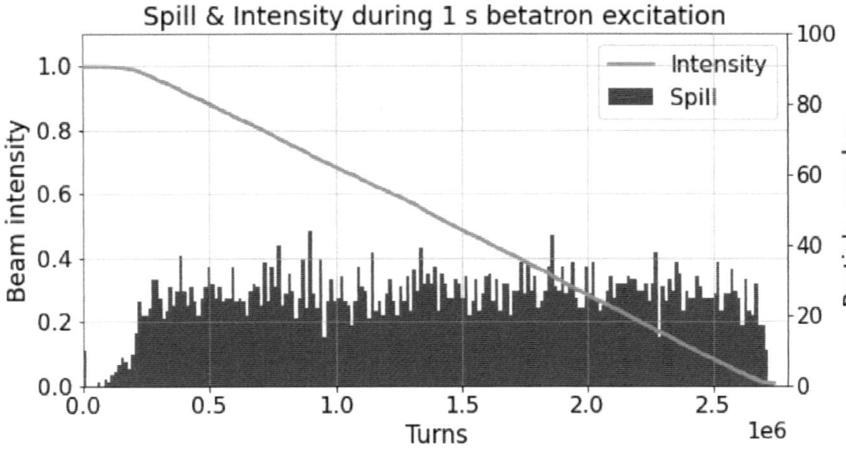

Figure 1. Beam and spill intensity as a function of turn number. 5,000 particles during 1 s of betatron excitation.

Figure 2 (top) in green shows the phase space and the Steinbach diagram [3] – tune vs amplitude – for a sample of 500 particles, cumulated over 25 turns, during which a particle is extracted. The large x_p distribution is due to the momentum increase and the cumulative visualization in a point with non-zero dispersion.

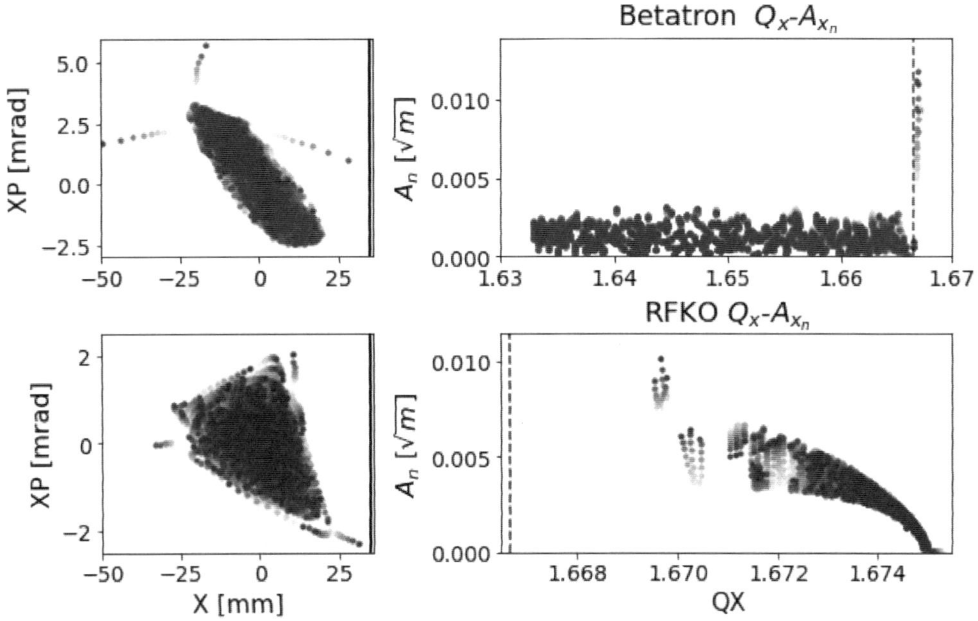

Figure 2. Cumulative phase-space and Steinbach diagram of 500 particles, 25 turns, for betatron (top - in green) and RFKO (bottom - in blue) extraction.

In figure 2 (bottom) in blue, the phase-space and tune evolution during RFKO excitation is shown. The left plot shows that the stable triangle phase-space is larger, and the right plot shows that when the particle's amplitude increases due to the $\delta x'$ kick, it experiences detuning, causing them to approach the resonance before extraction.

This RFKO excitation was then performed for 5,000 particles over 2,800,000 turns, and figure 3, in blue, shows the number of particles extracted per turn and the resulting beam intensity decrease, for an emittance of 1π mm mrad. The amplitude of the RFKO kick strength for this extraction spill is plotted as the dashed blue line in µrad, and varies between 1-2 µrad, which is compatible with previous RFKO simulations of PIMMS machines [19].

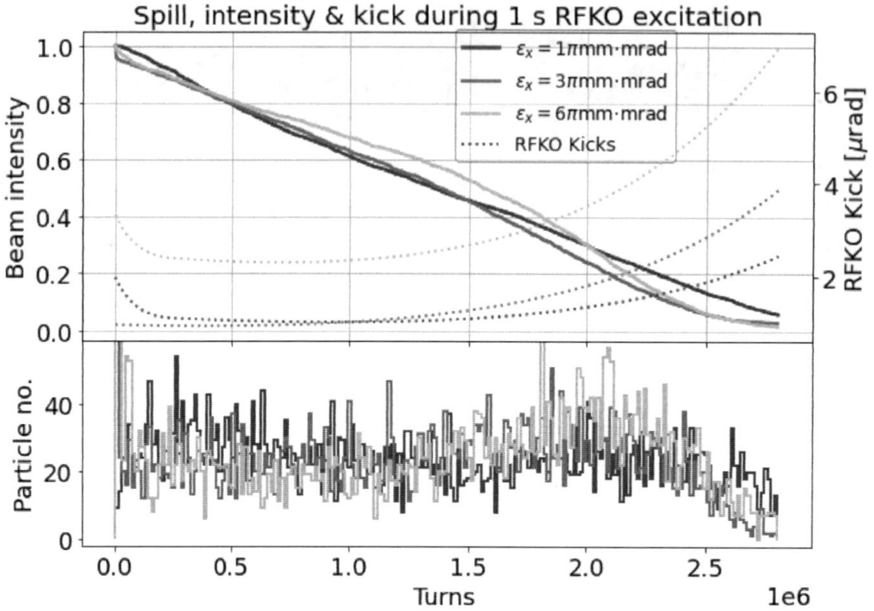

Figure 3. 5,000 particles extracted via RFKO over 1 s for three initial emittances, with optics settings as per table 1.

The effect of increasing the emittance to 3π mm mrad, is shown in figure 3 in red, and shows it is compatible with the nominal RFKO optics, but requires increases to the RFKO kick function for higher turns. Increasing emittance to 6π mm mrad requires moving the tune away from the resonance, as shown in table 1, else a substantial amount of the beam is extracted within the first few turns. The larger tune distance results in a larger required kick amplitude, between 1.5 to 3×, to extract the particles at the same rate.

To verify the validity of RFKO as a method for FLASH extraction, first the conditions required for a suitable spill timescale must be established, and then the feasibility of these conditions should be evaluated. Figure 4 shows that for a uniform spill totalling 10,000 particles over 280,000 turns (100 ms), the necessary RFKO kick varies between 2 µrad - 12 µrad, which is up to 6× higher than the kick used for 1 s of spill at the same emittance. From equation (1) for a CNAO-based exciter, this is equivalent to a voltage range from 700 V - 4.2 kV.

4. Conclusion

With MAD-X and Maptrack [10, 11] as a basis, flexible tracking and visualisation tools were developed for the slow extraction modelling of the PIMMS machine, with the intention of upgrading it to suit NIMMS advanced accelerator options. First the spill and behaviour of extraction was characterised for nominal betatron core excitation, and then the optics were

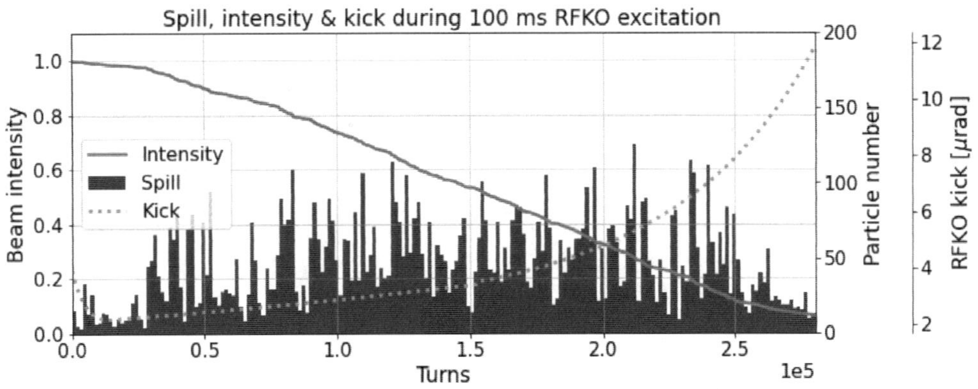

Figure 4. 10,000 particles extracted via RFKO over 100 ms.

adapted for RFKO excitation. Uniform spill could be produced by increasing the RFKO kick exponentially with time [18], and extracting higher emittance is possible by shifting the tune of the beam away from the resonance and providing a higher amplitude kick. Delivering a beam within 100 ms can be achieved by providing RFKO kicks on the order of 10 μrad, although the voltage required to provide kicks of this strength is on the order of kilovolts, which is difficult to achieve with existing hardware.

To gain experimental validation for both high emittance and high rate extraction, tests will need to be performed on an ion synchrotron to measure the voltage limits of existing RFKO exciters. It should be possible to observe how much of the beam can become extracted when the exciter is turned to its maximum strength. If this is not possible, alternative extraction techniques will need to be determined to meet FLASH conditions. The alternative FLASH techniques that are also compatible with bunched beam are based on optics changes. For example, tune sweep through the resonance can be done quickly using air-core quadrupoles or Constant Optics Slow Extraction (COSE) [11], which can potentially generate very fast extraction. Application of these two techniques alone or in a hybrid scheme with RFKO shall be investigated to determine the beam extraction scheme optimal for FLASH therapy.

Acknowledgments

This study was partially supported by the European Union's Horizon 2020 research and innovation programme under grant agreement no. 101008548 (HITRIplus).

References

[1] Vretenar M *et al.* 2021 *Proc. 12th Int. Particle Accelerator Conf. (IPAC'21)* (Campinas, Brazil) pp 1240–43
[2] Benedetto E *et al.* 2020 Comparison of Accelerator Designs for an Ion Therapy and Research Facility Tech. rep. CERN Geneva CERN-ACC-NOTE-2020-0068,NIMMS-Note-001
[3] Badano L *et al.* (CERN-TERA Foundation-MedAustron Oncology-2000 Collaboration) 1999 Proton-Ion Medical Machine Study (PIMMS), 1 Tech. rep. CERN Geneva
[4] Bryant P J *et al.* (Accelerator Complex Study Group) 2000 Proton-Ion Medical Machine Study (PIMMS), 2 Tech. rep. CERN Geneva CERN-PS-2000-007-DR
[5] Iwata Y *et al.* 2010 *Nucl. Instrum. Methods Phys. Res., Sect.* A **624**
[6] Favaudon V *et al.* 2014 *Sci. Transl. Med.* **6** 245ra93
[7] Jolly S, Owen H, Schippers M and Welsch C 2020 *Physica Med.* **78** 71–82
[8] Grote H and Schmidt F 2003 MAD-X: An Upgrade from MAD8 Tech. rep. CERN Geneva CERN-AB-2003-024-ABP
[9] Forest E, Schmidt F and McIntosh E 2002 Introduction to the polymorphic tracking code: Fibre bundles,

polymorphic Taylor types and "Exact tracking" Tech. rep. CERN Geneva CERN-SL-2002-044-AP, KEK-REPORT-2002-3

[10] Velotti F M maptrack `https://gitlab.cern.ch/abt-optics-and-code-repository/simulation-codes/maptrack`

[11] Sota A, Fraser M A and Gibson S 2020 *Optimisation of Slow Extraction and Beam Delivery from Synchrotrons* Master's thesis Royal Holloway, U. of London

[12] Taylor R Sloexlab `https://github.com/RebTaylor/sloexlab`

[13] Karastathis N Numerical analysis of fundamental frequencies `https://github.com/nkarast/PyNAFF`

[14] Furukawa T *et al. Proc. 8th Eur. Particle Accelerator Conf. (EPAC'02)* (Paris, France) pp 2739–41

[15] Peters A 2020 Review of spill structure enhancements at HIT and other ion therapy facilities

[16] Savazzi S *et al.* 2019 *Proc. 10th Int. Particle Accelerator Conf. (IPAC'19)* (Melbourne, Australia) pp 3469–71

[17] Kühteubl F 2020 *Design Study of Radio Frequency Knockout Slow Extraction for the MedAustron Synchrotron* Master's thesis Technische Universität Wien

[18] Schömers C 2013 *Entwicklung einer dynamischen Intensitätsregelung für das Heidelberger Ionenstrahl-Therapiesynchrotron* Ph.D. thesis Johann Wolfgang Goethe-Universität

[19] Carmignani N *et al.* 2010 *Proc. 1st Int. Particle Accelerator Conf. (IPAC'10)* (Kyoto, Japan) pp 3891–93

Beam optics study for a potential VHEE beam delivery system

C S Robertson[1], P N Burrows[1], M Dosanjh[1,2], A Latina[2] and A Gerbershagen[3,2]

[1]University of Oxford, Keble Road, Oxford OX1 3RH, UK
[2]CERN, Espl. des Particules 1, Meyrin 1211, Switzerland
[3]PARTREC, UMCG, University of Groningen, Groningen 9747 AA, Netherlands

E-mail: `cameron.robertson@physics.ox.ac.uk`

Abstract. VHEE (Very High Energy Electron) therapy can be superior to conventional radiotherapy for the treatment of deep seated tumours, whilst not necessarily requiring the space and cost of proton or heavy ion facilities. Developments in high gradient RF technology have allowed electrons to be accelerated to VHEE energies in a compact space, meaning that treatment could be possible with a shorter linac. A crucial component of VHEE treatment is the transfer of the beam from accelerator to patient. This is required to magnify the beam to cover the transverse extent of the tumour, whilst ensuring a uniform beam distribution. Two principle methodologies for the design of a compact transfer line are presented. The first of these is based upon a quadrupole lattice and optical magnification of beam size. A minimisation algorithm is used to enforce certain criteria on the beam distribution at the patient, defining the lattice through an automated routine. Separately, a dual scattering-foil based system is also presented, which uses similar algorithms for the optimisation of the foil geometry in order to achieve the desired beam shape at the patient location.

1. Introduction and background

1.1. Current modalities of radiotherapy

Radiotherapy (RT) is an essential component of cancer treatment, required by 50% of patients [1]. In RT, a treatment beam is used to damage the DNA of the tumour cells and cause cell death. The main goal of RT is to cause as much damage to the tumour as possible, whilst reducing dose to any surrounding healthy tissue. This is known as increasing the "Therapeutic Window". This is most critical when the tumour is seated deep within a patient, as organs particularly susceptible to radiation may be traversed. The dose deposition profile from an RT beam is dependent on the particle used for treatment [2]. Modern techniques in conventional (X-ray) RT have been developed to reduce dose to healthy tissues longitudinally and laterally [3]. Hadron beams can provide more precise treatments with less dose to healthy tissues than conventional RT due to the Bragg peak [2]. They can also be manipulated directly with magnets to allow scanning of the beam across the tumour. As such, they are very well suited for the treatment of deep-seated tumours. Hadron treatment facilities are however much larger and more expensive than conventional RT facilities [4]. This is due to the requirement for cyclotrons or synchrotrons for acceleration rather than the small LINACs used for conventional

RT, extended shielding requirements as well as the much larger gantries for bending the higher rigidity hadron beams.

1.2. VHEE
A promising modality for RT is VHEE (Very High Energy Electron) therapy. These are defined as electron beams with energies above 50 MeV, and would be capable of reaching deep-seated tumours [5]. There is evidence that VHEE beams would be less sensitive to inhomogeneities in the patient tissues than hadron beams. With the implementation of scanning and/or focusing, VHEE beams could also provide superior tumour conformity and avoidance of healthy tissues compared to conventional RT [6, 7]. Two crucial advantages that VHEE would have over hadron therapy are the required cost and space. A 200 MeV electron beam would require weaker magnets and/or a more compact gantry for delivering treatment. Developments in high gradient X-band RF-technology have also allowed the possibility of electrons being accelerated to VHEE energies in a very compact LINAC [8]. This would require less space than the circular accelerators used for hadron therapy.

1.3. Aim
The aim of this work is to design a transfer line for a 100 MeV VHEE beam from acceleration to the patient, using a simple model. This is an exploration and verification of the methodology for this design process, rather than a fully practical implementation. The beam should meet certain requirements for treatment whilst also being compact and efficiently designed to minimise associated costs.

1.4. Initial conditions and beamline geometry
The gantry was based upon a "Riesenrad" style gantry, consisting of a beamline with a single dipole [9]. This gantry would be rotated along with the patient couch to allow irradiation from multiple angles, shown in figure 1. The initial beam was assumed to have a uniform transverse distribution with a radius of 1 mm, an angular divergence of 1 mrad, and a momentum spread of 0.25%. It was required to be magnified to the radius of a large tumour - here taken to be between 50 mm and 100 mm. Furthermore, the beam was required have a uniform transverse distribution to ensure an even dose across the tumour. The study of scanned and focused beam delivery was left for future work. Achromaticity of the beamline whilst desirable for symmetry in both transverse axes was not a key aim of the design processes at this stage, but instead the physical beam profile was the most fundamental consideration.

2. Method and results
2.1. Magnification optics
The first method for developing the gantry was based around a lattice of quadrupoles. This was carried out using Nelder-Mead [10] (simplex) minimisation with RF-Track, a code developed at CERN [11]. The optimisation process was carried out using a merit function to characterise the transverse beam distribution at the patient. The merit function is composed of a sum of weighted terms, summarised by

$$M = \omega_1 M_{\text{mean}} + \omega_2 M_{\text{mag}} + \omega_3 M_{\text{kurtosis}} \tag{1}$$

for each transverse dimension. M_{mean} and M_{mag} are terms which are minimal when the distribution at the patient is centred and magnified to a defined radius, respectively. M_{kurtosis} is minimised when the beam has a constant transverse distribution across this radius, and is found using the kurtosis (tailedness) of the beam [12]. The weights in the merit function were

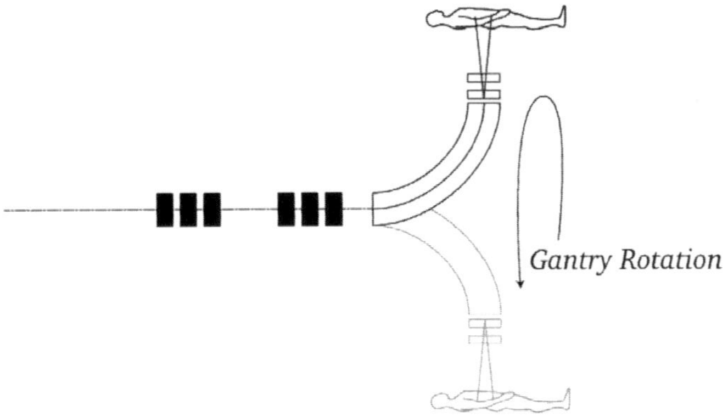

Figure 1. Riesenrad gantry concept [9].

determined through iterative investigation of optimisation results, attempting to balance these values for the desired final distribution.

The minimisation for this function was carried out using the gradients and positions of a number of quadrupole magnets as variables, along with the pole tip face angles of the single dipole. Limits were enforced on the strengths of the quadrupole magnets with respect to the required beam apertures to ensure magnetic fields below saturation for resistive magnets. At the end of the gantry, a 2 m long space was assumed to leave room for beam diagnostics and additional optics. The number of quadrupole magnets was to be kept to a minimum – particularly those downstream of the dipole as these would significantly add to the total cost of the gantry due to their required rotation. The optimisation was run a significant number of times with random seeds being used for input variables within the described bounds. This was necessary since the Nelder-Mead method is susceptible to returning local minima. The solution deemed most successful is presented here, with the gantry layout and quadrupole strengths as well as the resultant transverse distribution in the patient displayed in figure 2.

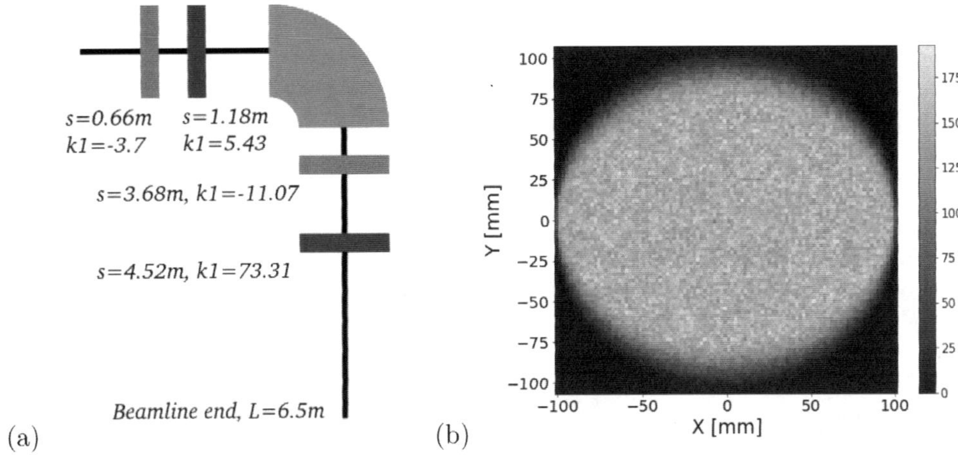

Figure 2. Magnification optics layout (a) and transverse beam distribution at patient location without collimation (b), colour represents density in arbitrary units.

Excellent consistency across the final beam out to 75 mm was achieved, with negligible large scale variation, corresponding to a kurtosis value of 2.000 in X and 2.004 in Y (the bending axis). After collimation out to 75 mm to retain only the flat region, 61% of the initial beam was retained. The rather large losses from collimation are due to the asymmetry between the transverse axes caused by dispersion.

2.2. Dual scattering foil

The second method of beam magnification proposed was the use of a dual scattering foil. The general principle of this is to firstly magnify a pencil beam by sending it through a flat scattering foil. The scattering process is random and thus the magnified beam will have an approximately

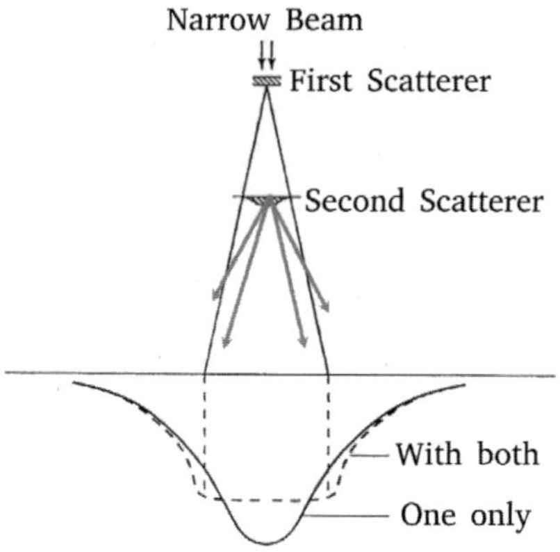

Figure 3. Dual scattering foil concept [13].

Gaussian transverse distribution. A 2nd scatterer with a more complex shape is then used to correct the Gaussian beam into a flat beam for treatment. This shape would be Gaussian to complement the beam, preferentially scattering the denser centre of the beam over the total beam radius. This concept is shown in figure 3.

To reduce particle production and associated energy losses as well as limiting divergence, a distance of 2.5 m between the 1st scatterer and the patient was enforced for this gantry. The two quadrupoles upstream of the dipole in the magnification optics design were retained for this setup in order to constrain the beam travelling through the dipole. The 1st scatterer was chosen to be composed of tantalum, as TOPAS [14, 15] simulations demonstrated a higher efficiency of scattering angle increase against photon production for high Z materials. Very high precision machining would be required for manufacturing a tantalum 2nd scatterer as it would be very thin. A much thicker, low Z material such as plastic would have more photon production however. Aluminium was chosen as a compromise between these two extremes. The same minimisation principle was utilised for optimisation of the foils. The geometry of the system was used as the variable in the optimisation. More precisely, the longitudinal thickness of the first scatterer, and the amplitude and standard deviation determining the shape of the 2nd, Gaussian scatterer were used as the input parameters along with the longitudinal position of the 2nd scatterer relative to the 1st. RF-Track and TOPAS were used in conjunction to carry out the simulations for this process. The same merit function as shown in Eq. 1 was again minimised from many random distributions of these variables.The optimum 2nd scatterer geometry is shown in figure 4, along with the resultant transverse beam distribution in the patient. A constant transverse beam distribution with a radius of 60 mm was achieved, with consistency of 94% in the transverse distribution density in this range. The corresponded to a kurtosis value of 2.021 in X and 1.985

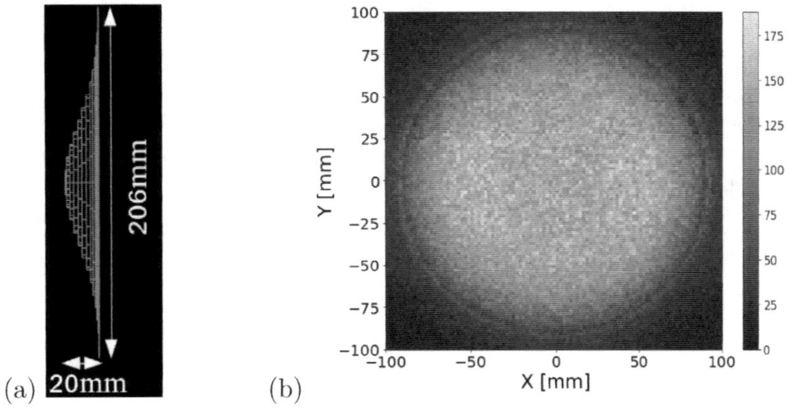

Figure 4. 2nd scatterer visualisation (a) and transverse beam density distribution at patient without collimation (b), colour represents density in arbitrary units.

in Y. This colimation to remove the more widely scattered particles was rather significant, and thus transmission through the dual scattering system with colimation was 55%.

3. Discussion

3.1. Comparison

Particularly in the case of the magnification optics method, the arrangement and strengths of the magnets would be crucial. Small alignment errors and beam jitter could have detrimental effects to the final beam. This method is also heavily dependent on the initial beam distribution, and one cannot assume that the same optics setup with a more realistic beam would provide a similar result. Even when minimised as far as possible as in this work, the beam at the patient still had a dose of X-rays that was comparable to that of the electrons for the dual scattering method. Enforcing the 2.5 m length between the 1st scatterer and the patient resulted in a large minimum gantry size. A solution to the above issues could involve making the first foil thicker and reducing the distance between it and the second foils, or moving the foils upstream of the single dipole; the concern for the latter would be that the scattering system increases the energy spread of the beam, and thus increases the effects of dispersion. This would also require a dipole magnet with a large aperture.

The dual scattering foil system while likely cheaper and easier to implement would always have the unavoidable issues of particle production compared to the pure VHEE beam from the magnification optics. However, the problems caused by misallignment and gantry rotation would likely be more severe for the magnification optics solution due to the increased weight and number of required beamline elements.

3.2. Future work

An valuable extension to this study would be a quantitative study of the effects of misallignment on the final distribution, for both methods. This could involve additional optimisations of a static setup to account for variations in the initial beam. The above magnification optics setup shown in figure 2 could be kept static and only the strengths of the magnets varied to be used as a transfer line for different distributions. One would expect that the foil system would be less dependent on the initial beam than the optics method, but this would also need to be quantified explicitly.

The impact of the X-ray production from the dual scattering foil on the patient dose distribution will require further study and comparison to the "pure" VHEE beam from the magnification optics. Due to the (relatively) high energy of the electrons, the details and method of collimation should also simulated for this study of dose in the patient. Monte-Carlo codes such as TOPAS and FLUKA [16, 17] could be used for this. Furthermore, the maximum achievable radius of the constant beam at the patient appeared to be somewhat limited by enforcing the strictly Gaussian shape of the 2nd scatterer. A more flexible approach allowing for non-Gaussian aberrations in the shape design could allow this to be solved.

4. Conclusion

Two general concepts were shown to explore methodologies for design of a VHEE treatment line, using an idealised initial beam distribution and Riesenrad treatment gantry. The first, using a lattice of quadrupoles, allowed a pure VHEE beam to be magnified and flattened with a relatively small gantry radius. The second, using a dual scattering foil system, also achieved this magnification and flattening, although this also resulted in unavoidable extra particle production from the beam-matter introduction. This was minimised by situating the system significantly upstream of the patient, although this led to a larger gantry radius. Both beamline designs were carried out using multiple simplex minimisation routines with random seeding. Future work will involve carrying out design studies with different initial beam distributions and studies of lattice imperfections.

References

[1] Borras J M, Lievens Y, Barton M, Corral J, Ferlay J, Bray F and Grau C 2016 *Radiotherapy and Oncology* **119** 5–11
[2] Yuan T Z, Zhan Z J and Qian C N 2019 *Cancer Communications* **39** 1–7
[3] Connell P P and Hellman S 2009 *Cancer Res.* **69** 383–92
[4] Amos R *et al.* 2018 *Clinical Oncology* **30** 280–4
[5] DesRosiers C, Moskvin V, Bielajew A F and Papiez L 2000 *Physics in Medicine & Biology* **45** 1781–1805
[6] Schüler E, Eriksson K, Hynning E, Hancock S L, Hiniker S M, Bazalova-Carter M, Wong T, Le Q T, Loo Jr B W and Maxim P G 2017 *Medical physics* **44** 2544–55
[7] Palma B, Bazalova-Carter M, Hårdemark B, Hynning E, Qu B, Loo B W and Maxim P G 2016 *Radiotherapy and Oncology* **119** 154–8
[8] Ronga M G, Cavallone M, Patriarca A, Leite A M, Loap P, Favaudon V, Créhange G and De Marzi L 2021 *Cancers* **13** 4942
[9] Owen H, Holder D, Alonso J and Mackay R 2013 *International Journal of Modern Physics A* **29** 14
[10] Olsson D M and Nelson L S 1975 *Technometrics* **17** 45–51
[11] Latina A 2020 Rf-track reference manual `https://zenodo.org/record/3887085#.YtAsp-xBxH4`
[12] Balanda K P and MacGillivray H L 1988 *The American Statistician* **42** 111–19
[13] LeBlanc J D 2012 *Design of electron dual foil scattering systems for Elekta Infinity radiotherapy accelerators* Master's thesis Louisiana State University
[14] Perl J, Shin J, Schümann J, Faddegon B and Paganetti H 2012 *Medical physics* **39** 6818–37
[15] Faddegon B, Ramos-Méndez J, Schuemann J, McNamara A, Shin J, Perl J and Paganetti H 2020 *Physica Medica* **72** 114–21
[16] Ahdida C *et al.* 2022 *Front. Phys.* **9**
[17] Battistoni G *et al.* 2015 *Annals of Nuclear Energy* **82** 10–18

Journal of Physics: Conference Series **2420** (2023) 012103 doi:10.1088/1742-6596/2420/1/012103

A compact synchrotron for advanced cancer therapy with helium and proton beams

M Vretenar[1], M E Angoletta1, G Bisoffi[1,3], J Borburgh[1], L Bottura[1], K Paļskis1[1,4], R Taylor[1], G Tranquille[1], E Benedetto[2] and M Sapinski[2]

[1] CERN, Geneva, Switzerland
[2] SEEIIST Association, Geneva, Switzerland
[3] INFN-LNL, Legnaro, Italy
[4] Riga Technical University, Riga, Latvia
[5] PSI, Villigen, Switzerland

Email: maurizio.vretenar@cern.ch

Abstract. Recent years have seen an increased interest in the use of helium for radiation therapy of cancer. Helium ions can be more precisely delivered to the tumour than protons or carbon ions, presently the only beams licensed for treatment, with a biological effectiveness between the two. The accelerator required for helium is considerably smaller than a standard carbon ion synchrotron. To exploit the potential of helium therapy and of other emerging particle therapy techniques, in the framework of the Next Ion Medical Machine Study (NIMMS) at CERN, the design of a compact synchrotron optimised for acceleration of proton and helium beams has been investigated. The synchrotron is based on a new magnet design, profits from a novel injector linac, and can provide both slow and fast extraction for conventional and FLASH therapy. Production of mini-beams, and operation with multiple ions for imaging and treatment are also considered. This accelerator is intended to become the main element of a facility devoted to a parallel programme of cancer research and treatment with proton and helium beams, to both cure patients and contribute to the assessment of helium beams as a new tool to fight cancer.

1. Cancer treatment with helium

The recent re-emergence of interest in the clinical use of helium ions for cancer treatment is mainly based on the underlying physical properties and corresponding biological effects - intermediate between the clinically approved proton and carbon ion beams.

Helium ion beams exhibit lower range straggling compared to proton beams, resulting in sharper Bragg peak and distal fall-off. Along with a reduction in multiple Coulomb scattering resulting in a decreased lateral penumbra, they can provide physical beam conformality comparable to carbon ion beams. Additionally, helium ions undergo less nuclear fragmentation processes than carbon ions, resulting in greatly reduced fragmentation tail and less complex mixed radiation beam, which in turn provides less uncertainties in biological effect estimations. The resulting secondary particle spectrum also exhibits decreased neutron production compared to carbon ions. The resulting neutron biological dose might be even lower than in proton beams [1], greatly reducing neutron dose associated risks in paediatric patients.

With linear energy transfer (LET) in the range of 4 to 40 keV/μm, helium ion beams exhibit an increased relative biological effectiveness (RBE) compared to protons, while also not reaching "overkill region" in the distal Bragg peak like high LET carbon beams [1]. With an increase in LET values, helium ions also provide reduction in oxygen enhancement ratio (OER) compared to proton beams, opening up certain possibilities for hypoxic tumour treatment.

Thanks to these characteristics, helium ion beams have potential to increase clinical efficacy for treatment sites in close proximity to vital organs and to even greater extent – i.e., in paediatrics. Along with better performance in ion radiography applications, pathways for in-vivo range verification and possibilities with treatment modalities as FLASH and mini-beams [2, 3], helium ion beam therapy holds a promising innovation position in cancer treatment.

2. Accelerator main parameters

The ideal accelerator to bring helium ions at cancer treatment energy is a compact synchrotron at a maximum magnetic rigidity of 4.5 T/m, corresponding to 220 MeV/u for ^{4}He ions with a penetration of 30 cm in water, sufficient to access all types of cancer under consideration. The synchrotron will allow for the acceleration of proton beams at the energies required for cancer treatment and above, for proton radiography.

The synchrotron can be the central element of a facility for cancer research and therapy as sketched in figure 1. A linear accelerator injects proton or helium beams in the synchrotron. Operating at higher duty cycle than required for synchrotron injection, it can send beam to a target for production of radioisotopes to be used for imaging or for cancer treatment with alpha-emitters [4]. The beams extracted from the synchrotron can go to the treatment rooms, one of which equipped with a rotating superconducting gantry of novel design [5], or to an experimental room.

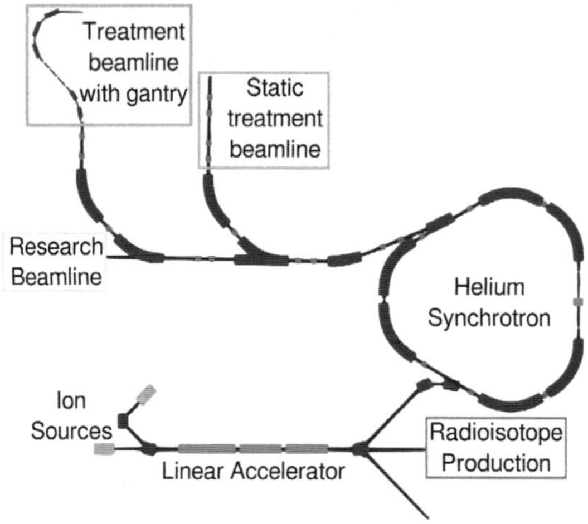

Figure 1. Layout of a compact cancer research and therapy facility with proton and helium beams.

The accelerator complex will use two (or more) ion sources, for protons and for ions. The helium source has to provide a $^{4}He^{2+}$ current of at least 2 mA to deliver 8 x 10^{10} ions at synchrotron via multi-turn injection, required to irradiate a 1 litre tumour with 2 Gy with a margin of a factor 2 to account for inefficiencies in the extraction process. This intensity is well within the capabilities of present ion sources. Additionally, the same source could provide lower intensity beams of 12C6+ for an experimental programme with heavier ions.

The helium beam out of the ion source is accelerated to the synchrotron injection energy of 5 MeV/u in a 352 MHz linac for q/m=1/2 made of a Radio Frequency Quadrupole followed by a Drift Tube Linac (DTL) tank of similar design to the CERN Linac4. An additional DTL tank designed for q/m=1 will bring exclusively the proton beam to its injection energy of 10 MeV. The second tank is left unpowered during the helium acceleration cycles.

For conventional therapy, the beam will be extracted from the synchrotron using slow extraction, requiring a horizontal tune near the third-integer. Radiofrequency knock-out (RF-KO) excitation will be used to provide uniform beam spill during treatment, which is an operational method in existing hadron therapy facilities. To upgrade the extraction to higher intensities and faster extraction rates to be compatible with FLASH modalities, the comparison of spill using high-voltage RF-KO exciters or alternative FLASH extraction modes, is presently under study [6].

3. RING layout and optics

For the lattice, a triangular shape was chosen for the sake of compactness and to ease the magnets manufacturing, following a similar approach to what was recently done for the design of a compact medical synchrotron for carbon-ions [7] with superconducting magnets.

A sketch of the synchrotron and preliminary optics are shown in figure 2; which still have to be optimized to properly accommodate the injection and extraction hardware.

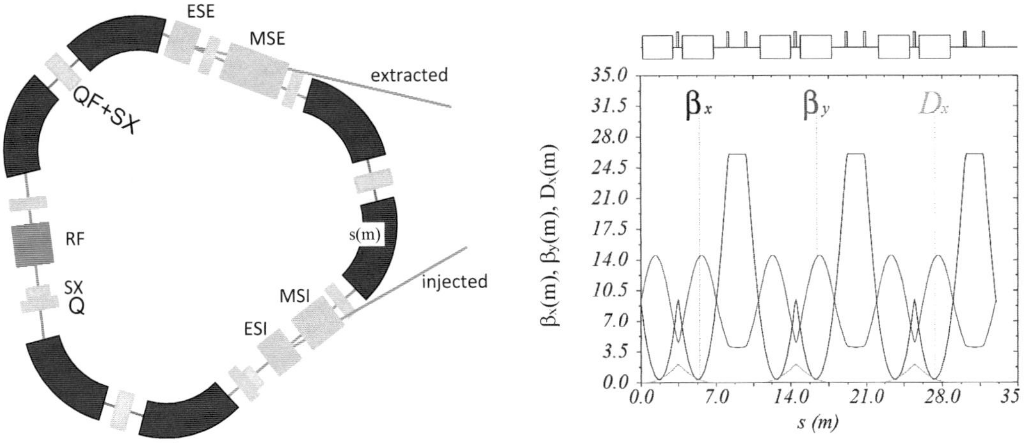

Figure 2. Sketch of the lattice layout and a preliminary optics, generated with MAD-X [8].

The lattice is made of three bending sections made of two 60° magnets with a strong quadrupole in between, to have zero dispersion in the three straight sections, which respectively host the injection and extraction hardware, and the RF cavities.

Because of the maximum beam rigidity of 4.5 Tm and the choice of the maximum field of 1.65 T, the resulting bending radius is 2.7 m. The straight sections are about 5 meters long, to accommodate the required hardware, which account for about 33 m total circumference length.

The 60° magnets have a small defocusing gradient to reduce the vertical beta functions in the bends and the quadrupoles in-between also carry sextupolar coils to control the chromaticity. Two additional quadrupoles are needed in each straight section for the optics and the working point adjustment, as well as sextupole(s) at the correct phase advance, to excite the 3rd order resonance for the slow extraction, and orbit correctors, which will also generate the extraction bump. These functions will be combined as much as possible in a single corrector magnet, carrying multiple components.

4. Magnet design

The synchrotron magnets were designed for compactness and simplicity, limiting power consumption, and keeping open the option of a later upgrade to higher energies and heavier ions. The main dipole magnet dominates the overall dimension, weight and cost of the ring. The baseline design has conservative field requirements, 1.65 T, and moderate current density in the water-cooled conductor, 13 A/mm^2. The aperture is 140 mm x 70 mm (H/V), and the dipole length is 2.56 m. The configuration chosen is a window-frame, achieving an excellent field linearity, which benefits operation. The dipole cross-section is reported in figure 3.

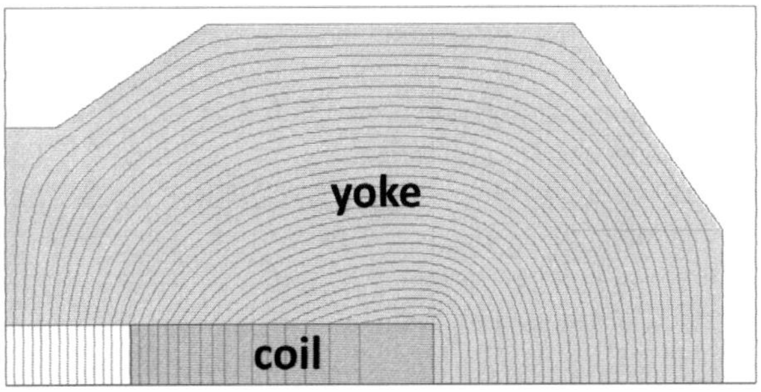

Figure 3. Dipole magnet cross section (one quarter).

Parametric studies show that in the range of dimensions considered, there is little benefit in increasing the bore field. Maintaining the conductor current density, hence the dissipated power (about 140 kW/dipole), an increase to 2 T (the upper practical end for this type of magnet) only yields a reduction of the machine radius by 20 cm. We finally remark that the coil space would be compatible with a superconducting coil. Though design details are not fully developed, this upgrade would allow the dipole to reach approximately 2.5 T bore field, i.e. as required to accelerate C ions to 430 MeV/u.

The corrector magnets are inspired by work performed for synchrotron light sources, where space is a premium [9]. The iron has sextupolar symmetry, but the coils can be powered independently and can generate normal dipole, skew quadrupole and normal sextupole corrections.

5. Injection and extraction

Historically, for ion treatment accelerators [10, 11], both injection and extraction use one electrostatic septum and several DC magnetic septa. Although the magnetic septa are not purely DC, since their strength is adjusted as a function of the particle species at injection and the extraction energy, their power consumption is not negligible and in the order of 10s of kW per device.

Assuming similar beam sizes and deflection angles for the magnetic injection septa as in medical ion treatment facilities [10, 11], today's technology will allow the reliable operation of pulsed, under vacuum, magnetic injection septa. The apparent septum thickness compared to the outside vacuum solution would be reduced. The power consumption of the pulsed septum solution would be a factor 1000 lower than a DC based solution. This also has an immediate beneficial effect on the cooling requirements of the facility. It is still to be evaluated if, by switching to this technology, the electrostatic septum would still be needed. For the extraction septa, the electrostatic septum is planned to be located either in the extraction region, if phase-advance and space allows, or in the straight section upstream, which would ease the requirements on the magnetic septa, but might require an increase of the bending magnets aperture. An attempt will be made to design the extraction trajectory such that it is passing through the back leg of the main dipole yoke. This will significantly reduce the required deflection angle to be provided by the magnetic septa, hence the required space in the straight section, allowing to minimize the accelerator footprint.

6. RF and diagnostics

The proposed RF system includes a digital Low-Level RF (LLRF) part that controls a wideband un-tunable High Level RF (HLRF) based on Finemet® alloy. These two RF building blocks rest on proven technologies, already operational in MedAustron and several proton or ion CERN machines [12]. The compact and modular HLRF is made of various 15 cm-long cells, each providing a peak voltage of 700 V. The LLRF can drive various harmonics, depending on the HLRF bandwidth and power, allowing different RF gymnastics and functional upgrades.

Because of the space constraints imposed by the compact design of the synchrotron, the instrumentation [13] will be limited to measure only the essential parameters of the circulating beam namely, intensity, position and tune (working point), and to transversely excite the beam for the RF-KO driven extraction. At the injection point, screens should be used to check the position and profile at that position, as well as after the first turn. A "semi-fast" transformer and a DC current transformer will provide the intensity measurement during multi-turn injection into the machine as well as along the accelerating cycle.

Electrostatic pickups which measure the closed orbit of the circulating beam can also monitor the trajectory at injection for the correction of the coherent oscillations of the injected beam. Horizontal and vertical pickups could be placed in the bending magnets, with a number of combined pickups installed in the straight sections inside the correcting elements (quadrupoles and dipoles).

The tune measurement system will be based on the Direct Diode Detection principle. In this scheme, electrode signals from a position pickup feed diode peak detectors, which down-mix the beam spectrum to the base-band. The system, despite its simplicity, is capable of measuring tunes of bunched and debunched ion beams with small beam kicks without changing the system gain.

7. Sustainability and operation

The energy consumption of the proposed facility was calculated. It is based on its use at 3 hours/day with protons, 6 hours/day with helium for therapy and 3 hours/day with helium for research. As to the synchrotron, once the actual electrical specifications of magnets and power converters (p.c.) are set, the power required by the components for the maximum beam energy, including the electrical efficiency of the p.c., must be scaled down (by ~40%) to the one related to the average use of the machine during the year, which is operated – in practice – at lower energy values than the maximum one. It is then further reduced by the duty cycle of the ramps (45% for protons and 50% for He) and by the total uptime of the machine, 37.5% with He and 12.5% with protons. As a result, the average value of the power required by all synchrotron magnets is ~170 kW. Assuming that the use of the beam lines is split in equal parts among SC gantry, horizontal treatment line, and two experimental lines, their overall power of 95 kW is dominated by the cryocoolers of the SC gantry (80 kW). The modern linac injector, equipped with Solid State Power Amplifiers, requires a power of ~40 kW. Minor contributions by vacuum pumps and beam instrumentation (10 kW), p.c. when in idle state (~17 kW) and cooling power (30% of the sum of the contributions above, i.e. 100 kW) must be added, for a grand total of ~430 kW.

8. Conclusions

The design of a compact facility for the production of proton and helium beams for cancer therapy has been presented. The technology is modern and innovative, including high intensity and FLASH-type extraction, but at the same time conservative and based on proven technologies. A path for a simple upgrade to carbon ion therapy has been defined, by upgrading the magnets with superconducting coils.

Figure 4 shows a preliminary layout of the synchrotron ring, which has a total circumference length of 33 m. The total surface covered by the complex will be in the range of only 1,600 m^2.

Figure 4. View of the 33 m circumference synchrotron.

Acknowledgements

This study was partially supported by the European Union H2020 research and innovation programme under GA 101008548 (HITRIplus).

References

[1] Mairani A *et al.* 2022 *Phys Med Biol.* **67** 15TR02
[2] Tessonnier T *et al.* 2021 *Int. J. Radiat. Oncol. Biol. Phys.* **111** 1011
[3] Schneider T, De Marzi L, Patriarca A and Prezado Y 2021 *Front. Phys.* **9** 595721
[4] Vretenar M, Mamaras A, Foka P and Bisoffi G 2022 *Proc. 13th Int. Particle Accelerator Conf. (IPAC'22)* (Bangkok, Thailand) pp 2996–99

[5] Pullia M G *et al.* 2022 *Proc. 13th Int. Particle Accelerator Conf. (IPAC'22)* (Bangkok, Thailand) pp 2966–69
[6] Taylor R L, Pasternak J, Benedetto E and Sapinski M 2022 *Proc. 13th Int. Particle Accelerator Conf. (IPAC'22)* (Bangkok, Thailand) pp 2988–91
[7] Benedetto E 2022 Carbon ion compact medical synchrotron: key parameters Tech. rep. CERN-ACC-NOTE-2022-0017, NIMMS-Note 008 CERN Geneva Switzerland
[8] Schmidt F and Grote H 2003 *Proc. Particle Accelerator Conf. (PAC'03)* (Portland, OR USA) pp 3497–99
[9] Milanese A, Huttel E and Shehab M M 2014 *Proc. 5th Int. Particle Accelerator Conf. (IPAC'14)* (Dresden, Germany) pp 1292–94
[10] Sermeus L, Borburgh J, Crescenti M, Fowler T, Hourican M and Metzmacher K D 2004 *Proc. 9th Eur. Particle Accelerators Conf. (EPAC'04)* (Lucerne, Switzerland) pp 1639–41
[11] Borburgh J *et al.* 2010 *Proc. 1st Int. Particle Accelerator Conf. (IPAC'10)* (Kyoto, Japan) pp 3954–56
[12] Angoletta M E *et al.* 2019 *Proc. 9th Low-Level RF Workshop (LLRF 2019)* (Chicago, US) (doi:10.48550/arxiv.1910.06643)
[13] Tranquille G 2013 *Proc. 2nd Int. Beam Instrumentation Conf. (IBIC'13)* (Oxford, UK) pp 664–66

Production of radioisotopes for cancer imaging and treatment with compact linear accelerators

M Vretenar[1], **A Mamaras**[1], **G Bisoffi**[1,2] **and P Foka**[3]

[1] CERN, Geneva, Switzerland
[2] INFN-LNL, Legnaro, Italy
[3] GSI, Darmstadt, Germany

Email: maurizio.vretenar@cern.ch

Abstract. Accelerator-produced radioisotopes are widely used in modern medicine, for imaging, for cancer therapy, and for combinations of therapy and diagnostics (theragnostics). Clinical trials are well advanced for several radioisotope-based treatments that might open the way to a strong request of specific accelerator systems dedicated to radioisotope production. While cyclotrons are the standard tool in this domain, we explore here alternative options using linear accelerators. Compared to cyclotrons, linacs have the advantage of modularity, compactness, and reduced beam loss with lower shielding requirements. Although in general more expensive than cyclotrons, linacs are competitive in cost for production of low-energy proton beams, or of intense beams of heavier particles. After a review of radioisotopes of potential interest, in particular produced with low-energy protons or helium, this paper presents two linac-based isotope production systems. The first is a compact RFQ-based system for PET (Positron Emission Tomography) isotopes, and the second is an alpha-particle linac for production of alpha-emitters. The accelerator systems are described, together with calculations of production yields for different targets.

1. Introduction

In the last decades, the use of radioisotopes for applications in medicine that include both diagnostics and therapy has shown a significant increase, generating a clear demand for novel solutions to produce standard and new radioisotopes with low-cost, easy-to-use, accelerators possibly located close to the hospitals where they are used.

At present, several hospitals are equipped with small proton cyclotrons at energy between 10 and 25 MeV, supplying radioisotopes which, via chemical purification and biological manipulations, are then transformed into radiopharmaceuticals and quickly provided to local patients.

Although cyclotrons are fairly small, the required massive shielding around accelerator and target imposes serious limitations to their installation in hospitals. Moreover, the costs related to the frequent and expensive maintenance required by commercial cyclotrons can be prohibitive for small hospitals. In this respect, linear accelerators (linacs) might provide a viable alternative to cyclotrons, because thanks to their well-defined beam optics up to the production target they are not affected by the large beam loss, particularly at extraction, leading to the sensitive activation levels of cyclotrons. In a linac, nearly all particles produced by the ion source impinge on the target, which is the only element that requires substantial radiation shielding: a linac can be hence operated in a simple radiation-controlled area with only the target placed inside a shielding casing. Linacs are lighter in weight than cyclotrons (they do not need large magnets), which makes them easier to handle and install and are very reliable, with few moving or high voltage components.

The first part of this paper reports calculations of the doses that can be produced at saturation with currents and energies achievable with a compact linac. In the second part, two linac configurations are presented, the first for production of Positron Emission Tomography (PET) imaging radiotracers ^{18}F or ^{11}C via low energy protons, and the second for production of ^{211}At for Targeted Alpha Therapy (TAT) via alpha particles.

2. Production yields

A numerical calculation method to estimate the production yields of two PET (^{18}F, ^{11}C) and one TAT (^{211}At) radioisotopes was developed [1], and after benchmarked to existing data [2-4], was applied for dose calculation at different conditions (energy, current), where no data were available. ^{18}F is produced via the nuclear reaction ^{18}O(p,n)^{18}F, ^{11}C via ^{14}N(p,α)^{11}C and ^{211}At via ^{209}Bi(α,2n)^{211}At. Yields were calculated for protons in the energy range 3–15 MeV and for alphas in the energy range 21–30 MeV.

2.1. Production yields for ^{18}F and ^{11}C

The ^{18}F saturation yields, calculated via the numerical calculation method, are reported in figure 1. The corresponding PET doses are also shown, knowing that an average dose of ^{18}F-FDG is ~400 MBq for a patient of 80 kg, based on literature data and routine experience of a hospital [1, 5, 6].

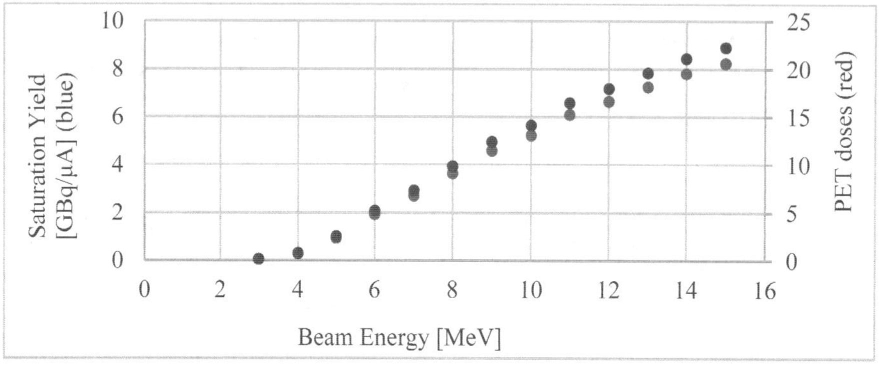

Figure 1. Saturation yield of ^{18}F (blue) and PET doses (red) as a function of E_b (saturation time of 3.5 $T_{1/2}$= 6.4 h).

For a 7 MeV final proton energy, the saturation yield is 2.71 GBq/μA, resulting, for 30 μA beam current, in ~80 GBq, which is equivalent to ~200 doses at saturation.

For ^{11}C production, a ^{14}N gas was taken as target material, which makes the extraction of ^{11}C from the produced carbon dioxide ($^{11}CO_2$) easy. Special care has to be taken on the selection of the appropriate conditions of temperature and pressure, as the gas chamber length should exceed the stopping range of the proton beam. The saturation yield (in GBq/μA) of the ^{11}C nuclear reaction was calculated for a proton energy up to 10 MeV onto a target at 20°C and 1 atm. The range of the 10 MeV proton beam turns out to be ~120 cm, calculated via the SRIM code [7]. The obtained ^{11}C production yield, after a saturation time of 3.5$T_{1/2}$ ≅ 1.2 hours, is 2.63 GBq/μA. ^{11}C saturation yields and PET doses vs E_b are reported in figure 2. In this case, an average dose of ^{11}C (to then compose ^{11}C(CH) – choline as the related radio molecule) is ~370÷740 MBq [8, 9]. With a 7 MeV, 20 μA beam current, the saturation yield is ~24 GBq, providing, at 350–750 MBq average dose per patient, 32–70 doses.

2.2. Production yields for TAT radioisotope ^{211}At

^{211}At is at an advanced stage of testing and expected to become the main component of new TAT radiopharmaceuticals. To calculate the most appropriate specifications for the TAT linac, we consider the reaction ^{209}Bi(α,2n)^{211}At, with the alpha beam impinging onto a solid ^{209}Bi target. The energy is limited in this case at 28 MeV (i.e. 7.5 MeV/u), set by the production of the undesirable ^{210}At (onset at ~28.5 MeV). Figure 3 shows the saturation yield (in GBq/μA) vs. E_b, between 20.7 MeV (lower threshold for the considered reaction) and 28 MeV.

The ^{211}At yields, with 28 MeV and 20 μA, an activity of 5.2 GBq after a saturation time of 3.5 $T_{1/2}$ ≅ 24 hours.

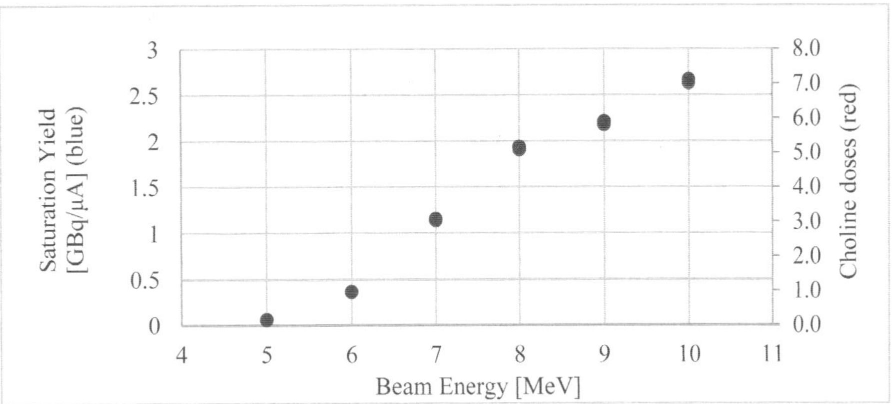

Figure 2. Saturation yield of ^{11}C (in GBq/μA) and Choline doses at saturation time of 3.5 $T_{1/2}$=1.2 h, versus E_b. The two series of points almost overlap in the graph.

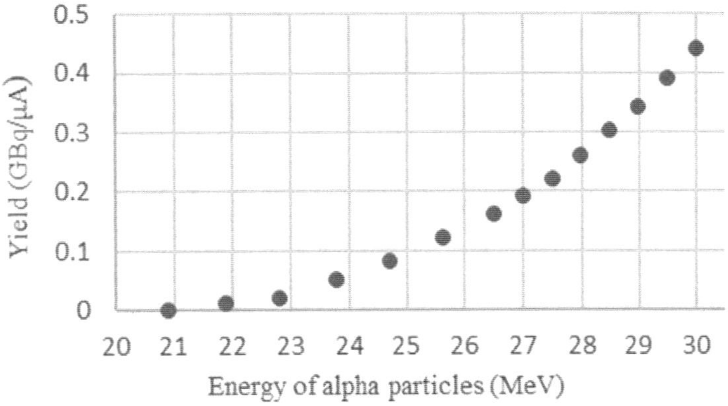

Figure 3. Saturation yield in GBq/μA of ^{211}At, at saturation time of 3.5 $T_{1/2}\cong$ 24 h, as a function of $E_{b,alpha}$.

Data available in the literature [10] allow calculating the equivalent doses required for treatment: 52–260 doses for post-surgery treatment of ovarian cancer (dose/patient: 20–100 MBq), and 15–74 doses for post-surgery brain tumour therapy (dose/patient: 70–350 MBq).

3. Compact linac designs

3.1. A compact linac for PET radioisotopes

Calculations in the previous section have shown that a useful number of doses per day of ^{18}F and ^{11}C can be generated by proton beams at relatively low energies.

The basic layout of the PET-isotopes production system is shown in figure 4, and its key parameters are presented in table 1. Following a general scheme already developed by the AccSys company in the 90's [11], this scheme differs in the use of an innovative compact RFQ design at 750 MHz recently developed at CERN [12] and in that it exploits the potential of modern solid-state RF technology. With respect to previous versions [13], this configuration is based on a simpler RFQ at lower energy (7 MeV) operating with higher average current.

The beam from the proton source is sent into a 2.5 m long 750 MHz RFQ made of a sequence of 0.5 m mechanical modules, which is the only accelerating structure of the linac. The RFQ RF amplifiers, feeding the 5 modules, consist of 100 kW solid-state elements.

The target system can house two interchangeable targets, to produce both ^{18}F and ^{11}C. A preliminary and conservative calculation of target shielding was made for E_b= 10 MeV assuming 2 μSv/h as maximum dose on the outside of the enclosure corresponding, in most legislations, to a simple radiation-controlled area accessible to personnel wearing a radiation dosimeter. Different configurations based on successive layers of iron, polyethylene and borated (5%) polyethylene were compared and progressively optimised for minimum overall

Journal of Physics: Conference Series **2420** (2023) 012104 doi:10.1088/1742-6596/2420/1/012104

thickness using genetic algorithms. The calculation indicates that a shielding with a total radius of 1 m can reduce the dose at contact during operation down to the required value [14].

The accelerator modules are water-cooled with the cooling circuit being connected to an external heat exchanger. The RFQ operates at 10% duty cycle, with a peak proton current of 300 μA generating a 210 W average beam power on target. The system weighs less than 5 tons for a floor space of less than 10 m^2, most of the weight being related to the local shielding of the target. The power from the mains is about 35 kVA.

It has been estimated that the 7 MeV isotope production system can be produced and commercialised at a 20% lower cost than equivalent cyclotron-based systems [15]. Additional savings would come from lower maintenance costs with respect to cyclotrons.

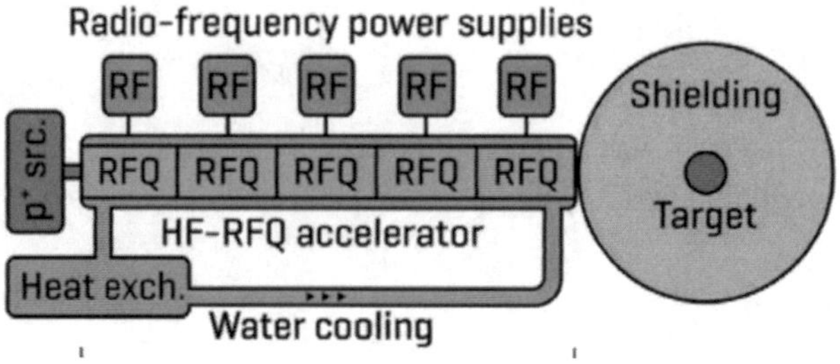

Figure 4. Scheme of the 7 MeV RFQ-based PET isotope production system.

Table 1. Parameters of the PET 7 MeV RFQ system.

Input / Output Energy	20 keV / 7 MeV
Accelerator length	2.5 m
Operating Frequency	750 MHz
Output peak current	300 μA
Duty cycle	10 %
Beam pulse repetition frequency	200 Hz
Beam pulse length	500 us
Output average current	30 μA
RF Power, peak	400 kW
RF Power, average	<30 kW
Surface (accelerator, source, target)	10 m^2
Surface (RF and equipment room)	10 m^2
Total weight of accelerator	350 kg
Total system weight	<5 tons
Mains power	35 kVA

3.2. A dedicated linac for ^{211}At production

The relatively low production energy and short half-life of ^{211}At make it an ideal therapeutic isotope to be produced by a compact linear accelerator close to a hospital. Cyclotrons for production of intense alpha beams would present unacceptable beam loss levels at extraction, giving a clear advantage to linacs for this specific application.

To achieve the required 28 MeV energy, a simple and cost-effective design consists in a newly designed 750 MHz RFQ for A/q=2 up to 5 MeV/u, followed by an A/q=2, 750 MHz Quasi-Alvarez DTL (QA-DTL) up to 7 MeV/u. The latter is a DTL where only one drift tube out of three contains a quadrupole [16]. The 15.5 cm diameter of the main drift tubes is sufficient to house a permanent quadrupole and the cooling channels required for 10% duty cycle.

The basic layout is shown in figure 5 and the accelerator parameters are presented in table 2. A commercial ECR source for alpha-particles produces a 0.5 mA current that is accelerated by the 750 MHz RFQ to 5 MeV/u. The latter is optimised for minimum length and RF power with a resulting beam transmission of 50%, following a design recently developed for carbon-ion therapy. The RFQ is followed by a QA-DTL with 27 cells and operating at an average accelerating field of 4 MV/m, bringing the energy to 7 MeV/u (28 MeV total). The accelerator operates at 10% duty cycle, producing an average current of 20 μA. Its total length, considering the ion source and the locally shielded production target, does not exceed 10 m. The overall RF power is < 800 kW, to be provided by 8 SSPA at 100 kW each. Assuming a 50% amplifier efficiency, the total mains power is about 160 kVA.

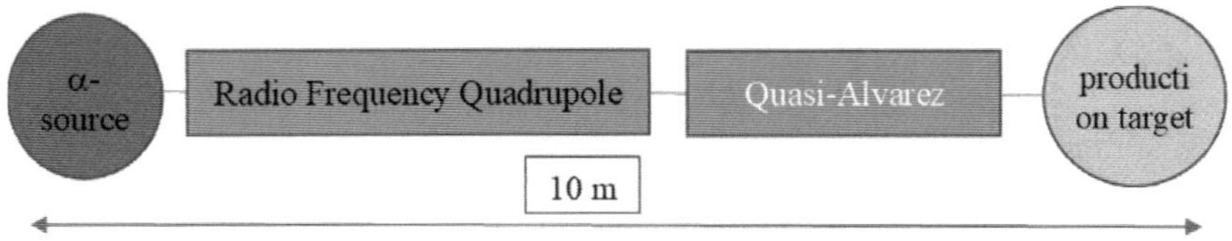

Figure 5. Scheme of the ^{211}At production system.

Table 2. Alpha accelerator parameters.

	RFQ	QA-DTL
f (MHz)	750	750
Input Energy (MeV/u)	0.015	5.06
Output Energy (MeV/u)	5.06	7
Length (m)	4.8	1.6
Peak current (μA)	200	200
Average current (μA)	20	20
Duty cycle (%)	10	10
Peak RF Power (kW)	600	171

Table 3. Accelerator-produced doses for patient treatment ($Doses_{treat}$), considering the preparation time (t_{prep}) after the End Of Bombardment (EOB).

Element	Half-life [min]	t_{PREP} [min]	$Doses_{EOB}$	$Doses_{TREAT}$
^{18}F	109,7	30	200	181
^{11}C	20,4	40	32÷70	8÷17
^{211}At - ovarian cancer	432	420	52÷260	27÷132
^{211}At - brain cancer	432	420	15÷74	8÷38

4. Conclusion

A compact linear accelerator is a very attractive option for producing an interesting selection of medical radioisotopes in a hospital site. With proton beams of 7 MeV and 30 μA protons, or alpha beams of 7 MeV/u and 20 μA, the production of radiopharmaceuticals containing respectively ^{18}F, ^{11}C and ^{211}At (the first two for PET imaging, the latter for therapy) is sufficient to cover the daily needs of a small-medium size hospital. Doses made available at the beginning of a therapy day by the two linac-based systems are shown in table 3. With respect to proton PET-cyclotrons of comparable beam specifications, compact proton linacs have smaller shielding and radiation concerns, lower cost, easier maintenance, lower weight. An α-linac to produce ^{211}At, necessarily limited to 7 MeV/u energy to avoid co-generation of ^{210}At, would be suited for targeted-alpha-therapy and might produce PET-imaging isotopes too, produced by an additional proton source.

Acknowledgements
This project has received funding from the European Union's H2020 R&I programme under Grant Agreement No 730871 (Project ARIES).

Journal of Physics: Conference Series **2420** (2023) 012104 doi:10.1088/1742-6596/2420/1/012104

References

[1] Mamaras A 2020 *Production of Radionuclides in Cyclotrons and Linear Accelerators for Medical Applications* BSc thesis Phys. Dept., Aristotle University, Thessaloniki, Greece
[2] Hess E *et al.* 2001 *Radiochimica Acta* **89** 357
[3] NDS-IAEA `https://www.nds.iaea.org/medical/n4p11c6.html`
[4] Alfarano A *et al.* 2005 *J. Phys.: Conf. Ser.* **41** 115
[5] Jain N *et al.* 2017 *J. Nucl. Med.* **58** 780
[6] Quinn B *et al.* 2016 *BMC Med. Imaging* **16** 41
[7] SRIM `http://www.srim.org/`
[8] Drugs `https://www.drugs.com/pro/choline-c-11.html`
[9] Seltzer M A *et al.* 2004 *J. Nucl. Med.* **45** 1233
[10] Guerard F *et al.* 2013 *Cancer Biother. Radiopharm.* **28** 1
[11] Hamm R W *et al.* 1986 *Proc. 1986 Linear Accelerator Conf. (LINAC'86)* (Stanford, CA USA) pp 141–43
[12] Vretenar M *et al.* 2014 *Proc. 27th Linear Accelerator Conf. (LINAC'14)* (Geneva, Switzerland) pp 935–38
[13] Vretenar M *et al.* 2016 *Proc. 28th Linear Accelerator Conf. (LINAC'16)* (East Lansing, MI USA) pp 704–09
[14] Vlachoudis V *et al.* 2017 *EPJ Web Conf.* **153** 06004
[15] Krawczyk P 2019 Modular Energies – Business Plan
[16] Warner D J 1988 *Proc. 1988 Linear Accelerator Conf. (LINAC'88)* (Newport News, VA USA) pp 109–11

Design of the 590 MeV proton beamline for the proposed TATTOOS isotope production target at PSI

M Hartmann[1,2], D Kiselev[1], D Reggiani[1], J Snuverink[1], H Zhang[1] and M Seidel[1,2]

[1] Paul Scherrer Institut, 5232 Villigen PSI, Switzerland
[2] École Polytechnique Fédérale de Lausanne (EPFL), Lausanne, Switzerland

E-mail: marco.hartmann@psi.ch

Abstract. IMPACT (Isotope and Muon Production with Advanced Cyclotron and Target Technologies) is a proposed initiative envisaged for the high-intensity proton accelerator facility (HIPA) at the Paul Scherrer Institute (PSI). As part of IMPACT, a radioisotope target station, TATTOOS (Targeted Alpha Tumour Therapy and Other Oncological Solutions) will allow the production of terbium radionuclides for therapeutic and diagnostic purposes. The proposed TATTOOS beamline and target will be located near the UCN (Ultra Cold Neutron source) target area, branching off from the main UCN beamline. In particular, the beamline is intended to operate at a beam intensity of 100 μA, requiring a continuous splitting of the main beam via an electrostatic splitter. Realistic beam loss simulations to verify safe operation have been performed and optimised using Beam Delivery Simulation (BDSIM), a Geant4 based tool enabling the simulation of beam transportation through magnets and particle passage through the accelerator. In this study, beam profiles, beam transmission and power deposits are generated and studied.

1. Introduction

The High Intensity Proton Accelerator (HIPA) at the Paul Scherrer Institute (PSI) is at the forefront of the high intensity frontier of particle accelerators delivering a 590 MeV continuous wave proton beam with currents of up to 2.4 mA (1.4 MW beam power) [1].

IMPACT (Isotope and Muon Production using Advanced Cyclotron and Target technologies) is a proposed initiative envisaged for HIPA [2]. As part of IMPACT, a radioisotope target station, TATTOOS (Targeted Alpha Tumour Therapy and Other Oncological Solutions) will allow the production of unprecedented quantities of terbium radionuclides for therapeutic and diagnostic purposes. The TATTOOS beamline is intended to operate at a beam intensity of 100 μA (60 kW beam power), requiring a continuous splitting of the high-powered beam via an electrostatic splitter [3].

A realistic model of the complete TATTOOS beamline from splitter to target was established for the first time in Beam Delivery Simulation (BDSIM) [4], a Geant4 [5] based tool enabling simulation of beam transportation through magnets and particle tracking in electromagnetic fields. BDSIM allows electromagnetic fields to be attached to user-built elements such as the splitter.

The splitter has to withstand significant power deposition and the beam losses due to scattering are a major aspect of the concept that must be evaluated. Indeed, in both the design and operational phases of high intensity accelerators it is critical to minimise beam losses to avoid damage and activation of components, while at the same time maintaining high transmission. In this paper, beam profiles, beam transmission to the target and power deposits are simulated and studied.

2. Proton beamline

2.1. Overview of the HIPA facility
An overview of the HIPA facility with the TATTOOS installation is illustrated in figure 1 [2].

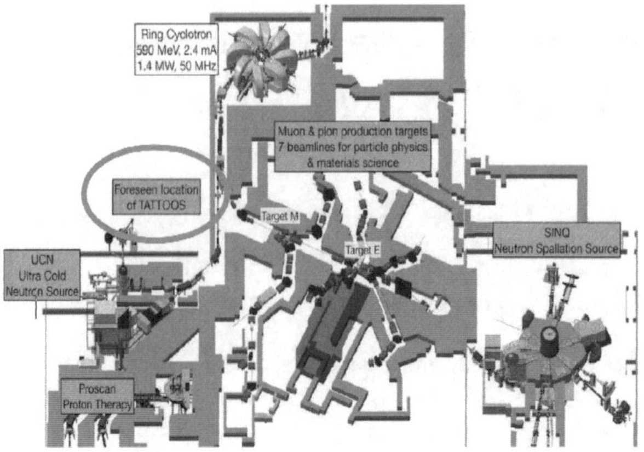

Figure 1. PSI's proton accelerator facility HIPA and the proposed TATTOOS installation circled in blue.

The proton beam is extracted from the Ring Cyclotron and transported to two target stations, Target M (TgM) and Target E (TgE) where secondary particles are produced for user experiments.

The proton beam also feeds two spallation sources, Swiss Spallation Neutron Source (SINQ) [6] and Ultra Cold Neutron source (UCN) [7], the latter running concurrently to the targets. A fast kicker magnet diverts the full intensity beam towards UCN via macro-pulsed kicks. Finally, the splitter located downstream from the kicker magnet can peel off a small portion of the main beam and send it to the UCN target. The elements of the TATTOOS beamline are described in detail in the following section.

2.2. TATTOOS beamline description
The simplest way of including the proposed TATTOOS beamline within the present framework is to split off part of the proton beam needed for TATTOOS by means of the splitter, use the existing magnetic septum magnet (ABS) to divert the peeled beam to the UCN beamline and finally steer it out of UCN and towards TATTOOS via a new bending magnet, specifically ramped up during TATTOOS operation.

In order to peel off such a high-power beam, the splitter possesses a distinct geometry (figure 2) consisting of 179 electrically grounded tungsten alloy strips with each strip tensioned by a pair of springs (all the strips have the same dimension and are 50 μm thick and 3 mm wide).

In total, the splitter amounts to a length of 1107 mm. Furthermore, the includes two cathodes, both at a negative voltage of -172 kV creating two electric fields on every side of the strips. The protons are thus pulled away from the strips at the end of the splitter for each beam [8]. A detailed electric field map has been obtained with ANSYS. Previous simulation studies [2] have demonstrated the viability of the splitter. In particular, the splitter can withstand the power deposition, with around 20 W being deposited on the first strip.

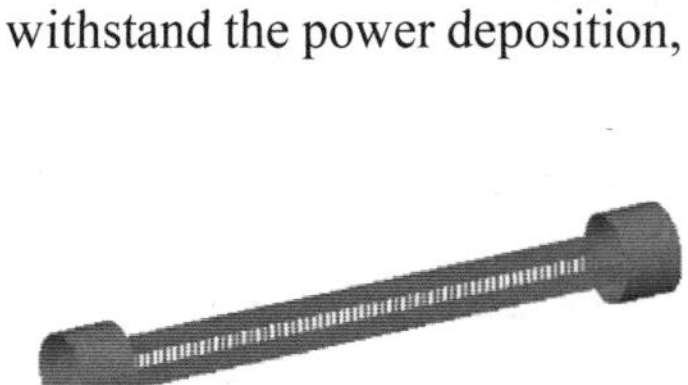

Figure 2. Splitter geometry implemented in BDSIM.

The proposed "alternating beam operation" consists in running TATTOOS between two UCN pulses (one UCN pulse lasts 8 s and the time between two pulses is 300 s). A dipole magnet, ABT (located downstream of the ABS), could be used to divert the beam from the UCN line into the new proposed TATTOOS beamline.

The combined effect of ABS and ABT would be a 13-degree bend. Furthermore, an additional dipole magnet, ADA will bend the beam a further 32 degrees for a total of 45 degrees towards the TATTOOS beamline. To make place for ADA, the existing QBB1 and QBB2 quadrupoles in the UCN beamline will be moved downstream by a few metres without having a detrimental effect on the UCN operation. Figure 3 illustrates the complete simulated BDSIM model from the beam splitter (left) to the target (right).

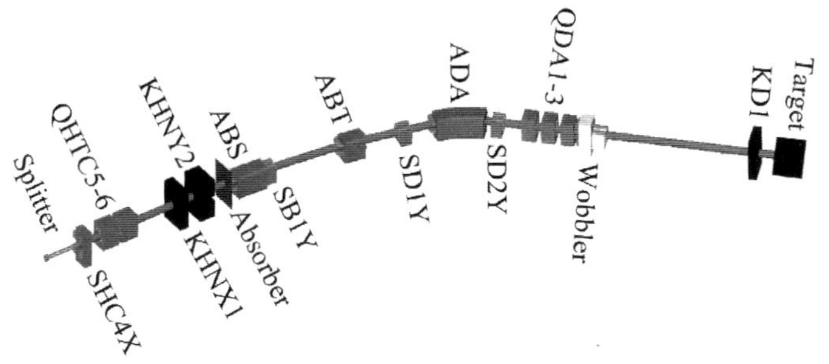

Figure 3. Complete TATTOOS beamline.

Several other components are required for the TATTOOS beamline. A quadrupole triplet (QDA1-3) will be used to shape the beam on target. A so-called "Wobbler" is used to flatten the beam footprint and achieve uniform target temperature. The Wobbler consists of two fast dipole magnets driven by AC current with a frequency of around 30 Hz [9].

Three dipole steering magnets (SB1Y, SD1Y and SD2Y) will be used to steer the beam in the vertical plane while steering in the horizontal plane is performed by the bending magnets.

In BDSIM, a fictious element called "Absorber" [10] was used to artificially remove the protons going into the SINQ line before the ABS septum, leaving only the TATTOOS protons.

A collimation system is also foreseen for TATTOOS. Existing collimators KHNX1 and KHNY2 are represented as jaw collimators in BDSIM, acting in the transverse x-y plane and cutting the beam in these two directions respectively. A last collimator, KD1 is placed at the end of the beamline to protect the target.

A comprehensive overview of all the elements in the TATTOOS beamline is given in ref. [2].

3. Simulations

3.1. Beamline optics validation

During the design of a beamline, it is of crucial importance to benchmark simulations against standard calculations or other dedicated simulation tools. The beamline lattice was designed in TRANSPORT [11] and the beam optics from BDSIM was benchmarked with MAD-X [12]. Proton tracking started 1.0054 m upstream of the upstream-end of the ABS for both codes and excludes the splitter since its implementation in conventional beam optics program such as MAD-X is complex. The TATTOOS line from the starting point to the target measures 21.479 m. Figure 4 illustrates the comparison of the beta-function between BDSIM and MAD-X. An excellent agreement is found between the two codes.

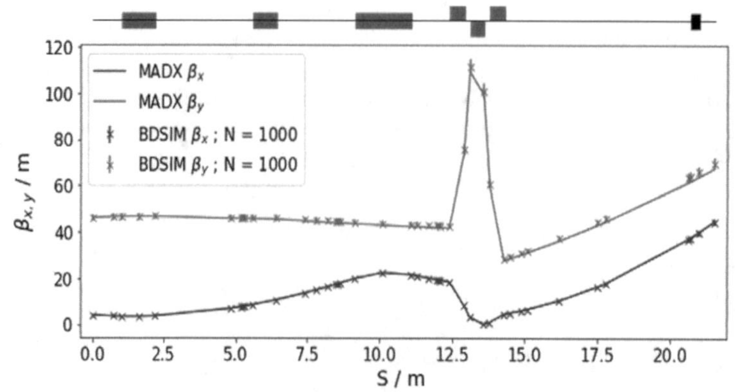

Figure 4. Benchmarking between BDSIM and MAD-X – beta-function validation.

3.2. Input file and horizontal beam profile

The initial beam is considered a six-dimensional Gaussian distribution representing a proton bunch in phase space. Protons are first tracked from splitter to target and an input file is then used to record the phase space coordinates after the splitter (at QHTC6) for each proton. By using the input file, CPU time is saved as the splitter field map does not have to be loaded for each simulation run. At the same time the properties of the split beam are preserved in the analysis. The horizontal beam profile is simulated with one million primary protons at QHTC6 as shown in figure 5.

A wide range of physics processes, including electromagnetic, hadronic and radioactive decay, are activated for the BDSIM simulation using the physics list physicsList=em ftfp_bert decay muon hadronic_elastic em_extra [10].

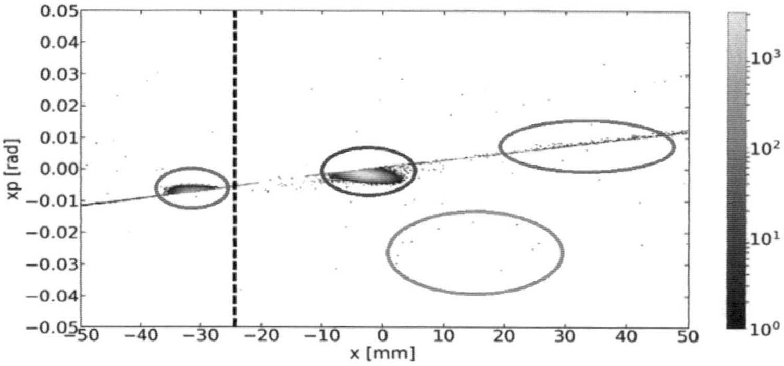

Figure 5. Horizontal phase space at QHTC6 approximately 1 m from the splitter.

From figure 5, the split and main beams can be clearly distinguished (red and blue circles respectively). Assuming a 2 mA beam, 100 µA is split. Off-momentum protons resulting from inelastic scattering with the splitter strips may be visible (green circle). Finally, a line of scattered protons due to the non-uniformity of the electric field acting on the strips may be noticed (grey circle).

3.3. Power deposition and transmission studies

The initial number of protons for the tracking simulations is taken from the split beam distribution before the collimators. From figure 5, only the protons with x < -25 mm (dashed line) are selected. These protons are then tracked along the beamline to the TATTOOS target. For transmission studies only the primary protons from this split beam are tracked, whereas for the power deposition studies, the entire beam (main and split) is tracked including the secondary particles.

In order to determine the collimator requirements for the TATTOOS beamline, the influence of the two collimators KHNX1 and KHNY2 is studied. The aim is to infer the aperture sizes suitable for minimising the energy losses. Parametric studies are performed whereby the aperture sizes of each collimator in x and y are varied. Several aperture settings are considered and compared to the case where there are no collimators. Figure 6 shows the power deposition from splitter to target for each KHNX1 aperture (KHNY2 is kept fixed at 0.080 m).

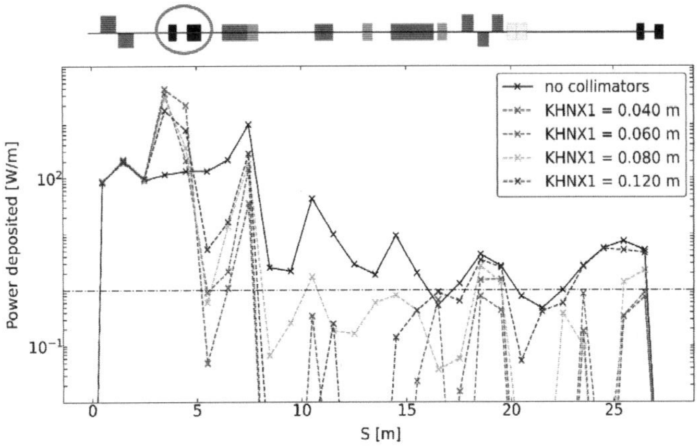

Figure 6. Power deposition from splitter to target.

From figure 6, the power deposited is highest at the collimators (red circle) but after the collimators, the power deposited is always lower when the collimators are present compared to when there are none. In general, the power deposited when the collimators are present is less than 1 W/m (dashed line in figure 6) which is a reference value for acceptable losses in high energy hadron accelerators [13].

Figure 7 illustrates the total power deposited after KHNY2 as a function of KHNY2 aperture (left) as well as transmission to the target (right).

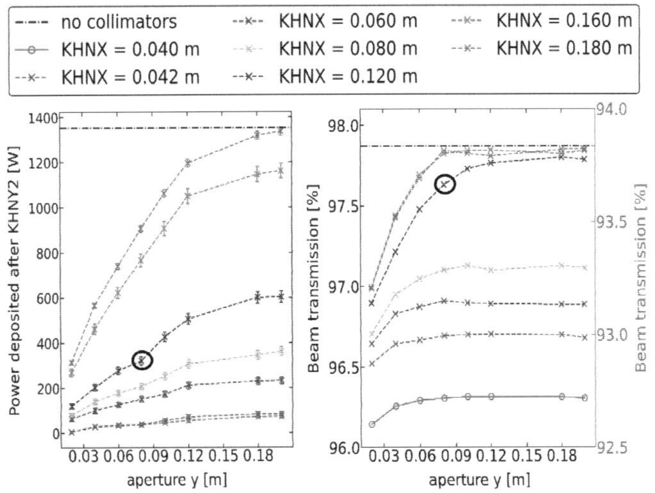

Figure 7. Power deposited and transmission to target as a function of KHNY2 aperture size for several KHNX1 aperture sizes.

As expected, the losses and transmission increase as the aperture size of the collimators is increased. The larger the aperture size, the more one approximates the case without collimators. Furthermore, as the KHNX1 aperture size is decreased to 0.040 m, the transmission is decreased by approximately 4%. This is because the edges of the collimator are cutting the core of the split beam, reducing the protons arriving at the target. Indeed, for a small aperture size, the beam is more sensitive to changes in position. An optimal collimator aperture could then be KHNX1 = 0.120 m, KHNY2 = 0.080 m (circled in figure 7) as the transmission here is high and the losses much lower than the case where there are no collimators.

4. Conclusion

In this paper, the TATTOOS lattice design was presented including the first realistic simulations of the beamline using BDSIM. With the current optimised collimator system, the simulation results show that the TATTOOS beamline may safely operate to produce radionuclides using a high-power beam that is split by a unique beam splitter without significant losses and with a high transmission. Furthermore, the splitter can withstand the power deposition and can hence be used as a viable component for TATTOOS. These simulation results will serve as model predictions for future TATTOOS beamline measurements.

References

[1] Kiselev D *et al.* 2021 *JPS Conf. Proc.* **33** 01104
[2] Eichler R *et al.* 2022 IMPACT conceptual design report Tech. rep. Paul Scherrer Institute Villigen PSI Switzerland PSI Rep. No.22-01
[3] Olivo M, Rohrer U, and Steiner E 1981 *IEEE Trans. Nucl. Sci.* **28** pp 3094-96
[4] Nevay L J *et al.* 2020 *Computer Physics Communications* **252** 107200
[5] Agostinelli S *et al.* 2003 *Nucl. Instrum. Methods Phys. Res., Sect.* A **506** pp 250-303
[6] Blau B *et al.*2009 *Neutron News* **20** pp 5-8
[7] Reggiani D *et al.* 2009 *Proc. 23rd Particle Accelerators Conf. (PAC'09)* (Vancouver, BC Canada) pp 1748-50
[8] Mariani E, Olivo M and Rossetti D 1998 *Proc. 6th Eur. Particle Accelerators Conf. (EPAC'98)* (Stockholm, Sweden) pp 2129-31
[9] Reggiani D *et al.* 2013 *Proc. 4th Int. Particle Accelerators Conf. (IPAC'13)* (Shanghai, China) pp 3954-56
[10] BDSIM Manual 2019 `http://pp.rhul.ac.uk/bdsim/manual`
[11] Brown K L, Rothaker F, Carey D C and Iselin C 1983 TRANSPORT: a computer program for designing charged particle beam transport systems Tech. rep. CERN, Geneva, Switzerland, Rep. CERN-80-04
[12] MAD-X 2016 `https://mad.web.cern.ch`
[13] Strasik I, Mustafin E and Steiner E 2010 *Phys. Rev. Spec. Top. Accel Beams* **13** 071004

Journal of Physics: Conference Series **2420** (2023) 012106 doi:10.1088/1742-6596/2420/1/012106

A novel intensity compensation method to achieve energy independent beam intensity at the patient location for cyclotron based proton therapy facilities

Vivek Maradia[1, 2], David Meer[1], Damian C Weber[1, 3, 4], Antony J Lomax[1, 2] and Serena Psoroulas[1]

[1] Paul Scherrer Institute, Villigen, Switzerland
[2] ETH Zurich, Zurich, Switzerland
[3] University Hospital Bern, Bern, Switzerland
[4] University Hospital Zurich, Zurich, Switzerland

Email: vivek.maradia@psi.ch

Abstract. In cyclotron-based proton therapy facilities, an energy selection system is typically used to lower the beam energy from the fixed value provided by the accelerator (250/230 MeV) to the one needed for the treatment (230-70 MeV). Such a system has the drawback of increase beam emittance and introducing an energy-dependent beam current at the patient location, resulting in energy dependent beam intensity ratios of about 10^3 between high and low energies. This complicates treatment delivery and challenges patient safety systems. As such, we propose the use of a dual energy degrader method that can reduce beam intensity for high-energy beams. The first degrader is made of high Z material and the second is made of low Z material and are placed next to each other. For high energies (190-230 MeV), we use only the first degrader to increase beam emittance after the degrader and thus lose intensity in the emittance selection collimators. For intermediate energy beams (110-190 MeV) we use the combination of both degraders, whereas for low energy beams (70-110 MeV), only the second degrader limits the increase in emittance. With this approach, energy-independent beam intensities at the patient location can be achieved, whilst localizing beam losses around the degrader.

1. Introduction

Proton therapy provides better dose distribution conformity and also better spares healthy tissues when com-pared with equivalent photon plans. Therefore, proton therapy has become a credible option in radiotherapy to treat certain types of cancers.

In proton therapy, based on the size and location of the tumor, proton beams with different energies are required to deliver the dose in the target volume. The energy required for patient treatments is typically in the range of 70-230 MeV.

Most proton therapy facilities use a cyclotron. Since a cyclotron produces beams of fixed energy (250 MeV for the PSI COMET cyclotron [1]), to modulate the energy of the beam, a degrader with an adjustable thickness is required. Unfortunately, beam scattering in the energy degrader increases the beam size, divergence, and energy spread beyond the beamline and gantry acceptance. It is unavoidable to use one or more collimator systems and an energy selection system (ESS) to cut these quantities to those that fit in the acceptance of the following beam transport system, to prevent unwanted beam losses along the beamline. This limited acceptance depends on the energy, the geometrical layout of the beam transport system, and the setting of the magnets in the beamline. At PSI, for example, for the highest energies (200-230 MeV), transmission through the beamline is about 30 %. However, for the lowest energies (70-100 MeV), transmission through the beamline is below 0.1 % [2-5].

The beam intensity at the patient would be strongly dependent on beam energy, due to this energy-dependent transmission. It may result in significant change in beam current at patient location and could have consequences for safety, due to a limitation of reaction times. Therefore, an intentional beam loss for higher energies is necessary to obtain the similar beam intensity for all energies.
This can be done by adjusting the intensity in the cyclotron, but one could also design a beamline setting with energy-dependent controlled beam losses at dedicated collimators in the beam-transport system.

In this study, we report a new way to do intensity compensation for high-energy beams using two degraders made of different materials. For the proof of principle simulation study, by using two degraders, we tried to achieve the similar transmission from the cyclotron to the end of the ESS beamline.

2. Methods and materials

2.1. Dual energy degrader method

To intentionally lose the intensities of the high energy beams, we designed an ESS beamline using two degraders (as shown in figure 1). We used two degraders made of different materials, Aluminium (D1) and Carbon (D2).

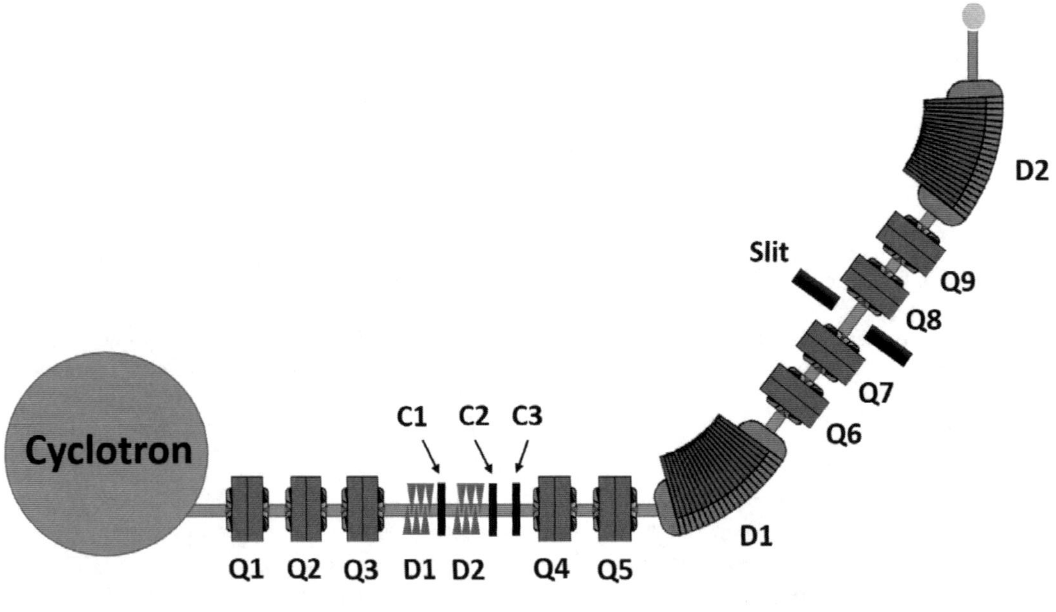

Figure 1. Schematic of the ESS beamline. (Q1-Q9: Quadrupole magnets, D1-D2: Dipole magnets, D1-D2: Degraders C1-C3 collimators).

To achieve efficient intensity compensation, we divide the energy degradation in three parts.

- High energy beams (190-230 MeV): we use degrader (D1) to choose the beam energy, collimator (C1) to choose beam size, and collimator (C2) to choose the beam divergence.
- Intermediate energy beams (110-190 MeV): we use the combination of degrader (D1) and degrader (D2) to achieve almost similar transmission at the end of ESS. Additionally, we use collimator (C2) to choose beam size and collimator (C3) to choose the bema divergence.
- Low energy beams (70-110 MeV): we use degrader (D2) to choose the beam energy, collimator (C2) to choose beam size, and collimator (C3) to choose the beam divergence.

2.2. Specification of degraders and collimators

The first degrader (D1) is made of Aluminium (Al) (assuming a density of 2.7 g/cm^3) and the second degrader (D2) is made of carbon (assuming a density of 1.7 g/cm^3), placed in the beam line as shown in figure 1. Due to the high atomic number, Aluminium provides higher scattering power compared to carbon. Table 1 summarizes the apertures of the collimators.

Table 1. Dimension of collimators.

	Radius in X-plane (mm)	**Radius in Y-plane (mm)**
C1	1.5	1.5
C2	4	4
C3	15	15

2.3. Beam optics and Monte Carlo simulation

The matrix formalism code TRANSPORT has been used to design new beam optics of the ESS beamline (see figure 2). The transmission of the beam was calculated with fast and light-weight Monte-Carlo beam optics code for the proton beamlines of the Paul Scherrer Institute, MinT [6]. MinT can compute the effects of beam degradation, multiple scattering, beam collimation and beam transmission. Additionally, it allows to do fitting.

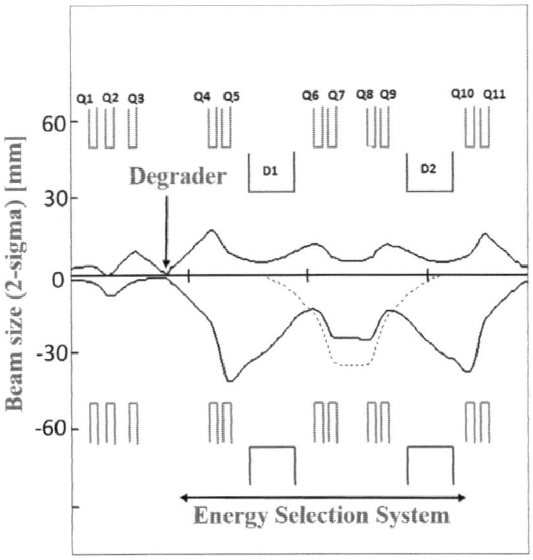

Figure 2. The new beam optics of ESS. The beam envelopes show the beam size in 2-sigma values and the dispersion (dashed line) along ESS beam line. The lower half of figure shows beam envelope in X-plane (bending plane) and the upper half shows envelope in Y-plane.

3. Results

3.1. Simulation results

For low energy beams (70 MeV), the transmission through the ESS is about 0.1 %. If we try to achieve the same 0.1 % transmission to all energy beams, the beam currents at the patient location will be very low. There-fore, to achieve reasonable beam currents at the isocenter, for the first proof of principle simulation study, we decided to do intensity compensation for energies higher than 110 MeV.

Figure 3 shows the transmission between cyclotron to end of ESS for different energy beams using the two de-grader method.

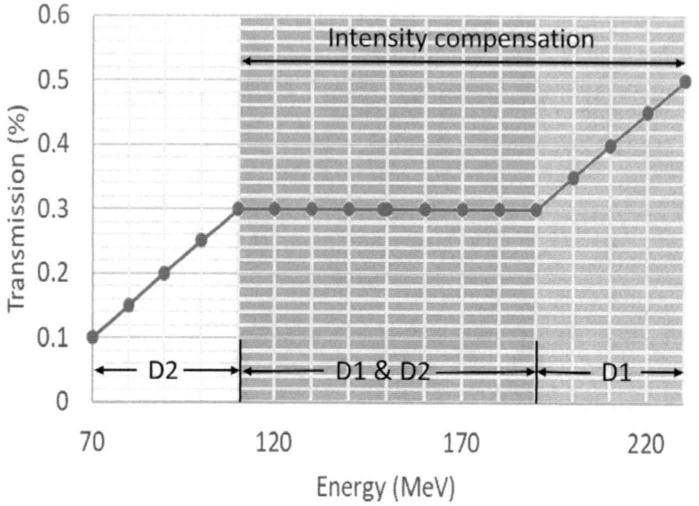

Figure 3. Beam transmission using two degrader method.

To achieve maximum beam loss for high energy beams (190-230 MeV beam) we used only the high Z degrader D1 made of Aluminium. Therefore, for high energy beams, the scattering through the degrader D1 was high and the emittance after the degrader was significantly high compared to low Z degrader material. Additionally, by using a smaller aperture of beam size selection collimator C1, we lose significant beam loss at C1 and achieve only 0.3 to 0.5 % transmission for high-energy beams.

For intermediate energy beams (110-190 MeV), our aim was to achieve almost the same transmission. Therefore, we used the combination of two degraders. We choose the thickness of both degraders by the iterative process using the fit function in MinT. We define the transmission at the end of ESS as a fitting constraint. The fitting function tries to find the thickness of two degraders to match the transmission value. The beam will lose part of the energy in degrader D1 and part of the energy in degrader D2 in a way that we achieve 0.3 % transmission at the end of ESS for all energies from 110 to 190 MeV.

For low energy beams, we used only degrader (D2) to achieve the maximum possible transmission through the ESS. As shown in figure 3, we achieved transmission of 0.1 to 0.3 for low 70 to 110 MeV beams.

4. Discussion

We have demonstrated that with the use of the two de-graders made of different materials, it is possible to achieve flat intensity compensation curve. However, for the proof of principle study, we tried to achieve the same transmission (from the cyclotron to the end of ESS) for 110 to 190 MeV beams. However, it is possible to achieve the same transmission from cyclotron to isocenter for all energy beams.

With this new intensity compensation method, we localized all the losses next to the degrader and avoid the need for an additional collimator along the beamline for intensity compensation. This helps in better shielding the design of the facility. Additionally, with this method, just by changing the degrader thickness of the two degraders, we can achieve any type of intensity compensation scheme based on the clinical need.

A few recent beam optics studies show that by using an asymmetric collimator or by transporting asymmetric emittance through the beamline with scattering foil, it is possible to achieve almost 1 % transmission for a 70 MeV beam. In this case, as the beam transmission for low energy is higher, by using two degrader methods, one could easily achieve the same 1 % transmission for all clinical energy range (70-230 MeV). With an asymmetric collimator, it is important to use two different apertures of the collimator (a larger aperture for low energy beams and a smaller aperture for high energy beams). However, the change of collimator during the treatment will increase the treatment delivery time. To avoid the use of different apertures of the collimator, one could defocus the beam before the degrader for high-energy beams.

One of the disadvantages of the method is that it will increase the length of the beam line by 0.7 to 1 m. How-ever it will reduce the shielding requirement in down-stream part of the beamline as there are no additional beam losses in coper collimators.

We did not look directly into the implement ability of this method. Further studies with different degrader materials would be essential to choosing the degrader material which could produce less long-lived radioactive isotopes.

5. Conclusion

In this proof of principle work, we have demonstrated that by using two degrader intensity compensation methods, it is possible to achieve almost the same transmission for all energy beams while localizing the beam losses of high energy beams around the degrader region. Additionally, it will reduce the shielding requirement for downstream beamline as there are no additional beam losses in downstream collimators for intensity compensation purposes.

References

[1] Schippers M, Duppich J, Goitein G, Hug E, Jermann M, Mezger A and Pedroni E for the PROSCAN team 2007 *Proc. 18th Int. Conf. on Cyclotrons and their Applications (Cyclotrons'07)* (Giardini-Naxos, Italy) pp 15-17

[2] Maradia V, Meer D, Weber D C, Lomax A J, Schippers J M and Psoroulas S 2021 *Med. Phys.* **48** pp 7713–22

[3] Maradia V *et al.* 2022 *Med. Phys.* **49** pp 0–10

[4] Maradia V *et al.* 2021 *Proc. 12th Int. Particle Accelerators Conf. (IPAC'21)* (Campinas, Brazil) pp 2477-79

[5] Maradia V *et al.* 2021 New Gantry Beam Optics Solution for Minimizing Treatment Time in Cyclotron- based Proton Therapy Facilities Presented at 59th Annual Particle Therapy Co-Operative Group Conference (PTCOG 2021) (Online) doi: 10.3929/ethz-b-000497120

[6] Baumgarten C 2022 MinT: A Fast Lightweight Envelope/Monte-Carlo Beam Optics Code for the Proton Beamlines of the Paul Scherrer Institute *Preprint* ArXiv:2202.07245

A novel method of emittance matching to increase beam transmission for cyclotron-based proton therapy facilities: simulation study

Vivek Maradia[1,2], David Meer[1], Damian C Weber[1,3,4], Antony J Lomax[1,2] Jacobus M Schippers[1] and Serena Psoroulas[1]

[1] Paul Scherrer Institute, Villigen, Switzerland
[2] ETH Zurich, Zurich, Switzerland
[3] University Hospital Bern, Bern, Switzerland
[4] University Hospital Zurich, Zurich, Switzerland

Email: vivek.maradia@psi.ch

Abstract. In proton therapy, high dose rates can reduce treatment delivery times, allowing for efficient mitigation of tumor motion and increased patient throughput. With cyclotrons however, high dose rates are difficult to achieve for low-energies as, typically, the emittance after the degrader is matched in both transversal planes using circular collimators, which does not provide an optimal matching to the acceptance of the following beamline. Transmission can however be substantially improved by transporting maximum acceptable emittances in both orthogonal planes, but at the cost of gantry angle-dependent beam shapes at isocenter. Here we demonstrate that equal emittances in both planes can be recovered at the gantry entrance using a thin scattering foil, thus ensuring gantry angle-independent beam shapes at the isocenter. We demonstrate in simulation that low-energy beam transmission can be increased by a factor of 3 using this approach compared to the currently used beam optics, whilst gantry angle-independent beam shapes are preserved. We expect that this universal approach could also bring a similar transmission improvement in other cyclotron-based proton therapy facilities.

1. Introduction

Brought into clinical practice at Paul Scherrer Institute (PSI) in the 1996, pencil beam scanning (PBS) is nowadays the standard beam delivery technique in proton therapy [1, 2]. However, current challenges of PBS particle therapy are the dosimetric uncertainties in treatment of moving targets and the relatively long treatment times involved. The dosimetric uncertainty can be minimized through the use of motion mitigation techniques, which aim to mitigate the interplay effect between the motions in the patient and the beam delivery; the most common motion mitigation techniques are breath-hold [3], rescanning [4], and gating [5]. For all these techniques, it is also desirable to have shorter treatment delivery times [6, 7]. One way to reduce the treatment delivery time for PBS proton therapy is to increase the intensity of the low-energy beams by improving the transmission of the beam from the cyclotron to the isocenter (patient position), thereby reducing beam-on time (the time required to deliver the dose) during treatment delivery.

Most of the proton therapy facilities use a cyclotron. Since a cyclotron produces beams of a fixed energy, to modulate the energy of the beam, an energy selection system (ESS), consisting of a degrader with an adjustable thickness followed by momentum selection, is required. However, due to scattering in the degrader, for low-energy beams, the emittance after the degrader is in the range of a few hundreds of π mm mrad. Therefore, to minimize beam losses in the beamline, it is necessary to use beam emittance selection collimators after the degrader to re-strict the emittance to the requirement of the following beamline or gantry. Currently, all cyclotron-based proton therapy facilities transport a maximum emittance of 30 π mm mrad through the beamline (in this work, beam sizes, divergences, and emittances are expressed as 2σ values), which limits the transmission of low-energy beams. At PSI for example, for the lowest energies (70-100 MeV), transmission through the beamline is below 0.1 % [8]. Such low transmission for these low energies causes an increase in beam-on time.

One way to achieve higher intensity beams at the isocenter is to transport a higher emittance through the following beamline and gantry [8-11]. At our facility, we can transport a maximum of ~65 π mm mrad in X-plane and ~139 π mm mrad in Y-plane. These will increase the beam transmission significantly compared to conventional 30 π mm mrad emittance transport in both planes, at the cost of an asymmetric emittance at the gantry entrance, leading to gantry angle dependent beam shapes at the isocenter. To achieve gantry angle independent beam shape at the isocenter, it is necessary to have same emittance at the entrances of the gantry.

In this study, we report on the use of a thin scattering foil, placed in the beamline between the ESS and gantry coupling point, to achieve equal emittances in both planes, whilst maintaining a high transmission through the beamline and gantry, a method also used in several synchrotron-based ion beam therapy facilities [12, 13]. In this work, all simulation investigations were performed with 70 MeV beam as our goal was to increase the trans-mission for low energy beams.

2. Mathods and Materials

2.1. Emittance matching with scattering foil

To increase the emittance in the X-plane to a similar value as the Y-plane emittance, but with minimally effect on the emittance in the Y-plane, the following boundary conditions have been applied: (as expressed schematically in figure 1).

- Beam waist is at the location of the scattering foil (R12 = 0 and R34 = 0).
- No dispersion at the scattering foil location (R16 =R26 = 0).
- Optimized focusing parameters based on twiss scattering formula for the thin scattering foil.
- The divergence ratio between X-plane and Y-plane must be smaller than the emittance ratio between X-plane and Y-plane at the entrance of scattering foil.

2.2. Specification of scattering foil

A tantalum (Ta) scattering foil 30 μm thick, assuming a density of 16.69 g/cm^3, is placed in the beam line as shown in figure 2.

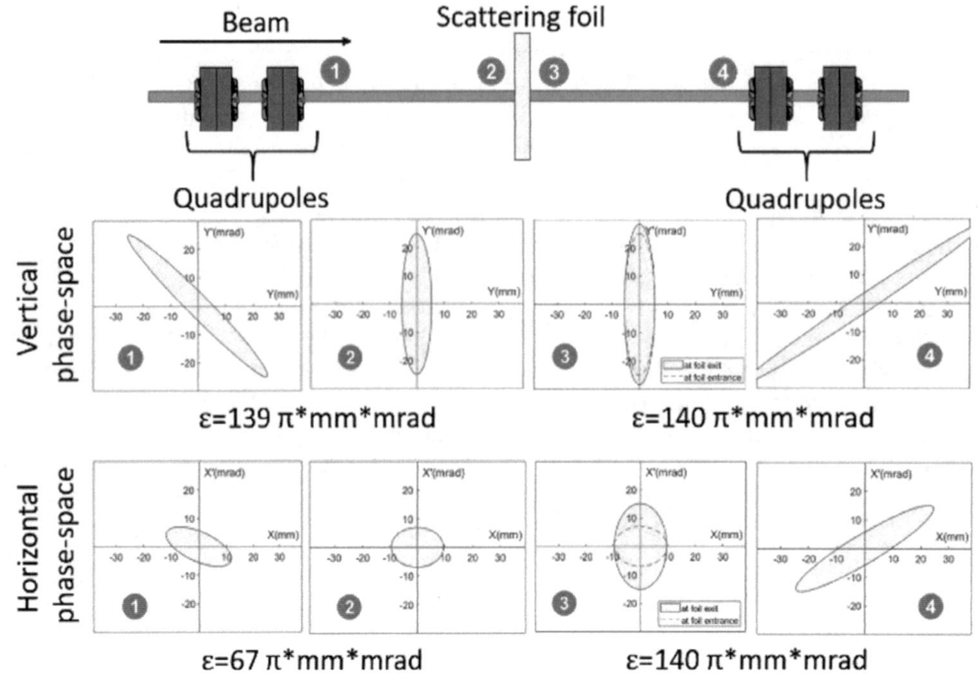

Figure 1. Schematic of the function of scattering foil. Both the horizontal and vertical phase-space ellipses and profiles at different location along the beamline (before and after the scattering foil).

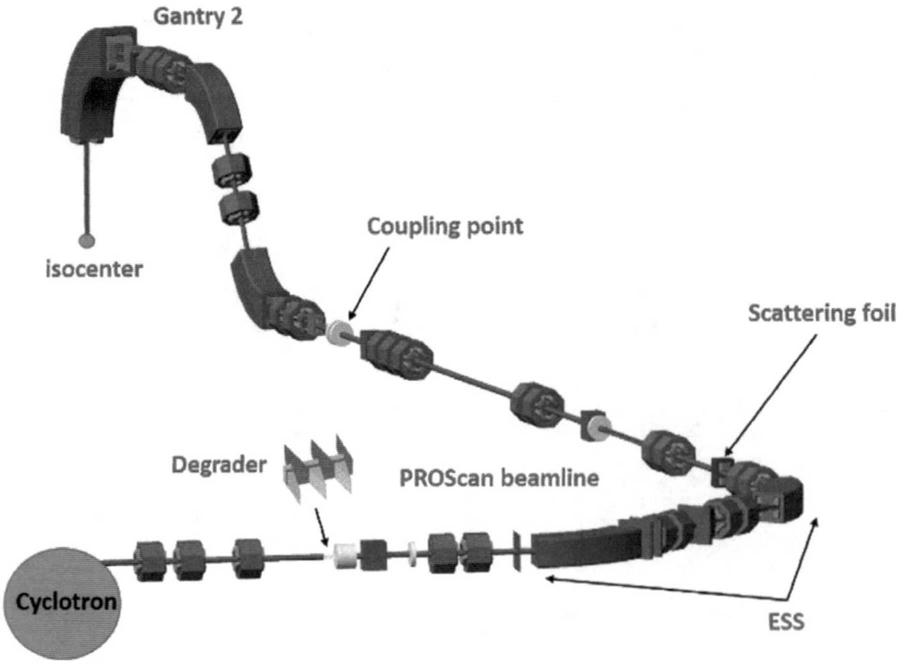

Figure 2. BDSIM model of PROSCAN beam line and Gantry 2.

2.3. Monte Carlo simulation with new beam-optics design

The matrix formalism code TRANSPORT has been used to design new beam optics to include the above described scattering foil. However, TRANSPORT cannot predict the scattering effect and beam losses along the beamline. To simulate the scattering effect and to calculate the transmission of the new beam optics and the beam size at isocenter, Monte Carlo simulations, based on beamline settings optimized with TRANSPORT, are re-quired. These have been performed using the BDSIM 1.4.133 Monte Carlo simulation toolkit [14]. Based on initial TRANSPORT beam optics, we did the BDSIM simulation up to scattering foil and calculated the beam parameters after the scattering foil. With these new parameters from BDSIM simulation, we redesign the beam optics in TRANSPORT from scattering foil to isocenter. After that, we did a BDSIM simulation for the full PROSCAN beamline and calculate the transmission along the beamline and beam size at the isocenter.

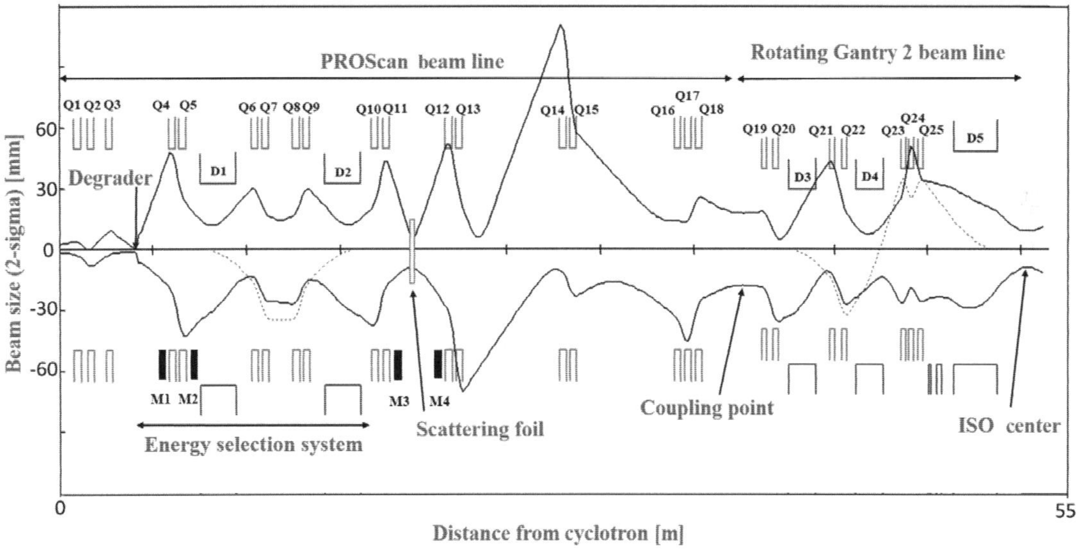

Figure 3. The new beam optics including a scattering foil transports 67 π mm mrad in X-plane and 139 π mm mrad in Y-plane up to scattering foil location and transports almost 140 π mm mrad (in both planes) from scattering foil to isocenter. The beam envelopes show the beam size in 2-sigma values and the dispersion (dashed line) along PSI's ESS beam line (The lower half of figure shows beam envelope in X-plane (bending plane) and the upper half shows envelope in Y-plane).

3. Results

3.1. Emittance simulation with scattering foil

The beam optics from cyclotron exit to the scattering foil location (position 2 in figure 1) have been designed such that we get a beam size of 9.5 mm and 7 mrad divergence in the X-plane, and 5.5 mm and 25 mrad in the Y-plane.

Table 1. Simulated beam parameters after scattering foil.

	Beam size (mm)	Beam divergence (mrad)
X-plane	9.5	14.7
Y-plane	5.5	25.5

With BDSIM simulation, we predict the beam parameters after the scattering foil. As shown in table 1, in X-plane, we achieved 9.5 mm beam size and 14.7 mrad divergence and in Y-plane, we achieved 5.5 mm beam size and 25.5 mrad divergence. After the scattering foil, we get 140 π mm mrad emittance in both planes (table 2).

Table 2. Simulated emittance value just before and after the scattering foil.

	Emittance before scattering foil (π mm mrad)	Emittance after scattering foil (π mm mrad)
X-plane	67	140
Y-plane	139	140

3.2. Transmission improvements with scattering foil

We have compared the transmission at different locations along the beamline for the reference beam optics transporting 30 π mm mrad (as in clinical use) and the new beam optics with scattering foil.

Table 3. Simulated transmission using reference beam optics and new beam optics with scattering foil. Transmission values are from cyclotron to different locations along the beam-line.

Location along beamline	Reference beam optics	New beam optics with scattering foil
X-plane	67	140
Y-plane	139	140

As we are transporting higher emittance, at the monitor M2 location (shown in figure 3), we get almost 5 % trans-mission for new beam optics while only 1.5 % transmission with reference beam optics. While passing through the ESS, we lose the beam in momentum selection slits in both cases. For reference beam optics, from the end of ESS to coupling, we do not lose the beam. However, when introducing the scattering foil, divergence increases in both planes, and the next quadrupole magnet is almost 2 m away. Therefore, losses between quadrupoles Q12 to Q15 are unavoidable, and another 25 % of the beam is therefore lost in the new beam optics. For the scattering foil case, the gantry beam optics was designed with 2:1 imaging which allows transporting high emittance through the gantry while having minimum losses [9]. Overall, then, with the use of asymmetric optics, 2:1 imaging in the gantry, and the introduction of the scattering foil, we predict an overall transmission of 0.42% from the cyclotron to the isocenter, which can be compared to the only 0.14 % transmission for the reference beam optics. However, transmission improvements come at the cost of an increased beam size. For the reference beam optics, the beam size at the isocenter is 10.5 mm whereas with the high transmission and scattering foil beam optics, this increases to 18.5 mm, representing a 76 % increase in beam size.

4. Conclusion

In this work, in simulation, we have demonstrated that by using a thin scattering foil placed in the beamline, we can match the emittance in both transverse planes by increasing the divergence in the low emittance-transporting plane (in our case X-plane). As this approach allows for transport of maximum acceptable emittance in both planes, this method substantially increases the low energy beam transmission through the beamline, while achieving gantry angle independent beam sizes at the isocenter. The obtained higher intensities (dose rates) could reduce treatment delivery times to aid motion mitigation techniques such as breath-hold, gating, and rescanning [15]. In addition, shorter treatment delivery times may increase patient throughput, lowering the cost of proton therapy treatment.

References

[1] Pedroni E *et al.* 1995 *Med. Phys.* **22** pp 37–53
[2] Lin S, Boehringer T, Coray A, Grossmann M and Pedroni E 2009 *Med. Phys.* **36** pp 5331–40
[3] Dueck J *et al.* 2016 *Int. J. Radiat. Oncol. Biol. Phys.* **95** pp 534–41
[4] Klimpki G *et al.* 2018 *Phys. Med. Biol.* **63** 145006
[5] Schätti A, Zakova M, Meer D and Lomax A J 2014 *Phys. Med. Biol.* **59** pp 3813–28
[6] Maradia V 2021 Intensity-modulated proton therapy treatment delivery of locally advanced non-small-cell lung cancers within a single breath-hold Presented at 4D Treatment Workshop for Particle Therapy (Delft, Netherlands) `doi: 10.3929/ethz-b-000515078`
[7] Maradia V *et al.* 2021 *Radiother. Oncol.* **161** pp S1488–S1489
[8] Maradia V, Meer D, Weber D C, Lomax A J, Schippers J M and Psoroulas S 2021 *Med. Phys.* **48** pp 7713–22
[9] Maradia V *et al.* 2022 *Med. Phys.* **49** pp 0–10
[10] Maradia V *et al.* 2021 *Proc. 12th Int. Particle Accelerators Conf. (IPAC'21)* (Campinas, Brazil) pp 2477-79
[11] Maradia V et al. 2021 New Gantry Beam Optics Solution for Minimizing Treatment Time in Cyclotron- based Proton Therapy Facilities Presented at 59th Annual Particle Therapy Co-Operative Group Conference (PTCOG 2021) (Online) `doi: 10.3929/ethz-b-000497120`
[12] Furukawa T and Noda K 2006 *Nucl. Instrum. Methods Phys. Res., Sect.* A **565** pp 430–38
[13] Furukawa T *et al.* 2006 *Proc. 10th Eur. Particle Accelerators Conf. (EPAC'06)* (Edinburgh, UK) pp 2322-24
[14] Nevay L J *et al.* 2016 *Proc. 7th Int. Particle Accelerators Conf. (IPAC'16)* (Busan, Korea) pp 3098-3100
[15] Maradia V, van de Water S, Meer D, Weber D C, Lomax A J and Psoroulas S 2022 *Radiother. Oncol.* **174** pp 23-29

Ripple pattern formation on silicon carbide surfaces by low-energy ion-beam erosion

Divya Gupta[1], G.R. Umapathy[2], Rahul Singhal[3] and Sanjeev Aggarwal[1]

[1] Ion Beam Centre, Department of Physics, Kurukshetra University, Kurukshetra 136119, India
[2] Inter University Accelerator Centre (IUAC), New Delhi 110001, India
[3] Malviya National Institute of Technology (MNIT), Jaipur, Rajasthan 302017, India

Email: divyagupta2017@kuk.ac.in

Abstract. A versatile experimental facility air insulated high current medium energy 200 kV Ion Accelerator, with the terminal voltage in the range of 30-200 kV has been running successfully at Ion Beam Centre, KUK for carry out multifarious experiments in material science and surface physics. This system offers single charge state, switching magnet with five exit ports and large area irradiation/implantation using hollow cathode ion source. Ion beam induced structures on the surfaces of semiconductors have potential applications in photonics, magnetic devices, photovoltaics, and surface-wetting tailoring etc. In this regard, silicon carbide (SiC) is a fascinating wide-band gap semiconductor for high-temperature, high-power and high-frequency applications. In this work, fabrication of ripple patterns is carried out on the SiC surfaces using 80 keV Ar^+ ion beam for different fluences at oblique incidence of 50^0. AFM study demonstrates that ripple wavelength and amplitude, ordering and homogeneity of these patterns vary linearly with argon ion fluence. The formation of such surface structures is attributed to the preferential sputtering of silicon in comparison to carbon. The evolution of high degrees of order is explained with the help of existing formalisms of coupling between surface topography and preferential sputtering.

1. Introduction

Oblique angle sputter erosion (OASE) is a versatile and cost-effective tool for patterning and structuring solid surfaces at the nanoscale level [1-3]. This is a scalable method that fabricates a myriad of patterns in a single technological step and is applicable to most solid materials, ranging from metals to semiconductors, and from organic to inorganic materials [1-9].

Considering the stochastic nature of OASE at micro- to nano-scopic scales, the spontaneous formation of patterns by this technique is a highly non-trivial process, and has long drawn attention due to its potential applications in surface technology and applied sciences [1-4].

In this regard, a versatile experimental facility 200 kV Ion Accelerator with ion beams having energy in the range of 30–200 keV is running successfully at Ion Beam Centre, KUK for carry out patterning and structuring the surfaces of different classes of materials. This system offers single charge state, switching magnet with five exit ports and large area irradiation/implantation using hollow cathode ion source [7].

For binary compounds, non-stoichiometric sputtering of one element in comparison to other occurs. This preferential sputtering determines the degree of patterning and structuring by oblique angle sputter erosion. Hence, it becomes extremely vital to have detailed knowledge of the composition modulation as a result of non-stoichiometric sputtering [8-9].

In this paper, we address the study of the temporal evolution of SiC surfaces with the aim of characterizing the features as a function of ion dose. This type of quantitative and qualitative analysis is required for designing large-scale ordered nanostructures on the sputtered surfaces.

2. Materials and methods

Silicon Carbide (SiC) thin films were deposited on Si(111) substrates employing RF sputtering method. These SiC thin films were then irradiated with 80 keV argon ions by 200 kV Ion Accelerator Facility available at Ion Beam Center, Kurukshetra University, Kurukshetra, India under a vacuum of 3.2 x 10^{-7} Torr. Argon ion fluence of 1×10^{18}, 2×10^{18} and 3×10^{18} ions cm^{-2} has been used at fixed incident angle of 50^0 [7].

Surface morphological evolution of these RF sputtered and argon irradiated SiC surfaces was profoundly evaluated by Atomic Force Microscopy (AFM) utilizing Bruker HR Mutlimode-8 available at Ion Beam Centre, Kurukshetra University, Kurukshetra. In order to get a better statistics, at least 5 AFM scans on each sample were performed.

3. Results and discussion

Figure 1 ((a)-(d)) shows the AFM images of RF sputtered SiC over Si(111) and SiC surfaces bombarded using 80 keV Ar^+ ions at fixed incident angle of 50^0 with fluences of 1×10^{18}, 2×10^{18} and 3×10^{18} ions cm^{-2} respectively. The corresponding Fast Fourier Transforms (FFT) is shown in insets of these micrographs.

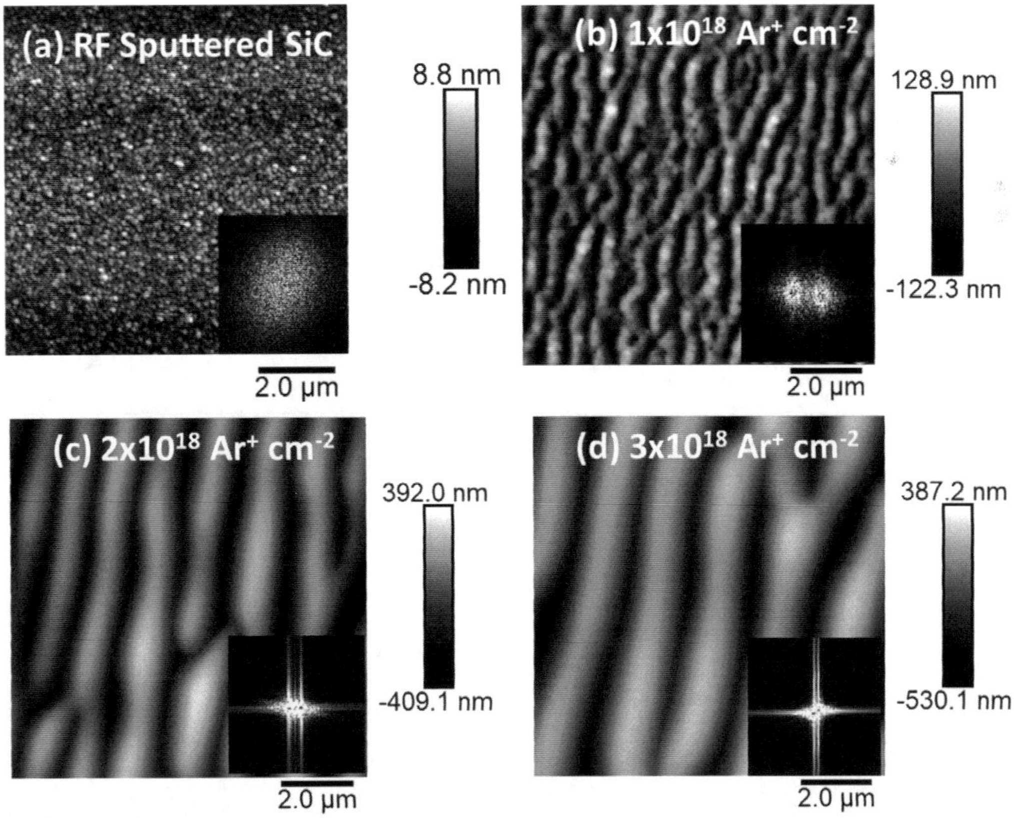

Figure 1.(a-d) AFM micrographs of as-deposited SiC thin films and Ar^+ sputtered SiC surfaces at various fluences of 1×10^{18}, 2×10^{18} and 3×10^{18} ions cm^{-2}.

Figure 1(a) reveals that globular type particles of SiC can be easily distinguished. These particles are uniform in size, with a lateral dimension of about 30-40 nm. FFT image depicts the random nature of the SiC particles after RF sputtering.

The temporal parameters i.e. wavelength and amplitude of evolved surface structures are calculated using the Nanoscope 1.8 software [8] and are plotted in figures 2 and 3, respectively.

For argon ion fluence of 1×10^{18} $Ar^{+}cm^{-2}$, periodic ripples of smaller dimensions have been observed (figure 1(b)). Ripples having wavelength of 588.4±0.87 nm are seen on the sample surface due to the ripening and coalescence (figure 2). From the FFT, we can see that these surface structures display square self-linking and homogeneity.

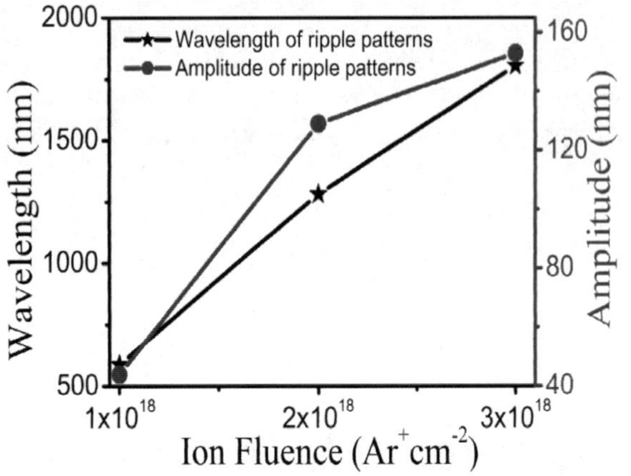

Figure 2. wavelength and amplitude of surface nano-structures over SiC exposed to 80 keV Ar^{+} ions at 50^0 oblique incidence for different fluences.

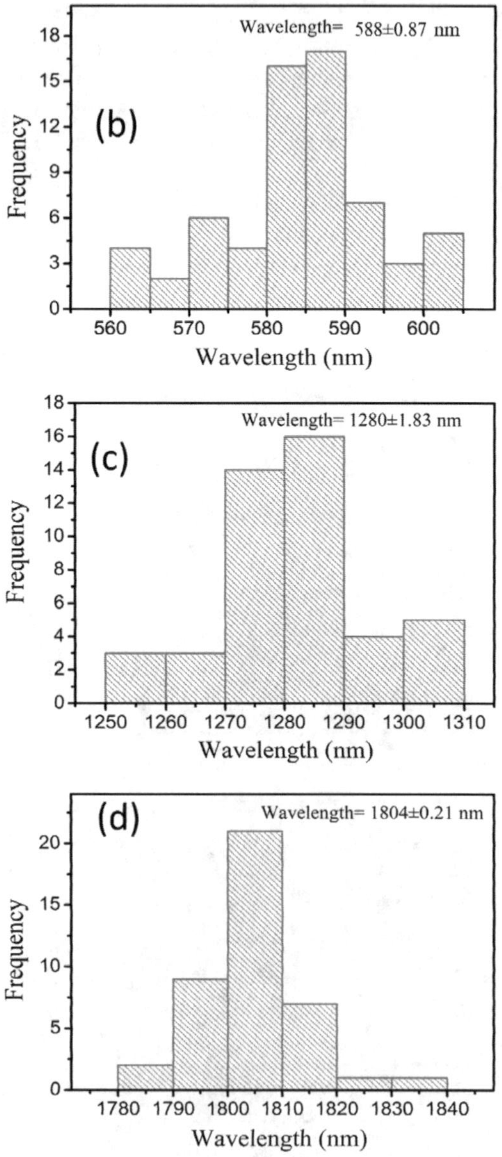

Figure 3. wavelength distribution of surface nano-structures over SiC exposed to 80 keV Ar^{+} ions at 50^0 oblique incidence.

Notably as the sputtering proceed to ion fluence of 2×10^{18} $Ar^{+}cm^{-2}$ (figure 1(c)), wavelength of these ripple pattern increases to 1280±1.83 nm. Close inspection of the FFT reveals that 2 fold symmetry increases significantly in comparison to earlier stage of sputtering. Moreover, the habit of these evolved patterns is sinusoidal.

Further increase in ion fluence to 3×10^{18} $Ar^{+}cm^{-2}$ leads to ordered ripple patterns with pronounced increase in wavelength and amplitude. At this stage of sputtering, the wavelength and amplitude is 1804±0.21 nm and 153 nm respectively as plotted in figure 2. Interestingly, the perfect square ordering among the evolved ripples can be easily inferred from the corresponding FFT image as shown in inset of figure 1(d). The variation of wavelength and amplitude of ripple patterns is shown in figure 2. Homogenous distribution of the wavelength observed for the sputtered SiC surfaces is plotted in the figure 3.

It can be inferred from figures 2 and 3 that both wavelength and amplitude of the ripple pattern increases with increase in argon ion fluence. Thus, SiC surfaces display ripple patterns with temporal parameters such as wavelength, self-assemble, spatial ordering under oblique angle sputtering. These temporal parameters exhibit dependence on the fluence of the bombarding ions.

This peculiar behaviour can be explained in terms of Sigmund's theory of sputtering for binary compounds [5]. The sputtering yield of silicon will be more in comparison to carbon due to significant difference in mass of both the elements in binary compound i.e. SiC. Thus, for all argon ion fluence, sputtered SiC surfaces will be rich in carbon because of preferential erosion of Si atoms from the surface region [2, 10-11].

4. Conclusion

The process of 80 keV argon ion irradiation induced sputtering at off-normal incidence of 50^0 is observed to be a promising tool for large area surface patterning of SiC. Ripple topography is developed on the sputtered surfaces, which becomes ordered with the increase in ion fluence. The specific role of sputtering time i.e. ion-fluence on argon sputtered SiC has been discussed. The formation of such surface structures is attributed to the preferential sputtering of silicon in comparison to carbon.

These structured SiC thin films; fabricated by oblique incidence of ions from particle accelerators; have been preferred for optoelectronics, MEMS, RF MEMS and TFT's due to relatively their high temperature resistance and slow degradation, which guarantees a larger compatibility with silicon-based technology.

Acknowledgements

The work has been supported by the Department of Science and Technology (DST) by funding major research project for utilizing 200 kV Ion Accelerator and related characterization facilities. Authors are highly grateful to the MHRD for RUSA 2.0 grants for Centre for Advanced Material Research (CAMR) to Kurukshetra University.

References

[1] Cuerno R and Kim J S 2020 *J. Appl. Phys.* **128** 180902
[2] Gupta D, Umapathy G R, Singhal R, Ojha S and Aggarwal S 2022 *Mater. Lett* **307** 131011
[3] Valbusa U, Boragno C and Mongeot B F 2002 *J. Phys.: Condens. Matter.* **14** 8153
[4] Gupta D *et al.* 2019 *Nanotechnology* **30** 385301
[5] Sigmund P 1981 Sputtering by ion bombardment: Theoretical concepts *Sputtering by Particle Bombardment I, Topics in Applied Physics* ed R Behrisch (Berlin: Springer-Verlag) pp 9-71
[6] Bradley R M and Harper J M E 1988 *J. Vac. Sci. Technol.* A **6** 2390
[7] Gupta D, Aggarwal S, Sharma A, Kumar S and Chopra S 2022 *Mater. Lett* **308** 131294
[8] `https://www.bruker.com/en/products-and-solutions/microscopes/materials-afm/multimode-8-hr-afm.html`
[9] Bradley R M and Shipman P D 2010 *Phys. Rev. Lett.* **105** 145501

[10] Gupta D, Chawla M, Singhal R and Aggarwal S 2019 *Sci. Rep* **9** 15531
[11] Zaporozchenko V I and Stepanova M G 1995 *Prog. Surf. Sci.* **49** 155

Journal of Physics: Conference Series **2420** (2023) 012109 doi:10.1088/1742-6596/2420/1/012109

Energy comparison of room temperature and superconducting synchrotrons for hadron therapy

G Bisoffi[1,2], E Benedetto[1,3], M Karppinen[1], M G Pullia[4], M Reza Khalvati[1], L Rossi[5], M Sapinski[6], M Sorbi[5], U Valente[7], R van Weelderen[1], G Venchi[4] and M Vretenar[1]

[1] CERN, Geneva, Switzerland
[2] INFN-LNL, Legnaro, Italy
[3] SEEIIST Association, Geneva, Switzerland
[4] CNAO Foundation, Pavia. Italy
[5] Universita' degli Studi di Milano and INFN, Milano, Italy
[6] PSI, Villigen, Switzerland
[7] Università La Sapienza, Rome, Italy

E-mail: giovanni.bisoffi@lnl.infn.it

Abstract. The yearly energy requirements of room temperature (RT) and superconducting (SC) magnet options of a new hadron therapy (HT) facility are compared. Special reference is made to the layouts considered for the proposed SEEIIST facility. Benchmarking with the RT CNAO HT centre in Pavia (Italy) was carried out. The energy comparison is centred on the different synchrotron solutions, assuming the same injector and lines in the designs. The beam current is 20 times higher than in present generation facilities: this allows efficient multi-energy extraction (MEE), which shortens the therapy treatment and is needed especially in the SC option, because of the slow magnet ramping time. Hence, power values of the facility in the traditional mode were converted into MEE ones, for a fair comparison between RT and SC magnets. Cryocoolers (c.c.) and a liquefier are also compared, for synchrotron refrigeration. This study shows that a RT facility in MEE mode requires the least average energy, followed by the SC synchrotron solution with a liquefier, while the most energy intensive solution is the SC one with c.c.

1. Introduction

Radiotherapy is a fundamental component of effective cancer treatment. Radiation therapy with protons and other ions, also called hadron therapy (HT), offers several advantages over the classical RT with X-rays. Nowadays, HT has reached the time of transitioning from a limited number of specialized institutions to many particle-therapy centres worldwide. In 2018, the conceptual study of a new ion therapy and research facility in Southeast Europe was proposed [1]. It consists in an advanced version of the PIMMS design [2], accelerating beams of p (up to 250 MeV), $^{4}He^{2+}$, $^{12}C^{6+}$ and heavier nuclei (up to 430 MeV/u), with 20 times more beam current than present generation machines. It will work in MEE mode [3], with the possibility of a single-turn extraction (10^{10} ions in ~1 μs) to perform FLASH therapy [4]. For the SEEIST facility, the gantry is assumed to be superconducting (SC) [5], while for the synchrotron both the room temperature (RT) and superconducting (SC) versions are studied. A SC synchrotron is certainly more compact than a RT one (see figure 1 and table 1). This paper provides a first evaluation of the average power required by the whole SEEIIST facility, comparing the RT and SC synchrotron cases.

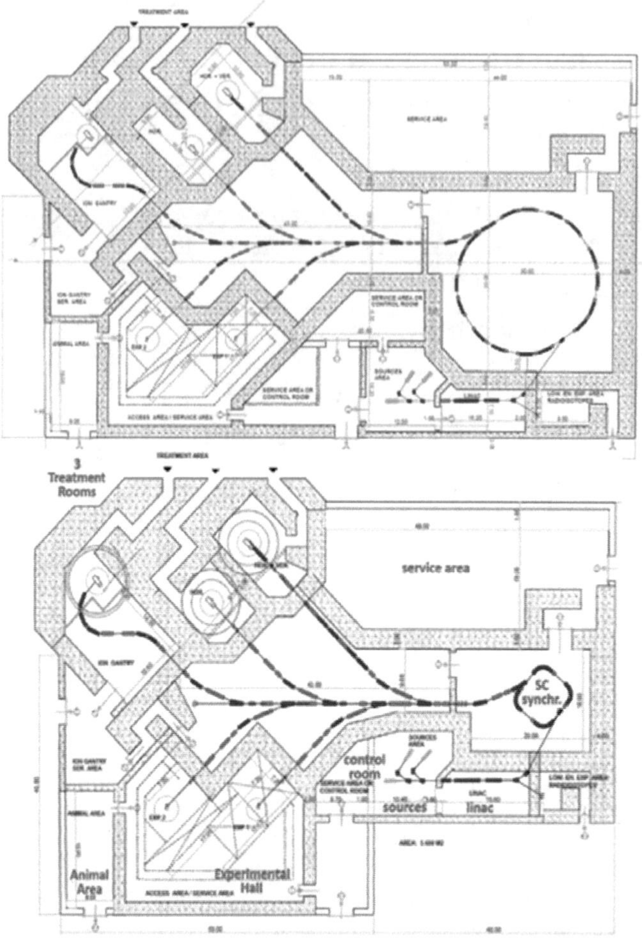

Figure 1. Layout of the SEEIST RT facility (16 dipole synchrotron, left) and the SC one (4 dipoles).

Table 1. Main specifications of the RT and SC synchrotrons (for $^{12}C^{6+}$).

	RT synchrotron	**SC synchrotron**
Energy	430 MeV/u	430 MeV/u
Circumference	70 m	27 m
$(B\rho)_{max}$	6.62 Tm	6.62 Tm
B_{max}	1.5 T	3.5 T
ρ	4.23 m	1.89 m
Ramp rate	2.4 T/s	1 T/s

The two options, differing by the synchrotron layout, have in common the RT linac injector, injection and extraction systems, the beam lines and the SC gantry.

2. The CNAO facility as benchmark

The calculation of the energy consumption of any accelerator facility is based on the electric characteristics of the most energy-intensive components, as well as on the setting corresponding to their average power and their operational duty cycle. The former can be derived from the machine specs, while the latter require explicit hypotheses on their realistic operation modes. The CNAO facility [6]. in Pavia ($E_{12\text{-}C}$ = 400 MeV/u) has technical specifications which are largely like the SEEIIST RT option: to benchmark the calculation method, we used them to calculate the consumption of the whole facility and compared a qualified portion of it with the measured energy supply at the technological station.

The CNAO injector, based on an RFQ and an IH-DTL (E_{fin} = 7 MeV/u), is kept on at a RF d.c. = 0.5 % for 93 % of the year (off only during long maintenance periods). Averaging the consumption over 1 year delivers an yearly-averaged power $P_{av,yr}$~66 kW. Waiting for an upgrade linac design to be completed, this value has been used for the RT and SC SEEIST layouts too.

The CNAO synchrotron, because of the beam current limitations from the injector, was not designed to be operated in the MEE mode: its magnets are ramped up and down for each of the energy values required by the treatment plan. Table 2 shows the synchrotron elements, with the electrical specifications required to calculate $P_{av,yr}$. $P_{max\text{-}T,all}$ values (electric power at the maximum current) are corrected upwards to include the power converter (p.c.) efficiency ($P_{maxT,all,plug}$), and downwards, considering the average value of the electrical power, consistent with the actual energy value at which the machine is operated ($P_{typ\text{-}T,all}$, see next paragraph). This value is further reduced by the use fraction of the machine for protons (10.8 %) and carbon ions (13.4 %) and by the cycle time fraction in which the machine is at the flattop. Overall, $P_{av,yr}$~90 kW for the synchrotron. Powering the extraction septa in pulsed mode has been proposed for the next machines [7].

With the same method adopted for the synchrotron, $P_{av,yr}$ for the three beam lines (2 horizontal, 1 horizontal+vertical) was calculated too, knowing that the field in the magnets does not cycle, but changes in steps – at the end of each machine cycle – according to the change in the beam energy. $P_{av,yr}$ ~90 kW for the beam lines, where the large vertical dipole counts for 1/3 of the total. Finally, consumption of the p.c. in idle state, vacuum pumps and beam instrumentations are added ($P_{av,yr}$~47 kW in total), with 30 % of the total for the cooling power. $P_{av,yr}$~346 kW is the total for the CNAO facility.

As to the operational data, those data provided for one year of operation at CNAO were assumed as a general reference in the energy comparison of the two SEEIIST layouts considered. Figure 2 shows the distribution of the beam energies, for p and ^{12}C, in one year at CNAO.

It is to be noted (as shown in table 2) that the effective total use of the synchrotron was only 24.2 %, a value confirmed experimentally and checked for consistency with another therapy centre.

Considering the increase of the maximum ^{12}C energy, the CNAO-statistics in the use of the facility (figure 2) has been applied to the RT and SC layouts (see next paragraph).

The electric cabin instrumentation can supply an approximate measurement of the $P_{av,yr}$ contributed by the time-variable part of the accelerator (~235±20 kW). The corresponding value, calculated as done above from the electrical specs and the duty cycle, is 193 kW. Given the approximation, this comparison confirms the adopted calculation approach rather well.

3. The SEEIIST RT facility

The RT SEEIIST facility, versus the CNAO one, offers: $E_{12\text{-}C,max}$= 430 instead of 400 MeV/u; 20 times higher beam current and MEE operation; larger fraction (65 vs 40 %) of the time planned for experimental tests; 5 beam lines, 1 of which equipped with a SC gantry.

The slightly larger final energy can be achieved by the same p.c. of CNAO, of which it can keep the electrical specifications, increasing the p.c. output current. The beam lines, realised with more modern p.c., have better specifications as to energy consumption. The MEE mode shortens the treatment time by eliminating large fractions of the dead time between successive energy steps, but keeps the synchrotron magnets at full field during the ~ 200 ms energy changes, delivering a modest 14 % $P_{av,yr}$ reduction.

Table 2. Calculation of the CNAO synchrotron average power from electrical specs and duty cycle.

	Synchrotron elements	N	R [Ω]	$I_{max\text{-}T}$ [A]	$P_{max\text{-}T,1}$ [kW]	$P_{max\text{-}T,all}$ [kW]	$\varepsilon_{p.c.}$	$P_{max\text{-}T,plug}$ [kW]	$P_{typ\text{-}T}$ [kW]	Yearly d.c.	CW-equiv. power	$P_{av,yr}$ [kW]
^{12}C	Dipoles (17 in 1 series)	1	0,08	2778	617,38	617,38	0,70	882,0	541,91	13,4%	47,2%	34,3
	Qpoles (24 in 3 families)	3	0,166	540	48,41	145,22	0,89	163,2	100,25	13,4%	47,2%	6,3
	Sext. 2x2 familes	2	0,067	500	16,75	33,50	0,90	37,2	22,87	13,4%	47,2%	1,4
	Resonance sextupole	1	0,039	500	9,75	9,75	0,70	13,9	8,56	13,4%	47,2%	0,5
	Inj. m. Septa (series of 2)	1	0,0044	3889	66,54	66,54	0,58	114,7	114,73	13,4%	100%	15,4
	Extr.m.s. type-1 (series of 3)	1	0,014	3479	169,45	169,45	0,77	220,1	135,21	13,4%	100%	18,1
p	Dipoles (17 in 1 series)	1	0,08	2778	617,38	617,38	0,70	882,0	57,41	10,8%	49,8%	3,1
	Qpoles (24 in 3 families)	3	0,166	540	48,41	145,22	0,89	163,2	10,62	10,8%	49,8%	0,6
	Sext. 2x2 familes	2	0,067	500	16,75	33,50	0,90	37,2	2,42	10,8%	49,8%	0,1
	Resonance sextupole	1	0,039	500	9,75	9,75	0,70	13,9	0,91	10,8%	49,8%	0,0
	Inj. m. Septa (series of 2)	1	0,0044	1745	13,40	13,40	0,58	23,1	23,10	10,8%	100%	2,5
	Extr.m.s. type-1 (series of 3)	1	0,014	1417	28,11	28,11	0,77	36,5	2,38	10,8%	100%	0,3
Others	Dipole washout for p											6,79
	Dipole washout for 12C											0,56
	RF cavity (Medaustron)	1				4,0	0,75	5,3	5,5%			0,30
TOTAL (Synchrotron)												90,3

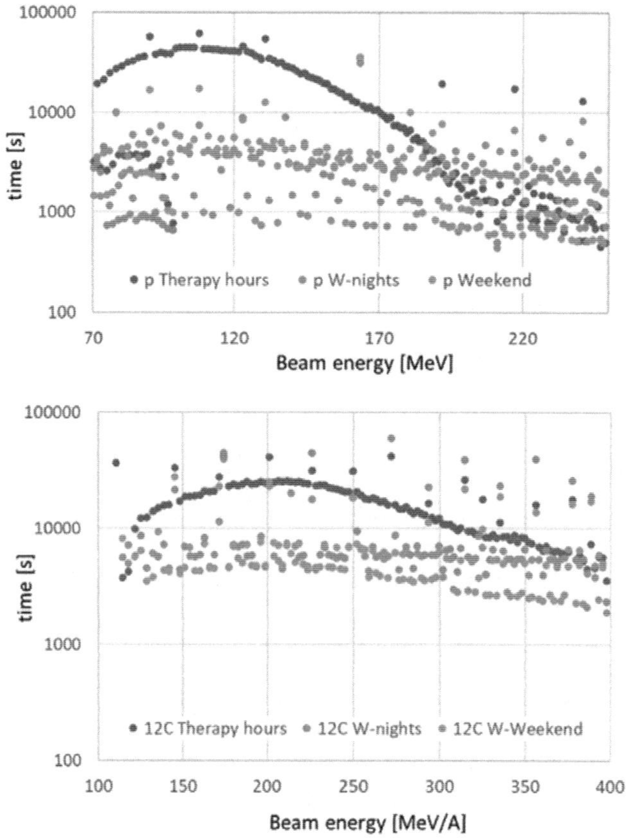

Figure 2. Time distribution vs. E_{beam} of proton and carbon beams during typical year 2020 of operation at CNAO.

The SC gantry [8], with 5 SC combined function dipoles (3 T field for the dipole and 3,5 T/m quadrupole gradient), 3 SC quadrupoles (40 T/m) and 7 room-T quadrupoles (25 T/m), features overall transient losses of 2 W/m at 4.2 K of the dipoles (7 m total length), for a total of 14 W. To this the contribution of the SC quadrupoles (4 W) must be added, together with the static consumption of the cryostats (2 W, with 0.3 W/m) and the contribution of the current leads (2 W, at ~0.1 W/kA). The total power of 22 W at 4.2 K can be removed by 11 double-stage c.c., consuming 7 kW at the plug each, delivering $P_{c.c.}$~77 kW in total, to which the $P_{av,yr}$ of the RT magnets (4 kW) must be added. In conclusion $P_{av,yr}$ ~81 kW for the gantry.

Calculations similar to those done for CNAO, applied to the RT version of the SEEIST facility of figure 1, deliver a total $P_{av,yr}$ ~379 kW.

4. The SEEIST SC facility

The SC option differs from the RT one in the synchrotron. Energy consumption occurs in the field ramps and is virtually zero in the plateaus. With MEE, it is necessary to sum the contributions of each individual field from the minimum field to the maximum magnetic field and backwards. A reference for the energy consumption of ramped SC magnets is the prototype realized in the DISCORAP experiment at INFN [9] (3.9 m long dipole, B_{max}=4.5 T, $\frac{dB}{dt}$=1 T/s). The largest contribution (two thirds) to the losses is the persistent current in the superconductor $P_{sp}=M_{pc}\dot{B}$, where M_{pc} is the magnetization. Interpolating the pc vs B in the conductor with a power function (in figure 3 the graph and the displayed equation), allows to integrate it between the field extremes 0.4 and 3.5 T:

$$U_{sp,ramp} = \int_{tmin}^{tmax} P_{sp} dt = \int_{t_{min}}^{t_{max}} P_{sp} \mathrm{dB} \left(\frac{dB}{dt}\right)^{-1}$$

A full ramp for C ions, covering the 0.4÷3.5 T range, requires $U_{sp,ramp}$~13 J/m at 4.2 K. The contribution of the coil ends, and of all other losses in %, were then added.

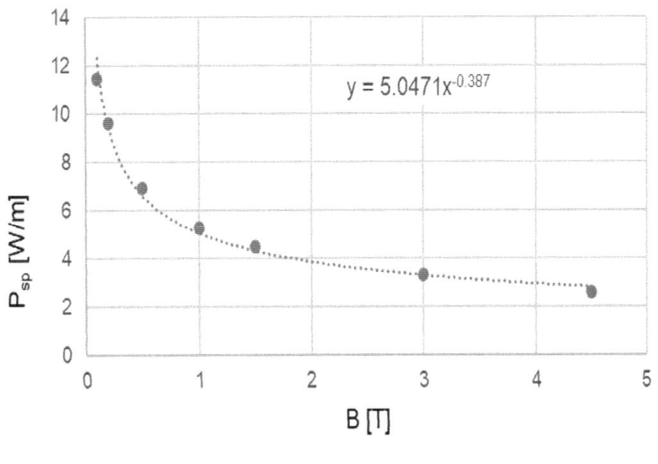

Figure 3. Power-per-meter vs. B (T) at 4.5 K required by the straight section of the DISCORAP magnet.

Knowing the energy of all treatments administered to patients in 2020, the corresponding SC dipole field and the energy required at 4.2 K is obtained, i.e. ~1 W. The static load of the cryostat (3.6 W), gives $P_{av,yr}$ (4.2 K) ~4.6 W.

Table 3. Yearly average power (in kW) of the 3 options.

Options	RT	SC (c.c.)	SC (refr.)
Linac	68,4	68,4	68,4
Synchrotron	76,0	192,0	97,0
Beam Lines	14,4	14,0	14,0
Others	80,3	80,3	80,3
SC Gantry	47,4	47,4	47,4
TOTAL	**286,3**	**402,1**	**307,1**
Cooling	85,9	120,3	91,8
Grand total	**372,2**	**522,4**	**398,9**

This would be the reference number if the dissipation were constant in time. However, dissipation is rather concentrated in the ramp-down to ramp-up periods of the dipoles, from one MEE to the next. It is mandatory not to exceed the critical temperature (T_C) of the magnet in those time spans. From CNAO statistics, the most critical case are the short ramps (achieving smaller B_{max} values, on the more dissipative part of figure 3), followed by short quasi-flattops for MEE. For each dipole, this corresponds to P = 14 W for a period dt = 2.2 s, with additional 0.15 W loss in the subsequent 50 s, before the next MEE starts. With c.c. able to remove 2 W each from a Cu-mass (c_p = 0.1 J/kgK being the Cu specific heat at 4.2 K), estimated in m = 400 kg, with ΔT = 0.3 K, to stay at least 1K away from T_C (5.5 K at 3.5 T), the number of needed c.c. (N_{cc}) is derived from

$$[P - N_{cc}(2\ W)] \leq mc_p\Delta T$$

giving N_{cc} = 5 per dipole. With 7 kW/cc, the total plug power for the SC dipoles would be 140 kW, a practical (ease of maintenance) but inefficient solution. As alternative, a commercial refrigerator [10] – of more demanding maintenance - may provide 130 W refrigeration power at 4 K (with liquid nitrogen precooling) with a plug power of 45 kW only.

For the SC solution, the energy contributions of the p.c., warm cables, busbars, etc. must be added, together with all the RT elements of a typical, albeit smaller, synchrotron.

$P_{av,yr}$ (SC facility) ~522 kW with c.c. and ~400 kW with a refrigerator. Last, it must be noted that future SC septa [11], would be cooled by the refrigerator, together with the magnets at no additional power (36 kW less in $P_{av,yr}$).

5. Conclusion

After benchmarking the calculation method on CNAO data, we calculated the $P_{av,yr}$ of a fully RT facility and a facility where the synchrotron is SC. The results are summarised in table 3. We note: that the overall average power is not huge ($P_{av,yr}$ ~0.5 MW), the RT solution being the one with the lower and the SC one (with c.c.) the one with the higher electricity bill. On the other hand, at equal energy consumption (RT and SC with liquefier being very close) the SC solution requires a smaller installed power.

References

[1] Amaldi U *et al.* 2021 *Front. Phys.* **8** 567466
[2] Bryant P J *et al.* 1999 Proton-Ion Medical Machine Study (PIMMS) Part 1 Tech. rep. CERN-PS-99-010-DI
[3] Iwata Y *et al.* 2010 *Proc. 1st Int. Particle Accelerators Conf. (IPAC'10)* (Kyoto, Japan) pp 79-81
[4] Norman H X Q *et al.* 2021 *Proc. 12th Int. Particle Accelerators Conf. (IPAC'21)* (Campinas, Brazil) pp 2465-68

[5] Iwata Y *et al.* 2018 *Proc. 9th Int. Particle Accelerators Conf. (IPAC'18)* (Vancouver, Canada) pp 1232-36
[6] Rossi S *et al.* 2011 Eur. Phys. J. Plus **126** 78
[7] Vretenar M *et al.* 2022 *Proc. 13th Int. Particle Accelerators Conf. (IPAC'22)* (Bangkok, Thailand) pp 811-14
[8] Amaldi U *et al.* 2021 SIGRUM – A superconducting ion gantry with Riboni's unconventional mechanics Tech. rep. CERN-ACC-NOTE-2021-0014 CERN-NIMMS-Note-002
[9] Volpini G *et al.* 2014 IEEE Trans. Appl. Supercond. **24** pp 1-5
[10] Refrigerator Datasheet, `https://www.linde-kryotechnik.ch/wp-content/uploads/2016/10/Datasheet_L70-LR70-D.pdf`
[11] Barna D and Novák M 2020 *Nucl. Instrum. Methods Phys. Res., Sect.* A **959** 163521

Study on construction of an additional beamline for a compact neutron source using a 30 MeV proton cyclotron

Y Kuriyama, M Hino, Y Iwashita, R Nakamura and H Tanaka

Institute for Integrated Radiation and Nuclear Science, Kyoto University, Kumatori, Japan

E-mail: kuriyama.yasutoshi.6n@kyoto-u.ac.jp

Abstract. The Institute for Integrated Radiation and Nuclear Science, Kyoto University (KURNS) has been actively using neutrons extracted from the research reactor (KUR) for collaborative research. Since the operation of KUR is scheduled to be terminated in 2026 according to the current reactor operation plan, the development of a general-purpose neutron source using the 30 MeV proton cyclotron (HM-30) installed at KURNS for Boron Neutron Capture Therapy (BNCT) research has been discussed as an alternative neutron source. In this presentation, we report on the conceptual design of an additional beamline for a compact neutron source using this cyclotron.

1. Introduction

At the Institute for Integrated Radiation and Nuclear Science, Kyoto University (KURNS), a medium-sized neutron source, neutrons extracted from a research reactor (KUR) with a maximum thermal power of 5 MW has been actively used for joint use. However, due to the deadline for the return of spent fuel, the KUR is not expected to continue operation after 2026, so there is an urgent need to construct a neutron source to replace the KUR.

Since 2007, Kyoto University has been developing a Boron Neutron Capture Therapy (BNCT) system using a proton cyclotron (HM-30), an average beam current of 1 mA and extraction beam energy of 30 MeV [1]. This system consists of HM-30 and a target station that moderates neutrons generated from Be targets to an energy suitable for BNCT [2, 3].

In 2019, HM-30 was transferred to Kyoto University, therefore, the consideration to use HM-30 as a proton accelerator for the KUR alternative neutron source had been started. Figure 1 shows KUR, Kyoto University Critical Assembly (KUCA), and Innovation Research Laboratory (Innovelab).

2. Accelerators installed in Innovelab at KURNS

Innovelab at KURNS is equipped with four ring accelerators and one linear accelerator. Among these accelerators, the 150MeV FFAG main-ring (MR) and its injector [4], the 11 MeV linear accelerator (linac) and HM-30 are currently in operation.

The Innovelab is divided into a medical building and an experimental building. HM-30 is located on the first floor of the medical building, and MR and linac are on the first floor of the experimental building. Figure 2 shows the layout of the accelerators in the Innovelab.

Figure 1. Appearance of KUR, KUCA and Innovelab in KURNS.

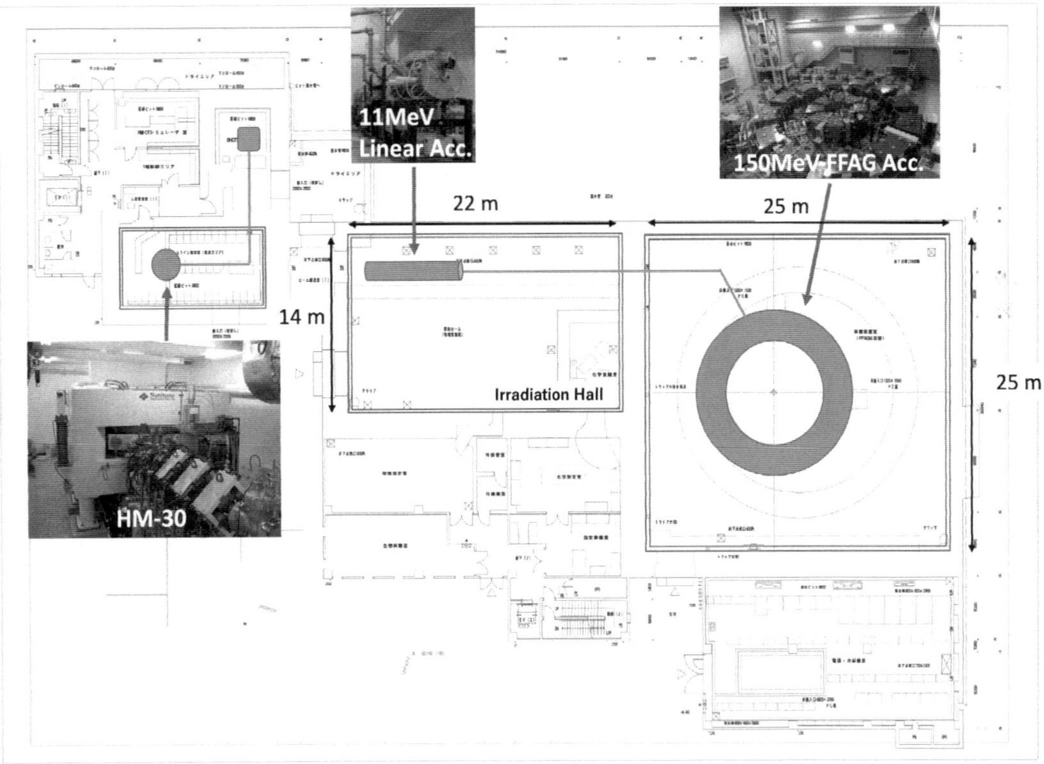

Figure 2. Layout of accelerators in Innovelab building at KURNS.

To make a room for a neutron generation target station in the irradiation hall (left-side of the experimental building), the linac has to move to the MR room (rigth side of the experimental building).

3. 30 MeV proton cyclotron (HM-30) at KURNS

30 MeV proton cyclotron (HM-30), manufactured by Sumitomo Heavy Industories, had been installed for BNCT research in Innovbelab medical building in 2008. Figure 3 shows HM-30, and table 1 shows the main specifications of the HM-30 [1].

Figure 3. HM-30 in Innovelab.

Table 1. Specifications of HM-30.

Particle	H^-
Injection Energy	30 keV
Extraction Energy	30 MeV
RF Frequency	73.1 MHz (h=4)
RF Voltage	200 kV/turn
Extraction Scheme	Foil
Extraction Beam Current	1 mA (DC)
Machine Size	W×D×H:3×1.6×1.7 m

For BNCT research, a 30 kW proton beam is extracted from the HM-30 and transported to the target station by 90-degree beam transport, as shown in figure 2.

Journal of Physics: Conference Series **2420** (2023) 012110 doi:10.1088/1742-6596/2420/1/012110

4. Additional beam transport for general purpose accelerator based neutron source

The target station for the general purpose accelerator based neutron source is being considered for installation in the irradiation hall in the Innovelab experimental building. Figure 4 shows the additional beam transport (BT2) under consideration in blue. The distance from the HM-30 extraction point to the neutron producing target is roughly 15∼20 m.

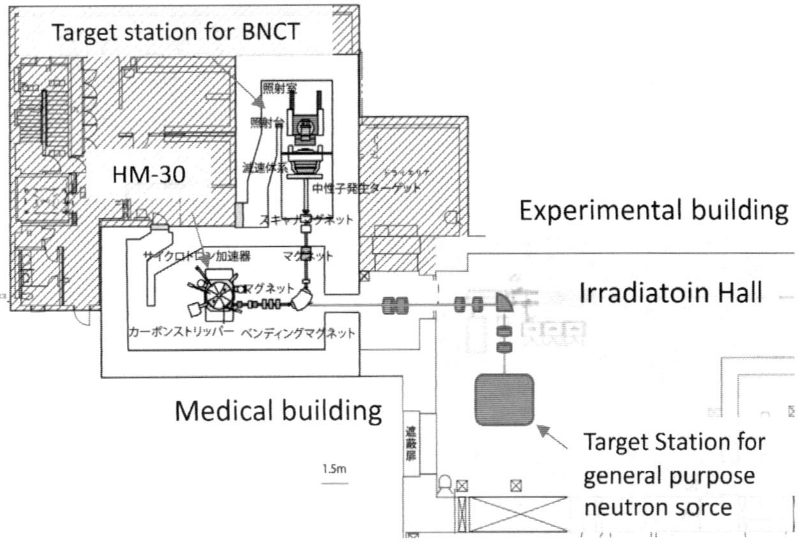

Figure 4. Additional beam transport for accelerator based neutron source. Blue objects indicate component of additional beam transport.

4.1. Construction of a calculation system using G4beamline

In order to perform beam transport calculations for the BT2 design, a calculation system using G4beamline (g4bl) [5] was developed. Figure 5 shows a 3D diagram of the calculation system constructed by g4bl.

The beam transport calculations for the HM-30 extraction beam are tentatively performed assuming gaussian distribution and equation (1) as the initial beam parameters.

$$\begin{aligned} \sigma_x &= \sigma_y = 3\text{mm} \\ \sigma_{x'} &= \sigma_y^{'} = 1\text{mrad} \\ \Delta P &= 1\% \end{aligned} \tag{1}$$

Figure 6 shows the initial beam size and the beam size calculated by g4bl at the neutron generation target located about 19 m from the HM-30 extraction position.

Figure 7 shows the beam size along beam transport, starting from the HM-30 extraction position.

The inner diameter of the beam pipe to be used for the additional beam transport will be 100 mm in diameter. Therefore, the beam size during beam transport must be kept under ±50 mm. Taking into account the heat load on the target, the beam size should be about the same as the target size.

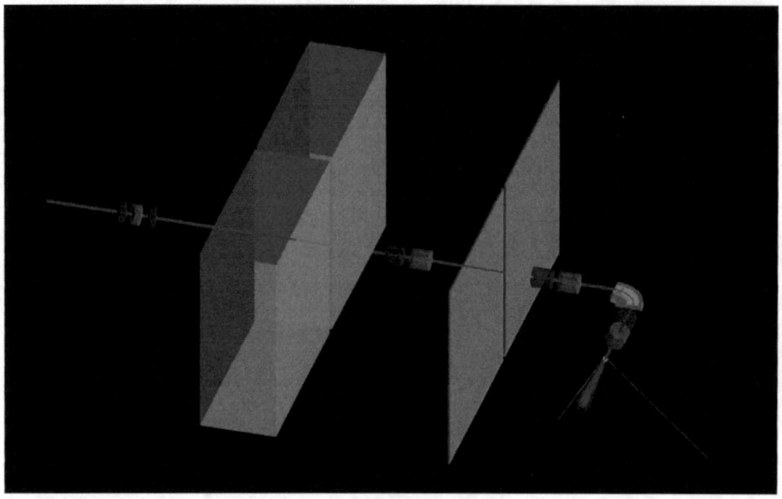

Figure 5. Computation geometry of additional beam transport for accelerator drive neutron source using G4beamline. Green object indicates bending magnet and blue and red objects indicate focus and defocus quadruople magnets, respectively. Be target is located at the endpoint.

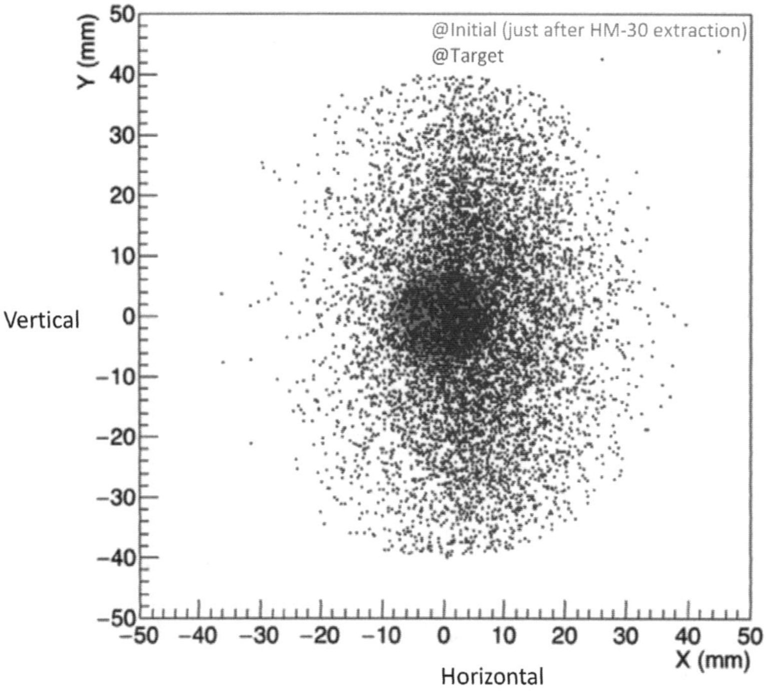

Figure 6. The comparison at initial condition (red dots) and after transport (blue dots) based on beam size using G4beamline.

5. Estimation of neutron generation by PHITS

5.1. Neutron generation targets

From the BNCT study, a 5.5 mm thick Be target is planned to be used as the neutron generation target. 5.5 mm thickness allows the proton range to be longer than the target thickness, so that

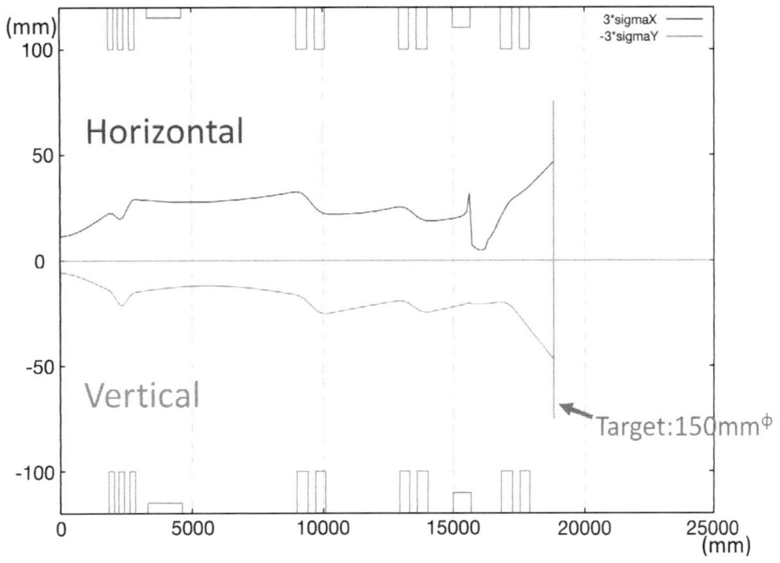

Figure 7. Beam tracking result of the additional beam transport (BT2) for general purpose accelerator neutron source using G4beamline.

the incident protons do not stop in the target but are injected into the cooling water. This will suppress the occurrence of blistering phenomena.

5.2. Estimation of neutron generation

PHITS (ver. 3.04) [6] was used to estimate the amount of neutrons produced when a uniform circular 30 MeV proton beam of 120 mm diameter is injected into a 5.5 mm thick Be target. The results of the PHITS calculations are shown in figure 8.

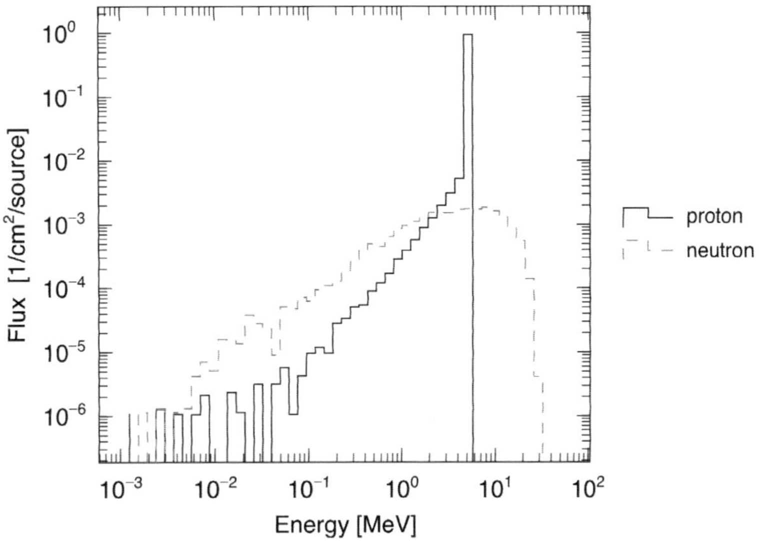

Figure 8. Calculated neutron and proton energy distribution from Be target using PHITS.

Based on the PHITS calculations, assuming an average beam current of 1 mA, the neutrons produced are expected to be 1.6E-2 neutrons/proton and the total neutrons produced are expected to be 1E14 neutrons/s.

6. Summary and future plans

Although the KUR has been actively used for joint use at KURNS, it is expected to be difficult to continue the operation of the KUR after 2026 due to the deadline for the return of spent fuel. Therefore, the development of a general-purpose neutron source using HM-30 has been discussed as an alternative neutron source. The conceptual design of the additional beam transport from HM-30 to the general purpose neutron generation target station has been carried out using g4bl. In addition, calculations using PHITS are being performed to estimate the neutron generation. Currently, we are at the stage of evaluation with tentative parameters, but we plan to perform beam transport calculations with more realistic beam parameters.

References

[1] Mitsumoto T *et al.* 2010 *Proc. 19th Int. Conf. on Cyclotrons and Their Applications (Cyclotrons'10)* (Lanzhou, China) pp 430–32
[2] Tanaka H *et al.* 2011 *Appl. Radiat. Isot.* **69** pp 1642–-45
[3] Tanaka H *et al.* 2009 *Nucl. Instrum. Methods Phys. Res., Sect.* B **267** pp 1970–77
[4] Mori Y *et al.* 2011 *Proc. 2nd Int. Particle Accelerator Conf. (IPAC'11)* (San Sebastian, Spain) pp 2685–87
[5] G4beamline `http://www.muonsinternal.com/muons3/G4beamline`
[6] Sato T *et al.* 2013 *J. Nucl. Sci. Technol.* **50** pp 913–23

Design of beam collection system with high transmission efficiency and tunablity

Y Yan[1,2], T Yang[1,2], D Y Li[1,2], H Cheng[1,2], Y Z Li[1,2], M J Wu[1,2], Y L Fang[1,2], F N Li[1,2], Z Guo[1,2], Z X Yuan[1,2,3], K Zhu[1,2,3], C Lin[*1,2,3] and X Q Yan[1,2,3]

[1]State Key Laboratory of Nuclear Physics and Technology, Peking University, Beijing, China
[2]Beijing Laser Acceleration Innovation Center, Huairou, Beijing, China
[3]Institute of Guangdong Laser Plasma Techology, Baiyun, Guangzhou, China

E-mail: lc0812@pku.edu.cn

Abstract. A beamline is required to focus proton beams produced by laser plasma accelerator to high charge density spots for practical applications. The Compact LAser Plasma Accelerator (CLAPA) coupled with an image-relaying beamline has been built at Peking University. While the collection section of this beamline is an electromagnetic quadrupoles (EMQs) triplet with certain aperture, more than 71 % of the protons are lost in the inlet end face or vacuum tube wall of triplet due to their divergence angles. Here we show the development of a high transmission efficiency and tunable collection system in the beamline by adding two permanent magnet quadrupoles (PMQs) with magnetic field strengths of about 200 and 120 T/m respectively in front of the triplet. As a result, by monitoring the relative position of the PMQs and the currents of the EMQs, the focusing energy is able to be adjusted ranging from 2 to 10 MeV, and the transmission efficiency is 1.4 to 3 times as high as only collected by EMQs triple.

1. Introduction

Laser plasma acceleration (LPA) with a accelerating gradient exceeding TV/m [1] has become a strong candidate for building future compact radiation sources [2] and high-energy particle accelerators [3, 4]. The applications of laser-accelerated ion beam , such as cancer treatment [5, 6], imaging [7], and warm-dense-matter generation [8], have received extensive attention. Under the current technical conditions of lasers and targets, Target Normal Sheath Acceleration (TNSA) [9] is the main direction to be considered for laser ion acceleration. TNSA proton beam has the characteristics of small source size (μm) [10], short pulse (ps) [11], wide energy spectrum (100 %), and large divergence angle ($\sim\pm150$ mrad). This requires a beamline to collect and focus TNSA protons for diverse applications. After passing through the beamline, most of the protons in the initial broad energy spectrum cannot be tranported, and only those with the aimed energy can reach the irradiation position. At the same time, the large divergence angle of the protons aggravates the beam loss, resulting a extremely low utilization rate. Therefore, the design of a beamline with high transmission efficiency is very important.

The mainstream collection elements used in beamlines are PMQs, EMQs and pulsed solenoids. PMQs have the advantages of high magnetic field gradient of the order of 100 T/m [12], small size, vacuum compatible, and relatively cheap; however, their invariable magnetic fields limit the regulated energy range. The magnetic field of EMQs can be easily adjusted by tuning the current, but they are large in size and weak in focusing ability. The pulsed solenoids can focus axisymmetrically and have large collection angles, but they require complex power supply systems, which makes them expensive and difficult to break through the high repetition frequency limit [13]. At present, a number of beam collection systems built by different research institutions have been reported [14, 15]. However, due to the shortcomings of each collection element, it is difficult for these systems constructed by a single type of collection elements to achieve high transmission efficiency, strong energy tuning capability and low cost simultaneously.

The Compact LAser Plasma Accelerator (CLAPA) coupled with a fully functional beamline has been built at Peking University [16]. EMQs triplet is used to collect and focus the proton beam and a dipole magnet is used for energy selection. This system has high stability and adjustability, but its transmission efficiency of the quasi-monoenergetic protons is relatively low at 20 %∼29 % [17]. This is because: (1) the protons with large divergence angle are blocked at the entrance of the beamline; (2) part of the protons that enter the beamline will further hit the vacuum tube wall due to the limitation of the aperture size and focusing ability of the EMQs triplet.

As mentioned above, PMQs have a high magnetic field gradient and can be very close to the point of interaction of the laser and the target. In this paper, we propose an improved scheme of the CLAPA beamline by adding a PMQs group in front of the existing EMQs triplet. Through optimized design, the beamline can achieve ±70 mrad collection angle of the proton beam in the energy range of 2∼10 MeV, whose transmission efficiency is 1.4 to 3 times as that only colleted by EMQs triplet.

2. Design of the high transmission efficiency collection system

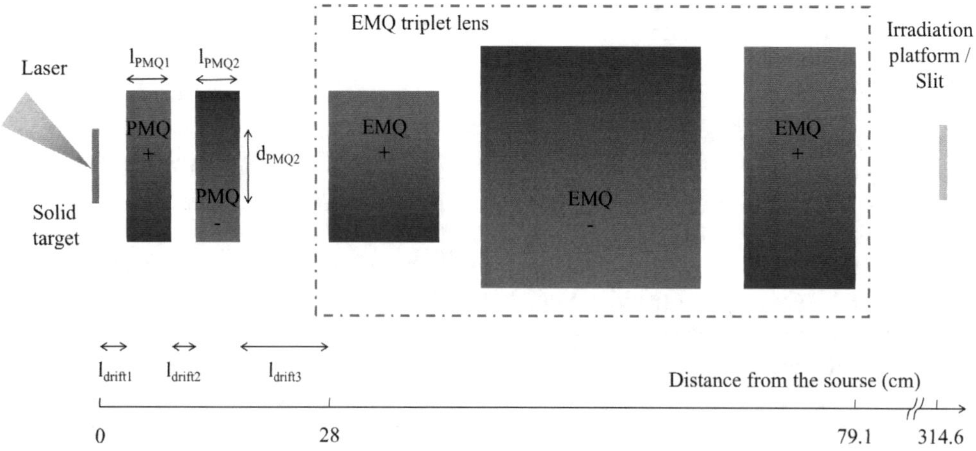

Figure 1. Schematic of the beam-collecting system.

The interaction between laser and solid target produces a high-current proton beam with an exponentially decaying energy distribution (cutoff energy is about 10 MeV) and a Gaussian divergence angle distribution (maximum divergence angle is about ±150 mrad) [18]. The proton beam is then injected into the beamline to be collected, focused and energy selected.

Table 1. Beam line simulation parameters.

Parameter	**Range**
maximum value of $k_{\mathrm{PMQ1,2}}$	25 KGs/cm
maximum value of $k_{\mathrm{EMQ1,2,3}}$	[5 KGs/cm, 2.5 KGs/cm, 2.5 KGs/cm]
$l_{\mathrm{PMQ1,2}}$	[10 mm, 50 mm]
l_{drift1}	$\geq$ 35 mm
$l_{\mathrm{drift2,3}}$	$\geq$ 25 mm

Transmission efficiency of the beamline is largely determined by the collection capacity of the collection system.

Figure 1 shows the schematic of the improved beam-collecting system. It consists of two PMQs and three EMQs. PMQ1, EMQ1 and EMQ3 are used for horizontal (x) focusing while PMQ2 and EMQ2 for vertical (y) focusing. The EMQs already exist and their dimensions and magnetic field parameters are in [16]. In order to select suitable and reliable PMQs parameters to match our experimental system and achieve the best experimental goal, we introduce genetic algorithm (GA) for analysis. For this work, protons with relatively high initial charge of 2-6 MeV are more worthy of attention as the current of protons decreases exponentially with energy increasing. The optimization goal was set such that: when 2, 3, 4, 5, 6 MeV is the center energy respectively, the collection efficiency satisfies equation (1), where η is defined by the ratio of the number of protons that are successfully transported to the irradiation platform to produce a beam spot with a radius of less than 1 cm to the number of protons that are initially accelerated by the laser.

$$x = max\left\{\sum \eta(E_i), E_i = 2, 3, 4, 5, 6\,\mathrm{MeV}\right\} \tag{1}$$

As illustrated by figure 1, the optimizer was allowed to vary the hardware parameters of the permanent magnet quadrupole lens including the lengths $l_{\mathrm{PMQ1,2}}$, inner diameters $d_{\mathrm{PMQ1,2}}$ and magnetic field gradients $k_{\mathrm{PMQ1,2}}$, and the adjustable parameters including the length of drifts $l_{\mathrm{drift1,2,3}}$ and the currents $I_{\mathrm{EMQ1,2,3}}$ of the electromagnetic quadrupole lens. There are some specific parameter constraints (see table 1): (1) the maximum magnetic field strengths of the magnets are limited by magnetization or circuits; (2) the volumes of the permanent magnet are limited by the space in the existing vacuum chamber (14 - 19 cm). That is, we have to design a proton collection system as compact as possible, and it must also be movable for replacement of other detection equipment; (3) simultaneously, due to the magnetic attraction force, the distance between two magnets should be larger than 25mm.

In all optimizitions, initial Twiss parameters of 50,000 protons were imported, and the population size was set to 50. The number of generations for each optimization run was typically hundreds of generations. After initial 50 generations, the progress of current optimal solution was monitored every few generations. Optimization was stopped when no significant improvement was found in the results. In order to avoid getting trapped in local optimum, the operation process was repeated several times, and the best results with the maximum $\sum \eta(E_i)$ were selected. The optimal parameters of two permanent magnet quadrupole magnets are shown in table 2.

3. Simulation verification of beam collection system design

We input the lattice design into a beam-transport simulation program TRACK for further verification. The radius of initial quasi-monoenergetic beam spot is set to 5 µm, and the half

Table 2. Theoretical optimal parameters of PMQs.

Parameter	Length	Inner Radius	Magnetic Field Gradient
PMQ1	40 mm	7.9 mm	195.3 T/m
PMQ2	40 mm	12 mm	112.5 T/m

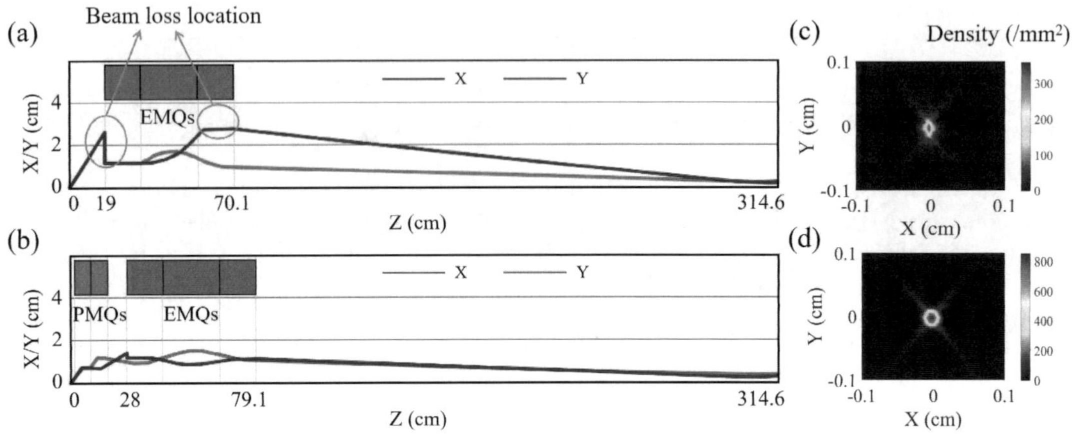

Figure 2. Simulated envelope evolution of 5 MeV protons transported by (a) only EMQs triplet and (b) the combination of PMQs and EMQs. (c) and (d) represent the corresponding transverse beam profile 3146 mm downstream the source. Colors represent the particle density.

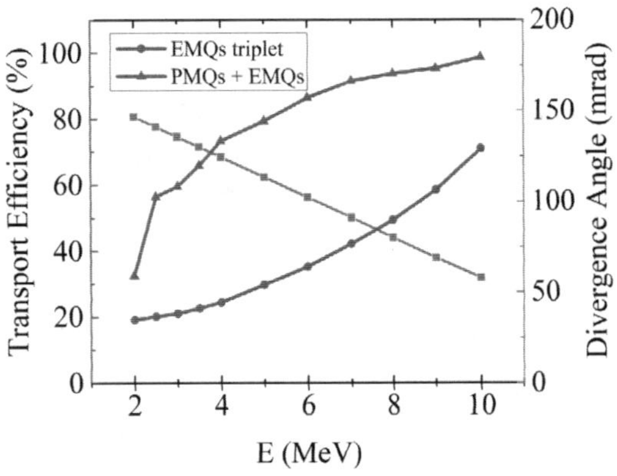

Figure 3. Transmission efficiency with (blue line with triangle points) and without (blue dotted line) PMQs' assisted collection for protons with different energies. Red line shows the corresponding initial half divergence angle in simulation.

divergence angle satisfies equation (2) according to our TNSA experiments [18].

$$\theta(\text{mrad}) = -11.1E(\text{MeV}) + 168.87 \tag{2}$$

As an example, the simulated envelope evolution of 5 MeV protons transported by only EMQs triplet and the combination of PMQs and EMQs are shown in figure 2(a) and (b) respectively. When only focused by the EMQs, the protons suffer a large amount of loss at the beamline entrance and tube wall due to the small geometric acceptance angle of the EMQs and the inability to restrain the proton divergence in time. However, the PMQs can be set closer to the target, and the strong focusing well constrains the divergence of protons. With the co-focusing of PMQs and EMQs, most of the protons within ± 70 mrad are successfully tranported to the focal plane. The corresponding transverse beam profile 314.6 cm downstream source in figure 2(c) and (d) demonstrate the protons have been focused into spots with an area of less than $2\times 2\,\mathrm{mm}^2$, and the maximum particle density has increased by a factor of 1.38 with PMQs' assisted collection. As can be seen in figure 3, transmission efficiency of the system for protons with different energies is 33%~99%, which is 1.4 to 3 times that of only the EMQs triplet.

4. Conclusion

In this paper, we propose to combine two PMQs and the EMQs triplet to realize the collection of highly divergent beams, so that the transport efficiency of the system for 2~10 MeV proton beams reaches 33%~99%. The optimal parameter sets corresponding to the collection of particles with different energies are obtained based on GA. By simultaneously adjusting the relative distance between the permanent magnet quadrupole lenses and the currents of EMQs, the beam collection system has a high ability to regulate the center energy.

References

[1] Maksimchuk A *et al.* 2000 *Phys. Rev. Lett.* **84** 4108
[2] Fuchs M *et al.* 2009 *Nat. Phys.* **5** 826–29
[3] Wagner F *et al.* 2016 *Phys. Rev. Lett.* **116** 205002
[4] Higginson A *et al.* 2018 *Nat. Commun.* **9** 724
[5] Kraft S D *et al.* 2010 *New J. Phys.* **12** 085003
[6] Yang G *et al.* 2021 *Front. Cell Dev. Biol.* **9** 1032
[7] Faenov A Y *et al.* 2009 *Appl. Phys. Lett.* **95** 101107
[8] Patel P K *et al.* 2003 *Phys. Rev. Lett.* **91** 125004
[9] Snavely R A *et al.* 2000 *Phys. Rev. Lett.* **85** 2945
[10] Borghesi M *et al.* 2004 *Phys. Rev. Lett.* **92** 055003
[11] Dromey B *et al.* 2016 *Phys. Rev. Lett.* **7** 10642
[12] Schollmeier M *et al.* 2008 *Phys. Rev. Lett.* **101** 055004
[13] Hofmann I *et al.* 2013 *Phys. Rev. Spec. Top. Accel Beams* **16** 041302
[14] Romano F *et al.* 2016 *Nucl. Instrum. Methods Phys. Res., Sect.* A **829** 153–58
[15] Kroll F *et al.* 2022 *Nature Physics* **18** 316-22
[16] Zhu J G *et al.* 2020 *Phys. Rev. Accel. Beams* **23** 121304
[17] Wu M J *et al.* 2020 *Nucl. Instrum. Methods Phys. Res., Sect.* A **955** 163249
[18] Geng Y X *et al* 2018 *Chin. Phys. Lett.* **35** 092901

J. Phys.: Conf. Ser. **2420** (2023) 012112 https://doi.org/10.1088/1742-6596/2420/1/012112

Retraction

Retraction: Investigation of polarized proton spin coherence time at storage rings (*J. Phys.: Conf. Ser.* 2420 012002)

Published 15 February 2023

This paper was retracted by IOP Publishing on 15th February 2023. The paper was published due to a technical error and should not have been published by this journal. The authors are free to publish this work elsewhere. IOP Publishing apologise for this error.

Retraction published: 15 February 2023

J. Phys.: Conf. Ser. **2420** (2023) 012113 https://doi.org/10.1088/1742-6596/2420/1/012113

Retraction

Retraction: Quasi-frozen spin concept of magneto-optical structure of NICA adapted to study the electric dipole moment of the deuteron and to search for the axion (*J. Phys.: Conf. Ser.* 2420 012052)

Published 15 February 2023

This paper was retracted by IOP Publishing on 15th February 2023. The paper was published due to a technical error and should not have been published by this journal. The authors are free to publish this work elsewhere. IOP Publishing apologise for this error.

Retraction published: 15 February 2023

Retraction

Retraction: Investigation of spin-decoherence in the NICA storage ring for the future EDM-measurement experiment (*J. Phys.: Conf. Ser.* **2420** 012041)

Published 15 February 2023

This paper was retracted by IOP Publishing on 15th February 2023. The paper was published due to a technical error and should not have been published by this journal. The authors are free to publish this work elsewhere. IOP Publishing apologise for this error.

Retraction published: 15 February 2023

AUTHOR INDEX

Ackermann, S ... 151
Aggarwal, Sanjeev ... 666
Akemoto, M ... 132
Aksentev, A E ... 253
Aksentyev, A ... 8, 318
Alesini, D ... 459, 531
Alexandre, P ... 352
Alves, M B ... 294
Angolettai, M E ... 637
Appleby, R B ... 580
Arakawa, D ... 132
Arakida, Y ... 132
Arduini, G ... 16
Arias, T ... 205
Arranz, F ... 489
Assenbaum, S ... 212
Auchettl, R ... 125
Avronsart, J ... 76
Bacha, B ... 436
Bahrdt, J ... 112
Bakos, J ... 112
Ballarino, A ... 118
Bando, Y ... 132
Banerjee, A ... 313
Barber, S ... 453
Barnes, M J ... 525
Barth, W ... 228
Baudrenghien, P ... 16
Bazrafshan, R ... 83
Behtouei, M ... 193, 358, 531
Bell, G R ... 199
Bellaveglia, M ... 459
Belokurov, V ... 513
Bencini, V ... 600
Benedetti, G ... 106
Benedetto, E ... 606, 612, 624, 637, 671
Benjamin, C ... 199
Berg, J S ... 51
Berg, J Scott ... 76
Berg, JS ... 61
Berger, C ... 453
Bernert, C ... 212
Bernhard, A ... 118
Bilgen, S ... 508
Bini, S ... 459
Bisoffi, G ... 637, 643, 671
Bisogni, M G ... 531
Blackler, I ... 56
Blaskiewicz, M ... 51, 56, 76
Boonpornprasert, P ... 176
Borburgh, J ... 637
Bortot, L ... 31
Bosco, F ... 193, 358, 365, 531
Bottura, L ... 618, 637
Brinker, Frank ... 163
Bruce, R ... 16
Bruni, Christelle ... 401
Brüning, O ... 16
Buda, S ... 436
Buffat, X ... 16
Buonomo, B ... 459
Burrows, P N ... 631
Burrows, Philip ... 413
Cai, M ... 169
Cai, Y ... 61
Calaga, R ... 16
Calvi, Marco ... 443
Calviani, M ... 573
Camacho, O ... 365
Cara, P ... 489
Cardelli, F ... 459
Carillo, M ... 193, 358, 365, 531
Carlà, M ... 106
Carmignani, N ... 138, 299
Carra, F ... 24
Carver, L R ... 138, 299
Castro, J Da Silva ... 352
Cayla, Jean-Noël ... 401
Cerutti, F ... 16
Chai, J S ... 483
Chaikovska, Iryna ... 401
Chance, Sophie ... 401
Chao, H-C ... 95
Chavanne, J ... 313
Chen, Kemin ... 89
Chen, Ye ... 163
Cheng, H ... 685
Cherif, A ... 544
Chiadroni, E ... 365
Chiggiato, P ... 538
Chrin, J ... 465
Chubar, O ... 313
Churn, H M ... 199
Cianciosi, F ... 299
Cirrone, P ... 531
Coleman, S J ... 453

Cook, N M ... 453
Corsini, Roberto ... 413
Costa, G ... 573
Cropp, Frederick ... 419
Curbis, F ... 151
Curry, C B ... 212
Cuttone, G ... 531
Danaeifard, A ... 145
Danneil, C ... 436
Dassa, L ... 612
De Arcangelis, D ... 531
De Gregorio, A ... 531
De Maria, R ... 16
De Matteis, E ... 612
De Sá, F H ... 294
Decking, Winfried ... 163
Delerue, N ... 544
Delerue, Nicolas ... 401
Della-Negra, S ... 508
Derbenev, A ... 436
Di Giulio, C ... 459
Di Martino, F ... 531
Di Paolo, F ... 193
Di Pasquale, E ... 459
Dilly, J ... 16
Diomede, M ... 459
Dollinger, G ... 600
Don, J ... 24
Donetti, M ... 612
Dönges, T ... 187, 265
Dosanjh, M ... 631
Dowd, R T ... 125
Droba, M ... 187, 259, 265, 306
Durfee, D ... 436
Dutheil, Y ... 519, 525, 618
Edelen, J P ... 453
Edelen, J ... 313
Efthymiopoulos, I ... 16
Ego, H ... 132, 234
Eichler, A ... 430
Einstein-Curtis, J ... 453
El Ajjouri, M ... 352
El Fekih, R Ben ... 352
Enomoto, Y ... 132
Erçağlar, V ... 465
Esposito, L S ... 573
Faillace, L ... 193, 358, 365, 459, 531
Fakhari, M ... 83
Falone, A ... 459
Fan, K ... 477, 502
Fang, Y L ... 685
Fantauzzi, S ... 193
Farabolini, Wilfrid ... 413
Favaudon, V ... 531
Felcini, E ... 606, 612
Feldmeier, E ... 277
Feng, C ... 169
Ferrari, E ... 151
Ferrari, Eugenio ... 443
Ficcadenti, L ... 358, 531
Fielder, R ... 95
Filippetto, Daniele ... 419
Fischer, W ... 51, 56
Flöttmann, K ... 83
Flowerdew, J A D ... 271
Foka, P ... 643
Francescone, D ... 531
Franciosini, G ... 531
Franzini, G ... 459
Frisella, G ... 606, 612
Fujii, H ... 407
Fujimoto, M ... 384
Fukasawa, A ... 365
Fukasaway, A ... 358
Fukuda, M ... 38, 45
Fukuma, H ... 378
Fukunishi, N ... 407
Funakoshi, Y ... 378
Furukawa, K ... 132
Gaebel, S ... 112
Gallo, A ... 459, 531
Gamage, B R ... 76
Gamage, BRP ... 61
Gamage, R ... 51
Gamelin, A ... 352
Ganter, Romain ... 443
García, E C Cortés ... 277
Gauthier, M ... 212
Gerbershagen, A ... 618, 631
Ghergherehchi, M ... 483
Giannetti, G ... 459
Giovannozzi, M ... 16
Giribono, A ... 365
Giuliano, L ... 358, 365, 531
Glenzer, S H ... 212
Gnacadja, E ... 240, 395, 586, 593
Göde, S ... 212
Goto, K ... 101
Gottschlich, S ... 112
Grenier, D ... 573
Grenier-Boley, E ... 573
Grimmer, S ... 112
Gu, D ... 169
Gu, X ... 51

Guler, Hayg 401
Guo, L 101, 384
Guo, Lei 89
Guo, Z 685
Gupta, Divya 666
Ha, D H 483
Ha, K 436
Haberer, Th 277
Hall, C 313
Hao, Y 51, 68, 76, 372
Hara, K 567
Harada, H 246
Harada, K 181
Hartmann, M 649
Hasegawa, K 567
Hatakeyama, Shuichiro 344
Hauer, V 489
Haziot, A 618
Heinrich, S 531
Hermes, P D 16
Hernalsteens, C 240, 395, 586, 593
Hidaka, Y 436
Higashi, N 181
Higo, T 132
Hillert, W 151
Himmerlich, M 538
Hino, M 678
Hoffstaetter, G 51
Hoffstaetter, GH 61
Hofler, A S 331
Holmes, D 68
Honma, H 132
Hosaka, Masahito 89
Hotchi, H 246
Hoummi, L 138, 299
Hu, Y 436
Huang, H 51
Hunt, J R 573
Iadarola, G 16
Iida, N 132
Iijima, T 234
Inami, K 234
Ishibashi, T 378
Ishii, K 477, 495, 502
Iwashita, Y 678
Iwata, S 477, 502
Jimbo, Kouichi 219
Jones, L B 199
Jones, R 16
Kakihara, K 132
Kallestrup, J 95
Kamitani, T 132
Karppinen, M 612, 671
Kärtner, F X 83
Kashiwagi, S 38, 45
Katagiri, H 132
Kato, R 181
Katoh, M. 384
Kawamura, M 132
Kelliher, D J 271
Kewisch, J 51, 61
Khachatryan, V 288
Khalvati, M Reza 671
Khan, A 338
Kiselev, D 649
Kittel, Christoph 443
Knaack, S 112
Kolokolchikov, S 1, 318
Kolthoff, I 24
Kondo, Y 234
Konno, S 38, 45
Korysko, Pierre 413
Kostoglou, S 16
Kotzian, G 471
Krafft, G A 331
Krainer, A M 573
Kramer, T 519, 525
Krasilnikov, M 176
Krogere, D 544
Krupa, Michal 413
Kuhn, C 112
Kurfürst, C 612
Kuriki, M 38, 45
Kuriyama, Y 678
Ladygin, V P 253
Ladygin, V 318
Laface, E 553, 560
Lagniel, J -M 283
Latina, A 631
Laube, F 112
Le Bec, G 313
Lechner, A 24
Lefevre, Thibaut 413
Leggieri, A 193
Letresor, A 352
Leung, Ka-Ngo 219
Li, D Y 685
Li, F N 685
Li, X 176
Li, Y Z 685
Li, Y 61, 205, 436
Liedl, A 459
Lin, C 685
Lindström, B 16

Liptak, Z ... 38, 45
Liuzzo, S M ... 138, 299
Loisch, G ... 513
Lomax, Antony J ... 655, 660
Longuevergne, D ... 508
Lovelace, H ... 51, 76
Lueangaramwong, A ... 176
Luo, Y ... 51, 56, 68, 76, 372
Machida, S ... 271
Maclean, E H ... 16
Maiano, C G ... 553, 560
Majernik, N ... 365
Mamaras, A ... 643
Mann, J I ... 205
Maradia, Vivek ... 655, 660
Marinoni, S ... 538
Mariotto, S ... 612
Marrese, F ... 193
Martin, I ... 95
Marusic, A ... 56
Marx, D ... 61
Matlis, N H ... 83
Matsumoto, H ... 477, 495, 502
Matsumoto, N ... 495, 502
Matsumoto, S ... 132
Matsumoto, T ... 132
Matsushita, H ... 132
Mayerhofer, M ... 600
Mazzoni, Stefano ... 413
Meer, David ... 655, 660
Melnikov, A A ... 253
Melnikov, A ... 1, 8, 318
Mercier, B ... 508
Mereghetti, A ... 606, 612
Meseck, A ... 112
Métral, E ... 16
Meusel, O ... 187, 259, 265, 306
Mibe, T ... 234
Migliorati, M ... 193, 358, 365, 531
Mikawa, K ... 132
Mirandola, A ... 612
Mirian, N ... 151
Miski-Oglu, M ... 228
Miura, T ... 132
Miyahara, F ... 132
Monard, Hugues ... 401
Montag, C ... 51, 56, 61, 68, 76, 372
Montañés, A A Del Barrio ... 519
Moriggi, D ... 459
Moriya, K ... 234
Morozov, V ... 51, 61, 76
Mostacci, A ... 193, 358, 365, 531
Motlagh, S Talebi ... 145
Mounet, N ... 16
Moutardier, Alexandre ... 401
Mucher, C Y ... 573
Muggli, P ... 390
Müller, A-S ... 118
Mumyapan, M ... 483
Muñoz, P Andreu ... 573
Nadolski, L S ... 352
Nagahashi, S ... 181
Nakajima, H ... 132
Nakamura, N ... 181
Nakamura, R ... 678
Nakamura, T ... 378
Nakanoya, T ... 246
Nakazawa, Y ... 234
Namgoong, H ... 483
Nangoi, J K ... 205
Nash, B ... 313
Natsui, T ... 132
Nechaeva, T ... 390
Nes, T H ... 118
Neupert, H ... 538
Nigorikawa, K ... 181
Niknejadi, P ... 151
Nikolaev, N ... 318
Nisson, E ... 51
Noakes, T C Q ... 199
Nogami, T ... 181
Nomen, O ... 489
Nomura, M ... 567
Norman, H XQ ... 580
Nosochkov, Y ... 61
Obier, F ... 513
Obina, T ... 181
Ogando, F ... 489
Ogawa, Y ... 132
Ohmi, K ... 378
Ohmori, C ... 567
Ohnishi, Y ... 378
Ohsawa, S ... 132
Okabe, K ... 246
Okayasu, Y ... 132
Okita, H ... 246, 567
Oliver, C ... 489
Ollier, R ... 352
Olsson, Teresia ... 157
Omeich, Maher ... 401
Omori, T ... 38, 45
Ondreka, D ... 265
Oogoe, T ... 132
Oponowicz, E ... 618

Otani, M 234
Otto, T 544
Padrazo, D 436
Pakuza, Collette 413
Paļskis, K 637
Palumbo, L 193, 358, 365, 531
Pannek, F 151
Paoluzzi, M M 567
Papaphilippou, Y 16
Paraskaki, G 151
Pasternak, J 624
Patera, V 531
Patriarca, A 531
Pauly, N 240, 395, 586, 593
Pedretti, M Foppa 544
Peggs, S 51
Pella, A 612
Pensavalle, J H 531
Perini, D 612
Perron, T 138, 299
Persson, T H B 16
Petit, V 538
Piacentini, L 612
Pierini, P 553
Piersanti, L 459
Pikurs, G 544
Pioli, S 459
Piquet, O 560
Pivi, M T F 606, 612
Plassard, F 436
Podadera, I 489
Podlech, H J 306
Podlech, H 187, 265
Pokharel, S 331
Pont, M 106
Pop, M 151
Porta, G Zevi Della 390
Pozzi, M 544
Prat, Eduard 443
Preston, Shaun 157
Prioli, M 612
Psoroulas, Serena 655, 660
Ptitsyn, V 51, 61
Pugnat, T 16
Pullia, M G 606, 612, 671
Qiang, J 51, 68, 372
Quaglia, S 459
Rahighi, J 145
Ramoisiaux, E 240, 395, 586, 593
Rasile, R 24
Rassool, R P 125
Ratkus, A 544, 612
Ravaioli, E 31
Redaelli, S 16
Reggiani, D 649
Rehaag, T J 199
Rehman, M A 132
Rehwald, M 212
Reiche, Sven 443
Reimann, S 306
Rethfeldt, C 112
Rial, E 112
Ribaud, I 508
Richter, S C 118
Richtmann, L 31
Robertson, C S 631
Roche, B 138
Rogosch-Opolka, A 112
Rohwer, T 83
Rosenzweig, J B 205
Rosenzweig, J 358, 365
Rossi, L 612, 671
Ruf, J 525
Sabbatini, L 459, 612
Sack, M 519, 525
Sacko, M 352
Saeidi, F 145
Sagan, D 61
Sagehashi, H 181
Saha, P K 246
Saito, N 234
Sakai, H 181
Salehi, E 384
Sammer, M 600
Samoilenko, D 151
Sánchez-Herranz, D 489
Sanmartí, M 489
Sapinski, M 624, 637, 671
Sarti, A 531
Satake, I 132
Satogata, T 51
Satoh, M 132
Sattonnay, G 508
Savazzi, S 606, 612
Scampati, M 459
Scarselletta, G 459
Schaper, L 151
Scheeler, U 228
Scheer, M 112
Scheinker, Alexander 419
Schietinger, T 465
Schietinger, Thomas 443
Schippers, Jacobus M 660
Schirmer, D 424

Schoenwaelder, C 212
Schoerling, D 118
Scholz, Matthias 163
Schramm, U 212
Schulte-Urlichs, K 187, 265
Seidel, M 649
Seimiya, Y 38, 45, 132
Senaj, V 519
Senes, Eugenio 413
Senichev, Y V 253
Senichev, Y 1, 8, 318
Shaftan, T 436
Sharma, S 338
Sharp, C J 573
Sheehy, S L 125, 223, 271, 580
Shibata, T 477, 495, 502
Shidara, T 132
Shimada, M 181
Shimada, T 567
Shirai, Toshiyuki 219
Shirakawa, A 132
Shobuda, Y 246
Shobuda, Yoshihiro 344
Shrey, T 56
Signorelli, MG 61
Singh, B K 489
Singh, B 95
Singh, Beni 157
Singhal, Rahul 666
Smaluk, V 338, 436
Snuverink, J 649
Söderén, M 471
Solvang, P S 471
Someya, H 132
Sorbi, M 671
Spataro, B 193, 358, 365, 531
Spiller, P 265
Steinberg, A F 580
Stella, A 459
Stephan, F 176
Sterbini, G 16
Stichelbaut, F 395
Stock, M 612
Stopjakova, V 471
Stopjakova, Viera 448
Sue, Y 234
Sugimoto, T 477, 495, 502
Sugiyama, Y 567
Sulc, A 430
Sumi, K 234
Suwada, T 132
Syresin, E M 253
Syresin, E 1, 8, 318
Taborelli, M 538
Taira, Y 384
Tajino, H 38, 45
Takahashi, T 38, 45
Takai, R 181
Takashima, Y 101
Takeuchi, Y 234
Tamura, F 246, 567
Tanaka, H 678
Tanaka, M 132
Tanaka, O 181
Tanimoto, Y 181
Taurines, S 352
Tavian, L 538
Taylor, G N 125
Taylor, R 624, 637
Tepikian, S 61, 76
Terui, S 378
Tesse, R 240, 395, 586, 593
Thielmann, M 544
Thoma, K I 187, 259, 265
Thoraud, S 352
Tian, Y 436
Timko, H 16
Tobiyama, M 378
Tocci, S 459
Tomás, R 16
Tommasini, D 612
Torims, T 544, 612
Torregrosa, C 573
Torrisi, G 193, 531
Toyama, Takeshi 344
Tranquille, G 637
Treffert, F 212
Tu, L J 169
Uberti, S 612
Uchiyama, T 181
Ueda, A 181
Ueki, R 378
Umapathy, G.R. 666
Unger, J 61
Urakawa, J 38, 45
Vaccarezza, C 365
Valente, U 671
Valetov, Eremey 325
Valletti, L 193
Valuch, D 471
Valuch, Daniel 448
Van Bibber, Karl 219
Van Der Veken, F 16
Van Tilborg, J 453

Van Weelderen, R 671
Vannozzi, A 531, 612
Vanwelde, M 240, 395, 586, 593
Vedani, M 544
Velloso, M M S 294
Venchi, G 671
Verra, L 390
Vilcans, J 612
Volpi, M 125
Volz, P 112
Vormann, H 228
Vossberg, M 228
Vretenar, M 544, 612, 637, 643, 671
Wagenblast, P 544
Wang, D 132
Wang, Faya 89
Wang, G 436
Wang, Guanliang 89
Wang, M 553
Wang, S T 288
Wang, Zhe 89
Weber, Damian C 655, 660
Wendt, Manfred 413
Wenninger, J 16
Werin, S 151
White, S 138, 299
Wiesner, C 24, 31
Willeke, F 51, 61, 68, 76, 372
Williams, S D 125
Williams, Scott David 401
Wilsken, T 430
Winkelmann, Lutz 163
Witte, H 76
Wollmann, D 24, 31
Wu, M J 685
Wu, Q 76
Xiao, B 76
Ximenes, R Franqueira 573
Xu, D 51, 56, 61, 68, 76, 372
Xu, Wei 89
Yamaguchi, H 101
Yamakita, M 407
Yamamoto, M 101, 567
Yan, X Q 685
Yan, Y 685
Yang, T 685
Yang, X 436
Yano, Y 132
Yap, J SL 580
Yaramyshev, S 228
Yasuda, H 234
Yokoya, K 38, 45
Yokoyama, K 132
Yoshida, M 132, 234
Yoshii, M 567
Yoshimoto, M 246
Yoshimoto, Masahiro 344
Yoshimoto, T 132
Yotsuzuka, M 234
Yu, L H 436
Yuan, Z X 685
Zamani, F 145
Zeil, K 212
Zelinotti, L 459
Zeng, R 560
Zerlauth, M 16
Zhang, H 649
Zhang, K Q 169
Zhang, R 132
Zhang, X 223
Zhao, Z T 169
Zhou, X 132
Zhu, K 685
Zhu, Z H 169
Zhu, Zihan 163